D1087332

Biochemistry of the Eye

PERSPECTIVES IN VISION RESEARCH

Series Editor: Colin Blakemore
University of Oxford
Oxford, England

Biochemistry of the Eye
Elaine R. Berman

Development of the Vertebrate Retina
Edited by Barbara L. Finlay and Dale R. Sengelaub

Parallel Processing in the Visual System
THE CLASSIFICATION OF RETINAL GANGLION CELLS AND ITS IMPACT ON THE NEUROBIOLOGY OF VISION
Jonathan Stone

Biochemistry of the Eye

Elaine R. Berman
Hadassah-Hebrew University Medical School
Jerusalem, Israel

Plenum Press • New York and London

Library of Congress Cataloging-in-Publication Data

Berman, Elaine R.
Biochemistry of the eye / Elaine R. Berman.
p. cm. -- (Perspectives in vision research)
ISBN 0-306-43633-7
1. Eye--Physiology. 2. Biochemistry. I. Title. II. Series.
[DNLM: 1. Eye--chemistry. WW 101 B516b]
QP475.B46 1991
612.8'4--dc20
DNLM/DLC
for Library of Congress 90-14353
CIP

ISBN 0-306-43633-7

A Division of Plenum Publishing Corporation
233 Spring Street, New York, N.Y. 10013

Printed in the United States of America

Dedicated to the memory of
Professor Isaac C. Michaelson

Preface

My first introduction to the eye came more than three decades ago when my close friend and mentor, the late Professor Isaac C. Michaelson, convinced me that studying the biochemistry of ocular tissues would be a rewarding pursuit. I hastened to explain that I knew nothing about the subject, since relatively few basic biochemical studies on ocular tissues had appeared in the world literature. Professor Michaelson assured me, however, that two books on eye biochemistry had already been written. One of them, a beautiful monograph by Arlington Krause (1934) of Johns Hopkins Hospital, is well worth reading even today for its historical perspective. The other, published 22 years later, was written by Antoinette Pirie and Ruth van Heyningen (1956), whose pioneering achievements in eye biochemistry at the Nuffield Laboratory of Ophthalmology in Oxford, England are known throughout the eye research community and beyond. To their credit are classical investigations on retinal, corneal, and lens biochemistry, beginning in the 1940s and continuing for many decades thereafter. Their important book written in 1956 on the *Biochemistry of the Eye* is a volume that stood out as a landmark in this field for many years.

In recent years, however, a spectacular amount of new information has been generated in ocular biochemistry. Moreover, there is increasing specialization among investigators in either a specific field of biochemistry or a particular ocular tissue. Therefore, subsequent books on the biochemistry of the eye have, of necessity, been multi-authored (Graymore, 1970; Anderson, 1983).

We have now in some ways come full circle. Notwithstanding the unique structures, functions, biochemical properties, and metabolic characteristics of individual ocular tissues, it is becoming increasingly evident that they also have many features in common. These include well-known pathways of glucose oxidation, energy production, and ion transport; in addition, newer areas of research have revealed a wide variety of other metabolic activities shared by many ocular tissues such as receptor-mediated membrane signal transduction systems, G proteins, defense mechanisms against light and oxygen toxicity, drug-metabolizing and -detoxifying systems, eicosanoid production, and many more. The scope of biochemistry has expanded considerably in recent years and now includes two major disciplines: cell biology and molecular biology. Recombinant DNA technology has been successfully applied to the lens for nearly a decade and is now having its impact on other ocular tissues such as the retina. The amino acid sequences of proteins can be deduced, gene structures studied, and gene localization determined by *in situ* hybridization. Increasing numbers of ocular proteins are being cloned and sequenced, and an understanding of disease processes at the molecular level in inherited disorders such as gyrate atrophy and blue cone monochromacy has already been achieved.

The emphasis in this volume is on work published during the 1980s, and it includes literature appearing until summer of 1989. Review articles summarizing earlier investigations are cited in appropriate sections. A major attempt has been made to cover material in depth and yet with maximum brevity, which is no simple task. Early responses to preliminary drafts of several chapters prompted the inclusion of an introductory chapter on selected topics in biochemistry relevant to the eye. The six chapters that follow are descriptions of individual ocular tissues beginning with the tear film and continuing posteriorly to the retina.

Many friends and colleagues have given of their time and patience at various stages during completion of the manuscript. Their comments, suggestions, permissions to use published illustrations, as well as access to manuscripts in press are gratefully acknowledged. I especially wish to thank Gene Anderson, Yogesh Awasthi, Wolfgang Baehr, Endre Balazs, Mike Berman, Tony Bron, Jerry Chader, Hugh Davson, Darlene Dartt, Ed Dratz, Lynette Feeney-Burns, Steve Fliesler, Ilene Gipson, Greg Hageman, John Harding, Paul Hargrave, Diane Hatchell, Carole Jelsema, Gordon Klintworth, Baruch Minke, Oded Meyuhas, Tom Mittag, Robert Molday, Beryl Ortwerth, David Papermaster, Alan Proia, John Scott, John Tiffany, Brenda and Ramesh Tripathi, Nicolaas van Haeringen, and Richard Young. In addition, a special note of thanks goes to Mrs. Bela Eidelman for expert secretarial assistance.

REFERENCES

Krause, A. C., 1934, *The Biochemistry of the Eye*, The Johns Hopkins Press, Baltimore.
Pirie, A., and van Heyningen, R., 1956, *Biochemistry of the Eye*, Charles C. Thomas, Springfield, IL.
Graymore, C. N. (ed.), 1970, *Biochemistry of the Eye*, Academic Press, London.
Anderson, R. E. (ed.), 1983, *Biochemistry of the Eye*, American Academy of Ophthalmology, San Francisco.

Elaine R. Berman

Introduction

GROSS ANATOMY AND STRUCTURE OF THE EYE

Diagrammatic horizontal cross-section of a vertebrate eye. The primate eye is approximately spherical in shape, measuring 24 mm in diameter in human adults. The posterior 85% of the globe is covered by the *sclera*, a dense, white, opaque protective coat that is not directly involved in the visual process. The *cornea*, which is lined by a thin (7- to 8-μm) *tear film*, covers the remaining anterior portion of the globe. It is a uniquely transparent tissue with high refractive power and unusual metabolic characteristics. Its posterior surface is bathed by the *aqueous humor* secreted by the *ciliary epithelium* into the posterior chamber. Light passes through both of these transparent media and reaches another transparent tissue, the *lens*, whose function is to focus incoming images onto the retina. Light then passes through the last of the transparent media, the *vitreous*. This viscous gel occupies about 90% of the total volume of the eye; it provides structural support for surrounding ocular tissues and also serves as a shock absorber against mechanical impact. All of these transparent media are secondary to the neural *retina*, the center of

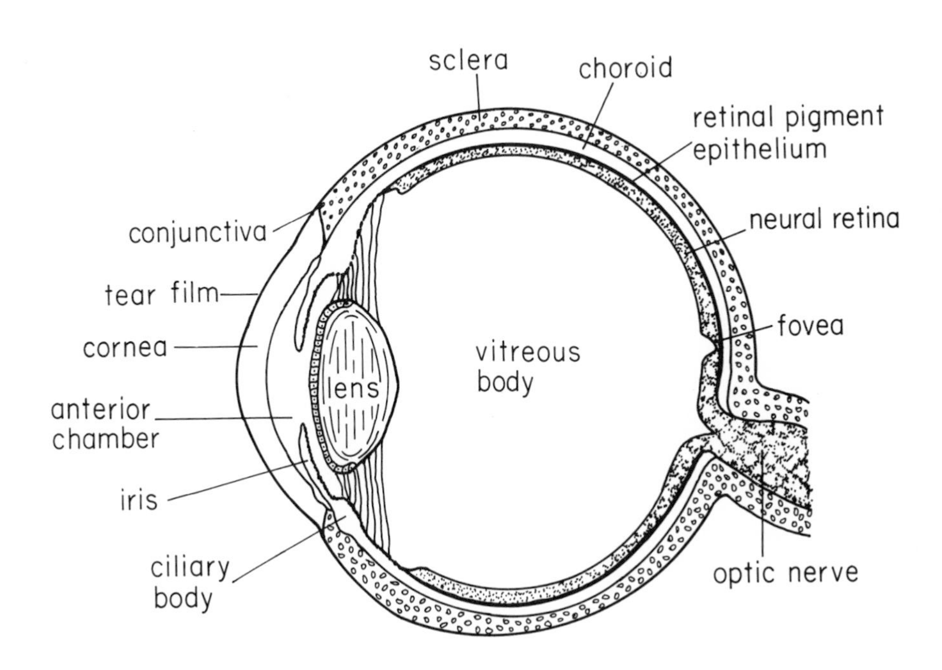

the visual process. It is here that light is absorbed by the photoreceptors (specialized organelles of the outermost layer of the neural retina) and converted into an electrical signal in a process called phototransduction. Light initiates a series of events that triggers hyperpolarization of photoreceptor plasma membranes. This signal reaches the photoreceptor synaptic region; further processing and integration take place in the secondary neurons, and the final signal is transmitted through the ganglion cell layer to the brain.

Contents

Selected Topics in Biochemistry Relevant to the Eye

1

The topics selected for inclusion in this chapter represent only a small fraction of the many fields of biochemistry and molecular biology. However, a brief introduction to general concepts as well as specific fields of research currently under active investigation in ocular tissues has been included to serve as background for Chapters 2 through 7. The material in the first section on cellular biochemistry is based almost entirely on several general texts (Lehninger, 1982; Darnell *et al.*, 1986; Stryer, 1988; Alberts *et al.*, 1989). For the remaining sections, additional source material is cited when relevant.

1.1. GENERAL CELLULAR BIOCHEMISTRY

1.1.1. Organization of the Cell

Biochemical reactions take place within the cell, on the plasma membrane and within specialized organelles (Fig. 1.1). Eukaryotic cells are not isolated; rather, they are in communication with adjacent cells and also receive signals originating from distant organs or tissues that modulate the activity of many key intracellular enzymes. These chemical signals are mediated by neurotransmitters, hormones, and a wide range of physiological agonists. The brief description of cellular organization given below highlights features common to many eukaryotic cells and is applicable to the majority of specialized cells in ocular tissues.

1.1.1.1. Cellular Organelles

The nucleus, which contains nearly all of the DNA of the cell, is surrounded by a nuclear envelope consisting of two paired membranes fused together at various intervals around openings termed nuclear pores. These openings function as channels, allowing free passage of various substances between the nucleus and the cytoplasm. The outer nuclear membrane is continuous with the rough endoplasmic reticulum in most eukaryotic cells.

It is here that DNA is transcribed by three different RNA polymerases, resulting in the formation of rRNA, tRNA, and mRNA. The essential feature of eukaryotic cells is that, whereas the genes that control protein synthesis are in the nucleus, the translation of

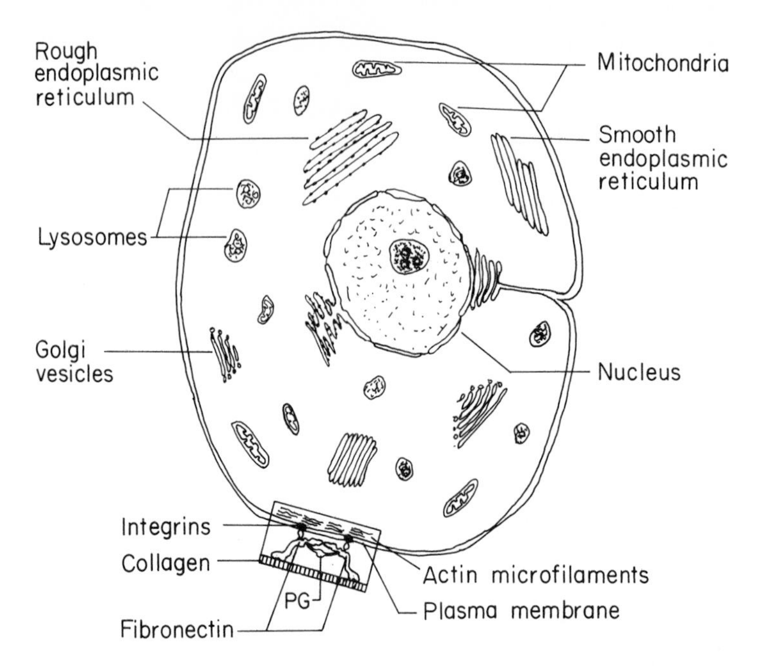

Figure 1.1. Schematic diagram of a eukaryotic cell showing the major cellular organelles. The inset illustrates the approximate form of transmembrane communication between actin filaments of the cytoskeleton and the extracellular matrix. Key components are integrins, 140-kDa transmembrane proteins that bind fibronectin on the extracytoplasmic face to actin filaments, possibly through vinculin. (For detailed reviews, see Burridge *et al.*, 1988; Ruoslahti, 1988a).

mRNA into protein occurs in the cytoplasm (see Section 1.1.2). The spatial and temporal separation of transcription and translation in eukaryotic cells allows a more intricate regulation of gene expression than is possible in prokaryotes.

Mitochondria, the centers of energy metabolism, are found in varying size, shape, and number throughout the cell. Measuring about 1 μm in diameter in most cells, they contain two membrane systems. The outer one, which completely surrounds the organelle, is smooth, whereas the inner membrane consists of multiple infoldings called cristae. The inner "core" of the mitochondrion is filled by a gel-like matrix.

The inner membranes, or cristae, and the matrix are the sites of oxidation of fatty acids and pyruvate to CO_2 and water through the tricarboxylic acid cycle. These reactions are coupled, at the same sites, to the synthesis of ATP from ADP and inorganic phosphate. Whereas glycolysis, which occurs in the cytoplasm, yields only 2 mol of ATP per mol of glucose, mitochondrial oxidation generates 36 mol (or 34 in some tissues). The essential step in the synthesis of ATP is the coupling of the flow of electrons from NADH and $FADH_2$ to the transport of protons across the inner mitochondrial membrane. This generates the proton-motive force, a combination of a proton concentration (pH) gradient and a membrane potential gradient. A multiprotein coupling factor, called ATP synthase (also known as H^+-ATPase or F_0F_1-ATPase) couples ATP synthesis to the electrochemical gradient of protons generated by the electron transport chain (see review by Futai *et al.*, 1989). The major components of mitochondrial electron transport are complexes of cytochromes b and c_1, cytochrome c, and cytochromes a and a_3, operating sequentially and utilizing interconversions of their Fe^{2+}- and Fe^{3+}-containing heme prosthetic groups for the transport of electrons. In the final step, a pair of electrons is transferred through cytochrome oxidase to molecular oxygen, the ultimate electron acceptor.

The majority of mitochondrial proteins are encoded by nuclear DNA, synthesized on cytoplasmic ribosomes, and then imported into the mitochondrion. However, it is now recognized that mitochondria also contain their own specific genes, and in fact the entire mitochondrial genome from many cell types has been cloned and sequenced. Mitochondrial DNA (mtDNA) is located in the matrix compartment of the organelle; this genome encodes 13 proteins, two rRNAs found in mitochondrial ribosomes, and 22 tRNAs that are used in the translation of mitochondrial mRNA. Mitochondrial gene products identified to date include certain components of electron transport located on the inner membrane: cytochrome b, three subunits of cytochrome c oxidase, and several subunits of ATP synthase (F_0F_1-ATPase) and of NADH-Q reductase.

The *endoplasmic reticulum* is a three-dimensional meshwork of membrane channels dispersed throughout the cytoplasm. The spaces within the channels, termed cisternae, function as "canals" for the transport of various substances throughout the cell as well as to the exterior. Of the two types of endoplasmic reticulum, smooth and rough, it is the latter that contains ribosomes on its surface; here the newly synthesized proteins pass into the lumen, where they are matured and sorted. These primary gene products are then transported to the Golgi complex where posttranslational modifications of secretory and membrane proteins, as well as glycoproteins and glycosaminoglycans, take place.

Golgi vesicles appear as a network of flattened membranous sacs, each surrounded by a single membrane. The vesicles located near the ends of the Golgi sacs are small and spherical. These regions, which appear as pinched-off edges of the larger Golgi vesicles, are the precursors of primary lysosomes. For its major role in glycoprotein synthesis, the Golgi stack is divided into three functionally distinct compartments, the cis, medial, and trans. Glycoproteins synthesized in the rough endoplasmic reticulum traverse all three compartments where sequential modifications, the final steps in complex oligosaccharide synthesis, take place (Kornfeld and Kornfeld, 1985).

Lysosomes are single membrane-bounded spherical organelles containing a broad spectrum of hydrolytic enzymes with acidic pH optima that degrade complex macromolecules such as proteins, nucleic acids, sphingolipids, and most glycoconjugates (glycosaminoglycans, proteoglycans, and glycoproteins). The substances to be degraded are brought into the lysosomes by endocytosis and by autophagy (as described below) and are then hydrolyzed. The end products of this degradative activity (sugars, amino acids, and other low-molecular-weight substances) diffuse into the cytoplasm and to other parts of the cell, where they may be either metabolized or reassembled into cellular macromolecules. In inherited storage disorders such as the mucopolysaccharidoses and sphingolipidoses (Berman, 1982), a defect in any one of the lysosomal hydrolases leads to an accumulation of undegraded material within the lysosome.

Primary lysosomes, which originate from the Golgi cisternae, fuse with endocytotic as well as autophagic vacuoles derived from extracellular and intracellular sources, respectively. The resulting digestive vacuoles are, by definition, secondary lysosomes. Small vesicles budding off from these secondary lysosomes may leave the cell by exocytosis through the plasma membrane. In many cells, substances that are not completely degraded by lysosomal enzymes accumulate and form residual bodies, some of which are autofluorescent. This is a prominent feature of retinal pigment epithelial cell metabolism and is discussed in Chapter 7, Section 7.4.5.3.

The various stages in the maturation of lysosomal enzymes and their transfer from the lumen of the endoplasmic reticulum (where they are discharged after synthesis) to the lysosome have recently been elucidated. In the first step, an N-linked oligosaccharide is attached to the lysosomal enzyme, and afterward some of the mannose residues in the oligosaccharide are phosphorylated, probably in the cis Golgi region. There is a receptor for mannose 6-phosphate on the luminal face of the Golgi complex that binds the phosphorylated protein. After transfer to acidic sorting vesicles, the phosphorylated lysosomal enzyme is dissociated from the receptor, and the phosphate group is hydrolyzed. The enzyme is then delivered to the lysosome, where it is further modified by proteolytic cleavage to an active form of lower molecular weight. There are also mannose 6-phosphate receptors on the plasma membranes of most cells that mediate in the binding and endocytosis of extracellular phosphorylated lysosomal enzymes. A defect in the phosphorylation of newly synthesized acid hydrolases forms the biochemical basis of two inherited disorders of lysosomal function, mucolipidosis II and III (see review by Neufeld and McKusick, 1983). Both are associated with varying degrees of corneal clouding.

Cells also contain *inclusion bodies*, cytoplasmic structures not bounded by a membrane. The principal ones are ribosomes, glycogen granules, and lipid droplets. In addition to being attached to the rough endoplasmic reticulum, *ribosomes* are also found in the free state in the cell cytoplasm. *Glycogen*, a high-molecular-weight polysaccharide, provides a reserve source of energy in the cell; it is a highly branched polymer of glucose residues joined in linear regions by α1,4 linkages and at branching points by α1,6 linkages. These granules are especially abundant in liver and muscle cells; in the eye, glycogen has been detected in the basal cell layers of the corneal epithelium and in Müller cells of the retina. In fat cells and in liver, the *lipid droplets* consist mainly of triglycerides. Vitamin A ester is also stored in the form of lipid droplets in liver and in the pigment epithelium of certain species.

All of the organelles and inclusion bodies are embedded in the cell *cytoplasm*, the structured "liquid" phase of the cell consisting of a viscous aqueous phase, the *cytosol*, and structural arrays of fibrous proteins called the *cytoskeleton* (see Section 1.1.1.3). The cytosol has a gel-like consistency and contains an exceedingly large number of dissolved solutes such as enzymes and enzyme systems (e.g., glycolytic enzymes), proteins that bind and transport a variety of metabolites, electrolytes, dissolved oxygen, organic solutes (e.g., glucose, amino acids, peptides, lactic acid), coenzymes, and a variety of nucleotides. Enzymes that oxidize glucose via the pentose phosphate pathway (hexose monophosphate shunt) are also localized in the cytosol. This pathway (see Figs. 3.2 and 5.3) has several important functions in many cells in general and in ocular tissues in particular. Through the oxidation and decarboxylation of glucose 6-phosphate, two reducing equivalents of NADPH are generated per mol of glucose. The NADPH is used for the synthesis of fatty acids and sterols by the cytochrome P450 system and in addition maintains glutathione (GSH) in the reduced state.

1.1.1.2. The Plasma Membrane

The plasma membrane is a highly selective semipermeable membrane that defines the outer limit of the cell. Its important and unique functions include (1) regulation of the molecular and ionic composition of the cell by means of proteins that serve as pumps, ion-gated channels, and enzymes, and (2) control of a continuous flow of information between the cells and their environment. The latter processes are modulated by specific receptors located on the cell surface that are coupled to membrane transduction systems. They play a major role in the generation of second messengers and the regulation of intracellular metabolism (see Section 1.3).

Special structural components of the plasma membrane, called gap junctions, allow short-range communication between cells. These junctions consist of composite plasma membranes from adjacent cells that are separated by a 2- to 3-nm space or gap. Their unique structure leads to the formation of 1- to 2-nm channels that allow ions and small molecules to pass between adjacent cells (see review by Evans, 1988). Gap junctions also provide a mechanism for propagation of electrical signals, coupled secretion, and other important cellular functions. The major channel-forming component of liver gap junctions is a 27- to 28-kDa polypeptide. Although gap junction protein is highly conserved, the molecular mass varies in different tissues; e.g., in heart it is a 45-kDa protein that shows some cross-reactivity with the liver protein (Revel *et al.*, 1987). Gap junction flow is modulated by hormone-induced phosphorylation.

The basic architecture of the plasma membrane, as well as of membranes surrounding most intracellular organelles, is the phospholipid bilayer, a sheet-like structure 6–10 nm in thickness and consisting of two layers of precisely aligned phospholipids. The polar (hydrophilic) heads of the phospholipids face both the outer and inner surfaces of the plasma membrane, while their fatty acyl chains form a continuous hydrophobic middle layer. Each layer, or each half of the phospholipid, is called a leaflet. Since all lipids and most proteins are able to move laterally in the plane of the membrane, the term fluid mosaic is often used to describe this universal membrane model. In general, the lower the ratio of cholesterol to unsaturated fatty acids, the greater is the fluidity of the membrane.

Membrane proteins consist broadly of two types: peripheral proteins, which are bound to the membrane by protein–protein interactions, and integral proteins, which span the membrane and interact directly with the fatty acyl chains of the interior hydrophobic layer. The majority of integral membrane proteins contain both hydrophilic and hydrophobic domains. The former lie outside of the membrane, either in the cytoplasm or on the exterior surface, whereas the latter are bound tightly to the membrane lipids in the middle layer of the plasma membrane. These integral, or transmembrane, proteins are insoluble in water and can only be extracted with the aid of detergents. Some integral proteins such as glycophorin in the erythrocyte cell membrane span the membrane only once, but others such as rhodopsin and the two major classes of hormone receptors, β-adrenergic and muscarinic cholinergic receptors, span the membrane seven times.

Apart from the phospholipid bilayer, another class of recently described membrane lipids, termed glycosyl-phosphatidylinositols, are found covalently linked to a diverse group of membrane proteins, of which at least 30 have been described to date (Low and Saltiel, 1988). The glycosyl-phosphatidylinositols located on the outer surface of the plasma membrane function to anchor these proteins to the cell surface, and at least one of them, on the inner surface, is thought to play an important role in insulin action. Hydrolytic enzymes have been identified that cleave this specific lipid–protein linkage and release the proteins. This suggests a potentially important mechanism for regulating the concentration and biological activity of cell-surface membrane proteins. Further details on the structure, linkages, and biological functions of these membrane lipids have been summarized in a recent review (Ferguson and Williams, 1988).

The asymmetric orientation of glycoconjugates in the lipid bilayer has been recognized for many years. The oligosaccharide side chains of the intrinsic glycoproteins are localized on the exterior surface of the cell and, together with the oligosaccharide moieties of membrane glycolipids and adsorbed glycoproteins and proteoglycans, form the cell coat or glycocalyx. It is the high content of sialic acid in the oligosaccharides of the glycocalyx that imparts a negative charge to most cells.

Two extensively studied glycoconjugates that form an integral part of the plasma membrane glycocalyx are (1) fibronectin and (2) heparan sulfate (or other proteoglycan).

Fibronectin. This asymmetric, V-shaped extracellular matrix and plasma glycoprotein has a subunit molecular mass of approximately 220 to 250 kDa. In the native state, it consists of a dimer joined by disulfide linkages at the C-terminal end of the molecule. It plays a central role in cell adhesion and, in addition, promotes the attachment of fibroblasts and other cells to collagen matrices. Its mode of attachment to the cell surface has been the focus of many recent investigations. Compelling evidence suggests that it is mediated by a specific receptor whose entire amino acid sequence has been deduced from cDNA clones (Argraves *et al.*, 1987). The fibronectin receptor is now recognized as a member of a superfamily of receptors (integrins) that mediate the attachment of many types of adhesive glycoproteins to the cell surface. The structure, molecular biology, and functions of fibronectin and its mode of interaction with the integrin family of receptors and with proteoglycans have been reviewed in detail (Ruoslahti, 1988a).

The native dimer of fibronectin has several different structural domains that bind specific types of macromolecules. One is a 32-kDa region located at the C-terminal end that binds heparan sulfate, heparin, and chondroitin sulfate proteoglycans. Another is a 40-kDa domain that binds several different types of collagen; this region of fibronectin is rich in N-linked oligosaccharides. The third, and largest, domain of fibronectin has a M_r of about 150,000 and appears to possess unique cell-spreading activity. An immunologically similar protein, called "soluble" or cold-insoluble globulin, is found in plasma (see review by Ruoslahti, 1988a).

Fibronectin has been detected in many extracellular matrices thoughout the body. It is secreted and deposited extracellularly by a wide variety of cells including cultured cells and is aligned to cytoplasmic actin filaments (see Fig. 1.1). There is evidence suggesting that this orientation is mediated through cytoskeletal components such as vinculin. Integrins, a recently described class of 140-kDa transmembrane proteins, link the extracellular matrix to the intracellular cytoskeleton (see reviews by Burridge *et al.*, 1988; Ruoslahti, 1988a). Immunofluorescent and immunoperoxidase staining techniques indicate not only a widespread distribution but also specific localizations of fibronectin in both human and rat ocular tissues (Sramek *et al.*, 1987; Kohno *et al.*, 1987). Fibronectin is thought to play an important, albeit controversial, role in the healing of experimentally induced corneal wounds (see Chapter 3, Section 3.8).

Heparan Sulfate. This proteoglycan is intimately associated with the plasma membranes of many cells; moreover, binding sites for heparan sulfate as well as other proteoglycans are present in fibronectin. Heparan sulfate and other cell-associated proteoglycans such as chondroitin sulfate/dermatan sulfate not only contribute to the unique structural organization of the glycocalyx but also have other important functions (see reviews by Fransson, 1987; Ruoslahti, 1988b). Apart from promoting cell adhesion, they play a direct role in matrix assembly. In addition, proteoglycans bind a large amount of water and hence occupy a large space relative to their weight. They thus serve a space-filling role, providing a carbohydrate–water milieu in the extracellular matrix.

1.1.1.3. The Cytoskeleton

The cytoskeleton of the cell consists of three major classes of fibrous proteins: microtubules, intermediate filaments, and actin microfilaments, measuring 25, 8–10, and 5–6 nm in diameter, respectively. The cytoskeleton is not a fixed or rigid framework; rather, it is a constantly changing, dynamic meshwork that not only gives the cell its characteristic shape and form but is also responsible for movements of cells and of subcellular organelles. As discussed above, another important function of the cytoskeleton is that of mediating the attachment of cells to their extracellular substratum. Actin filaments are joined through vinculin and transmembrane proteins (integrins) to cell surface glycoproteins such as fibronectin and laminin; they in turn are anchored to collagen present in the extracellular matrix.

Microtubules. These cytoskeletal components are long hollow structures whose walls are composed of 13 globular subunits, or protofilaments, of α- and β-tubulin. Each of the subunits has a M_r of 55,000, and together they form an αβ dimer or heteropolymer 8 nm in length. They are arranged in a "head-to-tail" orientation running parallel to the longitudinal axis of the tubule. Microtubules undergo reversible assembly–disassembly, i.e., polymerization and depolymerization, in the living cell, a process that can also be induced *in vitro* by alternate chilling and warming. *In vivo*, microtubules grow by addition of dimers to one end, the A end, and are lost preferentially at the opposite, D end. The plant alkaloid colchicine binds with high affinity to the αβ dimer and blocks the assembly of further dimers to the tubule; since microtubules are in a steady "balanced" state of assembly and disassembly, the net effect of colchicine is depolymerization of the microtubules. Two of the many known functions of microtubules are the formation of the mitotic spindle in dividing cells and the axonal transport of vesicles, proteins, and other substances in neurons.

Intermediate Filaments. The intermediate filaments (IFs) are composed of different classes of proteins specific for the cell type. Five major types of IFs have been recognized for many years: (1) keratins, which form the tonofilaments of epithelial cells; (2) desmin filaments, which anchor the Z disks in muscle cells; (3) neurofilaments, which are considered the slow-moving components of axonal transport; (4) vimentin fibers, prominent in fibroblasts, which maintain organelles in fixed positions; and (5) glial fibrillary acidic protein (GFAP), an IF expressed only in glial cells, whose precise function is not known.

However, recent studies on amino acid sequences and gene structures suggest a different classification of IFs based on five distinct types, the first four of which have been shown to belong to distinct gene families (see review by Steinert and Roop, 1988). Some 15 types of 44- to 60-kDa acidic keratins have been classified as type I IFs; another 15 different neutral or basic 50- to 70-kDa keratins are considered as type II IFs; 54-kDa vimentin, desmin, and glial fibrillary acidic protein (GFAP) are classified as type III; and three neurofilament proteins have been classified as type IV. Type V IFs now include lamins, which form the karyoskeleton on the inner surface of the nuclear membrane. The structures, patterns of gene organization and expression, and intracellular organization of the newly classified IFs have been discussed in detail by Steinert and Roop (1988).

Actin. Actin microfilaments (F-actin) measure 5–6 nm in diameter and are usually localized directly beneath the plasma membrane. They are polymers of G actin, a globular 42-kDa protein, whose filaments are arranged as two strands of monomers wound about each other in a right–handed helix. Actin fibrils, through their interactions with vinculin and/or integrins, serve as intracellular "anchoring structures" to the extracellular matrix (see Fig. 1.1). Monomeric actin contains tightly bound ATP or ADP. Hydrolysis of ATP causes the lengthening of the filament at one end by incorporation of actin-ATP monomers, with a concomitant shortening of the other end by the release of actin-ADP. Fungal products called cytochalasins block the polymerization of actin by binding to the free ends of elongating filaments. These substances inhibit cell movement in fibroblasts and also have other striking effects on the shape of many nonmuscle cells.

Myosin is a fibrous 200-kDa structural protein containing two heavy chains and two pairs of two different types of light 20-kDa chains. In muscle, these monomers form a bipolar aggregate containing about 300 molecules of myosin. Binding of myosin to actin is at the "headpiece" end of the molecule containing the light chains; this binding also activates the ATPase in the myosin molecule. The hydrolysis of ATP leads to muscle contraction, which is triggered by a rise in Ca^{2+} concentration. In smooth muscle, contraction is regulated by two second messengers, Ca^{2+} and cAMP, whose intracellular concentrations are modulated by hormones and a wide range of physiological agonists acting on cell membrane receptors. Actin–myosin microfilaments have been found in most ocular tissues.

1.1.1.4. Cell-to-Cell Communication

Communication within small groups of adjacent cells is through gap junctions, as discussed above. Signaling between cells over great distances is achieved through an elaborate system of chemical messengers. There are a large number and variety of hormones secreted by various endocrine cells that act on specific target cells to effect a biological response.

In general, small lipophilic hormones such as corticosteroids diffuse freely through the plasma membrane and interact with cytosolic and nuclear receptors. These substances are believed to alter the pattern of genomic expression; they modify DNA transcription and are also thought to affect the stability of certain mRNA molecules. The responses induced are slow and long-lasting; they probably play a crucial role in growth and differentiation.

Water-soluble hormones such as peptides and various catecholamines act as ligands that bind to cell surface receptors, change the properties of the receptor, and initiate membrane transduction (see Section 1.3). These substances mediate rapid intracellular short-term responses in a matter of minutes or even seconds. They usually modify the activity of enzymes already present in the cell, e.g., adenylate cyclase. They also induce the mobilization of second messengers such as Ca^{2+} from storage sites in the endoplasmic reticulum.

Of the general concepts that have been developed in this broad field, only a few can be discussed in this section. For example, the same receptor may occur in a variety of cells, but the binding of a hormone or other ligand may elicit entirely different responses in different cells. Acetylcholine receptors are found not only on striated muscle cells and in the iris (Chapter 4, Section 4.2.6.1), but also on acinar cells of the pancreas, in the lacrimal gland (Chapter 2, Section 2.1.1), and in the retina (Chapter 7, Section 7.7.2). The release of acetylcholine at neuromuscular junctions triggers contraction, whereas the release of acetylcholine adjacent to an acinar cell results in exocytosis of secretory granules.

A wide variety of chemical messengers, whose number is growing every day, can function as hormones. The most commonly used method for the detection of receptors in target tissues is by the use of radioactive hormones, a methodology of high specificty and sensitivity. The K_m values obtained from such studies are defined as the dissociation constants of the receptor–hormone complex. It is a precise measure of the affinity of the ligand for the receptor.

The activity, or amount, of functional receptor is regulated in part by the level of ligand or hormone to which it is exposed. However, the major mechanism in the regulation of receptor activity is through phosphorylation, which causes its inactivation. This reduction (down-regulation or desensitization) in the response of receptors to hormonal stimulation has been studied extensively for β-adrenergic agonists such as epinephrine (see Section 1.3.4) and is analogous to the deactivation of the light-bleached form of rhodopsin after phosphorylation.

1.1.1.5. Transport of Ions and Low-Molecular-Weight Solutes

That the plasma membrane of virtually all cells is highly selective in controlling the uptake and/or retention of ions, water, nutrients such as glucose, amino acids, and other low-molecular-weight solutes has been recognized for many years. The concentrations of these substances in the cell cytosol are very different from those in the interstitial and extracellular fluids. Gradients exist for nearly all ions and solutes; they are produced and maintained by both passive transport (simple diffusion and facilitated diffusion) and active transport (an energy-requiring process). In both cases the transport of substances across the cell membrane is mediated by transmembrane proteins (called transporters or permeases) that form conduits or channels in the membrane (Darnell *et al.*, 1986; Stryer, 1988). Active transport of ions and small molecules requires a coupled input of free energy, whereas passive transport does not.

Passive Diffusion. Simple diffusion is fairly nonspecific and requires no energy. Dissolved gases such as O_2 and CO_2, as well as small uncharged molecules such as urea and ethanol, traverse the phospholipid bilayer and enter the aqueous phase of the cell cytoplasm at a rate that is directly proportional to their concentration gradients across the membrane. Hydrophobic molecules diffuse across membranes at a rate proportional to their hydrophobicity, a consequence of their ability to dissolve in the lipid bilayer of the cell membrane.

Facilitated diffusion of ions and solutes such as glucose has the following characteristics. (1) The transport of the substance displays saturation kinetics (similar to enzyme-catalyzed reactions). The rate of uptake at low concentrations is proportional to the concentration of the transported molecule. The rate increases until a maximum, called the

V_{max}, is reached. Binding of the molecule to the "enzyme," i.e., the transporter or permease, is defined as the K_m, the concentration of the transported molecule that gives the half-maximal rate of transport. (2) The rate of transport of molecules is much faster than that of simple diffusion. (3) The process is highly specific in that each permease transports only a single substance or, at most, a group of closely related substances.

ATP-Dependent Active Transport and Cotransport. The most ubiquitous transport system in animal cells is the Na^+,K^+-ATPase pump, which transports three Na^+ out of the cell and two K^+ ions into it. This 270-kDa transmembrane protein is an $\alpha_2\beta_2$ tetrapolymer consisting of two large (112-kDa) α subunits and two small (50-kDa) glycosylated β subunits. The latter do not appear to be essential for ATPase or transport activity. The large α subunit contains six to eight transmembrane helices and has catalytic sites for ATPase activity as well as three high-affinity sites for the binding of Na^+ ions on its cytosolic surface. In addition, two high-affinity sites for K^+ ions and one for ouabain are present on its exterior surface. A thiol group close to the catalytic site of Na^+,K^+-ATPase is potentially vulnerable to oxidation and may play a role in cataractogenesis (see Chapter 5, Section 5.6.8.3). The primary structure of the large α subunit deduced from cDNA sequencing is highly conserved, showing 95% homology of amino acid sequences among all species examined, suggesting that they are derived from a common ancestor (Rossier *et al.*, 1987).

Three types of ion-motive ATPases are now recognized (Pedersen and Carafoli, 1987): the "P" type, which forms a covalent phosphorylated aspartate intermediate during ion transport; the "V" type, which is present in storage granules and Golgi vesicles; and the "F" type, such as F_0F_1-ATPase, which is found in the mitochondrial inner membrane (see Section 1.1.1.1). The "V" type of ATPase transports only H^+, whereas the "P" type transports a variety of cations by mechanisms summarized below.

The physiological concentrations of the transported cations, Na^+ and K^+, are such that they are pumped against their electrochemical gradients by the coupled hydrolysis of ATP and a concomitant phosphorylation of an aspartyl residue in the α subunit of Na^+,K^+-ATPase. It is the binding of Na^+ that triggers the phosphorylation, which in turn induces a conformational change in the enzyme, resulting in the outward transport of Na^+. Extracellular K^+ triggers dephosphorylation and moves inward. These two ions are thought to be transported in both directions through channels created by conformational changes in the ATPase molecule resulting from phosphorylation and dephosphorylation of aspartate. The enzyme exists in two reversible conformational states. In the E1 state, three Na^+ ions and ATP bind to their respective sites on the cytoplasmic side of the membrane. In the presence of tightly complexed Mg^{2+}, the ATP is hydrolyzed, and the phosphate forms a high-energy acyl phosphate bond with aspartate. The protein then changes to the E2 conformation, and the three Na^+ ions move out of the cell. Two K^+ ions then bind on the outer surface, the acyl phosphate is hydrolyzed, the enzyme reverts to the E1 conformation, and the two K^+ ions then enter the cell.

Two inhibitors of Na^+,K^+-ATPase are extensively used to locate and identify this enzyme in various cell membranes. One is the cardiac glycoside ouabain, which binds with high affinity to the outer surface of the large α subunits. Its action on the enzyme is that of inhibiting the dephosphorylation reaction and the conversion from E2 to E1 conformation; ouabain must be on the extracellular face of the membrane to exert its inhibitory effect. Another specific inhibitor of Na^+,K^+-ATPase is vanadate ion (V^{5+}). As a structural analogue of the phosphate group, it blocks the conformational changes induced by phosphorylation of the enzyme that are required for ion transport. Vanadate inhibits the ATPase and the pump only when it is inside the cell.

A Ca^{2+}-ATPase pump is present in a wide variety of membranes and has been most extensively investigated in the sarcoplasmic reticulum of skeletal muscle, where it is particularly abundant. Like the Na^+,K^+-ATPase pump, it is a transmembrane protein of high M_r; analyses of the cloned gene for its 110-kDa subunit show that its sequence is similar to that of the α subunit of Na^+,K^+-ATPase (see Stryer, 1988). Moreover, the mechanism for the transport of Ca^{2+} is strikingly similar to that of Na^+,K^+-ATPase. Conformational changes between E2 and E1 driven by the phosphorylation and dephosphorylation of a specific aspartate residue result in the outward transport of two Ca^{2+} for each ATP hydrolyzed.

Cotransport is another important type of membrane transport. It is not directly driven by the hydrolysis of ATP; rather, it utilizes energy stored in the transmembrane gradient of Na^+ generated by the Na^+,K^+-ATPase pump (see review by Wright *et al.*, 1986). Many cells are thus able to accumulate glucose and amino acids against concentration gradients by this energy-coupling mechanism. Glucose and Na^+ bind to the same transporter and enter the cell together. Transport of molecules in the same direction is called symport. As noted above, however, glucose can also enter the cell by facilitated diffusion, and it is probable that both transport mechanisms are operative in most cells. When ions are transported in opposite directions across cell membranes, the process is called antiport. An example of this is the Na^+/H^+ antiport that functions in controlling intracellular pH. Another example is the Na^+/Ca^{2+} antiporter, a transmembrane protein that has recently been characterized in a number of cell membranes including the photoreceptor plasma membrane (see Chapter 7, Section 7.2.1.1). This exchanger transports three Na^+ into the cell for each Ca^{2+} extruded.

1.1.2. Synthesis of Proteins and Nucleic Acids

The central aspect of protein synthesis is the continuous flow of coded information from nuclear DNA (where all the information needed for transcription is stored) to RNA (which carries the encoded information into the cytoplasm, where its translation into proteins takes place). RNA is a long polymer, predominantly single-stranded, composed of four ribonucleotides whose base units are adenine (A), cytosine (C), guanine (G), and uracil (U). The nucleotides are joined by phosphodiester bonds that link the 5′-hydroxyl of one ribose unit to the 3′-hydroxyl of the next. DNA is similar except that its sugar residue is deoxyribose instead of ribose, and thymine (T) replaces uracil (U). In the native state, DNA is a double helix of two antiparallel chains having complementary nucleotide sequences. The chains are held together by hydrogen bonds, which function to maintain a precise apposition of the bases in the two strands: A is paired with T, and G is paired with C. Because of the geometry of the double helix, purines (A and G) are always paired with pyrimidines (T and C).

Three different kinds of RNA are involved in the synthesis of proteins. Messenger RNA (mRNA), a copy of the genetic information encoded in DNA, carries the genetic code in the form of a base sequence that determines the amino acid sequence of the protein to be synthesized. The two other RNAs, tRNA (transfer RNA) and rRNA (ribosomal RNA), combine with specific proteins to form the protein synthesis machinery.

Table 1.1. The Genetic Code[a]

First position (5′ end)	Second position U	C	A	G	Third position (3′ end)
	Phe	Ser	Tyr	Cys	U
U	Phe	Ser	Tyr	Cys	C
	Leu	Ser	[b]	[b]	A
	Leu	Ser	[b]	Trp	G
	Leu	Pro	His	Arg	U
C	Leu	Pro	His	Arg	C
	Leu	Pro	Gln	Arg	A
	Leu	Pro	Gln	Arg	G
	Ile	Thr	Asn	Ser	U
A	Ile	Thr	Asn	Ser	C
	Ile	Thr	Lys	Arg	A’
	Met[c]	Thr	Lys	Arg	G
	Val	Ala	Asp	Gly	U
G	Val	Ala	Asp	Gly	C
	Val	Ala	Glu	Gly	A
	Val	Ala	Glu	Gly	G

[a]Bases are listed as ribonucleotides.
[b]Three termination codons: UAA, UAG, and UGA.
[c]AUG (Met) is the most common initiation signal; AUG also codes for internal methionine positions.

The discovery of how mRNA functions in the genetic code is one of the great scientific achievements of modern biochemistry. In this classical work, synthetic mRNA composed only of poly U residues was shown to yield peptides containing only phenylalanine (Nirenberg and Matthaei, 1961). This led to the elucidation of the genetic code: a “commaless” three-letter code consisting of three nucleotide bases that specifies the linear arrangement of 20 amino acids. Each of these triplets is called a codon; there are 61 (out of a possible 64) that are used for specifying the alignment of these amino acids. Many amino acids therefore have more than one codon, as shown in Table 1.1. Of the 64 possible codons, three (UAA, UAG, and UGA) are used for chain termination.

The processing of primary RNA transcripts from DNA into the functional products, mRNA, tRNA, and rRNA, is a very complex process. Precursor RNA molecules are produced in the nucleus and undergo several biochemical modifications before entering the cytoplasm. A schematic diagram of the overall steps in transcription and translation is shown in Fig. 5.6. RNA synthesis is carried out by three separate RNA polymerases. The precursors to 18 S, 5.8 S, and 28 S rRNAs are produced in the nucleolus of the cell by RNA polymerase I. The precursors to mRNAs and tRNAs are made by RNA polymerases II and III, respectively. Promoters for RNA polymerase II are located at the 5′ side of the start site for transcription and are composed of a diverse set of regulatory sequences. The one closest to the start site, called the TATA box, is present in many eukaryotic genes encoding mRNAs. A CAAT box and a GC-rich region are also common in many genes. Most of the primary products of RNA polymerase II undergo a series of covalent modifications converting them into functional mRNAs. As shown in Fig. 5.6, these consist of (1) addition of a “cap” at the 5′ end of the RNA molecule, a 7-methyl guanylate in 5′→5′ linkage to an initial nucleotide; (2) cleavage by an endonuclease followed by addition of a poly-A tail, a stretch of about 150 to 250 adenylate residues, to the 3′ end of the molecule; and (3) removal of intervening RNA sequences (introns) and the joining of remaining fragments (exons), a process of major importance called RNA splicing.

Even though polymerase II transcripts (the primary RNA transcripts) comprise close to half of the RNA synthesized by the cell, they are unstable and short-lived; hence, nascent RNA and the cytoplasmic mRNA derived from it comprise only a minor fraction of the total RNA present. Fortunately, however, the long chain of poly-A residues at the 3′ end of the molecule provides a simple way of separating the small amounts of poly(A)$^+$mRNA transcripts from the dominant tRNA and rRNA molecules. This is achieved by passing the mixture of newly synthesized RNAs through an affinity column containing either poly-U (polyuridylic acid) or poly-dT (polythymidylic acid). The poly(A)$^+$mRNAs hybridize with the poly-dT or poly-U and are thus retained on the column, whereas the rRNAs and tRNAs pass through in the effluents. The poly(A)$^+$mRNA is then eluted under nonhybridizing conditions such as low ion concentrations.

In the cell, the mature poly(A)$^+$mRNA, together with functional tRNA and rRNA, is quickly transferred to the cytoplasm, probably through the nuclear pores. The translation of the coded message from mRNA into protein is usually considered in three stages: initiation, elongation, and termination. The initiation site in mRNA is the codon AUG; it is here that Met-tRNA$_i^{Met}$ together with initiation factors, GTP, and a small ribosomal subunit form a complex. These are specific sites where the mRNA base sequences just preceding the codons show an unusually high affinity for ribosomes due to the presence of an mRNA "cap" (discussed above), which seals off the 5′ end of the mRNA. A large ribosomal subunit joins the complex, the bound GTP is hydrolyzed, the initiation factors are released, and Met-tRNA$_i^{Met}$ becomes bound to the ribosome.

Elongation begins when a second amino acid, correctly bound to its specific tRNA, is brought into proper position on the ribosome. The codons in the mRNA do not select amino acids directly; rather, another adapter molecule, aminoacyl-tRNA, binds to the codon in the mRNA strand and transfers the attached amino acid to the nascent polypeptide chain on the ribosome. The crucial steps in protein synthesis, i.e., attaching the correct aminoacyl-tRNAs depend on accurate codon–anticodon base pairing. The tRNA molecules are folded into three-dimensional structures having the form of a stem-loop resembling a cloverleaf. One of the loops contains the anticodons containing base pairs that are complementary, but not identical, to the triplets in the mRNA codon.

There are two types of tRNA binding sites: the A site accommodates incoming aminoacyl-tRNA, and the P site contains peptidyl-tRNA residues. The first amino acid, Met-tRNA$_i^{Met}$, is bound to the ribosome at the P site. The next amino acid to be added, an aminoacyl-tRNA, binds at the A position of the ribosome and forms a peptide bond with the carboxyl group of methionine. The peptidyl-tRNA of the second amino acid then leaves the A site and moves to the P site. The energy for this translocation is furnished by the hydrolysis of GTP bound to an activated protein complex. This cycle is repeated for the addition of each amino acid until the peptide fragment is completed.

Termination occurs when the ribosome reaches one of the three termination codons, UGA, UAG, or UAA (Table 1.1). These are recognized by termination factors, proteins that catalyze the hydrolysis of peptidyl-tRNA to tRNA and the newly synthesized peptide chain. The ribosome then separates from the mRNA and, after dividing into its two subunits (small and large), is ready for a new cycle of protein synthesis.

Nucleic acid synthesis, the correct assembly of the nucleotides in DNA or RNA chains, is a simpler process than that of protein synthesis. The elucidation of the copying mechanisms, and its impact on molecular biology, is discussed below in Section 1.1.3. Nucleic acids, both DNA and RNA, are replicated by copying the information coded in the preexisting duplex DNA strand, the template. The information encoded in the base pairs is preserved, but the copy is not identical; rather it is complementary. Individual nucleotides are linked by phosphate groups joining sugars at their 3′ and 5′ positions. The synthesis of new RNA and DNA proceeds from the 5′ to the 3′ end of the molecules; by convention, polynucleotide sequences are read in the 5′→3′ direction. In the synthesis of new nucleotide chains, the phosphate of an incoming nucleotide is attached to the 3′-hydroxyl of the sugar residue, forming a phosphodiester bond. The transcription of DNA to RNA is carried out by RNA polymerases, and in virtually all cells both strands of DNA are used for this transcription.

1.1.3. Molecular Biology, a Brief Introduction

Until rather recently, biochemistry, cell biology, and classical genetics have been considered as three separate scientific disciplines, each with its own methodology, terminology, and technology. However, over the past decade, molecular genetics has emerged as a major and powerful scientific discipline that encompasses all of these fields; genes that code for virtually any protein can be isolated, sequenced, altered, reintroduced into individual cells, and expressed there as proteins. A complete description of this vast field is beyond the scope of this chapter; it is hoped that the brief overview given below of the techniques, definitions, and terminology used in molecular biology will serve as an orientation for a better understanding of the regulatory proteins, enzymes, and structural proteins in ocular tissues. Most of the material that follows on nucleic acid sequencing and recombinant DNA technology is described in great detail in several excellent books (Darnell *et al.*, 1986; Stryer, 1988; Alberts *et al.*, 1989).

1.1.3.1. Sequencing of Nucleic Acids

Most if not all eukaryotic cells, with the exception of sperm and ova, are diploid; they contain two copies of each chromosome. Human cells contain 23 pairs of chromosomes. Nearly all of the chromosomal DNA is associated with five different nuclear proteins termed histones. Sequences of 150 to 180 base pairs along the DNA chains are bound to nucleosomes, which are made up of one molecule of histone H1, and two molecules of histones H2A, H2B, H3, and H4. The haploid human genome contains about 3.0×10^9 base pairs.

Cellular DNA is a two-stranded (double-helix) structure that can be separated (melted) by heating in dilute salt solution and subsequently reassociated (annealed) by lowering the temperature and raising the concentration of ions. Only complementary strands of either DNA or RNA reassociate in a process called molecular hybridization. There are several methods for detecting reassociated hybrid molecules, one of the most commonly used being the nitrocellulose filter-binding assay. The double-stranded DNA is melted and, while still single-stranded, is attached to the nitrocellulose matrix. Molecular hybrids are then formed by exposure to labeled RNA (or DNA) having sequences that are complementary to the bound nucleic acid. Hybridization is induced under appropriate conditions of ionic strength and temperature, and the unhybridized single strands of labeled nucleic acid are washed away. If labeled RNA had been used, subsequent digestion with a ribonuclease removes all unpaired (nonhybridized) RNA. This method allows the detection of as little as one part in 10^6 of hybridized RNA.

Restriction enzymes are endonucleases, bacterial in origin, that are used for partial mapping of DNA. These enzymes, of which several hundred are now known, cleave DNA at four to six base-pair restriction sites, generating precisely reproducible fragments that can be separated by gel electrophoresis. The use of two or more restriction enzymes allows the order of the restriction sites to be determined. Moreover, if the DNA is terminally labeled, it can be partially digested, and many sites can be located with a single restriction enzyme.

The next step is to determine to which restriction fragment(s) a probe (labeled nucleic acid with known features) will hybridize. Southern blots are widely used for this purpose (Southern, 1975). After electrophoresis, the denatured fragments are transferred, or blotted, onto nitrocellulose sheets, which are then exposed to the probe. The DNA fragments on the blot that are complementary to the probe will hybridize with them. If the probe is radiolabeled, detection is by autoradiography, which provides a high degree of sensitivity.

Another technique, termed northern blot analysis, is used to detect specific RNA molecules. In this case, the RNA is denatured and subjected to gel electrophoresis, blotting, and hybridization with a labeled nucleic acid (DNA or RNA that is complementary to the RNA molecule of interest). Autoradiography gives a quantitative estimation of the amount of the probed RNA, and its position in the gel allows an estimation of its size.

1.1.3.2. Recombinant DNA Technology, cDNA, mRNA

The techniques described above for the isolation of pure samples of specific DNA sequences are applicable mainly to simple systems (e.g., a virus). If the DNA of interest is from an animal cell, the complexity of the DNA is so great that until recently it was impossible to isolate a specific fragment of interest from a restriction digest since its concentration was too low.

To circumvent this problem, techniques known collectively as recombinant DNA are now available for isolating pure DNA samples from complex genomes. Two types of DNA preparations are grown in bacterial cells: one of them, a cDNA clone, is a cultured host cell containing a molecule of cDNA (complementary DNA) that was copied from mRNA; the other is a genomic clone, a cultured host cell containing a fragment of genomic DNA. The DNA of interest must be linked to a vector that can replicate independently within a bacterial host. Two of the most widely used vectors are bacteriophage lambda and plasmid, a small circular extrachromosomal DNA molecule that replicates independently in bacterial hosts such as *E. coli*.

A full set of genomic DNA or cDNA clones, called libraries, can then be produced. The cDNA clones are made by the use of reverse transcriptase enzymes. For preparation of libraries of genomic clones, which are copies of DNA from chromosomes, bacteriophage is used as the vector. The resulting genomic library is a collection of recombinant molecules containing all the DNA sequences of a particular tissue or organism. Purification of the gene of interest is carried out after its detection by hybridization. Once the pure DNA is obtained and is available in unlimited quantities, it can be altered (mutated) and reinserted into cells.

These techniques allow investigators to use the gene to identify a protein, or the reverse. It is thus possible, starting with cells from almost any organ or tissue, to isolate their total mRNA and to generate cDNA from it. By selecting and sequencing the cDNA clones, one can deduce protein sequences from the DNA sequence. Peptides can be produced and injected into animals to produce antibodies; the protein to which the antibody was directed can then be isolated. In the reverse process, a pure protein (or enzyme or hormone) is isolated, and a short peptide sequence prepared from it. Oligonucleotides coding for this amino acid sequence can be prepared and used to screen a genomic or cDNA library for the specific DNA sequence.

1.2. EXTRACELLULAR MATRIX

1.2.1. Collagen

1.2.1.1. General Characteristics

Collagen is the major protein of the extracellular matrix, accounting for about 30% of the total body protein in most vertebrates. Among the "classical" types I–III and V, the basic unit, tropocollagen, consists of a long asymmetric molecule approximately 300 nm in length and 1.5 nm in diameter with a M_r of 285,000 to 300,000 (Fig. 1.2). Each fibril is composed of three parallel polypeptide chains, called α chains, coiled in the form of a

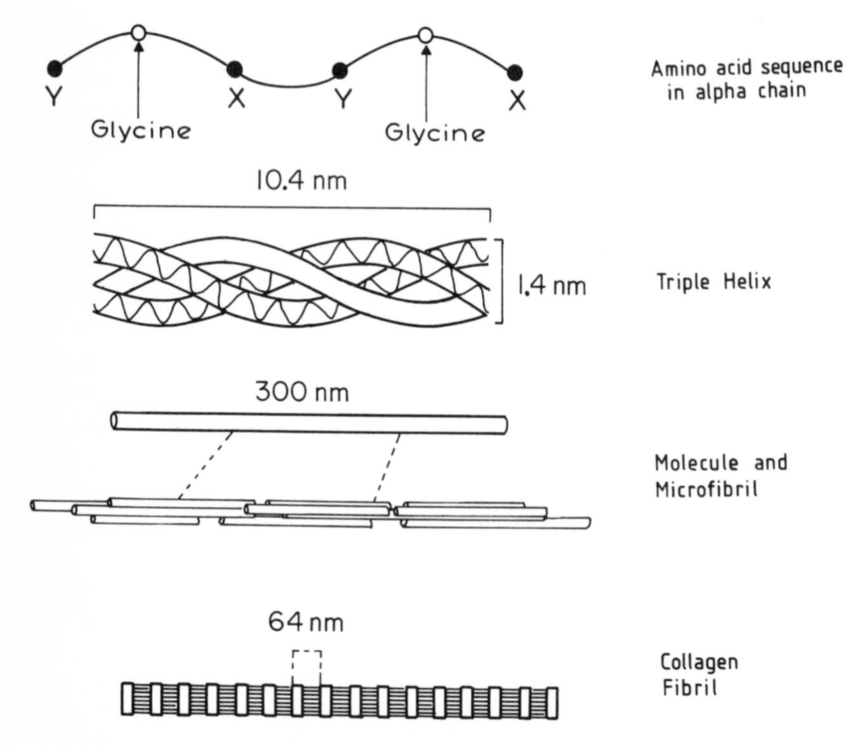

Figure 1.2. Diagrammatic representation of the assembly of a collagen fibril from polypeptide chains. This mode of assembly and quarter staggering apply mainly to types I, II, III, and V collagen.

Table 1.2. Structure and Localization of the Genetically Distinct Collagens[a]

Type	Chain composition	Localization	
		Ocular	Nonocular
I	$[\alpha 1(I)]_2\ \alpha 2(I)$	Cornea	Bone, skin
II	$[\alpha 1(II)]_3$	Vitreous	Cartilage
III	$[\alpha 1(III)]_3$	Cornea	Blood vessels; skin
IV	$[\alpha 1(IV)]_3$, $[\alpha 2(IV)]_3$	Lens capsule, cornea[b]	All basement membranes
V	$\alpha 1(V)$, $\alpha 2(V)$, $\alpha 3(V)$	Cornea, retina	Interstitial tissues
VI	$\alpha 1(VI)$, $\alpha 2(VI)$, $\alpha 3(VI)$	Cornea	Interstitial tissues
VII	$[\alpha 1(VII)]_3$	Cornea[c]	Anchoring fibrils
VIII	$\alpha 1(VIII)$	Descemet's membrane	Some endothelial cells
IX	$\alpha 1(IX)$, $\alpha 2(IX)$, $\alpha 3(IX)$	Vitreous	Cartilage
X	$\alpha 1(X)$		Cartilage
XI[d]	1α, 2α, 3α		Cartilage
XII[e]	$\alpha 1(XII)$		Tendon

[a]Adapted from Martin *et al.* (1985); Lindberg and Pinnell (1982); Pinnell and Murad (1983).
[b]Basal lamina of the epithelium; Descemet's membrane.
[c]Anchoring fibrils of the epithelial basal lamina.
[d]From Mayne and Burgeson (1987).
[e]From Dublet and van der Rest (1987).

triple helix. This conformation is made possible by its unusual amino acid sequence in which nearly every third amino acid is glycine. Two amino acids characteristic of collagen, 4-hydroxyproline and hydroxylysine, together account for 10–15% of the amino acids. The majority of polypeptide chains characterized to date are of two types, α1 and α2; a third type, α3, is present in type V collagen and in several other recently described collagens (Table 1.2). The three α chains are hydrogen-bonded to each other; the hydrogen donors are the peptide amino groups of glycine, and the acceptors are the peptide carboxyl groups of other residues along the chain. The hydroxyl groups of OH-proline also participate in hydrogen bonding, which further stabilizes the triple helix. In collagen types I–III and V, several tropocollagen molecules lie adjacent to one another in a staggered configuration; individual tropocollagen fibrils are displaced from one another by about one-quarter of their length. This quarter-staggering, which is characteristic of the collagens in this group, produces a striated effect, the resulting strands having a periodicity of 64 to 67 nm when viewed by electron microscopy.

This foregoing discussion is the classic view of banded collagens; however, with the development of immunologic probes and ultrastructural visualization following rotary shadowing with platinum or other metals, many previous notions are now being revised (see review by Burgeson, 1988). All collagens, not only the newly described types VI–XII but also the well-known types I–V contain globular domains in addition to triple-helical segments. The globular domains vary in size from 10 to 150 kDa per α chain. They are usually polypeptides but may also be glycoprotein, as in type VI collagen, or proteoglycan, as in type IX collagen. The globular domains are usually but not always located at the C- and N-terminal regions of the molecule. Thus, collagens are extremely diverse structurally. Twelve genetically distinct types are now recognized; their structures and localizations in ocular and nonocular tissues are shown in Table 1.2. A brief description of the newly characterized collagen types VI–XII is given below.

Type VI collagen is widely distributed in the extracellular matrix and is reported to be a major collagen of corneal stroma (Zimmermann *et al.*, 1986). It is characterized by extensive interchain disulfide cross-linking and large nonhelical globular domains (Jander *et al.*, 1984). The triple-helical domain comprises only about one-third of the molecule, with the remainder consisting of globular polypeptides at both the C- and N-terminal ends (Martin *et al.*, 1985; Burgeson, 1988). Other features also point to the unique molecular structure of type VI collagen. It is present in interstitial tissues as microfibrils assembled as dimers and tetramers; the latter are aggregated into multimers or microfilaments with an axial periodicity of about 110 nm (Engvall *et al.*, 1986). Native type VI collagen contains two tightly linked 140- and 260-kDa polypeptides that form an integral part of the molecule. At least one of them, the 140-kDa polypeptide, is a glycoprotein. Type VI collagen isolated from bovine uterus consists of two nonidentical 140-kDa subunits [α1(VI) and α2(VI)]; the α3(VI) chain has been identified as a 200-kDa subunit, which in some tissues may have arisen from a precursor 260-kDa chain (Trueb and Winterhalter, 1986).

Type VII collagen, first characterized in human amnion, is the major structural component of anchoring fibrils (Sakai *et al.*, 1986; Lunstrum *et al.*, 1986). In the eye, it is found in the basal lamina of the corneal epithelium (see Chapter 3, Section 3.2). This collagen contains three identical α chains but is unique in that it is approximately 1.5 times longer than other interstitial collagens (Bentz *et al.*, 1983). Moreover, it is the largest vertebrate collagen characterized to date. The biosynthetic form, procollagen type VII, contains a large and complex C-terminal 450-kDa globular domain, a central helical domain of approximately 510 kDa, and a smaller N-terminal globular region. The type VII collagen extracted from tissues is somewhat smaller and appears to be a dimer, stabilized by disulfide bonding and having a smaller N-terminal domain than the procollagen form (Lunstrum *et al.*, 1987). It has been proposed that the monomeric form in the tissue, a 960-kDa protein, should be called type VII collagen.

Type VIII collagen was originally designated EC (endothelial collagen) after its characterization in aortic and corneal endothelial cells. It has since been found as a synthetic product of other cell types as well. Two major 50-kDa peptides, representing the triple-helical domains of type VIII collagen, have been isolated by limited pepsin digestion from bovine Descemet's membrane (Kapoor *et al.*, 1986). Studies on type VIII collagen synthesized by rabbit corneal endothelial cells suggest a model consisting of three 61-kDa α1 chains arranged in a predominantly helical structure, with nonhelical pepsin-sensitive terminal domains of M_r 14,700 and 4000–5000 (Benya and Padilla, 1986).

Type IX collagen, recently detected in embryonic chick vitreous (see Chapter 6, Section 6.2.1.1), is unusual because it is the only collagen known to date that contains a covalently linked chondroitin and/or dermatan sulfate proteoglycan chain (Vaughan *et al.*, 1985; Bruckner *et al.*, 1985). Also known as proteoglycan lt, type M, or HMW-LMW collagen, type IX collagen is composed of three different disulfide-linked genetically distinct polypeptide chains, α1, α2, and α3 (van der Rest *et al.*, 1985). Data from cDNA and peptide sequencing show that the glycosaminoglycans are attached in the noncollagenous (NC3) domain of the α2(IX) chain and that the attachment region consists of an unusual oligosaccharide sequence of Gly-Ser-Ala-Asp (McCormick *et al.*, 1987).

Type X collagen has been identified in cartilage of many species. The triple helix comprises about two-thirds of the molecule, and, like other recently described collagens, it has nonhelical polypeptide extensions. The globular domains located at the N-terminal end of the molecule have been partially characterized by use of monoclonal antibodies and rotary shadowing techniques (Summers *et al.*, 1988).

The structures and functions of types XI and XII collagens are not as yet completely defined. Type XI collagen is heterotrimeric, and all of the chains show some nonhelical domains (Morris and Bachinger, 1987). The existence of type XII collagen has been inferred from a cDNA clone (pMG377) that was found to encode it. Some of the pepsin-derived fragments isolated from type XII collagen of chick embryo tendon have properties similar to those of type IX collagen (Dublet and van der Rest, 1987).

1.2.1.2. Biosynthesis and Degradation

Nascent collagen polypeptide chains are synthesized on the rough endoplasmic reticulum as procollagens, soluble precursor molecules that are about 50% longer than tropocollagen. The basic units, called pro-α chains, are each distinct gene products (Table 1.2). Each α chain contains a short hydrophobic N-terminal "signal" peptide as well as N- and C-terminal extension peptides at each end of the molecule. These peptide segments have noncollagen amino acid sequences and form globular (nonhelical) domains at both termini. As described below, they are removed after the procollagen is secreted extracellularly. The newly synthesized pro-α chains enter the lumen of the endoplasmic reticulum; the "signal" peptide is released, and three pro-α chains combine to form a triple helix. Certain lysine and proline residues are hydroxylated, and an N-linked high-mannose oligosaccharide is added.

The procollagen polypeptides are then transported to the Golgi complex, where several posttranslational modifications occur: glycosylation, sulfide bond formation, and stabilization of the procollagen triple helix. The molecule is then secreted extracellularly by exocytosis and converted to tropocollagen after proteolytic cleavage of the N- and C-terminal peptides. In the extracellular space, the fibrils undergo covalent cross-linking and extensive polymerization to form the final tropocollagen fibril, which for collagen types I, II, III, and V consists of a quarter-staggered array of four to eight tropocollagen molecules. These long asymmetric molecules are then aggregated into fibers of varying lengths and diameters specific for the wide range of extracellular matrices in which they are found. The final strands of types I, II, III, and V collagens have characteristic cross-striations spaced 68 nm apart. Other newly described collagens do not have this characteristic periodicity, with the possible exception of collagen type XI.

Degradation and remodeling of mammalian collagen is a normal physiological process, although the enzymatic basis of these changes is only partially understood (Harper, 1980). Vertebrate collagenase has been known for several decades, and although mammalian enzymes can be demonstrated in a number of tissues under normal conditions and during remodeling, most of our knowledge comes from studies of collagen degradation under pathological conditions. The most extensively studied model in the eye is in corneal wound healing after an alkali burn (see Chapter 3, Section 3.8.2).

Mammalian collagenase exists in precursor forms in many tissues. The biosynthesis and control of these zymogens, as well as activator systems that convert procollagenases to active enzyme, have been studied in several tissues: bovine gingiva, rat uterus, and synovial and skin fibroblasts (Harper, 1980). The initial attack involves proteolytic scission of each polypeptide of the collagen molecule, producing two reaction products: a large peptide (75% of the molecule) and a smaller one (25% of the collagen fiber). These peptides can then be degraded by a rather wide range of proteolytic enzymes such as neutral proteases, collagenolytic cathepsins, and collagen peptidases.

1.2.2. Basal Lamina (Basement Membrane): Type IV Collagen, Laminin, Fibronectin, and Proteoglycan

The basal lamina is a continuous, moderately dense specialized layer of extracellular matrix that underlies all epithelial cells. It consists of a fibrous network of collagen and glycoconjugates, and functions as a filter or barrier in limiting the access of macromolecules into epithelial cells. Basal lamina also plays an important, though not completely understood, role in differentiation and repair of epithelia by providing a scaffolding over which new cells can migrate.

All components of the basal lamina are synthesized by the epithelial cells that rest on it, and although the composition may vary from tissue to tissue or even from region to region within the same basal lamina, the major components that characterize all basal laminae are type IV collagen and laminin. Other oligosaccharides and/or glycoconjugates identified in a wide variety of basal laminae are fibronectin and proteoglycan (usually heparan sulfate). Fibronectin and cell surface proteoglycan have been described above (Section 1.1.1.2), since they are also major constituents of the plasma membrane glycocalyx. Type IV collagen and laminin, specific components of basal lamina, are discussed here.

Type IV collagen is unusual in many respects. First, it has a high content of hydroxylysine and 3-hydroxyproline, a very low content of alanine, and is almost fully glycosylated. Second, the pro-α chains have extra long extension peptides that, unlike types I–III and V collagen, are not cleaved after secretion. Hence, this collagen does not form typical fibrils with characteristic 68-nm striations; instead they form an open nonfibrillar network in which like ends of the molecules are joined to one another. Mammalian collagenases that degrade other collagens do not act on type IV collagen.

Laminin, the other major component of basal lamina, is a 400-kDa fibrous glycoprotein. It is formed from two subunits of 200 kDa each that are linked by disulfide bonds in the native state. Unlike fibronectin, which binds to all types of collagen, laminin binds specifically to type IV collagen.

1.2.3. Glycosaminoglycans and Proteoglycans

1.2.3.1. General Characteristics

Called mucopolysaccharides since their discovery in the 1930s, the carbohydrate chains of these high-molecular-weight polymers are now more appropriately termed glycosaminoglycans (GAGs). In the native state, with the exception of hyaluronic acid, the glycosaminoglycan chains are covalently linked to a core protein, the complex being termed a proteoglycan. There are several families of core proteins whose size, structure, and amino acid sequences are only now being deduced. With the development of cDNA clones for proteoglycans from different sources, the extent to which different members of these groups constitute distinct gene families should be clarified (Hassel *et al* ., 1986; Ruoslahti, 1988b).

Three types of carbohydrate–protein linkages are now recognized in all proteoglycans (or glycoconjugates).

The *O*-glycosidic linkage of β configuration between xylose and serine has been established for heparin, chondroitin sulfate, dermatan sulfate, and heparan sulfate proteoglycans (Roden, 1980). It is alkali-labile and does not appear to be present in any other type of glycoconjugate.

The N-glycosidic linkage of β configuration between N-acetylglucosamine and the amide group of asparagine is the characteristic carbohydrate–protein linkage in corneal keratan sulfate proteoglycan. This linkage is also found in most glycoproteins, e.g., fetuin, orosomucoid, and ovalbumin. Its formation, both in corneal keratan sulfate proteoglycan and in glycoproteins, is inhibited by tunicamycin, as discussed below.

The *O*-glycosidic linkage of α configuration between N-acetylgalactosamine and serine or threonine is found in cartilage proteoglycan as well as in the majority of mucins.

Proteoglycans are exceptionally diverse, due not only to the many different types of core proteins but also to the heterogeneity of their carbohydrate chains. The brief discussion that follows is based mainly on several recent reviews that describe in detail the biosynthesis, structure, and function of proteoglycans in a wide variety of cells and tissues (Kornfeld and Kornfeld, 1985; Evered and Whelan, 1986; Poole, 1986; Hassell *et al.*, 1986; Elbein, 1987; Hirschberg and Snider, 1987; Ruoslahti, 1988b).

A simple classification and nomenclature for proteoglycans, based on the chemical composition of the GAG chains, is given in Table 1.3. This represents the classical definitions of these macromolecules, but with the exception of hyaluronic acid, their actual chemical composition is far more complex. Hence, these definitions must be considered merely as a guideline for nomenclature; a complete classification scheme of all proteoglycans is a task for the future. The glycosaminoglycan (GAG) chains, at first thought to be homogeneous and of easily defined composition, are in fact an exceedingly diverse group of macromolecules. Most proteoglycans isolated from tissues under mild conditions are found to be polydisperse, having a wide range of molecular weights. Moreover, they are also heterogeneous as shown by highly varying compositions of the carbohydrate chains. These characteristic features of proteoglycans arise during post-

Table 1.3. Major Classes of Glycosaminoglycans

Glycosaminoglycan	Amino sugar[a]	Uronic acid[b]
Hyaluronic acid	Glucosamine	Glucuronic
Chondroitin 4(6)-sulfate	Galactosamine	Glucuronic
Dermatan sulfate	Galactosamine	Iduronic acid (glucuronic acid)
Keratan sulfate	Glucosamine	Galactose[c]
Heparan sulfate	Glucosamine	Glucuronic acid (iduronic acid)
Heparin	Glucosamine	Iduronic acid (glucuronic acid)

[a]Amino sugars of all glycosaminoglycans are acetylated except heparin, which contains mainly sulfated amino sugars.
[b]Carbohydrate chains contain both glucuronic acid and iduronic acid. The minor uronic acid is shown in parentheses.
[c]Keratan sulfate contains galactose in place of uronic acid in the disaccharide repeating unit.

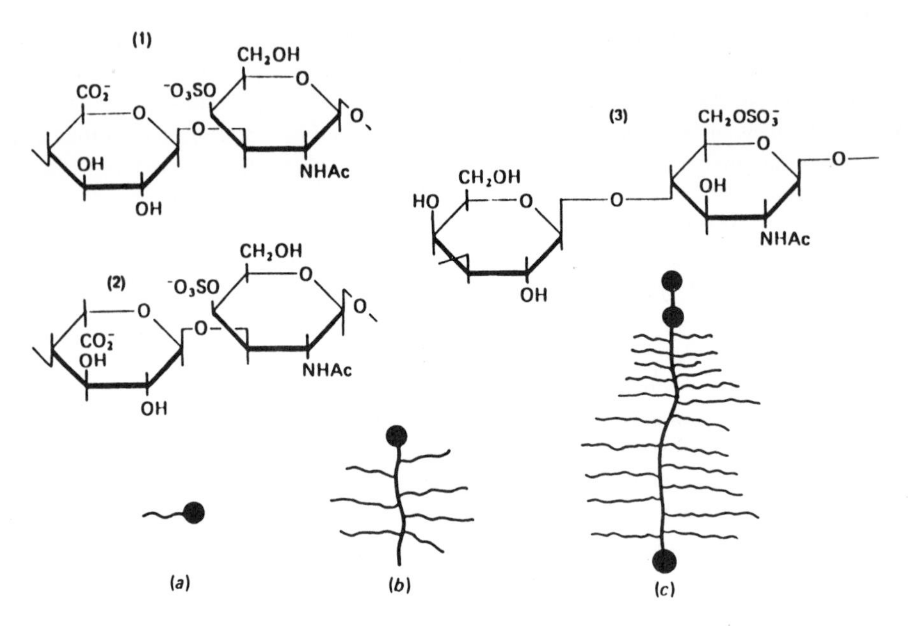

Figure 1.3. Structures of glycosaminoglycans and proteoglycans. The upper panel shows the structures of (1) chondroitin 4-sulfate, (2) dermatan sulfate, and (3) monosulfated keratan sulfate. The lower part of the diagram shows the physical appearance of three representative proteoglycans obtained by rotary shadowing and electron microscopy: (a) small (tadpole) proteoglycan, (b) large proteoglycan, and (c) very large proteoglycan. (From Scott, 1988.) (Reprinted by permission from *Biochem. J.* **252:** 313, © 1988 The Biochemical Society, London.)

translational processing, a complex multienzyme synthesis in the Golgi complex discussed below in Section 1.2.3.2.

The GAG side chains consist of disaccharide repeating units composed of alternating residues of uronic acid (either glucuronic or iduronic) and acetylated amino sugars (either N-acetylglucosamine or N-acetylgalactosamine) (Fig. 1.3). The notable exception to this is keratan sulfate (KS), which contains galactose in place of uronic acid. Each GAG, i.e., the oligosaccharide chain, is named according to the chemical composition of the dominant disaccharide repeating unit as well as the position of the sulfate. For example, chondroitin sulfate (CS) may be substituted at either the 4- or 6-hydroxyl group of N-acetylgalactosamine, giving rise to two different families of GAG molecules, chondroitin 4- or 6-sulfate, respectively. Heparan sulfate and heparin contain N-sulfated (amidosulfate) groups on the amino sugars in addition to O-ester linkages. Hyaluronic acid is the only GAG that contains no sulfate. Dermatan sulfate (DS) and heparan sulfate (HS) have other important structural features; namely, their oligosaccharide side chains contain regions of varying uronic acid composition as well as different degrees of sulfation of the amino sugar residues.

Much of our current understanding of the structure and chemical composition of proteoglycans is the result of highly improved techniques now available for their isolation and characterization (Hascall and Kimura, 1982). Proteoglycans are solubilized by dissociative extraction with the denaturing solvent guanidine HCl at a concentration of 4–6 M. Dialysis against solvent of lower ionic strength (associative conditions) in the presence of protease inhibitors causes the reaggregation of the proteoglycan. Fractionation and purification are achieved by a variety of methods: isopycnic CsCl density gradients, ion-exchange chromatography, rate zonal-velocity centrifugation, and molecular sieve chromatography. Controlled alkali, or alkali borohydride, hydrolysis and selective proteolytic digestion are used to identify the oligosaccharide groupings in the various GAG chains. Peptide mapping and, more recently, cDNA clones are used to identify the protein sequences.

Extensive investigations on cartilage proteoglycan have shown that it belongs to a group of chondroitin sulfate proteoglycans that form aggregates with hyaluronic acid. The core protein has a molecular mass of approximately 200 kDa, and the total M_r of the fully aggregated molecule is approximately 2.5×10^6. This form, which can be visualized *in situ* by electron microscopy, contains about 100 GAG chains of chondroitin sulfate and 30–60 GAG chains of keratan sulfate. The overall structure resembles a "bottle brush" in which the N terminus of each proteoglycan chain is bound to hyaluronic acid, the specific region being termed the hyaluronic acid binding region (HABR). The chains are stabilized by formation of a specific α-*O*-glycosidic linkage to the core protein.

Other major types of proteoglycans include large and small nonaggregating chondroitin sulfate proteoglycans found in skin and bone, respectively, and dermatan sulfate proteoglycans found in a wide variety of tissues, including the corneal stroma (see Chapter 3, Section 3.3.2.1). These proteoglycans vary in size from as low as 100 kDa in cornea to as high as 1×10^6 kDa in fibroblasts. The core protein constitutes about 30% to 60% of the molecule, and the proteoglycans contain varying proportions of chondroitin sulfate or dermatan sulfate chains, depending on the extent to which glucuronic acid is epimerized to iduronic acid during synthesis. Heparan sulfate proteoglycans are found on the cell surface of virtually all cells, where they mediate in attachment and spreading to the substratum through binding to fibronectin. Heparan sulfate also forms an integral part of virtually all basement membranes. However, the cell surface and basement membrane heparan sulfate proteoglycans differ greatly in composition and structure (Poole, 1986). Heparin, found mainly in mast cells, is also present as a proteoglycan.

Poole (1986) has proposed a general classification of proteoglycans based on their size and degree of aggregation. Large aggregating chondroitin sulfate proteoglycans are found in cartilage, aorta, and tendon, and large nonaggregating forms are present in skin and muscle. Dermatan sulfates of high iduronic acid content are found in skin, sclera, and tendon, whereas the dermatan sulfate proteoglycan in other tissues such as cornea has a relatively low iduronic acid content. Other possible classifications have also been suggested by Ruoslahti (1988b), who stressed that owing to their highly heterogeneous nature, it may not be possible in the future to deal with proteoglycans as a single group of substances. Precise classifications will be possible when the results of cloning and sequence analyses of cDNAs encoding the protein cores become available.

1.2.3.2. Biosynthesis and Degradation

Most investigations of proteoglycan biosynthesis have been carried out on cartilage chondrocytes from normal cells or from chondrosarcomas. Similar mechanisms probably exist in all extracellular matrices including the corneal stroma. The diversity of proteoglycans derives not only from the large number of different core proteins but also from the complex posttranslational modifications occurring in the final assembly of the proteoglycan.

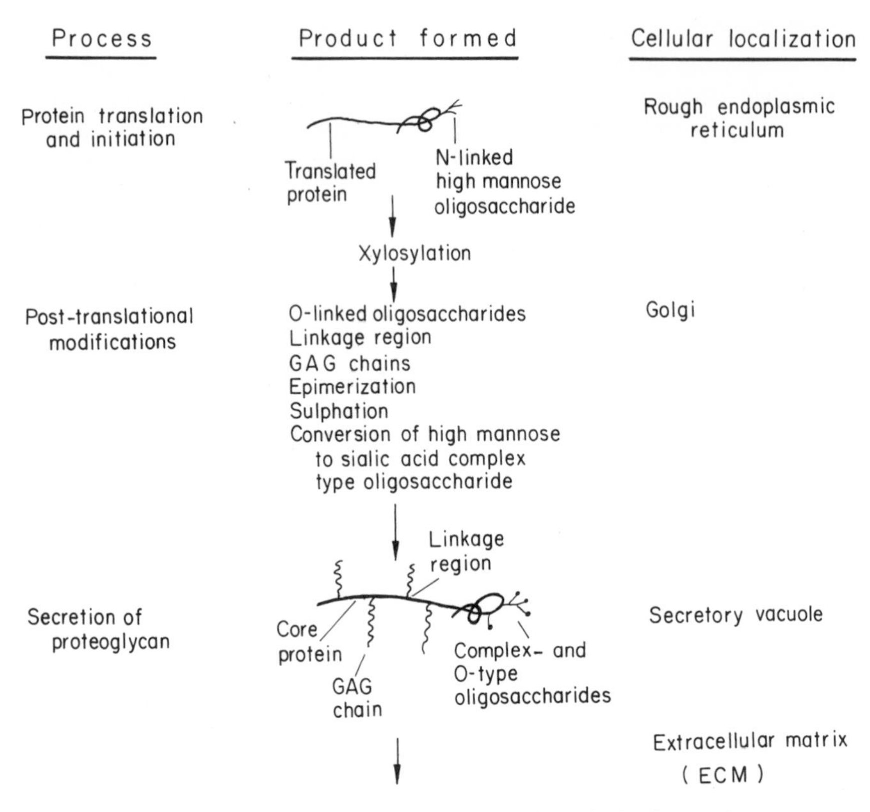

Figure 1.4. Diagram showing major steps in the biosynthesis of proteoglycans.

The functional core protein, i.e., the initial translation product, is assembled on the rough endoplasmic reticulum (Fig. 1.4). It then enters the lumen, where N-linked high-mannose-containing oligosaccharides are attached, a step mediated by dolichol-oligosaccharide intermediates. Dolichol is a very long unsaturated linear lipid composed of about 15–20 isoprene units. It is thought to span the phospholipid bilayer four or five times (Darnell *et al.*, 1986); its polar domain, a terminal phosphate, is the site of attachment of activated oligosaccharides (Stryer, 1988). The transfer of oligosaccharide from the dolichol carrier to asparagine residues in the core protein is a crucial step in the synthesis of the N-glycosidic bond. The antibiotic tunicamycin inhibits the synthesis of this linkage by blocking the formation of the initial dolichol-oligosaccharide intermediate, dolichol-P-P-N-acetylglucosamine. Specifically, tunicamycin inhibits the transfer of N-acetylglucosamine 1-phosphate from UDP-N-acetylglucosamine 1-phosphate to dolichol phosphate. Under normal conditions, in the absence of inhibitors, proteoglycan formation continues with the formation of the N-glycosidic bond by an oligosaccharide–protein transferase. The final synthetic step in the endoplasmic reticulum is xylosylation of the core protein.

Posttranslational modifications are rapid and occur during passage of the N-linked glycopeptide through the cis, medial, and trans Golgi complex. In a complex series of at least six sequential steps, the primary gene product, the core protein containing N-linked high-mannose oligosaccharides, is subjected to a variety of processing reactions resulting in the production of three types of oligosaccharides: high-mannose, hybrid, and complex. These involve mainly a reduction in mannose residues and the addition of specific sugar residues prior to secretion. Monensin, an ionophore that inhibits the transport of glycoproteins through the endoplasmic reticulum and Golgi complex, causes disruptions in the morphological integrity of the Golgi complex and prevents the release of newly synthesized glycoconjugates.

As shown in Fig. 1.4, the synthesis of GAG chains is carried out by specific glycosyltransferases through UDP-glucose and UDP-N-acetylglucosamine intermediates. The sugar moieties are inserted directly and afterward epimerized to the corresponding galactose sugars. Chain elongation is terminated by sulfation of terminal amino sugar residues. These GAG chains are then added to the appropriate amino acid residue of the core protein, either to serine (or threonine) residues for cartilage proteoglycan or to asparagine for corneal keratan sulfate proteoglycan. The O-linked oligosaccharides are also added at this stage; further processing steps include sulfation and epimerization of glucuronic acid to iduronic acid. In the final step before secretion, high-mannose-type oligosaccharides are converted to the complex type of linkage by addition of sialic acid, fucose, and galactose (Poole, 1986). The completed proteoglycan is then packaged into secretory vesicles and secreted into the extracellular matrix.

All extracellular matrices show a regular and normal turnover of proteoglycans. There is strong evidence suggesting extracellular cleavage of the core protein by proteinases, although the enzymes have not been fully characterized. However, with the exception of heparan sulfate, which is cleaved extracellularly by heparanases produced by platelets and activated lymphocytes (Poole, 1986), GAG chains are not extensively degraded extracellularly.

Most proteoglycans, either intact or after partial extracellular proteolytic degradation, are taken into the cell by endocytosis. After fusion with primary lysosomes to form secondary lysosomes, the proteoglycans are degraded by two types of enzymes, proteinases and glycosidases. Core proteins are degraded by the aspartate proteinase, cathepsin D, and by the cysteine proteases, cathepsins H, B, and L. The oligosaccharide (GAG) chains are degraded initially by endohexosaminidase or endoglucuronidase, and the resulting segments are then acted on sequentially by a series of specific exoglycosidases and sulfatases, starting at the nonreducing terminus of the oligosaccharide.

1.3. RECEPTOR-STIMULATED MEMBRANE SIGNAL TRANSDUCTION AND SECOND MESSENGERS

The binding of hormones, neurotransmitters, and other physiological agonists to specific receptors on the cell surface initiates a complex series of events in which G proteins (guanine nucleotide regulatory proteins) play a key role. These heterotrimeric proteins transduce extracellular messages of the endocrine and nervous systems by coupling membrane receptors to two major effector enzymes: adenylate cyclase and phospholipase C. The activation of these enzymes leads to the rapid production of second messengers that in turn generate a wide range of cellular responses, the most important being Ca^{2+} mobilization and the activation of protein kinases A and C. These kinases not only regulate the phosphorylation of key proteins but are themselves substrates for protein phosphorylation.

The three major components of transmembrane signaling in the adenylate cyclase system and in the receptor-stimulated breakdown of polyphosphoinositides (PI) are (1) the signal receivers, i.e., adrenergic and muscarinic cholinergic receptors; (2) the transducing elements, a family of G proteins; and (3) the effectors, enzymes whose activation results in the production of second messengers.

In addition to the G protein-linked adrenergic and muscarinic cholinergic receptors, two other major classes of plasma membrane receptors are now recognized; these differ from one another not only in structure and localization but also in the mechanisms used for transmembrane signaling (Huganir and Greengard, 1987). These receptor classes are (1) the nicotinic acetylcholine receptor (nAChR), a transmitter-dependent ion channel, and (2) the growth factor receptors for platelet-derived growth factor (PDGF), epidermal growth factor (EGF), insulin-like growth factor I (IGF-I), and insulin. The latter two receptors have recently been characterized in retinal photoreceptors (Chapter 7, Section 7.2.2.4). These two classes of receptors, as well as the G protein-linked receptors, all share one important characteristic: their activities are regulated by protein phosphorylation (Sibley *et al.*, 1987; Huganir and Greengard, 1987).

The membrane transduction systems in ocular tissues are similar in most respects to those under investigation in a wide variety of nonocular tissues. A complete description of this rapidly expanding field is beyond the scope of this chapter; hence, only an overview of the major transduction systems now being investigated in ocular tissues is included in this section. The first is the hormone-sensitive adenylate cyclase enzyme complex, which has been extensively investigated for several decades. The second is the receptor-stimulated breakdown of polyphosphoinositides with the production of two second messengers, inositol 1,4,5-trisphosphate (IP_3) and diacylglycerol (DG). A third recently described membrane receptor system involves atrial natriuretic factor (ANF), a peptide hormone that activates membrane-bound guanylate cyclase.

The two major classes of receptors, muscarinic cholinergic and β-adrenergic, have been purified to homogeneity, and the genes encoding them have been cloned and sequenced in many laboratories (see reviews by Dohlman *et al.*, 1987; Kerlavage *et al.*, 1987). Muscarinic receptors (Peralta *et al.*, 1987; Bonner *et al.*, 1987), like β-adrenergic receptors (Dixon *et al.*, 1986; Kobilka *et al.*, 1987) and rhodopsin (Applebury and Hargrave, 1986), form part of a multi-gene family. All of them show regions of homology in 20–28 hydrophobic amino acids that are located in the membrane-spanning hydrophobic segments. A "universal" model is now recognized. These (and possibly other) receptors contain seven transmembrane hydrophobic domains embedded in the phospholipid bilayer. The transmembrane helices are linked through hydrophilic amino acid sequences that form loops in the aqueous media on both the extracellular and the cytosolic sides of the membrane. The C terminus is located on the cytosolic face, while the N terminus is on the exterior surface. The degree of homology varies considerably among various species and tissues, but the most conserved structures are those in the hydrophobic transmembrane regions. In extensively studied muscarinic cholinergic receptors from brain and heart, the hydrophilic loops of the receptors located on the outer face of the membrane have limited domains for the formation of ligand-binding sites (Kerlavage *et al.*, 1987). Hence, as in the case of rhodopsin, their binding sites are probably located within the membrane, in "pockets" or clefts formed by the hydrophobic segments. Phosphorylation sites in β-adrenergic and muscarinic cholinergic receptors are located in serine residues near the C terminus (Dohlman *et al.*, 1987). Ligand-binding activity is regulated by phosphorylation and dephosphorylation of the receptors, now known to be the molecular basis of desensitization phenomena (see Section 1.3.4).

G proteins (guanine nucleotide regulatory proteins), which play the central role in membrane transduction, are heterotrimers composed of subunits designated as α, β, and γ, in decreasing order of molecular mass (Stryer and Bourne, 1986; Gilman, 1987; Graziano and Gilman, 1987; Stryer, 1988). The G proteins couple ligand-activated receptors to membrane-embedded effector proteins, acting as "on–off switches" in a wide variety of cells (Spiegel, 1987). Those that have been characterized to date are G_s and G_i, the stimulatory and inhibitory proteins, respectively, in the adenylate cyclase system; G_t, or transducin, the protein that couples the light activation of rhodopsin to activation of phosphodiesterase in photoreceptors; and G_p, the protein that mediates the ligand-induced breakdown of polyphosphoinositides by phospholipase C (Berridge, 1987; Cockcroft, 1987). Another G protein, G_o, has been isolated from bovine brain, but as yet little is known of its physiological role. The responses of effector molecules can also be induced in G proteins independently of receptor activation by nonhydrolyzable analogues of GTP, and by Fl^- plus Al^{3+}, the active ligand being $[AlF_4]^-$ (Gilman, 1987). These substances, as well as bacterial toxins such as pertussis toxin, interact directly with the α subunit of G proteins to exert their stimulatory effects.

Although the adenylate cyclase system and the receptor-stimulated breakdown of phosphoinositides constitute two separate and distinct membrane transduction systems, there are points of interaction between the two in many cell types (Nishizuka, 1986; Enna and Karbon, 1987). Protein secretion in the lacrimal gland is under the control of two, or possibly even three, such interacting signaling systems (Chapter 2, Section 2.1.1.1). These interactions may either counteract or potentiate cellular responses through common modes of operation (Taylor and Merritt, 1986). The main components in all cases consist of three plasma membrane-associated proteins: receptor, G protein, and effector. Both of the major systems modulate protein phosphorylation by activating protein kinases (A and C), and both of them, directly or indirectly, regulate the concentration of cytosolic Ca^{2+}, but by different mechanisms, as described below. There is also evidence that the two systems may be coupled through their G proteins to the activation of more than one effector by a single receptor subtype. This was shown by the simultaneous inhibition of adenylate cyclase and activation of phosphoinositide hydrolysis by a recombinant muscarinic cholinergic receptor (Ashkenazi *et al.*, 1987).

1.3.1. Adenylate Cyclase System

Adenylate cyclase is a multicomponent enzyme system whose three distinct interacting protein components were described several years ago by Rodbell (1980). They are all localized in the cell membrane, and the total receptor–enzyme complex extends from the outer cell surface to the inner cell cytoplasm. A schematic representation of the dual regulation of adenylate cyclase activity, based in part on several reviews (Gilman, 1984; Lefkowitz *et al.*, 1983; Levitzki, 1987), is shown in Fig. 1.5.

1.3.1.1. Hormone-Stimulated Adrenergic Receptors

The two types of receptors, stimulatory and inhibitory, are intrinsic membrane proteins whose hydrophobic regions form seven transmembrane domains. As discussed

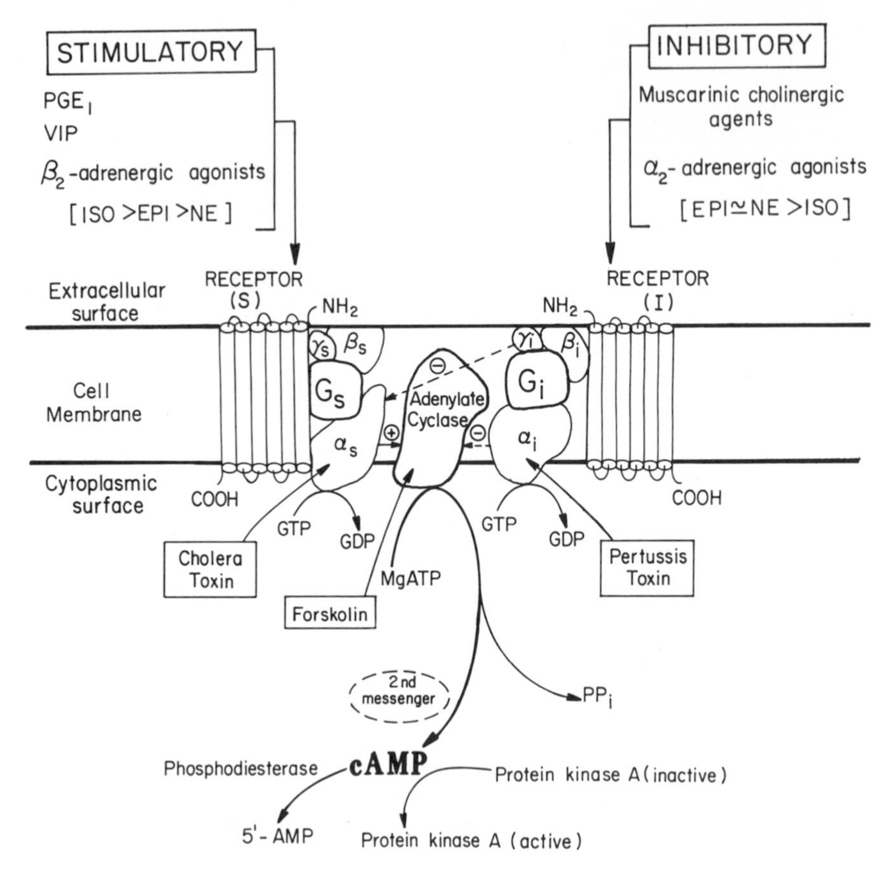

Figure 1.5. Dual control of the hormone-sensitive adenylate cyclase system. Drugs, hormones, and physiological agonists bind to membrane-spanning receptors that are coupled to specific G proteins, G_s or G_i. Transduction of one of these signals to the catalytic site of adenylate cyclase on the cytoplasmic surface of the membrane results in either its activation or its inhibition. Certain agents such as cholera toxin or pertussis toxin cause ribosylation of the stimulatory or inhibitory α subunits, respectively, of the G proteins. Forskolin activates adenylate cyclase directly without intervention of the regulatory G protein, G_s. Further details are given in the text. Abbreviations used: PGE_1, prostaglandin E_1; VIP, vasoactive intestinal peptide; ISO, isoproterenol; EPI, epinephrine; NE, norepinephrine.

above, these receptors show considerable homology and are part of a multi-gene family. Some of the hormone ligands that bind to the stimulatory β_2 receptor are shown in Fig. 1.5. Others include ACTH and glucagon. The functional activity of these receptors is modulated by phosphorylation, a key process in desensitization (see review by Benovic *et al.*, 1988 and Section 1.3.4). Also shown in Fig. 1.5 are some of the hormones, α_2-adrenergic agents, and other physiological agonists that bind to the inhibitory receptor. Each of these ligand–receptor complexes interacts with the closely associated α subunits of the G proteins, giving either an activating or an inhibitory signal to the enzyme adenylate cyclase.

1.3.1.2. G Proteins

These guanine nucleotide-binding proteins were originally called N or G/F proteins (Stiles *et al.*, 1984), but the term G protein is now universally accepted to define the heterotrimeric transducing proteins. The α subunits of the stimulatory (G_s) and inhibitory (G_i) proteins of the adenylate cyclase system have molecular masses of 46 and 41 kDa, respectively (Graziano and Gilman, 1987; Gilman, 1987); 52-kDa stimulatory α subunits have also been detected in some tissues. In all cases they show over 95% homology, but small regions of unique sequences confer functional differences in their ability to activate or inhibit adenylate cyclase. All α subunits have high-affinity binding sites for GTP, and they also show GTPase activity. In addition they have binding sites for cholera toxin and pertussis toxin (Fig. 1.5). The β and γ subunits form a tight complex that dissociates as a single unit from the α subunit in the presence of activated receptor. The β subunits have a molecular mass of 35 or 36 kDa, whereas the γ subunits are small 5-kDa peptides.

Molecular interactions between the receptor–ligand complex, the G protein subunits, and the enzyme in signal transduction have been studied extensively and are summarized in several reviews and texts (Graziano and Gilman, 1987; Gilman, 1987; Levitsky, 1987; Stryer, 1988). Adenylate cyclase is activated by the following sequence of events. Following binding of an agonist, the receptor–ligand complex interacts with the G protein in its inactive or resting state. In this state, GDP is bound to the α subunit. After binding, GDP is displaced from the α subunit and is replaced by GTP. The binding of GTP causes the dissociation of the G protein into its α and βγ subunits; there is a concomitant decrease in the affinity of the ligand for the receptor. The α subunit of G_s (with its bound GTP) is thought to undergo a conformational change that causes it to bind to and activate the adenylate cyclase. The activation is terminated by the hydrolysis of bound GTP to GDP. This causes reassociation of the α subunits with the βγ subunits, and the complexing of GDP to the α subunits returns G_s to its inactive state.

Mechanisms of inhibition of adenylate cyclase by G_i are not entirely clear (Fig. 1.5). The GTP–α subunit complex of G_i is a relatively weak inhibitor of adenylate cyclase. As an alternative, it has been suggested that free βγ subunits of G_i interact with the α subunits of G_s, and the formation of an αβγ complex "ties up" the α subunit of G_s making it unavailable for activation of adenylate cyclase (Gilman, 1984, 1987; Smigel *et al.*, 1985; Levitzki, 1987). This model has been modified by other investigators and should be considered somewhat hypothetical.

The α subunits of both G_s and G_i are specific substrates for NAD^+-dependent ADP-ribosylation catalyzed by bacterial toxins. Interaction with cholera toxin or pertussis toxin causes ribosylation of the α subunit of G_s or G_i, respectively. This ribosylation causes irreversible functional modifications in the regulatory proteins; in the former case, adenylate cyclase is "permanently" activated, and in the latter, "permanently" inhibited.

1.3.1.3. Adenylate Cyclase

This hydrophobic membrane-bound enzyme is a glycoprotein of M_r approximately 150,000 (Pfeuffer *et al.*, 1985; Smigel, 1986). Its catalytic site is located on the cytoplasmic face of the plasma membrane, and there is evidence suggesting that it may be a transmembrane protein with hydrophilic groups located on the outer surface of the plasma membrane. The enzyme catalyzes the conversion of MgATP to cAMP. As described above, the activity of adenylate cyclase is regulated by the G proteins after their interaction with stimulatory or inhibitory receptors. Isolated preparations of highly purified adenylate cyclase are generally inactive with their physiological substrate, MgATP, if G_s regulatory protein is not present.

The only agent known at present that can activate adenylate cyclase directly is forskolin, a diterpene that has been isolated from the root of an Indian medicinal plant, *Coleus forskohlii*. (Seamon and Daly, 1981). It was first discovered as a potent vasodilator and smooth muscle relaxant, and its structural formula was established as the 7-acetoxy-1α,6β,9α-trihydroxy derivative of manoyl oxide. Forskolin activates adenylate cyclase in intact tissues as well as in cell-free preparations in the absence of G_s, suggesting that it acts directly at the catalytic site of the enzyme. The activation is rapid and completely reversible; studies using radiolabeled terpene have revealed both low-affinity and high-affinity forskolin binding sites in membranes isolated from brain, platelets, liver, adipocytes, and other tissues (Seamon and Wetzel, 1984; Seamon and Daly, 1985; Nelson and Seamon, 1986; Shi *et al.*, 1986). These binding sites are closely associated with an activated complex of adenylate cyclase and the α subunit of G_s, possibly at the interface between these two proteins. Even though adenylate cyclase does not require G_s for its activation in the presence of forskolin, there is considerable potentiation of activity when both forskolin and a stimulatory hormone such as epinephrine are present together. This has also been observed in the ciliary processes (see Chapter 4, Section 4.2.6.4.).

The importance of cAMP in cellular physiology is its role as a second messenger in the activation of intracellular protein kinases (see review by Taylor, 1987). The most extensively studied of this group of enzymes is cAMP-dependent protein kinase (protein kinase A), which is composed of two pairs of subunits: two regulatory and two catalytic. The cAMP binds to the regulatory subunits and causes their dissociation from the catalytic subunits. The kinase activity of the catalytic subunits is then activated, and in this form the enzyme transfers the γ-phosphate of MgATP to serine, threonine, or tyrosine residues of a variety of substrate proteins. The production of cAMP and the activation of cAMP-dependent protein kinase(s) trigger a cascade of chemical reactions, resulting in a severalfold amplification of the initiating signal. Among the most extensively studied systems controlled by β-adrenergic stimulation of adenylate cyclase are the synthesis and degradation of liver glycogen and the regulation of several key enzymes in the Embden–Meyerhof glycolytic pathway. In ocular tissues, β-adrenergic receptors that modulate adenylate cyclase activity and hence regulate the production of cAMP are present in the lacrimal gland (Chapter 2, Section 2.1.1), corneal epithelium (Chapter 3, Section 3.1.5), ciliary processes (Chapter 4, Section 4.2.6.2) and retinal pigment epithelium (Chapter 7, Section 7.4.4.5).

The intracellular level of cAMP is also regulated by enzymes that control its hydrolysis. The principal ones are phosphodiesterases, many of which are activated by another second messenger, Ca^{2+}. One of the hormonal systems controlling the concentration of cytosolic Ca^{2+} is the receptor-stimulated breakdown of phosphatidylinositol 4,5-bisphosphate. Many of the regulatory functions of cytosolic Ca^{2+} are controlled by calmodulin, a ubiquitous cytosolic protein that forms a that Ca^{2+}/calmodulin complex binds to cAMP phosphodiesterase and activates it. Thus, β-adrenergic stimulation and activation of adenylate cyclase lead to a complex series of intracellular reactions containing finely tuned amplification and feedback mechanisms that assure rapid intracellular responses to external stimuli.

1.3.2. Polyphosphoinositide Breakdown

The agonist-stimulated breakdown of phosphatidylinositol 4,5-bisphosphate (PIP_2) is a major signal system for the generation of two intracellular second messengers: inositol

1,4,5-trisphosphate (IP_3) and diacylglycerol (DG) (Berridge and Irvine, 1984). It appears to be a central control mechanism in a wide variety of cells throughout the animal and plant kingdom. This membrane transduction system is also present in many ocular tissues: lacrimal gland (Chapter 2, Section 2.1.1), corneal epithelium (Chapter 3, Section 3.1.4), iris smooth muscle (Chapter 4, Section 4.2.6), ciliary processes (Chapter 4, Section 4.2.6.5), retina (Chapter 7, Sections 7.1.2.2, 7.2.2.2, and 7.8.3), and retinal pigment epithelium (Chapter 7, Section 7.4.4.5). The brief description given below of polyphosphoinositide (PI) turnover is based on several extensive review articles and texts (Berridge and Irvine, 1984; Takai *et al.*, 1984; Hirasawa and Nishizuka, 1985; Hokin, 1985; Abdel-Latif, 1986; Berridge, 1987; Majerus *et al.*, 1988). An overview of the phosphoinositide cascade (Stryer, 1988) is shown in Fig. 1.6.

Phosphatidylinositol (PI) generally comprises only a small fraction of the total phospholipids in cell membranes. It is known to be metabolized by pathways common to all phospholipids (e.g., hydrolysis to lyso derivatives by phospholipase A_1 or A_2). However, in addition, a small portion (about 10–20%) can be phosphorylated by ATP in two successive enzymatic steps by specific PI kinases, leading to the formation of PIP_2, the key intermediate in phosphoinositide-generated second messengers.

Over 25 receptor types are known to be associated with the phosphoinositide effect (Hokin, 1985), and the stimuli (or agonists) that elicit PIP_2 turnover consist of a wide variety of substances: hormones, growth factors, neurotransmitters, and many others. A few examples of these receptors (with the agonist given in parentheses) are: muscarinic cholinergic (acetylcholine), α_1-adrenergic (norepinephrine), serotonergic, dopaminergic, and peptidergic (substance P). Light also serves as an agonist in PI activation in vertebrate (Chapter 7, Section 7.2.2.2) and invertebrate photoreceptors (Chapter 7, Section 7.8.3). These agents activate membrane-bound phospholipase C (also called PIP_2·phosphodiesterase or phosphoinositidase), resulting in the production of two second messengers: inositol 1,4,5-trisphosphate (IP_3) and diacylglycerol (DG). Cytosolic forms of phospholipase C that may also be related to the phosphoinositide effect have been characterized in brain and may be present in other tissues (Ryu *et al.*, 1987a,b).

The signal transduction is mediated by a G protein named G_p to distinguish it from G_s and G_i that modulate the adenylate cyclase system. The G_p protein is not as well characterized as other G proteins, and there are many discrepancies in experimental data (Gilman, 1987; Spiegel, 1987; Berridge, 1987). In some cells, e.g., human neutrophils and turkey erythrocyte membranes, guanine nucleotides potentiate the agonist stimulation of phospholipase C (Harden *et al.*, 1987); moreover, in some but not all cell types, pertussis toxin blocks the agonist-stimulated breakdown of phosphoinositides (Paris and Pouyssegur, 1987). Like other G proteins, G_p is activated by NaF and $AlCl_3$ through formation of an $[AlF_4]^-$ complex (Berridge, 1987). There are, however, still some controversial aspects regarding the phospholipase C-coupled G protein, since it has been studied under a variety of experimental conditions in many different cell types and using a wide range of agonists and inhibitors. Some of the discrepant data may be explained by the fact that muscarinic cholinergic agents have opposite effects on the adenylate cyclase system and on PIP_2 breakdown. In the former case, these agonists inhibit adenylate cyclase (see Fig. 1.5), whereas in the phosphoinositide cascade, muscarinic agonists stimulate PIP_2 breakdown (Fig. 1.6). It has been suggested that there may be different G_p proteins in different tissues, or there may be an inhibitory G protein involved in phosphoinositide breakdown (Berridge, 1987).

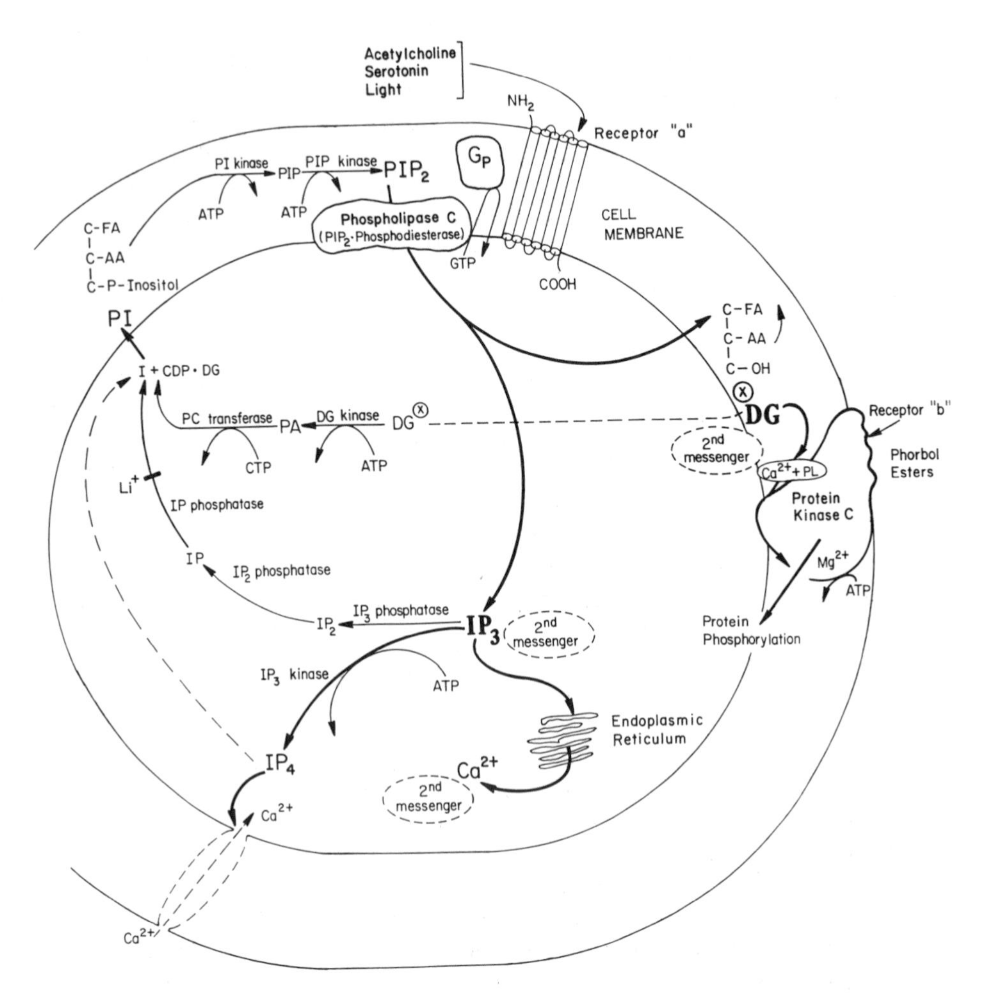

Figure 1.6. Polyphosphoinositide (PI) turnover. Diagrammatic representation of receptor-coupled activation of phospholipase C and formation of phosphoinositide-derived second messengers resulting from the hydrolysis of PIP_2. Only three of the many known agonists that activate this system are shown. The receptor ("a") has been well characterized (see text) as an integral membrane protein containing seven helical membrane-spanning domains. Transduction is mediated by a G protein, G_p, which activates phospholipase C and triggers the breakdown of phosphatidylinositol 4,5-bisphosphate (PIP_2) to inositol 1,4,5-trisphosphate (IP_3) and diacylglycerol (DG). Generation of these messengers leads to a variety of cellular responses, two of the major ones being the mobilization of Ca^{2+} by IP_3 and the activation of protein kinase C by DG in the presence of Ca^{2+} and phospholipid. Also shown is the activation of this kinase at a different receptor site ("b") by phorbol esters. This receptor has not been isolated or characterized but may be protein kinase C itself. Abbreviations used: PI, phosphatidylinositol; PIP, phosphatidylinositol 4-phosphate; PIP_2, phosphatidylinositol 4,5-bisphosphate; IP, inositol 1-phosphate; IP_2, inositol 1,4-bisphosphate; IP_3, inositol 1,4,5-trisphosphate; IP_4, inositol 1,3,4,5-tetrakisphosphate; I, inositol; DG, diacylglycerol; PA, phosphatidic acid; CDP-DG, cytidine diphosphodiacylglycerol; FA, fatty acid; AA, arachidonic acid; PIP_2-PDE, phosphoinositide phosphodiesterase (phospholipase C, phosphoinositidase); G_p, G protein that activates phospholipase C; PS, phosphatidylserine; PC transferase, CTP phosphatidate cytidyl transferase; DG kinase, diacylglycerol kinase.

Whatever the mechanism and role of G_p in activating phospholipase C, hydrolysis of PIP_2 results in the rapid (seconds or minutes) generation of two second messengers, IP_3, a soluble phosphorylated derivative of inositol, which diffuses into the cytosol, and diacylglycerol, which remains in the plasma membrane (Fig. 1.6). The IP_3 causes a rapid mobilization of Ca^{2+} from bound sites in the endoplasmic reticulum; receptors for IP_3 have been detected on the surface of the endoplasmic reticulum of certain cells. The release of Ca^{2+} as another second messenger leads to biological and physiological effects of major importance to the cell (Abdel-Latif, 1986). The brief initial response is modulated by calmodulin, whereas protein kinase C, discussed below, is responsible for the longer sustained effect (Rasmussen, 1986a,b). Inositol 1,4,5-trisphosphate (IP_3) may also be phosphorylated to inositol 1,3,4,5-tetrakisphosphate (IP_4) by a specific kinase first detected in brain (Irvine et al., 1986) but now known to have ubiquitous distribution. In contrast to the transient rise in intracellular Ca^{2+} triggered by IP_3, IP_4 is thought to mediate a prolonged elevation of intracellular Ca^{2+} by opening Ca^{2+} gates or ion channels in the plasma membrane and allowing an influx of extracellular Ca^{2+} (Houslay, 1987). The IP_4 is then hydrolyzed to inositol 1,3,4-trisphosphate, an isomer that has a slower turnover than inositol 1,4,5-trisphosphate. This isomer is also converted to inositol, but by a different pathway from that established for inositol 1,4,5-trisphosphate. Thus, another branching point involving two metabolic pathways of inositol 1,4,5-trisphosphate has been established; moreover, higher phosphorylated forms of inositol (IP_5 and IP_6) have been found in animal cells, suggesting the presence of a series of biologically important inositol kinases in many tissues (Irvine *et al.*, 1986; Berridge, 1987). Highly specific kinases and phosphatases are in fact involved in the intracellular metabolism of inositol phosphates (see review by Majerus *et al.*, 1988). Many of these enzymes have now been characterized, and the metabolic interrelationships among inositol and its phosphorylated derivatives are undoubtedly far more complex than those depicted in Fig. 1.6.

The other second messenger released by the receptor-stimulated hydrolysis of PIP_2 is diacylglycerol (DG), which is directly involved in the activation of the phospholipid- and Ca^{2+}-dependent enzyme, protein kinase C (Fig. 1.6). Activation of this enzyme is also coupled to the tumor-producing phorbol ester receptor (Ashendel, 1985). Protein kinase C is a ubiquitous enzyme with a multitude of biological activities (Nishizuka, 1986). There is a family of protein kinase C genes coding proteins with a high degree of homology but having structural differences that give rise to different biological responses (see review by Kikkawa *et al.*, 1989). Two or three distinct types of protein kinase C have been characterized in rat (Woodgett and Hunter, 1987) and rabbit (Jaken and Kiley, 1987) brain. The various forms of this enzyme are, however, ubiquitous, and several subspecies have been cloned. In its initial cellular response, protein kinase C is thought to act synergistically with Ca^{2+} in provoking a positive response; this may be immediately followed by a negative feedback response (down-regulation) effected by the phosphorylation of its own membrane receptor as well as others (see review by Kikkawa *et al.*, 1989). Most importantly, a network of signaling systems has been proposed (Nishizuka, 1986), suggesting that the two major classes of membrane receptors for signal transduction may counteract one another in bidirectional control systems and potentiate one another in unidirectional systems. Thus, protein kinase C is thought to play a pivotal role in modulating a myriad of essential biological activities. This enzyme, like protein kinase A, contains both regulatory and catalytic domains. Both can be autophosphorylated, and in this state the enzyme is fully activated (Mochly-Rosen and Koshland, 1987).

A second major role for DG involves prostaglandin (eicosanoid) biosynthesis. Arachidonic acid, usually located at the *sn*-2 position of the glycerol backbone, is liberated by the action of nonspecific lipases and/or specific ones. The arachidonic acid released is rapidly oxidized through cyclooxygenase, lipoxygenase, and cytochrome P450 pathways to a series of eicosanoids, which themselves have a broad range of biological activities.

Phosphatidylinositol (PI) is rapidly regenerated by two converging enzymatic pathways shown in Fig. 1.6. Inositol 1,4,5-trisphosphate is converted to inositol (I) by the consecutive action of three specific IP phosphatases. It may also be converted through the 1,3,4 isomer described above, although these steps have not been clearly established. Diacylglycerol (DG) is phosphorylated by a DG kinase to produce phosphatidic acid (PA). The final step in the regeneration of phosphatidylinositol (PI) occurs in the plasma membrane.

The two pathways in the phosphoinositide cascade resulting from hydrolysis of PIP_2 are not completely independent of one another; newer concepts suggest that there may in fact be considerable interaction and cooperation between the two messengers (Berridge, 1987). This may occur at several intracellular levels and involves subtle feedback controls in which calcium plays a pivotal role.

Finally, a general point regarding nomenclature concerns the use of inositol throughout the foregoing discussion. The active species of *myo*-inositol trisphosphate is known systematically as *D-myo*-inositol 1,4,5-trisphosphate (Parthasarathy and Eisenberg, 1986). The other inositol derivatives formed in the conversion of this substance to *myo*-inositol are of the same configuration and conformation. The stereochemistry of the newly identified 1,3,4,5-tetrakisphosphate has not yet been established. While some authors use the *myo* prefix, others do not. Nevertheless, until proven otherwise, it is implied that all derivatives of inositol generated during the ligand-induced breakdown of polyphosphoinositides are of the D-*myo* configuration.

1.3.3. Atrial Natriuretic Factor and Guanylate Cyclase

Atrial natriuretic factor (ANF) is one of a group of biologically active polypeptide hormones secreted by atrial cardiac myocytes in response to fluid overload. They have potent natriuretic and diuretic effects; in addition, they induce generalized vasodilation and have powerful vasorelaxant activity. Crude extracts of rat atria contain several forms of ANF, with molecular masses ranging from about 4 to 19 kDa. One of them, an ANF peptide designated cardionatrin I, was purified, sequenced, and characterized as a disulfide-bonded 28-amino acid peptide (Flynn *et al.*, 1983, 1985). Its structure is shown in Fig. 1.7. Athough it was at first thought to be synthesized solely in atrial granules (hence the name), the identification of ANF gene transcripts in the hypothalamus and other areas of the central nervous system of the rat suggest *in situ* synthesis in these, and possibly other, sites in the body (Gardner *et al.*, 1987).

Sequence analyses of natrial peptides isolated from rat plasma show that the major fraction in the circulation and the peptide with the highest biological activity is the 28-amino acid peptide cardionatrin I, or α-rat ANP (Fig. 1.7). Designated Ser-Leu-Arg-Arg-APIII (Schwartz *et al.*, 1985), it is present in human plasma at concentrations of about 25 to 100 pg/ml. The half-life of cardionatrin I in the circulation is of the order of minutes, and binding sites to specific receptors have been demonstrated in a number of tissues (discussed below) as well as in rabbit ciliary processes (Chapter 4, Section 4.2.8).

```
H-Ser-Leu-Arg-Arg-Ser-Ser-Cys-Phe-Gly-Gly-Arg-Ile-Asp-Arg-Ile-Gly
                           |
                           S——S
                              |
    Ala-Gln-Ser-Gly-Leu-Gly-Cys-Asn-Ser-Phe-(Arg)-Tyr-OH
```

Figure 1.7. Amino acid sequence of ANF peptide, cardionatrin I (also called α rat ANP).

High-affinity binding sites for a synthetic ANF (a 26-amino acid fragment, 8–33 C-terminal acid) are present in kidney and vascular endothelium (Napier *et al.*, 1984) as well as other tissues. Many of them have been purified to homogeneity and shown to consist of two identical 60- to 70-kDa subunits, probably linked in the native state by disulfide bonds (Schenk *et al.*, 1987; Shimonaka *et al.*, 1987). A more precise characterization of the ANF receptor has been obtained by translating kidney and lung poly(A)$^+$mRNA in a rabbit reticulocyte lysate system (Uchida *et al.*, 1987). The primary translation product is a 58-kDa protein that then undergoes posttranslational glycosylation to form a mature 70-kDa subunit. The native state in the membrane is that of a disulfide dimer containing two 70-kDa subunits. Of particular interest is recent evidence suggesting that the receptor is (covalently) linked to guanylate cyclase in the form of a 180-kDa protein complex (Paul *et al.*, 1987; Takayanagi *et al.*, 1987).

The activation of guanylate cyclase by ANF was demonstrated shortly after its isolation and characterization. Incubation of ANF with aortic ring segments (Winquist *et al.*, 1984) or with kidney homogenates (Waldman *et al.*, 1984) causes a rapid (within 2.5 min) and significant increase in cGMP levels in these tissue preparations. Homogenates from several other rat tissues showed a similar effect. Guanylate cyclase activity in rabbit cerebral microvessels is stimulated by a factor of more than 200% in the presence of rat ANP (rANP, the 1–28 Ser-Leu-Arg-Arg-atriopeptin III) (Steardo and Nathanson, 1987). Choroid plexus, as well as epithelial cells isolated from it, shows similar responses to the intact 1–28 rANP peptide. Moreover, there is a 30-fold increase in cGMP in the presence of phosphodiesterase inhibitors, theophylline and IBMX.

The synthetic ANF specifically activates the particulate form of guanylate cyclase in both a concentration- and time-dependent manner (Winquist *et al.*, 1984; Waldman *et al.*, 1984). It has also been shown that ANF concomitantly inhibits the production of cAMP, possibly by negative coupling to adenylate cyclase through G_i, the inhibitory G protein of the adenylate cyclase complex (Anand-Srivastava *et al.*, 1987). The mechanisms controlling the diverse responses of this unusual receptor system remain to be elucidated.

1.3.4. Desensitization

Desensitization, also known as adrenergic subsensitivity or down-regulation, is a general phenomenon describing the decrease in drug responsiveness after long-term treatment with epinephrine or other adrenergic drugs. It is not limited to the eye but is, in fact, ubiquitous in biological systems. The biochemical basis of desensitization involves the covalent modification of membrane receptors by phosphorylation, the primary mechanism in the regulation of receptor function (Sibley and Lefkowitz, 1985; Sibley *et al.*, 1987; Huganir and Greengard, 1987; Benovic *et al.*, 1988). Known for many years in the retina (Chapter 7, Section 7.8.2), the phosphorylation of bleached rhodopsin by rhodopsin kinase decreases the functional coupling of rhodopsin to its G protein, transducin. This system shows striking homology to the recently discovered β-adrenergic receptor kinase (βARK), an 80-kDa protein kinase that specifically and preferentially phosphorylates serine residues of agonist-occupied β-adrenergic receptors in a variety of cell types (Benovic *et al.*, 1986, 1987, 1988). Receptors coupled to PIP_2 breakdown (α_1-adrenergic and muscarinic cholinergic) also show agonist-induced desensitization, the mechanism in some cells being phosphorylation catalyzed by protein kinase C (Leeb-Lundberg *et al.*, 1985, 1987; Bouvier *et al.*, 1987). However, in other systems such as chick cardiac tissue, protein kinase C does not appear to play a direct role in desensitization (Kwatra *et al.*, 1987).

Two major types of desensitization are now recognized: homologous (or autoregulatory) and heterologous (see reviews by Sibley and Lefkowitz, 1985; Huganir and Greengard, 1987; Sibley *et al.*, 1987; Benovic *et al.*, 1988). The former are regulated by autophosphorylation and are considered to be agonist-specific, whereas in the latter type, phosphorylation can be catalyzed by other kinases and thus appears to be agonist-nonspecific. The homologous type requires agonist occupancy of the receptor and is mediated by βARK (β-adrenergic receptor kinase); heterologous desensitization is more general and in some cell types appears to be regulated by phosphorylation of the β-adrenergic receptor on at least two different sites by cAMP-dependent protein kinase A. In rat adipocytes, heterologous desensitization is expressed as a decrease in functional activity of the G_i α subunit, and concomitant increased responsiveness to stimulatory agents resulting in increased quantity and activity of the G_s α subunit (Parsons and Stiles, 1987). Desensitization has been observed in the iris–ciliary body (Chapter 4, Section 4.2.7). It appears to involve both α1-adrenergic and muscarinic cholinergic receptors, but whether it is of the homologous or heterologous type has not been established.

1.4. OXYGEN FREE RADICALS AND HYDROGEN PEROXIDE TOXICITY, LIPID PEROXIDATION, AND PHOSPHOLIPASE A_2

Many questions are commonly asked by those who may not be familiar with this relatively new and rapidly expanding field. What are free radicals? How are they generated? Are they normally present in cells or in biological fluids? How do they cause tissue injury?

All body tissues are potential targets of free radical injury, and it was predicted over a decade ago that ocular tissues may be particularly susceptible to oxygen toxicity because of their unique metabolic characteristics (Feeney and Berman, 1976). There is now a wide body of evidence showing that oxidative stress, light, and antioxidant-deficient diets do indeed induce biochemical and functional changes in many tissues, including those of the eye (Varma and Lerman, 1984; Wiegand *et al.*, 1984; Handelman and Dratz, 1986).

The extent to which tissue injury from free radicals occurs under normal conditions, in the absence of oxidative or other external stress, is not known, but it appears to be minimal. This is because aerobic cells have evolved protective mechanisms that limit the generation and/or accumulation not only of free radicals but also of other potentially toxic substances such as H_2O_2. These protective agents (enzymes, antioxidants, and free radical scavengers) are normal cellular components, and their role in protecting the cell from free radical injury is discussed below. Several general texts and reviews are available on all aspects of this broad topic and should be consulted for detailed information (Slater, 1984; Halliwell and Gutteridge, 1984a,b; Weiss, 1986; Fridovich, 1983, 1986, 1989; Fridovich and Freeman, 1986; Cadenas, 1989).

1.4.1. Generation of Free Radicals and H_2O_2

The majority of aerobic cells utilize the oxidation potential of oxygen to maximum advantage in the cytochrome system, where the complete tetravalent reduction of oxygen to water by the cytochrome system is used to generate ATP. This highly efficient process is the primary source of cellular energy and accounts for about 95% of the oxygen consumption in most tissues. However, the remaining 5% of oxygen consumed by aerobic cells is reduced to water in univalent steps that generate three types of reactive intermediates (Fig. 1.8).

Two of the products arising from the univalent reduction of oxygen are free radicals, defined as molecules containing an unpaired electron, and the third is H_2O_2. These three intermediates have varying degrees of reactivity and toxicity (Fridovich, 1983) and, once produced, have the potential of interacting in many unique ways in biological systems. Before discussing these aspects, it is important to consider the specific sources of free radicals in the cell.

The free radicals shown in Fig. 1.8 arise from many intracellular sources and by different mechanisms (Freeman, 1984; Fridovich and Freeman, 1986). Despite the technical difficulties involved in precise estimations of superoxide anion (O_2^-) production in specific cell fractions, their generation has been convincingly demonstrated in isolated preparations of mitochondria, endoplasmic reticulum, peroxisomes, and plasma membrane as well as the cytosol.

A major source of free radicals in mitochondria is the ubiquinone–cytochrome b region of the cytochrome carrier system. Since mitochondria are a rich source of superoxide dismutase (SOD), the O_2^- produced is rapidly converted to H_2O_2, which then diffuses into the cytosol. Endoplasmic reticulum and nuclear membranes generate O_2^- through autooxidation of flavoprotein-containing cytochrome reductases linked to the cytochrome P450 and b_5 detoxifying reactions (Freeman, 1984; Fridovich and Freeman, 1986).

Cytosol is also an important site of free radical production in the cell. It is probable that O_2^- is generated during the catalytic recycling of a number of oxygenases and dehydrogenases, the best characterized being xanthine oxidase (Freeman, 1984). In some,

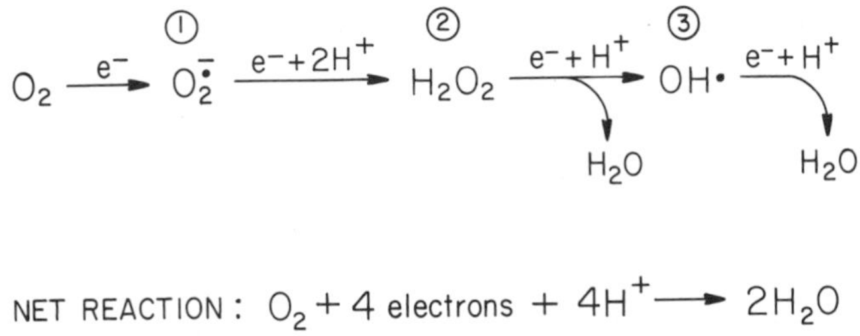

$$O_2 \xrightarrow{e^-} \overset{(1)}{O_2^{\dot{-}}} \xrightarrow{e^- + 2H^+} \overset{(2)}{H_2O_2} \xrightarrow{e^- + H^+} \overset{(3)}{OH\cdot} \xrightarrow{e^- + H^+} H_2O$$

(the $e^- + H^+$ steps from H_2O_2 and from $OH\cdot$ each release H_2O)

$$\text{NET REACTION: } O_2 + 4 \text{ electrons} + 4H^+ \longrightarrow 2H_2O$$

Figure 1.8. Reactive intermediates formed by the univalent reduction of molecular oxygen: (1) superoxide anion (O_2^-); (2) hydrogen peroxide; and (3) hydroxyl radical ($OH\cdot$). In the net reaction, one molecule of oxygen is reduced to two molecules of water.

but not all, tissues this enzyme can be formed from xanthine dehydrogenase during tissue ischemia, a condition that also leads to an accumulation of xanthine. Subsequent reintroduction of oxygen (reperfusion) generates active free radical species such as O_2^-, and secondary reactions lead to the production of H_2O_2 and hydroxyl radical (OH·) (McCord, 1986; Zweier, 1988). The cellular damage resulting from ischemia/reperfusion, which has been documented in many tissues, can be minimized or even prevented by prior administration of free radical scavengers (Granger *et al.*, 1986; Simpson *et al.*, 1987). Xanthine/xanthine oxidase is one of the most efficient free radical-generating systems known (Fridovich, 1972), and is widely used as a model system *in vitro*.

In addition to enzymes, there are low-molecular-weight cytosolic components that serve as potential sources of free radicals. Suitably chelated Fe^{3+} can be reduced to Fe^{2+} by naturally occurring reducing substances such as thiols and ascorbate; in the presence of H_2O_2, Fe^{2+} may catalyze the formation of OH· through a Fenton-type reaction. Thiols and ascorbate, as well as other low-molecular-weight cytosolic components such as hydroquinones, catecholamines, and flavins, can undergo autooxidation and form O_2^- and H_2O_2 (Fridovich and Freeman, 1986).

Plasma membranes of phagocytic cells such as polymorphonuclear neutrophils (PMNs) are a major source of three reactive species. Superoxide anion and H_2O_2 are generated directly, and OH· indirectly, by a membrane-bound NADPH oxidase during the metabolic burst following phagocytosis or in response to appropriate chemotactic stimuli. This is an important component of the inflammatory response of tissues and is thought to be one of the principal causes of tissue injury. Free radical production is also associated with the enzymatic oxidation of arachidonic acid leading to the production of biologically active substances such as eicosanoids and leukotrienes (Lands, 1985).

The free radicals produced at the various sites described above are short-lived. They undergo several different kinds of chemical and/or enzymatic transformations, and their steady-state concentration in cells, or in body fluids such as aqueous humor, is exceedingly low. Intracellular concentrations of O_2^- and H_2O_2 may be in the range of about 10^{-11} to 10^{-12} and 10^{-7} to 10^{-9}, respectively (see Freeman, 1984). These estimations, however, need further verification.

1.4.2. Free Radical Scavengers

Potentially toxic free radical species are generated continuously in the cell. However, under normal circumstances there is no substantial buildup in intracellular levels, due in part to their high rate of reactivity but mainly because of a wide variety of protective mechanisms in the cell (see review by Fridovich and Freeman, 1986). These include superoxide anion-degrading enzymes (SODs), antioxidants such as ascorbate, free radical scavengers such as α-tocopherol, β-carotene (Burton and Ingold, 1984), and H_2O_2-degrading enzymes (peroxidases, catalase, and glutathione peroxidase). Most or all of these agents have ubiquitous distribution in animal cells. In ocular tissues, catalase and SOD have been recognized for more than a decade (Hall and Hall, 1975; Bhuyan and Bhuyan, 1977, 1978; Crouch *et al.*, 1978). Of the two known types of SOD, mitochondrial Mn-containing SOD and cytoplasmic Cu- and Zn-containing SOD (McCord and Fridovich, 1969; Fridovich, 1975, 1989; Fridovich and Freeman, 1986), it is the latter type that has been detected by immunochemical methods in some, but not all, ocular tissues (Rao *et al.*, 1985).

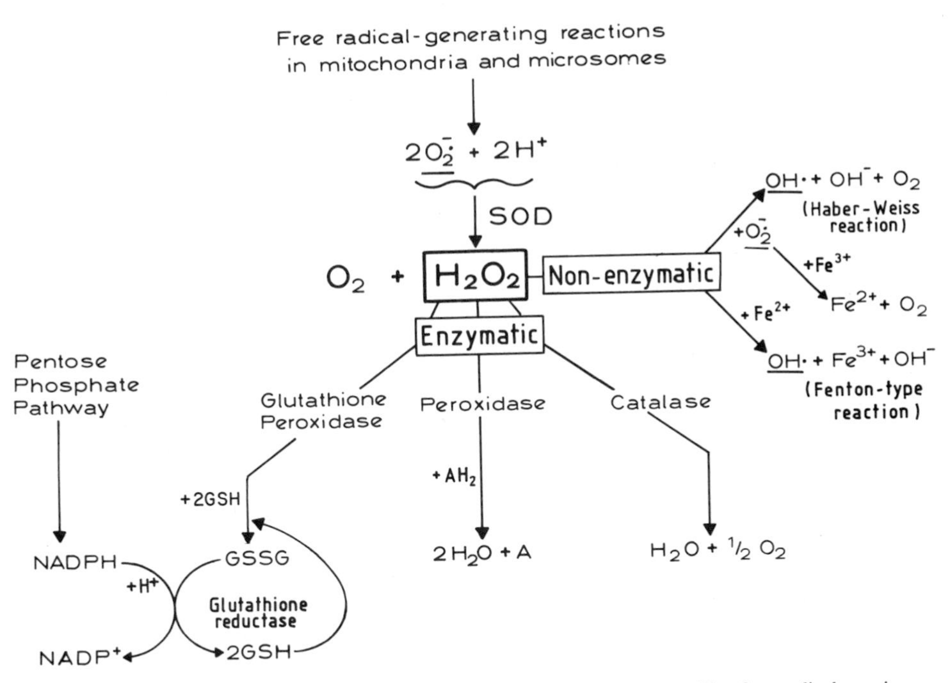

Figure 1.9. Degradation of H_2O_2 by enzymatic and nonenzymatic pathways. The free radical species are underlined. Maintenance of reducing conditions for the continuous generation of GSH depends on an adequate supply of NADPH from the pentose phosphate pathway.

An overview of the fate of O_2^- and H_2O_2 produced intracellularly is shown in Fig. 1.9. Two molecules of superoxide anion generated by the univalent reduction of molecular oxygen undergo dismutation, either spontaneous (which is relatively slow) or enzymatic. The latter pathway, through SOD, is probably the major mechanism used by respiring cells (McCord and Fridovich, 1969; Fridovich and Freeman, 1986). The H_2O_2 produced may then undergo either enzymatic or nonenzymatic degradation. In the enzymatic pathways, H_2O_2 is reduced to water by specific and/or nonspecific peroxidases, catalase, or glutathione peroxidase. The latter is the most extensively investigated pathway and is probably the most important one physiologically in most tissues. The reduction of H_2O_2 and of lipid hydroperoxides (discussed in Section 1.4.3) by glutathione (GSH) is catalyzed by glutathione peroxidase, a selenium enzyme; the reaction is coupled to glutathione reductase, which, in the presence of NADPH, effectively regenerates the GSH needed for continued reduction of H_2O_2. The maintenance of this pathway requires reducing conditions in the cell, i.e., an ample supply of NADPH. The major source of this nucleotide is the pentose phosphate pathway.

There is a large body of controversial, but often compelling, evidence for nonenzymatic transformations of H_2O_2 catalyzed by trace amounts of chelated metal ions (Fig. 1.9). Known as Haber–Weiss or Fenton-type reactions, they are considered to be a potential source of OH·, the most reactive of all oxygen free radicals and a key intermediate in primary cellular injury (Halliwell and Grootveld, 1987). In model systems of liposomes, peroxidation can be initiated by Fe^{2+} and H_2O_2; Fe^{3+} is required, and the rate of the reaction is determined by the ratio of Fe^{3+}: Fe^{2+} (Minotti and Aust, 1987). In other studies using the P388D$_1$ strain of transformed cells, exposure to H_2O_2 results in inhibition of ADP phosphorylation as well as inhibition of net glucose uptake and lactate production (Hyslop *et al.*, 1988). Glycolysis is specifically blocked at the glyceraldehyde 3-phosphate dehydrogenase step, partially through direct inactivation of the enzyme and partially through reduction in the redox potential of its nicotinamide cofactors. Exposure to the oxidant for 15 min results in a dose-dependent cell dysfunction and death.

The foregoing discussion, while describing briefly the major aspects of free radical production, did not consider certain problems inherent in the measurement of the free radical itself. These species are not only short-lived, but they are also present at very low concentrations. The methods used for their direct measurement, spin trapping and electron spin resonance (ESR), require sophisticated instrumentation and relatively large amounts of material. For practical purposes, most investigators use indirect methods to measure free radical production in biological systems (see reviews of methodology by Packer, 1984; Greenwald, 1985). The most commonly used assays are SOD-inhibitable reduction of ferricytochrome c and luminol (or lucigenin)-enhanced chemiluminescence. Both have been successfully applied to measurements of O_2^- production in activated PMNs and in certain other cells and tissues as well.

Free radical and H_2O_2 toxicity are thought by many to be important causes of certain pathological conditions in the eye, but the active species of free radicals are difficult to study directly for reasons discussed above. Two general approaches, albeit indirect, are commonly used and have yielded valuable information: (1) measurements of the protective systems in a particular tissue (SOD, GSH, glutathione peroxidase, ascorbic acid, α-tocopherol, or H_2O_2-degrading enzymes) and (2) creation of a stress (light, oxygen, interference with the NADPH-generating system, or dietary limitation of essential antioxidants) that overwhelms the protective mechanisms and creates measurable tissue insult. Both lines of investigation have been used for corneal endothelium (Chapter 3, Section 3.5.3), aqueous humor (Chapter 4, Section 4.1.2), trabecular meshwork (Chapter 4, Section 4.3.2.6), lens (Chapter 5, Section 5.3.2.2), and retina (Chapter 7, Section 7.2.3).

1.4.3. Lipid Peroxidation

A brief outline of the intracellular sources of free radicals generated by the univalent reduction of molecular oxygen, as well as cellular defense mechanisms, has been given in the preceding two sections. Very low concentrations of free radicals and of H_2O_2 may normally be present in cells; moreover they can be produced as a result of stress under a variety of physiological and/or pathological conditions. Hence, questions of how or where they exert their toxic effects should be addressed.

Important targets of free radical attack are cellular membranes, the most vulnerable molecules being the polyunsaturated fatty acids (PUFAs), most if not all of which are esterified to phospholipids (Halliwell and Gutteridge, 1984a,b; Slater, 1984; Sevanian and Hochstein, 1985; Weiss, 1986). In addition, however, a wide variety of structural proteins (both membrane-bound and cytosolic), as well as enzymes, and DNA and RNA are targets of direct attack by free radicals. These biomolecules can also be damaged or inactivated indirectly as a result of cross-linking with malonaldehyde generated by the breakdown of peroxidized PUFAs. The resulting structural changes in cellular membranes compromise many of their most important functions: selective permeability, enzyme activity, receptor availability and ion transport, to name but a few. Free radical-initiated damage to cellular membranes has been implicated in the etiology of a large number of degenerative and inflammatory diseases, in aging processes, and in cancer (Bulkley, 1983; Slater, 1984; Torrielli and Dianzani, 1984; Halliwell and Gutteridge, 1984a).

Lipid peroxidation can be defined as the free radical-induced nonenzymatic oxidation of long-chain polyunsaturated fatty acids (PUFAs). The two PUFAs present in esterified form in membrane phospholipids that are most vulnerable to attack are arachidonate (20:4) and docosahexaenoate (22:6). Peroxidation of these fatty acids is usually considered in three stages: initiation, propagation, and termination (Feeney and Berman, 1976; Feeney-Burns *et al.*, 1980; Slater, 1984; Sevanian and Hochstein, 1985). In the initial attack, an allylic hydrogen atom is abstracted from a PUFA (RH), leaving behind an unpaired electron which then forms a lipid free radical, R· (see Fig. 7.11). This intermediate undergoes molecular rearrangement to a conjugated diene, which rapidly reacts with O_2 to form a lipid peroxide radical, ROO·.

It is the lipid free radical, ROO·, that initiates the propagation phase of lipid peroxidation, since it has sufficient oxidizing potential to attack a neighboring PUFA (RH). This interaction results in the production of another lipid free radical, ROO·, and a lipid hydroperoxide, ROOH. The organic lipid hydroperoxide can be reduced enzymatically by both selenium-dependent glutathione peroxidase and glutathione S-transferase, a selenium-independent enzyme that displays glutathione peroxidase activity. However, in the presence of chelated transition metals, it may also be converted to peroxy or alkoxy free radicals, RO·. These active species further propagate the chain reaction in lipid peroxidation by abstracting additional hydrogen atoms.

The termination of lipid peroxidation, i.e., the interruption of the autooxidative chain reaction, is carried out by free radical scavengers. The two that have been most extensively studied are α-tocopherol (vitamin E), a membrane-bound lipid, and ascorbic acid (vitamin C), the ubiquitous water-soluble sugar acid. Interaction between the two to terminate free radical chain reactions is fairly well understood (see reviews by McCay, 1985; Weiss, 1986). Vitamin E reacts directly with the lipid free radical, ROO·, to form a relatively stable phenoxyl radical. This vitamin E-derived radical can be either oxidized by reacting with another lipid free radical or possibly reduced back to vitamin E by interaction with ascorbic acid. Whatever mechanism prevails for the regeneration of α-tocopherol, its presence in lipid membranes in general, and in rod outer segments in particular (Dilley and McConnell, 1970), has been established and suggests that it plays an important role in protection from free radical damage. Tissue degeneration is brought about by vitamin E deficiency, and the degeneration is amplified in most species by selenium deficiency (see Chapter 7, Section 7.2.3.4).

The difficulty of detecting free radicals directly was discussed above. Similarly, methods for measuring lipid peroxidation are far from satisfactory since the initial reaction products, hydroperoxides, are not only unstable but are also present in very low amounts; this makes quantitative estimations by conventional methods difficult (Pryor, 1987). However, the recent development of HPLC for separation of complex mixtures of phospholipid peroxides, combined with gas chromatography–mass spectrometry (GC–MS) for their identification, forms a sound basis for future methodological advances in this field (Hughes *et al.*, 1986; van Kuijk and Dratz, 1987; Smith and Anderson, 1987). This approach has excellent potential, but until the instrumentation becomes available at affordable prices, lipid peroxidation will still have to be detected by indirect methods in many laboratories, in spite of the pitfalls (Smith and Anderson, 1987). Most of these assays measure breakdown products, the most widely used being the somewhat nonspecific measurement of malondialdehyde (MDA) or, more accurately, thiobarbituric acid-reacting substances. A wide variety of other methods are also available, and the large number itself reflects the fact that each has its drawbacks (Packer, 1984; Smith and Anderson, 1987).

1.4.4. Phospholipase A_2

The PUFAs that undergo lipid peroxidation are located mainly in the *sn*-2 position of phospholipids, and there is evidence suggesting that phospholipase A_2 may play a role in the detoxification of these phospholipid hydroperoxides (van Kuijk *et al.*, 1987). In model membranes consisting of phospholipid-containing peroxidized liposomes, a preferential hydrolysis of peroxidized fatty acids by phospholipase A_2 has been demonstrated (Sevanian and Kim, 1985). About 30–60% of the total peroxides were released by this enzyme after a 15-min incubation, whereas under the same conditions only about 15% of the nonoxidized fatty acids are hydrolyzed. Moreover, peroxidized fatty acids esterified to membrane-bound phospholipids do not react with selenium-dependent glutathione peroxidase; prior hydrolysis by phospholipase A_2 is a necessary prerequisite for their reduction by this enzyme (van Kuijk *et al.*, 1985). A second type of selenium glutathione peroxidase has also been described that acts directly on peroxidized phospholipids in membranes (Ursini *et al.*, 1985). However, this phospholipid hydroperoxide glutathione peroxidase is a minor component, accounting for only about 2% of the major glutathione peroxidase activity. It has not as yet been reported in ocular tissues.

These interesting observations raise the possibility of peroxide "regulation" in membranes. The sequential action of phospholipase A_2 (activated by physiological agents such as Ca^{2+} and/or increased levels of peroxidized lipids) and selenium-dependent glutathione peroxidase selectively removes and destroys membrane-bound peroxidized fatty acids. This could control or even prevent the alternate pathway that prevails in intact peroxidized phospholipids, namely, their breakdown to reactive species that cause further production of lipid peroxides.

1.5. ARACHIDONIC ACID AND EICOSANOIDS (PROSTAGLANDINS, LEUKOTRIENES)

The *sn*-2 position of the glycerol backbone of phospholipids is esterified mainly to unsaturated long-chain fatty acids, one of the most abundant being arachidonic acid (AA), a 20-carbon fatty acid containing four double bonds (20:4). This is generally true for the majority of tissues except the retinal photoreceptors, which contain docosahexaenoate (22:6) as the principal unsaturated fatty acid (Chapter 7, Section 7.2.1.2). It has long been recognized that arachidonic acid (AA) in the free form, after hydrolytic release from phospholipids, is oxidized by a series of enzymatic reactions (the "arachidonic acid cascade") that lead to the production of prostaglandins and a large number of other biologically active metabolites now collectively termed eicosanoids.

An overview of AA metabolism—its production and subsequent oxidation by three major enzymatic pathways—is shown in Fig. 1.10. Tissues normally contain only trace amounts of free AA; however, it is released from phospholipids by two mechanisms,

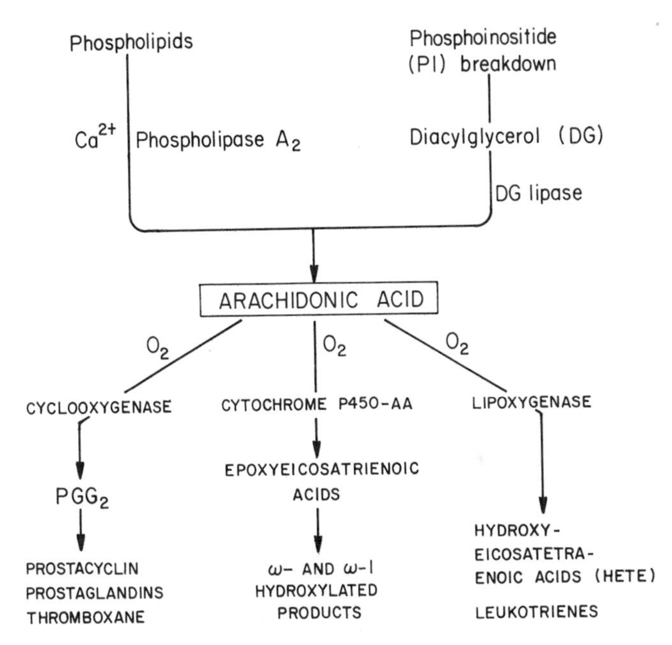

Figure 1.10. Schematic diagram showing the production of arachidonic acid and its oxidation by three major pathways.

hydrolytic cleavage from phospholipids by Ca^{2+}-activated phospholipase A_2, a relatively slow process, and receptor-stimulated polyphosphoinositide (PI) breakdown. In the latter pathway, the formation of diacylycerol (DG) and the hydrolytic release of AA by DG lipase are rapid; this hormone-stimulated mechanism is considered to be an important source of free AA and eicosanoids in tissues.

Prostaglandins are potent biologically active lipids of varying structures. They have one feature in common; namely, they are all 20-carbon carboxylic acids containing a cyclopentane ring; some representative members of this group are shown in Fig. 1.11. The first product formed by cyclooxygenase is PGG_2, which is rapidly converted to PGH_2. The latter is highly unstable and through a series of enzymatic conversions discussed below is converted to prostacyclin, prostaglandins, and thromboxane. In current nomenclature, the first two capital letters stand for prostaglandin, and the third defines the type of substituent found on the hydrocarbon chain. The numerical subscript indicates the number of double bonds outside of the ring. The subscripts α and β after the numerical subscripts in the F series indicate the orientation of the hydroxyl group at carbon 9 in the cyclopentane ring. The products of lipoxygenase activity are hydroxyeicosatetraenoic acids (HETEs) and leukotrienes.

Of the three pathways of AA oxidation shown in Fig. 1.10, two (cyclooxygenase and lipoxygenase) have been recognized for many years. The third, microsomal oxidation through an oxygen- and NADPH-dependent cytochrome P450 system, has only been established recently. Many of the enzymes as well as the diverse products formed through these pathways have been characterized in a variety of nonocular tissues as well as in corneal epithelium (Chapter 3, Section 3.1.6), iris–ciliary body (Chapter 4, Section

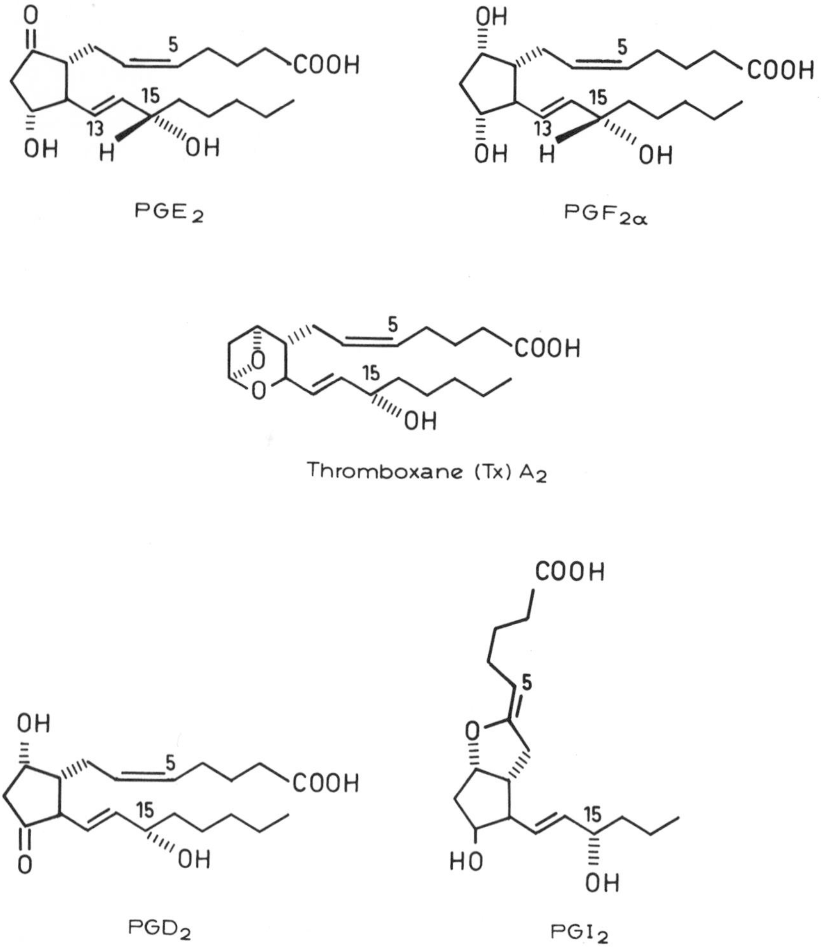

PGE_2

$PGF_{2\alpha}$

Thromboxane (Tx) A_2

PGD_2

PGI_2

Figure 1.11. Structures of prostaglandins (D_2, E_2, and $F_{2\alpha}$), prostacyclin (PGI_2), and thromboxane (TXA_2).

4.2.4) and other tissues in the anterior segment of the eye, and retina (Chapter 7, Section 7.1.2.2). The brief overview presented in this section is based in part on the comprehensive review by Needleman *et al.* (1986).

A summary of the well-known cyclooxygenase and lipoxygenase pathways of arachidonic acid oxidation is given in Fig. 1.12. Although the conversion of AA to PGH_2 appears as two steps, it is actually catalyzed by a single enzyme, prostaglandin endoperoxide synthase (PES) (Needleman *et al.*, 1986). The cyclooxygenase activity inserts two molecules of oxygen into AA to yield PGG_2, while the hydroperoxidase activity of PES reduces PGG_2 to PGH_2. Nonsteroidal antiinflammatory agents such as indomethacin and aspirin inhibit only the cyclooxygenase activity, i.e., the formation of PGG_2 from AA (Higgs *et al.*, 1987), and not the peroxidase activity of PES. The major classes of eicosanoids formed from PGH_2 are prostaglandins (PGD_2, PGE_2, $PGF_{2\alpha}$), prostacyclin (PGI_2), and thromboxanes (TXA_2 and TXB_2).

A series of lipoxygenase enzymes oxidize AA to produce 5-, 11, 12-, and 15-

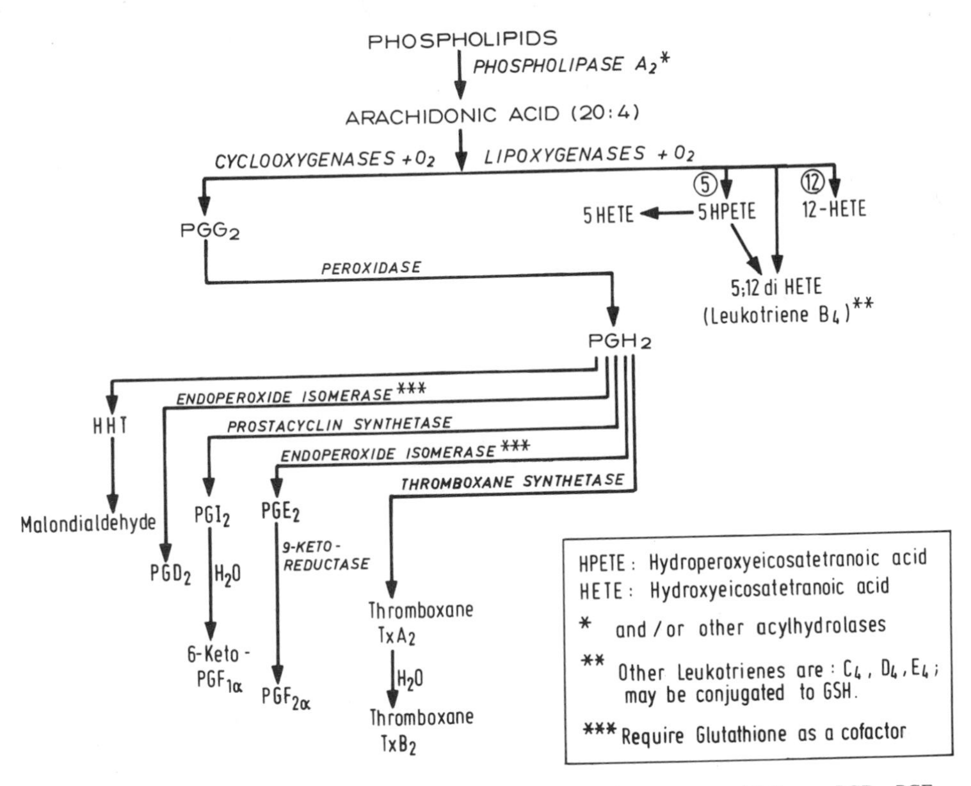

Figure 1.12. Metabolism of arachidonic acid. The final products of cyclooxygenase oxidation are PGD_2, PGE_2, $PGF_{2\alpha}$, 6-keto-$PGF_{1\alpha}$, and thromboxane (TXB_2). The final products of the lipoxygenase pathway are leukotrienes.

stereoisomers of hydroperoxyeicosatetraenoic acids (HPETEs), which are then converted to corresponding hydroxyeicosatetraenoic acids (HETEs) by peroxidatic reduction. There is great variability among tissues in their content of lipoxygenases, and only the 5- and 12-HETE stereoisomers are shown in Fig. 1.12. Leukotrienes, the most potent and biologically active of all the eicosanoids, are produced through the 5-lipoxygenase pathway, with leukotriene A_4 as the first intermediate. This enzyme requires Ca^{2+}, ATP, and other stimulatory factors for maximal activity (Rouzer and Samuelsson, 1987). Other leukotrienes (B_4, C_4, and D_4) are produced from the first, A_4, intermediate. The leukotrienes are involved in chemotaxis and inflammatory responses; one of them, leukotriene D_4, is the slow-reacting substance in anaphylaxis (SRS-A).

The third and most recently described pathway of AA oxidation is through microsomal cytochrome P450, the terminal acceptor for NADPH-dependent mixed-function oxidases (Schwartzman *et al.*, 1987a). The oxidation by NADPH-cytochrome P450 reductase shows an absolute requirement for both oxygen and NADPH. The first products formed from AA are epoxides, which are rapidly hydrolyzed and converted (by ω, ω-1, and ω-2 hydroxylations of eicosatrienoic intermediates) to a complex mixture of polar metabolites (Oliw *et al.*, 1982; Needleman *et al.*, 1986). This pathway of microsomal cytochrome P450-mediated oxidation of AA is inhibited by SKF-525A and carbon monoxide. In addition to the epoxidation of AA, some prostaglandins (PGA_1, PGE_1, and

PGE_2) can be hydroxylated by the microsomal cytochrome P450 system (Needleman *et al.*, 1986). Leukotriene B_4, which has a wide range of biological activities, is also metabolized (or inactivated) by ω- and ω-1 hydroxylation via cytochrome P450-dependent monooxygenases (Romano *et al.*, 1987). In rat liver (Capdevila *et al.*, 1986) and in corneal epithelium (Schwartzman *et al.*, 1987b), the microsomal cytochrome P450 system also produces several HETEs from AA, a predominant one being 12(*R*)-hydroxyeicosatetraenoic acid.

There are important and complex interrelationships between arachidonic acid metabolism and eicosanoid production, on the one hand, and free radicals, on the other. Although somewhat speculative and often contradictory, some of these interactions are thought to play important roles in mediating tissue injury and inflammation. For example, fatty acid hydroperoxides may react directly with the heme-containing cyclooxygenase, yielding a radical complex that abstracts the 13-*s*-hydrogen from AA (Mead, 1984). The free radical intermediate then cyclizes to yield PGG_2, the key precursor of eicosanoids via the cyclooxygenase pathway (Fig. 1.12). This reaction sequence may explain the interesting observation that lipid hydroperoxides greatly enhance eicosanoid production in biological systems (Lands, 1985).

Conversely, eicosanoids themselves appear to stimulate free radical formation (Torrielli and Dianzani, 1984). Prostaglandin PGG_2, the initial product formed by cyclooxygenase, is a substrate of PG hydroperoxidase and, under certain conditions, could be converted to hydroperoxide-forming free radicals such as OH·. There are also some paradoxical observations on the role of α-tocopherol and phospholipase A_2 in free radical production and arachidonic acid release. As pointed out above (Section 1.4.4), phospholipase A_2 may play a protective role in membrane damage by selectively hydrolyzing potentially harmful peroxidized phospholipid fatty acids (van Kuijk *et al.*, 1987). The well-known role of α-tocopherol as a potent free radical scavenger has been discussed above. The interesting observation that α-tocopherol is a potent inhibitor of phospholipase A_2 (Douglas *et al.*, 1986) suggests that it plays a dual role; it not only terminates free radical chain reactions, but it may also modulate the release of arachidonic acid in some tissues.

1.6. DRUG-METABOLIZING ENZYMES (CYTOCHROMES P450 AND MIXED-FUNCTION OXIDASES)

The cytochrome P450s comprise a multi-gene family of microsomal heme proteins that catalyze the monooxygenation of physiological substances such as steroids and fatty acids as well as foreign substances (xenobiotics) such as drugs and chemicals (see review by Hall, 1985). Each species of cytochrome P450 consists of a 30- to 60-kDa apoprotein together with a prosthetic group, protoheme. Their name is derived from the characteristic peak at 450 nm observed in the presence of carbon monoxide. Microsomal drug-metabolizing systems are present mainly in liver, the most extensively investigated tissue; however, in recent years they have also been detected in a variety of extrahepatic tissues as well as in several ocular tissues (Shichi, 1984). These include corneal epithelium (Chapter 3, Section 3.1.7.1), ciliary body (Chapter 4, Section 4.2.5), and retinal pigment epithelium (Chapter 7, Section 7.4.4.6). Their main function is to transform hydrophobic substrates into compounds that can either be excreted or detoxified further by conjugation to glutathione, a reaction catalyzed by glutathione S-transferase. Some cytochrome P450s

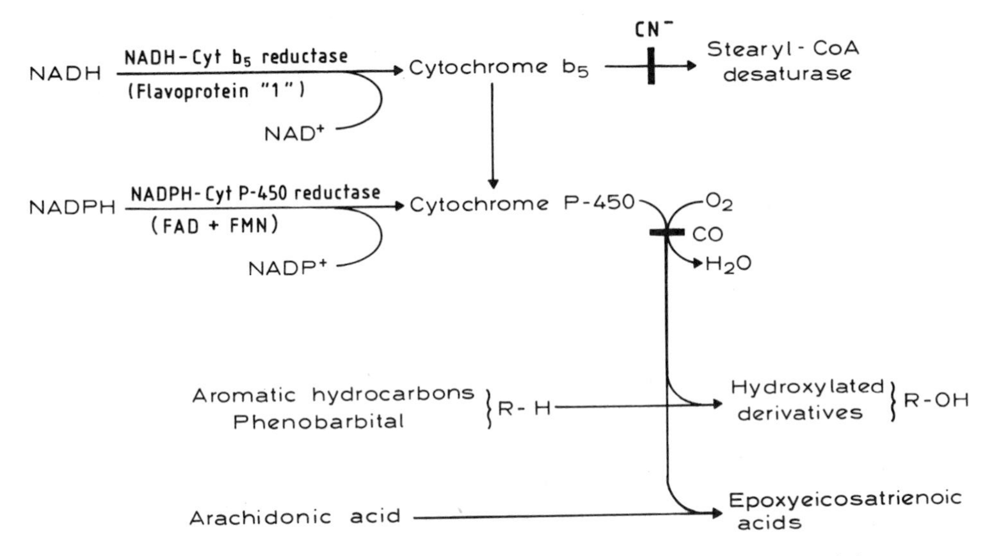

Figure 1.13. The NADPH-cytochrome P450- and NADH-cytochrome b_5-mediated microsomal electron transport system. Monooxygenases catalyze the introduction of one atom of oxygen into substances such as phenobarbital and aromatic hydrocarbons, converting them to hydroxylated compounds. This system also oxidizes arachidonic acid to epoxyeicosatrienoic acids (Schwartzman *et al.*, 1987a).

are endogenous; others can be induced by glucocorticoids (Schuetz *et al.*, 1984), phenobarbital, or 3-methylcholanthrene. The latter two agents give rise to at least six different heme proteins, but in actual fact, the number of inducible forms may be considerably greater (Shichi, 1984). Moreover, the inducibility of the various forms of cytochrome P450 in the eye (Shichi *et al.*, 1975) and in liver and other tissues (Nebert *et al.*, 1981) is under genetic control.

An essential component of the microsomal drug-metabolizing system is NADPH-cytochrome P450 reductase, a flavoprotein that contains both flavin adeneine nucleotide (FAD) and flavin mononucleotide (FMN). This enzyme catalyzes electron transfer from NADPH to the cytochromes P450 and the mixed-function oxidases, also called monooxygenases (Fig. 1.13). In this reaction one atom of oxygen from molecular oxygen is introduced into a substrate, while the other atom is reduced to water. Another electron transport chain consists of the flavoprotein "1"-containing NADH-cytochrome b_5 reductase. During hydroxylation reactions, electrons flow from both the cytochrome b_5 and the cytochrome P450 systems to the monooxygenases. The former is inhibited by cyanide and the latter by carbon monoxide.

Aryl hydrocarbon hydroxylase, which uses benzo(a)pyrene as substrate, is one of the most widely studied of the monooxygenases; others include 7-ethoxyresorufin-*o*-deethylase and benzphetamin demethylase. This system also catalyzes the hydroxylation of proline and lysine residues of pro-α collagen peptides.

1.7. RETINOL TRANSPORT, METABOLISM, AND FUNCTIONS

Vitamin A was discovered at the turn of the century, and its requirement for growth and differentiation of epithelia was established shortly afterward. Many years later, its role

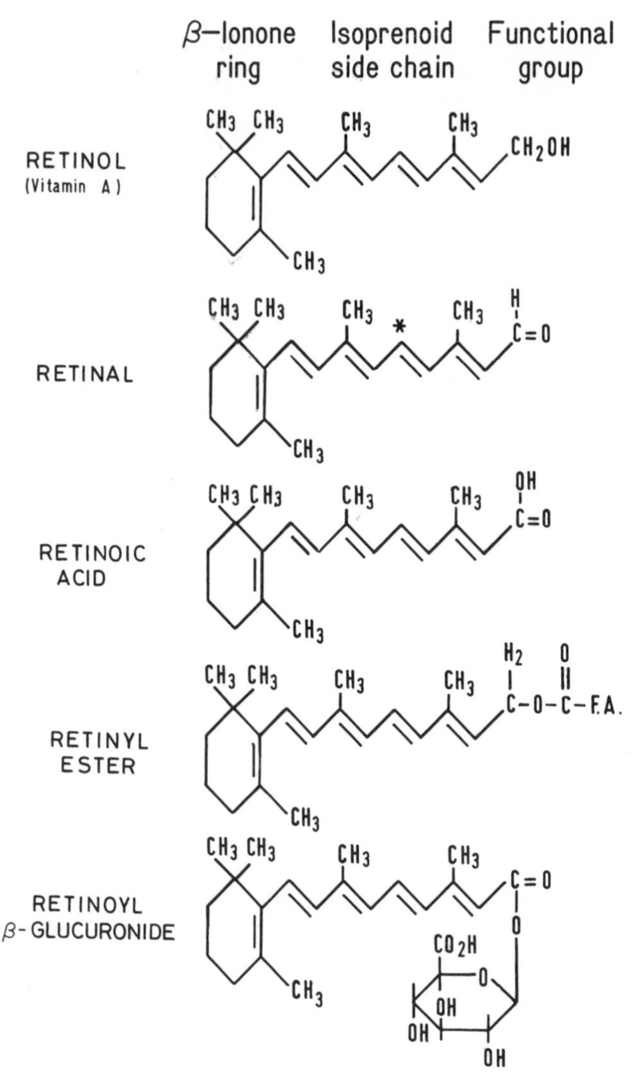

Figure 1.14. Structures of all-*trans* retinol (vitamin A) and other naturally occurring retinoids. The asterisk designates stereoisomeric forms. The "straight" form shown is all-*trans*, and the "bent" form, 11-*cis*, results from isomerization at carbon 11 (indicated by the asterisk).

in the visual process, as a precursor of retinaldehyde (11-*cis* retinal), was established by the classic work of Wald (1935). In current nomenclature, retinol (vitamin A) belongs to a class of compounds termed retinoids that have three structural features in common: a hydrophobic head (the β-ionone ring), a conjugated isoprenoid side chain, and terminal polar functional groups on carbon 15, which can be chemically or enzymatically modified (Fig. 1.14). Except for the 11-*cis* isomers found in the retina, the retinoids in virtually all body tissues are of all-*trans* configuration. The naturally occurring dietary precursor of retinoids is β-carotene, a symmetrical 40-carbon carotenoid consisting of two β-ionone rings joined by an isoprene chain. Several full descriptions of retinoids have appeared

(DeLuca and Shapiro, 1981; Chader, 1982, 1984; Olson *et al.*, 1983; Sporn *et al.*, 1984). These sources form the basis of the present discussion, which, owing to space limitations, covers only selected aspects of this field.

1.7.1. Absorption and Transport

The natural dietary sources of vitamin A are animal tissues such as liver that are rich in retinyl esters and leafy green vegetables, which contain precursor β-carotene(s). These water-insoluble substances are first "solubilized" in the intestine through formation of mixed micelles with bile salts and various lipolytic products of ingested fats (Fig. 1.15). Dietary retinyl esters are hydrolyzed in the intestinal lumen to free retinol. β-Carotene enters the mucosal cells and is converted to retinol in two enzymatic steps: the first involves cleavage at the central double bond by β-carotene 15,15′-dioxygenase, yielding

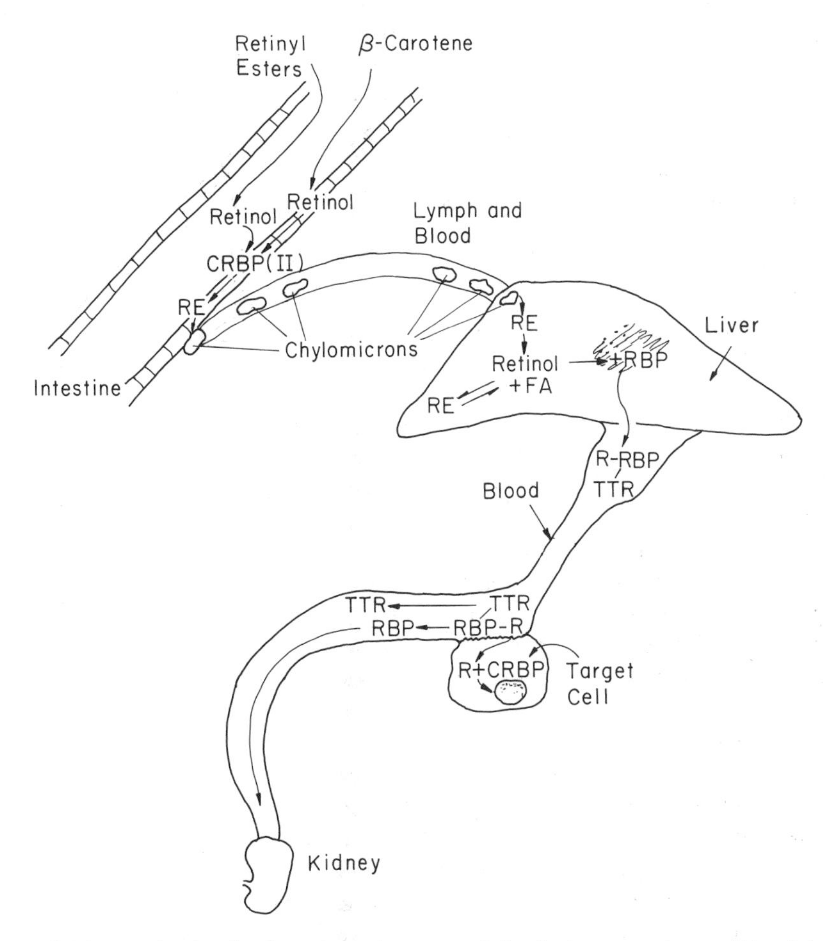

Figure 1.15. Diagram showing the absorption and transport of vitamin A to target tissues. Abbreviations are: CRBP(II), cytosol retinol-binding protein type II; RE, retinyl esters; FA, fatty acids; RBP, retinol-binding protein; R-RBP, retinol–retinol-binding protein complex; TTR, transthyretin (formerly prealbumin); CRBP, cellular retinol-binding protein.

two molecules of retinaldehyde; in the second step, the aldehyde is reduced to retinol. Retinol deriving from these two sources as well as retinaldehyde produced from β-carotene form high-affinity complexes with a specific retinol-binding protein of the intestinal mucosa designated cytosolic retinol-binding protein type II [CRBP(II)] (MacDonald and Ong, 1987). This binding protein plays an important role in facilitating the enzymatic esterification of retinol to retinyl esters (Ong *et al.*, 1987), which are subsequently incorporated into chylomicrons.

The chylomicrons are transported through the lymph and the general circulation, and they are partially metabolized in certain extrahepatic tissues. The chylomicron remnants, enriched in cholesterol and containing essentially all of the retinyl esters absorbed from the intestine, enter the liver. This organ is the major storage depot for vitamin A in the body. Close to 95% of the vitamin is in the esterified form; the remainder is present either as the free alcohol or bound to cellular retinol-binding protein (CRBP). Immunocytochemical studies have localized this binding protein to both parenchymal and fat-storing cells; the latter are visibly stained with antibody in normal rats and heavily stained in animals fed excess vitamin A (Kato *et al.*, 1984). Radioimmunoassays of CRBP in normal and vitamin A-deficient rats further support the view that liver stores of CRBP are regulated by nutritional status (Kato *et al.*, 1985).

Retinyl esters are stored mainly as lipid droplets in the cytoplasm of the parenchymal and stellate (fat-storing) cells of the liver, with very little being found in Kupffer cells (Olson and Gunning, 1983). These retinyl ester-rich globules are readily observed by electron microscopy and can be isolated from liver homogenates as a flotating layer after ultracentrifugation (Berman *et al.*, 1979). Retinol is mobilized from tissue stores for secretion after hydrolysis by hepatic retinyl palmitate hydrolase, an enzyme having many unusual properties: it requires a bile salt for activation and/or stimulation, it appears to be localized in the "nuclear" fraction of the cell, and it shows highly variable activity from one animal to the next when measured in rats (Harrison *et al.*, 1979; Prystowsky *et al.*, 1981). Moreover, the enzyme is inhibited *in vitro* by α-tocopherol (Napoli *et al.*, 1984).

Following hydrolysis, retinol forms a 1:1 complex with retinol-binding protein (RBP), a 21-kDa protein that is synthesized in the endoplasmic reticulum of the liver parenchymal cells and has one binding site for retinol (Goodman, 1984). It is the specific vehicle for the transport of vitamin A in the plasma to peripheral target tissues (Fig. 1.15). Neither the rate of synthesis of RBP nor the transcriptional levels of RBP mRNA appear to be dependent on liver stores of retinol (Soprano *et al.*, 1982). Further studies of RBP mRNA transcription have shown that liver is not the only site of RBP synthesis; extrahepatic tissues such as kidney contain about 5–10% of the level of RBP mRNA as that found in liver, whereas lung, spleen, brain, and other peripheral tissues contain smaller amounts (Soprano *et al.*, 1986). The RBP in these tissues could provide a vehicle for returning extrahepatic stores of retinol to the liver for recycling, or it may serve more specific function(s), as speculated for the lacrimal gland (Chapter 2, Section 2.1.1).

After secretion from the liver, the retinol–RBP complex is bound to plasma transthyretin (TTR; formerly called prealbumin), and in this form it circulates to target tissues such as lacrimal gland (Chapter 2, Section 2.1.1) and retinal pigment epithelium (Chapter 7, Section 7.4.4.2). In target cells such as the retinal pigment epithelium, the complex binds to specific receptors on the plasma membrane and then dissociates. The retinol enters the cell, and the RBP returns to the circulation; some of it is excreted in the kidney, but a large part is reabsorbed (Kato *et al.*, 1984) and returns to the parenchymal and stellate cells of the liver, where it is degraded (Gjoen *et al.*, 1987). Retinol from the circulating retinol–RBP–transthyretin complex is also taken up by these liver cells and recycled (Blomhoff *et al.*, 1985). It has been estimated that close to half of the retinol secreted from the liver as retinol–RBP complex returns to the liver within 50–60 h.

1.7.2. Metabolism

Retinoid metabolism in the neural retina and pigment epithelium is well understood, but for other tissues apart from liver, our information is at best fragmentary. It appears that in many nonocular tissues, retinoic acid is a key intermediate in the metabolism of retinol, and there is considerable speculation that either the acid or one of its polar metabolites is the biologically active form of vitamin A (Chytil, 1984). Retinoic acid supports the growth and epithelial differentiation functions of vitamin A but does not substitute in reproduction or in the visual process. Mammalian tissues oxidize retinol to retinoic acid and other polar metabolites (DeLuca *et al.*, 1981), but the reverse reaction does not seem to occur. Convincing evidence has accumulated from *in vivo* studies following injection of tritiated precursor that retinol is oxidized to retinoic acid in the rat. However, the low endogenous levels of retinoic acid under physiological conditions have not been accurately estimated until recently. A newly developed method utilizes improved HPLC procedures and internal retinoid standards for the detection of nanomolar amounts of retinyl palmitate and retinoic acid in the circulation and in liver (Cullum and Zile, 1986). In rats fed tritiated all-*trans* retinyl acetate for specified periods of time, levels of 13-*cis* and all-*trans* retinoic acid as low as 12.0 and 15.0 ng/g liver, respectively, could be accurately quantitated.

Many studies on the metabolic fate of retinoic acid have been carried out both *in vivo*, after injection of the tritiated acid into living animals, or *in vitro*, by incubation with subcellular fractions from liver and other tissues. Under these conditions, retinoic acid is metabolized to a series of polar substances, one of the major products formed being 5,6-epoxyretinoic acid (DeLuca *et al.*, 1981). However, the physiological significance of this metabolite is uncertain, since it is inactive in supporting growth of vitamin A-deficient animals. Tritiated all-*trans* retinoic acid is also isomerized to 13-*cis* retinoic acid when injected into rats; the major metabolites found under these conditions are conjugated forms excreted in the bile as a mixture of all-*trans* and 13-*cis* retinoyl-β-glucuronide (Zile *et al.*, 1982). The glucuronide has also been detected in liver under normal physiological conditions (Cullum and Zile, 1986), and in this form it is active in inhibiting the proliferation of leukemia HL-60 cells (Zile *et al.*, 1987).

Within the cell, cytosolic retinoid-binding proteins play important functional roles in retinoid metabolism. Binding proteins for both retinol (CRBP) and retinoic acid (CRABP) have been found in a wide variety of cells and tissues throughout the body. The CRBP and CRABP from liver, testes, and other nonocular tissues are both 14.6-kDa proteins (see review by Chytil and Ong, 1984). These binding proteins serve to stabilize the hydrophobic retinoids in the aqueous milieu of the cell; they also facilitate certain enzymatic conversions, for example, esterification of retinol in the intestinal mucosa and in other tissues. Most importantly, they appear to play a major role in translocation of retinol (and possibly retinoic acid) to the nucleus, which results in an alteration of gene expression. Although these proteins have many structural features in common, the two retinoid-binding proteins differ immunologically from one another and also show different binding characteristics toward their specific ligands (Chytil and Ong, 1984). Both human (Colantuoni *et al.*, 1985) and rat (Sundelin *et al.*, 1985; Sherman *et al.*, 1987) CRBP have been cloned and sequenced. The single polypeptide chain contains 134 amino acid residues and has a M_r of 14,600; although there is some homology with other proteins that bind hydrophobic ligands, no homology with plasma RBP could be found.

1.7.3. Functions

It is now recognized that vitamin A has multiple functions in the organism and in the cell. Its role in vision is well understood, but its effects on genome expression and other biological processes are only now being unfolded at the molecular level.

In the most general view, vitamin A has three major biological functions: somatic, reproductive, and visual (see review by Zile and Cullum, 1983). At the cellular level, they are both nuclear and extranuclear (Olson *et al.*, 1983). Many general functions can be supported by either retinol or retinoic acid (Dowling and Wald, 1960; Chader, 1982): growth, epithelial cell differentiation, and maintenance of the differentiated state. Specific functions in reproduction and in the visual process have an obligatory requirement for retinol. The present discussion is limited to the role of vitamin A in the synthesis of glycoconjugates as well as its hormone-like action and nuclear function in genomic expression.

A direct involvement of retinol in glycoconjugate synthesis was proposed more than a decade ago (DeLuca, 1977), and considerable experimental evidence has since accumulated supporting this view. Incorporation of mannose into liver glycoproteins is sharply diminished in vitamin A-deficient rats, and the synthesis of high-molecular-weight glycoconjugates in corneal epithelium is increased in retinoic acid-repleted rats (Hassell *et al.*, 1980). Retinol may be involved in lipid-linked pathways of glycoprotein biosynthesis analogous to those in the dolichol pathway. The formation of a key intermediate in glycopeptide biosynthesis in rat liver, mannosylretinyl phosphate, is stimulated by retinyl palmitate (Hassell et al., 1978). Whether these pathways are mediated directly by retinol, or indirectly through a retinoic acid intermediate, requires further clarification.

A hormone-like action analogous to steroids, operating at the level of the nucleus, has often been postulated as a major function of vitamin A (Omori and Chytil, 1982; Chader, 1982; Wolf *et al.*, 1983; Olson *et al.*, 1983). Some evidence for this action in retinoblastoma cells was reported over a decade ago (Wiggert *et al.*, 1977), and additional findings are summarized in Chapter 7, Section 7.10.1.

The action of retinol on the nucleus is mediated by CRBP, which has a specific and saturable ability to transfer cytosolic retinol to specific binding sites in nuclear chromatin and thus influence nuclear transcription (Liau *et al.*, 1981, 1985). The binding of retinol to chromatin is noncovalent and requires CRBP as the specific transfer agent. Nuclear binding sites for retinoic acid bound to CRABP have also been reported, suggesting a mechanism that may be similar to that of retinol in its ability to modify gene expression (Sani and Banerjee, 1983). Direct effects of both retinol and retinoic acid on genomic expression *in vivo* have been shown in vitamin A-deficient rats by analyzing a series of poly(A)$^+$RNA transcripts from testes, intestinal mucosa, and liver (Omori and Chytil, 1982). These retinoids influence gene expression *in vivo* by activation as well as suppression of the genome. In other investigations using cultured keratinocytes, mRNAs isolated from cells grown in the presence or absence of vitamin A direct the translation of entirely different keratins (Fuchs and Green, 1981). Retinoic acid suppresses the expression of terminal differentiation-specific large keratins and at the same time induces the synthesis of new keratins not normally expressed *in vivo* (Kopan *et al.*, 1987). Thus, a specific effect of retinoids on mRNAs coding for specific proteins—in this case, keratins—has been clearly established.

The foregoing description of the nuclear functions of vitamin A does not preclude possible cytoplasmic or membrane functions (Wolf *et al.*, 1983). Fibroblasts from 3T3 cell lines treated with tumor-promoting phorbol esters release fibronectin from their cell surface, an effect that can be blocked by retinoic acid, even in enucleated cells (Bolmer and Wolf, 1982). In rats, however, vitamin A may specifically modulate fibronectin synthesis, possibly through the genome. A threefold increase in plasma fibronectin levels has been found in vitamin A-deficient rats (Zerlauth *et al.*, 1984). Further studies using a cDNA clone specific for rat liver fibronectin revealed a two- to fourfold increase in liver fibronectin mRNA as well as a greatly increased transcription rate in vitamin A-deficient animals (Kim and Wolf, 1987). The effect could be reversed in these hepatocytes by addition of either retinyl acetate or retinoic acid. It thus appears that vitamin A regulates the synthesis of hepatic fibronectin at the level of its mRNA transcription.

1.8. REFERENCES

Abdel-Latif, A. A., 1986, Calcium-mobilizing receptors, polyphosphoinositides, and the generation of second messengers, *Pharmacol. Rev.* **38:**227–272.

Alberts, B., Bray, D., Lewis, J., Raff, M., Roberts, K., and Watson, J. D., 1989, *Molecular Biology of the Cell*, 2nd ed. Garland Publishing, New York.

Anand-Srivastava, M. B., Srivastava, A. K., and Cantin, M., 1987, Pertussis toxin attenuates atrial natriuretic factor-mediated inhibition of adenylate cyclase, *J. Biol. Chem.* **262:**4931–4934.

Applebury, M. L., and Hargrave, P. A., 1986, Molecular biology of the visual pigments, *Vision Res.* **26:**1881–1895.

Argraves, W. S., Suzuki, S., Arai, H., Thompson, K., Pierschbacher, M. D., and Ruoslahti, E., 1987, Amino acid sequence of the human fibronectin receptor, *J. Cell Biol.* **105:**1183–1190.

Ashendel, C. L., 1985, The phorbol ester receptor: A phospholipid-regulated protein kinase, *Biochim. Biophys. Acta* **822:**219–242.

Ashkenazi, A., Winslow, J. W., Peralta, E. G., Peterson, G. L., Schimerlik, M. I., Capon, D. J., and Ramachandran, J., 1987, An M2 muscarinic receptor subtype coupled to both adenylyl cyclase and phosphoinositide turnover, *Science* **238:**672–675.

Benovic, J. L., Strasser, R. H., Caron, M. G., and Lefkowitz, R. J., 1986, β-Adrenergic receptor kinase: Identification of a novel protein kinase that phosphorylates the agonist-occupied form of the receptor, *Proc. Natl. Acad. Sci. USA* **83:**2797–2801.

Benovic, J. L., Mayor, F., Jr., Staniszewski, C., Lefkowitz, R. J., and Caron, M. G., 1987, Purification and characterization of the β-adrenergic receptor kinase, *J. Biol. Chem.* **262:**9026–9032.

Benovic, J. L., Bouvier, M., Caron, M. G., and Lefkowitz, R. J., 1988, Regulation of adenylyl cyclase-coupled β-adrenergic receptors, *Annu. Rev. Cell Biol.* **4:**405–428.

Bentz, H., Morris, N. P., Murray, L. W., Sakai, L. Y., Hollister, D. W., and Burgeson, R. E., 1983, Isolation and partial characterization of a new human collagen with an extended triple-helical structural domain, *Proc. Natl. Acad. Sci. USA* **80:**3168–3172.

Benya, P. D., and Padilla, S. R., 1986, Isolation and characterization of type VIII collagen synthesized by cultured rabbit corneal endothelial cells, *J. Biol. Chem.* **261:**4160–4169.

Berman, E. R. 1982, Sphingolipidoses and neuronal ceroid-lipofuscinosis, in: *Pathobiology of Ocular Disease, Part B* (A. Garner and G. K. Klintworth, eds.), Marcel Dekker, New York, pp. 897–929.

Berman, E. R., Segal, N., and Feeney, L., 1979, Subcellular distribution of free and esterified forms of vitamin A in the pigment epithelium of the retina and in liver, *Biochim. Biophys. Acta* **572:**167–177.

Berridge, M. J., 1987, Inositol trisphosphate and diacylglycerol: Two interacting second messengers, *Annu. Rev. Biochem.* **56:**159–193.

Berridge, M. J., and Irvine, R. F., 1984, Inositol trisphosphate, a novel second messenger in cellular signal transduction, *Nature* **312:**315–321.

Bhuyan, K. C., and Bhuyan, D. K., 1977, Regulation of hydrogen peroxide in eye humors. Effect of 3-amino-1H-1,2,4-triazole on catalase and glutathione peroxidase of rabbit eye, *Biochim. Biophys. Acta* **497:**641–651.

Bhuyan, K. C., and Bhuyan, D. K., 1978, Superoxide dismutase of the eye. Relative functions of superoxide dismutase and catalase in protecting the ocular lens from oxidative damage, *Biochim. Biophys. Acta* **542:**28–38.

Blomhoff, R., Norum, K. R., and Berg, T., 1985, Hepatic uptake of [^{3}H]retinol bound to the serum retinol binding protein involves both parenchymal and perisinusoidal stellate cells, *J. Biol. Chem.* **260:**13571–13575.

Bolmer, S. D., and Wolf, G., 1982, Retinoids and phorbol esters alter release of fibronectin from enucleated cells, *Proc. Natl. Acad. Sci. USA* **79:**6541–6545.

Bonner, T. I., Buckley, N. J., Young, A. C., and Brann, M. R., 1987, Identification of a family of muscarinic acetylcholine receptor genes, *Science* **237:**527–532.

Bouvier, M., Leeb-Lundberg, L. M. F., Benovic, J. L., Caron, M. G., and Lefkowitz, R. J., 1987, Regulation of adrenergic receptor function by phosphorylation, *J. Biol. Chem.* **262:**3106–3113.

Bruckner, P., Vaughan, L., and Winterhalter, K. H., 1985, Type IX collagen from sternal cartilage of chicken embryo contains covalently bound glycosaminoglycans, *Proc. Natl. Acad. Sci. USA* **82:**2608–2612.

Bulkley, G. B., 1983, The role of oxygen free radicals in human disease processes, *Surgery* **94:**407–411.

Burgeson, R. E., 1988, New collagens, new concepts, *Annu. Rev. Cell Biol.* **4:**551–577.

Burridge, K., Fath, K., Kelly, T., Nuckolls, G., and Turner, C., 1988, Focal adhesions: Transmembrane junctions between the extracellular matrix and the cytoskeleton, *Annu. Rev. Cell Biol.* **4:**487–525.

Burton, G. W., and Ingold, K. U., 1984, β-Carotene: An unusual type of lipid antioxidant, *Science* **224:**569–573.

Cadenas, E., 1989, Biochemistry of oxygen toxicity, *Annu. Rev. Biochem.* **58:**79–110.

Capdevila, J., Yadagiri, P., Manna, S., and Falck, J. R., 1986, Absolute configuration of the hydroxyeicosatetraenoic acids (HETEs) formed during catalytic oxygenation of arachidonic acid by microsomal cytochrome P-450, *Biochem. Biophys. Res. Commun.* **141:**1007–1011.

Chader, G. J., 1982, Retinoids in ocular tissues: Binding proteins, transport, and mechanism of action, in: *Cell Biology of the Eye* (D. S. McDevitt, ed.), Academic Press, New York, pp. 377–433.

Chader, G. J., 1984, Vitamin A, in: *Pharmacology of the Eye*, *Handbook of Pharmacology*, Vol. 69 (M. L. Sears, ed.), Springer-Verlag, Berlin, pp. 367–384. Chytil, F., 1984, Retinoic acid: Biochemistry, pharmacology, toxicology, and therapeutic use, *Pharmacol. Rev.* **36:**93S–100S.

Chytil, F., and Ong, D. E., 1984, Cellular retinoid binding proteins, in: *The Retinoids*, Vol. 2 (M. B. Sporn, A. B. Roberts, and D. S. Goodman, eds.), Academic Press, New York, pp. 89–123.

Cockcroft, S., 1987, Polyphosphoinositide phosphodiesterase: Regulation by a novel guanine nucleotide binding protein, G_p, *Trends Biochem. Sci.* **12:**75–78.

Colantuoni, V., Cortese, R., Nilsson, M., Lundvall, J., Bavik, C.-O., Eriksson, U., Peterson, P. A., and Sundelin, J., 1985, Cloning and sequencing of a full length cDNA corresponding to human cellular retinol-binding protein, *Biochem. Biophys. Res. Commun.* **130:**431–439.

Crouch, R., Priest, D. G., and Duke, E. J., 1978, Superoxide dismutase activities of bovine ocular tissues, *Exp. Eye Res.* **27:**503–509.

Cullum, M. E., and Zile, M. H., 1986, Quantitation of biological retinoids by high-pressure liquid chromatography: Primary internal standardization using tritiated retinoids, *Anal. Biochem.* **153:**23–32.

Darnell, J., Lodish, H., and Baltimore, D., 1986, *Molecular Cell Biology*, Scientific American Books, W. H. Freeman, New York.

DeLuca, L. M., 1977, Vitamin A and glycosylation. The direct involvement of vitamin A in glycosyl transfer reactions of mammalian membranes, *Vitam. Horm.* **35:**1–57.

DeLuca, L. M., and Shapiro, S. S. (eds.), 1981, *Modulation of Cellular Interactions by Vitamin A and Derivatives (Retinoids)*, *Ann. N.Y. Acad. Sci*, **359:**1–430.

DeLuca, H. F., Zile, M., and Sietsma, W. K., 1981, The metabolism of retinoic acid to 5,6-epoxyretinoic acid, retinoyl-β-glucuronide, and other polar metabolites, *Ann. N.Y. Acad. Sci.* **369:**25–36.

Dilley, R. A., and McConnell, D. G., 1970, Alpha-tocopherol in retinal rod outer segments of bovine eyes, *J. Membr. Biol.* **2:**317–323.

Dixon, R. A. F., Kobilka, B. K., Strader, D. J., Benovic, J. L., Dohlman, H. G., Frielle, T., Bolanowski, M. A., Bennett, C. D., Rands, E., Diehl, R. E., Mumford, R. A., Slater, E. E., Sigal, I. S., Caron, M. G., Lefkowitz, R. J., and Strader, C. D., 1986, Cloning of the gene and cDNA for mammalian β-adrenergic receptor and homology with rhodopsin, *Nature* **321:**75–79.

Dohlman, H. G., Caron, M. G., and Lefkowitz, R. J., 1987, A family of receptors coupled to guanine nucleotide regulatory proteins, *Biochemistry* **26:**2657–2664.

Douglas, C. E., Chan, A. C., and Choy, P. C., 1986, Vitamin E inhibits platelet phospholipase A_2, *Biochim. Biophys. Acta* **876:**639–645.

Dowling, J., and Wald, G., 1960, The biological function of vitamin A acid, *Proc. Natl. Acad. Sci. USA* **46:**587–608.

Dublet, B., and van der Rest, M., 1987, Type XII collagen is expressed in embryonic chick tendons, *J. Biol. Chem.* **262:**17724–17727.

Elbein, A. D., 1987, Inhibitors of the biosynthesis and processing of N-linked oligosaccharide chains, *Annu. Rev. Biochem.* **56:**497–534.

Engvall, E., Hessle, H., and Klier, G., 1986, Molecular assembly, secretion, and matrix deposition of type VI collagen, *J. Cell Biol.* **102:**703–710.

Enna, S. J., and Karbon, E. W., 1987, Receptor regulation: Evidence for a relationship between phospholipid metabolism and neurotransmitter receptor-mediated cAMP formation in brain, *Trends Pharmacol. Sci.* **8:**21–24.

Evans, W. H., 1988, Gap junctions: Towards a molecular structure, *BioEssays* **8:**3–6.

Evered, D., and Whelan, J., (eds.), 1986, *Functions of the Proteoglycans, Ciba Foundation Symposium 124*, John Wiley & Sons, Chichester.

Feeney, L., and Berman, E. R., 1976, Oxygen toxicity: Membrane damage by free radicals, *Invest. Ophthalmol.* **15:**789–792.

Feeney-Burns, L., Berman, E. R., and Rothman, H., 1980, Lipofuscin of human retinal pigment epithelium, *Am. J. Ophthalmol.* **90:**783–791.

Ferguson, M. A. J., and Williams, A. F., 1988, Cell-surface anchoring of proteins via glycosyl-phosphatidylinositol structures, *Annu. Rev. Biochem.* **57:**285–320.

Flynn, T. G., de Bold, M. L., and de Bold, A. J., 1983, The amino acid sequence of an atrial peptide with potent diuretic and natriuretic properties, *Biochem. Biophys. Res. Commun.* **117:**859–865.

Flynn, T. G., Davies, P. L., Kennedy, B. P., de Bold, M. L., and de Bold, A. J., 1985, Alignment of rat cardionatrin sequences with the preprocardionatrin sequence from complementary DNA, *Science* **227:**323–325.

Fransson, L.-A., 1987, Structure and function of cell-associated proteoglycans, *Trends Biochem. Sci.* BB12:406–411.

Freeman, B. A., 1984, Biological sites and mechanisms of free radical production, in: *Free Radicals in Molecular Biology, Aging and Disease* (D. Armstrong, R. S. Sohal, R. G. Cutler, and T. F. Slater, eds.), Raven Press, New York, pp. 43–52.

Fridovich, I., 1972, Quantitative aspects of the production of superoxide anion radical by xanthine oxidase, *J. Biol. Chem.* **247:**4053–4057.

Fridovich, I., 1975, Superoxide dismutases, *Annu. Rev. Biochem.* **44:**147–159.

Fridovich, I., 1983, Superoxide radical: An endogenous toxicant, *Annu. Rev. Pharmacol. Toxicol.* **23:**239–257.

Fridovich, I., 1986, Biological effects of the superoxide radical, *Arch. Biochem. Biophys.* **247:**1–11.

Fridovich, I., 1989, Superoxide dismutases. An adaptation to a paramagnetic gas, *J. Biol. Chem.* **264:**7761–7764.

Fridovich, I., and Freeman, B., 1986, Antioxidant defenses in the lung, *Annu. Rev. Physiol.* **48:**693–702.

Fuchs, E., and Green, H., 1981, Regulation of terminal differentiation of cultured human keratinocytes by vitamin A, *Cell* **25:**617–625.

Futai, M., Noumi, T., and Maeda, M., 1989, ATP synthase (H^+-ATPase): Results by combined biochemical and molecular biological approaches, *Annu. Rev. Biochem.* **58:**111–136.

Gardner, D. G., Vlasuk, G. P., Baxter, J. D., Fiddes, J. C., and Lewicki, J. A., 1987, Identification of atrial natriuretic factor gene transcripts in the central nervous system of the rat, *Proc. Natl. Acad. Sci. USA* **84:**2175–2179.

Gilman, A. G., 1984, G proteins and dual control of adenylate cyclase, *Cell* **36:**577–579.

Gilman, A. G., 1987, G proteins: Transducers of receptor-generated signals, *Annu. Rev. Biochem.* **56:**615–649.

Gjoen, T., Bjerkelund, T., Blomhoff, H. K., Norum, K. R., Berg, I., and Blomhoff, R., 1987, Liver takes up retinol-binding protein from plasma, *J. Biol. Chem.* **262:**10926–10930.

Goodman, D. S., 1984, Plasma retinol-binding protein, in: *The Retinoids*, Vol. 2 (M. B. Sporn, A. B. Roberts, and D. S. Goodman, eds.), Academic Press, Orlando, FL, pp. 41–88.

Granger, D. N., Hollwarth, M. E., and Parks, D. A., 1986, Ischemia–reperfusion injury: Role of oxygen-derived free radicals, *Acta Physiol. Scand. [Suppl.]* **548:**47–63.

Graziano, M. P., and Gilman, A. G., 1987, Guanine nucleotide-binding regulatory proteins: Mediators of transmembrane signaling, *Trends Pharmacol. Sci.* **8:**478–481.

Greenwald, R. A. (ed.), 1985, *CRC Handbook of Methods for Oxygen R adical Research*, CRC Press, Boca Raton, FL.

Hall, M. O., and Hall, D. O., 1975, Superoxide dismutase of bovine and frog rod outer segments, *Biochem. Biophys. Res. Commmun.* **67:**1199–1204.

Hall, P. F., 1985, Role of cytochromes P-450 in the biosynthesis of steroid hormones, *Vitam. Horm.* **42:**315–368.

Halliwell, B., and Grootveld, M., 1987, The measurement of free radical reactions in humans, *FEBS Lett.* **213:**9–14.

Halliwell, B., and Gutteridge, J. M. C., 1984a, Oxygen toxicity, oxygen radicals, transition metals and disease, *Biochem. J.* **219:**1–14.

Halliwell, B., and Gutteridge, J. M. C., 1984b, Lipid peroxidation, oxygen radicals, cell damage, and antioxidant therapy, *Lancet* **1:**1396–1397.

Handelman, G. J., and Dratz, E. A., 1986, The role of antioxidants in the retina and retinal pigment epithelium and the nature of prooxidant-induced damage, *Adv. Free Radicals Biol. Med.* **2:**1–89.

Harden, T. K., Stephens, L., Hawkins, P. T., and Downes, C.P., 1987, Turkey erythrocyte membranes as a model for regulation of phospholipase C by guanine nucleotides, *J. Biol. Chem.* **262:**9057–9061.

Harper, E., 1980, Collagenases, *Annu. Rev. Biochem.* **49:**1063–1078.

Harrison, E. H., Smith, J. E., and Goodman, D. S., 1979, Unusual properties of retinyl palmitate hydrolase activity in rat liver, *J. Lipid Res.* **20:**760–771.

Hascall, V. C., and Kimura, J. H., 1982, Proteoglycans: Isolation and characterization, in: *Methods in Enzymology*, Vol. 82 (L. W. Cunningham and D. W. Frederiksen, eds.), Academic Press, New York, pp. 769–800.

Hassell, J., Silverman-Jones, C., and DeLuca, L., 1978, The *in vivo* stimulation of mannose incorporation into mannosylretinyl phosphate, dolichylmannosyl phosphate and specific glycopeptides of rat liver by high doses of retinyl palmitate, *J. Biol. Chem.* **253:**1627–1631.

Hassell, J. R., Newsome, D. A., and DeLuca, L. M., 1980, Increased biosynthesis of specific glycoconjugates in rat corneal epithelium following treatment with vitamin A, *Invest. Ophthalmol. Vis. Sci.* **16:**642–647.

Hassell, J. R., Kimura, J. H., and Hascall, V. C., 1986, Proteoglycan core protein families, *Annu. Rev. Biochem.* **55:**539–567.

Higgs, G. A., Salmon, J. A., Henderson, B., and Vane, J.R., 1987, Pharmacokinetics of aspirin and salicylate in relation to inhibition of arachidonate cyclooxygenase and antiinflammatory activity, *Proc. Natl. Acad. Sci USA* **84:**1417–1420.

Hirasawa, K., and Nishizuka, Y., 1985, Phosphatidylinositol turnover in receptor mechanism and signal transduction, *Annu. Rev. Pharmacol. Toxicol.* **25:**147–170.

Hirschberg, C. B., and Snider, M.D., 1987, Topography of glycosylation in the rough endoplasmic reticulum and Golgi apparatus, *Annu. Rev. Biochem.* **56:**63–87.

Hokin, L. E., 1985, Receptors and phosphoinositide-generated second messengers, *Annu. Rev. Biochem.* **54:**205–235.

Houslay, M. D., 1987, Egg activation unscrambles a potential role for IP_4, *Trends Biochem. Sci.* **12:**1–2.

Huganir, R. L., and Greengard, P., 1987, Regulation of receptor function by protein phosphorylation, *Trends Pharmacol. Sci.* 8:472–477.

Hughes, H., Smith, C. V., Tsokos-Kuhn, J. O., and Mitchell, J. R., 1986, Quantitation of lipid peroxidation products by gas chromatography-mass spectrometry, *Anal. Biochem.* **152:**107–112.

Hyslop, P. A., Hinshaw, D. B., Hasley, W. A. Jr., Schraufstatter, I. U., Sauerheber, R. D., Spragg, R. G., Jackson, J. H., and Cochrane, C. G., 1988, Mechanisms of oxidant-mediated cell injury. The glycolytic and mitochondrial pathways of ADP phosphorylation are major intracellular targets inactivated by hydrogen peroxide, *J. Biol. Chem.* **263:**1665–1675.

Irvine, R. F., Letcher, A. J., Heslop, J. P., and Berridge, M. J., 1986, The inositol tris/tetrakisphosphate pathway demonstration of Ins(1,4,5)P_3 3-kinase activity in animal tissues, *Nature* **320:**631–634.

Jaken, S., and Kiley, S. C., 1987, Purification and characterization of three types of protein kinase C from rabbit brain cytosol, *Proc. Natl. Acad. Sci. USA* **84:**4418–4422.

Jander, R., Troyer, D., and Rauterberg, J., 1984, A collagen-like glycoprotein of the extracellular matrix is the undegraded form of type VI collagen, *Biochemistry* **23:**3675–3681.

Kapoor, R., Bornstein, P., and Sage, E. H., 1986, Type VIII collagen from bovine Descemet's membrane: Structural characterization of a triple-helical domain, *Biochemistry* **25:**3930–3937.

Kato, M., Kato, K., and Goodman, D. S., 1984, Immunocytochemical studies on the localization of plasma and of cellular retinol-binding proteins and of transthyretin (prealbumin) in rat liver and kidney, *J. Cell. Biol.* **98:**1696–1704.

Kato, M., Blaner, W. S., Mertz, J. R., Das, K., Kato, K., and Goodman, D.S., 1985, Influence of retinoid nutritional status on cellular retinol- and cellular retinoic acid-binding protein concentrations in various rat tissues, *J. Biol. Chem.* **260:**4832–4838.

Kerlavage, A. R., Fraser, C. M., and Venter, J. C., 1987, Muscarinic cholinergic receptor structure: Molecular biological support for subtypes, *Trends Pharmacol. Sci.* **8:**426–431.

Kikkawa, U., Kishimoto, A., and Nishizuka, Y., 1989, The protein kinase C family: Heterogeneity and its implications, *Annu. Rev. Biochem.* **58:**31–44.

Kim, H.-Y., and Wolf, G., 1987, Vitamin A deficiency alters genomic expression for fibronectin in liver and hepatocytes, *J. Biol. Chem.* **262:**365–371.

Kobilka, B. K., Dixon, R. A. F., Frielle, T., Dohlman, H. G., Bolanowski, M. A., Sigal, I. S., Yang-Feng, T. L., Francke, U., Caron, M. G., and Lefkowitz, R. J., 1987, cDNA for the human β_2-adrenergic receptor: A protein with multiple membrane-spanning domains and encoded by a gene whose chromosomal location is shared with that of the receptor for platelet-derived growth factor, *Proc. Natl. Acad. Sci. USA* **84:**46–50.

Kohno, T., Sorgente, N., Ishibashi, T., Goodnight, R., and Ryan, S. J., 1987, Immunofluorescent studies of fibronectin and laminin in the human eye, *Invest. Ophthalmol. Vis. Sci.* 28:506–514.

Kopan, R., Traska, G., and Fuchs, E., 1987, Retinoids as important regulators of terminal differentiation: Examining keratin expression in individual epidermal cells at various stages of keratinization, *J. Cell Biol.* **105:**427–440.

Kornfeld, R., and Kornfeld, S., 1985, Assembly of asparagine-linked oligosaccharides, *Annu. Rev. Biochem.* **54:**631–664.

Kwatra, M. M., Leung, E., Maan, A. C., McMahon, K. K., Ptasienski, J., Green, R.D., and Hosey, M.M., 1987, Correlation of agonist-induced phosphorylation of chick heart muscarinic receptors with receptor desensitization, *J. Biol. Chem.* **262:**16314–16321.

Lands, W. E. M., 1985, Interactions of lipid hydroperoxides with eicosanoid biosynthesis, *J. Free Radicals Biol. Med.* **1:**97–101.

Leeb-Lundberg, L. M. F., Cotecchia, S., Lomasney, J. W., DeBernardis, J. F. Lefkowitz, R. J., and Caron, M. G., 1985, Phorbol esters promote α_1-adrenergic receptor phosphorylation and receptor uncoupling from inositol phospholipid metabolism, *Proc. Natl. Acad. Sci. USA* **82:**5651–5655.

Leeb-Lundberg, L. M. F., Cotecchia, S., DeBlasi, A., Caron, M. G., and Lefkowitz, R. J., 1987, Regulation of adreneregic receptor function by phosphorylation. I. Agonist-promoted desensitization and phosphorylation of α1-adrenergic receptors coupled to inositol phospholipid metabolism in DDT_1 MF-2 smooth muscle cells, *J. Biol. Chem.* **262:**3098–3105.

Lefkowitz, R. J., Stadel, J. M., and Caron, M. G., 1983, Adenylate cyclase-coupled beta-adrenergic receptors: Structure and mechanisms of activation and desensitization, *Annu. Rev. Biochem.* **52:**159–186.

Lehninger, A. L., 1982, *Principles of Biochemistry*, Worth, New York.

Levitzki, A., 1987, Regulation of hormone-sensitive adenylate cyclase, *Trends Pharmacol. Sci.* **8:**299–303.

Liau, G., Ong, D. E., and Chytil, F., 1981, Interaction of the retinol/cellular retinol-binding protein complex with isolated nuclei and nuclear components, *J. Cell Biol.* **91:**63–68.

Liau, G., Ong., D. E., and Chytil, F., 1985, Partial characterization of nuclear binding sites for retinol delivered by cellular retinol binding protein, *Arch. Biochem. Biophys.* **237:**354–360.

Lindberg, K. A., Jr., and Pinnell, S. R., 1982, Collagen and its disorders, in: *Pathobiology of Ocular Disease, Part B* (A. Garner and G. K. Klintworth, eds.), Marcel Dekker, New York, pp. 1009–1031.

Low, M. G., and Saltiel, A. R., 1988, Structural and functional roles of glycosyl-phosphatidylinositol in membranes, *Science* **239:**268–275.

Lunstrum, G. P., Sakai, L. Y., Keene, D. R., Morris, N. P., and Burgeson, R. E., 1986, Large complex globular domains of type VII procollagen contribute to the structure of anchoring fibrils, *J. Biol. Chem.* **261:**9042–9048.

Lunstrum, G. P., Kuo, H.-J., Rosenbaum, L. M., Keene, D. R., Glanville, R. W., Sakai, L. Y., and Burgeson,

R. E., 1987, Anchoring fibrils contain the carboxyl-terminal globular domain of type VII procollagen, but lack the amino-terminal globular domain, *J. Biol. Chem.* **262:**13706–13712.

MacDonald, P. N., and Ong, D. E., 1987, Binding specificities of cellular retinol-binding protein and cellular retinol-binding protein, type II, *J. Biol. Chem.* **262:**10550–10556.

Majerus, P. W., Connolly, T. M., Bansal, V. S., Inhorn, R. C., Ross, T. S., and Lips, D. L., 1988, Inositol phosphates: Synthesis and degradation, *J. Biol. Chem.* **263:**3051–3054.

Martin, G. R., Timpl, R., Muller, P. K., and Kuhn, K., 1985, The genetically distinct collagens, *Trends Biochem. Sci.* **10:**285–287.

Mayne, R., and Burgeson, R. E., 1987, *Structure and Function of Collagen Types*, Academic Press, Orlando, FL.

McCay, P. B., 1985, Vitamin E: Interactions with free radicals and ascorbate, *Annu. Rev. Nutr.* **5:**323–340.

McCord, J. M., 1986, Superoxide radical: A likely link between reperfusion injury and inflammation, *Adv. Free Radical Biol. Med.* **2:**325–345. McCord, J. M. and Fridovich, I., 1969, Superoxide dismutase: An enzymatic function for erythrocuprein (hemocuprein), *J. Biol. Chem.* **244:**6049–6055.

McCormick, D., van der Rest, M., Goodship, J., Lozano, G., Ninomiya, Y., and Olsen, B. R., 1987, Structure of the glycosaminoglycan domain in the type IX collagen-proteoglycan, *Proc. Natl. Acad. Sci. USA* **84:**4044–4048.

Mead, J. F., 1984, Free radical mechanisms in lipid peroxidation and prostaglandins, in: *Free Radicals in Molecular Biology, Aging and Disease* (D. Armstrong, R. S. Sohal, R. G. Cutler, and T. F. Slater, eds.), Raven Press, New York, pp. 53–66.

Minotti, G., and Aust, S. D., 1987, The requirement for iron(III) in the initiation of lipid peroxidation by iron(II) and hydrogen peroxide, *J. Biol. Chem.* **262:**1098–1104.

Mochly-Rosen, D., and Koshland, D. E., Jr., 1987, Domain structure and phosphorylation of protein kinase C, *J. Biol. Chem.* **262:**2291–2297.

Morris, N. P., and Bachinger, H. P., 1987, Type XI collagen is a heterotrimer with the composition (1α,2α,3α) retaining non-triple-helical domains, *J. Biol. Chem.* **262:**11345–11350.

Napier, M. A., Vandlen, R. L., Albers-Schonberg, G., Nutt, R. F., Brady, S., Lyle, T., Winquist, R., Faison, E. P., Heinel, L. A., and Blaine, E. H., 1984, Specific membrane receptors for atrial natriuretic factor in renal and vascular tissues, *Proc. Natl. Acad. Sci. USA* **81:**5946–5950.

Napoli, J. L., McCormick, A. M., O'Meara, B., and Dratz, E. A, 1984, Vitamin A metabolism: α-Tocopherol modulates tissue retinol levels *in vivo*, and retinyl palmitate hydrolysis *in vitro*, *Arch. Biochem. Biophys.* **230:**194–202.

Nebert, D. W., Eisen, H. J., Negishi, M., Land, M. A., and Hjelmeland, L. M., 1981, Genetic mechanisms controlling the induction of polysubstrate monooxygenase (P-450) activities, *Annu. Rev. Pharmacol. Toxicol.* **21:**431–462.

Needleman, P., Turk, J., Jakschik, B. A., Morrison, A. R., and Lefkowith, J.B., 1986, Arachidonic acid metabolism, *Annu. Rev. Biochem.* **55:**69–102.

Nelson, C. A., and Seamon, K. B., 1986, Binding of [^{3}H]forskolin to human platelet membranes, *J. Biol. Chem.* **261:**13469–13473.

Neufeld, E. F., and McKusick, V. A., 1983, Disorders of lysosomal enzyme synthesis and localization: I-cell disease and pseudo-Hurler polydystrophy, in: *The Metabolic Basis of Inherited Disease*, 5th ed. (J. B. Stanbury, J. B. Wyngaarden, D. S. Fredrickson, J. L. Goldstein, and M. S. Brown, eds.), McGraw-Hill, New York, pp. 778–787.

Nirenberg, M. W., and Matthaei, J. H., 1961, The dependence of cell-free protein synthesis in *E.Coli* upon naturally occurring or synthetic polyribonucleotides, *Proc. Natl. Acad. Sci. USA* **47:**1588–1602.

Nishizuka, Y., 1986, Studies and perspectives of protein kinase C, *Science* **233:**305–312.

Oliw, E. H., Guengerich, F. P., and Oates, J.A., 1982, Oxygenation of arachidonic acid by hepatic monooxygenases. Isolation and metabolism of four epoxide intermediates, *J. Biol. Chem.* **257:**3771–3781.

Olson, J. A., and Gunning, D., 1983, The storage form of vitamin A in rat liver cells, *J. Nutr.* **113:**2184–2191.

Olson, J. A., Bridges, C. D. B., Packer, L., Chytil, F., and Wolf, G., 1983, The function of vitamin A, *Fed. Proc.* **42:**2740–2746.

Omori, M., and Chytil, F., 1982, Mechanism of vitamin A action. Gene expression in retinol-deficient rats, *J. Biol. Chem.* **257:**14370–14374.

Ong, D. E., Kakkad, B., and MacDonald, P. N., 1987, Acyl-CoA- independent esterification of retinol bound to cellular retinol-binding protein (type II) by microsomes from rat small intestine, *J. Biol. Chem.* **262:**2729–2736.

Packer, L. (ed.), 1984, *Oxygen Radicals in Biological Systems, Methods in Enzymology*, Vol. 105, Academic Press, Orlando, FL.

Paris, S., and Pouyssegur, J., 1987, Further evidence for a phospholipase C-coupled G protein in hamster fibroblasts, *J. Biol. Chem.* **262:**1970–1976.

Parsons, W. J., and Stiles, G. L., 1987, Heterologous desensitization of the inhibitory A_1 adenosine receptor–adenylate cyclase system in rat adipocytes, *J. Biol. Chem.* **262:**841–847.

Parthasarathy, R., and Eisenberg, F., Jr., 1986, The inositol phospholipids: A stereochemical view of biological activity, *Biochem. J.* **235:**313–322.

Paul, A. K., Marala, R. B., Jaiswal, R. K., and Sharma, R. K., 1987, Coexistence of guanylate cyclase and atrial natriuretic factor receptor in a 180-kD protein, *Science* **235:**1224–1226.

Pedersen, P. L., and Carafoli, E., 1987, Ion motive ATPases. I. Ubiquity, properties, and significance to cell function. *Trends Biochem. Sci.* **12:**146–150.

Peralta, E. G., Winslow, J. W., Peterson, G. L., Smith, D. H., Ashkenazi, A., Ramachandran, J., Schimerlik, M. I., and Capon, D. J., 1987, Primary structure and biochemical properties of an M_2 muscarinic receptor, *Science* **236:**600–605.

Pfeuffer, E., Dreher, R.-M., Metzger, H., and Pfeuffer, T., 1985, Catalytic unit of adenylate cyclase: Purification and identification by affinity crosslinking, *Proc. Natl. Acad. Sci. USA* **82:**3086–3090.

Pinnell, S. R., and Murad, S., 1983, Disorders of collagen, in: *The Metabolic Basis of Inherited Disease*, 5th ed. (J. B. Stanbury, J. B. Wyngaarden, D. S. Fredrickson, J. L. Goldstein, and M. S. Brown, eds.), McGraw-Hill, New York, pp. 1425–1449.

Poole, A. R., 1986, Proteoglycans in health and disease: Structures and functions, *Biochem. J.* **236:**1–14.

Pryor, W. A., 1987, Detection of lipid hydroperoxides, *Free Radical Biol. Med.* **3:**317.

Prystowsky, J. H., Smith, J. E., and Goodman, D. S., 1981, Retinyl palmitate hydrolase activity in normal rat liver, *J. Biol. Chem.* **256:**4498–4503.

Rao, N. A., Thaete, L. G., Delmage, J. M., and Sevanian, A., 1985, Superoxide dismutase in ocular structures, *Invest. Ophthalmol. Vis. Sci.* **26:**1778–1781.

Rasmussen, H., 1986a, The calcium messenger system (first of two parts), *N. Engl. J. Med.* **314:**1094–1101.

Rasmussen, H., 1986b, The calcium messenger system (second of two parts), *N. Engl. J. Med.* **314:**1164–1170.

Revel, J. P., Yancey, S. B., Nicholson, B., and Hoh, J., 1987, Sequence diversity of gap junction proteins, in: *Junctional Complexes of Epithelial Cells, Ciba Foundation Symposium 125*, John Wiley & Sons, Chichester, pp. 108–127.

Rodbell, M., 1980, The role of hormone receptors and GTP-regulatory proteins in membrane transduction, *Nature* **284:**17–21.

Roden, L., 1980, Structure and metabolism of connective tissue proteoglycans, in: *The Biochemistry of Glycoproteins and Proteoglycans* (W. J. Lennarz, ed.), Plenum Press, New York, pp. 267–371.

Romano, M. C., Eckardt, R. D., Bender, P. E., Leonard, T. B., Straub, K. M., and Newton, J. F., 1987, Biochemical characterization of hepatic microsomal leukotriene B_4 hydroxylases, *J. Biol. Chem.* **262:**1590–1595.

Rossier, B. C., Geering, K., and Kraehenbuhl, J. P., 1987, Regulation of the sodium pump: How and why? *Trends Biochem. Sci.* **12:**483–487.

Rouzer, C. A., and Samuelsson, B., 1987, Reversible, calcium-dependent membrane association of human leukocyte 5-lipoxygenase, *Proc. Natl. Acad. Sci USA* **84:**7393–7397.

Ruoslahti, E., 1988a, Fibronectin and its receptors, *Annu. Rev. Biochem.* **57:**375–413.

Ruoslahti, E., 1988b, Structure and biology of proteoglycans, *Annu. Rev. Cell Biol.* **4:**229–255.

Ryu, S. H., Cho, K. S., Lee, K.-Y., Suh, P.-G., and Rhee, S. G., 1987a, Purification and characterization of two immunologically distinct phosphoinositide-specific phospholipase C from bovine brain, *J. Biol. Chem.* **262:**12511–12518.

Ryu, S. H., Suh, P.-G., Cho, K. S., Lee, K.-Y., and Rhee, S. G., 1987b, Bovine brain cytosol contains three immunologically distinct forms of inositolphospholipid-specific phospholipase C, *Proc. Natl. Acad. Sci. USA* **84:**6649–6653.

Sakai, L. Y., Keene, D. R., Morris, N. P., and Burgeson, R. E., 1986, Type VII collagen is a major structural component of anchoring fibrils, *J. Cell Biol.* **103:**1577–1586.

Sani, B. P. and Banerjee, C. K., 1983, Cellular receptor mediation of the action of retinoic acid, in: *Modulation and Mediation of Cancer by Vitamins* (F. L. Meyskens and K. N. Prasad, eds.), S. Karger, Basel, pp. 153–161.

Schenk, D. B., Phelps, M. N., Porter, J. G., Fuller, F., Cordell, B., and Lewicki, J. A., 1987, Purification and subunit composition of atrial natriuretic peptide receptor, *Proc. Natl. Acad. Sci. USA* **84:**1521–1525.

Schuetz, E. G., Wrighton, S. A., Barwick, J. L., and Guzelian, P. S., 1984, Induction of cytochrome P-450 by glucocorticoids in rat liver. I. Evidence that glucocorticoids and pregnenolone 16α-carbonitrile regulate *de novo* synthesis of a common form of cytochrome P-450 in cultures of adult rat hepatocytes and in the liver *in vivo*, *J. Biol. Chem.* **259:**1999–2006.

Schwartz, D., Geller, D. M., Manning, P. T., Siegel, N. R., Fok, K. F., Smith, C. E., and Needleman, P., 1985, Ser-Leu-Arg-Arg-atriopeptin III: The major circulating form of atrial peptide, *Science* **229:**397–400.

Schwartzman, M. L., Masferrer, J., Dunn, M. W., McGiff, J. C., and Abraham, N. G., 1987a, Cytochrome P450, drug metabolizing enzymes and arachidonic acid metabolism in bovine ocular tissues, *Curr. Eye Res.* **6:**623–630.

Schwartzman, M. L., Balazy, M., Masferrer, J., Abraham, N. G., McGiff, J. C., and Murphy, R. C., 1987b, 12(*R*)- Hydroxyeicosatetraenoic acid: A cytochrome P450-dependent arachidonate metabolite that inhibits Na^+,K^+-ATPase in the cornea, *Proc. Natl. Acad. Sci USA* **84:**8125–8129.

Scott, J. E., 1988, Proteoglycan–fibrillar collagen interactions, *Biochem. J.* **252:**313–323.

Seamon, K. B., and Daly, J. W., 1981, Forskolin: A unique diterpene activator of cyclic AMP-generating systems, *J. Cyclic Nucleotide Res.* **7:**201–224.

Seamon, K. B., and Daly, J. W., 1985, High-affinity binding of forskolin to rat brain membranes, *Adv. Cyclic Nucleotide Protein Phosphorylation Res.* **19:**125–135.

Seamon, K. B., and Wetzel, B., 1984, Interaction of forskolin with dually regulated adenylate cyclase, *Adv. Cyclic Nucleotide Protein Phosphorylation Res.* **17:**91–99.

Sevanian, A., and Hochstein, P., 1985, Mechanisms and consequences of lipid peroxidation in biological systems, *Annu. Rev. Nutr.* **5:**365–390.

Sevanian, A., and Kim, E., 1985, Phospholipase A_2 dependent release of fatty acids from peroxidized membranes, *J. Free Radicals Biol. Med.* **1:**263–271.

Sherman, D. R., Lloyd, R. S., and Chytil, F., 1987, Rat cellular retinol-binding protein: cDNA sequence and rapid retinol-dependent accumulation of mRNA, *Proc. Natl. Acad. Sci. USA* **84:**3209–3213.

Shi, Q.-H., Ruiz, J.A. and Ho, R.-J., 1986, Forms of adenylate cyclase, activation and/or potentiation by forskolin, *Arch. Biochem. Biophys.* **251:**156–165.

Shichi, H., 1984, Biotransformation and drug metabolism, in: *Pharmacology of the Eye, Handbook of Pharmacology*, Vol. 69 (M. L. Sears, ed.), Springer-Verlag, Berlin, pp. 117–148.

Shichi, H., Atlas, S. A., and Nebert, D. W., 1975, Genetically regulated aryl hydrocarbon hydroxylase induction in the eye: Functional significance of the drug-metabolizing system for the retinal pigmented epithelium–choroid, *Exp. Eye Res.* **21:**557–567.

Shimonaka, M., Saheki, T., Hagiwara, H., Ishido, M., Nogi, A., Fujita, T., Wakita, K.-i, Inada, Y., Kondo, J., and Hirose, S., 1987, Purification of atrial natriuretic peptide receptor from bovine lung. Evidence for a disulfide-linked subunit structure, *J. Biol. Chem.* **262:**5510–5514.

Sibley, D. R., and Lefkowitz, R. J., 1985, Molecular mechanisms of receptor desensitization using the β-adrenergic receptor-coupled adenylate cyclase system as a model, *Nature* **317:**124–129.

Sibley, D. R., Benovic, J. L., Caron, M. G., and Lefkowitz, R. J., 1987, Regulation of transmembrane signaling by receptor phosphorylation, *Cell* **48:**913–922.

Simpson, P. J., Mickelson, J. K., and Lucchesi, B. R., 1987, Free radical scavengers in myocardial ischemia, *Fed. Proc.* **46:**2413–2421.

Slater, T. F., 1984, Free-radical mechanisms in tissue injury, *Biochem. J.* **222:**1–15.

Smigel, M. D., 1986, Purification of the catalyst of adenylate cyclase, *J. Biol. Chem.* **261:**1976–1982.

Smigel, M. D., Ferguson, K. M., and Gilman, A. G., 1985, Control of adenylate cyclase activity by G proteins, *Adv. Cyclic Nucleotide Protein Phosphorylation Res.* **19:**103–111.

Smith, C. V., and Anderson, R. E., 1987, Methods for determination of lipid peroxidation in biological samples, *Free Radical Biol. Med.* **3:**341–344.

Soprano, D. R., Smith, J. E., and Goodman, D. S., 1982, Effect of retinol status on retinol-binding protein biosynthesis rate and translatable messenger RNA level in rat liver, *J. Biol. Chem.* **257:**7693–7697.

Soprano, D. R., Soprano, K. J., and Goodman, D. S., 1986, Retinol-binding protein messenger RNA levels in the liver and in extrahepatic tissues of the rat, *J. Lipid Res.* **27:**166–171.

Southern, E. M., 1975, Detection of specific sequences among DNA fragments separated by gel electrophoresis, *J. Mol. Biol.* **98:**503–517.

Spiegel, A. M., 1987, Signal transduction by guanine nucleotide binding proteins, *Mol. Cell. Endocrinol.* **49:**1–16.

Sporn, M. B., Roberts, A. B., and Goodman, D. S. (eds.), 1984, *The Retinoids*, Vols. 1 and 2, Academic Press, Orlando, FL.

Sramek, S. J., Wallow, I. H. L., Bindley, C., and Sterken, G., 1987, Fibronectin distribution in the rat eye, *Invest. Ophthalmol. Vis. Sci.* **28:**500–505.

Steardo, L., and Nathanson, J. A., 1987, Brain barrier tissues: End organs for atriopeptins, *Science* **235:**470–473.

Steinert, P. M., and Roop, D.R., 1988, Molecular and cellular biology of intermediate filaments, *Annu. Rev. Biochem.* **57:**593–625.

Stiles, G. L., Caron, M. G., and Lefkowitz, R. J., 1984, β-Adrenergic receptors: Biochemical mechanisms of physiological regulation, *Physiol. Rev.* **64:**661–743.

Stryer, L., 1988, *Biochemistry*, 3rd ed., W. H. Freeman, New York.

Stryer, L., and Bourne, H. R., 1986, G proteins: A family of signal transducers, *Annu. Rev. Cell Biol.* **2:**391–419.

Summers, T. A., Irwin, M. H., Mayne, R., and Balian, G., 1988, Monoclonal antibodies to type X collagen, *J. Biol. Chem.* **263:**581–587.

Sundelin, J., Anundi, H., Tragardh, L., Eriksson, U., Lind, P., Ronne, H., Peterson, P. A., and Rask, L., 1985, The primary structure of rat liver cellular retinol-binding protein, *J. Biol. Chem.* **260:**6488–6493.

Takai, Y., Kikkawa, U., Kaibuchi, K., and Nishizuka, Y., 1984, Membrane phospholipid metabolism and signal transduction for protein phosphorylation, *Adv. Cyclic Nucleotide Protein Phosphorylation Res.* **18:**119–158.

Takayanagi, R., Inagami, T., Snajdar, R. M., Imada, T., Tamura, M., and Misono, K. S., 1987, Two distinct forms of receptors for atrial natriuretic factor in bovine adrenocortical cells, *J. Biol. Chem.* **262:**12104–12113.

Taylor, C. W., and Merritt, J. E., 1986, Receptor coupling to polyphosphoinositide turnover: A parallel with the adenylate cyclase system, *Trends Pharmacol. Sci.* **7:**238–242.

Taylor, S. S., 1987, Protein kinases: A diverse family of related proteins, *BioEssays* **7:**24–29.

Torrielli, M. V., and Dianzani, M. U., 1984, Free radicals in inflammatory disease, in: *Free Radicals in Molecular Biology, Aging and Disease*, (D. Armstrong, R. S. Sohal, R. G. Cutler, and T. F. Slater, eds.), Raven Press, New York, pp. 355–379.

Trueb, B., and Winterhalter, K. H., 1986, Type VI collagen is composed of a 200 kD subunit and two 140 kD subunits, *EMBO J.* **5:**2815–2819.

Uchida, K., Shimonaka, M., Saheki, T., Ito, T., and Hirose, S., 1987, Identification of the primary translation product of atrial natriuretic peptide receptor mRNA in a cell-free system using anti-receptor antiserum, *J. Biol. Chem.* **262:**12401–12402.

Ursini, F., Maiorino, M., and Gregolin, C., 1985, The selenoenzyme phospholipid hydroperoxide glutathione peroxidase, *Biochem. Biophys. Acta* **839:**62–70.

van der Rest, M., Mayne, R., Ninomiya, Y., Seidah, N. G., Chretien, M., and Olsen, B. R., 1985, The structure of type IX collagen, *J. Biol. Chem.* **260:**220–225.

van Kuijk, F. J. G. M., and Dratz, E. A., 1987, Detection of phospholipid peroxides in biological samples, *Free Radical Biol. Med.* **3:**349–354.

van Kuijk, F. J. G. M., Handelman, G. J., and Dratz, E. A., 1985, Consecutive action of phospholipase A2 and glutathione peroxidase is required for reduction of phospholipid hydroperoxides and provides a convenient method to determine peroxide values in membranes, *Free Radical Biol. Med.* **1:**421–427.

van Kuijk, F. J. G. M., Sevanian, A., Handelman, G. J., and Dratz, E. A., 1987, A new role for phospholipase A_2: Protection of membranes from lipid peroxidation damage, *Trends Biochem. Sci.* **12:**31–34.

Varma, S. D., and Lerman, S. (eds.), 1984, Proceedings of the First International Symposium on Light and Oxygen Effects on the Eye, *Curr. Eye Res.* **3:**1–271.

Vaughan, L., Winterhalter, K. H., and Bruckner, P., 1985, Proteoglycan Lt from chicken embryo sternum identified as type IX collagen, *J. Biol. Chem.* **260:**4758–4763.

Wald, G., 1935, Carotenoids and the visual cycle, *J. Gen. Physiol.* **19:**351–371.

Waldman, S. A., Rapoport, R. M., and Murad, F., 1984, Atrial natriuretic factor selectively activates particulate guanylate cyclase and elevates cyclic GMP in rat tissues, *J. Biol. Chem.* **259:**14332–14334.

Weiss, S. J., 1986, Oxygen, ischemia and inflammation, *Acta Physiol. Scand. [Suppl.]* **548:**9–37.

Wiegand, R. D., Jose, J. G., Rapp, L. M., and Anderson, R. E., 1984, Free radicals and damage to ocular tissues, in: *Free Radicals in Molecular Biology, Aging, and Disease* (D. Armstrong, R. S. Sohal, R. G. Cutler, and T. F. Slater, eds.), Raven Press, New York, pp. 317–353.

Wiggert, B., Russell, P., Lewis, M., and Chader, G., 1977, Differential binding to soluble nuclear receptors and effects on cell viability of retinol and retinoic acid in cultured retinoblastoma cells, *Biochem. Biophys. Res. Commun.* **79:**218–225.

Winquist, R. J., Faison, E. P., Waldman, S. A., Schwartz, K., Murad, F., and Rapoport, R. M., 1984, Atrial natriuretic factor elicits an endothelium-independent relaxation and activates particulate guanylate cyclase in vascular smooth muscle, *Proc. Natl. Acad. Sci. USA* **81:**7661–7664.

Wolf, G., Levin, L. V., and Bolmer, S. D., 1983, Multiple functions of vitamin A: Nuclear and extranuclear, in: *Modulation and Mediation of Cancer by Vitamins* (F. L. Meyskens and K. N. Prasad, eds.), S. Karger, Basel, pp. 146–152.

Woodgett, J. R., and Hunter, T., 1987, Isolation and characterization of two distinct forms of protein kinase C, *J. Biol. Chem.* **262:**4836–4843.

Wright, J. K., Seckler, R., and Overath, P., 1986, Molecular aspects of sugar:ion cotransport, *Annu. Rev. Biochem.* **55:**225–248.

Zerlauth, G., Kim, S. Y., Winner, J. B., Kim, H.-Y., Bolmer, S. D., and Wolf, G., 1984, Vitamin A deficiency and serum or plasma fibronectin in the rat and in human subjects, *J. Nutr.* **114:**1169–1172.

Zile, M. H., and Cullum, M. E., 1983, The function of vitamin A: Current concepts, *Proc. Soc. Exp. Biol. Med.* **172:**139–152.

Zile, M. H., Inhorn, R. C., and DeLuca, H.F., 1982, Metabolism *in vivo* of all-*trans*-retinoic acid, *J. Biol. Chem.* **257:**3544–3550.

Zile, M. H., Cullum, M. E., Simpson, R. U., Barua, A. B., and Swartz, D. A., 1987, Induction of differentiation of human promyelocytic leukemia cell line HL-60 by retinoyl glucuronide, a biologically active metabolite of vitamin A, *Proc. Natl. Acad. Sci. USA* **84:**2208–2212.

Zimmermann, D. R., Trueb, B., Winterhalter, K. H., Witmer, R., and Fischer, R. W., 1986, Type VI collagen is a major component of the human cornea, *FEBS Lett.* **197:**55–58.

Zweier, J. L., 1988, Measurement of superoxide-derived free radicals in the reperfused heart, *J. Biol. Chem.* **263:**1353–1357.

Tears 2

2.1. INTRODUCTION AND GENERAL DESCRIPTION

The precorneal and conjunctival tear film forms an interface between the air and ocular tissues. Some of its important functions are (1) lubrication of the eyelids, (2) formation of a smooth and even layer over an otherwise irregular corneal surface, and (3) provision of antibacterial systems for the ocular surface and nutrients for the corneal epithelium (Bron, 1985; Lamberts, 1987a). The tear film also serves as a vehicle for the entry of PMNs (polymorphonuclear leukocytes) in case of injury and dilutes and washes away toxic irritants from the ocular surface.

Tear film has a trilaminar structure (Fig. 2.1), consisting of (1) a thin (0.1-μm) anterior lipid layer, (2) an intermediate (7-μm) aqueous layer, and (3) an innermost mucous layer whose component particles are loosely bound to the glycocalyx of the corneal and conjunctival epithelial surfaces. The depth of the mucous layer is usually estimated by histochemical staining reactions and in humans is thought to vary from approximately 0.02 to 0.04 μm. However, improved methods for stabilizing and preserving the mucus of guinea pig tear film for electron microscopic examination revealed that this layer is in fact much thicker (Nichols *et al.*, 1985). It measures 0.8–1.0 μm over the cornea and varies from 2.0 to 7.0 μm over the conjunctiva because of the greater irregularity of the conjunctival surface.

A large body of new information, deriving mainly from improved analytical techniques, has appeared since the two previous reviews of tear biochemistry (van Haeringen, 1981; Stanifer *et al.*, 1983). Despite many interesting studies reported on animal tear fluid, the emphasis in this chapter is on human tears, their normal composition, as well as biochemical changes occurring in certain pathological or disease states.

The lacrimal secretory system has two components: (1) the large orbital and smaller palpebral portions of the lacrimal gland, which together account for about 95% of the aqueous component of tears, and (2) the accessory lacrimal glands of Krause and Wolfring located in the conjunctival stroma. The lacrimal gland is innervated primarily by parasympathetic fibers; secretion is controlled mainly by the cholinergic fibers of this system, but in addition it is modulated by adrenergic sympathetic stimulation, peptidergic agents (e.g., vasoactive intestinal peptide, VIP), and humoral factors (see reviews by Lamberts, 1987a; Bron, 1988). It is now recognized that regulation of lacrimal gland secretion is in fact an exceedingly complex process. Recent investigations discussed in Section 2.1.1 have established the presence of adrenergic and cholinergic membrane transduction systems that modulate the levels of intracellular second messengers and control the secretory activities of the lacrimal gland. In contrast to the major advances made in elucidating the mechanisms of lacrimal gland secretion, little is known of the factors controlling secretion from the accessory glands.

The trilaminar tear film originates from three sources. The lacrimal glands secrete a

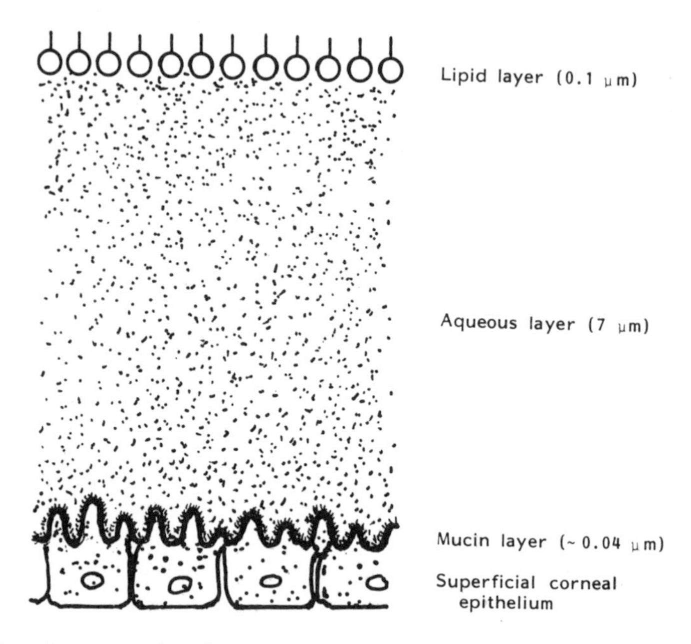

Figure 2.1. Schematic representation of precorneal tear film showing the trilaminar structure composed of (1) high-molecular-weight mucous glycoproteins originating mainly from conjunctival goblet cells; (2) water-soluble low-molecular-weight secretions of electrolytes, organic solutes, and proteins from the lacrimal glands; and (3) lipids originating from the meibomian glands.

variety of water-soluble substances that comprise the intermediate tear film layer: these include electrolytes, proteins, retinol, immunoglobulins, and enzymes. The thin anterior lipid layer, which contains both polar and nonpolar lipids, is secreted by the meibomian glands. The inner mucous layer is secreted mainly, if not entirely, by the conjunctival goblet cells. There are, however, reports that small amounts may also be derived from the lacrimal gland acinar cells and from subsurface secretory vesicles of the conjunctival and corneal epithelium (Dilly and Mackie, 1981; Greiner *et al.*, 1985; Dilly, 1985). This seems doubtful, since antibody against purified mucin fractions fails to label the lacrimal gland, and the substances thought to be mucin may have been glycoproteins instead (see Huang and Tseng, 1987, and review by Bron, 1988). Thus, the possible contribution of lacrimal acinar cells and conjunctival subsurface vesicles to the total ocular mucin requires further clarification. The goblet cells are not innervated but may respond to humoral agents such as serotonin and certain prostaglandins. The mucins secreted by the goblet cells are intimately attached to the infoldings of the superficial cornea and the conjunctival epithelium, creating a hydrophilic layer that ensures an even distribution of the tear film (Lamberts, 1987a). About 2–3 μl of mucins are produced daily (compared to 2–3 ml/day of aqueous tears).

2.1.1. Lacrimal Glands: Protein and Electrolyte Secretion

2.1.1.1. Protein Secretion

Lacrimal glands are typical exocrine glands that secrete proteins by exocytosis. The process is carried out by the acinar cells, which comprise about 80% of the gland (see review by Dartt, 1989). These cells are surrounded by nerve endings that release neurotransmitter substances after stimulation. Agonists, primarily neurotransmitters, diffuse to specific receptors on the basolateral membranes of the cell; their binding leads to activation of a variety of membrane-associated and cytosolic enzymes that modulate the intracellular levels of second messengers. Increases in their levels trigger the release of secretory granules containing stored proteins that had been synthesized in the endoplasmic reticulum–Golgi system. This type of regulated protein secretion in a variety of cell types has been reviewed in detail by Burgess and Kelly (1987). Stimulation of nerve endings in the lacrimal gland causes secretion of water and electrolytes across the apical membrane into the lumen, where it washes the protein into the ductal system and finally onto the ocular surface (see reviews by Dartt, 1989, and Mircheff, 1989). Thus, lacrimal gland secretion is now perceived in terms of release of neurotransmitters on the basolateral surface, where the nerve endings are located, and transduction of this signal through the membrane and across the cell to the apical surface, where regulated protein secretion occurs.

Experimental evidence supporting these recently developed concepts has come from a variety of *in vitro* and *in vivo* investigations in rabbits and rats. However, owing to space limitations only a few highlights of these important new findings can be summarized in this section. An early report (Friedman *et al.*, 1981) demonstrated β-adrenergic receptors in rat lacrimal gland that, on stimulation by agonists such as isoproterenol, activate adenylate cyclase and cause the secretion of peroxidase. Secretion is also stimulated by peptide hormones such as corticotropin and α-melanotropin, and by dibutyryl-cAMP (Jahn *et al.*, 1982). Moreover, secretory activity is accompanied by phosphorylation of particulate proteins, which were first detected using lacrimal gland homogenates (Dartt *et al.*, 1982) and later identified in purified basolateral plasma membranes of rat extraorbital lacrimal glands (Dartt *et al.*, 1988a). The phosphorylation of 52-kDa and 91-kDa peptides by a Ca^{2+}/calmodulin-dependent protein kinase and a cAMP-dependent protein kinase, respectively, may be crucial steps in the regulation of protein secretion.

Lacrimal stimulation by ACTH (adrenocorticotropic hormone) is potentiated by the cholinergic agonist carbachol, whereas a combination of ACTH and phenylephrine gives additive effects (Cripps *et al.*, 1987). These findings suggest independent activation of peptidergic as well as cholinergic and α-adrenergic receptor systems, with interaction between the intracellular pathways prior to exocytosis. Evidence supporting this view has been reviewed in detail by Dartt (1989). Two other agents that activate adenylate cyclase, vasoactive intestinal peptide (VIP) (Dartt *et al.*, 1984, 1988b) and forskolin (Mauduit *et al.*, 1983), also stimulate protein secretion in rat lacrimal gland. A recent study has shown that the cholinergic and VIP-dependent pathways interact and potentiate lacrimal gland protein secretion after an intracellular increase in either cAMP or Ca^{2+} (Dartt *et al.*, 1988c). In addition, stimulation of protein secretion by phorbol esters as well as muscarinic cholinergic and α_1-adrenergic agonists that activate protein kinase C shows clearly the direct involvement of phosphoinositide breakdown and Ca^{2+} mobilization in lacrimal gland protein secretion (Dartt *et al.*, 1988d). These findings provide compelling evidence that two membrane transduction systems play major roles in generating second messengers that modulate secretory activity in lacrimal gland acinar cells. A schematic representation of these interacting systems is shown in Fig. 2.2.

An excellent summary of current views on protein secretion in the lacrimal gland has been given by Dartt (1989), who suggests that not just two but possibly three transduction systems may be present. In one of them, cholinergic agents act by stimulating phos-

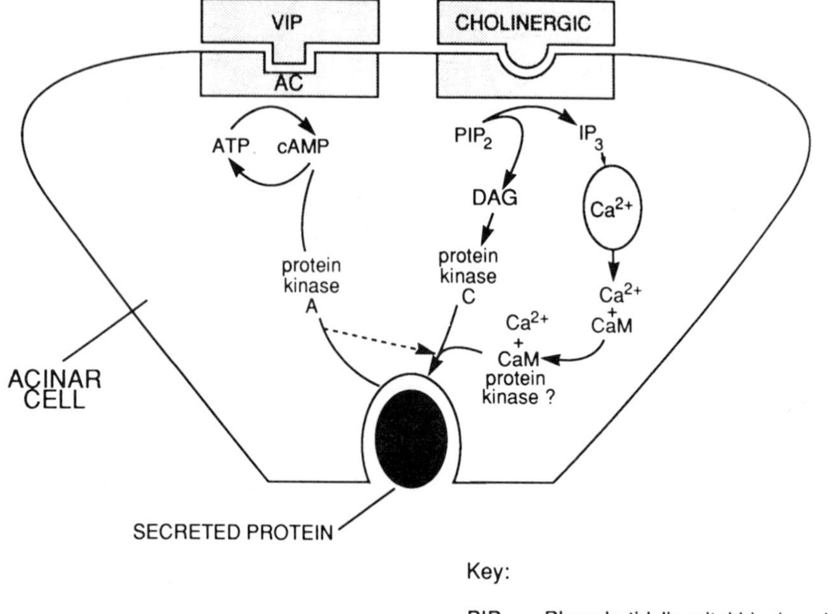

Figure 2.2. Model of cellular pathways in stimulus–secretion of lacrimal gland proteins by VIP (vasoactive intestinal peptide) and cholinergic agents. The two pathways are thought to interact at the sites indicated, potentiating lacrimal gland secretion after increases in intracellular levels of cAMP and Ca^{2+}. Substrate(s) phosphorylated by protein kinases A and C are thought to play key role(s) in triggering exocytosis. (From Dartt *et al.*, 1988c)

phoinositide turnover, resulting in the production of two second messengers, inositol 1,4,5-trisphosphate (IP_3) and diacylglycerol (DG) (see Chapter 1, Section 1.3.2 and Fig. 1.6). The Ca^{2+} mobilized from stores in the endoplasmic reticulum by IP_3, in conjunction with calmodulin, is thought to stimulate secretion by activation of protein kinase C. This enzyme is also activated by phorbol esters and by DG, and in all cases phosphorylation of endogenous proteins appears to be a key event. A second signal transduction system that modulates protein secretion is the stimulation of adenylate cyclase; this results in the production of cAMP and the activation of protein kinase A, an enzyme that is directly or perhaps indirectly involved in lacrimal protein secretion. In addition, α_1-adrenergic agonists also stimulate lacrimal gland secretion, but they use pathways different from those utilized by cholinergic agonists and VIP (Dartt, 1989). This possible third transduction system has not yet been characterized in detail.

2.1.1.2. Water and Electrolyte Secretion

Cholinergic agonists and VIP also stimulate fluid and electrolyte secretion in the lacrimal gland, and it is possible that they use the same transduction systems as those described for protein secretion (see review by Dartt, 1989). Two of the second messengers, Ca^{2+} and cAMP, directly activate ion movement and could stimulate secretion by increasing the activity of ion-gated channels. There may, however, be some differences since fluid secretion is a function of both acinar and duct cells, whereas protein secretion is functionally coupled only to the acinar cells.

The control of water and electrolyte secretion in the lacrimal gland has been investigated extensively in the past few years. As these studies are more physiological than biochemical, the topic can only be treated briefly. The reader is referred to a highly informative review by Mircheff (1989), which serves as the basis for the discussion that follows and in addition contains original literature citations that could not be included in this brief chapter.

Lacrimal gland fluid is produced in two stages: (1) secretion of a primary fluid resembling an isotonic ultrafiltrate of plasma in the acinus–early intercalated duct region and (2) secretion of a KCl-rich fluid in subsequent ductal segments. Electrolyte secretion by the lacrimal acinus is similar to that of other secretory cell types, and there is now direct evidence for the presence of Na^+,K^+-ATPase, K^+ channels, Cl^- channels, and Na^+Cl^--coupled transporters in lacrimal cells.

The presence of Ca^{2+}- and voltage-dependent K^+ channels has been demonstrated in lacrimal acinar cell basolateral membranes, and patch-clamp studies suggest that phosphorylation by a cAMP-dependent protein kinase prolongs the Ca^{2+}-dependent open time of the channel. Patch-clamp methodology has also demonstrated Ca^{2+}-dependent Cl^- channels that are activated by acetylcholine. These channels appear to be preferentially localized on the apical plasma membrane of the acinar cell. The presence of Na^+/H^+ antiporters has been indirectly deduced from early experiments showing that activation of either muscarinic cholinergic or α-adrenergic receptors accelerates the unidirectional flux of Na^+ into lacrimal acinar cells (Parod et al., 1980). Direct evidence for these antiporters has come from *in vitro* studies on isolated membrane vesicles from rat lacrimal gland (Mircheff *et al.*, 1987). An outwardly directed H^+ gradient was shown to be coupled to antiporter-mediated Na^+ influx localized in the basolateral lacrimal cell membrane. The Na^+/H^+ antiporters are thought to play an important role in regulating the pH of acinar cells. Recently Cl^-/HCO_3^- antiporters and $NaKCl_2$ symporters have also been detected in acinar cells (see review by Mircheff, 1989).

A carbachol-stimulated, ouabain-sensitive Na^+,K^+-ATPase was first described more than a decade ago in rat lacrimal acinar cells (Putney and van de Walle, 1979). This was confirmed in the rabbit, where autoradiographic studies of ouabain binding sites in slices of lacrimal gland showed preferential localization in the basolateral plasma membranes (Dartt *et al.*, 1981). However, this enzyme is not localized exclusively in the basolateral membrane, since later studies were able to detect immunoreactivity on the apical surface of acinar cells as well (see review by Mircheff, 1989). Nevertheless, the total contribution of the activity on the apical surface is thought to be small compared to that in the basolateral membrane.

Acinar electrolyte secretion is regulated by complex interactions among many, or all, of the components described above. It has been suggested by Mircheff (1989) that stimulation of either muscarinic cholinergic or α-adrenergic receptors leads directly to a series of responses mediated by changes in the intracellular levels of Ca^{2+}. They involve activation of Cl^-- selective channels on the apical plasma membrane and K^+ channels on the basolateral surface. An array of Na^+/H^+ and Cl^-/HCO_3^- antiporters on the basolateral plasma membrane are also activated following receptor stimulation. The precise sequence of events and the role of Na^{+},K^{+}-ATPase in electrolyte and water secretion under normal conditions are now beginning to be understood in terms of ion-gated channels and ionic fluxes located in specific regions of the acinar cell plasma membrane.

2.1.2. Physical Properties; Collection of Tears

Some of the physical characteristics of human tear fluid are summarized in Table 2.1. It is, however, recognized that precise estimations of the chemical composition, or the physical properties, of tear fluid are subject to large error owing to a number of factors. These include (1) the small samples; (2) evaporation during collection; (3) wide variability between individuals; (4) diurnal variations (Fullard and Carney, 1984); and, most importantly, (5) the method of collection (van Haeringen, 1981). Despite many attempts to "standardize" tear collection, there are inherent flaws that make the collection of perfectly reliable and reproducible samples for analysis very difficult.

The Schirmer test is a simple, inexpensive method for measuring tear volume. A strip of filter paper, usually 35 mm long and 5 mm wide, is inserted into the lower conjunctival sac. The tears are adsorbed by the strip, and the wetting of 5–6 mm of paper in 5 min is considered a normal value. Detailed studies on the capillary flow of tears in a series of human subjects indicate that the Schirmer filter strip method for tear collection is basically reliable under most conditions (Holly *et al.*, 1982/1983).

Table 2.1. Physical Properties of Human Tear Fluid[a]

Parameter	Value
Composition	98% water
	2% solids
Volume	6.2 ± 2.0 μl
Thickness	6.0–7.0 μm
Rate of secretion	
Unstimulated	1.2 μl/min
	(0.5–2.2 μl/min)
Anesthetized	
Schirmer	1.8 μl/min
Fluorophotometry	0.3 μl/min
Turnover rate	
Normal	12–16%/min
Stimulated	300%/min
Osmolarity	310–334 mOsm/k[b]
pH	7.5 ± 0.16[c]

[a]Adapted from Jordan and Baum (1980), Stanifer *et al.* (1983), and Bron (1985).
[b]With a freezing point depression method, a mean value of 318 mOsm/kg was found in 324 tear samples obtained from six healthy young adults (Benjamin and Hill, 1983).
[c]Mean value of unchallenged tears from six subjects (Carney *et al.*, 1989). Tears have substantial buffering capacity, especially in the acid range, mainly because of bicarbonate and partly from proteins or other components.

Filter paper strips are also used for collecting tear fluid for biochemical analyses. In this case the maximum amount of fluid is collected, and the tear fluid solutes are eluted and analyzed. However, because of unavoidable contact with the conjunctiva, the filter paper technique is not atraumatic, and some influx of serum components can be expected. Moreover, these samples may contain varying amounts of mucus. These problems can be circumvented by collecting tear fluid with a glass capillary tube. Although this is more time consuming, there is little conjunctival irritation, and fluid collected in this way contains mainly secretions from the lacrimal glands with little or no mucus contamination or influx of serum components. Comparisons of tear fluid composition using these two methods for collection show that although the concentrations of lacrimal proteins (lysozyme and lactoferrin) are similar, the concentrations of typical serum proteins (albumin, transferrin, and IgG) are significantly higher in tears collected with a Schirmer filter paper strip than by a capillary tube (Stuchell *et al.*, 1984).

Cellulose sponges are reported to be highly effective adsorbents for the collection and quantitative analysis of tear fluid (van Agtmaal *et al.*, 1987). The method appears to be atraumatic, although this has not been established with certainty. One of the principal advantages of the sponge compared to filter paper is that the adsorbed tears can be eluted by simple centrifugation, resulting in nearly quantitative recovery of undiluted tear fluid.

Tear flow can be artificially stimulated by a variety of agents in order to obtain larger samples, but the use of these techniques is questionable since proteins derived from the conjunctiva such as albumin, IgG, lactic dehydrogenase, and lysosomal enzymes show a "dilution effect" on stimulation. Hence, widely varying results for concentrations of certain proteins are obtained by inducing tear flow through either mechanical or chemical irritation or by the use of volatile lacrimating agents.

2.2. CHEMICAL COMPOSITION OF TEAR FLUID

2.2.1. Low-Molecular-Weight Solutes and Retinol

The concentrations of electrolytes and organic solutes in human tears are summarized in Table 2.2. Comparison with serum levels shows similar concentrations for Na^+, Cl^-, HCO_3^-, Mg^{2+}, and urea. However, other electrolytes are present at markedly different levels: K^+ is about six times higher, and Ca^{2+} about five times lower, in tears than in serum. Lactate (probably originating from the corneal epithelium) is about four to ten times higher in tears than in serum, while pyruvate levels are about the same in the two fluids. The concentrations of glucose (Daum and Hill, 1982) and protein are about ten and 30 times lower, respectively, in tear fluid than in serum.

Small amounts of all-*trans* retinol (see Chapter 1, Section 1.7) have been detected by high-pressure liquid chromatography (HPLC) in human and rabbit tear fluid (Ubels and MacRae, 1984; Speek *et al.*, 1986). The mean concentration in humans is 1.6 μg/dl, and the range of measured values is from trace amounts to as high as 1.06 (Speek *et al.*, 1986) or even 3.3 μg/dl (Ubels and MacRae, 1984). Taking into account these wide variations, the average concentration of retinol in human tears appears to be about 1/30 of that in serum (Table 2.2). The levels of retinol in rabbit tears are considerably higher than in humans, the mean concentration being 6.9 ± 0.9 μg/dl (Ubels and MacRae, 1984). Retinol levels in tear fluid reflect the nutritional status. Thus, tear fluid from vitamin

Table 2.2. Electrolytes and Organic Solutes in Human Tears and Serum[a]

Component	Tears	Serum
Electrolytes (mM)		
Na^+	120–165	130–145
Cl^-	118–135	95–125
HCO_3^-	20–26	24–30
Mg^{2+}	0.5–0.9	0.7–1.1
K^+	20–42	3.5–5
Ca^{2+}	0.4–1.1	2.0–2.6
Organic solutes (mM)		
Glucose	0.1–0.60	4–6
Urea	3.0–6.0	3.3–6.5
Lactate	2–5	0.5–0.8
Pyruvate	0.05–0.35	0.1–0.2
Ascorbate	0.008–0.04	0.04–0.06
All-*trans* retinol (μg/dl)	0.04–1.06[b]	30–60
	trace–3.3[c]	

[a]Adapted from van Haeringen (1981) and Stanifer *et al.* (1983).
[b]Range of retinol values in nine adult volunteers determined by HPLC (Speek *et al.*, 1986).
[c]The mean concentration of retinol in 11 human tear samples determined by HPLC is 1.6 ± 0.2 μg/dl (Ubels and MacRae, 1984).

A-deficient rabbits has no detectable retinol, and in marginally nourished Thai children the level of tear fluid retinol is significantly elevated 2 months after an oral dose of 110 mg of retinyl palmitate but returns to base-line levels 4 months afterward (van Agtmaal *et al.*, 1988).

Retinol is secreted into the tear fluid by the lacrimal gland, and the rate of secretion is enhanced, in a dose-dependent manner, by cholinergic drugs and by VIP (Ubels *et al.*, 1986). This suggests that both of the major receptor-stimulated membrane transduction systems recently characterized in the lacrimal gland (see Section 2.1.1) may modulate retinol secretion. However, a clear correlation or relationship between the secretion of retinol and protein was not established in this study. It is unlikely that retinol would be present in the free form in tear fluid because of its insolubility in aqueous media and its membranolytic properties. Hence, as in serum and in the cytosol of a variety of cell types, retinol is noncovalently complexed to retinol-binding proteins. This form, however, accounts for only a small fraction of the total intracellular stores; the major portion of retinol in tissues such as liver and retinal pigment epithelium is stored as long-chain fatty acyl esters either in oil droplets or in membranous fractions of the cell.

Similarly, in rat and rabbit lacrimal glands, retinol is also stored in the esterified form; the predominant esters in these species are retinyl palmitate and retinyl stearate, respectively (Ubels *et al.*, 1988). The concentration of retinyl palmitate in rabbit lacrimal gland is similar to that in rabbit epidermis, but is about 200 times lower than in testis or retina and 2500 times lower than in liver. The lacrimal gland retinyl esters did not arise as contamination from chylomicrons or chylomicron remnants present in blood, since similar values were obtained after removal of blood by perfusion. Further experiments with vitamin A-deficient rabbits showed that the retinol from orally administered labeled retinyl acetate is rapidly incorporated into labeled ester in the gland. Vitamin A does not appear to have a special function in the lacrimal gland itself; rather, this gland serves as a target tissue that functions in the transport of vitamin A from blood to tear fluid. Whether tear fluid is a physiological source of retinol for the avascular corneal epithelium is still an open question.

The finding that the major stores of retinol in lacrimal gland are esterified to long-chain fatty acid raises the question of how it is secreted into tear fluid where the only form detected to date is that of unesterified retinol. A reasonable mechanism suggests that lacrimal gland retinyl ester is hydrolyzed and then complexed to a carrier protein for secretion. A recent study strongly suggests that retinol and protein in rabbit lacrimal gland are secreted by the same mechanism (Ubels *et al.*, 1989). Both pilocarpine and VIP increase the rate of secretion of retinol and protein in a proportional manner, and under all conditions the retinol/protein ratio in lacrimal gland fluid in the rabbit is 3.3 ng retinol/mg protein.

To identify the lacrimal gland protein that is bound or complexed to retinol prior to secretion, vitamin A-deficient rabbits were administered oral doses of labeled retinyl acetate, and lacrimal gland fluid was analyzed for labeled retinol-bound protein(s) (Ubels *et al.*, 1989). Only one protein fraction in the 20-kDa range was found to carry labeled retinol. Although not identified, it could represent either RBP synthesized and expressed locally as in other tissues (see Chapter 1, Section 1.7.1) or tear-specific prealbumin, both of which have molecular weights in this range. The latter seems unlikely in spite of experimental evidence suggesting that a complex with unusual spectral characteristics is formed after prolonged incubation of labeled retinol with tear-specific prealbumin isolated from human ocular mucus (Chao and Butala, 1986). Further investigations are clearly needed to identify the physiological protein carrier for retinol in the lacrimal fluid.

2.2.2. Proteins

Tear fluid to be used for protein analysis is collected by either filter paper or glass capillary pipets; the samples are usually pooled, centrifuged, and then analyzed by a variety of techniques. These include one- or two-dimensional gel electrophoresis (Janssen and van Bijsterveld, 1981; Gachon *et al.*, 1980, 1982/1983), combined electrophoretic and immunologic methods (Gachon *et al.*, 1979; Janssen and van Bijsterveld, 1983a), and HPLC (Boonstra and Kijlstra, 1984). The most sensitive method developed to date utilizes size-exclusion HPLC combined with specific ELISAs for individual proteins (Fullard, 1988). All of the investigations reveal very complex patterns. As many as 50 to 60 proteins and/or enzymes have been detected in tear fluid, though not all have been identified.

The three principal tear proteins are lysozyme, lactoferrin, and tear-specific prealbumin (TSP); they originate solely from the lacrimal gland, and there is a high degree of correlation in the relative proportions of these proteins in separate analyses of individual tear samples (Janssen and van Bijsterveld, 1983a,b). A summary of most of the proteins identified to date in human tear fluid is given in Table 2.3.

Lysozyme, present at a concentration of 1 to 4 mg/ml, is a powerful bacteriolytic agent whose specific action is the dissolution of the cell walls of microorganisms. Its lytic activity toward *Micrococcus lysodeikticus* forms the basis of the widely used assay for

Table 2.3. Proteins in Human Tear Fluid[a]

Component	Concentration (mg/ml)
Total protein	5–9
Major proteins	
Lysozyme	2.4 ± 0.7
	0.7–1.1[b]
	4.6 ± 0.5[c]
Lactoferrin	1.5 ± 0.4
	2.4 ± 0.4[d]
	0.9–1.4[b]
Tear-specific prealbumin	0.5–1.5[e]
	1.2 ± 0.4[c]
Other proteins	
Albumin	0.054
Antiproteinases	
Secretory component (SC)	
Glycoproteins	
Orosomucoid	
a_1-glycoprotein	
Transferrin	
Ceruloplasmin	
Zn-α_2-glycoprotein	
Immunoglobulins	
sIgA (IgA)	0.41 ± 0.15
	0.4–0.6[b]
	0.3–0.5[c]
	0.1–0.3[f]
IgG	0.032
	trace–0.01[b]
	0.0005[c]
	0.002–0.017[f]
IgM	0.0003[c]
	0.002–0.015[f]

[a]Adapted from Gachon *et al.* (1982/1983).
[b]Measured by ELISA: high and low values represent younger and older individuals, respectively (Mackie and Seal, 1984).
[c]Analyzed by size-exclusion HPLC and ELISA (Fullard, 1988).
[d]ELISA assay (Kijlstra *et al.*, 1983).
[e]Generally estimated at 10–20% of the total proteins.
[f]Measured by ELISA (Coyle and Sibony, 1986).

lacrimal gland function, and its mode of action as a bacteriolytic agent is described in Section 2.3.

Lactoferrin, also present at a relatively high concentration in tear fluid, is produced entirely by the main and accessory lacrimal glands (Gillette and Allansmith, 1980). Its antimicrobial activity as well as its possible role in protecting the ocular surface from oxygen free radical injury are discussed below in Sections 2.3 and 2.5, respectively.

Tear-specific prealbumin (TSP) is immunologically distinct from serum prealbumin; it also differs chemically in amino acid composition, especially in its very low content of tryptophan (Chao and Butala, 1986). Although TSP comprises 10–20% of the total tear proteins, its precise functions in tear fluid are speculative. Apart from its reported bactericidal activity, it has a putative role in retinol transport in tear fluid (see Section 2.2.1).

Other proteins found in tears include small amounts of albumin, antiproteinases, glycoproteins, and immunoglobulins. Although not listed in Table 2.3, there was a report (Ford *et al.*, 1976) describing a bactericidal protein, β-lysin, in tear fluid. These findings have, however, been questioned by other investigators (Janssen *et al.*, 1984), and at present its existence in tear fluid seems doubtful.

2.2.3. Glycoconjugates

Precise chemical definitions of these substances in tears have been difficult owing to their relative inaccessibility and the paucity of starting material, not to mention the widely differing methodologies used for isolation and characterization. Added to this is the often unclear terminology. The terms mucin and mucous glycoprotein are used interchangeably to define the high-molecular-weight viscous glycoprotein(s) secreted by the conjunctival goblet cells that give ocular mucus its characteristic properties. However, other terms such as "mucin-like glycoproteins" and "mucoisolates" have recently been introduced; further work on their composition and origin is needed in order to clarify their relationship to ocular mucins and mucous glycoproteins.

2.2.3.1. Glycoproteins

Several of the soluble proteins secreted by the lacrimal gland into the aqueous phase of tear fluid are glycoproteins (Table 2.3). These include orosomucoid, α_1-glycoprotein, transferrin, ceruloplasmin, Zn-α_2-glycoprotein, and lactoferrin. In addition, mucin-like glycoproteins containing about 60% carbohydrate have been detected in saline and urea extracts of crude human ocular mucus obtained from the outer canthus (Chao *et al.*, 1983a,b). These macromolecules are linked to meibomian-type lipids. They have molecular weights as high as 10^5, and in some of the fractions, at least 22% of the serine (or threonine) residues are linked to N-acetylgalactosamine through O-glycosidic bonds. This distinguishes the ocular mucin-like glycoproteins from plasma glycoproteins, which have mainly N-glycosidic carbohydrate–protein linkages.

Crude saline extracts of ocular mucus also contain all of the major proteins present in tear fluid, as shown by immunologic analyses (Chao *et al.*, 1987). This would be expected, since in the normal eye, soluble components of the aqueous layer of tears such as proteins and glycoproteins are adsorbed onto, or become engulfed by, the mucus strands that are formed in the lower fornix and afterward accumulate in the inner canthus (Moore and Tiffany, 1979). A "mucoisolate" of complex structure containing rather substantial amounts of lipids, mucins, and glycosaminoglycans (GAGs) as well as small amounts of IgA accounts for about 12% of the saline-extractable protein of human ocular mucus (Chao *et al.*, 1987).

In other studies, examination of mucus threads by impression cytology (Tseng, 1985) revealed considerable cellular debris and other contaminants that could potentially degrade native ocular mucin. To obtain undegraded starting material, freshly secreted mucin from rabbit eyes was collected by bathing the ocular surface in buffered saline (Tseng *et al.*, 1987). Protease inhibitors were added to the crude samples of ocular mucin and a purified mucous glycoprotein (mucin) was then isolated after gel filtration and guanidine- and CsCl-gradient ultracentrifugation. This mucin contains 84% carbohydrate and has a M_r exceeding 10^6. Monoclonal antibodies against this rabbit ocular mucin reacted to epitopes present in both rabbit and human conjunctival goblet cells; they also bound to antigenic sites in tear film mucin (Huang and Tseng, 1987).

The water-soluble "mucoisolate" (Chao *et al.*, 1987) and the "native ocular mucin" (Tseng *et al.*, 1987) described above were isolated from saline extracts of ocular mucus and from buffer-soluble mucin discharged on the ocular surface, respectively. In addition to these studies using saline washes as starting material, the insoluble mucus threads have also been used as starting material for the isolation of mucous glycoproteins, as described below. Despite many detailed chemical analyses, the precise relationships of various mucins isolated from different types of starting material is not entirely clear.

2.2.3.2. Mucins of the Ocular Mucus

Ocular mucus has many of the characteristics of mucus found in other parts of the body such as the respiratory and gastrointestinal tracts. The thin layer of mucus covering the epithelial cells in these and other tissues is defined as a viscous, sticky, water-insoluble gel containing unique glycoproteins with $M_r \geq$ 1–2 million (Allen, 1983). Mucous glycoproteins differ both structurally and chemically from serum (or tissue) glycoproteins. The sugars in mucous glycoproteins, which usually comprise about 70% of the weight of the molecule, consist of hundreds of short polysaccharide chains attached to a central core. The overall structure resembles a "bottle brush" in which the carbohydrate chains represent the bristles and the protein core the central wire support. The sequence of sugars in the side chains is precise, and with few exceptions, the principal sugars in mucous glycoproteins are galactose, fucose, acetylated amino sugars, and sialic acid. The carbohydrate side chains are linked to the protein core by alkali-labile O-glycosidic bonds between the hydroxyl groups of serine or threonine and N-acetylgalactosamine at its reducing end.

Native-state mucous glycoproteins are hydrophilic anionic polymers of glycoprotein subunits joined together by disulfide bridges. The native macromolecule is difficult to analyze, but treatment with denaturing (reducing) agents cleaves the disulfide linkages and yields subunits that can be analyzed by SDS-PAGE.

It is the mucous glycoproteins that contribute to the highly viscous nature of ocular mucus. Human tears have a relative viscosity of about 2.8 at a shear rate of less than 4 s^{-1} (Bron, 1985; Kaura and Tiffany, 1986). The rheological behavior of mucous glycoproteins is non-Newtonian—i.e., the viscosity is dependent on the shear rate—and in such systems, the viscosity of the solution falls as the shear rate increases. The non-Newtonian behavior of tear fluid may have an advantage in lowering the viscosity during blinking, but between each blink, the viscosity would rise, thus stabilizing the tear film. Ocular mucous glycoproteins lower the surface tension of tears from about 70 dynes/cm to about half of that value, but other tear proteins contribute little to this effect.

Ocular mucous glycoproteins have been isolated from threads of mucus drawn from the inner canthus of human donors. The mucous strands were washed, and the residue solublized with and without reducing agents (Moore and Tiffany, 1979). Three principal fractions could be resolved by SDS-PAGE. Two of them (GP1 and GP2) had molecular weights greater than 1×10^6; a third component (GP3M) with a M_r of 200,000 represented a subunit of the GP1 glycoprotein. Antibodies raised against GP2 showed that it originates solely from the conjunctival goblet cells. The principal sugars in crude human ocular mucus are sialic acid, galactose, glucose, and N-acetylglucosamine (Moore and Tiffany, 1981). The carbohydrate side chains in the GP2 fraction are short, consisting of only one to four sugar residues; all of the serine and threonine residues are linked in O-glycosidic bonds to carbohydrate.

The studies described above were on pooled samples, but data are also available on individual mucus samples collected from the inferior fornix of normal individuals and from patients with mucin deficiency dry eye syndromes (Wells *et al.*, 1986). After washing of the samples to remove soluble proteins, the insoluble pellets were analyzed by SDS-PAGE under dissociating conditions. Four PAS-staining glycoproteins of molecular weights greater than 200,000 were consistently observed in individuals with normal conjunctiva. There were no great differences in the glycoprotein patterns in mucus samples from patients with various types of mucin deficiency syndromes (see Section 2.6.2).

2.2.4. Immunoglobulins

Immunoglobulin A (IgA), the principal immunoglobulin in tears, is normally found in unstimulated tears at concentrations ranging from about 0.1 to 0.6 mg/ml (Table 2.3) (Gachon *et al.*, 1982/1983). It is present in the form of an 11 S dimer consisting of two IgA molecules linked by a small protein, the J-chain of secretory component (SC). The whole complex, termed sIgA (secretory IgA), is secreted locally by plasma cells found in the lacrimal glands and the conjunctiva. Specific IgA may appear in tears in response to local or remote challenge (see review by Bron and Seal, 1986). Natural sIgA antibodies present in tears are induced and maintained through antigen ingestion and stimulation of the common mucosal immune system (CMIS) (Gregory and Allansmith, 1986). Orally administered antigen elicits IgA antibody responses in tears (Montgomery *et al.*, 1984). The mechanism for gastrointestinal stimulation of IgA antibody responses is thought to involve initial triggering and subsequent transport and seeding of distal secretory sites such as lacrimal gland by committed IgA precursor cells.

Immunoglobulin G (IgG) is normally present at very low concentrations. It is readily detected after trauma and is believed to originate as seepage from the conjunctival vessels. Immunoglobulin M (IgM) has only been identified recently in tears; its detection requires the use of glass capillaries for collecting the tear fluid and ELISA immunoassay for its quantitative measurement (Coyle and Sibony, 1986; Fullard, 1988). Immunoglobulins C and E may be present in trace amounts (or not at all) in tears collected atraumatically; IgE levels are increased in allergic eye disease.

2.2.5. Enzymes

A large number of glucose-metabolizing enzymes are present in tear fluid (Table 2.4), including several involved in anaerobic glycolysis. In addition, there are a number of oxidative (tricarboxylic acid cycle) enzymes. They are probably secreted normally in small amounts by the conjunctiva; their level in tears increases during mechanical irritation.

Antiproteases have been identified in tears, but at much lower levels than in plasma (Stanifer *et al.*, 1983). In infections of the eye or in corneal ulcerations, both α1-antitrypsin and α2-macroglobulin are found in increased amounts in tears. Both of these proteins can, under certain conditions, inhibit collagenase, but not necessarily corneal collagenase or plasmin. The origin of these two antiproteases is unclear; they may either be derived from the plasma or produced locally by the lacrimal glands. Proteases have also been detected in human tears (Tsung and Holly, 1981). Of six proteolytic enzymes assayed, those having the highest activity were cathepsins B and C; a trypsin-like enzyme activity has also been detected.

Table 2.4. Enzymes in Tear Fluid[a]

Enzyme	Source
Glycolytic	
Hexokinase	Conjunctiva
Pyruvate kinase	Conjunctiva
Lactic dehydrogenase (five isoenzymes)	Corneal epithelium
Tricarboxylic acid cycle	
Isocitric dehydrogenase	Conjunctiva
Pentose phosphate shunt	
Glucose 6-phosphate dehydrogenase (G6PD)	
Lysozyme	Lacrimal gland
Amylase	Lacrimal gland
Proteases	
Cathepsin B-like	
Cathepsin C-like	
Trypsin-like	
Antiproteases	
α1-antitrypsin	Lacrimal gland
α1-antichymotrypsin	Unknown
α2-macroglobulin	Unknown
Peroxidase	Lacrimal gland
Plasminogen activator	Lacrimal gland
Lysosomal acid hydrolases	Lacrimal gland[b]
Hexosaminidase A	
α-Galactosidase	
α-Fucosidase	
α-Mannosidase	
β-Galactosidase	
α-Glucosidase	
α-Iduronidase	
Sulfatases A and B	

[a]Adapted from Stanifer *et al.* (1983), van Haeringen (1981), and van Haeringen and Thorig (1984).
[b]Lysosomal acid hydrolases are also secreted by the conjunctiva.

Lysosomal acid hydrolases, originating mainly from the lacrimal glands, are present in tear fluid at levels two to ten times greater than in serum. Studies on the activities of ten lysosomal enzymes in human tears are consistent with the view that they originate from the lysosomes of the lacrimal gland (van Haeringen and Glasius, 1980). Tear fluid is an easily accessible, noninvasive source of lysosomal enzymes and has been used successfully in the diagnosis of several inherited lysosomal storage diseases (see Section 2.6.4).

Other enzymes detected in human tear fluid include peroxidase (van Haeringen *et al.*, 1979; Fullard, 1988), amylase, and plasminogen activator (PA) (Thorig *et al.*, 1983). The latter plays an important role in fibrinolysis, catalyzing the conversion of plasminogen to plasmin (Hayashi and Sueishi, 1988). Plasminogen activator is thought to arise from vascular endothelial cells of the conjunctival vessels and corneal epithelium in unstimulated tears and from the lacrimal gland in stimulated tears. Human tears contain two types of PA, urokinase PA (u-PA) and tissue PA (t-PA).

2.2.6. Lipids

The past decade has seen significant advances in our understanding of the meibomian gland and its lipid secretions, due in great part to the development of advanced analytical techniques by two groups of investigators, Nicolaides in Los Angeles and Tiffany at Oxford. The first and only comprehensive review of meibomian gland secretions by Tiffany (1987) is highly recommended for detailed information and literature citations on all aspects of this topic. The present discussion is of necessity limited to a general overview, with major emphasis placed on the composition of human meibomian lipids; data are also available on two other species, rabbit and cattle (Tiffany, 1979, 1987; Tiffany and Marsden, 1982).

The outermost portion of the tear film is a lipid layer measuring about 0.1 μm in depth and consisting of a complex mixture of polar and nonpolar lipids secreted by the meibomian glands. Owing to its low melting point (35°C), the superficial lipid layer of the preocular tear film is always fluid in the living eye despite the wide range of ambient temperatures to which the ocular surface is exposed. It is the unusual combination of unsaturated and branched-chain fatty acids and fatty alcohols, varying in length from eight to 32 carbon atoms, in both free and esterified forms, that accounts for the low melting point of this lipid secretion (Tiffany and Marsden, 1986). Meibomian secretion functions as a barrier to the inward movement of skin surface lipid; in addition, by making the lid margin hydrophobic, it prevents the spillover of tears (Tiffany, 1987). The lipid layer also has other important functions: it reduces the rate of evaporation from the open eye (Mishima and Maurice, 1961), provides lubrication for the eyelid/ocular interface, and contributes to the optical properties of the tear film.

The major lipid classes identified in meibomian gland samples obtained from human donors are wax esters, sterol esters (mainly cholesterol), and triglycerides (Tiffany, 1978). Hydrocarbons and free fatty acids, both straight and branched chain, as well as even and odd numbered, are also present in some individuals. These studies revealed great variability between individuals both in the composition and in the relative proportions of specific lipids. Hence, a typical lipid composition for human meibomian secretion is not easy to define. Later analytical data on the composition of meibomian lipids expressed directly from glands of postmortem eyes have been reported (Nicolaides *et al.*, 1981), and the range of values, based on the presence or absence of hydrocarbons, is summarized in Table 2.5. Detailed studies of the di- and triesters of human and cattle meibomian gland illustrate the complexity of structures that could result from random esterification of fatty acids and alcohols (Nicolaides and Santos, 1985). Some 2760 different species of wax monoesters could be found in human material, and perhaps hundreds of thousands of species could exist if all possible combinations of di- and triesters are present (Tiffany, 1987).

Free fatty acids (FFA) are a small but important component of meibomian secretions. In humans with no known pathological condition, the FFA comprise about 0.21% to 1.3% of the total meibomian lipids (Dougherty and McCulley, 1986a). The chain lengths range from 12 to 29 carbon atoms. Straight-chain fatty acids of 16 and 18 carbon atoms comprise about half of the total FFA, and isobranched and anteisobranched FFA account for another one-third. Meibomian lipids from human and cattle eyes also contain an unusual group of ω-hydroxy fatty acids ranging in size from 30 to 36 carbon atoms (Nicolaides and Ruth, 1982/1983). They are primarily monoenoic and are present as three homologous series. Whether they are in the free or esterified state could not be deter-

Table 2.5. Lipid Classes in Human Meibomian Secretions[a]

Component	Percentage of total
Wax esters	32–35
Sterol (mainly cholesterol) esters	27–29
Polar lipids	14.8–16
Hydrocarbons[b]	0–7
Diesters[c]	7.7–8.4
Triglycerides	3.7–4.0
Free sterols	1.6–1.8
Free fatty acids[d]	0.5–2.1

[a]Adapted from Nicolaides *et al.* (1981).
[b]Highest figures may be contamination from cosmetics or environment.
[c]Detailed analytical studies of diesters in human and bovine meibomian secretions have been reported by Nicolaides and Santos (1985).
[d]Mainly unsaturated, with chain lengths of 12 to 31 carbon atoms, some straight chain and some branched.

mined, and although their physiological function is not known, they may play a role in assisting the spread of other hydrophobic lipids (e.g., wax and sterol esters) over the tear film.

Metabolism and/or biosynthesis of meibomian lipids has only recently been investigated. Incubation of freshly excised bovine meibomian glands with labeled acetate results in the incorporation of radioactivity into all of the major lipid classes (Kolattukudy *et al.*, 1985). About 61% of the total label was incorporated into wax and sterol esters, and the radioactivity was almost evenly distributed between the alcohol and fatty acid residues. Further studies using microsomal membranes prepared from bovine meibomian glands demonstrated more clearly some of the enzymatic mechanisms involved in the biosynthesis of very-long-chain fatty acids (Anderson and Kolattukudy, 1985). An acyl-CoA synthetase is active in the synthesis of 83% and 8% of the C16 and C18 fatty acids, respectively.

The synthesis of very-long-chain fatty acids proceeds by elongation of C16 or C18 preformed fatty acids, using NADPH as the preferred cofactor. The same microsomal system is also active in the reduction of acyl-CoA species to corresponding alkanols, which is catalyzed by an acyl-CoA reductase that requires both NADH and NADPH (Kolattukudy and Rogers, 1986). This system is functionally coupled to esterifying enzymes that require exogenous acyl-CoA or ATP plus CoA. Thus, meibomian gland microsomes contain enzymes that actively synthesize wax esters according to the following possible sequence: fatty acids are elongated using malonyl-CoA and NADPH as substrates, these long-chain fatty acids are then reduced by the acyl-CoA reductase to alcohols, and in the final step the alcohols are esterified to wax esters.

The synthesis of unsaturated fatty acids (C16:1 and C18:1) from labeled acetate has been reported in rabbit meibomian glands *in vitro* (Tiffany, 1987). Some radioactivity is also present in long-chain alcohols, suggesting that reduction and chain elongation are important biosynthetic pathways in rabbit, as in bovine, meibomian gland. Studies with branched-chain primers show that the fatty acid synthetase preferentially utilizes α-methyl-branched structures for fatty acid chain initiation. Desaturase enzymes acting on initially formed saturated fatty acids to form families of monoenoic acids may precede elongation reactions at specific stages in the synthesis of long-chain fatty acids (see review by Tiffany, 1987).

2.3. BACTERICIDAL PROPERTIES

Tear fluid contains several systems having bactericidal or bacteriolytic activities. The principal one, and most extensively studied, is lysozyme, which accounts for 30–40% of the total protein in human tear fluid (see Table 2.3). This activity has been found in human and monkey tear fluid but not in lower species (van Haeringen and Thorig, 1984). The bacteriolytic properties of lysozyme derive from its ability to hydrolyze β1,4 glycosidic bonds in the cell walls of gram-positive organisms such as *Micrococcus lysodeikticus*. The products of this hydrolysis are disaccharides of N-acetylglucosamine and N-acetylmuramic acid, with some attached peptide side chains. The loss of this essential component of the bacterial cell wall results in its dissolution. Apart from its direct lytic effects in certain bacteria, the antibacterial action of lysozyme may be considerably expanded through its stimulation of phagocytosis of microorganisms by macrophages, monocytes, and neutrophils (see reviews by Bron and Seal, 1986; Bron, 1988). Moreover, it displays chitinase activity, which is thought to contribute antifungal activity, and it may also interact with lactoferrin in providing antibacterial defense. Thus, although lysozyme can act on its own as a bacteriolytic agent, its major action appears to be in conjunction with other agents present in tear fluid.

A heat-labile nonlysozyme antibacterial factor (NLAF), also known as β-lysin (Ford *et al.*, 1976), has been reported, but its existence has been challenged. One study has shown that even though fractionated human tears contains three bactericidal components, none of them is readily identifiable as NLAF or β-lysin (Selsted and Martinez, 1982). In other studies, purification of tear fluid proteins by ultrafiltration and affinity chromatography revealed that the only fraction with antibacterial activity is lysozyme (Janssen *et al.*, 1984). This protein was identified by immunologic, spectrophotometric, and agar diffusion assays. Taken together, these findings strongly suggest that even though platelets, serum, and certain other body fluids contain an antibacterial protein β-lysin, its presence in tear fluid is doubtful.

Lactoferrin also displays bacteriostatic properties, which are thought to be related to its efficient iron-binding capacity. It is generally assumed that this protein acts by inhibiting the growth of iron-dependent bacteria. The average concentration of lactoferrin in normal human tears is 2.2 mg/ml, accounting for about 25% of the tear protein (Kijlstra *et al.*, 1983). Apart from its role in nonspecific defense against invading microorganisms, lactoferrin also plays a role in preventing complement activation in inflammatory conditions, as described in Section 2.4.

Small amounts of peroxidase have been detected in human tear fluid (Fullard, 1988). It is immunologically similar to the enzyme found in human parotid saliva gland but is different from human leukocyte myeloperoxidase (Tenovuo *et al.*, 1984). Although no oxidizing reaction is detectable in human tears, oxidizable substrates such as Cl^- are present at high concentrations. As to the biological functions of peroxidase, it not only forms part of the broad spectrum of antimicrobial agents produced by the lacrimal gland, but it may also play a role in protecting the eye from potential hydrogen peroxide toxicity.

2.4. INFLAMMATORY AND ANTIINFLAMMATORY SYSTEMS

Prostaglandin F has been detected in trace quantities in tear fluid (Dhir *et al.*, 1979). Although its origin in tears has not been established, the level may be increased in inflammatory conditions such as vernal conjunctivitis and chronic trachoma. The presence of other eicosanoids in tear fluid is uncertain.

Complement C1 to C9, as well as factors that inhibit complement-mediated reactions such as lysis of antibody-coated erythrocytes, are present together in human tears (Yamamoto and Allansmith, 1979; Veerhuis and Kijlstra, 1982). The principal inhibitory factor identified to date is tear lactoferrin. This protein isolated from human tears exerts a dose-dependent anticomplement effect; at a concentration of 4.8 $\mu g/ml$, it causes a 50% inhibition of complement-mediated lysis of antibody-coated sheep erythrocytes. In other studies using a newly developed ELISA method, complement activation in human serum was measured by analyzing C3, C4, and C5 deposition on coated immune complexes (Kievits and Kijlstra, 1984). In this system, tears inhibit the classical C3 deposition after activation of C4 complement. The active factor appears to be tear lactoferrin, which may inhibit the formation of classical C3 convertase.

2.5. PROTECTION FROM OXYGEN FREE RADICAL TOXICITY

Little is known about protective mechanisms on the ocular surface against oxygen free radical toxicity (Chapter 1, Section 1.4), although there appear to be substances in human tear fluid that inhibit the formation of hydroxyl radical (Kuizenga *et al.*, 1987). In an *in vitro* system for generation of hydroxyl radicals with xanthine/xanthine oxidase (in the presence of trace amounts of Fe^{3+} and EDTA as a chelator), addition of human tears markedly inhibits the formation of this oxygen free radical. Two putative inhibitors were isolated from fractionated human tears and tentatively identified as Ca^{2+} and lactoferrin. No scavengers for either superoxide anion or H_2O_2 could be detected in human tears in these studies.

These findings were interpreted in terms of formation of hydroxyl radical in the presence of catalytic amounts of Fe^{3+}, in the form of an EDTA complex, by a Fenton-type or Haber–Weiss reaction (see Fig. 1.9). In an inflamed eye, PMNs can generate superoxide anion and H_2O_2, and as pointed out above, there do not appear to be protective systems against these substances. If not destroyed, they could be converted to hydroxyl radical in the presence of catalytic amounts of Fe^{3+}. Iron is only effective when chelated to EDTA, and it is interesting to point out that many eyedrop preparations use EDTA as a preservative. A protective effect for Ca^{2+} may be in its ability to chelate the EDTA, thus making it unavailable for forming the essential Fe^{3+}–EDTA catalytic complex. The action of lactoferrin in preventing hydroxyl radical formation may be related to its powerful iron-binding properties, which are maximally effective when it is not already saturated with iron. The latter may be applicable to tear lactoferrin (Kuizenga *et al.*, 1987).

2.6. CLINICAL DISORDERS

Many disorders of nonwettability of the tear film resulting from aqueous deficiency are known, the most common being keratoconjunctivitis sicca (KCS or dry eye); others, either acquired, congenital, or inherited, include Sjogren's syndrome, Riley–Day syndrome, and congenital alacrima (see reviews by Bron, 1985; Lamberts, 1987b). Mucin deficiency diseases include Stevens–Johnson syndrome, hypovitaminosis A (xerophthalmia), and cicatricial pemphigoid. Another common disorder is blepharitis, an inflammatory condition of the eyelid caused by meibomian gland dysfunction. Many types of blepharitis, both acute and chronic, are known, but a precise classification has been difficult to establish. Moreover, certain types of blepharitis are often associated with keratoconjunctivitis sicca. A full discussion of these complex clinical disorders is far beyond the scope of this chapter. Readers are referred to detailed reviews of the multiple causes of dry eye syndromes as well as current preventive and therapeutic possibilities (Bron, 1985, 1988; Lamberts, 1987b). In addition, a detailed review of meibomian gland lipid secretion by Tiffany (1987) provides considerable insight into possible causes of various types of blepharitis.

2.6.1. Keratoconjunctivitis Sicca (Dry Eye)

The broadest definition of dry eye is an insufficiency, qualitative or quantitative, of precorneal tear fluid resulting in nonwetting and instability of the tear film. It is an age-related aqueous deficiency syndrome that is more common in women than in men and usually has its onset during the fifth or six decade of life (Bron, 1985). It is most probably caused by infiltration or degeneration of the lacrimal glands.

In an interesting model proposed by Holly (1985), the properties of normal tear film and its behavior can be explained in physicochemical terms, and in diseased states its rupture is attributed to an influx of lipid contamination into the mucus layer. Focal areas of hydrophobicity are thus created in this hydrophilic layer, and the tear film ruptures over the hydrophobic spots. Unstable tear film is considered the underlying cause of all dry eye syndromes.

Methods available to assess nonwetting of the ocular surface have been reviewed in detail (Stanifer *et al.*, 1983; Bron, 1985; Lamberts, 1987b). The standard test used clinically is the Schirmer filter paper test described in Section 2.1.2. The aqueous-deficient eye has a considerably slower rate of fluid production than the accepted normal value of 5–6 mm of wetting in 5 min. A rise in osmolarity of the tear fluid in keratoconjunctivitis sicca (KCS) appears to be directly related to the lowered rate of fluid secretion measured by the Schirmer test strip (Gilbard *et al.*, 1978). Apart from these two diagnostic tests, tear instability or dysfunction can be measured as the break-up time (BUT). In this test, the time elapsing between the appearance of a dry spot in the tear film and the last blink after instillation of fluorescein is defined as the break-up time. The reliability of this test has however been questioned, since fluorescein itself can be considered as a provocative test, reducing the stability of the tear film to a greater degree in dry eye patients than in normal subjects (Bron, 1985). Several other tests are also available for assessing tear stability (see Bron, 1985; Lamberts, 1987b), and two of them, noninvasive break-up time (NIBUT) and surface tension measurements, may have certain advantages over other commonly used tests. In the NIBUT test, the break-up time is measured by viewing the specular image of a grid projected onto the precorneal film. Despite the wide range of values observed in normal subjects (Bron, 1985), a significantly reduced NIBUT was found in a group of 35 dry eye patients compared to 65 normal individuals (Tiffany *et al.*, 1989). Measurements of surface tension of small tear fluid samples from these two groups showed higher values in dry eye patients than in normal subjects. There is some correlation between surface tension and NIBUT for both dry eyes and normals, but only at moderate levels of significance. Nevertheless, these tests may provide additional parameters for evaluating the tear film stability of dry eye patients.

Tear fluids from KCS patients have greatly reduced levels of the three major lacrimal gland proteins: tear-specific prealbumin, lysozyme, and lactoferrin (Janssen and van Bijsterveld, 1981, 1982). Even though these findings support the widely held view of lacrimal gland dysfunction in KCS, their lowered concentrations in tear fluid may not be specific to KCS since these proteins are also lower than normal in such diverse conditions as smog irritation and herpes simplex infection (Lamberts, 1987b). Moreover, even in individuals with no known tear film disorder, the concentrations of lysozyme and lactoferrin decline with age (Mackie and Seal, 1984). Hence, comparisons of lacrimal gland protein concentrations in KCS patients must be made using age-matched controls. That there is in fact a decrease in lacrimal gland function with age has been demonstrated more directly using experimental animals. Carbachol-stimulated protein secretion declines from about 264 to about 175 μg/g tissue in lacrimal glands from 4- and 24-month-old rats, respectively (Bromberg and Welch, 1985). The age-related changes in lacrimal gland function in the rat appear to be somewhat similar to those observed in human subjects.

Increased levels of serum albumin (Janssen and van Bijsterveld, 1982) and ceruloplasmin (Mackie and Seal, 1984) are usually found in dry eye patients. The presence of elevated levels of these serum proteins probably reflects seepage from chronically inflamed conjunctival vessels (Mackie and Seal, 1984), suggesting a breakdown in the blood–conjunctival barrier (Janssen and van Bijsterveld, 1981).

2.6.2. Mucin Deficiency Diseases

Cicatricial pemphigoid and Stevens–Johnson syndrome are two disorders usually described as mucin-deficient dry eye syndromes. Surprisingly, SDS-PAGE analyses of mucin samples from a series of patients with these disorders showed no quantitative or qualitative differences in any of the four high-molecular-weight glycoprotein subunits found in mucin from normal individuals (Wells *et al.*, 1986). Other studies, however, suggest a small but significant decrease in mucin levels in cicatricial pemphigoid patients (Kinoshita *et al.*, 1983). Nevertheless, since rather substantial amounts of mucin-like glycoproteins were still detected in the preocular film of these patients, it seems likely that mucin deficiency *per se* is not the only cause of nonwetting associated with cicatricial pemphigoid (Lamberts, 1987b).

The classical example of a mucin deficiency caused by degeneration or loss of goblet cells is hypovitaminosis A. This can be induced experimentally in laboratory animals and is found endemically in developing countries (see Chapter 3, Section 3.1.3). Vitamin A is necessary for the maturation of goblet cells; it also plays an essential role in the biosynthesis of cell surface glycoconjugates. Nonwetting of the ocular surface, one of the earliest signs of vitamin A deficiency, is generally attributed to the loss of mucous glycoproteins. Given the importance of vitamin A to the ocular surface, its secretion by the lacrimal gland into the tear fluid has obvious physiological implications in the maturation and maintenance of normal goblet cell function.

2.6.3. Blepharitis

Meibomian glands are modified sebaceous glands located in the upper and lower eyelids; they are responsible for the secretion of the outermost lipid layer of the tear film (see Section 2.2.6). Functional abnormalities in the meibomian glands are believed to play a major role in the pathogenesis of blepharitis. Many types of blepharitis, both acute and chronic, are known, but a precise classification of this broad spectrum of disorders has yet to be established. The causes of blepharitis are not understood, but inflammation, hypertrophy, or keratinization of the epithelial cells lining the ducts and orifices is present in the majority of cases (see review by Tiffany, 1987). The normal outflow of lipid secretion is obstructed, and the retained lipids exacerbate the inflammation, which may ultimately lead to necrosis of the gland. It is also possible that modifications in the lipid composition lead to increased viscosity of the fluid, which could impede its secretion from the gland. The formation of lipid-laden nodules or lipogranulomas (chalazion) throughout the eyelid is characteristic of most if not all types of blepharitis; although these nodules are painless, they cause considerable discomfort.

Lipolytic enzymes secreted by microorganisms on the lid or conjunctiva may alter the fatty acid composition of the tear film, which in turn could adversely affect its physical characteristics. However, a bacteriological involvement in blepharitis is by no means clear. One study showed that no pathogenic flora could be detected in lid and conjunctival cultures of 88% of blepharitis cases (Seal *et al.*, 1985). Meibomian secretions in several clinically distinct types of chronic blepharitis have been examined, and the major lipid components, wax and sterol esters, appear to be normal (Dougherty and McCulley, 1986a). Similarly, the total amounts of free fatty acid (FFA) in these blepharitis patients were within normal limits. However, when compared to normal individuals, patients with various types of blepharitis have decreased amounts of C12:0, anteisobranched C15:0, and C23:0 free fatty acids, as well as increased levels of isobranched C22:0. These changes were thought to reflect abnormalities in the meibomian secretion itself, since the samples analyzed had been taken from freshly extruded meibomian secretion and not tear film lipids. However, since free fatty acids represent such a small fraction of the total meibomian secretion (see Table 2.5), the significance of these findings is unclear.

Further studies have attempted to identify microorganisms of the lid and conjunctiva of blepharitis patients and to correlate their lipid-degrading activities with specific types of blepharitis (Dougherty and McCulley, 1986b). The question of bacterial involvement in blepharitis is far from settled, but it is interesting to note that subclinical doses of tetracycline are reported to be beneficial in some patients. The possible effectiveness of this antibiotic is not through an inhibitory action on lid and conjunctival bacteria; rather, it may be inhibiting *de novo* lipid synthesis in the meibomian gland (Tiffany, 1987).

The duct epithelial cells of normal meibomian gland from a variety of species display ultrastructural characteristics suggestive of keratinization, and in humans there is compelling evidence that meibomian duct obstruction, probably related to abnormalities of hyperkeratinization, plays a role in the pathogenesis of meibomian gland dysfunction (Gutgesell *et al.*, 1982). Similar changes can be induced in the rabbit after long-term topical application of 2% epinephrine (Jester *et al.*, 1982). Immunofluorescent and immunobiochemical studies of normal human and rabbit meibomian gland show convincingly that the entire duct epithelial cell population is committed to keratin expression, specifically to the acidic 56.5- and basic 65- to 67-kDa keratin protein pair (Jester *et al.*, 1989a). These two keratin proteins are characteristic molecular markers expressed in keratinizing skin epithelia, and one of the earliest changes in epinephrine-induced meibomian gland dysfunction in the rabbit is full keratinization or hyperkeratinization of the meibomian gland duct epithelium (Jester *et al.*, 1989b). Whereas in normal meibomian gland, the characteristic 56.5/65- to 67-kDa protein pair is only detected in whole meibomian glands and not within the excreta, in experimentally induced meibomian gland dysfunction, these proteins are expressed within the excreta early in the development of the disease. These authors concluded that hyperkeratinization of the duct epithelium, which results in plugging of the meibomian gland, leads to gland dysfunction and is associated with decreased lipid secretion. This view is supported by the finding of increased lipid content in rabbits with clinically evident epinephrine-induced meibomian gland dysfunction, compared to normal controls (Nicolaides *et al.*, 1989). The lipids that accumulate in the glands of these rabbits were identified as free cholesterol and ceramides, and the analytical data raise the possibility that some hydrolysis of sterol esters had occurred. This was surmized by the eightfold increase in anteiso fatty acids with chain lengths longer than C20 that are normally esterified to sterol esters.

Hyperkeratinization of the gland ducts is thus a prominent feature of experimentally induced meibomian dysfunction, and similar pathological changes have also been observed in a recently described mouse model of the disease (Jester *et al.*, 1988). This animal, the rhino ($hr^{rh}hr^{rh}$) mouse, is a single-gene recessive mutant initially characterized by abnormal epidermal differentiation and maturation. Further studies demonstrated hyperkeratinization of the meibomian gland duct and marked plugging of the gland orifice. Loss of well-developed acini and atrophy of the gland occur by 1 year of age in this inherited model of meibomian gland dysfunction.

2.6.4. Diagnosis of Lysosomal Enzyme Storage Disorders

Although not related to abnormalities in the tear film, quantitative measurements of lysosomal acid hydrolases in tear fluid may be useful in the diagnosis of certain inherited lysosomal storage diseases. Tear fluid collected on a Schirmer strip provides a convenient noninvasive source of these enzymes, and whereas there are high and measurable amounts in normal tear fluid (van Haeringen and Glasius, 1980; van Haeringen, 1981), these hydrolases are virtually absent in the following disorders: Tay–Sachs disease (hexosaminidase A); Fabry's disease (α-galactosidase); fucosidosis (α-fucosidase); mannosidosis (α-mannosidase); Hurler and Scheie syndromes (α-iduronidase); and metachromatic leukodystrophy (sulfatase A).

2.7. REFERENCES

Allen, A., 1983, Mucus—A protective secretion of complexity, *Trends Biochem. Sci.* **8:**169–173.

Anderson, G. J., and Kolattukudy, P. E., 1985, Fatty acid chain elongation by microsomal enzymes from the bovine meibomian gland, *Arch. Biochem. Biophys.* **237:**177–185.

Benjamin, W. J., and Hill, R. M., 1983, Human tears: Osmotic characteristics, *Invest. Ophthalmol. Vis. Res.* **24:**1624–1626.

Boonstra, A., and Kijlstra, A., 1984, Separation of human tear proteins by high performance liquid chromatography, *Curr. Eye Res.* **3:**1461–1469.

Bromberg, B. B., and Welch, M. H., 1985, Lacrimal protein secretion: Comparison of young and old rats, *Exp. Eye Res.* **40:**313–320.

Bron, A. J., 1985, Prospects for the dry eye, *Trans. Ophthalmol. Soc. U.K.* **104:**801–826.

Bron, A. J., 1988, Eyelid secretions and the prevention and production of disease, *Eye* **2:**164–171.

Bron, A. J., and Seal, D. V., 1986, The defences of the ocular surface, *Trans. Ophthalmol. Soc. U.K.* **105:**18–25.

Burgess, T. L., and Kelly, R. B., 1987, Constitutive and regulated secretion of proteins, *Annu. Rev. Cell Biol.* **3:**243–293.

Carney, L. G., Mauger, T. F., and Hill, R.M., 1989, Buffering in human tears: pH responses to acid and base challenge, *Invest. Ophthalmol. Vis. Sci.* **30:**747–754.

Chao, C.-C. W., and Butala, S. M., 1986, Isolation and preliminary characterization of tear prealbumin from human ocular mucus, *Curr. Eye Res.* **5:**895–901.

Chao, C.-C. W., Vergnes, J.-P., and Brown, S. I., 1983a, Fractionation and partial characterization of macromolecular components from human ocular mucus, *Exp. Eye Res.* **36:**139–150.

Chao, C.-C. W., Vergnes, J.-P., and Brown, S. I., 1983b, O-Glycosidic linkage in glycoprotein isolates from human ocular mucus, *Exp. Eye Res.* **37:**533–541.

Chao, C.-C. W., Butala, S. M., Zaidman, G., and Brown, S.I., 1987, Immunological study of proteins and mucosubstance in saline soluble human ocular mucus, *Invest. Ophthalmol. Vis. Sci.* **28:**546–554.

Coyle, P. K., and Sibony, P. A., 1986, Tear immunoglobulins measured by ELISA, *Invest. Ophthalmol. Vis. Sci.* **27:**622–625.

Cripps, M. M., Bromberg, B. B., Patchen-Moor, K., and Welch, M. H., 1987, Adrenocorticotropic hormone stimulation of lacrimal peroxidase secretion, *Exp. Eye Res.* **45:**673–683.

Dartt, D. A., 1989, Signal transduction and control of lacrimal gland protein secretion: A review, *Curr. Eye Res.* **8:**619–636.

Dartt, D. A., Moller, M., and Poulsen, J. H., 1981, Lacrimal gland electrolyte and water secretion in the rabbit: Localization and role of (Na^+ + K^+)-activated ATPase, *J. Physiol.* **321:**557–569.

Dartt, D. A., Guerina, V. J., Donowitz, M., Taylor L., and Sharp, G. W. G., 1982, $Ca2^+$- and calmodulin-dependent protein phosphorylation in rat lacrimal gland, *Biochem. J.* **202:**799–802.

Dartt, D. A., Baker, A. K., Vaillant, C., and Rose, P. E., 1984, Vasoactive intestinal polypeptide stimulation of protein secretion from rat lacrimal gland acini, *Am. J. Physiol.* **247:**G502–G509.

Dartt, D. A., Rose, P. E., Joshi, V. M., Donowitz, M., and Sharp, G. W. G., 1985, Role of calcium in cholinergic stimulation of lacrimal gland protein secretion, *Curr. Eye Res.* **4:**475–483.

Dartt, D. A., Mircheff, A. K., Donowitz, M., and Sharp, G. W. G., 1988a, Ca^{2+}- and cAMP-induced protein phosphorylation in lacrimal gland basolateral membranes, *Am. J. Physiol.* **254:**G543–G551.

Dartt, D. A., Shulman, M., Gray, K. L., Rossi, S. R., Matkin, C., and Gilbard, J. P., 1988b, Stimulation of rabbit lacrimal gland secretion with biologically active peptides, *Am. J. Physiol.* **254:**G300–G306.

Dartt, D. A., Baker, A. K., Rose, P. E., Murphy, S. A., Ronco, L. V., and Unser, M. F., 1988c, Role of cyclic AMP and Ca^{2+} in potentiation of rat lacrimal gland protein secretion, *Invest. Ophthalmol. Vis. Sci.* **29:**1732–1738.

Dartt, D. A., Ronco, L. V., Murphy, S. A., and Unser, M. F., 1988d, Effect of phorbol esters on rat lacrimal gland protein secretion, *Invest. Ophthalmol. Vis. Sci.* **29:**1726–1731.

Daum, K. M. and Hill, R. M., 1982, Human tear glucose, *Invest. Ophthalmol. Vis. Sci.* **22:**509–514.

Dhir, S. P., Garg, S. K., Sharma, Y. R., and Lath, N. K., 1979, Prostaglandins in human tears, *Am. J. Ophthalmol.* **87:**403–404.

Dilly, P. N., 1985, Contribution of the epithelium to the stability of the tear film, *Trans. Ophthalmol. Soc. U.K.* **104:** 381–389.

Dilly, P. N., and Mackie, I. A., 1981, Surface changes in the anaesthetic conjunctiva in man, with special reference to the production of mucus from a non-goblet-cell source, *Br. J. Ophthalmol.* **65:**833–842.

Dougherty, J. M., and McCulley, J. P., 1986a, Analysis of the free fatty acid component of meibomian secretions in chronic blepharitis, *Invest. Ophthalmol. Vis. Sci.* **27:**52–56.

Dougherty, J. M., and McCulley, J. P., 1986b, Bacterial lipases and chronic blepharitis, *Invest. Ophthalmol. Vis. Sci.* **27:**486–491.

Ford, L. C., DeLange, R. J., and Petty, R. W., 1976, Identification of a nonlysozymal bactericidal factor (beta lysin) in human tears and aqueous humor, *Am. J. Ophthalmol.* **81:**30–33.

Friedman, Z. Y., Lowe, M., and Selinger, Z., 1981, β-Adrenergic receptor stimulated peroxidase secretion from rat lacrimal gland, *Biochim. Biophys. Acta* **675:**40–45.

Fullard, R. J., 1988, Identification of proteins in small tear volumes with and without size exclusion HPLC fractionation, *Curr. Eye Res.* **7:**163–179.

Fullard, R. J., and Carney, L. G., 1984, Diurnal variation in human tear enzymes, *Exp. Eye Res.* **38:**15–26.

Gachon, A. M., Verrelle, P., Betail, G., and Dastugue, B., 1979, Immunological and electrophoretic studies of human tear proteins, *Exp. Eye Res.* **29:**539–553.

Gachon, A. M., Lambin, P., and Dastugue, B., 1980, Human tears: Electrophoretic characteristics of specific proteins, *Ophthalmic Res.* **12:**277–285.

Gachon, A. M., Richard, J., and Dastugue, B., 1982/1983, Human tears: Normal protein pattern and individual protein determinations in adults, *Curr. Eye Res.* **2:**301–308.

Gilbard, J. P., Farris, R. L., and Santamaria, J., 1978, Osmolarity of tear microvolumes in keratoconjunctivitis sicca, *Arch. Ophthalmol.* **96:**677–681.

Gillette, T. E., and Allansmith, M. R., 1980, Lactoferrin in human ocular tissues, *Am. J. Ophthalmol.* **90:**30–37.

Gregory, R. L., and Allansmith, M. R., 1986, Naturally occurring IgA antibodies to ocular and oral microorganisms in tears, saliva and colostrum: Evidence for a common mucosal immune system and local immune response, *Exp. Eye Res.* **43:**739–749.

Greiner, J. V., Weidman, T. A., Korb, D. R., and Allansmith, M. R., 1985, Histochemical analysis of secretory vesicles in nongoblet conjunctival epithelial cells, *Acta Ophthalmol.* **63:**89–92.

Gutgesell, V. J., Stern, G. A., and Hood, C. I., 1982, Histopathology of meibomian gland dysfunction, *Am. J. Ophthalmol.* **94:**383–387.

Hayashi, K., and Sueishi, K., 1988, Fibrinolytic activity and species of plasminogen activator in human tears, *Exp. Eye Res.* **46:**131–137.

Holly, F. J., 1985, Physical chemistry of normal and disordered tear film, *Trans. Ophthalmol. Soc. U.K.* **104:**374–380.

Holly, F. J., Lamberts, D. W., and Esquivel, E. D., 1982/1983, Kinetics of capillary tear flow in the Schrimer strip, *Curr. Eye Res.* **2:**57–70.

Huang, A. J. W., and Tseng, S. C. G., 1987, Development of monoclonal antibodies to rabbit ocular mucin, *Invest. Ophthalmol. Vis. Sci.* **28:**1483–1491.

Jahn, R., Padel, U., Porsch, P.-H., and Soling, H.-D., 1982, Adrenocorticotropic hormone and alpha-melanocyte-stimulating hormone induce secretion and protein phosphorylation in the rat lacrimal gland by activation of a cAMP-dependent pathway, *Eur. J. Biochem.* **126:**623–629.

Janssen, P. T., and van Bijsterveld, O. P., 1981, Comparison of electrophoretic techniques for the analysis of human tear fluid proteins, *Clin. Chim. Acta* **114:**207–218.

Janssen, P. T., and van Bijsterveld, O. P., 1982, Immunochemical determination of human tear lysozyme (muramidase) in keratoconjunctivitis sicca, *Clin. Chim. Acta* **121:**251–260.

Janssen, P. T., and van Bijsterveld, O. P., 1983a, Origin and biosynthesis of human tear fluid proteins. *Invest. Ophthalmol. Vis. Sci.* **24:**623–630.

Janssen, P. T., and van Bijsterveld, O. P., 1983b, The relations between tear fluid concentrations of lysozyme, tear-specific prealbumin and lactoferrin, *Exp. Eye Res.* **36:**773–779.

Janssen, P. T., Muytjens, H. L., and van Bijsterveld, O. P., 1984, Nonlysozyme antibacterial factor in human tears. Fact or fiction? *Invest. Ophthalmol. Vis. Sci.* **25:**1156–1160.

Jester, J. V., Rife, L., Nii, D., Luttrull, J. K., Wilson, L., and Smith, R. E., 1982, *In vivo* biomicroscopy and photography of meibomian glands in a rabbit model of meibomian gland dysfunction, *Invest. Ophthalmol. Vis. Sci.* **22:**660–667.

Jester, J. V., Rajagopalan, S., and Rodrigues, M., 1988, Meibomian gland changes in the rhino (hr^{rh} hr^{rh}) mouse, *Invest. Ophthalmol. Vis. Sci.* **29:**1190–1194.

Jester, J. V., Nicolaides, N., and Smith, R. E., 1989a, Meibomian gland dysfunction. I. Keratin protein expression in normal human and rabbit meibomian glands, *Invest. Ophthalmol. Vis. Sci.* **30:**927–935.

Jester, J. V., Nicolaides, N., Kiss-Palvolgyi, I., and Smith, R. E., 1989b, Meibomian gland dysfunction. II. The role of keratinization in a rabbit model of MGD, *Invest. Ophthalmol. Vis. Sci.* **30:**936–945.

Jordan, A., and Baum, J., 1980, Basic tear flow. Does it exist? *Ophthalmology* **87:**920–930.

Kaura, R., and Tiffany, J. M., 1986, The role of mucous glycoproteins in the tear film, in: *The Preocular Tear*

Film (F. J. Holly, D. W. Lamberts, and D. L. MacKeen, eds.), Dry Eye Institute, Lubbock, TX, pp. 728–732.

Kievits, F., and Kijlstra, A., 1984, Complement regulation in tears, in: *Protides of the Biological Fluids* (H. Peeters, ed.), Pergamon Press, Oxford, pp. 31–34.

Kijlstra, A., Jeurissen, S. H. M. and Koning, K. M., 1983, Lactoferrin levels in normal human tears, *Br. J. Ophthalmol.* **67:**199–202.

Kinoshita, S. Kiorpes, T. C., Friend, J., and Thoft, R. A., 1983, Goblet cell density in ocular surface disease, a better indicator than tear mucin, *Arch. Ophthalmol.* **101:**1284–1287.

Kolattukudy, P. E., and Rogers, L., 1986, Acyl-CoA reductase and acyl-CoA:fatty alcohol acyl transferase in the microsomal preparation from the bovine meibomian gland, *J. Lipid Res.* **27:**404–411.

Kolattukudy, P. E., Rogers, L. M., and Nicolaides, N., 1985, Biosynthesis of lipids by bovine meibomian glands, *Lipids* **20:**468–474.

Kuizenga, A., van Haeringen, N. J., and Kijlstra, A., 1987, Inhibition of hydroxyl radical formation by human tears, *Invest. Ophthalmol. Vis. Sci.* **28:**305–313.

Lamberts, D. W., 1987a, Physiology of the tear film, in: *The Cornea, Scientific Foundations and Clinical Practice* (G. Smolin and R. A. Thoft, eds.), Little, Brown, Boston, pp. 38–52.

Lamberts, D. W., 1987b, Keratoconjunctivitis sicca, in: *The Cornea, Scientific Foundations and Clinical Practice* (G. Smolin and R. A. Thoft, eds.), Little, Brown, Boston, pp. 387–405.

Mackie, I. A., and Seal, D. V., 1984, Diagnostic implications of tear protein profiles, *Br. J. Ophthalmol.* **68:**321–324.

Mauduit, P., Herman, G., and Rossignol, B., 1983, Forskolin as a tool to study the β-adrenergic receptor-elicited, labeled protein secretion in rat lacrimal gland, *FEBS Lett.* **153:**21–24.

Mircheff, A. K., 1989, Lacrimal fluid and electrolyte secretion: A review, *Curr. Eye Res.* **8:**607–617.

Mircheff, A. K., Ingham, C. E., Lambert, R. W., Hales, K. L., Hensley, C. B., and Yiu, S. C., 1987, Na^+/H^+ antiporter in lacrimal acinar cell basal-lateral membranes, *Invest. Ophthalmol. Vis. Sci.* **28:**1726–1729.

Mishima, S., and Maurice, D. M., 1961, The oily layer of the tear film and evaporation from the corneal surface, *Exp. Eye Res.* **1:**39–45.

Montgomery, P. C., Rockey, J. H., Majumdar, A. S., Lemaitre-Coelho, I. M., Vaerman, J.-P., and Ayyildiz, A., 1984, Parameters influencing the expression of IgA antibodies in tears, *Invest. Ophthalmol. Vis. Sci.* **25:**369–384.

Moore, J. C., and Tiffany, J. M., 1979, Human ocular mucus. Origins and preliminary characterisation, *Exp. Eye Res.* **29:**291–301.

Moore, J. C., and Tiffany, J. M., 1981, Human ocular mucus. Chemical studies, *Exp. Eye Res.* **33:**203–212.

Nichols, B. A., Chiappino, M. L., and Dawson, C. R., 1985, Demonstration of the mucous layer of the tear film by electron microscopy, *Invest. Ophthalmol. Vis. Res.* **26:**464–473.

Nicolaides, N., and Ruth, E. C., 1982/1983, Unusual fatty acids in the lipids of steer and human meibomian gland excreta, *Curr. Eye Res.* **2:**93–98.

Nicolaides, N., and Santos, E. C., 1985, The di- and triesters of the lipids of steer and human meibomian glands, *Lipids* **20:**454–467.

Nicolaides, N., Kaitaranta, J. K., Rawdah, T. N., Macy, J. I., Boswell, F. M., III, and Smith, R. E., 1981, Meibomian gland studies: Comparison of steer and human lipids. *Invest. Ophthalmol. Vis. Sci.* **20:**522–547.

Nicolaides, N., Santos, E. C., Smith, R. E., and Jester, J. V., 1989, Meibomian gland dysfunction. III. Meibomian gland lipids, *Invest. Ophthalmol. Vis. Sci.* **30:**946–951.

Parod, R. J., Leslie, B. A., and Putney, J. W., Jr., 1980, Muscarinic and α-adrenergic stimulation of Na and Ca uptake by dispersed lacrimal cells, *Am. J. Physiol.* **239:**G99–G105.

Putney, J. W., Jr., and van de Walle, C. M., 1979, Effect of carbachol on ouabain-sensitive uptake of ^{86}Rb by dispersed lacrimal gland cells, *Life Sci.* **24:**1119–1124.

Seal, D. V., McGill, J. I., Jacobs, P., Liakos, G. M., and Goulding, N. J., 1985, Microbial and immunological investigations of chronic non-ulcerative blepharitis and meibomianitis, *Br. J. Ophthalmol.* **69:**604–611.

Selsted, M. E., and Martinez, R. J., 1982, Isolation and purification of bactericides from human tears, *Exp. Eye Res.* **34:**305–318.

Speek, A. J., van Agtmaal, E. J., Saowakontha, S., Schreurs, W. H. P., and van Haeringen, N. J., 1986, Fluorometric determination of retinol in human tear fluid using high-performance liquid chromatography, *Curr. Eye Res.* **5:**841–845.

Stanifer, R. M., Andrews, J. S., and Kretzer, F. L., 1983, Tear film, in: *Biochemistry of the Eye* (R. E.

Anderson, ed.), Manuals Program, American Academy of Ophthalmology, San Francisco, pp. 7–22.

Stuchell, R. N., Feldman, J. J., Farris, R. L., and Mandel, I. D., 1984, The effect of collection technique on tear composition, *Invest. Ophthalmol. Vis. Sci.* **25:**374–377.

Tenovuo, J., Soderling, E., and Sievers, G., 1984, The peroxidase system in human tears, in: *Protides of the Biological Fluids* (H. Peeters, ed.), Pergamon Press, Oxford, pp. 107–110.

Thorig, L., Wijngaards, G., and van Haeringen, N. J., 1983, Immunological characterization and possible origin of plasminogen activator in human tear fluid, *Ophthalmic Res.* **15:**268–276.

Tiffany, J. M., 1978, Individual variations in human meibomian lipid composition, *Exp. Eye Res.* **27:**289–300.

Tiffany, J. M., 1979, The meibomian lipids of the rabbit. I. Overall composition, *Exp. Eye Res.* **29:**195–202.

Tiffany, J. M., 1987, The lipid secretion of the meibomian glands, *Adv. Lipid Res.* **22:**1–62.

Tiffany, J. M., and Marsden, R. G., 1982, The meibomian lipids of rabbit. II. Detailed composition of the principal esters. *Exp. Eye Res.* **34:**601–608.

Tiffany, J. M., and Marsden, R. G., 1986, The influence of composition on physical properties of meibomian secretion, in: *The Preocular Tear Film* (F. J. Holly, D. W. Lamberts, and D. L. MacKeen, eds.), Dry Eye Institute, Lubbock, TX, pp. 597–608.

Tiffany, J. M., Winter, N., and Bliss, G., 1989, Tear film stability and tear surface tension, *Curr. Eye Res.* **8:**507–515.

Tseng, S. C. G., 1985, Staging of conjunctival squamous metaplasia by impression cytology, *Ophthalmology* **92:**728–733.

Tseng, S. C. G., Huang, A. J. W., and Sutter, D., 1987, Purification and characterization of rabbit ocular mucin, *Invest. Ophthalmol. Vis. Sci.* **28:**1473–1482.

Tsung, P.-K., and Holly, F. J., 1981, Protease activities in human tears, *Curr. Eye Res.* **1:**351–355.

Ubels, J. L. and MacRae, S. M., 1984, Vitamin A is present as retinol in the tears of humans and rabbits, *Curr. Eye Res.* **3:**815–822.

Ubels, J. L., Foley, K. M., and Rismondo, V., 1986, Retinol secretion by the lacrimal gland, *Invest. Ophthalmol. Vis. Sci.* **27:**1261–1268.

Ubels, J. L., Osgood, T. B., and Foley, K. M., 1988, Vitamin A is stored as fatty acyl esters of retinol in the lacrimal gland, *Curr. Eye Res.* **7:**1009–1016.

Ubels, J. L., Rismondo, V., and Osgood, T. B., 1989, The relationship between secretion of retinol and protein by the lacrimal gland, *Invest. Ophthalmol. Vis. Sci.* **30:**952–960.

van Agtmaal, E. J., van Haeringen, N. J., Bloem, M. W., Schreurs, W. H. P., and Saowakontha, S., 1987, Recovery of protein from tear fluid stored in cellulose sponges, *Curr. Eye Res.* **6:**585–588.

van Agtmaal, E.J., Bloem, M. W., Speek, A. J., Saowakontha, S., Schreurs, W. H. P., and van Haeringen, N. J., 1988, The effect of vitamin A supplementation on tear fluid retinol levels of marginally nourished preschool children, *Curr. Eye Res.* **7:**43–48.

van Haeringen, N. J., 1981, Clinical biochemistry of tears, *Surv. Ophthalmol.* **26:**84–96.

van Haeringen, N. J., and Glasius, E., 1980, Lysosomal hydrolases in tears and the lacrimal gland: Effect of acetylsalicylic acid on the release from the lacrimal gland, *Invest. Ophthalmol. Vis. Sci.* **19:**826–829.

van Haeringen, N. J., and Thorig, L., 1984, Enzymology of tear fluid, in: *Protides of the Biological Fluids* (H. Peeters, ed.), Pergamon Presss, Oxford, pp. 399–401.

van Haeringen, N. J., Ensink, F. T. E., and Glasius, E., 1979, The peroxidase–thiocyanate–hydrogenperoxide system in tear fluid and saliva of different species, *Exp. Eye Res.* **28:**343–347.

Veerhuis, R., and Kijlstra, A., 1982, Inhibition of hemolytic complement activity by lactoferrin in tears, *Exp. Eye Res.* **34:**257–265.

Wells, P. A., Ashur, M. L., and Foster, C. S., 1986, SDS-gradient polyacrylamide gel electrophoresis of individual ocular mucus samples from patients with normal and diseased conjunctiva, *Curr. Eye Res.* **5:**823–831.

Yamamoto, G. K., and Allansmith, M. R., 1979, Complement in tears from normal humans, *Am. J. Ophthalmol.* **88:**758–763.

Cornea 3

The cornea is composed of five distinct anatomic layers lying parallel to its surfaces: (1) an outermost epithelial layer that comprises about 8–10% of the total thickness; (2) the basal lamina (basement membrane) underlying the epithelium; (3) the stroma, which comprises about 90% of the total thickness; (4) Descemet's membrane; and (5) the innermost endothelium, a single layer of cells that accounts for about 1% of the total thickness. Being an avascular tissue, the nutritional needs of the cornea are met by the tear film (the source of atmospheric oxygen) and the aqueous humor (the major source of glucose) (Stanifer *et al.*, 1983; Maurice, 1984; Friend, 1987).

Normal cornea contains 78% water (Maurice, 1984); the major structural components are collagen and proteoglycans, which comprise about 12–15% and 1–3%, respectively, of the wet weight of the tissue. In addition, there are other noncollagenous structural proteins, soluble proteins and glycoproteins, and lipids. Localization in specific cellular or acellular layers of the cornea has been established for many of them. Although each is discussed in the appropriate section, an overview of the macromolecular and structural composition of the cornea is shown in tabular and diagrammatic forms in Table 3.1 and Fig. 3.1, respectively. Low-molecular-weight solutes and electrolytes are also present in the cornea, the most extensively studied being Na^+ and Cl^-. The transport of these ions anteriorly across the epithelium into the tears and posteriorly across the endothelium into the aqueous plays a major role in controlling the hydration of the tissue (Maurice, 1984).

3.1. EPITHELIUM

3.1.1. Oxidative Metabolism

Three types of epithelial cells can be distinguished histologically: (1) the superficial squamous cells (one to two layers), (2) the wing-shaped cells (two to three layers), and (3) the basal columnar cells (one layer). Together these five to seven layers of cells comprise the corneal epithelium, which is, biochemically, a major center of oxidative metabolism in the cornea. Oxygen penetrates the epithelium from the preocular tear film at a rate of 3.5–4.0 $\mu l/cm^2$ per h. Glucose, the principal substrate metabolized, reaches the epithelium mainly from the stroma; its overall oxidative metabolism is shown in outline form (excluding individual enzymatic steps) in Fig. 3.2.

The first step in glucose utilization is phosphorylation to glucose 6-phosphate; this key metabolite is, in turn, metabolized by three different pathways. The major one is glycolysis, which accounts for about 85% of the glucose metabolized by the corneal epithelium. The last step in glycolysis is the production of pyruvate. This substrate can be

Table 3.1. Macromolecular and Structural Components in Various Layers of the Cornea

Components	Epithelium	Basal lamina	Stroma	Descemet's membrane	Endothelium
Collagen					
Type I			■		
Type III			■		
Type IV		■		■	
Type V			■		
Type VI			■		
Type VII		■			
Type VIII				■	
Proteoglycans					
Keratan sulfate			■		
Dermatan/chondroitin sulfate			■		
Heparan sulfate		■			
Glycogen	■[a]				
Fibronectin	■[b]		■	■	
Keratin	■				
Vimentin[c]					■
Actin	■				
Laminin		■		■	
Fibrillin[d]			■	■	
Proteins[e]	■		■		

[a]In the basal cell layer of the epithelium.
[b]During development in rabbit cornea (Cintron *et al.*, 1984); presence in adult basal lamina depends on the species.
[c]Detected by immunofluorescence microscopy using a monoclonal antibody to human vimentin (Risen *et al.*, 1987).
[d]A nonsulfated 350-kDa glycoprotein identifiable as 10-nm microfibrils (Sakai *et al.*, 1986).
[e]Soluble proteins, mainly 54-kDa corneal antigen.

either reduced by lactic dehydrogenase to lactic acid or decarboxylated, with the resulting 2-carbon acetate fragment being oxidized through the tricarboxylic acid cycle. There is a net yield of 2 mol of ATP per mol of glucose through glycolysis, and 36 mol from the tricarboxylic acid oxidative pathway. The latter is clearly the major source of ATP in the corneal epithelium. A small amount of glucose 6-phosphate is converted to glycogen, which is localized mainly in the basal cell layer. Glycogen serves as an energy source either during periods of oxygen deprivation or in case of trauma or as a consequence of poorly fitting contact lenses; under these conditions the glycogen stores are rapidly depleted (Friend, 1987).

A physiologically important pathway of glucose 6-phosphate metabolism in corneal epithelium is the pentose phosphate pathway, often called the hexose monophosphate (HMP) shunt. Some estimations suggest that about 35% of the glucose metabolized by the corneal epithelium is through this "shunt" (Friend, 1987), but under certain conditions, it may be as high as 66% (Geroski *et al.*, 1978). One of the major functions of the pentose phosphate pathway, apart from providing a framework for the metabolism of pentoses in general, is the generation of NADPH. This is the principal source of reducing equivalents for maintaining substances such as glutathione and ascorbate in the reduced state. Maintenance of a reducing environment plays a role in protecting the corneal epithelium from potential oxidative damage by free radicals or H_2O_2 (see Section 3.1.7.3).

New noninvasive techniques are now available for studying carbohydrate metabolism

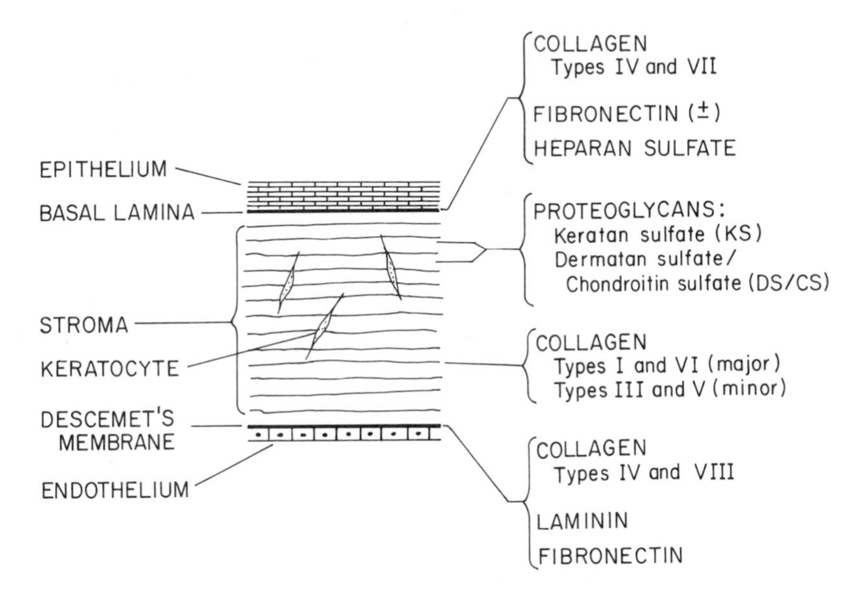

Figure 3.1. Diagrammatic representation of the structural organization of the cornea.

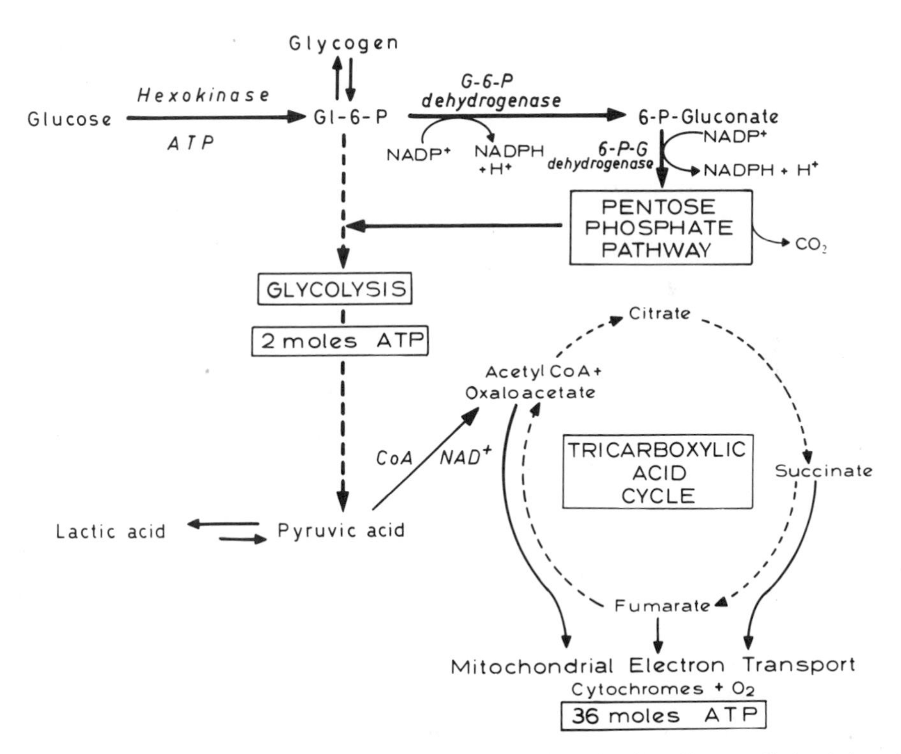

Figure 3.2. Principal pathways of glucose metabolism in the corneal epithelium. The overall glycolytic and oxidative pathways are similar to those found in nonocular tissues and in other ocular tissues such as the retina.

in the cornea. The ATP content of porcine epithelium has been calculated as 6.66 μmol/g wet weight by phosphorus-31 nuclear magnetic resonance (^{31}P-NMR) (Greiner *et al.*, 1985a), although in other animal species this value is somewhat lower. ATP represents about 30% of the total detectable phosphorus in the epithelium. This same technique is used to monitor rates of metabolic changes of specific metabolites (Greiner *et al.*, 1985b) and has been applied to calculations of the glycolytic activity of corneas excised from human donor eyes (Gottsch *et al.*, 1986). Lactate formation was found to be 0.50 μmol/h in intact corneas and about half of that value in deepithelialized corneas. The conversion of alanine to lactate in isolated rabbit cornea has been detected by the use of ^{13}C-NMR spectroscopy (Gottsch *et al.*, 1988). Another promising new technique, redox fluorometry, measures the fluorescence intensity of oxidized mitochondrial flavoproteins and reduced cytosolic and mitochondrial pyridine nucleotides (NADH and NADPH) (Masters, 1984a,b). With this method, an accurate assessment of cellular function is feasible at the mitochondrial level, and the effects of corneal hypoxia can be estimated with great accuracy.

3.1.2. Structural Components

3.1.2.1. Cytoskeleton

All of the major classes of cytoskeletal proteins (see Chapter 1, Section 1.1.1.3) have been characterized in corneal epithelium: intermediate filaments, represented by the keratins, actin microfilaments, and microtubules, the latter being present during mitosis.

Keratins are 10-nm intermediate filaments found in a wide variety of epithelia. They are a diverse multi-gene family of closely related polypeptides that can be divided into acidic (type I) and basic (type II) subfamilies. Biochemical, immunologic and cDNA analyses have established the existence of at least 20 human keratins (Moll *et al.*, 1982; Steinert *et al.*, 1985). Tissue distribution studies show that acidic and basic keratins of specific molecular size form keratin pairs that are frequently coexpressed (Cooper *et al.*, 1985); the appearance of a specific keratin pair provides an excellent marker for studying epithelial differentiation.

In the cornea, keratin filament composition was first studied in bovine epithelial cells (Gipson and Anderson, 1980). The keratins normally found in corneal epithelium consist of the 55-kDa acidic and 64-kDa basic types (often abbreviated as 55/64-kDa pairs); 40-kDa acidic and 50/58-kDa pairs have also been detected in normal epithelium. By using a monoclonal antibody specific for the 64-kDa corneal keratin, it has been possible to distinguish between the basal cells of the central corneal epithelium and those of the limbus on the basis of the keratin patterns expressed both *in vivo* and in cultured cells (Schermer *et al.*, 1986). Corneal epithelial stem cells are thought to be located in the limbus, a finding with important implications in resurfacing after epithelial cell abrasion (see Section 3.8.1).

Actin filaments measuring 5–7.5 nm in diameter are localized as networks in the apical surface cytoplasm of the two outermost superficial cell layers; they are also detectable in deeper layers, including the basal cells, by immunohistochemical techniques. During basal cell migration, or after epithelial debridement, the filaments are redistributed to the leading edges and basal portions of the mobile cells (Gipson and Anderson, 1977).

3.1.2.2. Extracellular Matrix Binding Proteins

Membrane receptors for extracellular matrix components such as laminin and fibronectin have been described in many cell types. Collagen-binding glycoproteins distinct from fibronectin are also associated with the plasma membrane; one such collagen-binding cell-surface 47-kDa glycoprotein has been named "colligin" (Kurkinen *et al.*, 1984).

Fluorescently labeled laminin, collagen, and fibronectin added to cultures of embryonic chick cornea bind to the basal surface of the cell, where specific binding sites for extracellular matrix components are thought to be present (Surgue and Hay, 1986). The binding causes striking changes in cell shape as well as reorganization of the intracellular cytoskeleton. Some of the collagen-binding glycoproteins from embryonic chick cornea have been isolated and shown to consist of a 47-kDa protein, which may be similar to the "colligin" described above, and a 70-kDa protein, which has many of the characteristics of an integral membrane protein (Surgue, 1987). The integrin family of transmembrane receptors shown in Fig. 1.1 may also be present in corneal epithelium.

3.1.3. Vitamin A, Retinoic Acid, and Xerophthalmia

In most populations throughout the world, the intake of vitamin A and/or carotene from dietary sources is more than adequate to meet all of the body's needs (Bondi and Sklan, 1984). Nevertheless, there are many areas of the world where malnutrition is endemic, and varying degrees of vitamin A deficiency may be present in large segments of the population. It has long been recognized that suboptimal dietary intake, especially in young children, affects mainly the epithelial cells of the gastrointestinal and respiratory tracts. In the eye, loss of conjunctival goblet cells and epithelial keratinization are prominent signs. In fact, of all the clinical signs associated with vitamin A deficiency, the earliest are usually found in the eye (see review by Sommer, 1983a). These consist of Bitot's spots in the conjunctiva and corneal xerosis; night blindness, which may not be recognized until a somewhat later stage, is also invariably present.

The cornea in experimentally induced vitamin A deficiency in the rat and in other laboratory animals has a lusterless appearance and irregular surface. Punctate epithelial erosions and keratinization are early signs of the disease. Exfoliating cells on the corneal surface coupled with an increase in density of keratofibrils and loss of glycogen deposits occur well before the onset of xerophthalmia in vitamin A-deficient rats (Carter-Dawson *et al.*, 1980). These changes probably represent a primary effect of vitamin A deficiency and not a secondary one resulting from infection, since the animals had been reared in a germ-free environment. In other studies, vitamin A-deficient rats were raised under standard laboratory conditions, and the animals were maintained until severe xerophthalmia (keratomalacia) developed (Leonard *et al.*, 1981). In this case, it was concluded that collagenase secreted from infiltrating leukocytes was the main cause of the corneal ulceration that occurs in the final stages of vitamin A deficiency and ultimately leads to complete corneal necrosis.

These studies, as well as many others over the years, leave little doubt that the cornea in general, and the epithelial cell layer in particular, is dependent on vitamin A for normal differentiation, development, and maintenance. Although there is now a rather considerable body of information on the metabolism and function of retinol in mammalian tissues and in special cell types (see Chapter 1, Section 1.7), relatively little is known about the cornea, even though it is a major target tissue for retinol.

The most likely mechanism for delivery of retinol to the cornea is by uptake from the circulation and/or from tears. Suspensions of epithelial cells from bovine cornea accumulate [^{3}H]retinol when incubated with [^{3}H]retinol–RBP (retinol-binding protein) (Rask *et al.*, 1980). This uptake is time dependent and substrate saturable, suggesting the presence of receptors on the epithelial cell membranes. Within the cell, vitamin A is bound to a 2 S (15-kDa) cytosolic retinol-binding protein (Wiggert *et al.*, 1977). The binding of [^{3}H]retinol to cellular RBP is greatly impaired in corneas of vitamin A-deficient rabbits (Wiggert *et al.*, 1982), possibly the consequence of either a decrease in its total amount in the tissue or a functional modification of its binding properties resulting from vitamin A deficiency.

Topically applied labeled retinol is metabolized by the cornea mainly to retinoic acid in both normal and vitamin A-deficient rabbits *in vivo* (Ubels and Edelhauser, 1985). Other more polar metabolites (that could not be identified) are also produced. In contrast to many rat tissues, which convert relatively small amounts of retinol to retinoic acid and polar metabolites, the rabbit cornea (in 2 h) oxidizes over 90% of the topically applied vitamin to retinoic acid and polar substances. Topically applied labeled all-*trans* retinoic acid is isomerized to 13-*cis* retinoic acid in normal rabbit corneas, but this isomerization does not occur in corneas of vitamin A-deficient animals.

Among the major functions of retinol or retinoic acid in corneal epithelium, two now appear to be clearly established: (1) the control of keratin expression and (2) the synthesis of glycoconjugates. Another important function, recently detected by ^{31}P nuclear magnetic resonance spectroscopy, is the activation and/or induction of creatine kinase (Hayashi *et al.*, 1989a).

Heavily keratinized corneal epithelia from vitamin A-deficient rabbits express the 56.5-kDa (acidic) and 65- to 67-kDa (basic) keratins; moreover, these specific keratins are also found in conjunctival and esophageal epithelia of vitamin A-deficient rabbits (Tseng *et al.*, 1984). The use of immunoblot techniques and subfamily-specific monoclonal antikeratin antibodies shows clearly that the 56.5/65- to 67-kDa keratins are expressed in keratinized epithelia induced by vitamin A deficiency and are found in addition to those normally present. There is also a marked decrease in 40-kDa keratin in vitamin A-deficient corneal epithelium.

Vitamin A appears to be required for the normal synthesis of cell surface glycoconjugates in the corneal epithelium. There is considerably greater incorporation of radioactive glucosamine into 220- to 250-kDa glycoproteins in corneal epithelium of vitamin A-deficient rats "repleted" with retinoic acid than in vitamin A-deficient animals (Hassell *et al.*, 1980a). Moreover, this is a dose-dependent phenomenon, since corneas of animals repleted with excess amounts of retinoic acid incorporate substantially more glucosamine than those receiving minimum amounts. Although the glycoconjugates were not identified in this study, the observations are interesting given the finding that fibronectin produced by retinoic acid-treated cultured chondrocytes has a higher proportion of complex-type carbohydrate chains than fibronectin isolated from untreated controls (Bernard *et al.*, 1984). Thus, retinoic acid appears to enhance glycosylation of cell surface glycoconjugates in many cell types, possibly by modifying genome expression (see Chapter 1, Section 1.7.3). Whether this holds true for the cornea remains to be established.

Xerophthalmia and its sequelae, keratomalacia, corneal xerosis or ulceration, and blindness, are not uncommon in developing countries. Systemic administration of vitamin A palmitate has generally been considered the most effective therapy, and if the corneal disease is not too advanced, there may be a certain degree of healing of the corneal lesions.

Pirie (1977) made the remarkable observation that topical application of retinoic acid was effective in reversing xerophthalmic lesions in vitamin A-deficient rats. Retinoic acid was afterward shown to reverse xerophthalmic lesions in human patients (Sommer and Emran, 1978) and in vitamin A-deficient rabbits (van Horn *et al.*, 1981). *In vitro* experiments have shown that in the normal rabbit, retinoic acid penetrates the cornea with an efficiency similar to that of commonly used ophthalmic drugs; however, its permeability is greatly decreased in xerophthalmic animals (Ubels and Edelhauser, 1982). These observations, and others, provided the basis for further testing of retinoic acid for therapeutic use. Retinoic acid enhances the healing rate of experimentally induced corneal epithelial wounds (Ubels *et al.*, 1983), and topical application to xerophthalmic patients (receiving vitamin A supplementation) speeds corneal healing in a larger proportion of cases than those receiving vitamin A alone (Sommer, 1983b). Tretinoin (all-*trans* retinoic acid) at a concentration of 0.1% reverses xerophthalmic lesions in vitamin A-deficient rabbits (Hatchell *et al.*, 1984), and in a petrolatum ointment, reversal of keratinization can be achieved at concentrations of 0.01% (Ubels *et al.*, 1985) or even as low as 0.0005% (Ubels *et al.*, 1987). The very low concentration of retinoic acid has no irritating side effects, although reversal of corneal keratinization in xerophthalmic rabbits is somewhat slower than at higher concentrations.

The rationale of using retinoic acid in place of, or in addition to, vitamin A in the management of xerophthalmia possibly lies in the fact that vitamin A requires a specific binding protein to reach its intracellular target site in the epithelial cells (Goodman, 1984). In malnourished individuals this protein may be absent or in limited supply; hence, the patient may be unable to utilize systemically administered vitamin A (Sommer and Muhilal, 1982). Most importantly, retinoic acid is a major metabolite of retinol in the cornea (Ubels and Edelhauser, 1985) and could be an important biologically active form of vitamin A in this tissue. Given this possibility, its dramatic therapeutic effects are not surprising. Nevertheless, retinoic acid in the form of ointments described above is still not used in areas of the world where xerophthalmia is endemic. Prevention of this disease by education in the use of dietary leafy vegetables as a natural source of vitamin A is a more practical approach to the problem, and when xerophthalmia does develop, treatment with massive doses of vitamin A palmitate is still highly effective.

3.1.4. Phosphoinositide Breakdown and Protein Kinase C

The receptor-stimulated hydrolysis of phosphatidylinositol 4,5-bisphosphate (PIP_2) is one of two major membrane signal transduction systems that generate second messengers in the cell (see Chapter 1, Section 1.3.2). Over 25 receptor types are known to be associated with the phosphoinositide (PI) effect; in corneal epithelium, the best characterized are the muscarinic cholinergic receptors, stimulated by acetylcholine and other cholinergic agents. In addition, PI turnover in rabbit corneal epithelium is stimulated through both serotonergic and α_1-adrenergic receptors. As shown in Fig. 1.6, the breakdown of PIP_2 generates two second messengers, inositol 1,4,5-trisphosphate (IP_3) and diacylglycerol (DG). The latter is directly involved in the activation of protein kinase C, a key enzyme that elicits a variety of intracellular responses. This enzyme, which is also activated directly by phorbol esters, may play a role in Cl- transport, as discussed in the following section. The present discussion summarizes current information on muscarinic cholinergic and serotonergic stimulation of PIP_2 turnover in corneal epithelium.

Corneal epithelium has an unusually high content of acetylcholine; fluorometric assays in calf, rabbit, and frog show concentrations of 16.9, 11.1, and 21.8 μg/g of tissue, respectively (Pesin and Candia, 1982). There are also active enzymes controlling its synthesis and degradation. Acetylcholinesterase, which is unrelated to the nerve supply of the tissue, appears to be produced *in situ* by corneal epithelial cells. During embryonic development of the chicken cornea there is a striking transient increase in the specific activity of this enzyme at day 15, followed by a rapid decrease before hatching (Sturges and Conrad, 1987). However, despite its widespread distribution, the functions of a cholinergic system in the corneal epithelium have been unclear until rather recently. It now appears that stimulation with cholinergic agonists such as carbachol enhances the activity of nuclear RNA and DNA polymerases both in normal epithelia (Colley *et al.*, 1985) and in resurfacing cells after an acid burn (Colley *et al.*, 1987). Most importantly, it is now recognized that cholinergic stimulation plays a major role in the agonist-induced hydrolysis of phosphatidylinositol 4,5-bisphosphate (PIP_2) in the corneal epithelium of rat (Chung *et al.*, 1985; Proia *et al.*, 1986), rabbit (Bazan *et al.*, 1985a), and probably other species as well.

Solubilized preparations of phospholipase C from rat cornea show preferential hydrolytic activity toward phosphatidylinositol 4,5-bisphosphate (PIP_2) compared to phosphatidylinositol (PI) (Chung *et al.*, 1985). The breakdown of PIP_2 is rapid, pointing to its direct involvement in cholinergic-stimulated phosphoinositide turnover. When the phospholipids of isolated rat cornea are labeled with [^{14}C]arachidonic acid (AA) and incubated with exogenous acetylcholine (ACh), the formation of labeled phosphatidic acid (PA) after 10 min is twice as great as that found in controls (Proia *et al.*, 1986). Since PA is an obligatory intermediate in the PI pathway (Fig. 1.6), these studies provide further evidence for active muscarinic cholinergic stimulation of PIP_2 breakdown in rat corneal epithelium. The activity of this pathway in the stroma/endothelium of the rat cannot be precisely assessed because these layers do not incorporate labeled AA in measurable amounts (Baratz *et al.*, 1987). In contrast, rat epithelium avidly incorporates labeled AA into phospholipids. A 37-s incubation of labeled rat corneas with ACh causes a 58% increase in labeled PA; after 5 min, this rises to 188%. Incubation of inositol-labeled rat epithelium with ACh for 5 min shows similar dramatic effects in the production of labeled inositol intermediates of PIP_2 hydrolysis.

An active PI membrane signal transduction system with the rapid production of IP_3 is also present in rabbit cornea (Bazan *et al.*, 1985a). The metabolism of inositol lipids is highest in the epithelium and lowest in the endothelium; stromal cells show intermediate values. Protein kinase C activity has been estimated in rabbit cornea by measuring the incorporation of labeled phosphate into histone (Bazan *et al.*, 1987). Here too, the activity is highest in the epithelium and is modulated by DG generated from the agonist-stimulated breakdown of PIP_2. Protein kinase C is also activated by phorbol esters (see Fig. 1.6); it shows maximal activity in the presence of 1.5 mM Ca^{2+} and 1 mM EGTA. Membrane-bound PI kinase is also very active in rabbit corneal epithelium, accounting for about half of the basal phosphorylation of the tissue.

Both serotonin (5-hydroxytryptamine; 5-HT) and its precursor 5-hydroxytryptophan

(5-HTP) are present in subepithelial or stromal nerve fibers of bovine cornea (Osborne, 1983) as well as in frog, pigeon, and guinea pig corneas (Osborne and Tobin, 1987). There are serotonin receptors in rabbit corneal epithelium that can be activated to stimulate the *in vitro* synthesis of cAMP in this tissue (Neufeld *et al.*, 1982).

Although α-adrenergic receptors are not thought to be present in rabbit cornea, the finding that norepinephrine stimulates the breakdown of both ^{32}P-labeled and inositol-labeled PIP_2 into DG and IP_3 strongly suggests the presence of α_1-adrenergic receptors in rabbit corneal epithelium (Akhtar, 1987). Serotonin also stimulates the phospholipase C-mediated hydrolysis of PIP_2, probably through activation of 5-HT_2 receptors. Incubation of permeabilized bovine epithelial cells with either GTPγS (a nonhydrolyzable analogue of GTP) or NaF results in enhanced breakdown of PIP_2 and increased formation of IP_3 (Akhtar, 1988). By analogy with other tissues, these findings provide indirect evidence that in corneal epithelium, a G protein may be involved in coupling α_1-adrenergic and 5-HT_2 receptors to the activation of phospholipase C.

3.1.5. Ion Transport

The epithelium has been described as a "tight" ion-transporting cell layer that functions both as a protective barrier and as an accessory fluid-secreting layer that augments the endothelial regulation of stromal hydration (Klyce and Crosson, 1985). The Cl^--dependent pump (described below) may account for as much as 15% of the water that leaves the stroma (Beekhuis and McCarey, 1986). The movement of Cl^- from the stroma and its secretion into the tear film appears to be modulated by several types of receptors: (1) β-adrenergic and serotonergic receptors, which are coupled to the activation of adenylate cyclase and the stimulation of cAMP synthesis, and (2) phorbol ester receptors, which activate protein kinase C. Dopamine and possibly muscarinic cholinergic agents that activate α_1-adrenergic receptors and mediate the breakdown of PIP_2 may also play a role in Cl^- transport, although these systems are only now beginning to be investigated in the corneal epithelium. Additionally, a Na^+,K^+-ATPase pump localized in the basolateral cell membranes of corneal epithelium and a Ca^{2+}-stimulated, Mg^{2+}-dependent ATPase in the plasma membranes (Reinach and Holmberg, 1987) play important roles in the transport of Na^+, K^+, and Ca^{2+}.

It has been recognized for many years that epinephrine and other β-adrenergic agonists stimulate the transport of Cl^- from the stromal side of the epithelium to the apical (tear) surface in isolated frog (Chalfie *et al.*, 1972), rabbit (Klyce *et al.*, 1973), and human (Fischer *et al.*, 1978) corneas. Other studies on rabbit corneal epithelium showed in addition a concomitant inward transport of Na^+ from the tear surface to the stroma (Klyce, 1975). The movement of Na^+ from tears into the epithelium is by passive diffusion, whereas the extrusion of this ion into the stroma is probably mediated by an active Na^+,K^+-ATPase pump localized in the basolateral membranes of the epithelium (Klyce and Crosson, 1985). The uptake of Cl- from the stroma is also coupled to this pump. Corneal epithelial Na^+,K^+-ATPase is strongly inhibited by a cytochrome P450-dependent metabolite of arachidonic acid metabolism that has been identified as 12(*R*)-HETE, or compound C (Schwartzman *et al.*, 1987a). An endogenous inhibitor of this type could play an important role in modulating the transport of ions across the basolateral membranes of the corneal epithelium.

The stromal thinning associated with the transport of Cl^- is enhanced in the presence of theophylline, a phosphodiesterase inhibitor that increases the intracellular levels of cAMP (Klyce, 1977). Stimulation of both serotonergic and adrenergic receptors (Klyce *et al.*, 1982; Neufeld *et al.*, 1983) causes enhanced cAMP production and an increase in the transport of Cl- in rabbit corneal epithelium. The adenylate cyclase system and cAMP-dependent protein kinase A (see Chapter 1, Section 1.3.1) are present in a variety of species. Apart from the extensively investigated rabbit and frog corneas (Reinach and Kirchberger, 1983), the enzyme has also been demonstrated in bovine corneal epithelium (Walkenbach *et al.*, 1981; Walkenbach and LeGrand, 1981). As to the mechanism of Cl^- secretion, it is well documented that intracellular changes in cAMP are involved in stimulus–secretion coupling. An increase in cellular cAMP elicits Cl^- secretion, and it has long been suspected that Ca^{2+} is the second messenger that mediates this response. Nevertheless, precise measurements of Ca^{2+} concentration in isolated bovine epithelial cells show no changes in response to either adrenergic or cholinergic agonists (Cork *et al.*, 1987). This rather unexpected finding may be explained in part by later studies using permeabilized cells (Reinach and Holmberg, 1989). It appears that Ca^{2+} may act as a negative feedback effector to inhibit a cAMP response and in this way could modulate the role of cAMP as a second messenger in receptor–effector coupling.

Active ion transport in rabbit corneal epithelium, as measured by changes in short-circuit current, is stimulated by tumor-promoting phorbol ester acetate (TPA), presumably through activation of Ca^{2+}- and phospholipid-dependent protein kinase C (Crosson *et al.*, 1986). The stimulation is Cl^--dependent, but neither β-adrenergic activation nor eicosanoids are required to produce the effect. Protein kinase C is also activated by DG produced in the agonist-induced breakdown of PIP_2. This membrane transduction system has been demonstrated in the cornea, and it could play a role in regulating ion movement through activation of protein kinase C.

3.1.6. Eicosanoids: Prostaglandins, Thromboxanes, HETEs, and Epoxyeicosatrienoic Acids

A large body of information has accumulated for more than a decade on eicosanoids in the anterior segment of the eye; most of this research has focused on the iris–ciliary body, where the arachidonic acid cascade is a prominent feature of inflammation. Eicosanoids are also produced in the cornea; as described below, all three of the cellular layers appear to be involved in the production of specific products. The general biochemistry and nomenclature of eicosanoids and some details of their generation from arachidonic acid (AA) by three major oxidative pathways are given in Chapter 1, Section 1.5.

The earliest study on eicosanoid production in the cornea was carried out on microsomal pellets isolated from homogenates of rabbit cornea (Kass and Holmberg, 1979). Incubation with labeled AA resulted in the production of small amounts of labeled eicosanoids generated through the cyclooxygenase pathway that were identified as prostaglandins $PGF_{2\alpha}$, PGE_2, and PGD_2, and thromboxane TXB_2. Further studies on the localization of eicosanoid production in specific cell layers of rabbit cornea show that epithelial cells and keratocytes produce mainly TXA_2 and PGE_2, respectively (Taylor *et al.*, 1982). Calf keratocytes also produce mainly PGE_2. These findings, as well as others described in the literature, and discussed below, are summarized in Table 3.2.

TABLE 3.2. Eicosanoid Production in the Cornea

Species	Cell layer/fraction	Detection	Major products	Reference
Rabbit	Whole tissue; microsomes	TLC	$PGF_{2\alpha}$; PGE_2 PGD_2; TXB_2	Kass and Holmberg (1979)
Rabbit	Epithelium keratocytes	TLC	TXB_2; $PGF_{2\alpha}$ PGE_2	Taylor *et al.* (1982)
Calf	Keratocytes	TLC	PGE_2; $PGF_{2\alpha}$	
Rabbit	Epithelium and endothelium	HPLC	HETE; TXB_2	Bazan *et al.* (1985b,c)
		HPLC	TXB_2	
	stroma	HPLC	PGE_2;12-HETE	
Rabbit	Whole cornea, stroma, stroma/epithelium, stroma/endothelium	TLC	12-HETE in all layers; PGE_2	Williams *et al.* (1985); Williams and Paterson (1986)
Calf	Epithelial microsomes	TLC	Epoxyeicosatrienoic acids	Schwartzman *et al.* (1985, 1987b)

Separate analyses of eicosanoid production in epithelium, stroma, and endothelium of rabbit cornea have largely corroborated the early investigations showing an active cyclooxygenase pathway; in addition, the lipoxygenase pathway has also been demonstrated in various layers of the cornea (Bazan *et al.*, 1985b; Williams *et al.*, 1985). Incubation of individual corneal layers with labeled AA shows that epithelium and endothelium produce mainly TXB_2 by the cyclooxygenase pathway, whereas in the stroma, the main product is PGE_2; 12-HETE is produced by the lipoxygenase pathway in both the stroma and epithelium. Arachidonic acid metabolism is markedly altered after a cryogenic injury (Bazan *et al.*, 1985b,c). There are large increases in cyclooxygenase products, particularly $PGF_{2\alpha}$ in the epithelium and endothelium, and increases of all eicosanoids, both cyclooxygenase and lipoxygenase products, in the stroma. Injection of labeled AA into the anterior chamber of the rabbit shows radioactivity distributed throughout the eye, with the posterior part of the eye containing 40% of the total radioactivity (Bazan, 1987). Most of the AA is incorporated into membrane lipids. Two hours after applying a cryogenic lesion, there is a large increase in eicosanoid metabolites originating from both the cyclooxygenase and lipoxygenase pathways.

Ascorbic acid inhibits 12-HETE production in rabbit corneal homogenates, although PGE_2 production is not affected (Williams and Paterson, 1986). The inhibitory action of ascorbic acid on the lipoxygenase pathway of AA metabolism appears to be related to the antioxidant properties of the vitamin. Still another modulation of eicosanoid production in the cornea may be through interleukin-1, which potentiates the production of TXB_2 in epithelial cell lines (Shams *et al.*, 1986).

Until rather recently, AA was believed to be metabolized only by two major pathways, cyclooxygenase and lipoxygenase. However, epoxidation of AA by a microsomal cytochrome P450-dependent mixed-function oxidase system (see Chapter 1, Section 1.5), first described in liver and kidney, has now been detected in several ocular tissues and is particularly active in corneal epithelium (Schwartzman *et al.*, 1985, 1987b). Microsomes from calf epithelium incubated with labeled AA in the presence of oxygen and an

NADPH-generating system convert about 60% of the substrate to several eicosanoids: 6-keto-$PGF_{1\alpha}$, $PGF_{2\alpha}$, PGE_2, and two specific cytochrome P450-epoxygenated metabolites. The formation of the first three metabolites is inhibited by indomethacin; hence, they are synthesized through the cyclooxygenase pathway. Production of cytochrome P450 metabolites is depressed in the presence of SKF-525A (which inhibits cytochrome P450-dependent enzymes by binding to hemoprotein) or carbon monoxide (which inhibits the oxygen-activating function of cytochrome P450). Compared to other ocular tissues, corneal epithelium has by far the highest capacity to oxidize AA via the cytochrome P450-dependent pathway; it is about 200 times more active in this tissue than in the ciliary body and appears to be the major pathway for AA oxidation in the corneal epithelium.

The major products of AA metabolism generated through the cytochrome P450 system are epoxyeicosatrienoic acids (see Fig. 1.10). At least two of them are now known to be biologically active. The earliest one isolated, compound C, is a potent inhibitor of corneal epithelial Na^+,K^+-ATPase. Another metabolite of AA generated through the cytochrome P450 system has been called compound D (Masferrer *et al.*, 1989). It is a potent vasodilator, causing marked vasodilation of conjunctival blood vessels when applied topically to the rabbit cornea. Compound D also has angiogenic properties, as shown by the appearance of new blood vessels in the cornea within 7 days after implantation of as little as 0.5 μg in the rabbit cornea.

3.1.7. Detoxification

3.1.7.1. Cytochrome P450 Drug-Metabolizing Enzymes

A general description of the microsomal cytochrome P450 monooxygenases (mixed-function oxidases) is given in Chapter 1, Section 1.6. Although the major activities of these enzymes are in the liver, they are also present in many extrahepatic tissues including ocular tissues. Active drug-metabolizing systems were detected many years ago in the ciliary body and retinal pigment epithelium (see review by Shichi, 1984), and two drug-metabolizing enzyme systems are now known to be present in corneal epithelium (Schwartzman *et al.*, 1987b). The activity of aryl hydrocarbon hydroxylase in microsomes of calf corneal epithelium is less than one-tenth of that in the ciliary body microsomes, although another drug-metabolizing enzyme, 7-ethoxycoumarin O-deethylase, is relatively more active. The latter has about one-half the activity found in ciliary body or retinal pigment epithelium.

Heme oxygenase has also been demonstrated in calf corneal epithelium (Schwartzman *et al.*, 1987b; Abraham *et al.*, 1987). This enzyme catalyzes the breakdown of heme, an integral component of cytochrome P450, and its activation by heavy metals causes a marked reduction in mixed-function oxygenase activity. Heme oxygenase activity in bovine corneal epithelium is about one-fourth of that found in retinal pigment epithelium or ciliary body. Of interest is that the level of heme oxygenase protein in human corneal epithelium is about ten times higher than that of bovine cornea. Immunologic studies have demonstrated considerable homology between heme oxygenase and NADPH cytochrome P450 (c) reductase proteins from human cornea and liver. The major importance of the hemoprotein regulatory heme oxygenase system and the cytochrome P450 enzymes in cornea is their potential role in local drug and hormone detoxification. This may be applicable to substances such as glucocorticoids, for which specific binding sites have been localized in the cytosol of bovine cornea (Lin *et al.*, 1984).

3.1.7.2. Glutathione S-Transferases

The glutathione S-transferases comprise a family of closely related enzymes present in multiple forms in many tissues. They have many functions, and two that appear to be important in ocular tissues are: (1) detoxification of xenobiotics by conjugating glutathione (GSH) with hydrophobic compounds having an electrophilic center and (2) protection of cellular integrity through a selenium-independent enzymatic reduction of organic hydroperoxides.

The initial step in mercapturic acid biosynthesis, conjugation of GSH to hydrophobic xenobiotics, is catalyzed by glutathione S-transferase. Subsequent reactions leading to the formation of mercapturic acid are carried out sequentially by γ-glutamyl transpeptidase, cysteinylglycinase, and N-acetyltransferase (see Fig. 4.2). Glutathione S-transferase activity as well as the other enzyme activities have been detected in extracts of bovine cornea (Saneto *et al.*, 1982a). Although individual cell layers of the cornea were not examined, it is reasonable to assume that the activities are present mainly in the epithelium. Cationic as well as anionic forms of glutathione S-transferase have been characterized in bovine (Saneto *et al.*, 1982b) and human cornea (Singh *et al.*, 1985). The glutathione S-transferases in both species display a wide range of substrate specificities toward a variety of xenobiotics that can be conjugated to GSH and afterward detoxified through the mercapturic acid pathway.

3.1.7.3. Protection from Free Radicals and H_2O_2

The corneal epithelium is well equipped to deal with oxidative stress, whatever its source may be. Two H_2O_2-degrading enzymes, catalase and glutathione peroxidase, as well as superoxide dismutase were detected more than a decade ago in rabbit corneal epithelium (Bhuyan and Bhuyan, 1977, 1978; Crouch *et al.*, 1978). Although there is little additional information on catalase and glutathione peroxidase activities of corneal epithelium, further studies on superoxide dismutase (SOD) have identified the copper–zinc cytosolic form of the enzyme in corneal epithelium and endothelium of rats, dogs, rabbits, and humans (Redmond *et al.*, 1984). Immunohistochemical methods have shown that in rat corneal epithelium, SOD is localized specifically in the basal and intermediate epithelial cells but not in the superficial cells (Rao *et al.*, 1985). Unlike the lens, corneal SOD does not undergo inactivation with age in the rat; the ratio of catalytically active enzyme to immunoreactive enzyme is close to unity in all age groups studied (Crouch *et al.*, 1984).

3.2. BASAL LAMINA (BASEMENT MEMBRANE)

The region defined morphologically as the epithelial basal lamina is secreted by the basal corneal epithelial cells. Bowman's layer (in the primate and avian cornea) lies directly posterior to the basal lamina, and together they comprise the basement membrane

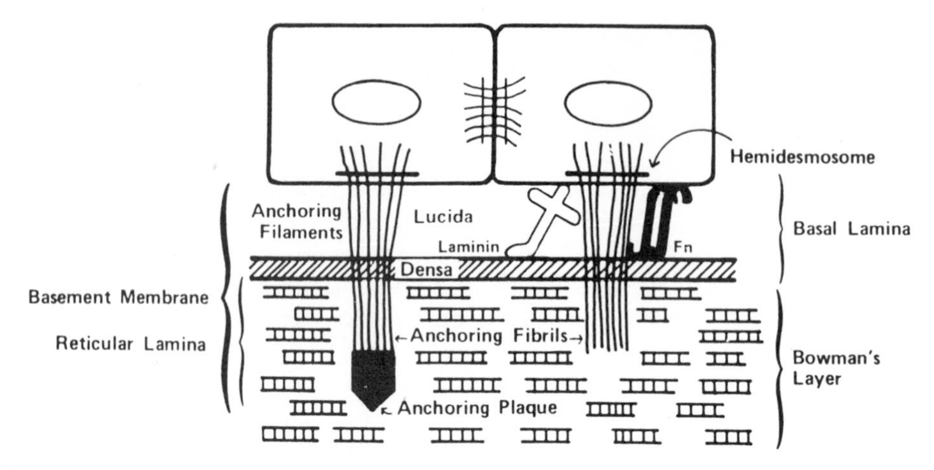

Figure 3.3. Schematic diagram of the basement membrane of human cornea showing attachment of the basal cells of the corneal epithelium to the basal lamina densa of the basement membrane by hemidesmosomes via anchoring filaments and anchoring plaques composed of type VII collagen. Other components mediating the attachment of basal epithelial cells to the basement membrane include two adhesive macromolecules, fibronectin (Fn) and laminin. (From Berman *et al.*, 1988)

of the corneal epithelium (Fig. 3.3). The attachment of the basal cells of the corneal epithelium to the basement membrane is mediated by hemidesmosomes through anchoring filaments and adhesive glycoproteins such as laminin and fibronectin, which are thought to promote the attachment of basal epithelial cells to two major integral components of the basal lamina, type IV collagen and heparan sulfate (Berman *et al.*, 1988).

Laminin, bullous pemphigoid antigen (localized intracellularly in the hemidesmosome plaque), and fibrin/fibrinogen have been identified in human corneal basement membrane by the use of immunofluorescent techniques (Millin *et al.*, 1986). This study also identified fibronectin in human basement membrane, and other independent investigations also suggest that fibronectin is present in the normal, uninjured basement membrane of human (Kohno *et al.*, 1987) as well as rat corneal epithelium (Sramek *et al.*, 1987). However, its presence in rabbit corneal basement membrane is controversial. Although fibronectin is found in developing stroma and Descemet's membrane of the rabbit (Cintron *et al.*, 1984), it may (Tervo *et al.*, 1986) or may not (Fujikawa *et al.*, 1981; Cintron *et al.*, 1984) be present in mature epithelial basement membrane. It is, however, an important adhesive macromolecule in human basal lamina (Berman *et al.*, 1988).

Two collagens, types IV and VII, have now been clearly identified in corneal basement membrane. The first, type IV collagen, was characterized several years ago in human epithelial basement membrane by immunofluorescent techniques (Newsome *et al.*, 1981) and more recently by the use of type-specific antibodies (Nakayasu *et al.*, 1986). Type VII collagen has only been detected and characterized rather recently. First described as a network of anchoring fibrils (Alvarado *et al.*, 1983), these structures were later shown to be composed of the newly described type VII collagen (Gipson *et al.*, 1987). They were characterized in human and rabbit basal lamina using monoclonal antibodies to the nonhelical domain of type VII collagen (see Chapter 1, Section 1.2.1.1). This complex network of fibrils, whose insertions in the basement membrane are directly opposite the hemidesmosomes, terminate as anchoring plaques in the anterior stroma (Gipson *et al.*,

1987). Studies on keratectomy wounds in the rabbit cornea show that the reassembly of adhesion structures composed of hemidesmosomes, anchoring filaments, and basement membrane reappear synchronously, but the normal unwounded state is not reached even by 12 months post-wounding (Gipson *et al.*, 1989).

3.3. STROMA

The corneal stroma is an extracellular matrix consisting of collagen and amorphous ground substances (proteoglycans, glycoproteins, and proteins), with a sparse population of specialized corneal fibrocytes (keratocytes) dispersed throughout the tissue. It comprises about 90% of the corneal thickness and plays a major role in corneal transparency, which is dependent on the degree of hydration and the orderly arrangement of the collagen fibers (see review by Maurice, 1984).

3.3.1. Collagen

3.3.1.1. General Characteristics

Collagen is the major structural component of the corneal stroma, accounting for 12–15% of its wet weight (or about 70% of its dry weight). The collagen fibers of mammalian stroma measure 25–30 nm in diameter and have typical 67-nm periodicity. They lie parallel to the corneal surface and extend continuously across the cornea. Parallel fibers in one lamella lie at oblique angles to those in the adjacent lamellae. The regular packing of the collagen fibers creates a lattice or three-dimensional diffraction grating, and the ability of the cornea to scatter 98% of the incoming light is thought to be the consequence of the equal spacing of the collagen fibers. According to the "lattice theory" of Maurice, scattered light waves interact in a perfectly ordered fashion, resulting in elimination of destructive interference (Maurice, 1957). However, this lattice need not be perfectly crystalline, as first proposed; although some ordering of the fibrils is necessary for transparency, the precise degree of regularity required is still somewhat uncertain (Maurice, 1984). Even though it has often been observed that loss of transparency in the swollen cornea is associated with changes in interfibrillar spacing of the collagen fibrils, the quantitative and biophysical aspects of this phenomenon have not yet been fully clarified.

The stromal collagens form part of the gene family of collagens, whose structure and composition are described in Chapter 1, Section 1.2.1.1. Of the seven types of collagen identified to date in the postnatal cornea, four are present in the stroma: types I, III, V, and VI (Table 3.3). The gene for human corneal type I procollagen has been assigned to chromosome 7, where the pro-α1 and possibly the pro-α2 gene(s) are encoded (Church *et al.*, 1981). The gene for human skin type I procollagen has been assigned to chromosome 17 (SundarRaj *et al.*, 1977). These findings imply that not only separate genes but also separate synthetic mechanisms may exist for these two type I procollagens.

The basic structural unit of types I, III, and V corneal collagen is tropocollagen, a long asymmetric molecule approximately 300 nm in length and 1.5 nm in diameter, with a molecular mass of 285 to 300 kDa. Each fibril is composed of three polypeptide chains, the α chains, coiled in the form of a triple helix. The other recently described major

Table 3.3. Collagen Types in Corneal Stroma

Type	Molecular composition	Relative content[a] (%)
Type I	$[\alpha 1(I)]_2$, $\alpha 2(I)$	50–55
Type III	$[\alpha 1(III)]_3$	≤1[b]
Type V	$\alpha 1(V)$, $\alpha 2(V)$, $\alpha 3(V)$	8–10
Type VI[c]	$\alpha 1(VI)$, $\alpha 2(VI)$, $\alpha 3(VI)$	25–30

[a]Estimations of normal adult human cornea, relative to dry weight of the tissue (Zimmermann *et al.*, 1988).
[b]The precise amount, or even its presence or absence, is still controversial, and detection depends on species, age, and physiological state of the tissue.
[c]Type VI collagen also contains extensive interchain disulfide linkages and large globular domains, one of which is a 140-kDa glycoprotein.

stromal collagen, type VI, is structurally unique. It has a high proportion of nonhelical globular polypeptides at both the C- and N-terminal ends of the helical matrix. A nonhelical 140-kDa glycoprotein is one of the characteristic components of type VI collagens.

Type I collagen is usually considered to be the major collagen of corneal stroma in most species. Small amounts of types III and V have also been been identified in pepsin-treated extracts of human corneas by standard biochemical techniques (Newsome *et al.*, 1981, 1982) and in intact human tissue using immunologic methods (Nakayasu *et al.*, 1986). In rabbit cornea too, the principal collagen is type I, which accounts for 93% of the total in newborn as well as adult rabbits (Lee and Davison, 1981). Type III collagen represents about 2% of the total at birth but is not detectable in adult cornea. Type V collagen forms 5% of the total collagen at birth and 9% in adult cornea. Pepsin-solubilized stromal extracts from bovine cornea show a similar pattern of collagen distribution (Lee and Davison, 1984). The major collagen, accounting for 89% of the total, is type I, whereas type III is detectable only in fetal bovine cornea, where it comprises about 1% of the total collagen.

The foregoing studies suggest that small amounts of type III collagen are present in the corneal stroma of some species, particularly—or perhaps only—in fetal life or at birth or in healing adult tissues. Its presence in unwounded adult cornea is still controversial. Direct biochemical extraction versus tissue culture techniques often give conflicting results, and marked species variation has been reported. Recent studies discussed below (Section 3.3.1.2) on organ-cultured rabbit corneas have partly resolved this controversy; sensitive radiolabeling and immunohistochemical techniques have been used to detect and characterize small amounts of type III collagen in neonatal and scar tissue in the rabbit cornea. On the other hand, type III collagen is not known to be present in chick cornea. The principal collagens of embryonic chick fibrils are types I and V; moreover, they are codistributed within the same collagen fibril (Birk *et al.*, 1988). Epithelial cells of primary embryonic chick cornea also synthesize collagen types II and IX early in development, but these specific types are not detectable in postnatal cornea.

The classical view that corneal stroma contains only types I, III, and V collagen has now been revised in the light of new findings showing that type VI is a major collagen in human (Zimmermann *et al.*, 1986) as well as bovine cornea (Alper, 1988). It may comprise as much as 25–30% of the dry weight of pepsin-solubilized tissue. This unusual collagen, which has been studied extensively in cartilage and other tissues, is a disulfide-linked high-molecular-weight protein whose helical domain accounts for only about one-third of the molecule (Ayad *et al.*, 1985). The remainder consists of globular domains located in the terminal regions of the molecule. The α1(VI) and α2(VI) helical chains consist of two nonidentical 140-kDa polypeptides, and the third, α3(VI) chain, is a 200-kDa peptide (Trueb and Winterhalter, 1986).

The 140-kDa glycoprotein domain of type VI collagen has been characterized in several nonocular tissues, and its presence in corneal stroma may be inferred indirectly from a number of studies. For example, an unusual disulfide-linked structural glycoprotein (SGP) containing collagen-derived amino acids was isolated from bovine corneal stroma before type VI collagen had been characterized (Alper, 1982/1983). More recent studies with an anti-SGP serum containing polyclonal antibody to the SGP complex show that the anti-SGP serum is specific for SGP and does not react with other stromal collagens or with glycoproteins such as fibronectin or laminin (Alper, 1988). Comparison of the properties of the anti-SGP serum with those of a polyclonal antibody specific for type VI collagen suggests that the SGP complex is probably derived from the tissue form of type VI collagen. This type of collagen is distinguishable from the noncollagenous microfibrillar bundles that form parallel and orthogonal arrays in the stroma (Bruns *et al.*, 1987) since the microfibrils appear to be associated with a 150-kDa glycoprotein rather than the 140-kDa glycoprotein characteristic of type VI collagen.

3.3.1.2. Metabolism

Specific enzymatic steps in the synthesis and/or turnover of corneal collagen have not been examined, but it is reasonable to assume that they are similar to those in nonocular tissues. Corneal collagen is formed during embryonic and early postnatal development; its turnover is probably minimal in normal uninjured adult cornea. Measurements of the incorporation of radioactive proline into collagen of newborn rabbits suggest a half-life of about 24–50 h (Lee and Davison, 1981). There are no equivalent biochemical data for adult cornea. A novel approach to understanding collagen turnover in corneal stroma involves its *in situ* labeling with thiocyanate and triazinyl chloride derivatives of fluorescent dyes (Davison and Galbavy, 1985). These studies suggest that while a certain portion of collagen synthesized in the growing cornea is turning over, there is also a stable fraction that persists for long periods of time.

Collagen synthesis has been examined *in vivo* in neonatal rabbit cornea and in scar tissue from adult animals following injection of labeled proline into the anterior chamber (Cintron *et al.*, 1981). The predominant collagens synthesized, albeit of low specific activity, were types I and V. Later studies by these investigators have utilized organ-cultured rabbit corneas incubated with labeled glycine under conditions that maintain the transparency of the tissue for 48 h. Improved sensitiviy and specificity provided evidence for the synthesis and deposition of type III (Cintron *et al.*, 1988) and type VI (Cintron and Hong, 1988) collagens, which were not detected previously using the *in vivo* labeling method. The synthesis of types I, III, and V collagen in both neonate and scar tissue from adult cornea was demonstrated in pepsin-digested formic acid extracts by Coomassie blue staining and fluorography of SDS-PAGE gels. Further evidence for the synthesis and deposition of type III collagen was obtained from immunohistochemical studies using type III collagen-specific monoclonal antibody. These authors concluded that type III collagen is synthesized by the endothelial cells, which deposit it within the growing

Descemet's membrane of neonatal cornea and in the posterior portion of scar tissue formed during wound healing. This is in contrast to the major collagens of corneal stroma, types I, V, and VI, which appear to be synthesized by stromal-derived fibroblasts and not by endothelial cells (Cintron *et al.*, 1988; Cintron and Hong, 1988). Type VI collagen has been demonstrated in normal adult cornea as well as in neonatal and healing corneas by the organ-culture technique described above. Owing to its unusual structure and its putative role as a space-filling structure, type VI collagen may be involved in a unique way in the development of transparency in neonatal cornea. It is deposited at a slower rate in corneal scars than in normal neonate tissue, and further studies on its turnover and precise localization in normal as well as scar tissue should clarify the physiological and structural role of this major corneal collagen.

3.3.2. Glycosaminoglycans and Proteoglycans

3.3.2.1. Chemical Composition

A brief historical introduction would seem appropriate before discussing current views on the structure and composition of the corneal proteoglycans. Termed mucopolysaccharides since their discovery in the 1930s, the carbohydrate chains of these high-molecular-weight polymers are now defined chemically as glycosaminoglycans (GAGs). In the native state, with the exception of hyaluronic acid, the glycosaminoglycan chains are covalently linked to a core protein, and the total complex is called a proteoglycan (PG) (see Chapter 1, Section 1.2.3.1).

The early investigations of Meyer and co-workers (1953) showed that corneal GAGs (mucopolysaccharides) consisted of about 65% keratan sulfate (KS) and 35% chondroitin sulfate (CS). These preparations did not, however, represent native PGs; rather, the polymers isolated consisted only of protein-free oligosaccharide (GAG) chains. The native PGs had been degraded because of extraction with NaOH and subsequent prolonged treatment with proteolytic enzymes. Later investigations using milder conditions for extraction and purification yielded PGs that were somewhat more intact. They contained varying amounts of covalently linked protein; moreover, the use of ion-exchange chromatography to fractionate and isolate the individual species showed considerable heterogeneity and polydispersity within the two major families of corneal PGs (Laurent and Anseth, 1961; Berman, 1970; Saliternik-Givant and Berman, 1970).

In the early 1970s, iduronic acid-containing dermatan sulfate (DS) proteoglycans were shown to comprise a significant proportion of the PGs of bovine cornea (Stuhlsatz *et al.*, 1972; Axelsson and Heinegard, 1975). These findings led to gradual revisions in the general definitions of corneal PGs, and the two major families now recognized in the cornea are keratan sulfate (KS) PGs and dermatan sulfate/chondroitin sulfate (DS/CS) PGs. The latter are also called simply dermatan sulfate (DS) PGs by some investigators. Gregory and co-workers (1982, 1988) have suggested the terms proteokeratan sulfates (PKS) and proteodermatan sulfates (PDS).

The techniques used today for the isolation and characterization of proteoglycans are based on dissociative extraction in 4–6 M guanidine as a crucial first step. Gel filtration, ion-exchange chromatography, and density-gradient centrifugation are then used to isolate and characterize the individual PGs present in the heterogeneous and polydisperse mixtures of native-state corneal PGs (Axelsson, 1984). A complete recovery and a "perfect"

Table 3.4. Composition and Structure of Corneal Proteoglycans[a]

Property	Proteoglycan: Keratan sulfate (KS)	Proteoglycan: Dermatan/chondroitin sulfate (DS/CS)
Disaccharide repeating unit	Galactose: N-acetylglucosamine	Iduronic/glucuronic acid: N-acetylgalactosamine
Oligosaccharides	High-mannose types linked via chitobiose[b] to asparagine	Mannose; galatose; N-acetylglucosamine
Position of sulfate	Carbon 6 in both sugars	Carbons 4 and 6
Other characteristics	Disulfide bonds	Polydisperse, heterogeneous
Average molecular mass	70–100 kDa	100–150 kDa
Core protein	32–40 kDa	40 kDa
GAG chain(s)	22kDa	55 kDa
Number of GAG chains	One to two	One or more
Carbohydrate-protein linkage regions	Fucosylated N-acetylglucosamine in complex type of N-glycosylamine linkage to asparagine; high mannose chains near linkage region	O-Glycosidic via galactose-galactose-xylose-serine (or threonine)
Percentage of total	60–70	30–40

[a]The data on linkage region and oligosaccharides of KS proteoglycan are for monkey cornea (Nilsson *et al.*, 1983). All other data are on bovine cornea (Axelsson and Heinegard, 1975, 1978, 1980; Axelsson, 1984; Poole, 1986).
[b]Chitobiose, N-acetylglucosamine-N-acetylglucosamine.

separation of *all* the PGs in corneal stroma has yet to be achieved. More emphasis is placed on biochemical characterization of the major families, analyses of their compositions and structures, identification of subpopulations, and isolation of individual PGs of sufficient purity to be used for producing monoclonal antibodies and ultimately for molecular cloning of their cDNAs and their genes.

A summary of the composition and structure of the two major PG families in bovine cornea (Axelsson and Heinegard, 1978, 1980) is given in Table 3.4. The weight-average M_r of the monomer KS proteoglycan is about 72,000. The corneal KS proteoglycans contain about 45% protein, 30–40% KS, and 10–12% mannose- and galactose-rich oligosaccharides. There may also be intrachain disulfide bonds in the peptide core. The DS/CS proteoglycans are especially heterogeneous and polydisperse. They have somewhat higher molecular weights than the KS proteoglycans and contain varying numbers of DS and CS side chains. The total DS/CS population of PGs accounts for about one-third of the proteoglycans of bovine cornea.

Two PG keratan sulfates (PKS-I and PKS-II) and two PG dermatan sulfates (PDS-I and PDS-II) have been isolated and purified from rabbit stroma (Gregory *et al.*, 1982). Although they resemble bovine PGs in some of their properties, there are nevertheless some important differences. The PKS-I (representing 21% of the total PGs extracted) and PKS-II (representing 43% of the total) contain 48% and 57% protein, respectively. Their amino acid compositions differ considerably; moreover, the oligosaccharides of PKS-II have a far higher content of sialic acid than those of PKS-I. The two dermatan sulfates,

PDS-I and PDS-II, of rabbit stroma contain 35% and 42% of their uronic acid as iduronic acid, respectively. The finding of two different KS PGs and DS PGs in rabbit stroma with marked differences in composition of their core proteins and in their covalently linked oligosaccharides has important implications in terms of specific binding sites to type I collagen (see Section 3.3.3).

As in other species, the proteoglycans of monkey cornea also display considerable size–charge heterogeneity (Nakazawa *et al.*, 1983b). The corneas used for these studies had been biosynthetically labeled with sulfate in combination with either mannose or serine. Two major classes of PGs were isolated, consisting of four subpopulations each of KS and DS proteoglycans. They differ from one another in GAG composition and in the number of GAG side chains as well as in the composition of the mannose-containing oligosaccharides linked to the core proteins.

The linkage regions have been characterized in KS proteoglycans isolated to a high state of purity from both bovine (Brekle and Mersmann, 1981) and monkey (Nakazawa *et al.*, 1983a; Nilsson *et al.*, 1983) corneas. All are of the N-glycosidic type between N-acetylglucosamine and asparagine, and two or possibly three branched structures have been identified. One of them contains a terminal fucose, two mannose residues, and one N-acetylglucosamine at the reducing end. A high-mannose oligosaccharide linkage-region glycopeptide isolated from monkey KS proteoglycan probably represents an intermediate in the normal dolichol pathway for the biosynthesis of N-glycosidic linkages. This is in accordance with the finding that tunicamycin, an inhibitor of the dolichol pathway of glycoprotein synthesis, also inhibits the biosynthesis of KS proteoglycan in embryonic chicken cornea (Hart and Lennarz, 1978). Structural analyses of other linkage-region glycopeptides suggest that there could also be alternate, probably minor, pathways for the formation of these linkages. An abbreviated summary of these findings is given in Table 3.4.

A large body of current evidence supports the view that PGs should now be classified into two major groups: aggregating and nonaggregating proteoglycans (Poole, 1986; Heinegard *et al.*, 1986). Cartilage PG is the best characterized of the aggregating proteoglycans. Two types of nonaggregating proteoglycans, broadly termed large and small, are also present in cartilage, skin, and other connective tissues. On the basis of immunologic properties, amino acid compositions of their core proteins, and other parameters, DS proteoglycan (and possibly KS proteoglycan) of cornea is now considered as belonging to the class of small, nonaggregating PGs whose M_r is less than 100,000–150,000 (Poole, 1986; Heinegard *et al.*, 1986). A diagram of corneal proteoglycan structures is shown in Fig. 3.4.

The production of polyclonal and monoclonal antibodies to KS proteoglycan has been reported from several laboratories. In one study, a highly purified KS proteoglycan isolated from bovine cornea contained nearly equimolar ratios of N-acetylglucosamine and galactose as well as considerable amounts of mannose and smaller amounts of fucose; it had 26% protein by weight (Conrad *et al.*, 1982). The absence of xylose indicated that it was free of CS proteoglycan. Antibodies raised in rabbits against this bovine KS proteoglycan reacted not only with the core protein but also with the native KS proteoglycan and even with the KS oligosaccharide side chains. The proteoglycan contains an antigenic 300-kDa component whose mobility is decreased to that of proteins corresponding to molecular weights of 55,000 and 40,000 after keratanase treatment.

Six monoclonal antibodies to rabbit KS proteoglycan have been prepared and shown

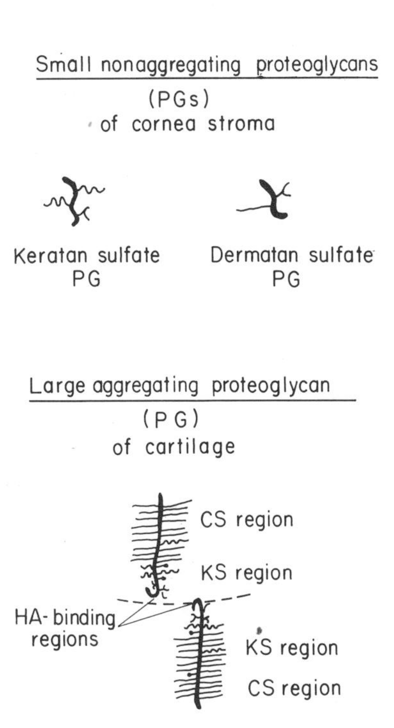

Figure 3.4. Schematic diagrams of corneal proteoglycan (PG) structures. Note the relatively small size of the nonaggregating corneal PGs in comparison with large aggregating cartilage PG. Other proteoglycan and glycosaminoglycan structures are given in Fig. 1.3.

to recognize a carbohydrate structure of the proteoglycan, probably the protein-linked oligosaccharide moiety, which comprises about 10% or more of the native KS proteoglycan (Funderburgh *et al.*, 1982/1983). Three of these antibodies were later shown to recognize antigenically related KS proteoglycans in a variety of noncorneal bovine and embryonic chicken tissues (Funderburgh *et al.*, 1987). Extracts from these tissues can bind both anti-KS monoclonal antibodies and anti-core-protein polyclonal antibody simultaneously. The binding is inhibited competitively by KS. These interesting findings suggest that the KS present in noncorneal tissues may be part of a family of proteoglycans containing antigenic sites both for corneal KS and for its core protein.

Monoclonal antibodies have been produced against carbohydrate portions of two KS proteoglycans isolated from rabbit corneal stroma, and their antigenic determinants have been identified (SundarRaj *et al.*, 1985). Twenty-eight antibodies were isolated; all of them reacted with both of the KS proteoglycans but not with scleral or conjunctival glycoconjugates. Keratanase digestion yielded protein cores that retain the linkage region and the closely associated mannose oligosaccharides. All but two of the monoclonal antibodies reacted with these protein cores. An interesting observation was that the monoclonals gave a positive immunohistochemical reaction with normal human cornea but not with cornea from patients with macular corneal dystrophy (see Section 3.10.2).

In summary, corneal proteoglycans are a complex and heterogeneous mixture of GAG-containing polyanionic macromolecules that contribute to the maintenance of the collagen organization and hence to the transparency of the cornea. More than half of the

PGs consist of KS proteoglycans, and the remainder are DS/CS proteoglycans. They belong to the group of small, nonaggregating proteoglycans found in a wide variety of extracellular matrices throughout the body. Greatly improved biochemical methodologies are now available for their isolation and characterization, and in recent years two additional techniques have been developed whose applications have already clarified many aspects of structure and composition of the PGs. One is the use of monoclonal antibodies to KS proteoglycans (and to DS/CS proteoglycans in the future), which provide structural information on proteoglycans through identification of their epitopes. They are also used for specific tissue localization in the normal and pathological cornea. The other is the molecular cloning of proteoglycan core proteins, a powerful tool now being developed that will define primary structures, identify preserved sequences, and allow assignment of the core proteins to specific gene families.

3.3.2.2. Metabolism

The enzymatic pathways and intracellular sites of biosynthesis, posttranslational modifications, and degradation of GAGs and proteoglycans in nonocular tissues are discussed in Chapter 1, Section 1.2.3.2. Our present understanding of these complex processes is based on investigations in a variety of cell types. It may be assumed that similar if not identical pathways are present in corneal cells, though they have not been examined directly.

Cultured keratocytes and epithelial cells from human and bovine cornea produce a wide spectrum of sulfated GAGs; each appears to synthesize a characteristic mixture of GAG molecules with varying proportions of HA (hyaluronic acid), CS, DS, and HS (heparan sulfate) (Bleckmann and Kresse, 1980). As described in the preceding section and shown in Table 3.4, keratan sulfate (KS) is the most abundant GAG in native corneal stroma, yet only small amounts of it are synthesized in culture either by keratocytes or by epithelial cells. The reasons for the striking change in expression of GAG biosynthesis in cultured corneal cells are not clear, but it is evident that the GAG profiles are very different from those found in native tissue.

On the other hand, organ cultures of rabbit cornea, maintained in media containing labeled glucosamine and sulfate, produce mainly DS and KS (Coster *et al.*, 1983). Thus, unlike cultured epithelial cells or keratocytes, organ-cultured corneas continue to synthesize the major endogenous stromal GAGs (KS and DS), albeit in somewhat different proportions from those found in the parent tissue. Moreover, the observation that the specific activity of the hexosamine in KS synthesized in organ-cultured cornea is only about one-half that in DS implies fundamental changes in synthetic rates of the major GAG chains *in vitro* compared to those prevailing *in vivo*.

Other studies using organ cultures of human cornea and confluent cultures of human skin and corneal fibroblasts showed marked differences in the patterns of GAG synthesis among the different cell types (Klintworth and Smith, 1981). As in rabbit organ cultures, KS forms a large fraction of the GAGs synthesized from radioactive precursors in cultured human corneas. In contrast, but similar to bovine cells, fibroblasts from human cornea appear to have lost most of their capacity to synthesize KS, even though some of the individual GAG-metabolizing characteristics of skin and corneal fibroblasts are retained in culture. In this context, recent studies by Scott and Haigh (1988a) address the interesting question of whether "oxygen lack" may control the rate of KS biosynthesis, and indirect but compelling evidence suggests that KS is characteristically produced when there is a limited supply of nutrient, particularly oxygen. This interesting hypothesis may account for the reciprocal relationship between the DS and KS contents of thick corneas (e.g., in cattle and rabbit) compared to thin ones, as in the mouse. The "oxygen lack" hypothesis may in addition provide a theoretical basis for determining the optimum conditions for studying GAG metabolism in general and KS biosynthesis in particular *in vitro* under conditions comparable to those prevailing *in vivo*.

The foregoing studies of GAG metabolism *in vitro* have centered on the biosynthesis of the polysaccharide side chains but not the total molecules, the proteoglycans (PGs). Developmental changes in PGs of embryonic chick cornea have been studied by immunohistochemical methods (Funderburgh *et al.*, 1986), and more recently direct biochemical methods have been applied to this problem in mammalian (rabbit) cornea (Gregory *et al.*, 1988). The major PG in fetal rabbit cornea is DS proteoglycan of high charge density; KS proteoglycan of low sulfate content is present, but only in small amounts. Postnatally, both the charge density and the total amount of KS proteoglycan increase dramatically, reaching adult levels between the second and eighth postnatal weeks. This corresponds to the time that the eyes have opened and the corneas are transparent. Thus, an increase in both the amount and degree of sulfation of KS proteoglycan, together with changes in the relative proportions of the two major families of corneal PGs, are major events during corneal maturation and the development of transparency.

Wounded corneas have been used as experimental models for monitoring GAG biosynthesis in normal and healing cornea. Such studies provide considerable insight on the role of proteoglycans in promoting and/or maintaining corneal transparency, since direct comparisons can be made between transparent and opaque scar tissue. One study showed that following a penetrating wound in rabbit cornea, the PGs isolated from 2-week-old opaque scars were of abnormally large size, even though the GAG chains appeared to be of normal size (Hassell *et al.*, 1983). Dermatan/chondroitin sulfate is the major GAG synthesized by the scar tissue; some hyaluronic acid is also present, but little or no keratan sulfate could be detected in the 2-week-old opaque scar. In 1-year-old scars, the normal pattern of proteoglycan synthesis was restored, and the interfibrillar spacing of the collagen fibrils was found to be similar to that in transparent cornea. The PGs in 1-year-old scars were the normal small nonaggregating type, implying that it was the synthesis of abnormally large PGs that had led to the formation of opaque, light-scattering scar tissue. Rather similar changes occur after partial-thickness radial nonperforating scalpel wounds in the rabbit cornea (Funderburgh and Chandler, 1989). In these experiments the PGs were labeled *in vivo* with radioactive glucosamine and sulfate, and comparisons were made between wounded and nonwounded corneas. Labeled PGs from wounded corneas are of larger molecular size than those isolated from controls. There is also a general reduction in PG content as well as reduced sulfation of KS proteoglycan and a concomitant accumulation of high-sulfated, high iduronic acid-containing DS proteoglycan. Thus, independent studies corroborate at least two major changes in scar tissue formed after corneal wounding: the PGs are of larger size than normal, and there is an accumulation of an altered form of iduronic acid-containing DS proteoglycan.

3.3.3. Proteoglycan–Collagen Interaction

Binding and/or structural interactions between the proteoglycans and collagen fibers of the corneal stroma have long been suspected, but only recently, through the development of an elegant electron-optical staining technique employing cupromeronic blue, has

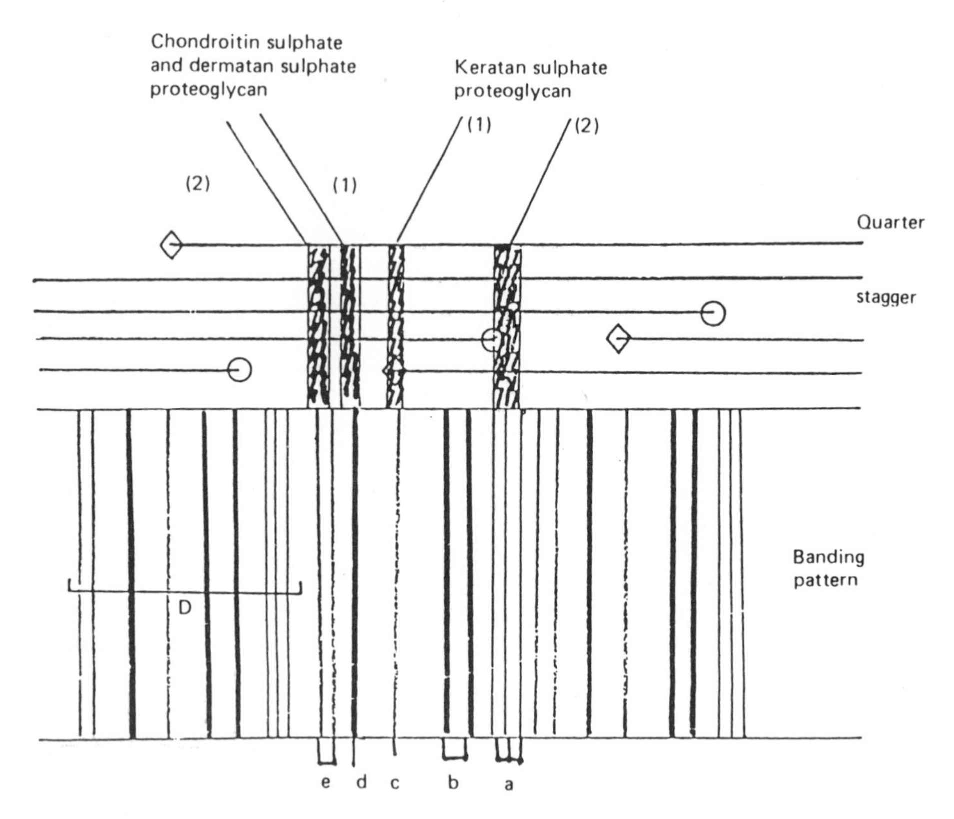

Figure 3.5. Diagrammatic representation of proteoglycan binding sites within the D period of type I collagen of rabbit cornea. The association of four proteoglycans of rabbit cornea (DS-PG I and II and KS-PG I and II) characterized by Gregory and co-workers (1982) at specific binding sites is shown in the upper portion of the diagram. The a–e banding pattern is shown in the lower portion of the diagram against the arrangement of collagen molecules in quarter stagger. The locations of the proteoglycans in the upper part of the diagram correspond to the a, c, d, and e banding patterns. (From Scott, 1988). (From Scott, 1988.) (Reprinted by permission from *Biochem. J.* **252:** 313, © 1988 The Biochemical Society, London.)

this been clearly demonstrated (Scott, 1985, 1986, 1988; Scott and Haigh, 1985). Incubation of 10-μm frozen sections of freshly excised rabbit cornea with cupromeronic blue in the presence of $MgCl_2$ at a "critical electrolyte concentration" of 0.3 M results in the formation of electron-opaque complexes with all sulfated GAGs. Individual corneal GAGs are identified by pretreatment with specific enzymes, keratanase or chondroitinase, that degrade KS or CS/DS, respectively. After incubation of the tissue sections, with or without enzymatic digestion, the preparations are rinsed, processed, and counterstained for visualization of the collagen fibrils. Striking electron micrographs are obtained showing an orthogonal arrangement of proteoglycans on the collagen fibrils. In this way, four separate and specific proteoglycan binding sites within the D period of type I collagen have been identified in rabbit cornea (Fig. 3.5). The KS proteoglycans associate at the a or c "step" bands, and DS proteoglycans are found at the d or e "gap zone." The PGs of rabbit cornea have been shown to consist of two major families of small nonaggregating KS and DS proteoglycans with differing core proteins (Gregory *et al.*, 1982; Poole, 1986). The four binding sites on type I collagen detected with the cupromeronic blue technique are thought to represent the specific sites of attachment for the four major proteoglycans of rabbit cornea. Whether the core proteins play a role in proteoglycan–collagen interaction requires further clarification.

X-ray diffraction mass distribution patterns of cupromeronic blue-stained tissue sections show that approximately 12% of the stromal proteoglycans are in an ordered arrangement associated with the specific binding sites shown in Fig. 3.5 (Meek *et al.*, 1986). The remainder appear to be diffusely distributed throughout the interfibrillar space. The various determinants for binding and interaction of PGs with collagen fibrils have been discussed in detail (Scott, 1988; Scott and Haigh, 1988a). Semiquantitative data suggest that the percentage occupancy of a fibril binding site reflects the percentage concentration of the particular PG in the tissue; there may be an equilibrium between bound and free proteoglycans.

Compelling evidence for the specificity of the binding sites has come from investigations of mouse cornea, which lacks chemically detectable KS proteoglycan. In this species, virtually no PG filaments could be detected at the a and c bands; hence, it appears that these sites are specific for KS proteoglycan (Scott and Haigh, 1988a). Other studies on the relative frequency of occupancy of bands d and e in rabbit cornea led to the assignment of DS-I to band d, and DS-II to band e; KS-I and KS-II have been assigned to bands c and a, respectively. A one-PG–one-binding site hypothesis proposed by Scott and co-workers (Scott and Haigh, 1985; Scott, 1986, 1988) has received further support from CEC ("critical electrolyte concentration") studies of bovine cornea (Scott and Haigh, 1988b). These investigations revealed very different binding behavior of a-band-associated and c-band-associated PGs with changing $MgCl_2$ molarity. Moreover, there are striking interspecies variations in KS proteoglycan binding patterns, which depend in part on the amount and type(s) of KS proteoglycans present. Interactions between DS proteoglycan and collagen fibrils are relatively more constant. In summary, the unique and highly specific technique developed by Scott and co-workers allows the "mapping" of PG binding sites on collagen fibrils of corneal stroma and to date has been applied to at least four species: rabbit, rat, mouse, and cattle.

3.4. DESCEMET'S MEMBRANE

Descemet's membrane is a morphologically unique basement membrane produced partially, or possibly entirely, by the endothelial cells (Labermeier and Kenney, 1983; Benya and Padilla, 1986; Kapoor *et al.*, 1986). Its thickness varies with age and species, being about 3 μm in young humans and 9–10 μm in adults (Murphy *et al.*, 1984). Bovine Descemet's membrane is 17 μm in thickness (Lee and Davison, 1984). In all species examined to date, both collagenous and noncollagenous substances are present. Laminin, a characteristic component of basement membranes that is closely associated with type IV collagen in extracellular matrices throughout the body, has been detected in human Descemet's membrane by immunofluorescent techniques (Kohno *et al.*, 1987). Fibronectin is also present in human (Kohno *et al.*, 1987) and rat (Sramek *et al.*, 1987) Descemet's membrane.

Three anatomic zones can be distinguished by immunofluorescent techniques (Newsome *et al.*, 1981): granular basement membrane-like material at the stromal interface, wide-spaced collagen in the middle zone, and basement membrane-like material again at the endothelial interface. Type IV collagen has been identified by immunofluorescent staining of the tissue as well as by amino acid analyses of α chains isolated from morphologically pure membrane preparations (Labermeier *et al.*, 1983). However, the principal collagen of Descemet's membrane appears to be type VIII, or "EC" (endothelial cell) collagen (Labermeier and Kenney, 1983). This recently characterized collagen (see Chapter 1, Section 1.2.1.1) was isolated from bovine Descemet's membrane after limited pepsinization of the tissue. Type IV and type VIII collagens can be selectively extracted according to their specific solubilities: type VIII collagen (identifiable by its 50-kDa $\alpha 1$ chains) is extracted by limited pepsinization, whereas type IV collagen can be solubilized only by phenol extraction. Further analyses of type VIII collagen from bovine Descemet's membrane show that its major helical components consist of two nonidentical 50-kDa polypeptides (Kapoor *et al.*, 1986). Type VIII collagen of rabbit Descemet's membrane is composed of three 61-kDa $\alpha 1$ chains, arranged mainly in a helical structure, with nonhelical domains at both ends (Benya and Padilla, 1986). Two pepsin-sensitive terminal peptides of M_r 14,700 and 4000–5000, representing the nonhelical domains of the molecule, have been isolated and partially characterized.

In summary, there is strong evidence that the major and minor collagens of native Descemet's membrane are types VIII and IV, respectively. Cultured rabbit endothelial cells actively synthesize these collagens as well as type I collagen under certain conditions.

3.5. ENDOTHELIUM

The corneal endothelium is a single layer of hexagonally shaped cells resting on Descemet's membrane and lining the posterior surface of the cornea. Its central role in normal corneal physiology has been the subject of several reviews (Riley, 1982, 1985; Waring *et al.*, 1982; Fischbarg and Lim, 1984; Maurice, 1984; Fischbarg *et al.*, 1985). This cell layer has two major functions. First, by continuously pumping fluid and ions out of the stroma into the aqueous, it is the principal mechanism responsible for maintaining corneal dehydration and transparency. These are energy-requiring processes in which the Na^+,K^+-ATPase pump plays a central role. The second function of the corneal endothelium is that of a physical barrier controlling the entry of fluid and dissolved solutes into the stroma from the aqueous humor. Even though maculae occludentes (focal tight junctions) have been observed in corneal endothelium, these structures do not constitute a barrier against the penetration of ions and small molecules from the aqueous through the intercellular spaces. The finding that larger molecules such as horseradish peroxidase and lanthanum can also flow across these junctional complexes has led to the concept of a "leaky barrier" for the corneal endothelium. The regulation of fluid and electrolyte flux in both directions through the endothelium is now considered in terms of a pump–leak model in which the metabolic pump offsets the fluid entering the stroma through this "leaky barrier." These two balanced processes form the physical and biochemical basis of corneal dehydration.

3.5.1. Carbohydrate Metabolism

The overall pattern of carbohydrate metabolism is similar to that of the epithelium (see Fig. 3.2), although the endothelium is considerably more active. Calculations of respiration (O_2 uptake), based on either unit volume of the endothelial cell or dry weight of tissue, show that the oxidative activity in the endothelium is five to six times greater than that in the epithelium (Riley, 1982; Maurice, 1984). This is not surprising, given the abundance of mitochondria and the great energy demands made on this single cell layer to maintain the metabolic pumps. It has been estimated that 93% of the cellular glucose 6-phosphate is converted to pyruvate by anaerobic glycolysis; 70% of the pyruvate is converted to lactic acid by lactic dehydrogenase, and the remainder is oxidized through the tricarboxylic acid cycle to produce ATP (Riley, 1982). Rabbit corneal endothelium contains an active pentose phosphate pathway, accounting for about 37% of the glucose metabolized by the cell (Geroski *et al.*, 1978). Stimulation of this pathway by adenosine has been shown in homogenates of cultured bovine endothelial cells (Zagrod and Whikehart, 1985). By following the fate of radioactively labeled sugar phosphates, significant increases in the formation of ribose 1- and 5-phosphate as well as sedoheptulose 7-phosphate in the presence of adenosine were detected. Although the mechanism of this effect is not clear, it appears that exogenous adenosine may be serving as a glucose substitute, entering the pentose phosphate pathway (after deamination), and stimulating the production of ATP.

Apart from the biochemical studies summarized above, there are relatively little quantitative data on glucose metabolism in the corneal endothelium. Now, however, new and promising noninvasive approaches for assessing glycolytic activity in whole and deepithelialized cornea have been reported utilizing nuclear magnetic resonance (NMR) spectroscopy (Greiner *et al.*, 1984; Gottsch *et al.*, 1986). Preliminary results on postmortem human eyes show that lactate formation from glucose in intact and deepithelialized cornea is approximately 0.50 and 0.20 μmol/h, respectively. Further development of these methods could lead to more precise estimations of endothelial metabolism and function.

3.5.2. Fluid and Ion Transport: Metabolic Pumps

The pump–leak model for corneal transparency is now envisaged as a two-directional flow of fluid and electrolytes; the energy-requiring pumps in the endothelium remove the fluid that enters the stroma through the "leaky barrier" of the endothelium. The existence of an energy-requiring pump for the transport of fluid from the stroma across the endothelium was established many years ago (Maurice, 1972). This finding, together with subsequent investigations, has lead to the concept of energy-coupled transport of fluid and electrolytes across the endothelium from the stroma to the aqueous. Supporting evidence for this notion has been summarized in several reviews (Riley, 1982; Fischbarg and Lim, 1984; Maurice, 1984).

There is a net flux of both Na^+ and HCO_3^- across the endothelium from the stroma to the aqueous, but no net flux has been established for either Cl^- or K^+ (Fischbarg and Lim, 1984). The energy-dependent transport of Na^+ and HCO_3^- is mediated by (1) Na^+,K^+-ATPase and (2) HCO_3^--dependent ATPase, respectively (Riley, 1982). Carbonic anhydrase may also be involved indirectly in the transport of HCO_3^-.

The *Na^+,K^+-ATPase* is a ubiquitous ouabain-sensitive enzyme whose mode of action has been discussed in Chapter 1, Section 1.1.1.5. It is localized in the lateral (and/or basolateral) plasma membranes of the corneal endothelium and plays a major role in fluid and Na^+ transport. Ouabain, a specific inhibitor of Na^+,K^+-ATPase in a variety of tissues including the corneal endothelium, causes corneal edema *in vivo* (e.g., when injected into the aqueous humor) and *in vitro* (e.g., after perfusing the isolated cornea from the endothelial surface) (Riley, 1982; Fischbarg and Lim, 1984). Both human and rabbit corneas, when mounted in an appropriate perfusion chamber, swell in a dose-dependent manner in the presence of ouabain at concentrations ranging from 10^{-8} to 10^{-5} M (Geroski *et al.*, 1984). This causes not only swelling of the stroma but also a dose-dependent endothelial edema. However, as shown by electron microscopy, the endothelial apical junctions remain intact. Thus, ouabain-induced swelling of the endothelial cells is a result of pump inhibition and is not related to impairment of the barrier function of the endothelium.

The specific activity of endothelial Na^+,K^+-ATPase of human cornea is 4.0 μmol P_i/mg protein per h (Ruf and Ebel, 1976), and similar values have been reported for rabbit and bovine endothelium. Comparative studies of the saturation kinetics of Na^+,K^+-ATPase in fresh and cultured bovine endothelial cells reveal a substantial drop in activity in second-passage cultures (Whikehart *et al.*, 1987). In the presence of sodium as the activating ion, the calculated V_{max} values for fresh and cultured cells are 5.6 and 2.0 μmol P_i/mg protein per 30 min, respectively.

The number of Na^+,K^+-ATPase sites per cell in two species (cattle and rabbit) has been quantitated by two different methods, and the results are in good agreement. Perfusion of whole mounted rabbit corneas with radioactive ouabain shows that it binds specifically, and with high affinity, to the endothelial pump sites in the cell membranes, mainly the lateral ones (Geroski and Edelhauser, 1984). This method shows 3.0 x 10^6 sites per cell. In another approach, measurements of labeled ouabain and $^{86}Rb^+$ uptake in cultured bovine endothelium show that the number of sites varies with the density of the cells (Savion and Farzame, 1986). Values ranging from 0.8 to 2.2 x 10^6 pump sites per cell have been calculated, with sparse cultures showing a larger number of binding sites than confluent ones.

An *HCO_3^--dependent ATPase* has been detected in rabbit corneal endothelium (Riley, 1977), where its specific activity is six times greater than in the epithelium. These and other findings suggested that this enzyme could play a role in generating the net HCO_3^- flux across the endothelium. Subsequent studies on the localization of HCO_3^--dependent ATPase in subcellular fractions of bovine endothelium have shown that the highest specific activities are in the mitochondrial fractions. Relatively little HCO_3^--dependent ATPase activity is detectable in the plasma membrane fractions, where Na^+,K^+-ATPase was shown to be specifically localized (Riley and Peters, 1981; Whikehart and Soppet, 1981). These investigators reasoned that since HCO_3^--dependent-ATPase is localized in the mitochondria and not on the plasma membrane, this enzyme does not play a direct role in the transport of HCO_3^-. However, the mitochondrial enzyme may function indirectly in ion transport in the endothelium through the production of ATP required by the plasma membrane-localized Na^+,K^+-ATPase pump.

Some investigators have suggested a role for carbonic anhydrase in contolling the net flux of HCO_3^- since inhibitors of this enzyme decrease the net movement of fluid from the stroma to the aqueous. However, the evidence for its involvement in ion transport is indirect and controversial (Riley, 1982). Carbonic anhydrase is not coupled to a source of free energy; therefore, it would not be expected to contribute to energy-requiring pump activity. Nevertheless, as described below, carbonic anhydrase may play an indirect role in HCO_3^- transport.

Newer experimental approaches and methodologies suggest that other models may be used to explain ion transport in general and HCO_3^- transport in particular in the corneal endothelium. The mechanisms proposed are based on the activities of electrogenic Na^+–HCO_3^- coupling (or cotransport) that is indirectly driven by a Na^+/H^+ exchanger (or antiport) (Jentsch *et al.*, 1985; Fischbarg *et al.*, 1985). The Na^+/H^+ antiport is thought to be located in the basolateral (stromal) plasma membrane; as in other cells throughout the body, this antiport is essential for the regulation of intracellular pH. In the corneal endothelium, it uses the inwardly directed Na^+ gradient (maintained by the active Na^+,K^+-ATPase localized in the basolateral membrane) to extrude H^+ from the cell (Jentsch *et al.*, 1985). The extracellular H^+ may react with HCO_3^- to form carbonic acid and CO_2, the latter reentering the cell by simple diffusion, where it is reconverted to HCO_3^- by carbonic anhydrase. This Na^+/H^+ exchanger is inhibitable by high concentrations of amiloride. An electrogenic Na^+–HCO_3^- cotransport is thought to mediate the exit of HCO_3^- at the apical plasma membrane of the endothelium. Cotransport implies that both of these anions move in the same direction, and hence this model is an appealing one that adds considerably to an understanding of the molecular basis of the net flux of both Na^+ and HCO_3^- from the stromal to the aqueous side of corneal endothelium.

3.5.3. Redox Systems

Oxidative stress, as it affects the endothelium, may play an important role in controlling corneal hydration. Both H_2O_2 and free radical intermediates such as superoxide anion are potential sources of oxidative damage in the anterior segment of the eye. Such injury does not normally occur in the endothelium owing to the presence of free radical scavengers (Bhuyan and Bhuyan, 1977, 1978; Redmond *et al.*, 1984; Rao *et al.*, 1985) as well as enzymatic and nonenzymatic pathways for the degradation of H_2O_2 (see Fig. 1.9). An important component of protection from oxidative stress in the endothelium is a redox system involving glutathione (GSH), glutathione peroxidase, and glutathione reductase. This system is maintained by the NADPH generated through the pentose phosphate pathway and has now been established as an essential part of endothelial metabolism (Ng and Riley, 1980).

Glutathione, ascorbate, and H_2O_2 are all known to be present in aqueous humor. The latter is found at concentrations of 20–30 μM, but this value may be higher in some patients undergoing cataract surgery (Spector and Garner, 1981). Perfusion of the endothelial surface of isolated rabbit cornea with 50 μM H_2O_2 in the absence of glucose results in considerable corneal swelling, which can be partly delayed by addition of glucose to the medium (Riley and Giblin, 1982/1983). Glutathione levels in the endothelium drop in the absence of glucose, and of the amount remaining, the proportion in the oxidized state increases. The H_2O_2 is detoxified by reaction with GSH and GSH peroxidase, and the oxidized glutathione (GSSG) that is generated can be recycled to GSH in the presence of NADPH and GSH reductase, as shown in Fig. 1.9. Fluid transport in the cornea is greatly compromised when intracellular levels of GSH are reduced to about one-third of their normal *in vivo* values (Ng and Riley, 1980).

Not only H_2O_2 but also 1,3-bis(2-chloroethyl)-1-nitrosourea (BCNU), a potent inhibitor of GSH reductase, causes pronounced swelling when added to the perfusion fluid bathing the endothelial cells of the isolated cornea (Riley, 1984). This can be prevented by GSH, suggesting that the endothelium requires a constant supply of NADPH to maintain sufficient levels of reduced glutathione to protect the cells from potential damage by elevated levels of H_2O_2. Studies with labeled mannitol and inulin show that the cellular injury caused by H_2O_2 and BCNU affects mainly the permeability of the endothelial cells rather than the transport processes controlled by the Na^+,K^+-ATPase pump (Riley, 1985).

High concentrations of ascorbate are found in the aqueous humor of most species, and the following reaction may explain the origin of aqueous humor H_2O_2 (Giblin *et al.*, 1984):

$$\text{Ascorbic acid} + O_2 \rightarrow \text{Dehydroascorbate} + H_2O_2$$

Therefore, it is not surprising that, similar to H_2O_2, ascorbate in the absence of glucose also causes swelling in the perfused cornea (Riley *et al.*, 1986). However, it is not ascorbate itself; rather, as the reaction shown above suggests, it is the rapid oxidation of ascorbate in the presence of O_2 that generates H_2O_2, a substance that is potentially toxic to endothelial cells if present at high enough concentrations. These three substances (ascorbate, H_2O_2, and O_2) may be in a dynamic equilibrium in the aqueous humor, the relative concentrations of each being modulated by the levels of GSH (Riley *et al.*, 1986).

3.5.4. Collagen Synthesis in Culture

Primary cultures of rabbit endothelial cells synthesize type IV collagen, one of the two major types characterized in Descemet's membrane (Kay *et al.*, 1982). However, in the presence of polymorphonuclear leukocyte (PMN)-conditioned medium, a minor fraction of cultured endothelial cells acquire a fibroblast-like morphology (Kay *et al.*, 1985). After serial passage, the morphologically modified cells become the dominant cell type, and a concomitant change in expression of collagen phenotype occurs. Almost no type IV collagen is produced; instead, the dominant type synthesized is type I, the interstitial collagen characteristic of cultured fibroblasts. Small amounts of type III collagen are also produced as the number of serial passages increases. Further studies of subcultured rabbit corneal endothelium without added PMN-conditioned medium reveal morphological changes, not to fibroblasts but rather to enlarged and flattened cells. These cultures continue to express type IV collagen as the major detectable species, but by the fourth passage, the rate of synthesis is only 22% of that found in primary cultures (Kay *et al.*, 1984). In contrast to type IV collagen produced by cultured rabbit endothelial cells, bovine endothelia synthesize type III as the major collagen (Kay and Oh, 1988). It accounts for about 60% of the total collagen synthesized in primary culture, and its expression is further enhanced in eighth-passage cells.

The investigations by Kay and co-workers cited above show that normal rabbit corneal endothelial cells do not synthesize type I collagen; moreover, there is no detectable production of type IV collagen in PMN-modulated endothelial cells. The phenotypic switch from type IV to type I collagen in PMN-modulated endothelial cells was thought to be an authentic response to factors released by the PMNs. The reasons for this change in expression are not clear, but it is known from studies on other cell types that collagen gene expression is regulated at many levels. Further studies on the molecular regulation of these two collagen types in both normal and PMN-modulated corneal endothelium have utilized cDNA probes to examine changes in the steady-state levels of their mRNAs (Kay, 1989). Somewhat unexpectedly it was found that the amounts of type I and type IV collagen mRNAs were similar, if not identical, in both cell types. Hence, collagen synthesis does not correlate with the steady-state mRNA levels; rather, there appear to be high levels of untranslated messages in both cell types. Another interesting finding was that type IV collagen mRNA is smaller in modulated cells than in normal ones. Thus, collagen gene expression in cultured rabbit endothelial cells may be regulated, at least in part, at the posttranscriptional level.

Type VIII collagen was not detected in the studies reported by Kay and co-workers (1982, 1984, 1985; Kay and Oh, 1988; Kaye, 1989), although a 50-kDa peptide characteristic of type VIII collagen may have been present in the preparations. When precautions are taken to avoid losses resulting from adsorption of type VIII collagen on glassware and other surfaces during isolation, type VIII collagen is readily identified in confluent secondary monolayers of rabbit endothelial cells (Benya and Padilla, 1986). In these studies, positive identification was achieved by two-dimensional mapping of cyanogen bromide (CNBr) peptides, and it was calculated that type VIII collagen represents about 60% of the collagenous proteins synthesized by cultured rabbit endothelial cells. Moreover, type VIII appears to be the major collagen of bovine corneal endothelium. Not only is it the most abundant collagen in native bovine Descemet's membrane (Labermeier and Kenney, 1983), but it is also the major species synthesized by cultured bovine endothelial cells (Kapoor *et al.*, 1986).

3.5.5. Phagocytic Properties and Cell-Surface Glycoproteins

Apical junctional complexes of the corneal endothelium provide a barrier to the passage of a wide variety of substances that would otherwise enter the endothelial cells from the aqueous humor. However, it has been observed that certain macromolecules do nevertheless enter the endothelium under a variety of conditions, presumably by phagocytosis, endocytosis, or both.

Confluent cultures of human corneal endothelial cells can internalize carmine particles and latex spheres (Tripathi and Tripathi, 1982). However, phagocytosis of latex spheres 0.5 μm in diameter results in a generalized disruption of the endothelial membranes and degeneration of the cells with a concomitant release of lysosomal enzymes (Hara *et al.*, 1985).

Other substances such as low-density lipoprotein (LDL) are not internalized by the strongly adherent, flattened endothelial cells characteristic of confluent monolayers of cultured cells (Goldminz *et al.*, 1979). In contrast, LDL is avidly taken up in preparations of sparsely cultured cells that are less strongly apposed and are without a well-developed barrier system. Formation of a barrier to substances such as LDL in confluent cultures is thought to be associated with the production of an extracellular fibronectin meshwork, a characteristic component of extracellular matrix (ECM) of a wide variety of cells and of corneal endothelium as well.

Extracellular matrix has other functions; for example, fibronectin-rich matrix derived from a variety of cell sources promotes cell growth in cultured corneal endothelium and also influences their morphology (Hsieh and Baum, 1985). Fibronectin arranged as arrays of long fibers induces a fibroblast-like appearance of the endothelial cells, whereas discrete clusters of short fibronectin fibers (such as those produced by corneal endothelial cells) promote the formation of typical polygonally shaped corneal endothelia.

Laminin is another high-molecular-weight glycoprotein of the endothelial extracellular matrix. Its secretion by cultured bovine endothelial cells is a function of cell density and cell growth (Gospodarowicz *et al.*, 1981). However, the two glycoproteins (laminin and fibronectin) have different extracellular localizations in cultured endothelial cells. Whereas fibronectin is detectable on both the apical and basal cell surfaces, laminin is associated almost entirely with the extracellular matrix produced beneath the cell monolayer and is in close proximity to type IV collagen. Thus, similar to other cells, the extracellular matrix of the corneal endothelium is characterized by a close structural association of type IV collagen and laminin.

In addition to fibronectin and laminin, other cell surface glycoproteins are also associated with the plasma membranes of cultured rabbit endothelial cells (Panjwani and Baum, 1985). Periodate oxidation of the membranes followed by reduction with NaB^3H_4 produces labeled sialoglycoproteins. After SDS-PAGE electrophoresis, at least eight labeled glycoproteins can be detected, ranging in size from 44 to 220 kDa. Further characterization could clarify their function(s) in endothelial plasma membranes.

3.5.6. Epidermal Growth Factor and Eicosanoids

Epidermal growth factor (EGF), a 6-kDa acidic protein found in mouse salivary gland and other tissues, has been studied extensively. It is mitogenic for a variety of cell types, including bovine (Gospodarowicz *et al.*, 1977), human (Fabricant *et al.*, 1981), and rabbit corneal endothelial cells (Raymond *et al.*, 1986).

The mitotic rate of rabbit endothelial cells in culture is stimulated 70% in the presence of EGF, and there is a striking change in morphology from polygonal to spindle-shaped cells, an appearance similar to that of migrating cells closing a wound (Raymond *et al.*, 1986). The cell elongation of cultured rabbit endothelial cells caused by EGF is greatly enhanced in the presence of indomethacin, but if prostaglandin E_2 (PGE_2) is also present, the cells retain their normal polygonal shape (Neufeld *et al.*, 1986). Corneal endothelium synthesizes PGE_2 from labeled arachidonic acid *in vivo* (Bazan *et al.*, 1985c) and *in vitro*, in incubated and/or cultured cells (Bazan *et al.*, 1985b; Gerritsen *et al.*, 1989). Other eicosanoids are also produced by corneal endothelium, as discussed below. Indomethacin, an inhibitor of the cyclooxygenase pathway of arachidonic acid metabolism, causes the complete inhibition of PGE_2 synthesis in endothelial cell cultures, implying that maintenance of the differentiated polygonal shape of corneal endothelial cells is dependent on the endogenous synthesis of PGE_2 (Raymond *et al.*, 1986; Neufeld *et al.*, 1986).

More recently, Jumblatt and co-workers (1988) have found that eicosanoid synthesis—and regulation of cell shape, mitotic activity, and state of differentiation—is modulated by the two major receptor-stimulated membrane transduction systems described in Chapter 1, Section 1.3. In one of them, EGF may stimulate the breakdown of PIP_2 (see Fig. 1.6), producing inositol 1,4,5-trisphosphate (IP_3) and diacylglycerol (DG), the latter causing the activation of protein kinase C. Alternatively, epidermal growth factor (like phorbol esters) may stimulate protein kinase C directly and thus promote cell elongation and increased mitosis. A second pathway in which PGE_2 may be involved is mediated indirectly through the adenylate cyclase system (see Fig. 1.5). There is compelling evidence from these interesting studies (Jumblatt *et al.*, 1988) that activation of both protein kinases A and C through separate though interacting pathways may play key roles in regulating endothelial cell mitosis and cell shape.

Further investigations are needed to define more precisely the roles of EGF, arachidonic acid metabolism, and prostaglandin E_2 (PGE_2) synthesis in endothelial differentiation. An early study on arachidonic acid metabolism in rabbit cornea showed that PGE_2 is a major product of the cyclooxygenase pathway in the endothelium and possibly also in the stroma; smaller amounts of 6-keto-$PGF_{2\alpha}$, PGD_2, and TXB_2 were also detected in these *in vitro* experiments (Bazan *et al.*, 1985b). As in rabbit, PGE_2 is also the major metabolite of arachidonic acid in bovine corneal endothelium; however, there was no evidence for TXB_2 synthesis in bovine endothelium (Gerritsen *et al.*, 1989). These studies also confirmed the findings of Schwartzman and co-workers (1987b) on the absence of a microsomal cytochrome P450-dependent mixed-function oxidase system for the metabolism of arachidonic acid in corneal endothelium.

3.6. LIPIDS

The major lipids of human (Feldman, 1967) and rabbit cornea (Reddy *et al.*, 1987) are triglycerides, cholesterol esters, and cholesterol; together they account for about three-quarters of the total lipids of the tissue. The remainder consist of sphingomyelin and four other phospholipids (PE, PC, PS, and PI). As shown in Table 3.5, phosphatidylcholine (PC) is the major phospholipid of whole rabbit cornea, and PE is the next most abundant. Although PI comprises about 10% of total lipids, other phosphoinositides are present in only trace amounts. Recent analyses of membrane lipids in cultured bovine endothelial cells give entirely different values for the major phospholipids: sphingomyelin is the most abundant phospholipid, and PC and PE comprise a somewhat smaller proportion of the total (Gerritsen *et al.*, 1989).

The fatty acid composition of membrane phospholipids from the epithelium, stroma, and endothelium of rabbit cornea is given in Table 3.6. Oleic acid (18:1) accounts for about half of the fatty acids esterified to phospholipids in all three cell layers, and palmitic acid (16:0) is the second most abundant. Endothelial phospholipids have the highest content of arachidonic acid (20:4), and epithelium the lowest. Small amounts of other unsaturated fatty acids are also present in corneal phospholipids.

TABLE 3.5. Phospholipid Composition of Rabbit Cornea[a]

Phospholipid	Percentage of total lipid P
PC	38.7
lyso-PC	6.8
PE	13.1
lyso-PE	10.3
PI	8.2
PS	8.3
SPH	9.4

[a]Adapted from Akhtar (1987). The relatively large amounts of lyso-derivatives may be artifacts arising from deacylation during extraction. Abbreviations used are: PC, phosphatidylcholine; PE, phosphatidylethanolamine; PI, phosphatidylinositol; PS, phosphatidylserine; SPH, sphingomyelin.

Table 3.6. Fatty Acid Composition of Corneal Phospholipids[a]

Fatty acid	Epithelium	Stroma	Endothelium
16:0	18.7	16	17
16:1	4.5	2.7	1.0
18:0	4.2	11.2	10.2
18:1	57.4	42	48
18:2	tr	1.1	2.3
20:1	2.6	1.2	3.1
20:4 (n–6)[b]	1.6	8.6	11.0

[a]Adapted from Bazan and Bazan (1984); data are expressed as mol%. Only the major fatty acids are shown, and the values do not include standard errors.
[b]n-6 designates the linoleic acid series.

Studies on the biosynthesis of corneal lipids in the rabbit show active incorporation of labeled acetate (Culp *et al.*, 1970) and arachidonic acid (Bazan and Bazan, 1984) into neutral lipids and phospholipids. Analyses of whole rabbit cornea after incubation with labeled arachidonate indicate that phosphatidylcholine (PC) and triglycerides have the most active rate of turnover of this unsaturated fatty acid. In the rat, labeled arachidonic acid is incorporated into phospholipids in all of the cell layers, with the epithelium showing the highest rate of activation and esterification (Proia *et al.*, 1986; Baratz *et al.*, 1987).

3.7. PROTEINS

The structural proteins of the cornea, consisting mainly of collagen, have been described in Section 3.3. Of the many soluble proteins, glycoproteins, and enzymes known to exist in the cornea, only three have been isolated and characterized (to varying degrees): the 54-kDa antigen, a 30-kDa glycoprotein, and an aminopeptidase that resembles aminopeptidase III of lens.

A 54-kDa antigen was first detected as the major soluble protein in bovine corneal extracts; it constituted 30% of the total soluble protein and was named BCP 54 (bovine corneal protein 54) (Alexander *et al.*, 1981; Silverman *et al.*, 1981). Although it is present throughout the cornea, its highest concentration is in the epithelial layer; it has not been detected in any other ocular tissue except lens epithelium. Proteins immunologically related to BCP 54 are present in several other mammalian corneas but not in nonmammalian species. The same protein has been isolated from serum of a patient with corneal melting disease; it was detected by its cross-reaction with the corneal protein and renamed 54-kDa antigen (Kruit *et al.*, 1986). This antigen is also abundant in rat cornea, and although antibody production is easily triggered, the antibodies do not reach the corneal epithelium (Eype *et al.*, 1987).

A 30-kDa glycoprotein has been isolated from corneas of several animal species as well as from human tissue (Leonardy *et al.*, 1985). This glycoprotein is identifiable by its behavior on Sepharose CL-4B, where it appears as the last of four peaks eluting after gel filtration chromatography of labeled extracts of corneal proteoglycans and glycoproteins. Its synthesis is dependent on epithelial and/or endothelial attachment to the stroma.

Bovine ocular tissues contain a number of aminopeptidases that are thought to play a role in the inactivation of biologically active neuropeptides (Stratford and Lee, 1985). The two tissues having the highest peptidase activity in the rabbit eye are corneal epithelium and iris–ciliary body. Bovine cornea contains an exopeptidase that is concentrated mainly in the endothelial cell layer (Sharma and Ortwerth, 1987). The purified enzyme has a broad substrate specificity pattern as well as many other properties similar to an aminopeptidase III isolated from the lens.

3.8. WOUND HEALING

3.8.1. Epithelial Abrasions and Keratectomy Wounds

The first step in wound closure after partial or total debridement is the migration of the epithelium as a sheet over the exposed stromal surface. The sheet travels along as a stratified, albeit flattened, epithelium. Mitosis ceases at the edges of the wound, and the area is filled by epithelial cells, strongly adherent to one another, sliding over the wounded area. Once the wound is closed, cells reform hemidesmosomal attachments to the basal lamina. Mitosis then resumes and continues until the epithelium reaches its full thickness of five to seven layers, 1–2 weeks after wounding. A variety of experimental approaches, both *in vivo* and *in vitro*, have been used to elucidate the biological and biochemical changes occurring during this unique example of tissue remodeling.

Actin filaments (see Chapter 1, Section 1.1.1.3) in nonmigrating corneal epithelia are present as an apical network in the superficial cell layers, but in cells migrating to cover an epithelial abrasion, they are concentrated in the basal regions of the cell, as networks of filaments in the leading edges (Gipson and Anderson, 1977). Actin may play an important role in epithelial cell migration, since cytochalasins B and D, known antagonists of actin polymerization, inhibit cell movement *in vitro* in the rat (Gipson *et al.*, 1982). Measurements of epithelial cell migration on glass or on collagen substrata confirmed the inhibitory effect of cytochalasin B in rat as well as in rabbit epithelial cells (Soong and Cintron, 1985). Calmodulin, a multifunctional calcium-binding protein, is also known to mediate cell motility in a variety of cells. Calmodulin inhibitors, which cause a redistribution of F-actin fibers, inhibit cell migration in rat epithelium but not in migrating epithelia of rabbit, which are devoid of stress fibers.

Two additional sources of epithelia for resurfacing the exposed stroma after debridement are now known: (1) the narrow zone between corneal and bulbar conjunctival epithelium, the limbal epithelium, and (2) conjunctival epithelia (Kinoshita *et al.*, 1982). These cell types are distinguishable from one another on the basis of glycogen content and protein profiles on SDS-PAGE.

Changes in keratin profiles have been demonstrated in regenerating corneal and conjunctival epithelial cells (Kinoshita *et al.*, 1983). Although in normal cornea, the two cell types have different keratin profiles, during the first 3 months after debridement, when conjunctival epithelia grow into the cornea, the keratin patterns of the conjunctival cells resemble those of corneal epithelia. A prominent 58-kDa band not detectable in normal corneal epithelia is expressed in migrating corneal epithelial cells. Moreover, a 51-

kDa band is also prominent during conjunctival migration. These interesting studies also showed that 3 months after healing, regenerated conjunctival epithelia on the corneal surface have keratin profiles similar, though not identical, to those of epithelial cells. The appearance of a 58-kDa keratin-like protein has also been noted in rabbit corneal epithelium in response to a keratectomy wound (Zieske *et al.*, 1987). Other studies, based on immunofluorescent staining characteristics of monoclonal antibodies to human epithelial keratin, also support the view that new keratins are expressed in regenerating corneal epithelium (Jester *et al.*, 1985). In rabbits receiving either full-thickness wounds or transcorneal freeze injury, the AE1 monoclonal antibody, which is specific for superficial cells of the corneal epithelium, was shown to react with cells at the leading edge of the migrating epithelium. In contrast, undamaged corneal epithelium showed normal fluorescence limited to the superficial cells. The change in keratin expression persisted until wound healing was complete. Two weeks after a freeze injury or one and one-half months after a penetrating wound, all of the epithelial cells expressed a normal AE1 immunofluorescent staining pattern.

Fibronectin is a high-molecular-weight plasma and extracellular matrix (ECM) glycoprotein that functions in cell-to-cell adhesion as well as in promoting the attachment of fibroblasts and other cells to the ECM. Although it is present in the basement membrane of rat and human cornea, most studies except those of Tervo *et al.* (1986) fail to detect fibronectin in normal uninjured adult rabbit epithelial basement membrane. The role of fibronectin–fibrin as a temporary matrix or substratum for the attachment of migrating corneal epithelial cells has remained controversial. In the rabbit, after a superficial epithelial scrape wound, fibronectin becomes deposited as a continuous layer on the denuded corneal basement membrane within a few hours after the injury (Fujikawa *et al.*, 1981). Fibrin is also deposited and, in fact, appears at an even earlier stage than fibronectin. The epithelium begins to migrate over the newly deposited fibronectin–fibrin layer about 22 h after wounding and completely covers the surface by 52 h. Both fibronectin and fibrin maintain peak levels for about 36–72 h; when the abraded surface is completely covered, they disappear to basal (nondetectable) levels.

Similar observations have been reported using *in vivo* as well as organ-culture models (Suda *et al.*, 1981/1982; Nishida *et al.*, 1983a; Tervo *et al.*, 1986, Watanabe *et al.*, 1987). In rabbit organ cultures, where no fibronectin is present initially on the corneal surface, the source of fibronectin in response to a keratectomy wound appears to be the corneal keratocytes (Phan *et al.*, 1986; Zieske *et al.*, 1987). Fibronectin is prominent in the stroma, but it is not detectable under migrating epithelium after superficial scrape wounds *in vitro* except at the leading edges of the wound.

Many studies have shown that the layer of fibronectin (or fibronectin–fibrin) that forms on the wounded surface facilitates and/or accelerates epithelial cell migration. However, few have addressed the question of whether fibronectin–fibrin *per se* is essential for epithelial cell adhesion and migration following wounding. In a direct approach to this question, three different experimental techniques have been developed to examine the influence of fibronectin–fibrin matrix on the healing rates of epithelial scrape wounds in rabbit and guinea pig corneas (Phan *et al.*, 1989a). The results show clearly that neither the inhibition of fibronectin or of fibrin deposition nor the presence of a natural fibronectin–fibrin matrix affects the rate of epithelial wound healing. Laminin is a normal component of the corneal basement membrane and functions as an attachment glycoprotein for epithelial cells to type IV collagen in the basement membrane. Hence, in superficial wounds, where the lamina densa remains intact and laminin is not removed, a provisional scaffolding of fibronectin–fibrin, although present, is not essential for epithelial cell migration.

Even in the case of superficial keratectomies in the rabbit, where the basement membrane and the anterior third of the stroma are surgically removed, addition of exogenous fibronectin does not accelerate epithelial cell migration (Phan *et al.*, 1989b). This is an apparent contradiction to the findings of Nishida and co-workers (1983a) and probably reflects different experimental techniques. Studies on superficial keratectomies by Phan and co-workers (1989b) show convincingly that an exogenously supplied fibronectin scaffolding is not essential for healing of a superficial keratectomy wound. Under these conditions, the substratum, probably fibronectin, is produced endogenously by the biochemically normal and active corneal cells at the wound edges.

The fibronectin–fibrin matrix, though not essential for cell migration in the rabbit, may nevertheless have other important biological effects. Its degradation (e.g., by plasminogen/plasmin following alkali burns) leads to an epithelial defect possibly caused by the generation of biologically active "uncoupled domains" of fibronectin (Berman *et al.*, 1988). Exogenous fibronectin may be beneficial in preventing corneal ulceration following an alkali burn in the cornea, where the fibrin–fibronectin scaffolding has been degraded by plasminogen activator released from injured epithelial cells (see Section 3.8.2).

The role of proteins, especially glycoproteins, in epithelial cell migration and wound healing has been studied in rats using a novel organ-culture technique (Gipson and Kiorpes, 1982). Epithelial abrasions are produced after death, and the corneas are removed and cultured. Controls without abrasions are treated the same way. When incubated in the presence of radiolabeled leucine or glucosamine, there is a dramatic increase in protein and glycoprotein synthesis during epithelial cell migration in the early stages of wound repair. Tunicamycin, a specific inhibitor of the asparagine-linked N-glycosidic bonds in glycoproteins, slows the rate of epithelial wound healing; in its presence, the incorporation of labeled glucosamine into epithelial glycoproteins is only 14% of that in control (unwounded) cultures (Gipson *et al.*, 1984). The incorporation of labeled leucine is essentially normal, suggesting that the synthesis of specific asparagine-linked glycoproteins may be essential for sustained migration of epithelial cells and for complete wound closure.

The glycoproteins that appear to be involved in cell migration are located on the epithelial cell surface, as shown by the greatly increased binding of concanavalin A (Con A) and wheat germ agglutinin (WGA) in cells at the leading edge of the migrating sheet (Gipson *et al.*, 1983; Zieske *et al.*, 1986). Moreover, specifically in the migrating cells and not in others, there is a nearly fourfold increase in 70- and 155-kDa glycoproteins. There are also greatly increased amounts of a 110-kDa protein detectable 16 h after wounding and during cell migration (Zieske and Gipson, 1986). This component has many characteristics of vinculin, a cytoskeletal protein recently identified in focal regions that mediate cell-to-substrate attachment in cultured rat epithelium (Soong, 1987). The 110-kDa protein is absent in control (unwounded) epithelium and disappears 24 h after the closure of debrided epithelium (Zieske and Gipson, 1986).

In summary, of the many factors involved in wound healing after superficial epithelial injuries, two appear to play major roles. The first is a matrix or substratum on the exposed basal lamina whose adhesive components promote cell migration by mediating the attachment of newly formed migrating cells. Fibronectin and/or fibrin can be formed as a response to superficial injuries but need not be added exogenously in order to enhance cell migration or wound healing. Other receptor complexes, not yet completely identified, can also provide the scaffolding required for attachment of migrating cells to the basal membrane. A second important factor is an increased synthesis of glycoproteins on the surface of migrating cells.

3.8.2. Alkali Burns

The devastating effects of burns (alkali, thermal, or other) of the cornea have been recognized for decades, both clinically and experimentally. The most extensively investigated have been alkali burns, which are produced experimentally by brief application of 2–4 N NaOH to the corneal surface. This results in the immediate death of nearly all corneal cells, and although the epithelium ultimately regenerates, the cornea develops a sterile ulcer, thought to be caused by a persistent epithelial defect. Many of the steps in the cascade of events following alkali burn have been elucidated and are now known to involve internal (biochemical and enzymatic) changes in the epithelium and stroma as well as extraocular events, most notably the infiltration of PMNs (polymorphonuclear leukocytes). Experimental findings reported below support the view that epithelial defect and stromal ulcerations result from the "trapping" of normal healing in a stage of proteolysis (Berman *et al.*, 1988).

The production of collagenase by cellular components of the cornea and/or by infiltrating PMNs, and with it the breakdown of stromal collagen, was proposed many years ago as the principal cause of corneal ulceration following alkali wounds (Brown *et al.*, 1970). More recently it has been shown that breakdown products generated by alkaline treatment of either Sigma type I collagen or bovine corneal collagen may act as chemotactic agents for stimulating the entry of PMNs, now considered to be the prime effectors of the inflammatory response (Pfister *et al.*, 1987). Other studies attempting to identify leukocyte chemoattractants in the alkali-burned cornea corroborate these observations and point to high-molecular-weight substance(s), possibly protein(s) or other breakdown products of collagen, as the principal chemotactic factors (Elgebaly *et al.*, 1987). Well-known low-molecular-weight chemotactic factors such as C5a, interleukin-1, and leukotriene B_4 do not appear to be involved.

Despite many investigations over the years, the precise source of collagenase in ulcerating alkali-burned corneas—or even its existence in the form of an inactive or precursor "procollagenase" in the normal uninjured cornea—has been difficult to establish (Berman *et al.*, 1977, 1979). This question has been partially resolved by the use of immunocytochemical techniques employing a specific antiserum to human skin collagenase (Gordon *et al.*, 1980). Collagenase was clearly detectable in the stroma of human ulcerating corneas but was not present in either the epithelium or stroma of nonulcerating corneas, even pathological ones (e.g., keratoconus or Fuchs' endothelial dystrophy), or in healed ulcers.

It is now recognized that collagenases isolated from various tissues are distinct gene products; the various types present in ulcerating corneas, originating from stromal cells or from infiltrating PMNs and often referred to as latent collagenase, have been difficult to identify. However, monoclonal antibodies have recently been raised against collagenase by immunizing mice with an enzyme purified from involuting rabbit uterus (Kao *et al.*, 1986). This makes it possible to distinguish between collagenases produced by PMNs after alkali burn and those produced by stromal fibroblasts. Immunofluorescent staining of ulcerating corneas localized these antibodies specifically to the PMNs on the superficial anterior surface of the tissue. There was no reaction either with PMNs within the stroma or with stromal cells (keratocytes). These findings suggest that most of the collagenase accumulating after an alkali burn is the result of discharge of the enzyme on the corneal surface by infiltrating PMNs (Kao *et al.*, 1986). Nevertheless, there is compelling evidence that corneal type I collagenase originating from stromal fibroblasts also plays an important role in stromal ulceration after an alkali wound (Berman *et al.*, 1988).

Thus, the PMNs observed in the extracellular matrix of the stroma as ulceration develops may (or may not) be secreting collagenase; however, the release of other degradative enzymes from PMN phagolysosomes cannot be ruled out. Whatever the role of PMNs may be in causing stromal destruction, if they are prevented from entering the cornea after alkali burn by applying glued-on methyl methacrylate lenses, stromal ulceration is greatly diminished (Kenyon *et al.*, 1979).

Normal corneal epithelium in culture is capable of stimulating corneal fibroblasts to produce and/or secrete latent type I collagenase (Johnson-Muller and Gross, 1978). Possibly present as a 40-kDa zymogen, latent collagenase from ulcerating corneas, presumably originating from stromal fibroblasts, can be converted to an active 23-kDa form by trypsin (Berman *et al.*, 1977) as well as by plasmin, as discussed below (Berman *et al.*, 1980).

A major factor elucidated in recent years on the pathogenesis of ulceration after alkali burns is plasminogen activator (PA), one of a group of serine protease enzymes found in a variety of tissues whose function is to modulate fibrinolysis. The two main types are urokinase-like PA (u-PA), first found in urine but also known to be present in diverse cell types, and tissue PA (t-PA), also present in many tissues, including cornea. First detected in the subepithelial membrane, PA activity is now known to be localized in all layers of the cornea: epithelium, stroma, and endothelium (Geanon *et al.*, 1987). In ulcerating corneas, PA appears to be released by epithelial cells and keratocytes as well as by infiltrating PMNs (Berman *et al.*, 1980). On the basis of its molecular size and immunoreactivity, corneal PA appears to be mainly u-PA-like.

Urokinase-like PA of rabbit cornea is a 46-kDa protein, although in many other tissues it appears to be of larger size (approximately 55 kDa). It acts (by limited proteolysis) on plasminogen, cleaving an arginine–valine bond to convert plasminogen into two chains, heavy and light, held together by two disulfide bonds (Pandolfi and Lantz, 1979). Plasminogen, a 90-kDa protein, is one of the globulins present in circulating blood. It enters tissues, including the cornea, in this (latent) form, and once inside the tissue, it can be acted on by PA to produce plasmin, a substance involved in a variety of inflammatory processes. Some of the known biological actions of plasmin include (1) lysis of fibrin clots, (2) cleavage of a specific component of complement C3 which generates a chemotactic factor for PMNs, (3) generation of vasoactive kinins, and (4) activation of latent collagenase in ulcerating cornea.

A PA–plasmin system has been found in ulcerating corneas; the plasmin released by PA activates latent collagenase by proteolytic cleavage of the 40-kDa latent form to an active 23-kDa species (Berman *et al.*, 1980). Addition of plasminogen to cultures of ulcerating corneas increases the levels of PA, collagenase, and collagen degradation products. It has been suggested that added plasminogen is activated by the endogenous PA and that the plasmin produced causes corneal cells to secrete additional PA as well as

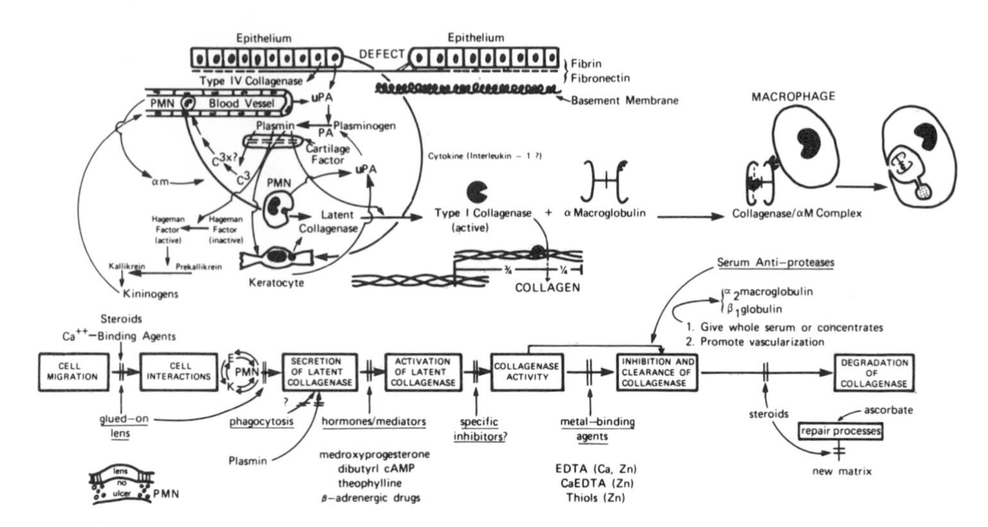

Figure 3.6. Tentative scheme showing the cascade of biochemical events following alkali burns in the rabbit cornea. (From Berman *et al.*, 1988.)

collagenase. Thus, there appears to be a cascading chain of events in ulcerating corneas; the plasmin generated causes the secretion of additional activator, which in turn results in further collagen degradation and destruction of the stromal matrix. A tentative scheme showing the development of stromal ulcerations following alkali burns is given in Fig. 3.6.

In addition to activating latent collagenase, the PA–plasmin system also causes the degradation of the newly formed fibrin–fibronectin layer deposited after an alkali burn (Berman *et al.*, 1983). As discussed above in studies on regeneration after epithelial abrasion, a fibrin–fibronectin layer forms on the anterior stromal surface shortly after simple epithelial debridement, remains until the wound closes, and then disappears (Fujikawa *et al.*, 1981). However, in the case of chemical trauma (alkali), in contrast to mechanical trauma (abrasion), the newly formed epithelium chronically secretes a protease, specifically u-PA. By remaining on the leading epithelial edge, the u-PA continues to produce plasmin, which in turn hastens the resorption of fibrin and fibronectin from the stromal surface (Berman *et al.*, 1983). Whereas in mechanical debridement, fibrin and fibronectin appear shortly after injury and are ultimately resorbed several days later in a regulated manner, this is not so in alkali burns, where these substances are resorbed quickly, in a matter of hours.

Fibronectin mediates the attachment of many cell types, including corneal epithelial cells, to the substratum. One peptide region in particular, localized in the cell-binding domain of fibronectin, is thought to be required for attachment to, and spreading of, epithelial cells. It is a pentapeptide of glycine-arginine-glycine-aspartic acid-serine (GRGDS). Several studies have shown that the free pentapeptide, by competing with intact fibronectin, inhibits the normal epithelial cell interactions between fibronectin and the extracellular matrix (ECM) (Nishida *et al.*, 1986, 1988; Berman *et al.*, 1988). Studies using synthetic GRGDS show that this pentapeptide inhibits the attachment of corneal epithelial cells to fibronectin-coated culture plates in a dose-dependent manner. As discussed above and shown in Fig. 3.6, persistent epithelial defect after alkali burns of rabbit corneas is associated with increased levels of active u-PA and accelerated breakdown of the subepithelial fibronectin–fibrin layer. Although a direct association has not been established between the release of a putative inhibitory pentapeptide following alkali burns and an uncoupling of cell and ECM binding domains, it is tempting to speculate that such events could take place (Berman *et al.*, 1988).

Still another factor in the proteolytic chain of events leading to stromal ulceration is the presence of both latent and active PA in alkali-burned corneas (Wang *et al.*, 1985). In cultured epithelial cells and in normal cornea, PA is present mainly in latent forms consisting of 46-kDa u-PA together with some 72-kDa u-PA. The latent PA in normal cornea can be activated by either trypsin or plasmin. By contrast, in ulcerating corneas, only active PA is present in early cultures, whereas in older ones, the latent form predominates. Cultures of ulcerating corneas contain an additional 35-kDa PA, which is thought to be derived from 46-kDa u-PA species by proteolytic degradation. The extent of ulceration is correlated with the conversion of latent to active PA, which in turn causes further degradation of the corneal structure.

3.8.3. Vitamin A and Retinoic Acid

Whereas an epithelial abrasion results in essentially complete wound closure in animals on normal diets, vitamin A-deprived animals develop keratomalacia-like stromal ulcerations after simple debridement (Sendele *et al.*, 1982). Moreover, there is a greater inflammatory response, characterized by infiltration of PMNs and release of collagenase, after mild thermal burns in vitamin A-deficient animals than in pair-fed controls (Seng *et al.*, 1982). The necrotic lesions in ulcerating xerophthalmic corneas in experimental animals show morphological evidence of proteolytic degradation, most of the damage being associated with the extracellular matrix (Twining *et al.*, 1985). Cathepsin B- and D-like activities, although present in normal corneas, are found at greatly elevated levels in ulcerating xerophthalmic corneas. Cathepsin D is an aspartate protease with optimal activity at pH 3.3, and cathepsin B is an endopeptidase active at pH 6. Together they are able to cleave a wide variety of peptide linkages in proteins, proteoglycans, and fragments produced by the action of collagenase. Additionally, cathepsin B may activate many types of zymogens, thus increasing the spectrum of potential degradative activities. The major sources of these enzymes in ulcerating xerophthalmic corneas are probably PMNs and macrophages.

Other studies have shown that the responses of ocular epithelia to corneal and conjunctival wounding in vitamin A-deficient rabbits are significantly different from those of pair-fed controls (El-Ghorab *et al.*, 1988). Vitamin A deficiency appears to result in delayed healing of conjunctival wounds in spite of increased mitotic rates. Epithelial wound healing is also delayed compared to pair-fed controls. Curiously, goblet cells are present in normal numbers in vitamin A-deficient rabbits even 5 to 8 weeks after liver stores had been depleted. It is also known that xerophthalmic rabbits are more susceptible to corneal infection than pair-fed control animals, and together these findings suggest that the integrity of the epithelial barrier is severely compromised in vitamin A-deficient rabbits.

Although a precise role for vitamin A in corneal wound healing is still unclear, it has long been recognized that healing of superficial or chemically induced wounds is grossly impaired in vitamin A-deficient animals. Its action may be at the level of synthesis of membrane glycoconjugates. The incorporation of radiolabeled sugars into epithelial glycoproteins is greatly diminished in tissues of vitamin A-deficient animals (De Luca, 1977). Administration of retinoic acid to vitamin A-deficient rats increases the incorporation of sugars into high-M_r epithelial glycoconjugates (Hassell *et al.*, 1980a). The tissue stores and secretion of retinol-binding protein in the liver are regulated by nutritional status (see Chapter 1, Section 1.7), and binding of labeled retinol to cellular retinol-binding protein of the cornea is impaired in vitamin A-deficient rabbits (Wiggert *et al.*, 1982). Which of these are primary and which are secondary effects of vitamin A action in the cornea are questions to be resolved in the future.

Alkali burns of rabbit cornea result in persistent (or recurring) epithelial defect and stromal ulceration, possibly due to resorption of the fibronectin–fibrin layer on the corneal surface by the plasminogen activator (PA)–plasmin system. Retinoids are thought to modulate PA synthesis in certain nonocular tissues, and this may also be true in the rat cornea. A recent study on reepithelialization after scalpel debridement in vitamin A-deficient animals shows that although PA is present in the epithelium, it is not detectable in the scrape-debrided wound area (Hayashi,*et al.*, 1988). A pseudomembrane composed of PMNs, cell debris, and fibrous exudate forms in the wounded area of vitamin A-deficient rat corneas but not in pair-fed controls. The migrating edge of regenerating epithelium overlays this membrane and is therefore not in contact with the stromal surface. The formation of this membrane, which appears to delay reepithelialization, was thought to have resulted from the absence of PA in the defect region. Thus, both overactivity (from an alkali burn) and underactivity (from vitamin A deficiency) are thought to perturb normal repair processes after corneal injury.

Other recent studies have shed further light on the possible role of fibronectin in the impaired epithelial wound healing in vitamin A deficiency. Reepithelialization after a mild debridement is delayed in severely vitamin A-deficient rats, and, unlike pair-fed controls, no fibronectin appears on the denuded corneal surface (Frangieh *et al.*, 1989). Delayed epithelial migration is associated with an inflammatory cell layer and occurs in the absence of fibronectin. These observations suggest possible disturbances in deposition and/or enhanced breakdown of fibronectin in response to scrape wounds in vitamin A-deficient rat corneas.

Transforming growth factor β (TGF-β), a 25-kDa polypeptide with a wide range of biological activities, is a potent modulator of growth and differentiation in many cell types (Sporn *et al.*, 1986, 1987). In general, TGF-β stimulates the proliferation of cells of mesenchymal origin, inhibits cells of epithelial origin, and significantly accelerates the wound-healing response. It has been suggested that retinoic acid and TGF-β share the role of modulating agent(s) for cell growth and transformation (van Zoelen *et al.*, 1986); moreover, retinoic acid selectively modulates the mitogenic effect of TGF-β in fibroblasts (Roberts *et al.*, 1985). Recent studies on the expression of TGF-β in rat corneas after mild epithelial abrasion show an acute inflammatory reaction and strong immunohistochemical staining for TGF-β in the stroma of vitamin A-deficient animals but not in pair-fed controls (Hayashi *et al.*, 1989b). As pointed out above, extensive inflammatory response to tissue injury is characteristic of vitamin A-deficient corneas (Seng *et al.*, 1982; Sendele *et al.*, 1982; Twining *et al.*, 1985). Shortly after injury, TGF-β is identifiable in inflammatory cells, suggesting that it plays a role in the acute inflammatory response and may mediate the influx of fibroblasts and macrophages into the inflammatory focus. An *in vitro* study on *myc*-transfected fibroblasts has shown that the absence of retinoic acid enhances the modulating effects of TGF-β (Roberts *et al.*, 1985); if this is also true *in vivo*, then the overall expression of TGF-β in wound healing in vitamin A-deficient corneas may be more prominent than in normal corneas.

3.8.4. Therapeutic Possibilities

The experimental findings described above provide some theoretical basis for treating alkali burns, trauma, and certain clinical conditions. In theory, collagenase inhibitors should be effective in preventing, or possibly minimizing, the ulceration resulting from alkali burns. Collagenase requires Ca^{2+} and Zn^{2+} for enzymatic activity, and the use of metal-chelating agents such as cysteine or Ca^{2+}-EDTA has been shown to prevent ulceration in alkali-burned rabbit corneas. Nevertheless, their usefulness in humans after an alkali burn remains to be established (Berman, 1978). Inhibition of collagenase (and other proteolytic enzymes) by serum antiproteases such as α_2-macroglobulin has been successfully demonstrated using type I collagenase from organ cultures of ulcerating corneas. However, when administered *in vivo* either by topical administration or by intrastromal injection, α_2-macroglobulin did not prevent ulceration.

Ascorbic acid, which functions as a cofactor in the hydroxylation of proline, is thought to promote the synthesis and secretion of collagen of high tensile strength during wound healing in skin and other tissues. Its usefulness in corneal burns is, however, questionable since parenterally administered ascorbate had no effect on the breaking strength of wounds caused by severe alkali burns in normal rabbits (Pfister *et al.*, 1981). It is only in animals with low levels of aqueous humor ascorbate that subcutaneously injected ascorbate appears to increase the breaking strength of the wounds.

Applying a glued-on methyl methacrylate contact lens to alkali-burned rabbit cornea shortly after inducing an injury prevents reepithelialization as well as PMN infiltration and ulceration (Kenyon *et al.*, 1979). The glued-on lens also arrests further ulceration of injured corneas that had not responded to other treatment. Although use of the lens itself has fallen into disfavor clinically, topically applied glue alone is commonly used to prevent or arrest stromal ulceration. A simpler procedure for reducing PMN accumulation in alkali-burned rabbit corneas consists of topical application of 10% citrate immediately after the injury (Haddox *et al.*, 1989). Continued use for 35 days appears to block both the early (0–7 days) and late (6–35 days) PMN infiltration and greatly reduces the incidence and severity of corneal ulceration.

Since corneal epithelial cells are known to have β-adrenergic receptors, it seems possible that *in vivo* "first messengers" such as β-adrenergic agonists could elevate the intracellular levels of cAMP and influence wound-healing processes. In fact, under certain experimental conditions, cAMP appears, paradoxically, to enhance the rate of epithelial wound healing (Jumblatt and Neufeld, 1981). A novel assay developed to measure epithelial cell migration *in vitro* uses confluent cultures of cryo-wounded rabbit epithelium (Jumblatt and Neufeld, 1986). β-Adrenergic agonists that increase intracellular levels of cAMP inhibit wound closure but at the same time promote cell–substrate adhesion. Other substances such as laminin, epidermal growth factor, and fibronectin had no effect on wound healing in this system.

Fibronectin has been purified from patients' autologous plasma (Nishida *et al.*, 1982) and used in the form of eye drops to treat corneal trophic ulcer and persistent epithelial defect (Nishida *et al.*, 1983b). There were no undesirable side effects, and some clinical improvement was noted. Fibronectin may also be effective in healing corneal epithelial defects in cases of herpetic keratitis that had not responded to conventional therapy (Nishida *et al.*, 1985). Six additional patients with refractory corneal epithelial defects who had not improved with standard therapy have been treated with topical fibronectin purified from autologous plasma (Phan *et al.*, 1987). The results of this phase I pilot trial were encouraging, with five out of six patients showing complete reepithelialization; in the 3- to 18-month follow-up period, there were two recurrences. Taken together, these observations point to a need for randomized large-scale clinical trials to evaluate the feasibility and usefulness of fibronectin therapy in specific disorders of the corneal epithelium.

Topically applied solutions of hyaluronate have a beneficial effect on epithelial resurfacing after alkali burns in the rabbit cornea (Chung *et al.*, 1989). Hyaluronate is extensively used in surgery of the anterior segment, and its unique viscoelastic properties suggest many as yet unknown applications to disorders of the cornea. Its mode of action appears to be nonspecific, possibly providing mechanical protection for the resurfacing epithelium.

The use of retinoic acid in experimentally induced endothelial injury has been described. Morphological studies on regeneration of the endothelium after gentle scraping of a small area (without damage to Descemet's membrane) demonstrate that after 24–48 h, the wounded area is covered completely by large irregularly shaped cells. As cells migrate to close the wound they elongate, but when mitosis ceases, the cells assume their normal hexagonal shape (Matsuda *et al.*, 1985). The healing rate of endothelial injuries is significantly hastened by topical application of 0.1% retinoic acid in petrolatum ointment (Matsuda *et al.*, 1986).

One of the principal known biological effects of epidermal growth factor (EGF) is enhancement of endothelial mitosis, and although its mode of action in epithelial or stromal wound healing is not understood, the availability of genetically engineered human EGF should stimulate further study of its potential usefulness in corneal wound healing. Stromal as well as epithelial wounds in *Macaca* monkeys have been effectively treated by topical application of biosynthetic human epidermal growth factor (h-EGF) in combination with a synthetic steroid (Brightwell *et al.*, 1985). In the rabbit, EGF alone enhances epithelial wound healing in alkali-burned corneas but does not prevent recurrent ulcerations and secondary breakdown in the corneal epithelial surface (Singh and Foster, 1987). A combination of therapies such as EGF, fibronectin, and corticosteroids may be more effective than either agent alone in the management of alkali burns.

3.9. KERATOCONUS

Keratoconus is a slowly progressive dystrophy characterized by the gradual appearance of bilateral cone-shaped protrusions of the central cornea caused by thinning and weakening of the stromal layer. This disorder is more heterogeneous than previously thought (Klintworth, 1982a; Yue *et al.*, 1984), and some cases may be familial (see review by Krachmer *et al.*, 1984).

That the stromal weakening is caused by abnormalities in the synthesis, turnover, or composition of the stromal collagen has long been suspected. However, the precise nature of these abnormalities remains obscure. Seven collagen types have been characterized to date in the cornea (Table 3.1). Types I, III, V, and VI are stromal components; Descemet's membrane contains types IV and VIII, and types IV and VII have been identified in the basal lamina.

Studies using radiolabeled proline or type-specific antibodies have revealed no gross abnormalities in either the composition or the distribution of collagen types I, III, IV, and V in keratoconus corneas (Newsome *et al.*, 1981; Nakayasu *et al.*, 1986). Moreover, cultured stromal keratocytes from keratoconus patients and from normal controls produce essentially identical amounts of types I, III, and V collagens, and the ratio of $\alpha1/\alpha2$ polypeptide chains in the type I collagens synthesized in keratoconus and control corneas is similar (Ihme *et al.*, 1983). A subtle difference between normal and keratoconus corneas may be an increase in type V collagen produced by cultured keratocytes from keratoconus corneas (Yue *et al.*, 1983). This, however, could not be confirmed in a later comparative study of collagen types I, III, IV, V, VI, and VII in keratoconus and normal corneas (Zimmermann *et al.*, 1988). Amino acid analyses and SDS electrophoresis of pepsin-solubilized collagen, together with specific antibodies to identify collagens by immunofluorescent techniques in tissue sections, revealed no differences in either composition or distribution between keratoconus and normal corneas. Moreover, no differences in cross-linking patterns have been detected in keratoconus corneas compared to normal controls (Oxlund and Simonsen, 1985; Critchfield *et al.*, 1988).

Investigations using purified type-specific antibodies and immunofluorescent techniques show rather extensive degenerative changes in the basal lamina of keratoconus corneas (Newsome *et al.*, 1981). Destruction, lysis, and disruption of the type IV collagen of the basement membrane has been confirmed by other investigators (Nakayasu *et al.*, 1986). In addition, there is a decrease in immunofluorescent staining for fibrin–fibrinogen in the basement membrane zone of keratoconus corneas compared to controls, although other components appear to be normal (Millin *et al.*, 1986).

In spite of normal collagen patterns, it has been found that organ cultures of keratoconus corneas produce substantially more collagenase and gelatinase than normal corneas (Kao *et al.*, 1982). Studies along similar lines have shown increases in collagenolytic activity toward type I and type IV collagen in primary cultures of keratoconus corneas (Ihalainen *et al.*, 1986). There appears to be not only increased breakdown but also enhanced synthesis of collagen in keratoconus corneas, as shown by increased incorporation of radioactive proline (Rehany and Shoshan, 1984) and elevated prolyl-4-hydroxylase activity (Ihalainen *et al.*, 1986). An accelerated breakdown of collagen, compensated by enhanced rates of synthesis, in keratoconus corneas implies faulty regulation of collagen turnover, an interesting speculation that requires further investigation.

Considerable attention has been focused on the chemistry and metabolism of corneal collagen in keratoconus, but a clear picture of the major biochemical defect has not yet emerged, possibly owing to the heterogeneity of the disorder (Yue *et al.*, 1984). New analytical data based on histochemical (Yue *et al.*, 1988) and biochemical (Critchfield *et al.*, 1988) analyses point to significant accumulations of glycoconjugates and decreases in total protein in keratoconus corneas compared to normal controls. The glycoconjugates have not been identified, but they are polyanionic in character; 4 M guanidine extracts of keratoconus corneas contain 25 times more neutral sugar and uronic acid than normal cornea. Whether abnormal accumulations of glycoconjugates can indirectly influence collagen metabolism and turnover remains to be established, but the recent elucidation of specific proteoglycan–collagen interactions in corneal stroma (see Section 3.3.3) opens other possibilities for studying the pathogenesis of keratoconus.

3.10. CORNEAL DYSTROPHIES

3.10.1. Fuchs' Endothelial Dystrophy

Fuchs' endothelial dystrophy is an autosomal dominant disorder with more severe penetrance in women than in men. It is characterized clinically by stromal and epithelial edema; the most prominent ultrastructural changes are abnormal corneal endothelial cells that undergo fibroblast-like transformation and produce a collagen-like layer posterior to the normal Descemet's membrane (Rodrigues and Waring, 1982). The stromal and epithelial edema are the consequence of either a breakdown in the endothelial barrier or a malfunctioning of the energy-dependent endothelial pump. In the end stages of the disease, both the barrier and the pump functions may be affected.

Histochemical assessment of cytochrome oxidase in surgically removed corneas from patients with Fuchs' endothelial dystrophy shows decreased activity in the central region of the dystrophic endothelium compared to normally functioning endothelium (Tuberville *et al.*, 1986). Other regional differences in energy metabolism were noted that could be related to decreased numbers of mitochondria and/or synthesis of abnormal Descemet's membrane components. Another abnormality in this disorder is a drastic reduction in Na^+,K^+-ATPase pump sites on the lateral membranes of the central region of the corneal endothelium compared to non-dystrophic controls (McCartney *et al.*, 1987a). This was demonstrated by the use of wheat germ agglutinin, a lectin that binds specifically to sugar residues of the ATPase subunit. A quantitative reduction in pump site density in Fuchs' endothelial dystrophy has also been shown using labeled ouabain as a specific probe for Na^+,K^+-ATPase (McCartney *et al.*, 1987b). These findings have been corroborated by immunohistochemical studies using a rabbit Na^+,K^+-ATPase antibody (McCartney *et al.*, 1987c). However, the reduction in pump sites is not specific for Fuchs' endothelial dystrophy; it has also been observed in other endothelial dystophies, thus confirming the essential role of the endothelial pump in maintaining normal hydration of the tissue.

3.10.2. Macular Corneal Dystrophy

This autosomal recessive disorder is characterized by deposits of Alcian blue-staining substances in the keratocytes and endothelial cells; extracellular deposits of abnormal substances are also observed in the stroma, frequently in the subepithelial regions (Klintworth, 1982b). There is a concomitant accumulation of single-membrane-limited cytoplasmic vacuoles containing fibrillar storage bodies resembling those found in the systemic mucopolysaccharidoses. Macular corneal dystrophy (MCD) was initially thought to be a localized storage disease, possibly involving keratan sulfate (KS) (Klintworth and Smith, 1980; Klintworth, 1982c). However, recent observations (discussed below) suggest a more complex pathogenesis, implicating a systemic involvement of skeletal tissues (Klintworth *et al.*, 1986) and also stromal accumulation of abnormal intermediate filaments associated with, but different from, vimentin (SundarRaj *et al.*, 1987).

The Alcian blue-staining material that accumulates in MCD has many properties resembling keratan sulfate (KS), especially its behavior toward specific GAG-degrading enzymes such as keratanase and chondroitin ABC lyase (Klintworth, 1982b). Nevertheless, organ cultures of MCD corneas show little or no synthesis of KS from radiolabeled precursors (sulfate and glucosamine) when compared to normal controls (Klintworth and Smith, 1980). Instead, an abnormal glycoprotein of lower molecular weight than KS proteoglycan accumulates in organ cultures (Klintworth and Smith, 1983) and may be related to an "immature KS proteoglycan" detected in other studies (Hassell *et al.*, 1980b). This glycoprotein, not present in normal cornea, was thought to represent an unprocessed precursor of mature KS proteoglycan.

Use of an antibody to monkey KS proteoglycan confirmed, as described in the foregoing discussion, that MCD corneas do not synthesize a normal KS proteoglycan (Nakazawa *et al.*, 1984). Instead, they produce an immunoreactive glycoprotein of lower sulfate content than KS proteoglycan from normal cornea; moreover, even though the oligosaccharides in this glycoconjugate are larger than normal, their carbohydrate composition is similar to that of KS proteoglycan present in normal human stroma (Nakazawa *et al.*, 1984). The MCD corneas do not in fact synthesize KS proteoglycan; rather, they produce an abnormal carbohydrate-rich molecule or glycoconjugate with a protein core that reacts positively to KS proteoglycan antibodies but is resistant to degradation by keratanase.

With the development of a sensitive enzyme-linked immunosorbent assay (ELISA) using a monoclonal antibody directed against sulfated epitopes in the carbohydrate moieties of both corneal and cartilage KS, very small amounts of KS can now be measured in blood and tissues of both normal individuals and MCD patients (Klintworth *et al.*, 1986; Thonar *et al.*, 1986). Using this assay, it was found that whereas detectable amounts of KS proteoglycan are present in extracts of normal cornea, virtually none is detectable in corneas of MCD patients. Of particular importance is the use of this method for measuring serum levels, which revealed average concentrations of 251 and <2 ng/ml in healthy individuals and in MCD patients, respectively. The average level of KS in serum of patients with corneal diseases other than MCD is 273 ng/ml.

Most of the serum KS proteoglycan is thought to represent degradation products of cartilage proteoglycans, a contention based on the finding that MCD patients had only about 1% of the normal serum level of KS proteoglycan despite the fact that more than half of them had undergone keratoplasty. This observation suggests that degradation products of corneal KS proteoglycan do not contribute substantially to the pool of circulating KS proteoglycan. These findings on antigenic similarities between corneal and cartilage KS proteoglycan corroborate the work described in Section 3.3.2.1 on the widespread occurrence in bovine and embryonic chick tissues of proteoglycans with antigenic and physical properties similar to corneal KS proteoglycan (Funderburgh *et al.*, 1987). It has been suggested that keratan sulfates in many noncorneal tissues form proteoglycan families containing both protein and KS antigenic sites related to corneal KS proteoglycan.

Although the chemical nature of the abnormal deposits in corneas of MCD patients has not been established, there is little doubt that the major product is an undersulfated or "immature" KS proteoglycan having many characteristics of a glycoprotein. There may, however, be other abnormal substances accumulating in MCD corneas, as shown in a study using monoclonal antibodies developed against antigens of corneal keratocytes (SundarRaj *et al.*, 1987). Of the large number that were screened for reactivity toward the abnormal deposits in MCD corneas, one of them (8F1-3) showed a particularly strong reaction, and it could be shown that in normal cornea, this monoclonal antibody recognizes an antigen associated with the intermediate-type filament, vimentin (see Chapter 1, Section 1.1.1.3). This antigen accumulates intracellularly in MCD cornea and is also prominent in subepithelial regions. It has been suggested that the antigen recognized by the 8F1-3 monoclonal antibody is a new intermediate filament-associated antigen and not a breakdown product of vimentin. The relationship of this antigen to the abnormal KS proteoglycan synthesized by MCD corneas requires further clarification.

3.11. MUCOPOLYSACCHARIDOSES AND MUCOLIPIDOSES

Corneal clouding is an important feature of most, though not all, of the inherited disorders of glycosaminoglycan (GAG) catabolism, the mucopolysaccharidoses. These are storage diseases caused by deficiencies of specific lysosomal enzymes, either exoglycosidases or sulfatases, required for the degradation of dermatan sulfate (DS), heparan sulfate (HS), and keratan sulfate (KS). All of the malfunctioning lysosomal enzymes are acid hydrolases with one exception: in the Sanfilippo C syndrome, the mutant enzyme is an acyltransferase. Incompletely degraded GAGs accumulate in tissues throughout the body, including the cornea. Details of the clinical signs, ultrastructural pathologies, and enzymatic defects have appeared in two extensive reviews (Klintworth, 1982b; McKusick and Neufeld, 1983).

All of the mucopolysaccharidoses are transmitted as autosomal recessive traits with the exception of the Hunter syndrome, which is X-linked. A summary of the specific enzyme deficiencies, as well as the presence or absence of corneal clouding, in this group of disorders is given in Table 3.7.

In those mucopolysaccharidoses associated with corneal clouding (MPS IH, MPS IS, MPS IV, and MPS VI), undegraded GAGs accumulate in the stroma. Such deposits are usually not present in the mucopolysaccharidoses having no corneal clouding (MPS II and MPS III). The highly specific nature of the enzymatic defects in this group of disorders, each affecting the cleavage of a particular glycosidic or sulfated bond in the GAG side chains, may determine the extent to which undegraded GAGs accumulate in the stroma (Klintworth, 1982b). Further characterization of the stromal proteoglycans, and especially the specific linkages in the complex mixture of GAGs present in this tissue, is required before a sound biochemical basis for corneal clouding in these disorders can be formulated.

In contrast to the mucopolysaccharidoses, the mucolipidoses comprise a highly heterogeneous group of inherited disorders, initially observed in patients having a broad spectrum of clinical signs resembling both lipid and mucopolysaccharide storage diseases. Varying degrees of corneal clouding are present in the majority of these patients. With the elucidation of the enzymatic defect in many of the mucolipidoses, these disorders have now been reclassified. Two reviews are available on the clinical and biochemical findings in some of the mucolipidoses (Berman, 1982; Neufeld and McKusick, 1983).

Table 3.7. Mucopolysaccharide Storage Diseases

Name	Eponym	Corneal clouding	Enzymatic defect
MPS I H	Hurler	Present	α-L-Iduronidase
MPS I S	Scheie	Present	α-L-Iduronidase
MPS I H/S	Hurler–Scheie	Present	α-L-Iduronidase
MPS II			
Severe	Hunter	Absent	Iduronate sulfatase
Mild	Hunter	Late in life	Iduronate sulfatase
MPS III A	Sanfilippo A	Absent	Heparan N-sulfatase
MPS III B	Sanfilippo B	Absent	N-Acetyl-α-D-glucosaminidase
MPS III C	Sanfilippo C	Absent	Acetyl Co A: α-glucosaminide-N-Acetyl transferase
MPS III D	Sanfilippo D	Absent	N-Acetyl glucosaminide 6-sulfatase
MPS IV A	Morquio A	Present	Galactosamine 6-sulfate sulfatase
MPS IV B	Morquio B	Present	β-Galactosidase
MPS V	Formerly Scheie		
MPS VI	Maroteaux–Lamy		
Severe		Present	Arylsulfatase B
Intermediate		Present	Arylsulfatase B
Mild		Present	Arylsulfatase B
MPS VII	Sly	Variable	β-Glucuronidase

One of the mucolipidoses, ML IV, has particularly prominent and characteristic ocular signs (severe corneal clouding and strabismus) that appear during the first year of life. Although few corneal transplants have been reported, one case of a conjunctival transplantation resulted in improved corneal clarity that persisted for more than a year (Dangel *et al.*, 1985). Although this could be useful in the management of ML IV, it is known that retinal degeneration and pigmentary retinopathy develop in later years, together with severe psychomotor retardation (Newell *et al.*, 1975). Light and transmission electron microscopy of a whole eye removed at autopsy from a patient with ML IV revealed that the accumulation of storage material is not confined to the cornea and conjunctiva; evidence for the storage of both glycosaminoglycans and phospholipids was found in macrophages, plasma cells, ciliary epithelial cells, Schwann cells, retinal ganglion cells, and vascular endothelial cells (Riedel *et al.*, 1985).

Since ML IV was first delineated as a specific disease entity (Berman *et al.*, 1974), at least 40 additional cases have been reported in the world literature (Crandall *et al.*, 1982; Amir *et al.*, 1987). Although there is a high frequency in Ashkenazi Jewish children, other ethnic groups also express the gene, and it now appears that the disorder may be more heterogeneous than first suspected (Amir *et al.*, 1987). Mucolipidosis IV is an autosomal recessive storage disorder, and the principal substances accumulating are gangliosides, phospholipids, and glycosaminoglycans. A partial deficiency of plasma membrane-associated ganglioside sialidase has been detected in cultured skin fibroblasts from ML IV patients (Bach *et al.*, 1979; Caimi *et al.*, 1982; Zeigler and Bach, 1985), but the biochemical basis of the specific ocular effects, especially the pronounced corneal clouding, remains unknown.

3.12. REFERENCES

Abraham, N. G., Lin, J. H.-C., Dunn, M. W., and Schwartzman, M. L., 1987, Presence of heme oxygenase and NADPH cytochrome P-450 (c) reductase in human corneal epithelium, *Invest. Ophthalmol. Vis. Sci.* **28:**1464–1472.

Akhtar, R. A., 1987, Effects of norepinephrine and 5-hydroxytryptamine on phosphoinositide-PO_4 turnover in rabbit cornea, *Exp. Eye Res.* **44:**849–862.

Akhtar, R. A., 1988, Guanosine 5′-O-thiotriphosphate and NaF stimulation of phosphatidylinositol 4,5-bisphosphate hydrolysis in bovine corneal epithelium, *Curr. Eye Res.* **7:**487–496.

Alexander, R. J., Silverman, B., and Henley, W. L., 1981, Isolation and characterization of BCP 54, the major soluble protein of bovine cornea, *Exp. Eye Res.* **32:**205–216.

Alper, R., 1982/1983, Isolation and preliminary characterization of a structural glycoprotein complex from bovine corneal stroma, *Curr. Eye Res.* **2:**479–487.

Alper, R., 1988, The bovine corneal SGP-complex is related to the tissue form of type VI collagen, *Curr. Eye Res.* **7:**31–42.

Alvarado, J., Murphy, C., and Juster, R., 1983, Age-related changes in the basement membrane of the human corneal epithelium, *Invest. Ophthalmol. Vis. Sci.* **24:**1015–1028.

Amir, N., Zlotogora, J., and Bach, G., 1987, Mucolipidosis type IV: Clinical spectrum and natural history, *Pediatrics* **79:**953–959.

Axelsson, I., 1984, Heterogeneity, polydispersity, and physiologic role of corneal proteoglycans. *Acta Ophthalmol.* **62:**25–38.

Axelsson, I., and Heinegard, D., 1975, Fractionation of proteoglycans from bovine corneal stroma, *Biochem. J.* **145:**491–500.

Axelsson, I., and Heinegard, D., 1978, Characterization of the keratan sulphate proteoglycans from bovine corneal stroma, *Biochem. J.* **169:**517–530.

Axelsson, I., and Heinegard, D., 1980, Characterization of chondroitin sulfate-rich proteoglycans from bovine corneal stroma, *Exp. Eye Res.* **31:**57–66.

Ayad, S., Chambers, C. A., Shuttleworth, C. A., and Grant, M. E., 1985, Isolation from bovine elastic tissues of collagen type VI and characterization of its form *in vivo*, *Biochem. J.* **230:**465–474.

Bach, G., Zeigler, M., and Schaap, T., 1979, Mucolipidosis type IV: Ganglioside sialidase deficiency, *Biochem. Biophys. Res. Commun.* **90:**1341–1347.

Baratz, K. H., Proia, A. D., Klintworth, G. K., and Lapetina, E. G., 1987, Cholinergic stimulation of phosphatidylinositol hydrolysis by rat corneal epithelium *in vitro*, *Curr. Eye Res.* **6:**691–701.

Bazan, H. E. P., 1987, Corneal injury alters eicosanoid formation in the rabbit anterior segment *in vivo*, *Invest. Ophthalmol.Vis. Sci.* **28:**314–319.

Bazan, H. E. P., and Bazan, N. G., 1984, Composition of phospholipids and free fatty acids and incorporation of labeled arachidonic acid in rabbit cornea. Comparison of epithelium, stroma and endothelium, *Curr. Eye Res.* **3:** 1313–1319.

Bazan, H. E. P., King, W. D., and Rossowska, M., 1985a, Metabolism of phosphoinositides and inositol polyphosphates in rabbit corneal epithelium, *Curr. Eye Res.* **4:**793–801.

Bazan, H. E. P., Birkle, D. L., Beuerman, R., and Bazan, N. G., 1985b, Cryogenic lesion alters the metabolism of arachidonic acid in rabbit cornea layers, *Invest. Ophthalmol. Vis. Sci.* **26:**474–480.

Bazan, H. E. P., Birkle, D. L., Beuerman, R. W., and Bazan, N. G., 1985c, Inflammation-induced stimulation of the synthesis of prostaglandins and lipoxygenase-reaction products in rabbit cornea, *Curr. Eye Res.* **4:**175–179.

Bazan, H. E. P., Dobard, P., and Reddy, S. T. K., 1987, Calcium- and phospholipid-dependent protein kinase C and phosphatidylinositol kinase: Two major phosphorylation systems in the cornea, *Curr. Eye Res.* **6:**667–673.

Beekhuis, W. H., and McCarey, B. E., 1986, Corneal epithelial Cl-dependent pump quantified, *Exp. Eye Res.* **43:**707–711.

Benya, P. D., and Padilla, S. R., 1986, Isolation and characterization of type VIII collagen synthesized by cultured rabbit corneal endothelial cells, *J. Biol. Chem.* **261:**4160–4169.

Berman, E. R., 1970, Proteoglycans of bovine corneal stroma, in: *Chemistry and Molecular Biology of the Intercellular Matrix*, Vol. 2 (E. A. Balazs, ed.), Academic Press, New York, pp. 879–886.

Berman, E. R., 1982, Mucolipidoses, in: *Pathobiology of Ocular Disease* (A. Garner and G. K. Klintworth, eds.), Marcel Dekker, New York, pp. 931–946.

Berman, E. R., Livni, N., Shapira, E., Merin, S., and Levij, I. S., 1974, Congenital corneal clouding with abnormal systemic storage bodies: A new variant of mucolipidosis, *J. Pediatr.* **84:**519–526.

Berman, M., 1978, Regulation of collagenase. Therapeutic considerations, *Trans. Ophthalmol. Soc. U.K.* **98:**397–405.

Berman, M., Leary, R., and Gage, J., 1977, Latent collagenase in the ulcerating rabbit cornea, *Exp. Eye Res.* **25:**435–445.

Berman, M., Leary, R., and Gage, J., 1979, Collagenase from corneal cell cultures and its modulation by phagocytosis, *Invest. Ophthalmol. Vis. Sci.* **18:**588–601.

Berman, M., Leary, R., and Gage, J., 1980, Evidence for a role of the plasminogen activator-plasmin system in corneal ulceration, *Invest. Ophthalmol. Vis. Sci.* **19:**1204–1221.

Berman, M., Manseau, E., Law, M., and Aiken, D., 1983, Ulceration is correlated with degradation of fibrin and fibronectin at the corneal surface, *Invest. Ophthalmol. Vis. Sci.* **24:**1358–1366.

Berman, M., Kenyon, K., Hayashi, K., and L'Hernault, N., 1988, The pathogenesis of epithelial defects and stromal ulceration, in: *The Cornea: Transactions of the World Congress on the Cornea III* (H. D. Cavanagh, ed.), Raven Press, New York, pp. 35–43.

Bernard, B. A., De Luca, L. M., Hassell, J. R., Yamada, K. M., and Olden, K., 1984, Retinoic acid alters the proportion of high mannose to complex type oligosaccharides on fibronectin secreted by cultured chondrocytes, *J. Biol. Chem.* **259:**5310–5315.

Bhuyan, K. C., and Bhuyan, D. K., 1977, Regulation of hydrogen peroxide in eye humors. Effect of 3-amino-1H-1,2,4-triazole on catalase and glutathione peroxidase of rabbit eye, *Biochim. Biophys. Acta* **497:**641–651.

Bhuyan, K. C., and Bhuyan, D. K., 1978, Superoxide dismutase of the eye. Relative functions of superoxide dismutase and catalase in protecting the ocular lens from oxidative damage, *Biochim. Biophys. Acta* **542:**28–38.

Birk, D. E., Fitch, J. M., Babiarz, J. P., and Linsenmayer, T. F., 1988, Collagen type I and type V are present in the same fibril in the avian corneal stroma, *J. Cell Biol.* **106:**999–1008.

Bleckmann, H., and Kresse, H., 1980, Glycosaminoglycan metabolism of cultured cornea cells derived from bovine and human stroma and from bovine epithelium, *Exp. Eye Res.* **30:**469–479.

Bondi, A., and Sklan, D., 1984, Vitamin A and carotene in animal nutrition, *Prog. Food Nutr. Sci.* **8:**165–191.

Brekle, A., and Mersmann, G., 1981, The carbohydrate-protein binding region in keratan sulfate from bovine cornea: Structure of a major binding region oligosaccharide, *Biochim. Biophys. Acta* **675:**322–327.

Brightwell, J. R., Riddle, S. L., Eiferman, R. A., Valenzuela, P., Barr, P. J., Merryweather, J. P., and Schultz, G.S., 1985, Biosynthetic human EGF accelerates healing of neodecadron-treated primate corneas, *Invest. Ophthalmol. Vis. Sci.* **26:**105–110.

Brown, S. I., Weller, C. A., and Akiya, S., 1970, Pathogenesis of ulcers of the alkali-burned cornea, *Arch. Ophthalmol.* **83:** 205–208.

Bruns, R. R., Press, W., and Gross, J., 1987, A large-scale, orthogonal network of microfibril bundles in the corneal stroma, *Invest. Ophthalmol. Vis. Sci.* **28:**1939–1946.

Caimi, L., Tettamanti, G., and Berra, B., 1982, Mucolipidosis IV, a sialolipidosis due to ganglioside sialidase deficiency, *J. Inherit. Metab. Dis.* **5:**218–224.

Carter-Dawson, L., Tanaka, M., Kuwabara, T., and Bieri, J. G., 1980, Early corneal changes in vitamin A deficient rats, *Exp. Eye Res.* **30:**261–268.

Chalfie, M., Neufeld, A. H., and Zadunaisky, J.A., 1972, Action of epinephrine and other cyclic AMP-mediated agents on the chloride transport of the frog cornea, *Invest. Ophthalmol.* **11:**644–650.

Chung, J.-H., Fagerholm, P., and Lindstrom, B., 1989, Hyaluronate in healing of corneal alkali wound in the rabbit, *Exp. Eye Res.* **48:**569–576.

Chung, S. M., Proia, A. D., Klintworth, G. K., Watson, S. P., and Lapetina, E. G., 1985, Deoxycholate induces the preferential hydrolysis of polyphosphoinositides by human platelet and rat corneal phospholipase C, *Biochem. Biophys. Res. Commun.* **129:**411–416.

Church, R. L., SundarRaj, N., and Rohrbach, D. H., 1981, Gene mapping of human ocular connective tissue proteins. Assignment of the strutural gene for corneal type I procollagen to human chromosome 7 in human corneal stroma-mouse fibroblast somatic cell hybrids, *Invest. Ophthalmol. Vis. Sci.* **21:**73–79.

Cintron, C., and Hong, B.-S., 1988, Heterogeneity of collagens in rabbit cornea: Type VI collagen, *Invest. Ophthalmol. Vis. Sci.* **29:**760–766.

Cintron, C., Hong, B.-S., and Kublin, C. L., 1981, Quantitative analysis of collagen from normal developing corneas and corneal scars, *Curr. Eye Res.* **1:**1–8.

Cintron, C., Fujikawa, L. S., Covington, H., Foster, C. S., and Colvin, R. B., 1984, Fibronectin in developing rabbit cornea, *Curr. Eye Res.* **3:**489–499.

Cintron, C., Hong, B.-S., Covington, H. I., and Macarak, E. J., 1988, Heterogeneity of collagens in rabbit cornea: Type III collagen, *Invest. Ophthalmol. Vis. Sci.* **29:**767–775.

Colley, A. M., Cavanagh, H. D., Drake, L. A., and Law, M. L., 1985, Cyclic nucleotides in muscarinic regulation of DNA and RNA polymerase activity in cultured corneal epithelial cells of the rabbit, *Curr. Eye Res.* **4:**941–950.

Colley, A. M., Law, M. L., Drake, L. A., and Cavanagh, H. D. 1987, Activity of DNA and RNA polymerases in resurfacing rabbit corneal epithelium, *Curr. Eye Res.* **6:**477–487.

Conrad, G. W., Ager-Johnson, P., and Woo, M.-L., 1982, Antibodies against the predominant glycosaminoglycan of the mammalian cornea, keratan sulfate-I, *J. Biol. Chem.* **257:**464–471.

Cooper, D., Schermer, A., and Sun, T.-T., 1985, Biology of Disease. Classification of human epithelia and their neoplasms using monoclonal antibodies to keratins: Strategies, applications, and limitations, *Lab. Invest.* **52:**243–256.

Cork, R. J., Reinach, P., Moses, J., and Robinson, K. R., 1987, Calcium does not act as a second messenger for adrenergic and cholinergic agonists in corneal epithelial cells, *Curr. Eye Res.* **6:**1309–1317.

Coster, L., Cintron, C., Damle, S. P., and Gregory, J. D., 1983, Proteoglycans of rabbit cornea: Labelling in organ culture and *in vivo*, *Exp. Eye Res.* **36:**517–530.

Crandall, B. F., Philippart, M., Brown, W. J., and Bluestone, D. A., 1982, Review article: Mucolipidosis IV, *Am. J. Med. Genet.* **12:**301–308.

Critchfield, J. W., Calandra, A. J., Nesburn, A. B., and Kenney, M. C., 1988, Keratoconus: I. Biochemical studies, *Exp. Eye Res.* **46:**953–963.

Crosson, C. E., Klyce, S. D., Bazan, H. E. P., and Bazan, N. G., 1986, The effect of phorbol esters on the chloride secreting epithelium of the rabbit cornea, *Curr. Eye Res.* **5:**535–541.

Crouch, R., Priest, D. G., and Duke, E. J., 1978, Superoxide dismutase activities of bovine ocular tissues, *Exp. Eye Res.* **27:**503–509.

Crouch, R. K., Patrick, J., Goosey, J., and Coles, W. H., 1984, The effect of age on corneal and lens superoxide dismutase, *Curr. Eye Res.* **3:**1119–1123.

Culp, T. W., Cunningham, R. D., Tucker, P. W., Jeter, J., and Deiterman, L. H., Jr., 1970, *In vivo* synthesis of lipids in rabbit iris, cornea and lens tissues, *Exp. Eye Res.* **9:**98–105.

Dangel, M. E., Bremer, D. L., and Rogers, G. L., 1985, Treatment of corneal opacification in mucolipidosis IV with conjunctival transplantation, *Am. J. Ophthalmol.* **99:**137–141.

Davison, P. F., and Galbavy, E. J., 1985, Fluorescent dyes demonstrate the uniform expansion of the growing rabbit cornea, *Invest. Ophthalmol. Vis. Sci.* **26:**1202–1209.

De Luca, L. M., 1977, The direct involvement of vitamin A in glycosyl transfer reactions of mammalian membranes, *Vitam. Horm.* **35:**1–57.

Elgebaly, S. A., Downes, R. T., Bohr, M., Forouhar, F., O'Rourke, J., and Kreutzer, D. L., 1987, Inflammatory mediators in alkali-burned corneas: Preliminary characterization, *Curr. Eye Res.* **6:**1263–1274.

El-Ghorab, M., Capone, A., Jr., Underwood, B. A., Hatchell, D., Friend, J., and Thoft, R. A., 1988, Response of ocular surface epithelium to corneal wounding in retinol-deficient rabbits, *Invest. Ophthalmol. Vis. Sci.* **29:**1671–1676.

Eype, A. A., Kruit, P. J., Gaag, R. v. d., Neuteboom, G. H. G., Broersma, L., and Kijlstra, A., 1987, Autoimmunity against corneal antigens. II. Accessibility of the 54 kD corneal antigen for circulating antibodies, *Curr. Eye Res.* **6:**467–475.

Fabricant, R. N., Alpar, A. J., Centifanto, Y. M., and Kaufman, H. E., 1981, Epidermal growth factor receptors on corneal endothelium, *Arch. Ophthalmol.* **99:**305–308.

Feldman, G. L., 1967, Human ocular lipids: Their analysis and distribution, *Surv. Ophthalmol.* **12:**207–243.

Fischbarg, J., and Lim. J. J., 1984, Fluid and electrolyte transports across corneal endothelium, in: *Current Topics in Eye Research*, Vol. 4 (J. A. Zadunaisky and H. Davson, eds.), Academic Press, Orlando, FL, pp. 201–223.

Fischbarg, J., Hernandez, J., Liebovitch, L. S., and Koniarek, J. P., 1985, The mechanism of fluid and electrolyte transport across corneal endothelium: Critical revision and update of a model, *Curr. Eye Res.* **4:**351–360.

Fischer, F. H., Schmitz, L., Hoff, W., Schartl, S., Liegl, O., and Wiederholt, M., 1978, Sodium and chloride transport in the isolated human cornea, *Pflugers Arch.* **373:**179–188.

Frangieh, G. T., Hayashi, K., Teekhasaenee, C., Wolf, G., Colvin, R. B., Gipson, I. K., and Kenyon, K. R., 1989, Fibronectin and corneal epithelial wound healing in the vitamin A-deficient rat, *Arch. Ophthalmol.* **107:**567–571.

Friend, J., 1987, Physiology of the cornea: Metabolism and biochemistry, in: *The Cornea. Scientific Foundations and Clinical Practice*, 2nd ed. (G. Smolin and R. A. Thoft, eds.), Little, Brown, Boston, pp. 16-38.

Fujikawa, L. S., Foster, C. S., Harrist, T. J., Lanigan, J. M., and Colvin, R. B., 1981, Fibronectin in healing rabbit corneal wounds, *Lab. Invest.* **45:**120–129.

Funderburgh, J. L., and Chandler, J. W., 1989, Proteoglycans of rabbit corneas with nonperforating wounds, *Invest. Ophthalmol. Vis. Sci.* **30:**435–442.

Funderburgh, J. L., Stenzel-Johnson, P. R., and Chandler, J. W., 1982/1983, Monoclonal antibodies to rabbit corneal keratan sulfate proteoglycan, *Curr. Eye Res.* **2:**769–775.

Funderburgh, J. L., Caterson, B., and Conrad, G. W., 1986, Keratan sulfate proteoglycan during embryonic development of the chicken cornea, *Dev. Biol.* **116:**267–277.

Funderburgh, J. L., Caterson, B., and Conrad, G. W., 1987, Distribution of proteoglycans antigenically related to corneal keratan sulfate proteoglycan, *J. Biol. Chem.* **262:**11634–11640.

Geanon, J. D., Tripathi, B. J., Tripathi, R. C., and Barlow, G. H., 1987, Tissue plasminogen activator in avascular tissues of the eye: A quantitative study of its activity in the cornea, lens, and aqueous and vitreous humors of dog, calf, and monkey, *Exp. Eye Res.* **44:**55–63.

Geroski, D. H., and Edelhauser, H. F., 1984, Quantitation of Na/K ATPase pump sites in the rabbit corneal endothelium, *Invest. Ophthalmol. Vis. Sci.* **25:**1056–1060.

Geroski, D. H., Edelhauser, H. F., and O'Brien, W. J., 1978, Hexose-monophosphate shunt response to diamide in the component layers of the cornea, *Exp. Eye Res.* **26:**611–619.

Geroski, D. H., Kies, J. C., and Edelhauser, H. F., 1984, The effects of ouabain on endothelial function in human and rabbit corneas, *Curr. Eye Res.* **3:**331–338.

Gerritsen, M. E., Rimarachin, J., Perry, C. A., and Weinstein, B. I., 1989, Arachidonic acid metabolism by cultured bovine corneal endothelial cells, *Invest. Ophthalmol. Vis. Sci.* **30:**698–705.

Giblin, F. J., McCready, J. P.. Kodama, T., and Reddy, V. N., 1984, A direct correlation between the levels of ascorbic acid and H_2O_2 in aqueous humor, *Exp. Eye Res.* **38:**87–93.

Gipson, I. K., and Anderson, R. A., 1977, Actin filaments in normal and migrating corneal epithelial cells, *Invest. Ophthalmol. Vis. Sci.* **16:**161–166.

Gipson, I. K., and Anderson, R. A., 1980, Comparison of 10 nm filaments from three bovine tissues, *Exp. Cell Res.* **128:**395–406.

Gipson, I. K., and Kiorpes, T. C., 1982, Epithelial sheet movement: Protein and glycoprotein synthesis, *Dev. Biol.* **92:** 259–262.

Gipson, I. K., Westcott, M. J., and Brooksby, N. G., 1982, Effects of cytochalasins B and D and colchicine on migration of the corneal epithelium, *Invest. Ophthalmol. Vis. Sci.* **22:**633–642.

Gipson, I. K., Riddle, C. V., Kiorpes, T. C., and Spurr, S. J., 1983, Lectin binding to cell surfaces: Comparisons between normal and migrating corneal epithelium, *Dev. Biol.* **96:**337–345.

Gipson, I. K., Kiorpes, T. C., and Brennan, S. J., 1984, Epithelial sheet movement: Effects of tunicamycin on migration and glycoprotein synthesis, *Dev. Biol.* **101:**212–220.

Gipson, I. K., Spurr-Michaud, S. J., and Tisdale, A. S., 1987, Anchoring fibrils form a complex network in human and rabbit cornea, *Invest. Ophthalmol. Vis. Sci.* **28:**212–220.

Gipson, I. K., Spurr-Michaud, S., Tisdale, A., and Keough, M., 1989, Reassembly of the anchoring structures of the corneal epithelium during wound repair in the rabbit, *Invest. Ophthalmol. Vis. Sci.* **30:**425–434.

Goldminz, D., Vlodavsky, I., Johnson, L. K., and Gospodarowicz, D., 1979, Contact inhibition and the regulation of endocytosis in the corneal endothelium: Correlation with a restricted surface receptor lateral mobility and the appearance of a fibronectin meshwork, *Exp. Eye. Res.* **29:**331–351.

Goodman, D. S., 1984, Plasma retinol-binding protein, in: *The Retinoids*, Vol. 2 (M. B. Sporn, A. B. Roberts, and D. S. Goodman, eds.), Academic Press, Orlando, FL, pp. 41–88.

Gordon, J. M., Bauer, E. A., and Eisen, A. Z., 1980, Collagenase in human cornea, *Arch. Ophthalmol.* **98:**341–345.

Gospodarowicz, D., Mescher, A. L., and Birdwell, C. R., 1977, Stimulation of corneal endothelial cell proliferation *in vitro* by fibroblast and epidermal growth factors, *Exp. Eye Res.* **25:**75–89.

Gospodarowicz, D., Greenburg, G., Foidart, J. M., and Savion, N., 1981, The production and localization of laminin in cultured vascular and corneal endothelial cells, *J. Cell. Physiol.* **107:**171–183.

Gottsch, J. D., Chen, C.-H., Aguayo, J. B., Cousins, J. P, Strahlman, E. R., and Stark, W. J., 1986, Glycolytic activity in the human cornea monitored with nuclear magnetic resonance spectroscopy, *Arch. Ophthalmol.* **104:**886–889.

Gottsch, J. D., Hairston, R. J., Chen, C.-H., Graham, C. R., Jr., and Stark, W. J., 1988, Corneal alanine metabolism demonstrated by NMR spectroscopy, *Curr. Eye Res.* **7:**253–256.

Gregory, J. D., Coster, L., and Damle, S. P., 1982, Proteoglycans of rabbit corneal stroma. Isolation and partial characterization, *J. Biol. Chem.* **257:**6965–6970.

Gregory, J. D., Damle, S. P., Covington, H. I., and Cintron, C., 1988, Developmental changes in proteoglycans of rabbit corneal stroma, *Invest. Ophthalmol. Vis. Sci.* **29:**1413–1417.

Greiner, J. V., Lass, J. H., and Glonek, T,. 1984, *Ex vivo* metabolic analysis of eye bank corneas using phosphorus nuclear magnetic resonance, *Arch. Ophthalmol.* **102:**1171–1173.

Greiner, J. V., Braude, L. S., and Glonek, T., 1985a, Distribution of phosphatic metabolites in the porcine cornea using phosphorus-31 nuclear magnetic resonance, *Exp. Eye Res.* **40:** 335–342.

Greiner, J. V., Kopp, S. J., and Glonek, T., 1985b, Phosphorus nuclear magnetic resonance and ocular metabolism, *Surv. Ophthalmol.* **30:**189–202.

Haddox, J. L., Pfister, R. R., and Yuille-Barr, D., 1989, The efficacy of topical citrate after alkali injury is dependent on the period of time it is administered, *Invest. Ophthalmol. Vis. Sci.* **30:**1062–1068.

Hara, S., Ishiguro, S., and Mizuno, K., 1985, Phagocytosis of polystyrene spheres in the rabbit corneal endothelium: Contribution of lysosomal enzymes to the endothelial degeneration, *Invest. Ophthalmol. Vis. Sci.* **26:**1631–1634.

Hart, G. W., and Lennarz, W.J., 1978, Effects of tunicamycin on the biosynthesis of glycosaminoglycans by embryonic chick cornea, *J. Biol. Chem.* **253:**5795–5801.

Hassell, J. R., Newsome, D. A., and De Luca, L. M., 1980a, Increased biosynthesis of specific glycoconjugates in rat corneal epithelium following treatment with vitamin A, *Invest. Ophthalmol. Vis. Sci.* **19:**642–647.

Hassell, J. R., Newsome, D. A., Krachmer, J. H., and Rodrigues, M. M., 1980b, Macular corneal dystrophy: Failure to synthesize a mature keratan sulfate proteoglycan, *Proc. Natl. Acad. Sci. USA* **77:**3705–3709.

Hassell, J. R., Cintron, C., Kublin, C. and Newsome, D. A., 1983, Proteoglycan changes during restoration of transparency in corneal scars, *Arch. Biochem. Biophys.* **222:**362–369.

Hatchell, D. L., Faculjak, M., and Kubicek, D., 1984, Treatment of xerophthalmia with retinol, tretinoin, and etretinate, *Arch. Ophthalmol.* **102:**926–927.

Hayashi, K., Frangieh, G., Kenyon, K. R., Berman, M., and Wolf, G., 1988, Plasminogen activator activity in vitamin A-deficient rat corneas, *Invest. Ophthalmol. Vis. Sci.* **29:**1810–1819.

Hayashi, K., Cheng, H.-M., Xiong, J., Xiong, H., and Kenyon, K. R., 1989a, Metabolic changes in the cornea of vitamin A-deficient rats, *Invest. Ophthalmol. Vis. Sci.* **30:**769–772.

Hayashi, K., Frangieh, G., Wolf, G., and Kenyon, K. R., 1989b, Expression of transforming growth factor-β in wound healing of vitamin A-deficient rat corneas, *Invest. Ophthalmol. Vis. Sci.* **30:**239-247.

Heinegard, D., Franzen, A., Hedbom, E., and Sommarin, Y., 1986, Common structures of the core proteins of interstitial proteoglycans, in: *Functions of the Proteoglycans, Ciba Foundation Symposium 124* (D. Evered and J. Whelan, eds.), John Wiley & Sons, Chichester, pp. 69–82.

Hsieh, P., and Baum, J., 1985, Effects of fibroblastic and endothelial extracellular matrices on corneal endothelial cells, *Invest. Ophthalmol. Vis. Sci.* **26:**457–463.

Ihalainen, A., Salo, T., Forsius, H., and Peltonen, L., 1986, Increase in type I and type IV collagenolytic activity in primary cultures of keratoconus cornea, *Eur. J. Clin. Invest.* **16:**78–84.

Ihme, A., Krieg, T., Muller, R. K. and Wollensak, J., 1983, Biochemical investigation of cells from keratoconus and normal cornea, *Exp. Eye Res.* **36:**625–631.

Jentsch, T. J., Keller, S.K., and Wiederholt, M., 1985, Ion transport mechanisms in cultured bovine corneal endothelial cells, *Curr. Eye Res.* **4:**361–369.

Jester, J. V., Rodrigues, M. M., and Sun, T.-T., 1985, Change in epithelial keratin expression during healing of rabbit corneal wounds, *Invest. Ophthalmol. Vis. Sci.* **26:**828–837.

Johnson-Muller, B., and Gross, J., 1978, Regulation of corneal collagenase production: Epithelial–stromal cell interactions, *Proc. Natl. Acad. Sci. USA* **75:**4417–4421.

Jumblatt, M. M., and Neufeld, A. H., 1981, Characterization of cyclic AMP-mediated wound closure of the rabbit corneal epithelium, *Curr. Eye Res.* **1**:189–195.

Jumblatt, M. M., and Neufeld, A. H., 1986, A tissue culture assay of corneal epithelial wound closure, *Invest. Ophthalmol. Vis. Sci.* **27**:8–13.

Jumblatt, M. M., Matkin, E. D., and Neufeld, A. H., 1988, Pharmacological regulation of morphology and mitosis in cultured rabbit corneal endothelium, *Invest. Ophthalmol. Vis. Sci.* **29**:586–593.

Kao, W. W.-Y., Vergnes, J.-P., Ebert, J., Sundar-Raj, C. V., and Brown, S. I., 1982, Increased collagenase and gelatinase activities in keratoconus, *Biochem. Biophys. Res. Commun.* **107**:929–936.

Kao, W. W.-Y., Ebert, J., Kao, C. W.-C., Covington, H., and Cintron, C., 1986, Development of monoclonal antibodies recognizing collagenase from rabbit PMN; the presence of this enzyme in ulcerating corneas, *Curr. Eye Res.* **5**:801–815.

Kapoor, R., Bornstein, P., and Sage, E. H., 1986, Type VIII collagen from bovine Descemet's membrane: Structural characterization of a triple-helical domain, *Biochemistry* **25**:3930–3937.

Kass, M. A., and Holmberg, N. J., 1979, Prostaglandin and thromboxane synthesis by microsomes of rabbit ocular tissues, *Invest. Ophthalmol. Vis. Sci.* **18**:166–171.

Kay, E. P., 1989, Expression of types I and IV collagen genes in normal and in modulated corneal endothelial cells, *Invest. Ophthalmol. Vis. Sci.* **30**:260–268.

Kay, E. P., and Oh, S., 1988, Modulation of type III collagen synthesis in bovine corneal endothelial cells, *Invest. Ophthalmol. Vis. Sci.* **29**:200–207.

Kay, E. P., Smith, R. E., and Nimni, M. E., 1982, Basement membrane collagen synthesis by rabbit corneal endothelial cells in culture. Evidence for an α chain derived from a larger biosynthetic precursor, *J. Biol. Chem.* **257**:7116–7121.

Kay, E. P., Nimni, M. E., and Smith, R. E., 1984, Stability of collagen phenotype in morphologically modulated rabbit corneal endothelial cells, *Invest. Ophthalmol. Vis. Sci.* **25**:495–501.

Kay, E. P., Smith, R. E., and Nimni, M. E., 1985, Type I collagen synthesis by corneal endothelial cells modulated by polymorphonuclear leukocytes, *J. Biol. Chem.* **260**:5139–5146.

Kenyon, K. R., Berman, M., Rose, J., and Gage, J., 1979, Prevention of stromal ulceration in the alkali-burned rabbit cornea by glued-on contact lens. Evidence for the role of polymorphonuclear leukocytes in collagen degradation, *Invest. Ophthalmol. Vis. Sci.* **18**:570–587.

Kinoshita, S., Kiorpes, T. C., Friend, J., and Thoft, R. A., 1982, Limbal epithelium in ocular surface wound healing, *Invest. Ophthalmol. Vis. Sci.* **23**:73–80.

Kinoshita, S., Friend, J., Kiorpes, T. C., and Thoft, R. A., 1983, Keratin-like proteins in corneal and conjunctival epithelium are different, *Invest. Ophthalmol. Vis. Sci.* **24**:577–581.

Klintworth, G. K., 1982a, Degenerations, depositions, and miscellaneous reactions of the cornea, conjunctiva, and sclera, in: *Pathobiology of Ocular Disease, Part B* (A. Garner and G. K. Klintworth, eds.), Marcel Dekker, New York, pp. 1431–1475.

Klintworth, G. K., 1982b, Disorders of glycosaminoglycans (mucopolysaccharides) and proteoglycans, in: *Pathobiology of Ocular Disease, Part B* (A. Garner and G. K. Klintworth, eds.), Marcel Dekker, New York, pp. 863–895.

Klintworth, G. K., 1982c, Current concept of macular corneal dystrophy, in: *Genetic Eye Diseases, Birth Defects: Original Article Series*, Vol. 18 (E. Cotlier, I. H. Maumenee, and E. R. Berman, eds.), Alan R. Liss, New York, pp. 463–477.

Klintworth, G. K., and Smith, C. F., 1980, Abnormal product of corneal explants from patients with macular corneal dystrophy, *Am. J. Pathol.* **101**:143–157.

Klintworth, G. K., and Smith, C. F., 1981, Difference between the glycosaminoglycans synthesized by corneal and cutaneous fibroblasts in culture, *Lab. Invest.* **44**:553–559.

Klintworth, G. K., and Smith, C. F., 1983, Abnormalities of proteoglycans and glycoproteins synthesized by corneal organ cultures derived from patients with macular corneal dystrophy, *Lab. Invest.* **48**:603–612.

Klintworth, G. K., Meyer, R., Dennis, R., Hewitt, A. T., Stock, E. L., Lenz, M. E., Hassell, J. R., Stark, W. J. Jr., Kuettner, K. E., and Thonar, E. J.-M. A., 1986, Macular corneal dystrophy. Lack of keratan sulfate in serum and cornea, *Ophthalmic Paediatr. Genet.* **7**:139–143.

Klyce, S. D., 1975, Transport of Na, Cl, and water by the rabbit corneal epithelium at resting potential, *Am. J. Physiol.* **228**:1446–1452.

Klyce, S. D., 1977, Enhancing fluid secretion by the corneal epithelium, *Invest. Ophthalmol. Vis. Sci.* **16**:968–973.

Klyce, S. D., and Crosson, C. E., 1985, Transport processes across the rabbit corneal epithelium: A review, *Curr. Eye Res.* **4:**323–331.

Klyce, S. D., Neufeld, A. H., and Zadunaisky, J. A., 1973, The activation of chloride transport by epinephrine and Db cyclic-AMP in the cornea of the rabbit, *Invest. Ophthalmol.* **12:**127–139.

Klyce, S. D., Palkama, K. A., Harkonen, M., Marshall, W. S., Huhtaniitty, S., Mann, K. P., and Neufeld, A. H., 1982, Neural serotonin stimulates chloride transport in the rabbit corneal epithelium, *Invest. Ophthalmol. Vis. Sci.* **23:**181–192.

Kohno, T., Sorgente, N., Ishibashi, T., Goodnight, R., and Ryan, S. J., 1987, Immunofluorescent studies of fibronectin and laminin in the human eye, *Invest. Ophthalmol. Vis. Sci.* **28:**506–514.

Krachmer, H., Feder, R. S., and Belin, M. W., 1984, Keratoconus and related noninflammatory corneal thinning disorders, *Surv. Ophthalmol.* **28:**293–322.

Kruit, P. J., van der Gaag, R., Broersma, L., and Kijlstra, A., 1986, Autoimmunity against corneal antigens. I. Isolation of a soluble 54 Kd corneal epithelium antigen, *Curr. Eye Res.* **5:**313–320.

Kurkinen, M., Taylor, A., Garrels, J. I., and Hogan, B. L. M., 1984, Cell surface-associated proteins which bind native type IV collagen or gelatin, *J. Biol. Chem.* **259:**5915–5922.

Labermeier, U., and Kenney, M. C., 1983, The presence of EC collagen and type IV collagen in bovine Descemet's membranes, *Biochem. Biophys. Res. Commun.* **116:**619–625.

Labermeier, U., Demlow, T. A., and Kenney, M. C., 1983, Identification of collagens isolated from bovine Descemet's membrane, *Exp. Eye Res.* **37:**225–237.

Laurent, T. C., and Anseth, A., 1961, Studies on corneal polysaccharides. II. Characterization, *Exp. Eye Res.* **1:**99–105.

Lee, R. E., and Davison, P. F., 1981, Collagen composition and turnover in ocular tissues of the rabbit, *Exp. Eye Res.* **32:**737-745.

Lee, R. E., and Davison P. F., 1984, The collagens of the developing bovine cornea, *Exp. Eye Res.* **39:**639–652.

Leonard, M. C., Maddison, L. K., and Pirie, A., 1981, A comparison between the enzymes in the cornea of the vitamin-A deficient rat and those of rat leucocytes, *Exp. Eye Res.* **33:**479–495.

Leonardy, N. J., Smith, C. F., Brown, C. F., and Klintworth, G. K., 1985, Intercellular relationships in the synthesis of macromolecules by organ cultures of corneas, *Invest. Ophthalmol. Vis. Sci.* **26:**1216–1222.

Lin, M. T., Eiferman, R. A., and Wittliff, J. L., 1984, Demonstration of specific glucocorticoid binding sites in bovine cornea, *Exp. Eye Res.* **38:**333–339.

Masferrer, J. L., Murphy, R. C., Pagano, P. J., Dunn, M. W., and Laniado-Schwartzman, M., 1989, Ocular effects of a novel cytochrome P-450-dependent arachidonic acid metabolite, *Invest. Ophthalmol. Vis. Sci.* **30:**454–460.

Masters, B. R., 1984a, Nonivasive redox fluorometry: How light can be used to monitor alterations of corneal mitochondrial function, *Curr. Eye Res.* **3:**23–26.

Masters, B. R., 1984b, Noninvasive corneal redox fluorometry, in: *Current Topics in Eye Research*, Vol. 4 (J. A. Zadunaisky, and H. Davson, eds.), Academic Press, Orlando, FL, pp. 139–200.

Matsuda, M., Sawa, M., Edelhauser, H. F., Bartels, S. P., Neufeld, A. H., and Kenyon, K. R., 1985, Cellular migration and morphology in corneal endothelial wound repair, *Invest. Ophthalmol. Vis. Sci.* **26:**443–449.

Matsuda, M., Ubels, J. L., and Edelhauser, H. F., 1986, Corneal endothelial healing rate and the effect of topical retinoic acid, *Invest. Ophthalmol. Vis. Sci.* **27:**1193–1198.

Maurice, D. M., 1957, The structure and transparency of the cornea, *J. Physiol.(Lond.)* **136:**263–286.

Maurice, D. M., 1972, The location of the fluid pump in the cornea, *J. Physiol. (Lond.)* **221:**43–54.

Maurice, D. M., 1984, The cornea and sclera, in: *The Eye*, 3rd ed. (H. Davson, ed.), Academic Press, New York, pp. 1–158.

McCartney, M. D., Wood, T. O., and McLaughlin, B. J., 1987a, Freeze-fracture label of functional and dysfunctional human corneal endothelium, *Curr. Eye Res.* **6:**589–597.

McCartney, M. D., Robertson, D. P., Wood, T. O., and McLaughlin, B. J., 1987b, ATPase pump site density in human dysfunctional corneal endothelium, *Invest. Ophthalmol. Vis. Sci.* **28:**1955–1962.

McCartney, M. D., Wood, T. O., and McLaughlin, B. J., 1987c, Immunohistochemical localization of ATPase in human dysfunctional corneal endothelium, *Curr. Eye Res.* **6:**1479–1486.

McKusick, V. A., and Neufeld, E. F., 1983, The mucopolysaccharide storage diseases, in: *The Metabolic Basis of Inherited Disease*, 5th ed. (J. B. Stanbury, J. B. Wyngaarden, D. S. Fredrickson, J. L. Goldstein, and M. S. Brown, eds.), McGraw-Hill, New York, pp. 751–777.

Meek, K. M., Elliott, G. F., and Nave, C., 1986, A synchrotron x-ray diffraction study of bovine cornea stained with cupromeronic blue, *Collagen Relat. Res.* **6:**203–218.

Meyer, K., Linker, A., Davidson, E. A., and Weissman, B., 1953, The mucopolysaccharides of bovine cornea, *J. Biol. Chem.* **205:**611–616.

Millin, J. A., Golub, B. M., and Foster, C. S., 1986, Human basement membrane components of keratoconus and normal corneas, *Invest. Ophthalmol. Vis. Sci.* **27:**604–607.

Moll, R., Franke, W. W., Schiller, D. L., Geiger, B., and Krepler, R., 1982, The catalog of human cytokeratins: Patterns of expression in normal epithelia, tumors and cultured cells, *Cell* **31:**11–24.

Murphy, C., Alvarado, J., and Juster, R., 1984, Prenatal and postnatal growth of the human Descemet's membrane, *Invest. Ophthalmol. Vis. Sci.* **25:**1402–1415.

Nakayasu, K., Tanaka, M., Konomi, H., and Hayashi, T., 1986, Distribution of types I, II, III, IV and V collagen in normal and keratoconus corneas, *Ophthalmic Res.* **18:**1–10.

Nakazawa, K., Newsome, D. A., Nilsson, B., Hascall, V. C., and Hassell, J. R., 1983a, Purification of keratan sulfate proteoglycan from monkey cornea, *J. Biol. Chem.* **258:**6051–6055.

Nakazawa, K., Hassell, J. R., Hascall, V. C., and Newsome, D. A., 1983b, Heterogeneity of proteoglycans in monkey corneal stroma, *Arch. Biochem. Biophys.* **222:**105–116.

Nakazawa, K., Hassell, J. R., Hascall, V. C., Lohmander, L. S., Newsome D. A., and Krachmer, J., 1984, Defective processing of keratan sulfate in macular corneal dystrophy, *J. Biol. Chem.* **259:**13751–13757.

Neufeld, A. H., Ledgard, S. E., Jumblatt, M. M., and Klyce, S. D., 1982, Serotonin-stimulated cyclic AMP synthesis in the rabbit corneal epithelium, *Invest. Ophthalmol. Vis. Sci.* **23:** 193–198.

Neufeld, A. H., Ledgard, S. E., and Yoza, B. K., 1983, Changes in responsiveness of the β-adrenergic and serotonergic pathways of the rabbit corneal epithelium, *Invest. Ophthalmol. Vis. Sci.* **24:**527–534.

Neufeld, A. H., Jumblatt, M. M., Matkin, E. D., and Raymond, G.M., 1986, Maintenance of corneal endothelial cell shape by prostaglandin E_2: Effects of EGF and indomethacin, *Invest. Ophthalmol. Vis. Sci.* **27:**1437–1442.

Neufeld, E. F., and McKusick, V. A., 1983, Disorders of lysosomal enzyme synthesis and localization: I-cell disease and pseudo-Hurler polydystrophy, in: *The Metabolic Basis of Inherited Disease*, 5th ed.(J. B. Stanbury, J. B. Wyngaarden, D. S. Fredrickson, J. L. Goldstein, and M. S. Brown, eds.), McGraw-Hill, pp. 778–802.

Newell, F. W., Matalon, R., and Meyer, S., 1975, A new mucolipidosis with psychomotor retardation, corneal clouding, and retinal degeneration, *Am. J. Ophthalmol.* **80:**440–449.

Newsome, D. A., Foidart, J.-M., Hassell, J. R., Krachmer, J. H., Rodrigues, M. M., and Katz, S. I., 1981, Detection of specific collagen types in normal and keratoconus corneas, *Invest. Ophthalmol. Vis. Sci.* **20:**738–750.

Newsome, D. A., Gross, J., and Hassell, J. R., 1982, Human corneal stroma contains three distinct collagens, *Invest. Ophthalmol. Vis. Sci.* **22:**376–381.

Ng, M. C., and Riley, M. V., 1980, Relation of intracellular levels and redox state of glutathione to endothelial function in the rabbit cornea, *Exp. Eye Res.* **30:**511–517.

Nilsson, B., Nakazawa, K., Hassell, J. R., Newsome, D.A., and Hascall, V.C., 1983, Structure of oligosaccharides and the linkage region between keratan sulfate and the core protein on proteoglycans from monkey cornea, *J. Biol. Chem.* **258:**6056–6063.

Nishida, T., Nakagawa, S., Awata, T., Nishibayashi, C., and Manabe, R., 1982, Rapid preparation of purified autologous fibronectin eyedrops from patient's plasma, *Jpn. J. Ophthalmol.* **26:**416–424.

Nishida, T., Nakagawa, S., Awata, T., Ohashi, Y., Watanabe, K., and Manabe, R., 1983a, Fibronectin promotes epithelial migration of cultured rabbit cornea *in situ*, *J. Cell Biol.* **97:**1653–1657.

Nishida, T., Ohashi, Y., Awata, T., and Manabe, R., 1983b, Fibronectin. A new therapy for corneal trophic ulcer, *Arch. Ophthalmol.* **101:**1046–1048.

Nishida, T., Nakagawa, S., and Manabe, R., 1985, Clinical evaluation of fibronectin eyedrops on epithelial disorders after herpetic keratitis, *Ophthalmology* **92:**213–216.

Nishida, T., Nakagawa, S., Watanabe, K., Yamada, K., McDonald, J., Otori, T., and Berman, M., 1986, Pathobiology of epithelial defects: Peptide (GRGDS) of fibronectin cell-binding domain inhibits corneal epithelial attachment and spreading on plasma fibronectin, *Invest. Ophthalmol. Vis. Sci. [Suppl.]* **27:**53.

Nishida, T., Nakagawa, S., Watanabe, K., Yamada, K. M., Otori, T., and Berman, M. B., 1988, A peptide from fibronectin cell-binding domain inhibits attachment of epithelial cells, *Invest. Ophthalmol. Vis. Sci.* **29:**1820–1825.

Osborne, N. N., 1983, The occurrence of serotonergic nerves in the bovine cornea, *Neurosci. Lett.* **35:**15–18.

Osborne, N. N., and Tobin, A. B., 1987, Serotonin-accumulating cells in the iris–ciliary body and cornea of various species, *Exp. Eye Res.* **44:**731–746.

Oxlund, H., and Simonsen, A. H., 1985, Biochemical studies of normal and keratoconus corneas, *Acta Ophthalmol.* **63:**666–669.

Pandolfi, M., and Lantz, E., 1979, Partial purification and characterization of keratokinase, the fibrinolytic activator of the cornea, *Exp. Eye Res.* **29:**563–571.

Panjwani, N., and Baum, J., 1985, Rabbit corneal endothelial cell surface glycoproteins, *Invest. Ophthalmol. Vis. Sci.* **26:**450–456.

Pesin, S. R., and Candia, O.A., 1982, Acetylcholine concentration and its role in ionic transport by the corneal epithelium, *Invest. Ophthalmol. Vis. Sci.* **22:**651–659.

Pfister, R. R., Hayes, S. A., and Paterson, C. A., 1981, The influence of parenteral ascorbate on the strength of corneal wounds, *Invest. Ophthalmol. Vis. Sci.* **21:**80–86.

Pfister, R. R., Haddox, J. L., Dodson, R. W., and Harkins, L. E., 1987, Alkali-burned collagen produces a locomotory and metabolic stimulant to neutrophils, *Invest. Ophthalmol. Vis. Sci.* **28:**295–304.

Phan, T.-M. M., Gipson, I. K., Foster, C. S., Zagachin, L., and Colvin, R. B., 1986, Endogenous production of fibronectin in corneal stromal wounds: An organ culture cross-species transplant study, *Invest. Ophthalmol. Vis. Sci. [Suppl.]* **27:**52.

Phan, T.-M. M., Foster, C. S., Boruchoff, S. A., Zagachin, L. M., and Colvin, R. B., 1987, Topical fibronectin in the treatment of persistent corneal epithelial defects and trophic ulcers, *Am. J. Ophthalmol.* **104:**494–501.

Phan, T.-M. M., Foster, C. S., Wasson, P. J., Fujikawa, L. S., Zagachin, L. M., and Colvin, R. B., 1989a, Role of fibronectin and fibrinogen in healing of corneal epithelial scrape wounds, *Invest. Ophthalmol. Vis. Sci.* **30:**377–385.

Phan, T.-M. M., Foster, C. S., Zagachin, L. M., and Colvin, R. B., 1989b, Role of fibronectin in the healing of superficial keratectomies *in vitro*, *Invest. Ophthalmol. Vis. Sci.* **30:**386–391.

Pirie, A., 1977, Effects of locally applied retinoic acid on corneal xerophthalmia in the rat, *Exp. Eye Res.* **25:**297–302.

Poole, A. R., 1986, Proteoglycans in health and disease: Structures and functions, *Biochem. J.* **236:**1–14.

Proia, A. D., Chung, S. M., Klintworth, G. K., and Lapetina, E. G., 1986, Cholinergic stimulation of phosphatidic acid formation by rat cornea *in vitro*, *Invest. Ophthalmol. Vis. Sci.* **27:**905–908.

Rao, N. A., Thaete, L. G., Delmage, J. M., and Sevanian, A., 1985, Superoxide dismutase in ocular structures, *Invest. Ophthalmol. Vis. Sci.* **26:**1778–1781.

Rask, L., Geijer, C., Bill, A., and Peterson, P. A., 1980, Vitamin A supply of the cornea, *Exp. Eye Res.* **31:**201–211.

Raymond, G. M., Jumblatt, M. M., Bartels, S. P., and Neufeld, A.H., 1986, Rabbit corneal endothelial cells *in vitro*: Effects of EGF, *Invest. Ophthalmol. Vis. Sci.* **27:**474–479.

Reddy, C., Stock, E. L., Mendelsohn, A. D., Nguyen, H. S., Roth, S. I., and Ghosh, S., 1987, Pathogenesis of experimental lipid keratopathy: Corneal and plasma lipids, *Invest. Ophthalmol. Vis. Sci.* **28:**1492–1496.

Redmond, T. M., Duke, E. J., Coles, W. H., Simson, J. A. V., and Crouch, R. K., 1984, Localization of corneal superoxide dismutase by biochemical and histocytochemical techniques, *Exp. Eye Res.* **38:**369–378.

Rehany, U., and Shoshan, S., 1984, *In vitro* incorporation of proline into keratoconic human corneas, *Invest. Ophthalmol. Vis. Sci.* **25:**1254–1257.

Reinach, P., and Holmberg, N., 1987, Ca-stimulated Mg dependent ATPase activity in a plasma membrane enriched fraction of bovine corneal epithelium, *Curr. Eye Res.* **6:**399–405.

Reinach, P., and Holmberg, N., 1989, Inhibition by calcium of beta adrenoceptor mediated cAMP responses in isolated bovine corneal epithelial cells, *Curr. Eye Res.* **8:**85–90.

Reinach, P. S., and Kirchberger, M. A., 1983, Evidence for catecholamine-stimulated adenylate cyclase activity in frog and rabbit corneal epithelium and cyclic AMP-dependent protein kinase and its protein substrates in frog corneal epithelium, *Exp. Eye Res.* **37:**327–335.

Riedel, K. G., Zwaan, J., Kenyon, K. R., Kolodny, E. H., Hanninen, L., and Albert, D. M., 1985, Ocular abnormalities in mucolipidosis IV, *Am. J. Ophthalmol.* **99:**125–136.

Riley, M. V., 1977, Anion-sensitive ATPase in rabbit corneal endothelium and its relation to corneal hydration, *Exp. Eye Res.* **25:**483–494.

Riley, M. V., 1982, Transport of ions and metabolites across the corneal endothelium, in: *Cell Biology of the Eye* (D. S. McDevitt, ed.), Academic Press, New York, pp. 53–95.

Riley, M. V., 1984, A role for glutathione and glutathione reductase in control of corneal hydration, *Exp. Eye Res.* **39:**751–758.

Riley, M. V., 1985, Pump and leak in regulation of fluid transport in rabbit cornea, *Curr. Eye Res.* **4:**371–376.

Riley, M. V., and Giblin, F. J., 1982/1983, Toxic effects of hydrogen peroxide on corneal endothelium, *Curr. Eye Res.* **2:**451–458.

Riley, M. V., and Peters, M. I., 1981, The localization of the anion-sensitive ATPase activity in corneal endothelium and its relation to corneal hydration, *Biochim. Biophys. Acta* **644:**251–256.

Riley, M. V., Schwartz, C. A., and Peters, M. I., 1986, Interactions of ascorbate and H_2O_2: Implications for *in vitro* studies of lens and cornea, *Curr. Eye Res.* **5:**207–216.

Risen, L. A., Binder, P. S., and Nayak, S. K., 1987, Intermediate filaments and their organization in human corneal endothelium, *Invest. Ophthalmol. Vis. Sci.* **28:**1933–1938.

Roberts, A. B., Roche, N. S., and Sporn, M. B., 1985, Selective inhibition of the anchorage-independent growth of *myc*-transfected fibroblasts by retinoic acid, *Nature* **315:**237–239.

Rodrigues, M. M., and Waring, G. O., III, 1982, Anterior and posterior corneal dystrophies, in: *Pathobiology of Ocular Disease, Part B* (A. Garner and G. K. Klintworth, eds.), Marcel Dekker, New York, pp. 1153–1166.

Ruf, W., and Ebel, H., 1976, (Na^+K^+)-Activated ATPase in human cornea, *Pflugers Arch.* **366:**203–210.

Sakai, L. Y., Keene, D. R., and Engvall, E., 1986, Fibrillin, a new 350-kD glycoprotein, is a component of extracellular microfibrils, *J. Cell Biol.* **103:**2499–2509.

Saliternik-Givant, S., and Berman, E. R., 1970, Biochemical heterogeneity of the corneal glycosaminoglycans, *Ophthalmic Res.* **1:**94–108.

Saneto, R. P., Awasthi, Y. C., and Srivastava, S. K., 1982a, Mercapturic acid pathway enzymes in bovine ocular lens, cornea, retina and retinal pigmented epithelium, *Exp. Eye Res.* **34:**107–111.

Saneto, R. P., Awasthi, Y. C., and Srivastava, S. K., 1982b, Purification and characterization of glutathione S-transferases from the bovine cornea, *Exp. Eye Res.* **35:**279–286.

Savion, N., and Farzame, N., 1986, Characterization of the Na, K-ATPase pump in cultured bovine corneal endothelial cells, *Exp. Eye Res.* **43:**355–363.

Schermer, A., Galvin, S., and Sun, T.-T., 1986, Differentiation-related expression of a major 64K corneal keratin *in vivo* and in culture suggests limbal location of corneal epithelial stem cells, *J. Cell Biol.* **103:**49–62.

Schwartzman, M. L., Abraham, N. G., Masferrer, J., Dunn, M. W., and McGiff, J. C., 1985, Cytochrome P450 dependent metabolism of arachidonic acid in bovine corneal epithelium, *Biochem. Biophys. Res. Commun.* **132:**343–351.

Schwartzman, M. L., Balazy, M., Masferrer, J., Abraham, N. G., McGiff, J. C., and Murphy, R. C., 1987a, 12(*R*)- hydroxyicosatetraenoic acid: A cytochrome P450-dependent arachidonate metabolite that inhibits Na^+,K^+-ATPase in the cornea, *Proc. Natl. Acad. Sci. USA* **84:**8125–8129.

Schwartzman, M. L., Masferrer, J., Dunn, M. W., McGiff, J. C., and Abraham, N. G., 1987b, Cytochrome P450, drug metabolizing enzymes and arachidonic acid metabolism in bovine ocular tissues, *Curr. Eye Res.* **6:**623–630

Scott, J. E., 1985, Proteoglycan histochemistry—A valuable tool for connective tissue biochemists, *Collagen Relat. Res.* **5:**541–575.

Scott, J. E., 1986, Proteoglycan-collagen interactions, in: *Functions of the Proteoglycans, Ciba Foundation Symposium 124* (D. Evered and J. Whelan, eds.), John Wiley & Sons, Chichester, pp. 104–124.

Scott, J. E., 1988, Proteoglycan-fibrillar collagen interactions, *Biochem. J.* **252:**313–323.

Scott, J. E., and Haigh, M., 1985, "Small"-proteoglycan: Collagen interactions: Keratan sulphate proteoglycan associates with rabbit corneal collagen fibrils at the "a" and "c" bands, *Biosci. Rep.* **5:**765–774.

Scott, J. E., and Haigh, M., 1988a, Keratan sulphate and the ultrastructure of cornea and cartilage: A "stand-in" for chondroitin sulphate in conditions of oxygen lack? *J. Anat.* **158:**95–108.

Scott, J. E., and Haigh, M., 1988b, Identification of specific binding sites for keratan sulphate proteoglycans and chondroitin-dermatan sulphate proteoglycans on collagen fibrils in cornea by the use of cupromeronic blue in "critical-electrolyte-concentration" techniques, *Biochem. J.* **253:**607–610.

Sendele, D. D., Kenyon, K. R., Wolf, G., and Hanninen, L. A., 1982, Epithelial abrasion precipitates stromal ulceration in the vitamin A-deficient rat cornea, *Invest. Ophthalmol. Vis. Sci.* **23:**64–72.

Seng, W. L., Kenyon, K. R., and Wolf, G., 1982, Studies on the source and release of collagenase in thermally burned corneas of vitamin A-deficient and control rats, *Invest. Ophthalmol. Vis. Sci.* **22:**62–72.

Shams, N. B. K., Sigel, M. M., Davis, J. F., and Ferguson, J. G., 1986, Corneal epithelial cells produce thromboxane in response to interleukin 1 (IL-1), *Invest. Ophthalmol. Vis. Sci.* **27:**1543–1545.

Sharma, K. K., and Ortwerth, B. J., 1987, Purification and characterization of an aminopeptidase from bovine cornea, *Exp. Eye Res.* **45:**117–126.

Shichi, H., 1984, Biotransformation and drug metabolism, in: *Pharmacology of the eye, Handbook of Pharmacology*, Vol. 69 (M. L. Sears, ed.), Springer-Verlag, Berlin, pp. 117–148.

Silverman, B., Alexander, R. J., and Henley, W. L., 1981, Tissue and species specificity of BCP 54, the major soluble protein of bovine cornea, *Exp. Eye Res.* **33:**19–29.

Singh, G., and Foster, C. S., 1987, Epidermal growth factor in alkali-burned corneal epithelial wound healing, *Am. J. Ophthalmol.* **103:**802–807.

Singh, S. V., Hong, T. D., Srivastava, S. K., and Awasthi, Y. C., 1985, Characterization of glutathione S-transferases of human cornea, *Exp. Eye Res.* **40:**431–437.

Sommer, A., 1983a, Effects of vitamin A deficiency on the ocular surface, *Ophthalmology* **90:**592–600.

Sommer, A., 1983b, Treatment of corneal xerophthalmia with topical retinoic acid, *Am. J. Ophthalmol.* **95:**349–352.

Sommer, A., and Emran, N., 1978, Topical retinoic acid in the treatment of corneal xerophthalmia, *Am. J. Ophthalmol.* **86:**615–617.

Sommer, A., and Muhilal, H., 1982, Nutritional factors in corneal xerophthalmia and keratomalacia, *Arch. Ophthalmol.* **100:**399–403.

Soong, H. K., 1987, Vinculin in focal cell-to-substrate attachments of spreading corneal epithelial cells, *Arch. Ophthalmol.* **105:**1129–1132.

Soong, H. K., and Cintron, C., 1985, Different corneal epithelial healing mechanisms in rat and rabbit: Role of actin and calmodulin, *Invest. Ophthalmol. Vis. Sci.* **26:**838–848.

Spector, A., and Garner, W. H., 1981, Hydrogen peroxide and human cataract, *Exp. Eye Res.* **33:**673–681.

Sporn, M. B., Roberts, A. B., Wakefield, L. M., and Assoian, R. K., 1986, Transforming growth factor-β: Biological function and chemical structure, *Science* **233:**532–534.

Sporn, M. B., Roberts, A. B., Wakefield, L. M., and de Crombrugghe, B., 1987, Some recent advances in the chemistry and biology of transforming growth factor-beta, *J. Cell Biol.* **105:**1039–1045.

Sramek, S. J., Wallow, I. H. L., Bindley, C., and Sterken, G., 1987, Fibronectin distribution in the rat eye, *Invest. Ophthalmol. Vis. Sci.* **28:**500–505.

Stanifer, R. M., Snyder, R. K., and Kretzer, F. L., 1983, Cornea, in: *Biochemistry of the Eye* (R. E. Anderson, ed.), American Academy of Ophthalmology, San Francisco, pp. 23–47.

Steinert, P. M., Steven, A. C., and Roop, D. R., 1985, The molecular biology of intermediate filaments, *Cell* **42:**411–419.

Stratford, R. E., Jr., and Lee, V. H. L., 1985, Ocular aminopeptidase activity and distribution in the albino rabbit, *Curr. Eye Res.* **4:**995–999.

Stuhlsatz, H. W., Muthiah, P. L., and Greiling, H., 1972, Occurrence of dermatan sulfate in calf cornea, *Scand. J. Clin. Lab. Invest.* **29** (Suppl. 123):31.

Sturges, S. A., and Conrad, G. W., 1987, Acetylcholinesterase activity in the cornea of the developing chick embryo, *Invest. Ophthalmol. Vis. Sci.* **28:**850–858.

Suda, T., Nishida, T., Ohashi, Y., Nakagawa, S., and Manabe, R., 1981/1982, Fibronectin appears at the site of corneal stromal wound in rabbits, *Curr. Eye Res.* **1:**553–556.

SundarRaj, C. V., Church, R. L., Klobutcher, L. A., and Ruddle, F. H., 1977, Genetics of the connective tissue proteins: Assignment of the gene for human type I procollagen to chromosome 17 by analysis of cell hybrids and microcell hybrids, *Proc. Natl. Acad. Sci. USA* **74:**4444–4448.

SundarRaj, N., Willson, J., Gregory, J. D., and Damle, S. P., 1985, Monoclonal antibodies to proteokeratan sulfate of rabbit corneal stroma, *Curr. Eye Res.* **4:**49–54.

SundarRaj, N., Barbacci-Tobin, E., Howe, W. E., Robertson, S. M., and Limetti, G., 1987, Macular corneal dystrophy: Immunochemical characterization using monoclonal antibodies, *Invest. Ophthalmol. Vis. Sci.* **28:**1678–1686.

Surgue, S. P., 1987, Isolation of collagen binding proteins from embryonic chicken corneal epithelial cells, *J. Biol. Chem.* **262:**3338–3343.

Surgue, S. P., and Hay, E. D., 1986, The identification of extracellular matrix (ECM) binding sites on the basal surface of embryonic corneal epithelium and the effect of ECM binding on epithelial collagen production, *J. Cell Biol.* **102:**1907–1916.

Taylor, L., Menconi, M., Leibowitz, H. M., and Polgar, P., 1982, The effect of ascorbate, hydroperoxides, and bradykinin on prostaglandin production by corneal and lens cells, *Invest. Ophthalmol. Vis. Sci.* **23:**378–382.

Tervo, T., Sulonen, J., Valtonen, S., Vannas, A., and Virtanen, I., 1986, Distribution of fibronectin in human and rabbit corneas, *Exp. Eye Res.* **42:**399–406.

Thonar, E. J.-M. A., Meyer, R. F., Dennis. R. F., Lenz, M. E., Maldonado, B., Hassell, J. R., Hewitt, A. T., Stark, W. J., Stock, E. L., Kuettner, K. E., and Klintworth, G. K., 1986, Absence of normal keratan sulfate in the blood of patients with macular corneal dystrophy, *Am. J. Ophthalmol.* **102:**561–569.

Tripathi, R. C., and Tripathi, B. J., 1982, Human trabecular endothelium, corneal endothelium, keratocytes, and scleral fibroblasts in primary cell culture. A comparative study of growth characteristics, morphology, and phagocytic activity by light and scanning electron microscopy, *Exp. Eye Res.* **35:**611–624.

Trueb, B., and Winterhalter, K. H., 1986, Type VI collagen is composed of a 200 kd subunit and two 140 kd subunits, *EMBO J.* **5:**2815–2819.

Tseng, S. C. G., Hatchell, D., Tierney, N., Huang, A. J.-W., and Sun, T.-T., 1984, Expression of specific keratin markers by rabbit corneal, conjunctival, and esophageal epithelia during vitamin A deficiency, *J. Cell Biol.* **99:**2279–2286.

Tuberville, A. W., Wood, T. O., and McLaughlin, B. J., 1986, Cytochrome oxidase activity of Fuchs' endothelial dystrophy, *Curr. Eye Res.* **5:**939–947.

Twining, S. S., Hatchell, D. L., Hyndiuk, R. A., and Nassif, K. F., 1985, Acid proteases and histologic correlations in experimental ulceration in vitamin A deficient rabbit corneas, *Invest. Ophthalmol. Vis. Sci.* **26:**31–44.

Ubels, J. L., and Edelhauser, H. F., 1982, Retinoid permeability and uptake in corneas of normal and vitamin A-deficient rabbits, *Arch. Ophthalmol.* **100:**1828–1831.

Ubels, J. L., and Edelhauser, H. F., 1985, *In vivo* metabolism of topically applied retinol and all-*trans* retinoic acid by the rabbit cornea, *Biochem. Biophys. Res. Commun.* **131:**320–327.

Ubels, J. L., Edelhauser, H. F., and Austin, K. H., 1983, Healing of experimental corneal wounds treated with topically applied retinoids, *Am. J. Ophthalmol.* **95:**353–358.

Ubels, J. L., Edelhauser, H. F., Foley, K. M., Liao, J. C., and Gressel, P., 1985, The efficacy of retinoic acid ointment for treatment of xerophthalmia and corneal epithelial wounds, *Curr. Eye Res.* **4:**1049–1057.

Ubels, J. L., Rismondo, V., and Edelhauser, H. F., 1987, Treatment of corneal xerophthalmia in rabbits with micromolar doses of topical retinoic acid, *Curr. Eye Res.* **6:**735–737.

van Horn, D. L., DeCarlo, J. D., Schutten, W. H., and Hyndiuk, R. A., 1981, Topical retinoic acid in the treatment of experimental xerophthalmia in the rabbit, *Arch. Ophthalmol.* **99:**317–321.

van Zoelen, E. J. J., van Oostwaard, T. M. J., and de Laat, S. W., 1986, Transforming growth factor-β and retinoic acid modulate phenotypic transformation of normal rat kidney cells induced by epidermal growth factor and platelet-derived growth factor, *J. Biol. Chem.* **261:**5003–5009.

Walkenbach, R. J., and LeGrand, R. D., 1981, Regulation of cyclic AMP-dependent protein kinase and glycogen synthase by cyclic AMP in the bovine cornea, *Exp. Eye Res.* **33:**111–120.

Walkenbach, R. J., LeGrand, R. D., and Barr, R. E., 1981, Distribution of cyclic AMP-dependent protein kinase in the bovine cornea, *Exp. Eye Res.* **32:**451–459.

Wang, H.-M., Berman, M., and Law, M., 1985, Latent and active plasminogen activator in corneal ulceration, *Invest. Ophthalmol. Vis. Sci.* **26:**511–524.

Waring, G. O., III, Bourne, W. M., Edelhauser, H. F., and Kenyon, K. R., 1982, The corneal endothelium. Normal and pathologic structure and function, *Ophthalmology* **89:**531–590.

Watanabe, K., Nakagawa, S., and Nishida, T., 1987, Stimulatory effects of fibronectin and EGF on migration of corneal epithelial cells, *Invest. Ophthalmol. Vis. Sci.* **28:**205–211.

Whikehart, D. R., and Soppet, D. R., 1981, Activities of transport enzymes located in the plasma membranes of corneal endothelial cells, *Invest. Ophthalmol. Vis. Sci.* **21:**819–825.

Whikehart, D. R., Montgomery, B., and Hafer, L. M., 1987, Sodium and potassium saturation kinetics of Na^+K^+-ATPase in plasma membranes from corneal endothelium: Fresh tissue vs. tissue culture, *Curr. Eye Res.* **6:**709–717.

Wiggert, B., Bergsma, D. R., Helmsen, R. J., Alligood, J., Lewis, M., and Chader, G. J., 1977, Retinol receptors in corneal epithelium, stroma and endothelium, *Biochim. Biophys. Acta* **491:**104–113.

Wiggert, B., Van Horn, D. L., and Fish, B. L., 1982, Effects of vitamin A deficiency on [^{3}H]retinoid binding to cellular retinoid-binding proteins in rabbit cornea and conjunctiva, *Exp. Eye Res.* **34:**695–702.

Williams, R. N., and Paterson, C. A., 1986, Modulation of corneal lipoxygenase by ascorbic acid, *Exp. Eye Res.* **43:**7–13.

Williams, R. N., Delamere, N. A., and Paterson, C. A., 1985, Generation of lipoxygenase products in the avascular tissues of the eye, *Exp. Eye Res.* **41:**733–738.

Yue, B. Y. J. T., Baum, J. L., and Smith, B.D., 1983, Identification of collagens synthesized by cultures of normal human corneal and keratoconus stromal cells, *Biochim. Biophys. Acta* **755:**318–325.

Yue, B. Y. J. T., Sugar, J., and Benveniste, K., 1984, Heterogeneity in keratoconus: Possible biochemical basis, *Proc. Soc. Exp. Biol. Med.* **175:**336–341.

Yue, B. Y. J. T., Sugar, J., and Schrode, K., 1988, Histochemical studies of keratoconus, *Curr. Eye Res.* **7:**81–86.

Zagrod, M. E., and Whikehart, D. R., 1985, Adenosine-stimulated production of sugar-phosphates in bovine corneal endothelium, *Invest. Ophthalmol. Vis. Sci.* **26:**1475–1483.

Zeigler, M., and Bach, G., 1985, Ganglioside sialidase distribution in mucolipidosis type IV cultured fibroblasts, *Arch. Biochem. Biophys.* **241:**602–607.

Zieske, J. D., and Gipson, I. K., 1986, Protein synthesis during corneal epithelial wound healing, *Invest. Ophthalmol. Vis. Sci.* **27:**1–7.

Zieske, J. D., Higashijima, S. C., and Gipson, I. K., 1986, Con A- and WGA-binding glycoproteins of stationary and migratory corneal epithelium, *Invest. Ophthalmol. Vis. Sci.* **27:**1205–1210.

Zieske, J. D., Higashijima, S. C., Spurr-Michaud, S. J., and Gipson, I. K ., 1987, Biosynthetic responses of the rabbit cornea to a keratectomy wound, *Invest. Ophthalmol. Vis. Sci.* **28:**1668–1677.

Zimmermann, D. R., Trueb, B., Winterhalter, K. H., Witmer, R., and Fischer, R. W., 1986, Type VI collagen is a major component of the human cornea, *FEBS Lett.* **197:**55–58.

Zimmermann, D. R., Fischer, R. W., Winterhalter, K. H., Witmer, R., and Vaughan, L., 1988, Comparative studies of collagens in normal and keratoconus corneas, *Exp. Eye Res.* **46:**431–442.

Aqueous, Iris–Ciliary Body, and Trabeculum 4

4.1. AQUEOUS HUMOR

4.1.1. Chemical Composition

The chemical components of aqueous humor derive from many sources, the principal ones being plasma (by passive diffusion) and the ciliary epithelium (mainly by active secretion) (Cole, 1984). Specific substances also enter the aqueous humor by diffusion (or secretion) from surrounding tissues: corneal endothelium, lens, and vitreous. These tissues also utilize many nutrients present in the aqueous; e.g., the principal source of glucose for the cornea and lens is the aqueous humor.

Depending on the primary source and/or rate of utilization of a particular substance, there may be significant differences in concentration between the anterior and posterior chambers (Riley, 1983). However, many of the chemical analyses reported in the literature have been performed on samples in which the source (anterior or posterior aqueous) was not precisely defined; hence, these estimations represent "average" levels of a particular substance in the total aqueous humor. Despite this shortcoming, analyses of aqueous humor are extremely valuable in understanding pathological processes of the anterior segment, particularly glaucoma. The unique composition of aqueous humor with respect to electrolytes, proteins, biologically active substances, and organic solutes such as GSH and ascorbate suggest that it is not a simple filtrate or secretory product; rather, it is an intraocular fluid whose production is carefully controlled and whose individual components are in a state of rapid turnover. The complex composition of the aqueous is modulated by many factors, and changes and/or imbalances in its carefully tuned chemical composition are thought to be both the cause and the consequence of pathological conditions in the anterior segment.

4.1.1.1. Ions and Low-Molecular-Weight Solutes

The steady-state concentrations of low-molecular-weight solutes in the aqueous reflect a dynamic equilibrium in which substances are continuously entering (from the plasma or surrounding tissues) and leaving (through the trabeculum and uveoscleral drainage). Details on the dynamics of these exchanges are beyond the scope of this chapter, but lactic acid provides an example of the complexities involved (Riley, 1983). The concentration of lactic acid is about six to nine times greater in aqueous humor than in plasma (Table 4.1). A small part of it enters the posterior aqueous from the plasma;

Table 4.1. Concentrations of Electrolytes and Low-Molecular-Weight Solutes in Human Aqueous Humor and Plasma[a]

Component	Aqueous humor	Plasma
Electrolytes (mM)		
Na^+	142	130–145
Cl^-	131	92–125
HCO_3^-	20	24–30
Mg^{2+}	1	0.7–1.1
K^+	4	3.5–5.0
Ca^{2+}	1.2	2.0–2.6
Organic solutes (mM)		
Ascorbate	1.1	0.04–0.06
H_2O_2	0.024–0.069[b]	0.02–0.10
Glutathione	$1–10 \times 10^{-3}$	[c]
Lactate	4.5	0.5–0.8
Citrate	0.1	0.1
Glucose	2.7–3.9	5.6–6.4
Urea	4.1	3.3–6.5
Amino acids	[d]	[d]

[a]Adapted from Riley (1983).
[b]The lower and higher values are mean levels in normal and cataractous individuals, respectively (Spector and Garner, 1981).
[c]Most of the glutathione in whole blood is in the erythrocytes, whose level is approximately 1000 times higher than that in plasma.
[d]Twenty-two amino acids have been identified and quantitated in rabbit, monkey, and human aqueous humor (see Riley, 1983). Their concentrations relative to those of plasma are discussed in the text.

however, most of the lactic acid is derived from the ciliary body and lens as the end product of anaerobic glycolysis. Additional lactate diffuses in posteriorly from the vitreous humor and anteriorly from the corneal endothelium. Some of these fluxes have been measured in the rabbit eye and are discussed in greater detail by Riley (1983).

The concentrations of low-molecular-weight solutes in human aqueous humor and plasma are shown in Table 4.1. Cations such as Na^+, K^+ and Mg^{2+} are present at similar concentrations in plasma and aqueous, whereas Ca^{2+} is considerably lower in the aqueous than in plasma. The two principal anions in aqueous humor, Cl^- and HCO_3^-, are also present at nearly the same concentrations as in plasma. This specific combination of ionic species is thought to maintain the electrical neutrality of the aqueous humor and to provide the buffering capacity needed by the surrounding tissues (see reviews by Riley, 1983; Cole, 1984).

Ascorbate levels are about 20 times higher in aqueous humor than in plasma. Only a small part is transported fron the plasma; in the majority of animal species it is secreted into the aqueous (in the reduced form) by the ciliary epithelium (Riley, 1983). A wide range of H_2O_2 concentrations have been found in aqueous humor, the mean levels in cataractous patients being more than double those found in normal controls (Spector and Garner, 1981). The direct correlation between ascorbate and H_2O_2 levels found in aqueous humor of rabbits and guinea pigs supports the view that the latter arises by nonenzymatic oxidation of ascorbic acid in the presence of molecular oxygen (Pirie, 1965; Giblin *et al.*,

1984). Glutathione, although present in very low concentrations in aqueous humor, nevertheless plays an important role in protecting the anterior segment from the potentially toxic effects of H_2O_2 as well as from oxidative damage by free radicals (see Chapter 1, Section 1.4). Over 99% of the glutathione in human aqueous humor is in the reduced (GSH) state (Riley *et al.*, 1980). A detoxifying mechanism involving the nonenzymatic oxidation and reduction of glutathione, ascorbate, and H_2O_2 has been proposed by Riley (1983) and is discussed in Section 4.1.2.

Glucose concentrations in the aqueous are about 76% of those in the plasma of young individuals, but this proportion decreases to 63% in older persons (Pohjola, 1966). Glucose is thought to reach the posterior aqueous from the ciliary epithelium by facilitated diffusion, and its steady-state concentration is determined by the rate of inward flux from the ciliary body and utilization by the corneal endothelium and lens.

Free amino acids are present at a wide range of concentrations in human, rabbit, and monkey aqueous humor; the levels reported in these species have been summarized in tabular form by Riley (1983). Amino acids found in human aqueous humor at concentrations greater than those in plasma are probably secreted by the ciliary epithelium. These include arg, leu, ile, met, phe, ser, thr, tyr, and val. Amino acids whose aqueous/plasma ratio is <1 include ala, glu, his, lys, and taurine. Three amino acids are present in only trace amounts in the aqueous: cys, pro, and gly. It appears as if greatly differing rates of diffusion from the plasma and/or secretion by the ciliary epithelium, on the one hand, and varying rates of utilization by lens and corneal endothelium, on the other, account for the wide variations in aqueous/plasma ratios of individual amino acids (Riley, 1983).

4.1.1.2. Catecholamines, Prostaglandins, and Hormones

Catecholamines, previously not thought to be present in aqueous humor, have recently been detected by the use of radioenzymatic techniques sensitive enough to measure picogram quantities with reasonable accuracy (Table 4.2). Norepinephrine, the most abundant of the catecholamines identified, is present at a concentration of approximately 1 ng/ml in patients undergoing cataract surgery (Cooper *et al.*, 1984; Trope and Rumley, 1985). The level of epinephrine is approximately ten times lower than that of norepinephrine in some, but not all, individuals. Whether dopamine is present endogenously in aqueous humor is controversial. In one study dopamine was found in only two out of nine individuals examined (Cooper *et al.*, 1984), but in a later study it could not be detected in any of the 14 cataract patients examined (Trope and Rumley, 1985). The source of catecholamines in aqueous humor is uncertain, but there is evidence suggesting that they may be derived from either the ciliary epithelium or the retina (Cooper *et al.*, 1984). Their level remains low because of efficient mechanisms of inactivation, which are thought to include uptake by adjacent tissues and, to a lesser extent, enzymatic degradation by dopamine β-hydroxylase (Gual *et al.*, 1983).

Steroid hormones are present in aqueous humor at levels of 10–25% of those found in plasma (Riley, 1983). They probably enter the aqueous by diffusion from plasma. Insulin has also been detected in rabbit aqueous.

Prostaglandins, another important group of biologically active substances, have been detected at concentrations of 2 ng/ml or less in the aqueous humor of untraumatized rabbit and human eye. Their concentrations can rise 50-fold after trauma or inflammation (Riley, 1983). These substances are synthesized in the iris–ciliary body and anterior uvea in

Table 4.2. Biologically Active Substances in Aqueous Humor

Component	Aqueous humor (ng/ml)	Plasma (ng/ml)
Prostaglandins	2	—
Cyclic AMP	8	
Catecholamines		
Norepinephrine	0.8[a]; 1.14[b]	0.311
Epinephrine	0–0.13	0–0.097
Dopamine[c]	0.12; 0.32	0.037
Steroid hormones		
Testosterone		
Estrogen		
Corticosterone		
Aldosterone		
Basic fibroblast growth factor[d]		

[a]Data of Cooper *et al.* (1984).
[b]Data of Trope and Rumley (1985).
[c]Small but measurable quantities of dopamine were found in two out of nine cataract patients examined (Cooper *et al.*, 1984), but in another series of patients, no dopamine could be detected (Trope and Rumley, 1985).
[d]This 17-kDa polypeptide was identified immunologically in aqueous samples obtained from human subjects under going cataract surgery (Tripathi *et al.*, 1988).

response to injury and are rapidly secreted into the aqueous humor. An overview of prostglandin, prostacyclin, and leukotriene production from arachidonic acid through three major enzymatic pathways is given in Chapter 1, Section 1.5, and further details on their production in the iris–ciliary body are found in Section 4.2.4. Of interest is the fact that not only are prostaglandins rapidly produced in response to injury, but they are effectively removed by active transport through the ciliary body. Hence, their lifetime in the aqueous humor is extremely short.

4.1.1.3. Proteins and Lipids

The total protein concentration in human, monkey, and rabbit aqueous humor is extremely low, and in humans is only about 1/500 of the plasma protein levels (Table 4.3.). The major aqueous humor protein is albumin, accounting for about one-half of the total protein when analyzed either by HPGFC (high-performance gel filtration chromatography) (Saari *et al.*, 1983) or by crossed immunoelectrophoresis (Inada *et al.*, 1984) or SDS-PAGE (Tripathi *et al.*, 1989). Owing to its relatively low molecular weight and small diffusion radius, albumin is able to pass through the tight junctions of the nonpigmented ciliary epithelium more readily than other plasma proteins (see review by Riley, 1983). Measurable amounts of transferrin and trace amounts of other proteins such as orosomucoid, α_1-acid glycoprotein, and immunoglobulins IgG and IgE are also normally present in the aqueous. The latter two are relatively large molecules and may be produced locally, as discussed below. Virtually all other plasma proteins are completely excluded by the molecular sieve in the ciliary epithelium (Riley, 1983; Cole, 1984). Insulin, of M_r 5700, has been found in rabbit aqueous humor, where it is present at about

Table 4.3. Protein Composition of Aqueous Humor and Plasma

Component	Aqueous humor (mg/dl)	Plasma (mg/dl)
Protein	12.4 ± 2.0[a]	7000
Albumin	5.5–6.5[b]	3400
Transferrin	1.3–1.7[b]	
Prealbumin	0.3–0.4[b]	
Orosomucoid		
α_1-acid glycoprotein		
Enzymes		
Carbonic anhydrase		
Lysozyme		
Diamine oxidase		
Plasminogen activator[c]		
Fibronectin	0.25	29
Immunoglobulins		
IgG	3.0	1270
IgE (Iu/ml)	<0.75	16–218
Complement C2–C7[d]		

[a]Mean protein values on aqueous samples carefully drawn from 20 human subjects prior to cataract surgery (Tripathi *et al.*, 1989).
[b]Calculated from crossed immunoelectrophoresis data (Inada *et al.*, 1984).
[c]Detected at only very low levels in calf and monkey aqueous by an ELISA method (Geanon *et al.*, 1987).
[d]Measured by radial immunodiffusion in aqueous samples removed from patients undergoing cataract surgery (Mondino and Rao, 1983a,b). Mean values are higher in patients with inflamed aqueous than in uninflamed aqueous.

3% of the plasma level. No information is available on insulin levels in other animal species, or in humans.

A frequent estimation of 20 mg/dl for the "average" protein concentration in the aqueous is based on analyses of samples obtained by a variety of procedures not all of which were free from artifacts. When care is taken to avoid trauma and blood contamination in a standardized procedure involving rapid withdrawal of samples with a 30-gauge needle, the protein concentrations in patients with widely varying medical histories undergoing cataract surgery appear to be remarkably similar (Tripathi *et al.*, 1989). The mean concentration was found to be 12.4 ± 2.0 mg/dl in 25 human eyes examined. Twelve major proteins ranging in size from 9 to 140 kDa were detected in these samples by SDS-PAGE. One of them, a 17-kDa polypeptide, has been characterized by immunologic methods as basic fibroblast growth factor (Tripathi *et al.*, 1988). Other proteins identified by their affinity for specific antisera, lectin binding, and Western blots include 67-kDa albumin, 25-kDa light chains of IgG, and 80-kDa transferrin (R. C. Tripathi and B. J. Tripathi, personal communication).

It has been reported that certain proteins in bovine and primate aqueous humor are able to obstruct microporous filters with pore dimensions similar to those of juxtacanalicular tissue of the trabecular meshwork (Johnson *et al.*, 1986). A recent study of calf aqueous undertaken to characterize these putative filter-binding proteins has led to the tentative identification, by two-dimensional gel electrophoresis, of four polypeptides whose unique properties may be responsible for hydrophobic interactions with microporous filters (Pavao *et al.*, 1989). Two of these were 28- and 48-kDa polypeptides not found in serum; two other proteins, also not detected in serum, appeared to have additionally charged, possibly adsorbed, components, which were not further characterized. That one or more of these (or other) proteins may arise from glycosylation of already existing polypeptides raises the interesting possibility of structural and/or functional changes in aqueous humor proteins after their secretion.

The concentrations of two aqueous immunoglobulins, IgG and IgE, have been quantitated by immunoassay in aqueous humor of cataract surgery patients (Stur *et al.*, 1983; Dernouchamps *et al.*, 1985). Their high molecular weights would preclude passage from the plasma through the tight junctions of the ciliary epithelium, and although their origin in aqueous humor is not known, it has been suggested that they are produced locally, possibly by IgE-producing lymphocytes that infiltrate the uvea (Dernouchamps *et al.*, 1985).

Fibronectin is present in bovine aqueous humor at 1/100 of the concentration found in human plasma (Reid *et al.*, 1982). The molecular weight, electrophoretic mobility, and amino acid composition of fibronectins isolated from these two sources are similar. Fibronectin levels in rabbit aqueous humor are significantly decreased after transcorneal freezing, and there is a concomitant deposition of a thickened collagenous matrix on the posterior surface of the cornea (Kenney *et al.*, 1986). These findings suggest that aqueous humor fibronectin levels may modulate collagen metabolism in the corneal endothelium.

There are few enzymes in aqueous humor, and those present are found in only trace amounts. Diamine oxidase has been detected in bovine aqueous; it probably originates from plasma, and because one of the products of the oxidative deamination of diamine substrates is H_2O_2, this enzyme has been postulated to play a role in the production of H_2O_2 in aqueous humor (Crabbe, 1985). However, since a native substrate for diamine oxidase, either an aromatic or aliphatic diamine, has not yet been detected in aqueous humor, this mechanism as a source of H_2O_2 is still uncertain. Plasminogen activator is detectable by an ELISA method at very low levels in calf and monkey aqueous humor (Geanon *et al.*, 1987). The amounts are very small compared to those found in the iris–ciliary body and corneal endothelium, hence, it seems unlikely that they originate from these tissues. Carbonic anhydrase is present in trace amounts in the aqueous and may play a role in maintaining the physiological pH of 7.5–7.6 of aqueous humor by controlling the equilibrium between HCO_3^-, dissolved CO_2, and water. Lysozyme has been detected in rabbit aqueous, but no data are available on its possible role as a bactericidal agent.

The aqueous/plasma ratio for total lipids is approximately 0.3 in rabbits, and some measurements have been made on monkey aqueous (see Riley, 1983). Fat-soluble substances would be expected to cross the cell membranes of the ciliary epithelium quite readily. However, most if not all of the plasma lipids are bound to high-molecular-weight lipoproteins, and molecules of this size would not be expected to pass through the molecular sieve in the ciliary epithelium. Nevertheless, small amounts of these substances have been detected in human aqueous humor samples obtained from postmortem eyes (R. C. Tripathi and B. J. Tripathi, personal communication).

4.1.1.4. Hyaluronic Acid and Hyaluronidase

Hyaluronic acid (HA) was detected in the aqueous humor over 50 years ago (Meyer and Smyth, 1938), but its possible physiological significance was not appreciated until

Table 4.4. Hyaluronic Acid Content of Aqueous Humor and Vitreous in Various Species[a]

Species	Aqueous humor[b] (μg/ml)	Vitreous (μg/ml)
Cattle	4.0	342–775
Sheep	2.9	75–328
Pig	1.2	124
Human	1.1	140–338
Rabbit	1.1	14–52
Opossum	0.2	0.2
Rat	0.2	30

[a]Adapted from Laurent (1983).
[b]Mean values.

Barany and co-workers (Barany, 1953/1954, 1956; Barany and Woodin, 1955) made the remarkable observation that perfusion of hyaluronidase into the aqueous humor of rabbits caused a marked drop in intraocular pressure. These observations lay dormant for nearly 30 years, but the current (almost routine) use of HA during anterior segment surgery has stimulated a search for further information on endogenous aqueous humor HA, its concentration, and its source.

Early estimations of the concentration of HA in aqueous humor were probably too high because it was measured as hexosamine, an amino sugar found in both glycoproteins and HA (Meyer and Smyth, 1938). However, more accurate values, based on a specific and sensitive radioassay, are now available (Laurent, 1983). As shown in Table 4.4, the concentration of HA in the aqueous humor is about 1/100 of that in the vitreous in all species examined except the opposum, where the concentrations are similar. The estimations of HA in aqueous humor were made as soon as possible after death because of large postmortem increases (in cattle eyes, and presumably in other species) resulting from diffusion from the vitreous (Laurent, 1983). The concentrations of HA in glaucomatous individuals were found to be normal, observations that cast some doubt on the commonly held view that a hyaluronidase-sensitive material in the aqueous may be one of the causes of impaired outflow facility in glaucoma. This would not, however, exclude a possible accumulation of HA or other glycosaminoglycan(s) in the trabecular meshwork, not necessarily originating from the aqueous, as a factor in impaired outflow facility (see Sections 4.3.1.1 and 4.3.2.3).

Some understanding of the origin of HA in aqueous humor may be gleaned from comparisons of its molecular weight in the aqueous and in the vitreous (Table 4.5). Detailed studies on rabbit and cattle have shown that in both sources, HA is polydisperse; i.e., it shows a broad range of molecular weight values (Laurent and Granath, 1983). In rabbit, the HA of aqueous humor has a weight-average M_r higher than that of vitreous; moreover, the turnover of HA in the vitreous is only 15% of that in the aqueous humor, the latter estimated as 3 μg/day (Laurent and Fraser, 1983). These findings imply that in the rabbit, aqueous humor HA is not a simple degradation product of vitreous HA; some, or possibly most, of it probably originates in the anterior segment of the eye. In adult cattle as in rabbit, a large proportion of the HA of aqueous humor is of higher M_r than that in the

Table 4.5. Molecular Weight of Hyaluronic Acid of Aqueous Humor and Vitreous[a]

Species	Aqueous humor[b]	Vitreous
Rabbit	5.1×10^6	$2–3 \times 10^6$
Cattle[b]	$0.53–2.8 \times 10^6$	$5–8 \times 10^5$

[a]Adapted from Laurent and Granath (1983). As described in the text, all samples are polydisperse; the values shown are the weight-average molecular weights.
[b]Molecular weights are for 3-month-old to 10-year-old cattle; values for newborns are not included, but they are considerably higher.

vitreous. However, unlike rabbit, cattle aqueous also contains HA of lower M_r that could have originated from the vitreous by diffusion.

Other studies using explants of human and monkey ciliary body as well as carefully dissected tips of the ciliary processes support the view that these tissues may secrete HA into the aqueous humor (Rohen *et al.*, 1984; Schachtschabel *et al.*, 1984). Radioactive glucosamine was incorporated into several glycosaminoglycans, the most prominent one being HA. A large portion of the HA was released into the medium, and by analogy with an *in vivo* situation, it seems possible that the ciliary body, at least in human and monkey, could be a source of aqueous humor HA.

Hyaluronidase has never been detected in native aqueous humor; all of the studies on the dramatic effect of this enzyme on increasing the outflow facility have in fact been done using exogenous (perfused) enzyme. There is, however, indirect evidence for hyaluronic acid-degrading activity in surrounding ocular tissues. Hyaluronic acid injected into the anterior chamber reaches a high level after 2 h, but is no longer detectable after 12 h (Iwata *et al.*, 1984; Miyauchi and Iwata, 1984). Its rapid elimination was attributed to enzymatic breakdown after diffusion into the iris–ciliary body and the trabeculum.

4.1.2. Protection from Oxidative Damage

All oxygen-consuming tissues produce small amounts of highly reactive free radicals by univalent reduction of oxygen, an alternative to the major pathway of oxygen reduction through the cytochrome system. However, virtually all cells have evolved protective mechanisms that limit the formation and/or accumulation not only of free radicals but also of potentially toxic substances such as H_2O_2 (see Chapter 1, Section 1.4 and Fig. 1.9). These protective mechanisms are of two types. The first consists of enzymes that destroy these substances almost as quickly as they are formed: (1) superoxide dismutase catalyzes the dismutation of two molecules of superoxide anion, which results in the production of H_2O_2 and O_2; (2) catalase degrades H_2O_2 to H_2O and O_2; and (3) selenium-dependent peroxidase, using biological reductants such as GSH (glutathione), reduces H_2O_2 or various hydroperoxides to water or hydroxy fatty acids, respectively. These enzymes are found mainly in the cytosolic compartment of the cell (although at least one type of Mn superoxide dismutase is a mitochondrial enzyme). The second mechanism protecting oxidative cells from membrane damage involves the free radical scavenger α-tocopherol (vitamin E), a membrane-bound antioxidant that quenches free radicals and effectively terminates oxidative chain reactions involved in lipid peroxidation.

Thus, two key substances play a central role in potential oxidative damage and cellular toxicity: free radicals and H_2O_2. Whether there are endogenous free radicals in aqueous humor is not known, since their transient nature would preclude accurate measurement in this fluid. Hydrogen peroxide is, however, known to be present at concentrations ranging from about 0.025 to 0.070 mM (Table 4.1). Early reports of low but detectable levels of H_2O_2 in bovine aqueous (Pirie, 1965), and of catalase and glutathione peroxidase activities in rabbit aqueous (Bhuyan and Bhuyan, 1977), were not fully appreciated at the time. However, recent interest in the possible role of H_2O_2 in cataract formation (Spector and Garner, 1981) and its potential toxicity to other ocular tissues have stimulated an intensive search for protective mechanisms in the anterior eye.

The intracellular enzymes described above (superoxide dismutase, catalase, and glutathione peroxidase) that protect most cells from oxygen toxicity have never been detected in aqueous humor. Instead, a nonenzymatic extracellular oxidoreduction system has been postulated as a major mechanism for protecting the anterior eye from free radical damage and H_2O_2 toxicity (Riley, 1983). As postulated by Pirie (1965), H_2O_2 in the aqueous arises mainly from a nonenzymatic reaction between ascorbic acid and molecular oxygen. Reduced ascorbic acid, the dominant form in aqueous humor (Riley, 1983), reacts nonenzymatically with molecular oxygen, producing oxidized (dehydro) ascorbic acid and H_2O_2. There is, moreover, a statistically significant and direct relationship between the levels of ascorbic acid and H_2O_2 in the aqueous humor of both rabbit and guinea pig (Giblin *et al.*, 1984).

Thus, in rabbit and guinea pig, potentially toxic H_2O_2 is generated constantly in the aqueous humor in a physiological manner by a nonenzymatic pathway. It is effectively degraded to harmless water by reaction with reduced glutathione (GSH), present at a concentration of about $1–10 \times 10^{-3}$ mM in the aqueous humor. This detoxification scheme requires a continuous supply of reducing substance, GSH, which enters the aqueous humor mainly by active secretion from the ciliary epithelium (Riley *et al.*, 1980; Riley, 1983). Reduced glutathione may also diffuse into the aqueous from the corneal endothelium or lens. Whatever its source, this system could effectively control the formation of free radicals and also ensure a low "steady-state" concentration of H_2O_2 in the aqueous humor. This scheme also requires a constant supply of ascorbic acid, which, like GSH, enters the aqueous humor by secretion from the ciliary epithelium.

Whether a redox cycle is present in the aqueous humor of other animal species or in humans remains to be established. A direct relationship between ascorbate and H_2O_2 could not be detected in human aqueous (Spector and Garner, 1981). Moreover, reduced ascorbate does not react appreciably with H_2O_2, although the oxidized form does. Since it is mainly the reduced form that is present in aqueous humor, the role of ascorbic acid in this proposed redox system requires further clarification.

Studies in other biological systems suggest that superoxide anion generated from the xanthine/xanthine oxidase system can oxidize the reduced form of ascorbic acid. Although this free radical has never been detected in aqueous humor, there is indirect evidence suggesting the presence of an endogenous substrate in the aqueous capable of generating superoxide anion (Mittag, 1984; Mittag *et al.*, 1985a). Addition of xanthine oxidase to aqueous humor *in vitro* results in a measurable decrease in ascorbic acid content, suggesting—though not directly proving—generation of superoxide anion from an endogenous substrate. Injection of xanthine oxidase into the anterior chamber of rabbits, or intravitreally into rats, results in infiltration of leukocytes into the aqueous humor. There is no enhancement of this response if xanthine is administered simultaneously. Leukocyte infiltration could not be elicited with boiled enzyme, and, consistent with the experimental findings on ascorbic acid, it has been inferred that an endogenous substrate for the xanthine/xanthine oxidase free radical-generating system is present in aqueous humor. The leukocyte infiltration may be interpreted in terms of a putative chemotactic factor generated by the interaction of superoxide anion with endogenous chemotactic precursor(s).

4.2. IRIS–CILIARY BODY

The iris and ciliary body are two separate tissues, highly specialized both anatomically and functionally. They are, however, difficult to separate mechanically in the majority of experimental animals including rabbit, the most extensively used animal model. The ciliary body overlaps the iris and is tightly bound to it. Hence, both intact iris and intact ciliary body usually cannot be dissected out from the same tissue preparation free of cross contamination. Many chemical analyses and biochemical investigations have been carried out on either the combined tissues or on pooled preparations of either iris or ciliary body. The specific tissue(s) examined, iris–ciliary body, iris, or ciliary body, is indicated in the discussion that follows. In recent years, new microsurgical techniques have been developed to obtain intact ciliary processes free of adhering tissues and highly suitable for a wide variety of biochemical and metabolic studies.

This section emphasizes biochemical aspects of the iris–ciliary body such as general metabolism, transport processes, eicosanoid production, H_2O_2 detoxification, cytochrome P450 drug-metabolizing enzymes, and membrane signal transduction and second messenger systems. Discussions on the pharmacology and physiology of aqueous humor dynamics and the control of intraocular pressure are beyond the scope of this chapter but are mentioned when relevant.

4.2.1. Aqueous Humor Formation

Aqueous humor is formed by the ciliary processes and reaches the posterior chamber by several mechanisms: (1) passive diffusion; (2) active secretion, a process localized in the nonpigmented epithelium and controlled primarily by active transport involving plasma membrane-associated Na^+,K^+-ATPase (Riley and Kishida, 1986); (3) carbonic anhydrase type II activity; and (4) ultrafiltration, a passive, nonenzymatic process regulated by both the differential hydrostatic pressure in the blood and the osmotic pressure of the ciliary body (Abdel-Latif, 1983).

Although the precise contribution of each of the above components to aqueous humor formation has not been clearly established, active secretion by the inner nonpigmented cell layer appears to play a major role (Cole, 1984; Riley and Kishida, 1986). The ciliary epithelium consists of two closely apposed cell layers, the nonpigmented epithelium and the pigmented epithelium (Fig. 4.1). The nonpigmented cells are rich in mitochondria, and it is reasonable to assume that sufficient ATP is generated through the tricarboxylic acid cycle to support the energy requirements of active secretion. The active transport of Na^+ and K^+ is mediated by a ouabain-sensitive Na^+,K^+-ATPase pump localized on the basal and lateral cell borders of the nonpigmented ciliary epithelial cells (see review by Cole, 1984). Autoradiographic studies with [^{3}H]ouabain on single isolated rabbit ciliary

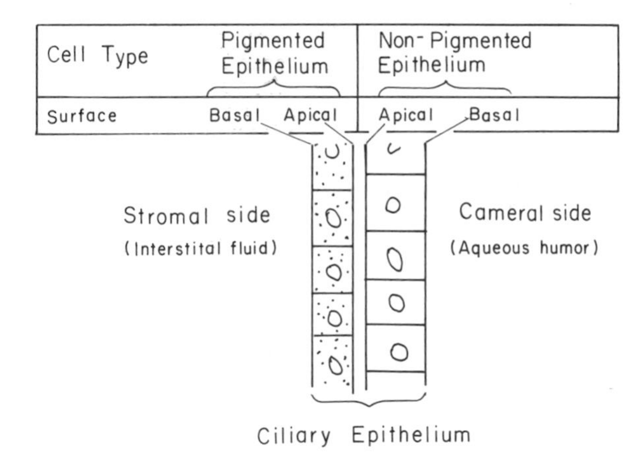

Figure 4.1. Diagram showing the orientation of the two cell layers of the ciliary epithelium. (From Cole, 1984).

processes show the specific localization of this enzyme within narrow infoldings of the basal and lateral interdigitations of this membrane (Usukura *et al.*, 1988). Other biochemical investigations on fractionated preparations of nonpigmented cells from cattle eyes confirm the localization of Na^+,K^+-ATPase specifically in plasma membrane-enriched fractions (Riley and Kishida, 1986).

In contrast, anion-stimulated ATPase is associated mainly with mitochondria-enriched fractions and probably does not play a direct role in either ion transport or secretion in the ciliary epithelium (Riley and Kishida, 1986). Other ions and low-molecular-weight solutes such as amino acids, glucose, inositol, and ascorbic acid are also secreted by the nonpigmented ciliary epithelium, but the mechanism of their transport across the ciliary epithelium into the posterior chamber requires clarification.

That carbonic anhydrase (CA) also plays an important role in aqueous humor formation has long been recognized. Parenterally administered sulfonamides such as acetazolamide and methazolamide, potent inhibitors of CA, reduce aqueous humor formation and lower the intraocular pressure (IOP) in rabbits (the most extensively investigated animal model) and cats as well as in humans (Friedland and Maren, 1984). Four isoenzymes of CA have been characterized in a variety of tissues: (1) CA I, the low-activity type; (2) CA II, the high-activity type; (3) CA III, the soluble, sulfonamide-resistant low-activity type; and (4) CA IV, the membrane-bound hydrophilic type (Wistrand, 1984; Wistrand *et al.*, 1986). Human ciliary body contains only CA II; it is found in both the pigmented and nonpigmented epithelium. Studies with a variety of CA inhibitors have shown that they act directly on the CA type II enzyme in the ciliary processes, causing a reduction in the rate of entry of sodium and bicarbonate ions into the posterior aqueous and a corresponding 50% reduction of aqueous flow. The mechanism may be similar to that present in corneal endothelium (see Chapter 3, Section 3.5.2), where electrochemical gradients generated by the Na^+,K^+-ATPase pumps are utilized in Na^+/H^+ exchange reactions and in Na^+/HCO_3^- cotransport.

Carbonic anhydrase inhibitors are administered parenterally because the drugs do not penetrate the cornea at concentrations high enough to inhibit the enzyme. The commonly used drugs are only effective in lowering IOP when 99% of the CA activity in the ciliary body is inhibited (Friedland and Maren, 1984). However, a newly developed CA inhibitor, trifluormethazolamide, appears to penetrate the cornea effectively and lower IOP in rabbits. This drug could be potentially useful in humans as well (Maren *et al.*, 1983).

4.2.2. Blood–Aqueous Barrier

First considered as a concept and later characterized as a structural entity, the presence of an anatomic barrier in the ciliary epithelium that is impermeable to the passage of macromolecules, mainly proteins, is now firmly established. The blood–aqueous barrier consists of tight junctions (zonulae occludentes) that appear as fusion points, sealing the cell membranes between adjacent nonpigmented ciliary epithelial cells (Raviola, 1977). They are located specifically at the apical end of the lateral plasma membranes of these cells. Freeze–fracture studies of glutaraldehyde-fixed ciliary epithelia suggest that the barrier is not merely a simple "system of pores" but rather consists structurally of a series of fibrillar networks that form more or less discontinuous ridges (Hirsch *et al.*, 1985). This type of structural organization could effectively block the passage of large molecules while at the same time allowing a flux of ions and low-molecular-weight solutes through channels within the junction. A breakdown in the barrier by injury, trauma, paracentesis, or other causes leads to a large influx of protein into the aqueous humor, producing what is usually termed "secondary aqueous." The protein concentration may increase by as much as 100-fold under these conditions (Riley, 1983).

4.2.3. Lipids

The iris–ciliary body is not a lipid-storing tissue, although it has a very active lipid metabolism. Analytical data are available on the lipid composition of rabbit (Abdel-Latif and Smith, 1979), bovine, and human iris (Anderson *et al.*, 1970; Culp *et al.*, 1970a). Phospholipids, including sphingomyelin, are the major lipids in all species examined and in bovine iris account for about 69% of the total lipids extracted (Culp *et al.*, 1970a). Cholesterol and neutral lipids comprise about 14% and 5%, respectively, of the total. The major phospholipids are phosphatidylcholine (PC), phosphatidylethanolamine (PE), phosphatidylserine (PS), and sphingomyelin (SPH), which together account for 85–90% of the phosphoglycerides in bovine (Anderson *et al.*, 1970) and rabbit iris (Table 4.6). Other phosphoglycerides are also present, accounting for about 7% of the total. Neutral lipids (di- and triacylglycerol) have been found in bovine (Anderson *et al.*, 1970) as well as rabbit iris (Abdel-Latif and Smith, 1979).

The fatty acid composition of iris lipids from various sources (rabbit, bovine, and human) is generally similar. All are esterified and consist mainly of long-chain saturated (palmitic, 16:0, and stearic, 18:0) and unsaturated (oleic, 18:1, linoleic, 18:2, and arachidonic, 20:4) fatty acids. The most important fatty acid physiologically is arachidonic acid, which is concentrated mainly in PE, PS, and their lyso derivatives (Abdel-Latif and Smith, 1979). Substantial amounts of arachidonic acid are also present in phosphatidylinositol (PI), a major precursor of eicosanoids produced by the receptor-stimulated breakdown of PIP_2.

The phospholipids and neutral lipids of the iris are in a state of rapid metabolic turnover (Culp *et al.*, 1970b; Abdel-Latif and Smith, 1982). Radioactive precursors such as acetate, fatty acids, glycerol, phosphate, choline, and arachidonic acid are incorporated into lipids both *in vivo* and *in vitro*. The turnover of arachidonic acid plays a key role in

Table 4.6. Phospholipid Composition of Rabbit Iris[a]

Phospholipid	Lipid phosphorus	
	μg/g tissue	Percent total
Phosphoglycerides		
Phosphatidylcholine[b]	6.34	46.41
Phosphatidylethanolamine[b]	3.15	23.10
Phosphatidylserine	1.03	7.54
Phosphatidylinositol	0.54	3.95
Lysophosphoglycerides		
Lysophosphatidylcholine	0.16	1.17
Lysophosphatidylethanolamine	0.20	1.46
Sphingomyelin	1.05	7.70
Other phosphoglycerides[c]	0.16–0.29	6–8

[a]Neutral chloroform–methanol extracts, adapted from the data of Abdel-Latif and Smith (1979).
[b]Includes plasmalogens.
[c]These include phosphatidic acid, cardiolipin, and di- and triphosphoinositides.

the production of prostaglandins and other eicosanoids in the iris–ciliary body. *In vitro* studies on rabbit iris show that the major part of labeled arachidonic acid is incorporated into neutral lipids, mainly triacylglycerol, as well as phosphatidylcholine (PC) (Abdel-Latif and Smith, 1982). Although PE and lyso-PE have the highest content of arachidonic acid, this fatty acid appears to be more rapidly metabolized than PC, a phospholipid that contains only about 13% arachidonic acid.

4.2.4. Eicosanoids: Prostaglandins and Leukotrienes

Eicosanoids have diverse and potent biological effects; they are directly involved in inflammatory reactions resulting from ocular trauma and, in addition, are thought to play a role in controlling intraocular pressure. Hence, eicosanoid metabolism in the anterior segment of the eye is an area of active investigation. Some of the known biological effects of prostaglandins and leukotrienes are summarized in Table 4.7.

The biochemical and enzymatic pathways of eicosanoid production from arachidonic acid (AA) in nonocular tissues are described in Chapter 1, Section 1.5. Although AA is present at barely detectable levels in the free form in the rabbit iris–ciliary body, it is rapidly released from esterified sites in phospholipids by a variety of stimuli such as trauma, irritation, or even simple mechanical agitation (Abdel-Latif, 1983). A major source of AA in rabbit iris sphincter muscle is the receptor-stimulated breakdown of phosphatidylinositol 4,5,-bisphosphate (PIP_2) induced by substance P (Yousufzai *et al.*, 1986). The AA released through this pathway is rapidly oxidized to prostaglandin PGE_2 and possibly other products, although they were not measured. This implicates the cyclooxygenase pathway in AA oxidation in rabbit sphincter muscle. In bovine ciliary body microsomes as well, cyclooxygenase is the major pathway of AA oxidation (Schwartzman *et al.*, 1987). It is about 30 times more active than the cytochrome P450-linked monooxygenase system.

The cyclooxygenase pathway of AA metabolism was in fact the first to be detected in

Table 4.7. Biological Responses to Prostaglandins and Leukotrienes

Substance	Effect
PGI_2	Vasodilation
PGE_1	Stimulates adenylate cyclase
TXA_2	Vasoconstriction
PGE_2	Miosis
$PGF_{2\alpha}$	Ocular hypertension; inflammation[a]
	Lowers IOP (low doses)[a]
	Increases aqueous protein by breakdown in blood–aqueous barrier
5, 12-HETE	Chemotactic for polymorphonuclear
12-HETE	leukocytes (PMNs)
Leukotrienes C_4, D_4, E_4	Anaphylaxis

[a]In rhesus monkeys and cats (Stern and Bito, 1982).

early studies on rabbit iris–ciliary body. Incubation of minced tissue or microsomes isolated from intact tissue with labeled AA for 15 to 30 min leads to the production of measurable amounts of cyclooxygenase products: $PGF_{2\alpha}$, PGD_2, TXB_2, PGI_2, and 6-keto-$PGF_{1\alpha}$ (Kass and Holmberg, 1979; Bhattacherjee *et al.*, 1979; Kass *et al.*, 1981). Tissues from inflamed eyes generate considerably greater amounts than control eyes, confirming the commonly held view that prostaglandin production is enhanced in inflammatory conditions. Both indomethacin (Kass and Holmberg, 1979; Bhattacherjee *et al.*, 1979) and aspirin (Abdel-Latif and Smith, 1982) inhibit cyclooxygenase activity and completely abolish the synthesis of prostaglandins. Imidazole inhibits the synthesis of TXB_2 but has no effect on other eicosanoids produced by the iris–ciliary body.

Steroids appear to exert their antiinflammatory effect by indirectly inhibiting phospholipase A_2, thus limiting the release of free AA. Steroids such as dexamethasone combine with cytosolic receptor protein, forming a complex that is then translocated to the nucleus. The hormone is thought to induce the synthesis of a phospholipase A_2 inhibitor, and at low substrate levels of AA, prostaglandin synthesis is effectively suppressed in the presence of glucocorticoids. As in other tissues, translocation of cytosolic glucocorticoid receptor to the nucleus has also been demonstrated in the rabbit iris–ciliary body and in adjacent corneoscleral tissue after topical application of labeled dexamethasone (Southren *et al.*, 1983a). Eicosanoid production was not measured, but by analogy with other tissues it is reasonable to assume that their production was suppressed due to inhibition of phospholipase A_2.

Inhibitors of eicosanoid production are thus of two types: those that prevent the release of AA (corticosteroids) and those that block the cyclooxygenase pathway of arachidonate oxidation (nonsteroidal antiinflammatory drugs such as indomethacin). Some of these effects are summarized in Table 4.8. Prostaglandins may play a role in mediating the disruption of the blood–aqueous barrier in the rabbit eye (Hoyng *et al.*, 1986), and although paracentesis-induced breakdown of the barrier can be only partially prevented by corticosteroids, it is almost completely blocked by indomethacin (van Delft *et al.*, 1987). Since both of these agents block eicosanoid production (by different mecha-

Table 4.8. Inhibitors of Prostaglandin Synthesis in Rabbit Iris–Ciliary Body

Drug	Enzyme inhibited
Aspirin	Cyclooxygenase
Indomethacin[a]	Cyclooxygenase
Imidazole	Thromboxane synthetase[b]
BW 755	Cyclooxygenase and lipoxygenase
Corticosteroids	Phospholipase A_2

[a]In some tissues indomethacin may also stimulate the synthesis of lipoxygenase products that are chemotactic for PMNs.
[b]Inhibition of TXB_2 synthesis may be through this enzymatic pathway (Kass and Holmberg, 1979).

nisms), these observations point to the complex biological effects of eicosanoids in ocular inflammation.

The foregoing discussion has considered mainly the cyclooxygenase pathway of AA metabolism; however, iris–ciliary body also oxidizes this fatty acid via the lipoxygenase pathway (see Figs. 1.10 and 1.12). The major product formed is 12-HETE, with small amounts of 5- and 5,12-diHETE also being produced in minced iris tissue from monkey, dog, cat, guinea pig, rat, and Dutch (pigmented) rabbits (Williams *et al.*, 1983). Somewhat unexpectedly, this pathway appears to be absent in the commonly used albino rabbit. The drug BW 755 inhibits both cyclooxygenase and lipoxygenase pathways, and although it resembles the antiinflammatory actions of the corticosteroids, the mechanisms of suppressing eicosanoid production are different.

The inflammatory responses of the anterior segment are not confined to the iris–ciliary body, since conjunctiva, anterior uvea, and eyelids also possess very active cyclooxygenase and lipoxygenase pathways for AA oxidation (Kass and Holmberg, 1979; Bhattacherjee *et al.*, 1979; Kulkarni and Srinivasan, 1983; Kulkarni *et al.*, 1984a,b, 1987). Labeled arachidonic acid is metabolized by these tissues *in vitro* through both pathways in a variety of animal species and in humans as well. The lipoxygenase products, 5- and 12-HETE as well as certain leukotrienes are chemotactic agents for polymorphonuclear leukocytes, and mixtures of leukotrienes C_4 and D_4 are known to be the slow-reacting substance (SRS) of guinea pig anaphylaxis.

4.2.5. Drug-Metabolizing Systems and Glutathione

The microsomal cytochrome P450 hemoproteins in liver and in nonhepatic tissues serve as terminal acceptors for the NADPH-dependent mixed-function oxidases (monooxygenases) (see Chapter 1, Section 1.6). This system catalyzes the oxidation of endogenous substrates (e.g., fatty acids and steroids) and foreign substances (xenobiotics) by introducing one atom of oxygen from molecular oxygen into a substrate while reducing the other atom to water. In this initial detoxification, considered as phase I in drug metabolism (see review by Shichi, 1984), the introduction of oxygen atoms to form hydroxylated derivatives converts drugs and other substances into products that are often far more reactive than the parent compound and hence potentially more toxic to the cell. Many of the electrophilic derivatives generated by the cytochrome P450 mixed-function oxidases then undergo a second stage of detoxification through at least two conjugation systems. One is the formation of glucuronides by UDP-glucuronyl transferase, an enzyme localized in the microsomes of liver and several other tissues including the ciliary body (Das and Shichi, 1981). A second important pathway is conjugation of electrophilic drugs (or their derivatives) with GSH (glutathione), reactions catalyzed by glutathione S-transferases. These enzymes comprise a multi-gene family of proteins whose tissue distribution, subunit composition, catalytic mechanism, and gene structure have been studied extensively (see review by Pickett and Lu, 1989). This pathway, conjugation with GSH and further metabolism to mercapturic acids, plays an important role in drug detoxification in the ciliary body, as described below.

The cytochrome P450 monooxygenase system (phase I reactions), when assayed as aryl hydrocarbon hydroxylase activity, is 20 times more active in homogenates of bovine ciliary body than in any other ocular tissue examined, including the iris (Das and Shichi, 1981; Shichi, 1984). This monooxygenase is present endogenously, but its activity is also inducible and appears to be under the same genetic regulation in the eye as in liver (Shichi *et al.*, 1975). A two- to tenfold increase in aryl hydrocarbon hydroxylase activity in several ocular tissues has been noted after injection of polycyclic hydrocarbons (β-naphthoflavone and 3-methylcholanthrene) into C57BL/6N mice. Two additional monooxygenase activities, 7-ethoxycoumarin-*o*-deethylase and benzphetamine demethylase, are present in bovine ciliary body microsomes at activities similar to those in retinal pigment epithelium (Schwartzman *et al.*, 1987). The complete cytochrome P450 drug-metabolizing system, including two of the monooxygenases described above, has been demonstrated in microsomal preparations isolated from ciliary processes of bovine ciliary body (Kishida *et al.*, 1986). The preparations showed characteristic difference spectra for both the cytochrome P450 and b_5 electron transport components; the calculated content of cytochrome P450 was 32 pmol/mg protein, which is about 3–4% of that found in rat liver. Two electron transfer systems, NADH– and NADPH–cytochrome c reductase, had activities of 268 and 18 nmol/min per mg protein, respectively.

Phase II detoxification of electrophilic xenobiotics is also extremely active in the ciliary body. A schematic diagram showing the initial conjugation with GSH, catalyzed by glutathione S-transferases, and subsequent enzymatic steps leading to the formation of mercapturic acid is shown in Fig. 4.2. Glutathione S-transferase activities have been detected in microsomal as well as cytosolic fractions of bovine ciliary body (Das and Shichi, 1981). In later studies, an isoenzyme with a pI of 5.8 was isolated and purified from 100,000 × *g* supernatants of bovine ciliary body homogenates (Shichi and O'Meara, 1986). This anionic isoenzyme accounts for about 25% of the total soluble enzyme in the tissue. Its substrate specificity, pI, amino acid composition, and peptide maps are similar, if not identical, to the μ class of glutathione S-transferases characterized in lens. Other studies have identified five forms of glutathione S-transferase belonging to the three major enzyme classes (α, μ, and π); these enzymes were found in carefully dissected preparations of bovine iris and ciliary body (Ahmad *et al.*, 1989). However, for their isolation and purification, combined iris–ciliary body preparations were used. The three classes of glutathione S-transferases were characterized by isoelectric focusing and Western blotting.

Three other enzymes in mercapturic acid synthesis, cystine aminopeptidase, N-acetyltransferase, and γ-glutamyl transpeptidase, have been localized to the microsomal fractions of bovine ciliary body (Das and Shichi, 1979, 1981). The γ-glutamyl transpeptidase

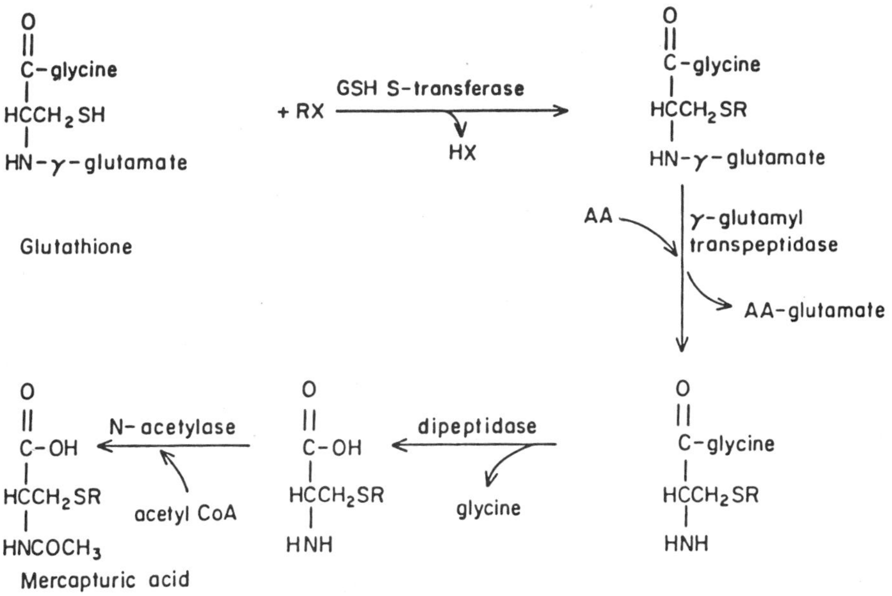

O
||
C–glycine
|
$HCCH_2SH$ + RX —GSH S-transferase→ (HX)
|
HN–γ–glutamate

Glutathione

O
||
C–glycine
|
$HCCH_2SR$
|
HN–γ–glutamate

AA → γ-glutamyl transpeptidase → AA-glutamate

O
||
C–glycine
|
$HCCH_2SR$
|
HNH

—dipeptidase→ (glycine)

O
||
C–OH
|
$HCCH_2SR$
|
HNH

—N-acetylase→ (acetyl CoA)

O
||
C–OH
|
$HCCH_2SR$
|
$HNCOCH_3$
Mercapturic acid

Figure 4.2. Proposed mechanism for the conjugation of xenobiotics and other substances with glutathione and the production of mercapturic acids. RX is a hydrophobic compound having an electrophilic group; AA is an acceptor amino acid. (From Awasthi *et al.*, 1980).

was purified 32-fold and found to be a glycoprotein with a pH optimum of 8.2 (Das and Shichi, 1979). The precise cellular localization of this enzyme has been achieved by the successful culturing of separate populations of pigmented and nonpigmented cells obtained after pronase digestion of bovine ciliary processes (Ng *et al.*, 1988). These studies showed that γ-glutamyl transpeptidase activity is associated almost exclusively with the nonpigmented cells of the ciliary epithelium.

The ciliary body has the potential of generating superoxide anion and hydroxyl radicals by one-electron reductions of oxygen (see Chapter 1, Section 1.4), but it is reasonable to assume that this tissue has effective mechanisms for the detoxification of active oxygen species. Data reported more than a decade ago on the rabbit eye showed that, compared to all other ocular tissues, the iris and ciliary body had the highest activities of catalase, superoxide dismutase, and glutathione (GSH) peroxidase (Bhuyan and Bhuyan, 1977, 1978). More recently, both of the two known types of GSH peroxidase activity have been characterized in bovine iris–ciliary body: GSH peroxidase I, the selenium-dependent enzyme, and GSH peroxidase II, the selenium-independent enzyme expressed by the glutathione S-transferase enzymes (Ahmad *et al.*, 1989). Examination of separately cultured pigmented and nonpigmented ciliary epithelial cells showed that catalase and superoxide dismutase are located predominantly in the nonpigmented cells (Ng *et al.*, 1988). Similarly, GSH peroxidase I was found to be significantly more active in nonpigmented than in pigmented cells, but selenium-independent GSH peroxidase II was more active in pigmented cells.

Selenium-dependent GSH peroxidase I is coupled to glutathione reductase in the majority of cells and tissues, including the ciliary body. Two glutathione reductases, GR-I and GR-II, have been isolated and purified to apparent homogeneity from bovine ciliary body (Ng and Shichi, 1986). The two enzymes have essentially similar amino acid composition and peptide patterns, but they differ in molecular weight and structure; GR-I has a M_r of about 140,000 and is composed of two identical subunits, whereas GR-II exists as aggregates of M_r over 670,000. Calculations of the activities and turnover numbers of these enzymes show that they play a major role in catalyzing the rapid reduction of oxidized glutathione (GSSG) produced in the tissue by the detoxification of peroxides. One bovine ciliary body has the potential of reducing 0.25 μmol of GSSG in less than 2 s.

There is little doubt that glutathione plays an important role in detoxification in the ciliary body, and a novel technique involving perfusion of the bovine eye provides a potentially useful tool for examining this function more precisely (Kishida *et al.*, 1985). Depletion of the endogenous stores of glutathione with *t*-butylhydroperoxide and the addition of nitrofurantoin (a glutathione reductase inhibitor) to the perfusion medium causes a marked decrease in glutathione levels, particularly reduced glutathione. There is strong evidence in the corneal endothelium (Chapter 3, Section 3.1.7.3) and in the aqueous humor (Section 4.1.2) for a redox cycle involving glutathione that protects these ocular tissues from oxidative stress. A similar mechanism may be operating in the ciliary body as well.

4.2.6. Cholinergic and Adrenergic Receptors

Nerve impulses are communicated across synapses by chemical transmitters (neurotransmitters), small diffusible molecules such as acetylcholine (ACh) and norepinephrine. The iris–ciliary body contains both cholinergic (parasympathetic) and adrenergic (sympathetic) innervation; these neurons are responsible for the synthesis, storage, and release of neurotransmitters. They also contain the enzymes necessary for their inactivation. A detailed description of neurotransmitters in the iris–ciliary body is beyond the scope of this chapter, but both acetylcholine and norepinephrine have been identified and studied in several animal species and in humans (Abdel-Latif, 1983). More recently, serotonin has also been detected in the iris–ciliary body of frog, pigeon, goldfish, and guinea pig by HPLC analysis (Osborne and Tobin, 1987), and studies with labeled serotonin agonists have revealed 5-HT_{1A} serotonin receptors in rabbit iris–ciliary body (Mallorga and Sugrue, 1987).

The receptors in human iris sphincter muscle are of the muscarinic cholinergic type (Hutchins and Hollyfield, 1984), whereas those of the iris dilator muscle are mainly the noradrenergic type, although muscarinic cholinergic receptors are probably also present. The emphasis in this section is on the role of cholinergic and adrenergic receptors in transmembrane signaling and the production of second messengers. Pharmacological activities and responses are considered only in those instances where they are correlated with biochemical data on the production of second messengers. For example, contraction of the rabbit iris sphincter muscle *in vitro* is mediated by substance P, an undecapeptide neurotransmitter known for many years as a strong miotic (Yousufzai *et al.*, 1986). Sphincter muscle contraction induced by substance P is accompanied by receptor-mediated breakdown of phosphatidylinositol 4,5-bisphosphate (PIP_2), leading to the rapid formation of two second messengers: inositol trisphosphate (IP_3) and diacylglycerol (DG). Since IP_3 plays a major role in the mobilization of Ca^{2+} from bound sites in the endoplasmic reticulum, this mechanism has been invoked as an explanation for the striking pharmacological actions of substance P on the iris sphincter muscle.

Adrenergic receptors (both α and β) in the ciliary epithelium, with their positive and negative coupling to the adenylate cyclase system, play a central role in aqueous humor production and the control of IOP. This dual control of the adenylate cyclase system is discussed in detail in Section 4.2.6.3. Activation of hormone receptor-coupled adenylate cyclase stimulates the phosphorylation of endogenous (Coca-Prados, 1985) and exogenous (Yoshimura *et al.*, 1987) proteins of rabbit and human ciliary processes. Protein phosphorylation in the ciliary processes is mediated by both cAMP-dependent protein kinases and protein kinase C, the latter shown by the enhanced phosphorylation of H1 histone by phorbol myristate acetate (PMA) (Mittag *et al.*, 1987c). Muscarinic receptors coupled to PIP_2 hydrolysis have also been demonstrated in human nonpigmented ciliary epithelial cells transfected with simian virus 40 (SV-40) (Wax and Coca-Prados, 1989). Thus, two membrane transduction systems responsible for the activation of cAMP-dependent protein kinase A and protein kinase C are present in the ciliary processes. Phosphorylation of endogenous substrates such as 57- to 58-kDa vimentin by these protein kinases and its role in water and electrolyte transport in the ciliary processes is discussed in Section 4.2.6.5.

4.2.6.1. Cholinergic Neurons and Receptors

The cholinergic neurons of the ciliary body have been recognized for many years through extensive pharmacological studies in a variety of animal species (for review see Abdel-Latif, 1983). A schematic representation of a cholinergic synapse is shown in Fig. 4.3. The end of the presynaptic axon is filled with synaptic vesicles containing acetylcholine (ACh). This neurotransmitter is synthesized near the presynaptic end of the axon from acetyl CoA and choline in a reaction catalyzed by choline acetyltransferase. It is then transported against a concentration gradient into the synaptic vesicles. An increase in intracellular Ca^{2+} triggers the release of ACh in quantum packets, each containing about 10^4 molecules, into the synaptic cleft. The contents of the synaptic vesicles are released

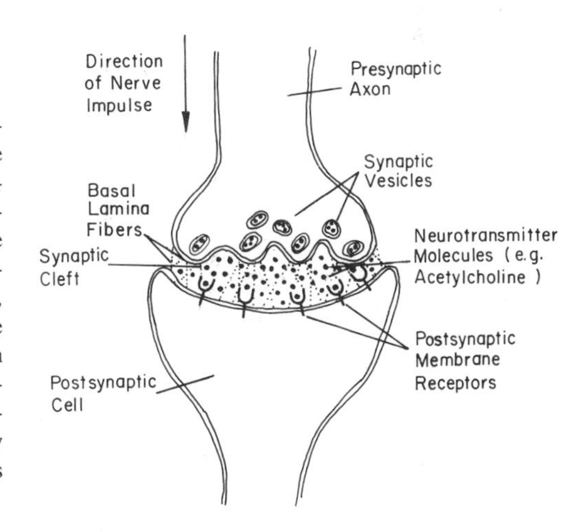

Figure 4.3. Schematic diagram of chemical transmission at a synapse. Neurotransmitter substance (e.g., acetylcholine) is synthesized in the presynaptic axon and stored in synaptic vesicles located at the end of the axon. A nerve impulse triggers the release of quantum packets of neurotransmitter molecules into the synaptic cleft, where they diffuse to the postsynaptic membrane and bind to specific receptor sites. This results in opening of cation gates in the postsynaptic membrane, depolarization, and excitation of the membrane. Removal of neurotransmitter substance by either uptake or enzymatic degradation restores the polarization of the postsynaptic membrane.

by exocytosis, and although the precise mechanism is not known, it is thought to be mediated by cAMP-dependent and/or Ca^{2+}- and calmodulin-dependent protein kinases (Darnell *et al.*, 1986). The vesicle contents (ACh) diffuse across the synaptic cleft and bind to specific postsynaptic membrane receptors, an interaction that produces a transient increase in the permeability of the postsynaptic membrane to both Na^+ and K^+. The immediate result is a large inward current of Na^+ and a smaller outward current of K^+. It is mainly the inward flow of Na^+ that causes depolarization of the postsynaptic membrane, which in turn generates an action potential in the adjacent axon. It is also known that increased K^+ permeability can result in a hyperpolarization response in this system. The ACh is rapidly hydrolyzed by a specific acetylcholine esterase, an extensively studied enzyme located in the postsynaptic cleft, where it is bound to a network of collagen-forming basal laminae. The hydrolysis of ACh causes an immediate repolarization of the membrane, which restores its excitability.

There are both muscarinic cholinergic and nicotinic cholinergic receptors in iris–ciliary body; the former can be blocked by atropine, and the latter by curare and α-bungarotoxin. As shown in studies with carbachol, muscarinic cholinergic receptors in the iris (as well as α_1-adrenergic receptors, discussed below) play an important role in the production of polyphosphoinositide-generated second messengers. Incubation of inositol-labeled rabbit iris smooth muscle with carbachol causes the production, within minutes, of inositol-labeled breakdown products of phosphatidylinositol 4,5-bisphosphate (PIP_2) (Akhtar and Abdel-Latif, 1984). These findings demonstrate the presence of the major membrane transduction system, receptor-stimulated polyphosphoinositide (PI) hydrolysis, in the iris smooth muscle.

4.2.6.2. Adrenergic Receptors

Norepinephrine, epinephrine, and dopamine are the principal neurotransmitters of adrenergic nerve fibers and smooth muscle junctions. These transmitters are synthesized from tyrosine, both locally at the sympathetic nerve terminals and systemically in the adrenal gland or within specific tissues by paracrine-like cells. The presence of both α- and β-adrenergic receptors in the iris–ciliary body of several animal species and in humans is well established.

In recent years, both subclasses of receptors have been divided into subtypes: numerical subscripts $_1$ and $_2$ are sometimes used to designate post- and presynaptic localization, respectively. In addition, the numerical subscripts are used to define the relative potencies or affinities of agonists and antagonists, regardless of their localization. Thus, four subtypes of adrenergic receptors are now recognized: α_1 or α_2 and β_1 or β_2 (Table 4.9). The use of radioactive ligands specific for each subtype has shown that in rabbit iris–ciliary body about 20–25% of the total adrenergic receptors are of the β_2 subtype, and the majority of α receptors are of the α_2 subtype (Mittag and Tormay, 1985a). In a novel technique for identifying the receptors in human iris–ciliary body, tissue obtained from postmortem eyes was homogenized and the plasma membranes isolated by isopycnic centrifugation (Wax and Molinoff, 1987). These preparations were used to study the potencies of a series of specific agonists and antagonists in their ability to displace the high-affinity binding of [^{125}I]iodopindolol. The results showed clearly that about 90% of the β-adrenergic receptors in human iris–ciliary body are of the β_2 subtype.

Table 4.9. Adrenergic Receptor Subtypes[a]

Receptor	Adrenergic agent: Agonist[b]	Adrenergic agent: Specific antagonist	Agonist-mediated effect
α_1	EPI ≥ NE > ISO Specific agonists: methoxamine, phenylephrine	Prazosin	Stimulate polyphosphoinositide (PI) turnover
α_2	EPI ≥ NE > ISO Specific agonists: clonidine, tramazoline	Yohimbine Rauwolscine	Inhibit adenylate cyclase
β_1	ISO > NE = EPI	Atenolol Betaxolol Metoprolol Practolol	Stimulate adenylate cyclase[c]
β_2	ISO > EPI > NE	Propanolol Butoxamine I-hydroxy benzylpindolol	Stimulate adenylate cyclase

[a]Modified from Lefkowitz *et al.* (1983).
[b]Abbreviations used: EPI, epinephrine; NE, norepinephrine; ISO, isoproterenol.
[c]Only in some systems but probably not in the ciliary body.

α-Adrenergic Receptors. Pharmacological and physiological studies on the role of phenylephrine, a specific α_1-adrenergic agonist, in the control of IOP or in the production of aqueous humor have been somewhat inconclusive. In fact, until recently a role for α-adrenergic receptors in aqueous humor formation by the ciliary epithelium had not been clearly established. However, functional α-adrenergic receptors, mainly of the α_2 subtype, have been characterized in membrane preparations of rabbit iris–ciliary body (Mittag *et al.*, 1985b). Moreover, it is now recognized that α-adrenergic receptors, specifically the α_2 subtype, play an important role in aqueous humor formation since they are linked in an inhibitory manner to the adenylate cyclase system (Mittag and Tormay, 1985b).

Epinephrine causes enhanced synthesis of prostaglandins from labeled arachidonic acid in rabbit iris–ciliary body *in vitro*, but only in the presence of relatively high concentrations of α-adrenergic agent (Bhattacherjee *et al.*, 1979). This stimulation was shown more convincingly using low pharmacological concentrations of phenylephrine, an α_1-adrenergic agonist (Engstrom and Dunham, 1982). The enhanced production of prostaglandins, particularly PGE_2 and $PGF_{2\alpha}$, from radiolabeled arachidonic acid in the presence of phenylephrine could be abolished by the α-adrenergic antagonist phenoxybenzamine. These and other findings raise the possibility that stimulation of α-adrenergic receptors in the iris–ciliary body augments prostaglandin synthesis either directly, by stimulation of the cyclooxygenase pathway of arachidonic acid oxidation, which seems unlikely, or indirectly, through agonist-stimulated breakdown of phosphatidylinositol 4,5-bisphosphate (PIP_2) and subsequent release of DG as a second messenger. The latter, after hydrolysis by DG lipase, would seem to be the the most probable physiological source of arachidonic acid, given the active receptor-stimulated phosphoinositide (PI) hydrolysis present in rabbit iris (Akhtar and Abdel-Latif, 1984).

β-Adrenergic Receptors and Adenylate Cyclase. The discovery of adenylate cyclase in rabbit ciliary processes and its activation by catecholamines provided the first evidence that this enzyme may play a role in aqueous humor formation (Waitzman and Woods, 1971). This initial observation has been confirmed independently in many laboratories. The modulation of aqueous humor production by adenylate cyclase was postulated from other lines of evidence; one was the observation that cholera toxin-induced activation of adenylate cyclase stimulates cAMP production and causes a drop in intraocular pressure (IOP) in the rabbit eye (Sears *et al.*, 1981; Gregory *et al.*, 1981). Further investigations in both rabbits (Nathanson, 1980) and humans (Nathanson, 1981) supported the view that the control of IOP was mediated through catecholamine-coupled β-adrenergic receptors, specifically of the β_2 subtype (see Fig. 1.5 and Table 4.9). Although these early observations are still valid, it must be borne in mind that other agonists such as vasoactive intestinal peptide (VIP) are also coupled to adenylate cyclase in the ciliary processes, as described below.

Subsequent investigations in a variety of nonocular tissues, and in the ciliary processes during the past few years, have led to the concept of dual control of adenylate cyclase mediated by stimulatory (β_2) and inhibitory (α_2) receptors (Mittag and Tormay, 1985b). Even more recently, aqueous humor production has been linked to phosphorylation of endogenous proteins by hormone-stimulated activation of protein kinases A and C through two major membrane transduction systems (see Section 4.2.6.5).

4.2.6.3. Adrenergic Agents, Aqueous Production, and Control of IOP

With the availability of highly specific radioligands, it is now possible to characterize more precisely than was possible before the various types of adrenergic receptors in the ciliary body. Moreover, the use of highly purified *in vitro* preparations of ciliary processes, uncontaminated by other cell types, allows both the identification and localization of the receptors that modulate adenylate cyclase activity in this tissue.

The *in vivo* localization of β-adrenergic receptors specifically in the ciliary processes (separate from those in the iris and uveal tissue) was first demonstrated in the rat by use of a fluorescent analogue of propranolol (Lahav *et al.*, 1978) and shortly afterward, *in vitro*, in purified homogenates of rabbit ciliary processes (Bromberg *et al.*, 1980). In the latter studies, it was shown that [^{125}I]hydroxybenzylpindolol [^{125}I]HYP, a β-adrenergic antagonist, appears in the same particulate fractions isolated by sucrose density gradient centrifugation as adenylate cyclase. The high-affinity binding sites for [^{125}I]HYP were tentatively identified as the physiologically active β-adrenergic receptors of the ciliary processes.

The adrenergic receptors in isolated human and rabbit ciliary processes are of the β_2 type, as shown by the stimulation of adenylate cyclase by isoproterenol, a β-adrenergic agonist, as well as by the 12-fold greater stimulation of the enzyme by epinephrine than by norepinephrine (Nathanson, 1980, 1981). The isoproterenol-stimulated adenylate cyclase activity is inhibited by β-adrenergic antagonists such as propranolol and timolol.

Studies with cholera toxin also provide evidence that adrenergic receptors, through modulation of adenylate cyclase activity, play a direct role in regulating aqueous humor formation and lowering IOP (Gregory *et al.*, 1981). As described in Chapter 1, Section 1.3.1.2., this toxin causes a "permanent" activation of adenylate cyclase by irreversible ribosylation of the α_s subunit of the G_s guanine nucleotide-binding regulatory protein. Intravitreal injection of cholera toxin decreases net aqueous flow and lowers IOP. However, these preparations may have been contaminated with bacterial endotoxin, which itself is a potent hypotensive agent. Regardless of this possible artifact, the production of cAMP in the ciliary processes is about seven times higher in cholera toxin-treated rabbits than in control animals, which is consistent with the role of cholera toxin in causing "permanent" activation of adenylate cyclase.

As the concept of dual adrenergic control of adenylate cyclase became firmly established in a variety of cell types, it was also found applicable to the ciliary body. It is now recognized that both α and β adrenergic receptors in the ciliary body modulate the activity of adenylate cyclase through the stimulatory and inhibitory G proteins G_s and G_i, respectively (see Chapter 1, Section 1.3.1 and Fig. 1.5). This has been shown in whole iris–ciliary body preparations (Mittag and Tormay, 1985b) and in intact excised ciliary processes of the rabbit (Bausher *et al.*, 1987). The use of radiolabeled adenine by Mittag and Tormay (1985b) allowed the measurement of *changes* in cAMP formation rather than its absolute concentration. In this way, not only increases but also—most importantly—decreases in adenylate cyclase activity can be estimated with considerable precision. These studies show that clonidine and α-methylnorepinephrine, specific α_2-adrenergic agonists, block the stimulation of adenylate cyclase by isoproterenol. Clonidine decreases the secretion of aqueous humor in ciliary processes, possibly by lowering the concentration of cAMP or possibly by other receptor-mediated mechanisms.

The isoproterenol-stimulated cAMP production in isolated rabbit ciliary processes, mediated by β_2-adrenergic receptors, is dose-dependent at agonist concentrations ranging from 0.1 to 1.0 μM (Bausher *et al.*, 1987). At higher concentrations, this stimulation is decreased, an effect that can be blocked by the α_2-adrenergic antagonist yohimbine. Cervical ganglionectomy did not abolish this effect, suggesting that the α_2 and β_2 receptors are postsynaptic. Moreover, these studies suggest interaction between β_2- and α_2-adrenergic receptors, a possibilty that was further explored using vasoactive intestinal peptide (VIP) (Bausher *et al.*, 1989). The α_2-adrenergic receptors in excised rabbit ciliary processes appear to be linked to the VIP receptors in such a way that stimulation of α_2-adrenergic receptors inhibits the VIP-induced stimulation of cAMP production. Thus, catecholamines may be regulating aqueous humor formation by inhibiting cAMP production via postjunctional α_2-adrenergic receptors.

Dose–response curves for catecholamine-stimulated cAMP production are best interpreted in terms of the relative affinities of specific agonists for either the β_2 stimulatory receptors linked to G_s protein or the α_2 inhibitory receptors linked to G_i protein. The negative coupling of the α_2-adrenergic receptors in rabbit ciliary processes appears to be mediated by an NaCl-dependent G protein (Kintz *et al.*, 1988). By blocking the β_2 receptors with propranolol, adenylate cyclase activity could be measured and evaluated as a function of α_2 occupancy. In these studies, α_2 selectivity was established in terms of the ability of α_2 agonists to inhibit the production of cAMP.

Despite the importance of cAMP in the regulation of aqueous humor production, there are few biochemical studies on the properties of the membrane-bound adenylate cyclase of the ciliary epithelium. Divalent cations are required for its activity, and Mg^{2+} specifically is essential for the formation of the MgATP substrate complex. However, other inorganic ions also influence the activity of this enzyme in particulate fractions of rabbit ciliary processes (Mittag *et al.*, 1987a). Investigations on the effects of four ions were carried out on three different activity states of adenylate cyclase: basal, G_s-stimulated through isoproterenol activation of β-adrenergic receptors, and forskolin-activated enzyme. The most striking effects were found for Mn^{2+}, which potentiated all three activity states of adenylate cyclase. Three other ions, Ca^{2+}, VO_3^-, and F^-, caused mixed responses, suggesting several possible modes of regulation for ciliary process adenylate cyclase. These findings differ substantially from the effects of Mn^{2+} in other cell types such as turkey erythrocytes, where this ion inhibits adenylate cyclase activity at the level of G_s protein. The stimulatory effect of Mn^{2+} on ciliary process adenylate cyclase appears to be localized at divalent cation-binding sites associated with the catalytic region of the enzyme; this effect is expressed by the extent of synergism between hormone-activated G_s and forskolin in stimulating adenylate cyclase activity (Mittag *et al.*, 1988). Whereas Mg^{2+} promotes the synergistic effect of the two modes of adenylate cyclase stimulation, the effect of Mn^{2+} is additive rather than synergistic.

4.2.6.4. Forskolin

Forskolin is a diterpene found in ethanolic extracts of the root of an Indian medicinal plant, *Coleus forskohlii*. It was first shown to be pharmacologically active as a vasodilator and smooth muscle relaxant, and further studies revealed that it is a potent activator of adenylate cyclase. Forskolin is a unique substance in that it acts directly at the catalytic site of the enzyme, rather than through the stimulatory G_s protein. A synergistic effect can be demonstrated in certain tissues, as well as in ciliary processes (as described above), when both forskolin and a β_2-adrenergic agonist are present together.

This unusual drug has become an important tool for elucidating the direct role of adenylate cyclase and cAMP on the production of aqueous humor and the control of IOP. It has a potent effect on lowering IOP and causing a reduction in the net aqueous flow in rabbits, monkeys, and humans (Caprioli and Sears, 1983; Smith *et al.*, 1984; Burstein *et al.*, 1984). Certain species, e.g., cynomolgus monkeys, may, however, build up a tolerance to this drug, since its hypotensive properties were essentially lost after 3 days of treatment (Lee *et al.*, 1984). Nevertheless, in rabbits, forskolin was found to be effective for 15 days (Caprioli *et al.*, 1984). A combination of acetazolamide and forskolin causes a reduction in IOP greater than either of the drugs alone, which is not surprising since they have different sites of action in the ciliary body; the former is an inhibitor of carbonic anhydrase, whereas the latter is an activator of the catalytic site of adenylate cyclase (Caprioli and Sears, 1984).

Forskolin has limited water solubility and is usually prepared as a suspension, with appropriate stabilizers, for topical application. A single dose of 1% forskolin administered to human volunteers causes a 35% reduction in the average IOP; there is a similar drop in net aqueous flow, but no change in outflow facility (Sears, 1985). The mechanism of this dramatic effect is thought to involve the ability of activated adenylate cyclase to alter the permeability of the nonpigmented epithelium to salt and water in such a way as to cause a net flow of fluid out of the posterior chamber and into the ciliary body. Other mechanisms have also been proposed to explain the hypotensive effects of forskolin; for example, by activation of presynaptic adenylate cyclase and raising of intracellular cAMP levels, it may enhance the secretion of norepinephrine from nerve endings in the iris–ciliary body (Jumblatt and North, 1986).

That cAMP may play a role in ion transport has long been suspected, but the molecular mechanism is unclear. This question has been addressed by measuring the short-circuit current of isolated rabbit iris–ciliary body using a specially designed chamber in which the composition of the bathing fluids on either the aqueous side or the blood side can be carefully controlled (Chu *et al.*, 1986). When added to the aqueous side, forskolin stimulates the short-circuit current by 37.5%. However, it has no effect when added to the blood side or if HCO_3^- is absent from the bathing media. These findings support the view that cAMP stimulates HCO_3^- transport in the ciliary processes. There is also evidence suggesting that the reduction in aqueous flow by forskolin in rabbits *in vivo* results from a small but significant increase in permeability of the blood–aqueous barrier (Bartels *et al.*, 1987). Moreover, forskolin potentiates the effectiveness of epinephrine in lowering IOP. This is best interpreted in terms of stimulation of adenylate cyclase by two mechanisms, one directly at the catalytic site by forskolin and the other through the β_2-receptor stimulation of G_s protein.

4.2.6.5. Protein Kinases

Phosphorylation of vimentin-type intermediate filaments has been demonstrated in both human and rabbit ciliary processes and in cultured ciliary epithelial cells (Coca-Prados, 1985). This 57- to 58-kDa protein was identified by both biochemical and immunologic methods, and its phosphorylation is stimulated severalfold by agents that activate adenylate cyclase. These include β_2-adrenergic agonists (isoproterenol and epinephrine) as well as agents such as cholera toxin and forskolin that act on the G_S regulatory protein or on the catalytic site of the enzyme, respectively (see Fig. 1.5). The enhanced phosphorylation can also be induced by 8-bromo-cAMP, a known activator of cAMP-dependent protein kinase(s). Thus, phosphorylation of vimentin is directly related to an elevation of intracellular levels of cAMP resulting from activation of adenylate cyclase. Other studies using soluble and particulate fractions from rabbit ciliary processes show a Ca^{2+}/calmodulin-dependent phosphorylation of vimentin that is especially active in the particulate fractions (Yoshimura *et al.*, 1989a). It was also noted that dephosphorylation, which can occur at incubation times greater than 1 min, may contribute substantially to the phosphorylation kinetics observed in crude fractions. The Ca^{2+}/calmodulin phosphorylation of vimentin appears to be more efficient than that of cAMP-dependent protein kinase(s), but the possibility of interaction between these systems must be kept in mind. Other endogenous proteins ranging in molecular size from about 18 to 205 kDa are also phosphorylated in a Ca^{2+}/calmodulin-dependent manner.

In addition to phosphorylation of endogenous proteins such as vimentin by cAMP-activated protein kinase(s), exogenous proteins are phosphorylated by a very active protein kinase C as well as other protein kinases characterized for the first time in both soluble and particulate fractions of rabbit ciliary processes (Yoshimura *et al.*, 1987).

Protein kinase C, a 90-kDa protein, was the major kinase detected in the ciliary processes; approximately one-quarter of the total enzyme activity of the tissue is membrane bound. There is compelling evidence in several nonocular tissues supporting the view that protein kinase C may regulate the activity of adenylate cyclase and modulate the level of cAMP. This may arise through interactions, or "crossovers," between the two major membrane transduction systems, the adenylate cyclase system and agonist-stimulated polyphosphoinositide (PI) breakdown. It is the latter that results in the production of diacylglycerol (DG), a second messenger that activates the phospholipid Ca^{2+}-dependent protein kinase C (see Fig. 1.6).

The activation of ciliary process protein kinase C is also coupled to phorbol esters such as phorbol myristate acetate (4β-PMA), and there is evidence suggesting that the phorbol ester receptor may be protein kinase C itself. Protein kinase C present in membrane fractions prepared from rabbit ciliary processes is activated by 4β-PMA and Ca^{2+} when H1 histone is used as the exogenous substrate (Mittag *et al.*, 1987b). However, this phorbol ester does not directly affect either adenylate cyclase or the cAMP-dependent protein kinase(s) in isolated ciliary process membranes. Only by addition of soluble protein kinase C from rat brain in the presence of agents that activate the α subunits of the stimulatory G_s protein (isoproterenol, VIP, and aluminum fluoride) could enhanced adenylate cyclase activation be demonstrated. The potentiation of adenylate cyclase activity by protein kinase C may also be explained in terms of phosphorylation of the α subunit of the inhibitory G_i protein; the resulting block of G_i would allow maximum activity for the stimulatory G_s protein. Whatever the basic mechanism, the observation that intravitreal injections of low doses of 4β-PMA cause a substantial drop in IOP in the rabbit eye suggests that the action is mediated by protein kinase C possibly acting through enhancement of adenylate cyclase responsiveness.

Muscarinic cholinergic stimulation of polyphosphoinositide (PI) hydrolysis has been demonstrated in human nonpigmented ciliary epithelial cells transfected with simian virus 40 (SV-40) (Wax and Coca-Prados, 1989). Muscarinic receptors of the M_3 subtype were identified. Also of interest was the observation that pertussis toxin reduced PI hydrolysis by 40% of control values. A 41-kDa polypeptide substrate of pertussis toxin bearing some resemblance the α subunit of G_i coupled to the inhibitory adenylase cyclase receptor was also isolated. The possible agonist-induced hydrolysis of PIP_2 with release of DG and activation of protein kinase C was not examined in this study.

Three other protein kinases (in addition to protein kinase C) have been characterized in both cytosolic and particulate membrane fractions of rabbit ciliary processes (Yoshimura *et al.*, 1987). All were identified and quantitated using H1 histone, casein, and myosin light chain as exogenous substrates. The three protein kinases (cAMP-dependent protein kinase type II, casein kinase type II, and protein kinase M) have considerably lower activity than protein kinase C. Neither cGMP-dependent nor calmodulin-dependent protein kinase activities could be detected in the preparations used for these studies. Elucidation of the endogenous substrates for these protein kinases could shed considerable light on their possible role in aqueous humor formation. Later studies by Yoshimura *et al.* (1989b) showed that the particulate fractions of rabbit ciliary processes contain a complete functional membrane transduction system for adenylate cyclase. Activation of receptors by β_2 agonists such as VIP in the presence of MgATP leads to the rapid phosphorylation of a series of endogenous proteins mediated by a membrane-bound cAMP-dependent protein kinase.

4.2.7. Desensitization

Desensitization, also known as down-regulation or adrenergic subsensitivity, is recognized clinically as a decrease in drug responsiveness after long-term treatment, either topically or systemically, with epinephrine or other adrenergic drugs. It is not limited to the eye but is in fact a widespread biological phenomenon. Recent studies on the two major receptor-coupled membrane transduction systems, hormone-stimulated adenylate cyclase and polyphosphoinositide (PI) hydrolysis, show that both are directly involved in desensitization in the iris–ciliary body. As discussed in Chapter 1, Section 1.3.4, two types of desensitization are now recognized: homologous and heterologous. The biochemical basis of desensitization is receptor inactivation through phosphorylation by protein kinases such as cAMP-dependent protein kinase, protein kinase C, and the recently discovered β-adrenergic receptor kinase (βARK). Agonist-occupied receptors are substrates for these protein kinases, and phosphorylation uncouples the receptors and impairs their interaction with the guanine regulatory proteins, the G proteins.

A loss of effectiveness of epinephrine (desensitization) in the rabbit eye is associated with a drug-induced decrease in the density of β-adrenergic receptors (Neufeld *et al.*, 1978), and even before the biochemical basis of desensitization was firmly established, a functional uncoupling of β-adrenergic receptors from their G proteins had been observed in membrane preparations of rabbit iris–ciliary body (Mittag and Tormay, 1981). Three successive topical doses of 2% epinephrine to the rabbit eye cause a marked decrease in the activation of adenylate cyclase by GTP or by the GTP analogue GppNHp.

Not only β-adrenergic receptors are affected; α_1-adrenergic-mediated phosphoinositide turnover can also become desensitized in the iris–ciliary body. The latter is manifested by decreased turnover of phosphatidylinositol 4,5-bisphosphate (PIP_2) in epinephrine-treated rabbits (Yousufzai and Abdel-Latif, 1987). Desensitization in this case was induced, as described above, by topical application of 2% epinephrine in three successive doses to the rabbit eye. The animals were sacrificed, and iris–ciliary bodies dissected out and incubated with ^{32}Pi to label the phosphoinositides. Comparison of desensitized tissues with untreated controls showed that in the epinephrine-sensitized iris–ciliary body there are (1) a decrease in the accumulation of second messengers IP_3 and DG, (2) decreased production of arachidonic acid and prostaglandins, and (3) attenuation of epinephrine-induced contraction of the iris muscle. By analogy with other cells (e,g, MF-2 smooth muscle cells), the mechanism of this adrenergic deactivation of phosphoinositide hydrolysis is probably through phosphorylation of the α_1 receptors by protein kinase C (Leeb-Lundberg *et al.*, 1987). In the rabbit iris, pilocarpine-induced subsensitivity to activation of muscarinic cholinergic receptors with carbachol is associated with decreased phosphorylation of myosin light chain (Yousufzai *et al.*, 1987). The desensitized tissue also showed decreases in carbachol-induced breakdown of PIP_2 as well as IP_3 accumulation and arachidonic acid release. These effects are specific to cholinergic receptors in the iris and are in accord with previous studies on α_1-adrenergic desensitization of the phosphoinositide response.

4.2.8. Atrial Natriuretic Factor

Atrial natriuretic factor (ANF) is a biologically active peptide synthesized in cardiac atrial tissue; when secreted into the blood, it has potent vasodilatory, natriuretic, and diuretic effects. The peptide has been cloned, sequenced, and synthesized (see Chapter 1, Section 1.3.3), and putative receptors have been identified in a variety of tissues. The receptors are coupled to the activation of membrane-associated guanylate cyclase, and they are the only hormone or neurotransmitter receptors known to be coupled to this enzyme (Nathanson, 1987). This activation is distinct from the muscarinic cholinergic stimulation of cytosolic guanylate cyclase, where the stimulation is directly related to the mobilization of Ca^{2+}.

The iris–ciliary body of rat contains 31 ng of immunoreactive atrial natriuretic protein (ANP)/g of tissue, concentrations that are severalfold higher than in rat plasma (Stone and Glembotski, 1986). The major form detected in extracts of iris–ciliary body by gel filtration is a 2.4-kDa peptide, similar in size to the peptide isolated from hypothalamus but smaller than the cardiac form. This difference in molecular size of ANP from the two tissues could also be demonstrated by reversed-phase high-pressure liquid chromatography. High-affinity receptors for synthetic ANF are localized on the plasma membrane of the basal infoldings of the "pigmented" ciliary process of the rat (Bianchi *et al.*, 1986).

Incubation of rat atrial natriuretic peptide 1–28 (rANP) with rabbit ciliary processes results in a 24–337% stimulation of basal guanylate cyclase activity (Nathanson, 1987). The rANP-stimulated guanylate cyclase activity is concentrated mainly in the ciliary processes and is selective for the complete peptide; the enzyme is only partially stimulated by rANP fragment 13–28, and the 1–11 fragment shows essentially no stimulatory effect. Other studies on rabbit ciliary processes have demonstrated maximum basal activity of guanylate cyclase in the presence of Mn^{2+} but only modest (about 30%) and somewhat variable stimulation by ANP obtained from commercial suppliers (Mittag *et al.*, 1987c).

A small inhibitory effect of ANP on forskolin-activated adenylate cyclase has been reported in isolated rabbit ciliary processes (Bianchi *et al.*, 1986), suggesting that ANF may be negatively coupled to adenylate cyclase. This, however, could not be confirmed in other *in vitro* studies carried out under similar, though not identical, experimental conditions (Mittag *et al.*, 1987c). The discrepancy may be explained by the finding that cGMP in the ciliary processes (as in many other tissues) regulates the activity of both the particulate phosphodiesterase and the soluble Ca^{2+}/calmodulin EGTA-inhibitable enzyme. The crude membrane preparations used in these studies contained not only the ANP receptor–guanylate cyclase complex but also cGMP-activated phosphodiesterase and the adenylate cyclase system. Therefore, stimulation of cGMP production by exogenous ANP would activate the particulate phosphodiesterase and lead to increased breakdown of cAMP. Whereas the preparations used by Bianchi *et al.* (1986) contained GTP and had no phosphodiesterase inhibitor, those used by Mittag *et al.* (1987c) had no GTP and contained 4 mM theophylline, a potent phosphodiesterase inhibitor. Under these conditions, no inbitory effect of ANP on the adenylate cyclase system could be detected in rabbit ciliary processes.

Although ANP appears to be effective in lowering intraocular pressure (IOP) in the rabbit, there are some discrepancies on the magnitude and persistence of the hypotensive effect. Intravitreal injection of 2–4 μg/eye of synthetic rat ANP (M_r 3,063) in the albino rabbit causes a small decrease in intraocular pressure after 16–24 h, but after 40 h the pressure returns to normal levels (Mittag *et al.*, 1987c). Other investigations (Nathanson, 1987) using rANP 1–28 show a dramatic decrease in IOP commencing 3 h after intravitreal injection in the rabbit and persisting for more than 48 h. Different doses administered and varying potencies of commercial preparations may explain some of these apparent discrepancies.

4.3. TRABECULAR MESHWORK

More than 80% of the aqueous humor leaves the eye through the chamber angle tissue into Schlemm's canal, and it is generally recognized that a rise in resistance through this outflow route is the principal cause of primary open-angle glaucoma (POAG). The aqueous humor first passes through the relatively large openings of the uveal network and afterward traverses the irregular spaces of the corneoscleral trabecular meshwork (TM) and the juxtacanalicular meshwork. From here the aqueous flows through the endothelial wall and Schlemm's canal, finally leaving the eye through the aqueous veins. The exact sites of flow resistance have not been established, but it is generally thought that the major site is in the juxtacanalicular tissue.

A large body of histochemical and ultrastructural information on the TM has been available for many years, but with the successful culturing of TM endothelial cells from monkey (Schachtschabel *et al.*, 1977), human (Polansky *et al.*, 1979), and other species, biochemical investigations are now feasible. Although most investigators use the term TM endothelial cells, the simpler term, trabecular cells, may be more appropriate because of their putative neural crest origin (R. C. Tripathi and B. J. Tripathi, personal communication). This section considers first the chemical composition of the trabecular meshwork, second, the biochemistry and metabolism of TM tissue and cultured trabecular cells, and finally, biochemical changes found in aging and in glaucoma.

4.3.1. Chemical Composition

The major chemical components of this structurally and functionally unique meshwork are type I collagen and elastin fibers. In addition, basement membrane-like structures consisting in part of type IV collagen have been identified in human TM by immunofluorescence and immunoperoxidase techniques (Rodrigues *et al.*, 1980). The TM is rich in glycosaminoglycans (GAGs) and also contains fibronectin and laminin, as described below.

4.3.1.1. Glycosaminoglycans and Proteoglycans

The classic work of Barany and co-workers (Barany, 1953/1954, 1956; Barany and Woodin, 1955) showing that perfusion of the aqueous outflow system with testicular hyaluronidase causes an increased outflow facility has been confirmed in many laboratories. This enzyme, an endoglycosidase, degrades hyaluronic acid (HA) and certain other GAGs such as chondroitin 4-(6)-sulfate to low-molecular-weight oligosaccharides. Numerous histochemical studies, mainly using Alcian blue or colloidal iron, combined with electron microscopy, have demonstrated GAGs in the TM in a variety of animal species and in humans (Zimmerman, 1957; Armaly and Wang, 1975; Grierson and Lee, 1975; Mizokami, 1977). They are localized on the surface of the TM cells lining Schlemm's canal and are in close association with the connective tissue elements and extracellular spaces of the TM. The combined use of ruthenium red stain and proteolytic digestion of

GAG–protein complexes suggests that in the cat, the connective tissue GAGs are complexed to protein, whereas those associated with endothelial cells are not (Richardson, 1982).

Because of the paucity of tissue in TM, it has been especially difficult to identify accurately the endogenous GAGs (see Chapter 1, Section 1.2.3.1). However, improved methodologies for their separation and the availability of highly purified enzymes of known specificity provide important tools for this purpose. The general procedure used for isolating the GAGs from the TM involves delipidation, proteolytic digestion, and precipitation of the crude ("protein-free") mixture of GAGs with ethanol. After electrophoresis on cellulose acetate membranes, Alcian blue staining usually reveals three broad bands. Alternatively, the GAGs may be resolved by gel filtration chromatography. Sequential degradation of the GAGs with the enzymes shown in Table 4.10 followed by zone electrophoresis allows a reasonably accurate characterization of the major classes of GAGs in the TM but does not provide information on the native-state proteoglycan composition.

Analyses of pooled TM from rabbit (Knepper *et al.*, 1981) and from human and monkey (Acott *et al.*, 1985) tissue suggest a broad spectrum of GAGs, with hyaluronic acid (HA) appearing as the most abundant species in human TM. Nevertheless, there is great variability in the relative proportions of individual GAGs in all species studied to date (Table 4.11). This is probably the consequence of differences in analytical techniques as well as species variation. The data on rabbit GAGs are qualitative only, but in other species, attempts have been made to measure the concentrations of the individual GAGs present in these complex mixtures. There are widely varying values for all of the components; for example, keratan sulfate (KS) comprises only 2% of the total GAG population

Table 4.10. Glycosaminoglycan-Degrading Enzymes

	Glycosaminoglycan[a]				
Enzyme	HA	CS	DS	KS	HS
Endoglycosidases					
Testicular hyaluronidase (EC 3.2.1.35)	+	+	+/−	−	−
Hyaluronate lyase[b] (EC 4.2.99.1)	+	−	−	−	−
Chondroitin AC lyase (EC 4.2.2.5)	+	+	−	−	−
Chondroitin ABC lyase (EC 4.2.2.4)	+	+	+	−	−
Heparitinase lyase	+	−	−	−	+
Keratanase	+	−	−	+	−
Exoglycosidases					
β-Galactosidase					
β-Glucuronidase					
β-N-Acetylhexosaminidase					

[a]The glycosaminoglycan that is degraded or depolymerized by the specific enzymes listed. Abbreviations used: HA, hyaluronic acid; CS, chondroitin 4- or 6-sulfate; DS, dermatan sulfate; KS, keratan sulfate; HS, heparatan sulfate.
[b]Isolated from *Streptomyces hyalurolyticus*.

Table 4.11. Endogenous Glycosaminoglycans of Trabecular Network

Glycosamino-glycan[a]	Calf[b] (% total GAG)	Rabbit[c] (Qualitative identification)	Monkey[d] (% total GAG)	Human[d] (% total GAG)
HA	15	+	9.9	24.2
CS	14		12.2	13.6
DS	43		13.1	20.9
KS	2	+	28.1	15.3
HS	24	+	16.4	17.8
Unidentified			19.8	9.5

[a]Abbreviations are as given in the footnote to Table 4.10.
[b]Adapted from Crean *et al.* (1986b).
[c]Adapted from Knepper *et al.* (1981).
[d]Adapted from Acott *et al.* (1985).

in calf TM, but in the monkey, it is the major species, accounting for 28% of the total GAGs.

Studies on hyaluronidase-infused rabbit eyes suggest that HA is an important component of the aqueous outflow resistance (Knepper *et al.*, 1984). This was shown by comparing the effect of testicular hyaluronidase (which partially degrades not only HA but also CS and possibly DS) with hyaluronate lyase (which degrades only HA, specifically and completely). The latter was found to be more effective than testicular hyaluronidase in degrading HA of the TM and in reducing aqueous outflow resistance. Hyaluronate lyase has a wider pH range than testicular hyaluronidase and, at physiological pH, completely removes all of the HA without affecting the other GAGs present in the TM. These findings suggest that HA, more than the other GAGs, may play a role in controlling outflow facility.

4.3.1.2. Fibronectin, Laminin, and Type IV Collagen

Fibronectin, laminin, and collagen have been identified in human TM by immunofluorescence and immunoperoxidase techniques (Rodrigues *et al.*, 1980; Polansky *et al.*, 1984; Sramek *et al.*, 1987). These characteristic components of the extracellular matrix are also produced by cultured TM cells, as discussed in Section 4.3.2.3. In a study of 13 postmortem eyes from elderly nonglaucomatous individuals, fibronectin was identified by means of a newly developed sensitive immunoassay utilizing avidin–biotinylated enzyme complex (ABC) for its detection (Floyd *et al.*, 1985). In these normotensive eyes, fibronectin, as observed in light microscopic studies of the stain, was localized mainly, if not entirely, in the endothelial region of Schlemm's canal and collecting channels as well as in the aqueous veins. Little or no fibronectin stain was detected in the trabecular meshwork. The distribution pattern of fibronectin appears to be similar to that of GAGs, although further work is needed to clarify this point.

Apart from *in situ* detection in postmortem tissue, laminin and type IV collagen have also been identified in extracellular matrices of cultured human TM by standard immunofluorescent techniques (Hernandez *et al.*, 1987). They are present only after the cells have reached confluence and display stable morphology. In addition to fibronectin, a positive identification of laminin and type IV collagen provides an important tool for identifying TM tissue in culture, since neither of these typical basement membrane components is present in fibroblasts originating from adjacent sclera.

4.3.1.3. Structural Proteins

Collagen and elastin are the principal structural components of the TM. Elastin has been characterized by electron microscopy but has not been analyzed chemically. Amino acid analyses of pepsin-solubilized samples of pooled human TM suggest that about 50–80% of the protein can be accounted for as type I collagen (Horstmann *et al.*, 1983). This is also true for bovine TM, which has, in addition, trace amounts of type V collagen (Grierson *et al.*, 1985a). However, immunofluorescent studies of human TM failed to reveal type I collagen, possibly because of the masking of antigenic determinants (Rodrigues *et al.*, 1980). Instead, electron-dense basement membrane-like material, consisting in part of type IV collagen, was abundantly distributed in the peripheral (cortical) portion of the trabecular beams adjacent to the endothelial cells as well as in the juxtacanalicular meshwork (Rodrigues *et al.*, 1980). Cattle eye TM does not have such a cortical zone, nor is there evidence for significant amounts of basement membrane in bovine TM. It is probably for this reason that type IV collagen is absent in this species (Grierson *et al.*, 1985b).

4.3.2. Endothelial Cells

A large portion of the outflow pathway is lined by endothelial cells of the TM. The establishment of successful lines of trabecular endothelial cells in culture from monkey (Schachtschabel *et al.*, 1977), human (Polansky *et al.*, 1979; Alvarado *et al.*, 1982; Schachtschabel *et al.*, 1982; Tripathi and Tripathi, 1982), cattle (Grierson *et al.*, 1985b), and calf eyes (Crean *et al.*, 1986a) now offers almost unlimited possibilities for studying the biochemistry and metabolism of this tissue. The cells from all species differentiate in culture and can be propagated through many passages, maintaining their morphological characteristics and functional properties. They are actively phagocytic, synthesize and secrete a variety of extracellular matrix components, and even respond to injury in a specific and physiological manner (Polansky *et al.*, 1984). Cultured TM cells provide a powerful tool for studying normal biochemistry as well as changes in aging and in primary open-angle glaucoma (POAG).

4.3.2.1. Cytoskeleton

The cytoskeletal elements of postmortem human TM cells (Gipson and Anderson, 1979) as well as cultured monkey (Ryder and Weinreb, 1986; Weinreb *et al.*, 1986) and bovine cells (Grierson *et al.*, 1986a) consist of a complex three-dimensional network of actin filaments, microtubules, and intermediate filaments. Of the three components of the cytoskeletal system, actin appears to be the most prominent and is thought to play a major role in maintaining both the shape and the intracellular organization of cultured monkey endothelial cells (Ryder and Weinreb, 1986; Ryder *et al.*, 1988).

Cytochalasin B is a potent actin-disrupting drug, and in cultured monkey TM cells at a concentration of 10^{-5} M, it causes dramatic changes not only in the structure of the actin filaments but also in the intracellular organization of the other two major cytoskeletal systems, the microtubules and the intermediate filaments (Weinreb *et al.*, 1986). Exposure to cytochalasin B also causes prominent changes in cell shape. In contrast, treatment of TM cells with colchicine, a microtubule-depolymerizing drug, causes only minimal changes in microtubule structure and has essentially no effect on either the organization of the cytoskeleton or on cell shape.

Despite some differences in the organization of actin filaments and microtubules as well as in cell shape, the overall appearance of human and monkey cultured TM cells in transmission electron microscopy is similar (Ryder *et al.*, 1988). In human TM cells, as in the monkey, actin filaments play a major role in maintaining the spatial relationship of the other two cytoskeletal components, microtubules and intermediate filaments. The latter were identified, using a mouse monoclonal antibody, as vimentin. The responses of human cultured TM cells to cytochalasin B and to colchicine are similar to those of monkey cells.

Independent immunocytochemical investigations using antibodies to four types of intermediate filaments show that two of them, vimentin and desmin, are present in human cultured TM cells (Iwamoto and Tamura, 1988). Although these studies did not show whether vimentin and desmin coexist in the same cell, the finding that both are present in human TM cells is an interesting observation. Certain similarities in structure and function may exist between TM cells and vascular smooth muscle cells, where these two intermediate filaments are also known to coexist.

Treatment of bovine meshwork cells with Triton X-100 or other nonionic detergent under carefully controlled conditions results in the extraction of the plasma membrane, cytosol, and many of the cellular organelles but leaves the nucleus, cytoskeletal filaments, and actin microfilaments preserved (Grierson *et al.*, 1986a). Immunohistochemical examination of detergent-extracted cells exposed to MgATP shows a dose-dependent decrease in surface area over a period of 1 h, accompanied by a disassembly of cytoskeletal stress fibers. Such a "contractile event" could be related to the phagocytic activities of these cells as well as to the maintenance of their normal shape and their potential to respond to hormonal agents.

The functions of actin, though not well understood, are thought to be mediated through adrenergic receptors, a view supported by the striking effects of epinephrine on human TM endothelial cells (Tripathi and Tripathi, 1984). Continuous exposure of cultured human TM to 10^{-5} M epinephrine causes marked impairement of cytokinetic cell movements and inhibition of phagocytic and mitotic activities within 3 days; after 4–5 days, considerable cell degeneration is observed. The effects of epinephrine are reversible if the drug is withdrawn within 3 days after exposure. These interesting observations are best interpreted in terms of recently identified β-adrenergic receptors, mainly of the β_2 subtype, in human TM (Jampel *et al.*, 1987). Studies using cultured human TM suggest that at least some of the receptors could be associated with the endothelial cells *in vivo*. These findings raise the possibility that β-adrenergic agents may modulate outflow facility by direct interaction with receptors in the trabecular meshwork. Other hormonal effects on TM cells are discussed in Section 4.3.2.5.

4.3.2.2. Carbohydrate Metabolism; Energy Production

Classical histochemical studies suggest that both glycolytic and oxidative pathways of glucose metabolism are present in TM or, more specifically, in the endothelial cells of the tissue. More recently, direct biochemical assays of glucose-metabolizing enzymes have been carried out using calf TM, which, because of its large size, provides sufficient tissue for a variety of enzymatic analyses. After homogenization, a high-speed (27,000 × *g*) supernatant was found to have active glycolytic activity, producing lactate from both glucose and fructose (Anderson *et al.*, 1980). In addition, mitochondria-enriched fractions oxidized three tricarboxylic acid substrates: succinate, malate, and glutamate. Calculations of the relative activities of oxidative and glycolytic pathways indicate that most of the ATP in the TM is generated from glycolysis.

Two glycolytic regulatory enzymes, hexokinase (Anderson *et al.*, 1984a) and phosphofructokinase (Anderson *et al.*, 1984b), have been detected in high-speed supernatants of calf TM homogenates. Isoenzymes of hexokinase as well as adult and fetal subtypes of hexokinase I were isolated, and some of their kinetic properties examined. Taking into account the endogenous glucose levels in aqueous humor, which may reflect the glucose concentration in TM, hexokinase may be under feedback inhibition by glucose 6-phosphate, a commonly used mechanism in many cells for controlling the entry of glucose into general cellular metabolism. Phosphofructokinase is another primary regulatory enzyme of glycolysis, whose activity in the TM, as in many other cell types, is regulated by NH_4^+ in a concentration-dependent manner. Given the relatively high concentration of NH_4^+ in aqueous humor (0.18 mM), this enzyme may under most conditions be operating at maximum velocity in the TM.

An active pentose phosphate pathway has been detected in calf TM homogenates, as shown by the preferential oxidation of [^{14}C-1]glucose over [^{14}C-6]glucose (Kahn *et al.*, 1983). Of the two key enzymes in this pathway, glucose 6-phosphate dehydrogenase (G6PD) and 6-phosphogluconate dehydrogenase (6-PGD), the G6PD has been examined in some detail (Nguyen *et al.*, 1986). Studies on the 450-fold purified enzyme show that it is activated by divalent cations and inhibited by sulfhydryl reagents such as *p*-chloromercuribenzoate. Its activity is regulated by the ratio of NADPH/NADP$^+$, being negligible when this ratio is as high as 10 and increasing as the ratio is lowered. The G6PD of calf TM displays an isoenzyme pattern similar to that of calf retina, and, as in most tissues, generation of NADPH through the pentose phosphate pathway may be important in providing reducing equivalents for the maintenance of GSH.

4.3.2.3. Metabolism of the Extracellular Matrix

The ability of cultured TM endothelial cells from monkey eyes to synthesize GAGs from labeled precursors was first reported more than a decade ago (Schachtschabel *et al.*, 1977). Primary cultures incubated with labeled sulfate or glucosamine synthesize mainly hyaluronic acid (HA), which accounts for 60–80% of the total GAGs produced. The remaining GAG was identified as chondroitin 4-sulfate. Cell lines cultured for as long as 3 years produce approximately the same amount of HA, but the sulfated GAG identified under these conditions appears to be dermatan sulfate (DS). Other studies comparing monkey TM cells in tissue culture and in organ culture have shown that although both systems produce HA and other sulfated GAGs, there are some differences in GAG profiles (Yue and Elvart, 1987). For example, the ratio of chondroitin 6- to chondroitin 4-sulfate is about four times higher in organ cultures of monkey TM than in tissue cultures, and greater amounts of HA are found in the medium of organ-cultured TM than in tissue cultures.

Organ explants (Rohen *et al.*, 1984) and cultured endothelial cells from human postmortem eyes (Polansky *et al.*, 1984) also produce GAGs from labeled precursors, the principal one being HA, as in the monkey eye. By using labeled acetate and sulfate to study the types of GAGs synthesized and their fate, it was found that most of the HA synthesized by the cells is released into the medium. In contrast, most of the sulfated GAGs synthesized are retained on the cell surface. In organ cultures of human TM, about 42% of the labeled GAGs are HA, and 60% of the GAGs released into the medium are HA (Polansky *et al.*, 1984). The ratio of "excreted" or "secreted" GAGs in organ culture is influenced by the presence (or absence) of serum in the incubating medium (Schachtschabel *et al.*, 1984). In general, HA, more than sulfated GAGs, appears to be preferentially "secreted" from the cell after *in vitro* synthesis. Whether this also occurs *in vivo*, into the trabecular spaces, remains to be established. It is interesting to note that exogenous HA added to human cultured TM cells grown in serum-free medium causes a marked stimulation in the synthesis of HA and other GAGs from labeled glucosamine precursor (Binninger *et al.*, 1987). About 90% of the newly synthesized GAGs are released into the medium under these conditions.

Glycosaminoglycan synthesis has also been studied in primary cultures of calf TM, and in this species the GAG profiles vary with the age and/or cell density of the culture (Crean *et al.*, 1986b). Consistent with previous observations, young cultures of calf TM produce mainly HA. However, as the cells reach confluency, sulfated GAGs become more prominent. Once the critical cell density is reached, the GAG profile more closely resembles the parent tissue. These findings imply that confluent cultures of TM more accurately reflect "native-state" metabolic activity than growth-phase cultures.

A novel *in vitro* technique for studying GAG synthesis in the anterior segment in general and in the TM in particular consists of infusion of labeled GAG precursors through the aqueous of a freshly enucleated rabbit eye (Knepper *et al.*, 1983). In the presence of radioactive sulfate and glucosamine, a nearly linear increase in GAG biosynthesis in the TM was observed for up to 2 h. Although the rate of synthesis in TM is relatively slow compared to cornea and iris–ciliary body, there is sufficient incorporation of radioactive precursors into the GAGs of TM to isolate and identify the individual components. This promising technique circumvents the potential artifacts of tissue culture and has many advantages over autoradiography for studying GAG metabolism in the anterior segment.

Corneoscleral explant organ culture in serum-free medium is another newly described system for studying GAG metabolism in human TM (Acott *et al.*, 1988). Postmortem eyes are preequilibrated in culture medium for 7 days prior to initiation of the experiments, and GAG metabolism is studied for periods of 1 to 2 weeks. The TM is structurally well preserved, and the pattern of the GAGs synthesized is distinctly different from that produced by either the corneal or the scleral tissue present in the explant. With labeled glucosamine as a precursor, the distribution and relative proportion of GAGs produced were: hyaluronic acid (22%), chondroitin sulfate (28%), dermatan sulfate (21%), and keratan sulfate (6%). This profile compares favorably with that of native human TM shown in Table 4.11 and does not show the excess proportion of hyaluronic acid produced by organ explants and cultures described above.

Although many studies have been reported on biosynthesis, very little is known about the degradation of GAGs in the TM. It is likely that, as in other tissues, lysosomal enzymes, both endoglycosidases and exoglycosidases, are responsible for their turnover in this tissue. As shown in Table 4.10, these enzymes isolated from a variety of sources are used exogenously for the selective degradation of GAGs, and the information gained is an important tool for their identification. However, only a few of these enzymes have been identified *in situ* in TM. Endogenous hyaluronidase activity (approximately 0.2 μg N-acetylglucosamine cleaved/h per mg protein) has been detected in cultured human trabecular cells (Polansky *et al.*, 1984). With maximum activity at pH 3.7, it is in all probability lysosomal in origin. This finding suggests a possible mechanism for regulating the metabolic turnover of GAGs in the TM, and especially in removing HA from outflow channels. Three exoglycosidases have been measured in cultures of human TM cells: N-acetylglucosaminidase, β-glucuronidase, and acid phosphatase. Their activities were 50, 0.3, and 5 μm/h per mg protein, respectively. Other lysosomal enzymes detected in cultured TM that appear to play a role in the phagocytic activities of these cells are discussed in Section 4.3.2.4.

The *in vitro* production of two other major structural components of TM, fibronectin and collagen, has been demonstrated in human (Worthen and Cleveland, 1982; Polansky *et al.*, 1984) and bovine cell cultures (Grierson *et al.*, 1985a). Use of a highly sensitive competitive radioimmunoassay to detect fibronectin shows that it is secreted continuously over a period of 96 h by cells subcultured up to seven passages (Worthen and Cleveland, 1982). Laminin is also detectable in some of these cultures. Bovine TM cells elaborate a fibronectin network similar to that produced by human cells (Grierson *et al.*, 1985a). Incubation of these cells with labeled proline, followed by extraction and precipitation with ammonium sulfate, has revealed α chains of type I collagen and traces of type V collagen. No other collagen types could be detected.

Human TM cells in culture are capable of producing significant amounts of tissue plasminogen activator (t-PA), especially when stimulated by thrombin or calcium ionophore (Polansky *et al.*, 1984). Quantitative measurements show that cultured TM synthesizes substantially more t-PA than vascular endothelial cells, the primary source of circulating t-PA (Shuman *et al.*, 1988). Moreover, cultured human TM cells produce only barely detectable levels of t-PA inhibitor. Hence, the balance between plasminogen activation and inhibition is strongly in favor of activation. Quantitative measurements of t-PA using a [^{125}I]fibrin-coated well assay show levels of 0.2 and 0.5 IU/mg protein in TM of dog and monkey, respectively (Park *et al.*, 1987). Similar values were found using an ELISA technique. Among the major roles postulated for t-PA in trabecular meshwork cells, two seem especially important: (1) as a fibrinolytic agent, it would prevent the formation of fibrin clots in the TM, and (2) it could activate the plasminogen–plasmin system, which is thought to play a role in modulating aqueous outflow resistance.

4.3.2.4. Phagocytosis

Phagocytic properties of the cells lining the outflow system have been recognized for many years. Acting as a "biological filter," the endothelial cells engulf and remove debris entering the TM from the aqueous humor, thus preventing blockage of the narrow outflow channels. One of the first quantitative assessments of the phagocytic activity of TM showed that cultured human TM cells are able to internalize latex spheres 1 μm in diameter as well as carmine particles added to the culture medium (Tripathi and Tripathi, 1982). Primary cultures of human TM have greater phagocytic capabilities than other cell types such as corneal endothelium, keratocytes, or scleral fibroblasts. Phagocytosis of cultured human TM is markedly inhibited in the presence of 10^{-5} M epinephrine (Tripathi and Tripathi, 1984).

Other studies on cultured human TM cells using radioactively labeled polystyrene latex microspheres have revealed that they are rapidly internalized and, once within the cells, occasionally fuse with organelles presumably belonging to the phagolysosomal system (Polansky *et al.*, 1984). The particles are positive for acid phosphatase, suggesting lysosomal involvement in the fusion. The use of a newly developed double fluorescence technique allows the quantitation and accurate estimation of the time sequence involved in adsorption and phagocytosis of particles in monkey TM cell cultures (Barak *et al.*, 1988). Considerable particle adsorption but relatively little phagocytosis takes place during the first hour of incubation with polystyrene beads, suggesting possible saturation of binding sites by the particles on the cell surfaces. After about 1 h, and continuing for 6 h, a linear internalization takes place accompanied by a "steady state" on the cell surface, possibly an equilibrium between adsorption and phagocytosis.

Cultured bovine TM cells phagocytize both biotic (γ-globulin-coated microspheres) and abiotic (uncoated microspheres) particles (Grierson *et al.*, 1986b). This lack of selectivity was also shown in other studies in which a wide variety of particles such as fibrin, erythrocytes, and gold particles were actively taken up by trabecular cells. Preconfluent cells are more avidly phagocytic than 2-week postconfluent cultures.

Phagocytic activity is a well-known property of normal trabecular endothelial cells and is more pronounced after certain types of injury (surgery or laser treatment). The phagocytic process itself affects certain biological properties of cultured bovine TM cells; for example, there is an increased uptake of tritiated glucosamine in cells undergoing active phagocytosis, and the fibronectin scaffolding is more extensive in cells preincubated with latex microspheres (Day *et al.*, 1986). There is also a marked increase in endogenous levels of lysosomal enzymes in cultured bovine TM cells following phagocytic challenge (Yue *et al.*, 1987). Histochemical evaluations and biochemical measurements of acid lipase, phosphatase, and acid esterase activities show statistically significant increases of lysosomal acid hydrolase activities following a 24-h incubation with either latex microspheres or zymosan particles.

4.3.2.5. Hormonal Effects and Prostaglandins

Steroid-induced glaucoma, usually defined as the increase in intraocular pressure (IOP) following either brief or prolonged steroid administration, has been recognized clinically for many years. Although the TM is thought to be the primary target of steroid action, other possible cellular sites in the anterior segment cannot be excluded (see review by Polansky and Weinreb, 1984). Glucocorticoid receptors have been characterized in a variety of target cells throughout the body. After entering the cell, glucocorticoid hormones exert their biological action by first binding to receptor molecules in the cytoplasm of target cells. The hormone–receptor complex is then translocated to the nucleus, where it alters the expression of specific genomic mRNAs. Only certain mRNA species are thought to be modulated by these steroid hormones; hence, only a limited number of proteins whose synthesis they direct would be affected by glucocorticoid hormones.

Functional high-affinity steroid receptors have been detected in human cultured TM cells (Weinreb *et al.*, 1981). The binding affinity of dexamethasone, as measured in fourth-passage cell cultures, is 5 nM. After 60 min incubation with excess labeled dexamethasone, nearly two-thirds of the specific binding is present in the nuclear fraction, and the remainder is in the cytosol. This is the characteristic pattern of intracellular translocation for glucocorticoid receptors in the majority of target cells. A similar sequence has been demonstrated in rabbit iris–ciliary body and adjacent corneoscleral tissue (Southren *et al.*, 1983a). As described above, glucocorticoids are thought to mediate their biological effects either by altering the transcriptional activity of genes coding for specific proteins or, alternatively, they may influence protein synthesis at the level of translation of specific mRNAs. This question has been addressed in a recent study undertaken to identify both the cellular and the secreted proteins whose synthesis is regulated by glucocorticoids (Partridge *et al.*, 1989). Exposure of human cultured TM cells to 10^{-7} M dexamethasone for 16 h results in a 60% inhibition of prostaglandin production, which would be expected, and in addition induces the expression of cellular proteins of M_r 35, 65, and 70 k, and secreted proteins of M_r 40, 90 and 100 k. The 70-kDa cellular protein is also expressed in scleral fibroblasts exposed to dexamethasone, leading these investigators to suggest that this specific 70-kDa cellular protein may be responsible for some of the known biological responses to glucocorticoids.

The antiinflammatory actions of steroids are accompanied by a myriad of physiological, immunologic, pharmacological, and biochemical responses throughout the body (see review by Polansky and Weinreb, 1984). Among the many possible causes of steroid-induced elevation in IOP, changes in cellular levels of prostaglandins and other eicosanoids may play an important role through their putative regulation of the aqueous outflow pathway. Eicosanoid metabolism in nonocular tissues is discussed in Chapter 1, Section 1.5, and in the first report of their production in human cultured TM, third-passage cells were shown to synthesize high levels of PGE_2 and moderate amounts of $PGF_{2\alpha}$ as well as prostacyclin 6-keto-$PGF_{1\alpha}$ (Weinreb *et al.*, 1983). Dexamethasone at concentrations as low as 10^{-8} M caused a marked inhibition of basal eicosanoid production. Later studies on eicosanoid production from labeled arachidonic acid showed that 80–90% of this eicosanoid precursor is incorporated into membrane lipids after a 2-h incubation (Gerritsen *et al.*, 1986). The principal eicosanoids produced were PGE_2 and 6-keto-$PGF_{1\alpha}$, the former being the predominant species. The conversion of labeled arachidonic acid to PGE_2 in human TM cells is remarkably high, ranging from 10–70% of the initial radioactivity added. This is in contrast to most other cell types, which generally oxidize only about 1–2% of arachidonic acid substrate to cyclooxygenase and/or lipoxygenase products. Production of the major prostaglandins in human cultured TM cells is inhibited by brief pretreatment with indomethacin or by prolonged (4- to 24-h) incubation with dexamethasone. The former is known to inhibit the cyclooxygenase pathway of AA metabolism, and dexamethasone under these conditions may be inducing the synthesis of phospholipase A_2 inhibitory protein(s).

Human cultured TM cells also metabolize labeled arachidonic acid through the lipoxygenase pathway, producing substantial amounts of 12- and 15-HETEs as well as leukotriene B_4 (Weinreb *et al.*, 1988). The production of lipoxygenase products was markedly inhibited by BW755c and partially inhibited by dexamethasone. These studies also confirmed that PGE_2 is the major cyclooxygenase product synthesized by human cultured TM cells, and $PGF_{2\alpha}$ appears to be derived from it through slow enzymatic reduction by 9-keto-reductase; the level of $PGF_{2\alpha}$ reaches a maximum only after 18 h of incubation.

A direct role for cyclic nucleotides in aqueous outflow facility has long been suspected but not firmly established. Perfusion of the anterior chamber of the monkey eye with cAMP analogues, whose action mimics that of isoproterenol or epinephrine, causes a marked increase in outflow facility (Neufeld, 1978). On the other hand, later investigations in the rabbit showed no temporal relationship between an increased production of cAMP following topical epinephrine and a drop in IOP (Boas *et al.*, 1981). The levels of cyclic nucleotide increased at first and then returned to base-line levels several hours before any drop in IOP was observed. In cultured human TM cells, epinephrine causes a marked inhibition of mitotic and phagocytic activity after 3 days' exposure, and after 5 days, irreversible degenerative changes are observed (Tripathi and Tripathi, 1984). The presence of β_2-adrenergic receptors in human TM endothelial cells (Jampel *et al.*, 1987) suggests a possible role for hormonal modulation of some of the biological activities of this tissue. For example, three β-adrenergic agonists were found to be especially effective in raising the basal levels of cAMP in cultured monkey TM cells: vasoactive intestinal peptide (VIP), L-isoproterenol, and prostaglandin E_1 (Koh and Yue, 1988). Other neuroendocrine agents such as glucagon, substance P, and dopamine appear to be inactive.

4.3.2.6. Detoxifying Activity and Glutathione

The TM is continually exposed to (at least) two potentially toxic substances that, in the absence of protective mechanisms, could cause serious membrane damage (see Chapter 1, Section 1.4). One is the free radical superoxide anion, which is undoubtedly generated as a byproduct of oxidative metabolism in the TM, and the other is H_2O_2, which could arise from two sources: (1) entry from the aqueous humor, where it is present at a concentration of approximately 25 μM, and (2) local production resulting from enzymatic dismutation of superoxide anion by superoxide dismutase (SOD). Several defense mechanisms are known to be present in most tissues, including the TM, that prevent potential damage from free radicals and H_2O_2.

The supernatants obtained from homogenates of calf TM after centrifugation at 10,000 × *g* contain high levels of both SOD and catalase; their specific activities are 0.184 and 0.884 U/mg tissue, respectively (Freedman *et al.*, 1985). The activities are similar in magnitude to those in iris and retina, and are considerably higher than in the lens. Catalase exists as a single enzyme, whereas three isoenzymes of SOD can be detected by gel electrophoresis of TM extracts. Two are the Cu- and Zn-containing cytosolic species, and the third, accounting for 10% to 20% of the total activity, has the properties of Mn-containing mitochondrial SOD.

In addition to catalase, a second major enzymatic pathway for the rapid decomposition of H_2O_2 is through the redox cycling of glutathione (GSH), as shown in Fig. 1.9. Two enzymes are involved in this coupled system: (1) selenium-dependent glutathione peroxidase, which catalyzes the reduction of H_2O_2 to water, producing oxidized glutathione (GSSH), and (2) glutathione reductase, which catalyzes the interconversion of oxidized and reduced forms of glutathione in the presence of NADPH or $NADP^+$, respectively. Studies on glutathione peroxidase in homogenates of calf TM have been reported using three substrates: H_2O_2, *tert*-butylhydroperoxide (tBHP), and glutathione (Scott *et al.*, 1984). The activities found for H_2O_2 and tBHP were 596 and 680 nmol/min per g tissue. Glutathione is present at a concentration of 0.4 μmol/g tissue in TM; all of it is in the reduced (GSH) form (Kahn *et al.*, 1983). Calculations of the H_2O_2-degrading potential of calf TM show that the glutathione peroxidase in this tissue has the capacity of catalyzing the decomposition of H_2O_2 over a wide range of concentrations, thus preventing any buildup under normal physiological conditions. Glutathione reductase has been characterized and purified 200-fold from homogenates of calf TM (Nguyen *et al.*, 1985). The K_m of this enzyme for GSSG is 78 μM, which is lower than the K_m of glutathione peroxidase for GSH by a factor of more than 100. The high affinity of glutathione reductase for GSSG could explain in part the common finding that in many tissues, including TM, glutathione is mainly in the reduced (GSH) form. Another factor maintaining the excess of the reduced form is the constant supply of NADPH through the pentose phosphate pathway (Kahn *et al.*, 1983; Nguyen *et al.*, 1986).

4.3.3. Biochemical Changes in Aging and in Glaucoma

Considerable attention has recently been focused on age-related biochemical changes in the TM and their possible role in the pathogenesis of primary open-angle glaucoma (POAG). This disorder, recognized clinically by marked elevations in IOP resulting from increased outflow resistance, is known to occur only in humans. It is more prevalent in older populations than in younger ones, but it has been difficult to study experimentally because there are no suitable animal models available. The majority of biochemical studies described below have utilized human postmortem trabeculum when available, as well as cells cultured from postmortem tissue.

Quantitative evaluations of TM cells in postmortem tissue from a series of patients ranging in age from newborn to 81 years show decreases of 58% and 47% in cellularity and cell numbers, respectively, with advancing age (Alvarado *et al.*, 1981). This loss of cellularity is not accompanied by changes in tissue area. The possibility of subtle changes in the metabolism of TM cells or in the chemical composition of the TM extracellular matrix resulting from a depleted cell population has been considered by Polansky *et al.* (1984) as a factor that could lead to altered functional capacity of the TM.

Several biochemical studies on age-related changes in the protein composition of human TM have been reported. Analyses of cyanogen bromide peptides of acetic acid-insoluble TM residues of human postmortem eyes show that up to 40 years of age, the major collagen of human TM is type I (Horstmann *et al.*, 1983). However, in individuals beyond the age of 40, the TM appears to be resistant to cyanogen bromide and pepsin digestion. Type IV collagen comprises about 1–8% of the total TM collagen, and there appear to be moderate increases with age. Large increases in the amount of protein-bound methionine oxidized to methionine sulfoxide were observed during certain stages of the aging process. Other studies of postmortem TM tissue from individuals ranging in age from 5 months to 87 years also suggest a possible age-related increase in type IV collagen (Millard *et al.*, 1987). There is a noticeable increase in staining intensity in older individuals of 140- and 160-kDa peptides after SDS-PAGE, bands that probably correspond to the α1 and α2 chains of type IV collagen. This would be consistent with the increased thickening of TM basal laminae found in older individuals. Another age-related change noted by Millard *et al.* (1987) was a decrease in staining intensity of both the 42-kDa G-actin band and a 58-kDa polypeptide. A recent study comparing protein profiles in TM from young and old cattle eyes showed the acetic acid-soluble protein fractions in both age groups to be similar to those of aged human TM (Russell *et al.*, 1989). One of the prominent changes in bovine TM was an age-related decrease in a 68-kDa polypeptide, which may be similar to the 58-kDa polypeptide component in human TM that also decreases with age. Oxidation and/or aggregation of TM proteins such as actin has been postulated as a potentially important age-related change in human and bovine TM (Russell *et al.*, 1989).

There is indirect evidence suggesting that certain types of glaucoma could result from blockage in the outflow facility by excess protein derived from the aqueous humor. This possibility has received support from studies on enucleated human eyes (Epstein *et al.*, 1978) as well as on a small number of patients with so-called protein glaucoma (Zirm, 1982). The notion of "protein glaucoma" has been further explored in a model system using microporous filters having flow dimensions similar to those found within the juxtacanalicular meshwork (Johnson *et al.*, 1986). The viscosity of aqueous humor is similar if not identical to that of isotonic saline, since the protein content of normal aqueous is too low to influence its viscosity. Yet bovine and primate aqueous humor have a greater flow resistance through microporous filters than isotonic saline. The filter-obstructing properties of aqueous humor could be abolished by papain treatment but not by hyaluronidase. Later studies using calf vitreous showed that the filter-blocking process involves two interacting classes of blocking components, arbitrarily called "A" and "B"; when present simultaneously in aqueous humor, they bind hydrophobically to the filter surface (Ethier *et al.*, 1989). Although the substances were not identified, one of the "class A" members could be albumin, the dominant protein of the aqueous. Filter blocking also requires a "class B" component, which is not necessarily a protein. The filter-blocking activity can be effectively disrupted with Triton X, a nonionic detergent that cannot be used experimentally; however, the possibility that other surfactants could be equally effective requires further study.

Alterations in cortisol metabolism have been detected in cultured cells from glaucomatous patients compared to cells from nonglaucomatous individuals. One is a 100-fold increase in Δ^4-reductase activity, and the other is a fourfold decrease in 3-oxidoreductase activity (Southren *et al.*, 1983b). Further studies using homogenates from cell cultures of glaucomatous and nonglaucomatous patients revealed that the defects in steroid metabolism arise from altered activities of the enzymes and not from availability of cofactors or endogenous inhibitors or activators (Weinstein *et al.*, 1985). As a result of these defects in cortisol-metabolizing enzymes (which may be either genetic or acquired), there is an accumulation in glaucomatous TM of both 5α- and 5β-dihydrocortisol, intermediates that are not found in nonglaucomatous cells. Independent experiments in rabbits show that 5β-dihydrocortisol potentiates the action of dexamethasone, causing a rise of IOP of 7 to 10 mm Hg (Southren *et al.*, 1985). These findings raise the interesting possibility that the accumulation of 5β-dihydrocortisol in the TM of glaucomatous patients could potentiate the action of endogenous glucocorticoids, resulting in the ocular hypertension that characterizes POAG.

Qualitative (histochemical) measurements of sialic acid in Schlemm's canal using polycationic ferritin and colloidal iron have been made on postmortem trabeculum from normal human and monkey eyes and on surgical specimens from five patients with POAG (Tripathi *et al.*, 1987). These studies showed a greater number of neuraminidase-sensitive binding sites in trabelular tissue from glaucoma patients than in normal controls, implying abnormalities in membrane receptors in the trabecular endothelial cells in POAG. These and other findings (Robey *et al.*, 1988) point to altered patterns of glycoconjugate biosynthesis in the TM of POAG patients. Freshly excised surgical specimens of glaucomatous TM display greatly reduced incorporation of labeled sulfate and glucosamine into proteoglycans compared to normal control tissue. Whether this is a consequence of the decrease in cellularity of glaucomatous TM (Alvarado *et al.*, 1981; Polansky *et al.*, 1984) or is a reflection of a specific enzymatic defect in glycoconjugate biosynthesis associated with glaucoma remains to be established.

Two new and promising lines of investigation could help to clarify many of the puzzling aspects of glaucoma. One is the development of quantitative methods to evaluate parameters such as binding affinities, dose–response relationships, and identification of receptors in all of the cell types in the anterior segment (Polansky and Alvarado, 1985). This should lead to a more systematic method for screening of drugs used in the treatment of the various forms of glaucoma. The other, at a more basic level, is the availability for the first time of monoclonal antibodies to human trabecular meshwork (Tripathi *et al.*, 1986). Definition of the antigenic determinants that delineate the TM endothelial cells provides a powerful tool not only for studying their embryological origin but also for investigating the epitopes expressed in aging and in glaucomatous eyes.

4.4. REFERENCES

Abdel-Latif, A. A., 1983, The iris–ciliary body, in: *Biochemistry of the Eye* (R. E. Anderson, ed.), American Academy of Ophthalmology, San Francisco, pp. 48–78.

Abdel-Latif, A. A., and Smith, J. P., 1979, Distribution of arachidonic acid and other fatty acids in glycerolipids of the rabbit iris, *Exp. Eye Res.* **29:**131–140.

Abdel-Latif, A. A., and Smith, J. P., 1982, Studies on the incorporation of [1-^{14}C]arachidonic acid into glycerolipids and its conversion into prostaglandins by rabbit iris. Effects of anti-inflammatory drugs and phospholipase A_2 inhibitors, *Biochim. Biophys. Acta* **711:**478–489.

Acott, T. S., Westcott, M., Passo, M. S., and van Buskirk, E. M., 1985, Trabecular meshwork glycosaminoglycans in human and cynomolgus monkey eye, *Invest. Ophthalmol. Vis. Sci.* **26:**1320–1329.

Acott, T. S., Kingsley, P. D., Samples, J. R., and van Buskirk, E. M., 1988, Human trabecular meshwork organ culture: Morphology and glycosaminoglycan synthesis, *Invest. Ophthalmol. Vis. Sci.* **29:**90–100.

Ahmad, H., Singh, S. V., Srivastava, S. K., and Awasthi, Y. C., 1989, Glutathione S-transferases of bovine iris and ciliary body: Characterization of isoenzymes, *Curr. Eye Res.* **8:**175–184.

Akhtar, R. A., and Abdel-Latif, A. A., 1984, Carbachol causes rapid phosphodiesteratic cleavage of phosphatidylinositol 4,5-bisphosphate and accumulation of inositol phosphates in rabbit iris smooth muscle; prazosin inhibits noradrenaline- and ionophore A23187-stimulated accumulation of inositol phosphates, *Biochem. J.* **224:**291–300.

Alvarado, J., Murphy, C., Polansky, J., and Juster, R., 1981, Age-related changes in trabecular meshwork cellularity, *Invest. Ophthalmol. Vis. Sci.* **21:**714–727.

Alvarado, J. A., Wood, I., and Polansky, J. R., 1982, Human trabecular cells. II. Growth pattern and ultrastructural characteristics, *Invest. Ophthalmol. Vis. Sci.* **23:**464–478.

Anderson, P. J., Wang, J., and Epstein, D. L., 1980, Metabolism of calf trabecular (reticular) meshwork, *Invest. Ophthalmol. Vis. Sci.* **19:**13–20.

Anderson, P. J., Karageuzian, L. N., Cheng, H.-M., and Epstein, D. L., 1984a, Hexokinase of calf trabecular meshwork, *Invest. Ophthalmol. Vis. Sci.* **25:**1258–1261.

Anderson, P. J., Karageuzian, L. N., and Epstein, D. L., 1984b, Phosphofructokinase of calf trabecular meshwork, *Invest. Ophthalmol. Vis. Sci.* **25:**1262–1266.

Anderson, R. E., Maude, M. B., and Feldman, G. L., 1970, Lipids of ocular tissues. III. The phospholipids of mature bovine iris, *Exp. Eye Res.* **9:**281–284.

Armaly, M. F., and Wang, Y., 1975, Demonstration of acid mucopolysaccharides in the trabecular meshwork of the rhesus monkey, *Invest. Ophthalmol. Vis. Sci.* **14:**507–516.

Awasthi, Y. C., Saneto, R. P., and Srivastava, S. K., 1980, Purification and properties of bovine lens glutathione S-transferase, *Exp. Eye Res.* **30:**29–39.

Barak, M. H., Weinreb, R. N., and Ryder, M. I., 1988, Quantitative assessment of cynomolgus monkey trabecular cell phagocytosis and adsorption, *Curr. Eye Res.* **7:**445–448.

Barany, E. H., 1953/1954, *In vitro* studies of the resistance to flow through the angle of the anterior chamber, *Acta Soc. Med. Upsaliensis* **54:**260–276.

Barany, E. H., 1956, The action of different kinds of hyaluronidase on the resistance of flow through the angle of the anterior chamber, *Acta Ophthalmol.* **34:**397–403.

Barany, E. H., and Woodin, A. M., 1955, Hyaluronic acid and hyaluronidase in the aqueous humour and the angle of the anterior chamber, *Acta Physiol. Scand.* **33:**257–290.

Bartels, S. P., Lee, S. R., and Neufeld, A. H., 1987, The effects of forskolin on cyclic AMP, intraocular pressure and aqueous humor formation in rabbits, *Curr. Eye Res.* 6:307–320.

Bausher, L. P., Gregory D. S., and Sears, M. L., 1987, Interaction between alpha$_2$- and beta$_2$-adrenergic receptors in rabbit ciliary processes, *Curr. Eye Res.* **6:**497–505.

Bausher, L. P., Gregory, D. S., and Sears, M. L., 1989, Alpha$_2$-adrenergic and VIP receptors in rabbit ciliary processes interact, *Curr. Eye Res.* **8:**47–54.

Bhattacherjee, P., Kulkarni, P. S., and Eakins, K. E., 1979, Metabolism of arachidonic acid in rabbit ocular tissues, *Invest. Ophthalmol. Vis. Sci.* **18:**172–178.

Bhuyan, K. C., and Bhuyan, D. K., 1977, Regulation of hydrogen peroxide in eye humours. Effect of 3-amino-1H-1,2,4-triazole on catalase and glutathione peroxidase of rabbit eye, *Biochim. Biophys. Acta* **497:**641–651.

Bhuyan, K. C., and Bhuyan, D. K., 1978, Superoxide dismutase of the eye, *Biochim. Biophys. Acta* **542:**28–38.

Bianchi, C., Anand-Srivastava, M. B., De Lean, A., Gutkowska, J., Forthomme, D., Genest, J., and Cantin, M., 1986, Localization and characterization of specific receptors for atrial natriuretic factor in the ciliary processes of the eye, *Curr. Eye Res.* **5:**283–293.

Binninger, E. A., Schachtschabel, D. O., and Rohen, J. W., 1987, Exogenous glycosaminoglycans stimulate hyaluronic acid synthesis by cultured human trabecular-meshwork cells, *Exp. Eye Res.* **45:**169–177.

Boas, R. S., Messenger, M. J., Mittag, T. W., and Podos, S. M., 1981, The effects of topically applied epinephrine and timolol on intraocular pressure and aqueous humor cyclic-AMP in the rabbit, *Exp. Eye Res.* **32:**681–690.

Bromberg, B. B., Gregory, D. S., and Sears, M. L., 1980, Beta-adrenergic receptors in ciliary processes of the rabbit, *Invest. Ophthalmol. Vis. Sci.* **19:**203–207.

Burstein, N. L., Sears, M. L., and Mead, A., 1984, Aqueous flow in human eyes is reduced by forskolin, a potent adenylate cyclase activator, *Exp. Eye Res.* **39:**745–749.

Caprioli, J., and Sears, M., 1983, Forskolin lowers intraocular pressure in rabbits, monkeys and man, *Lancet* **1:** 958–960.

Caprioli, J., and Sears, M., 1984, Combined effect of forskolin and acetazolamide on intraocular pressure and aqueous flow in rabbit eyes, *Exp. Eye Res.* **39:**47–50.

Caprioli, J., Sears, M., Bausher, L., Gregory, D., and Mead, A., 1984, Forskolin lowers intraocular pressure by reducing aqueous inflow, *Invest. Ophthalmol. Vis. Sci.* **25:**268–277.

Chu, T.-C., Candia, O. A., and Iizuka, S., 1986, Effects of forskolin. prostaglandin $F_{2\alpha}$, and Ba^{2+} on the short-circuit current of the isolated rabbit iris-ciliary body, *Curr. Eye Res.* **5:**511–516.

Coca-Prados, M., 1985, Regulation of protein phosphorylation of the intermediate-sized filament vimentin in the ciliary epithelium of the mammalian eye, *J. Biol. Chem.* **260:**10332–10338.

Cole, D. F., 1984, Ocular fluids, in: *The Eye* (H. Davson, ed.), Academic Press, New York, pp. 269–390.

Cooper, R. L., Constable, I. J., and Davidson, L., 1984, Aqueous humor catecholamines, *Curr. Eye Res.* **3:**809–813.

Crabbe, M. J. C., 1985, Ocular diamine oxidase activity, *Exp. Eye Res.* **41:**777–778.

Crean, E. V., Sherwood, M. E., Casey, R., Miller, M. W., and Richardson, T. M., 1986a, Establishment of calf trabecular meshwork cell cultures, *Exp. Eye Res.* **43:**503–517.

Crean, E. V., Tyson, S. L., and Richardson, T. M., 1986b, Factors influencing glycosaminoglycan synthesis by calf trabecular meshwork cell cultures, *Exp. Eye Res.* **43:**365–374.

Culp, T. W., Tucker, P. W., Ratliff, C. R., and Hall, F. F., 1970a, Chromatographic analysis of ocular lipids. I. Bovine and human iris tissue, *Biochim. Biophys. Acta* **218:**259–268.

Culp, T. W., Cunningham, R. D., Tucker, P. W., Jeter, J., and Deiterman, L. H., Jr., 1970b, *In vivo* synthesis of lipids in rabbit iris, cornea and lens tissues, *Exp. Eye Res.* **9:**98–105.

Darnell, J., Lodish, H. and Baltimore, D., 1986, *Molecular Cell Biology*, W. H. Freeman, New York, pp. 715–769.

Das, N. D., and Shichi, H., 1979, Gamma-glutamyl transpeptidase of bovine ciliary body: Purification and properties, *Exp. Eye Res.* **29**:109–121.

Das, N. D., and Shichi, H., 1981, Enzymes of mercapturate synthesis and other drug-metabolizing reactions—Specific localization in the eye, *Exp. Eye Res.* **33**:525–533.

Day, J., Grierson, I., Unger, W. G., and Robins, E., 1986, Some effects of phagocytosis on bovine meshwork cells in culture, *Exp. Eye Res.* **43**:1077–1087.

Dernouchamps, J. P., Magnusson, C. G. M., Michiels, J., and Masson, P. L., 1985, Immunoglobulin E in aqueous humour, *Exp. Eye Res.* **40**:321–325.

Engstrom, P., and Dunham, E. W., 1982, Alpha-adrenergic stimulation of prostaglandin release from rabbit iris-ciliary body in vitro, *Invest. Ophthalmol. Vis. Sci.* **22**:757–767.

Epstein, D. L., Jedziniak, J. A., and Grant, W. M., 1978, Obstruction of aqueous outflow by lens particles and by heavy-molecular-weight soluble lens proteins, *Invest. Ophthalmol. Vis. Sci.* **17**:272–277.

Ethier, C. R., Kamm, R. D., Johnson, M., Pavao, A. F., and Anderson, P. J., 1989, Further studies on the flow of aqueous humor through microporous filters, *Invest. Ophthalmol. Vis. Sci.* **30**:739–746.

Floyd, B. B., Cleveland, P. H., and Worthen, D. M., 1985, Fibronectin in human trabecular drainage channels, *Invest. Ophthalmol. Vis. Sci.* **26**:797–804.

Freedman, S. F., Anderson, P. J., and Epstein, D. L., 1985, Superoxide dismutase and catalase of calf trabecular meshwork, *Invest. Ophthalmol. Vis. Sci.* **26**:1330–1335.

Friedland, B. R., and Maren, T. H., 1984, Carbonic anhydrase: Pharmacology of inhibitors and treatment of glaucoma, in: *Pharmacology of the Eye, Handbook of Pharmacology*, Vol. 69 (M. L. Sears, ed.), Springer-Verlag, Berlin, pp. 279–309.

Geanon, J. D., Tripathi, B. J., Tripathi, R. C., and Barlow, G. H., 1987, Tissue plasminogen activator in avascular tissues of the eye: A quantitative study of its activity in the cornea, lens, and aqueous and vitreous humors of dog, calf, and monkey, *Exp. Eye Res.* **44**:55–63.

Gerritsen, M. E., Weinstein, B. I., Gordon, G. G., and Southren, A. L, 1986, Prostaglandin synthesis and release from cultured human trabecular-meshwork cells and scleral fibroblasts, *Exp. Eye. Res.* **43**:1089–1102.

Giblin, F. J., McCready, J. P., Kodama, T., and Reddy, V. N., 1984, A direct correlation between the levels of ascorbic acid and H_2O_2 in aqueous humor, *Exp. Eye Res.* **38**:87–93.

Gipson, I. K., and Anderson, R. A., 1979, Actin filaments in cells of human trabecular meshwork and Schlemm's canal, *Invest. Ophthalmol. Vis. Sci.* **18**:547–561.

Gregory, D., Sears, M., Bausher, L., Mishima, H., and Mead, A, 1981, Intraocular pressure and aqueous flow are decreased by cholera toxin, *Invest. Ophthalmol. Vis. Sci.* **20**:371–381.

Grierson, I., and Lee, W. R., 1975, Acid mucopolysaccharides in the outflow apparatus, *Exp. Eye Res.* **21**:417–431.

Grierson, I., Kissun, R., Ayad, S., Phylactos, A., Ahmed, S., Unger, W. G., and Day, J. E., 1985a, The morphological features of bovine meshwork cells *in vitro* and their synthetic activities, *Albrecht von Graefes Arch. Klin. Exp. Ophthalmol.* **223**:225–236.

Grierson, I., Robins, E., Unger, W., Millar, L., and Ahmed, A, 1985b, The cells of the bovine outflow system in tissue culture, *Exp. Eye Res.* **40**:35–46.

Grierson, I., Millar, L., Yong J. D., Day, J., McKechnie, N. M., Hitchins, C., and Boulton, M., 1986a, Investigations of cytoskeletal elements in cultured bovine meshwork cells, *Invest. Ophthalmol. Vis. Sci.* **27**:1318–1330.

Grierson, I., Day, J., Unger, W. G., and Ahmed, A., 1986b, Phagocytosis of latex microspheres by bovine meshwork cells in culture, *Albrecht von Graefes Arch. Klin. Exp. Ophthalmol.* **224**:536–544.

Gual, A., Blanco, J., Belmonte, C., and Garcia, A. G., 1983, Dopamine β-hydroxylase activity in human aqueous humor, *Exp. Eye Res.* **37**:99–102.

Hernandez, M. R., Weinstein, B. I., Schwartz, J., Ritch, R., Gordon, G. G., and Southren, A. L., 1987, Human trabecular meshwork cells in culture: Morphology and extracellular matrix components, *Invest. Ophthalmol. Vis. Sci.* **28**:1655–1660.

Hirsch, M., Montcourrier, P., Arguillere, P., and Keller, N., 1985, The structure of tight junctions in the ciliary epithelium, *Curr. Eye Res.* **4**:493–501.

Horstmann, H.-J., Rohen, J. W., and Sames, K., 1983, Age-related changes in the composition of proteins in the trabecular meshwork of the human eye, *Mech. Ageing Dev.* **21**:121–136.

Hoyng, P. F. J., Verbey, N., Thorig, L., and van Haeringen, N. J., 1986, Topical prostaglandins inhibit trauma -induced inflammation in the rabbit eye, *Invest. Ophthalmol. Vis. Sci.* **27**:1217–1225.

Hutchins, J. B., and Hollyfield, J. G., 1984, Autoradiographic identification of muscarinic receptors in human iris smooth muscle, *Exp. Eye Res.* **38:**515–521.

Inada, K., Baba, H., and Okamura, R., 1984, Quantitative determination of human aqueous proteins by crossed immunoelectrophoresis, *Jpn. J. Ophthalmol.* **28:**1–8.

Iwamoto, Y., and Tamura, M., 1988, Immunocytochemical study of intermediate filaments in cultured human trabecular cells, *Invest. Ophthalmol. Vis. Sci.* **29:**244–250.

Iwata, S., Miyauchi, S., and Takehana, M., 1984, Biochemical studies on the use of sodium hyaluronate in the anterior eye segment. I. Variation of protein and ascorbic acid concentration in rabbit aqueous humor, *Curr. Eye Res.* **3:**605–610.

Jampel, H. D., Lynch, M. G., Brown, R. H., Kuhar, M. J., and De Souza, E. B., 1987, β-Adrenergic receptors in human trabecular meshwork, *Invest. Ophthalmol. Vis. Sci.* **28:**72–779.

Johnson, M., Ethier, C. R., Kamm, R. D., Grant, W. M., Epstein, D. L., and Gaasterland, D., 1986, The flow of aqueous humor through micro-porous filters, *Invest. Ophthalmol. Vis. Sci.* **27:**92–97.

Jumblatt, J. E., and North, G. T., 1986, Potentiation of sympathetic neurosecretion by forskolin and cyclic AMP in the rabbit iris–ciliary body, *Curr. Eye Res.* **5:**495–502.

Kahn, M. G., Giblin, F. J., and Epstein, D. L., 1983, Glutathione in calf trabecular meshwork and its relation to aqueous humor outflow facility, *Invest. Ophthalmol. Vis. Sci.* **24:**1283–1287.

Kass, M. A., and Holmberg, N. J., 1979, Prostaglandin and thromboxane synthesis by microsomes of rabbit ocular tissues, *Invest. Ophthalmol. Vis. Sci.* **18:**166–171.

Kass, M. A., Holmberg, N. J., and Smith, M. E., 1981, Prostaglandin and thromboxane synthesis by microsomes of inflamed rabbit ciliary body-iris, *Invest. Ophthalmol. Vis. Sci.* **20:**442–449.

Kenney, M. C., Lewis, W., Redding, J., and Waring, G. O., 1986, Decreased fibronectin levels in aqueous humor after corneal injury, *Ophthalmic Res.* **18:**165–171.

Kintz, P., Himber, J., de Burlet, G., and Andermann, G., 1988, Characterization of alpha$_2$-adrenergic receptors, negatively coupled to adenylate cyclase, in rabbit ciliary processes, *Curr. Eye Res.* **7:**287–292.

Kishida, K., Kodama, T., O'Meara. P. D., and Shichi, H., 1985, Glutathione depletion and oxidative stress: Study with perfused bovine eye, *J. Ocular Pharmacol.* **1:**85–99.

Kishida, K., Matsumoto, K., Manabe, R., and Sugiyama, T., 1986, Cytochrome P-450 and related components of the microsomal electron transport system in the bovine ciliary body, *Curr. Eye Res.* **5:**529–533.

Knepper, P. A., Farbman, A. I., and Telser, A. G., 1981, Aqueous outflow pathway glycosaminoglycans, *Exp. Eye Res.* **32:**265–277.

Knepper, P. A., Collins, J. A., Weinstein, H. G., and Breen, M., 1983, Aqueous outflow pathway complex carbohydrate synthesis in vitro, *Invest. Ophthalmol. Vis. Sci.* **24:**1546–1551.

Knepper, P. A., Farbman, A. I., and Telser, A. G., 1984, Exogenous hyaluronidases and degradation of hyaluronic acid in the rabbit eye, *Invest. Ophthalmol. Vis. Sci.* **25:**286–293.

Koh, S.-W. M., and Yue, B. Y. J. T., 1988, Effects of agonists on the intracellular cyclic AMP concentration in monkey trabecular meshwork cells, *Curr. Eye Res.* **7:**75–80.

Kulkarni, P. S., and Srinivasan, B. D., 1983, Synthesis of slow reacting substance-like activity in rabbit conjunctiva and anterior uvea, *Invest. Ophthalmol. Vis. Sci.* **24:**1079–1085.

Kulkarni, P. S., Fleisher, L., and Srinivasan, B. D., 1984a, The synthesis of cyclooxygenase products in ocular tissues of various species, *Curr. Eye Res.* **3:**447–452.

Kulkarni, P. S., Rodriguez, A. V., and Srinivasan, B. D., 1984b, Human anterior uvea synthesizes lipoxygenase products from arachidonic acid, *Invest. Ophthalmol. Vis. Sci.* **25:**221–223.

Kulkarni, P. S., Kaufman, P. L., and Srinivasan, B. D., 1987, Cyclo-oxygenase and lipoxygenase pathways in cynomolgus and rhesus monkey conjunctiva, anterior uvea and eyelids, *Curr. Eye Res.* **6:**801–808.

Lahav, M., Melamed, E., Dafna, Z., and Atlas, D., 1978, Localization of beta receptors in the anterior segment of the rat eye by a flurescent analogue of propranolol, *Invest. Ophthalmol. Vis. Sci.* **17:**645–651.

Laurent, U. B. G., 1983, Hyaluronate in human aqueous humor, *Arch. Ophthalmol.* **101:**129–130.

Laurent, U. B. G., and Fraser, J. R. E., 1983, Turnover of hyaluronate in the aqueous humour and vitreous body of the rabbit, *Exp. Eye Res.* **36:**493–504.

Laurent, U. B. G., and Granath, K. A., 1983, The molecular weight of hyaluronate in the aqueous humour and vitreous body of rabbit and cattle eyes, *Exp. Eye Res.* **36:**481–492.

Lee, P.-Y., Podos, S. M., Mittag, T., and Severin, C., 1984, Effect of topically applied forskolin on aqueous humor dynamics in cynomolgus monkey, *Invest. Ophthalmol. Vis. Sci.* **25:**1206–1209.

Leeb-Lundberg, L. M. F., Cotecchia, S., DeBlasi, A., Caron, M. G., and Lefkowitz, R. J., 1987, Regulation of adrenergic receptor function by phosphorylation. I. Agonist-promoted desensitization and phosphorylation of α_1-adrenergic receptors coupled to inositol phospholipid metabolism in DDT_1, MF-2 smooth muscle cells, *J. Biol. Chem.* **262:**3098–3105.

Lefkowitz, R. J., Stadel, J. M., and Caron, M. G., 1983, Adenylate cyclase-coupled beta-adrenergic receptors: Structure and mechanisms of activation and desensitization, *Annu. Rev. Biochem.* **52:**159–186.

Mallorga, P., and Sugrue, M. F., 1987, Characterization of serotonin receptors in the iris + ciliary body of the albino rabbit, *Curr. Eye Res.* **6:**527–532.

Maren, T. H., Jankowska, L., Sanyal, G,, and Edelhauser, H. F., 1983, The transcorneal permeability of sulfonamide carbonic anhydrase inhibitors and their effect on aqueous humor secretion, *Exp. Eye Res.* **36:**457–480.

Meyer, K., and Smyth, E. M., 1938, On the nature of the ocular fluids. II. The hexosamine content, *Am. J. Ophthalmol.* **21:**1083–1090.

Millard, C. B., Tripathi, B. J., and Tripathi, R. C., 1987, Age-related changes in protein profiles of the normal human trabecular meshwork, *Exp. Eye Res.* **45:**623–631.

Mittag, T., 1984, Role of oxygen radicals in ocular inflammation and cellular damage, *Exp. Eye Res.* **39:**759–769.

Mittag, T., and Tormay, A., 1981, Desensitization of the β-adrenergic receptor–adenylate cyclase complex in rabbit iris–ciliary body induced by topical epinephrine, *Exp. Eye Res.* **33:**497–503.

Mittag, T. W., and Tormay, A., 1985a, Adrenergic receptor subtypes in rabbit iris–ciliary body membranes: Classification by radioligand studies, *Exp. Eye Res.* **40:**239–249.

Mittag, T. W., and Tormay, A., 1985b, Drug responses of adenylate cyclase in iris–ciliary body determined by adenine labelling, *Invest. Ophthalmol. Vis. Sci.* **26:**396–399.

Mittag, T. W., Hammond, B. R., Eakins, K. E., and Bhattacherjee, P., 1985a, Ocular responses to superoxide generated by intraocular injection of xanthine oxidase, *Exp. Eye Res.* **40:**411–419.

Mittag, T. W., Tormay, A., Severin, C., and Podos, S. M., 1985b, Alpha-adrenergic antagonists: Correlation of the effect on intraocular pressure and on α_2-adrenergic receptor binding specificity in the rabbit eye, *Exp. Eye Res.* **40:**591–599.

Mittag, T. W., Tormay, A., Ortega, M., and Podos, S. M., 1987a, Effects of inorganic ions on rabbit ciliary process adenylate cyclase, *Invest. Ophthalmol. Vis. Sci.* **28:**2049–2056.

Mittag, T. W., Yoshimura, N., and Podos, S. M., 1987b, Phorbol ester: Effect on intraocular pressure, adenylate cyclase, and protein kinase in the rabbit eye, *Invest. Ophthamol. Vis. Sci.* **28:**2057–2066.

Mittag, T. W., Tormay, A., Ortega, M., and Severin, C., 1987c, Atrial natriuretic peptide (ANP), guanylate cyclase, and intraocular pressure in the rabbit eye, *Curr. Eye Res.* **6:**1189–1196.

Mittag, T. W., Tormay, A., and Podos, S. M., 1988, Manganous chloride stimulation of adenylate cyclase responsiveness in ocular ciliary process membranes, *Exp. Eye Res.* **46:**841–851.

Miyauchi, S., and Iwata, S., 1984, Biochemical studies on the use of sodium hyaluronate in the anterior eye segment. II. The molecular behavior of sodium hyaluronate injected into anterior chamber of rabbits, *Curr. Eye Res.* **3:**611–617.

Mizokami, K., 1977, Demonstration of masked acidic glycosaminoglycans in the normal human trabecular meshwork, *Jpn. J. Ophthalmol.* **21:**57–63.

Mondino, B. J., and Rao, H., 1983a, Complement levels in normal and inflamed aqueous humor, *Invest. Ophthalmol. Vis. Sci.* **24:**380–384.

Mondino, B. J., and Rao, H., 1983b, Hemolytic complement activity in aqueous humor, *Arch. Ophthalmol.* **10:**465–468.

Nathanson, J. A., 1980, Adrenergic regulation of intraocular pressure: Identification of β_2-adrenergic-stimulated adenylate cyclase in ciliary process epithelium, *Proc. Natl. Acad. Sci. USA* **77:**7420–7424.

Nathanson, J. A., 1981, Human ciliary process adrenergic receptor: Pharmacological characterization, *Invest. Ophthalmol. Vis. Sci.* **21:**798–804.

Nathanson, J. A., 1987, Atriopeptin-activated guanylate cyclase in the anterior segment, *Invest. Ophtalmol. Vis. Sci.* **28:**1357–1364.

Neufeld, A. H., 1978, Influences of cyclic nucleotides on outflow facility in the vervet monkey, *Exp. Eye Res.* **27:**387–397.

Neufeld, A. H., Zawistowski, K. A., Page, E. D., and Bromberg, B. B., 1978, Influences on the density of β-adrenergic receptors in the cornea and iris–ciliary body of the rabbit, *Invest. Ophthalmol. Vis. Sci.* **17:**1069–1075.

Ng, M. C., and Shichi, H., 1986, Purification and properties of glutathione reductases from bovine ciliary body, *Exp. Eye Res.* **43:**477–489.

Ng, M. C., Susan, S. R., and Shichi, H., 1988, Bovine non-pigmented and pigmented ciliary epithelial cells in culture: Comparison of catalase, superoxide dismutase and glutathione peroxidase activities, *Exp. Eye Res.* **46:**919–928.

Nguyen, K. P. V., Weiss, H., Karageuzian, L. N., Anderson, P. J., and Epstein, D. L., 1985, Glutathione reductase of calf trabecular meshwork, *Invest. Ophthalmol. Vis. Sci.* **26:**887–890.

Nguyen, K., Lee, D. A., Anderson, P. J., and Epstein, D. L., 1986, Glucose 6-phosphate dehydrogenase of calf trabecular meshwork, *Invest. Ophthalmol. Vis. Sci.* **27:**992–997.

Osborne, N. N., and Tobin, A. B., 1987, Serotonin-accumulating cells in the iris–ciliary body and cornea of various species, *Exp. Eye Res.* **44:**731–746.

Park, J. K., Tripathi, R. C., Tripathi, B. J., and Barlow, G. H., 1987, Tissue plasminogen activator in the trabecular endothelium, *Invest. Ophtalmol. Vis. Sci.* **28:**1341–1345.

Partridge, C. A., Weinstein, B. I., Southren, A. L., and Gerritsen, M. E., 1989, Dexamethasone induces specific proteins in human trabecular meshwork cells, *Invest. Ophthalmol. Vis. Sci.* **30:**1843–1847.

Pavao, A. F., Lee, D. A., Ethier, C. R., Johnson, M. C., Anderson, P. J., and Epstein, D. L., 1989, Two-dimensional gel electrophoresis of calf aqueous humor, serum, and filter-bound proteins, *Invest. Ophthalmol. Vis. Sci.* **30:**731–738.

Pickett, C. B., and Lu, A. Y. H., 1989, Glutathione S-transferases: Gene structure, regulation, and biological function, *Annu. Rev. Biochem.* **58:**743–764.

Pirie, A., 1965, Glutathione peroxidase in lens and a source of hydrogen peroxide in aqueous humour, *Biochem. J.* **96:**244–253.

Pohjola, S., 1966, The glucose content of the aqueous humor in man, *Acta Ophthalmol. Suppl.* **28:**11–80.

Polansky, J. R., and Alvarado, J. A., 1985, Isolation and evaluation of target cells in glaucoma research: Hormone receptors and drug responses, *Curr. Eye Res.* **4:**267–279.

Polansky, J. R., and Weinreb, R. N., 1984, Anti-inflammatory agents. Steroids as anti-inflammatory agents, in: *Pharmacology of the Eye, Handbook of Pharmacology,* Vol. 69 (M. L. Sears, ed.), Springer-Verlag, Berlin, pp. 459–538.

Polansky, J. R., Weinreb, R. N., Baxter, J. D., and Alvarado, J., 1979, Human trabecular cells. I. Establishment in tissue culture and growth characteristics, *Invest. Ophthalmol. Vis. Sci.* **18:**1043–1049.

Polansky, J. R., Wood, I. S., Maglio, M. T., and Alvarado, J. A., 1984, Trabecular meshwork cell culture in glaucoma research: Evaluation of biological activity and structural properties of human trabecular cells in vitro, *Ophthalmology* **91:**580–595.

Raviola, G., 1977, The structural basis of the blood–ocular barriers, *Exp. Eye Res. Suppl.* **25:**27–63.

Reid, T., Kenney, M. C., and Waring, G. O., 1982, Isolation and characterization of fibronectin from bovine aqueous humor, *Invest. Ophthalmol. Vis. Sci.* **22:**57–61.

Richardson, T. M., 1982, Distribution of glycosaminoglycans in the aqueous outflow system of the cat, *Invest. Ophthalmol. Vis. Sci.* **22:**319–329.

Riley, M. V., 1983, The chemistry of the aqueous humor, in: *Biochemistry of the Eye* (R. E. Anderson, ed.), American Academy of Ophthalmology, San Francisco, pp. 79–95.

Riley, M. V., and Kishida, K., 1986, ATPases of ciliary epithelium: Cellular and subcellular distribution and probable role in secretion of aqueous humor, *Exp. Eye Res.* **42:**559–568.

Riley, M. V., Meyer, R. F., and Yates, E. M., 1980, Glutathione in the aqueous humor of human and other species, *Invest. Ophthalmol. Vis. Sci.* **19:**94–96.

Robey, P. G., Kirshner, J. A., Cummins, C. E. III, Ballintine, E. J., Rodrigues, M. M., and Gaasterland, D.E., 1988, Synthesis of glycoconjugates by trabecular meshwork of glaucomatous corneoscleral explants, *Exp. Eye Res.* **46:**111–115.

Rodrigues, M. M., Katz, S. I., Foidart, J.-M., and Spaeth, G. L., 1980, Collagen, factor VIII antigen, and immunoglobulins in the human aqueous drainage channels, *Ophthalmology* **87:**337–345.

Rohen, J. W., Schachtschabel, D. O., and Berghoff, K., 1984, Histoautoradiographic and biochemical studies on human and monkey trabecular meshwork and ciliary body in short-term explant culture, *Albrecht von Graefes Arch. Klin. Exp. Ophthalmol.* **221:**199–206.

Russell, P., Garland, D., and Epstein, D. L., 1989, Analysis of the proteins of calf and cow trabecular meshwork: Development of a model system to study aging effects and glaucoma, *Exp. Eye Res.* **48:**251–260.

Ryder, M. I., and Weinreb, R. N., 1986, The cytoskeleton of the cynomolgus monkey trabecular cell. I. General considerations, *Invest. Ophthalmol. Vis. Sci.* **27:**1305–1311.

Ryder, M. I., Weinreb, R. N., Alvarado, J., and Polansky, J., 1988, The cytoskeleton of the cultured human trabecular cell, *Invest. Ophthalmol. Vis. Sci.* **29:**251–260.

Saari, K. M., Aine, E., and Parviainen, M. T., 1983, Determination of protein content in aqueous humour by high-performance gel filtration chromatography, *Acta Ophthalmol.* **81:**611–617.

Schachtschabel, D. O., Bigalke, B., and Rohen, J. W., 1977, Production of glycosaminoglycans by cell cultures of the trabecular meshwork of the primate eye, *Exp. Eye Res.* **24:**71–80.

Schachtschabel, D. O., Rohen, J. W., Wever, J., and Sames, K., 1982, Synthesis and composition of glycosaminoglycans by cultured human trabecular meshwork cells, *Albrecht von Graefes Arch. Klin. Exp. Ophthalmol.* **218:**113–117.

Schachtschabel, D. O., Berghoff, K., and Rohen, J. W., 1984, Synthesis and composition of glycosaminoglycans by explant cultures of human ciliary body and ciliary processes in serum-containing and serum-free defined media, *Albrecht von Graefes Arch. Klin. Exp. Ophthalmol.* **221:**207–209

Schwartzman, M. L., Masferrer, J., Dunn, M. W., McGiff, J. C., and Abraham, N. G., 1987, Cytochrome P450, drug metabolizing enzymes and arachidonic acid metabolism in bovine ocular tissues, *Curr. Eye Res.* **6:**623–630.

Scott, D. R., Karageuzian, L. N., Anderson, P. J., and Epstein, D. L., 1984, Glutathione peroxidase of calf trabecular meshwork, *Invest. Ophthalmol. Vis. Sci.* **25:**599–602.

Sears, M. L., 1985, Regulation of aqueous flow by the adenylate cyclase receptor complex in the ciliary epithelium, *Am. J. Ophthalmol.* **100:**194–198.

Sears, M., Gregory, D., Bausher, L., Mishima, H., and Stjernschantz, J., 1981, A receptor for aqueous humor formation, in: *New Directions in Ophthalmic Research* (M. L. Sears, ed.), Yale University Press, New Haven, pp.163–183.

Shichi, H., 1984, Biotransformation and drug metabolism, in: *Pharmacology of the Eye*, *Handbook of Pharmacology,* Vol. 69 (M. L. Sears. ed.), Springer-Verlag, Berlin, pp. 117–148.

Shichi, H., and O'Meara, P. D., 1986, Purification and properties of anionic glutathione S-transferase from bovine ciliary body, *Biochem. J.* **237:**365–371.

Shichi, H., Atlas, S. A., and Nebert, D. W., 1975, Genetically regulated aryl hydrocarbon hydroxylase induction in the eye: Possible significance of the drug-metabolizing enzyme system for the retinal pigmented epithelium-choroid, *Exp. Eye Res.* **21:**557–567.

Shuman, M. A., Polansky, J. R., Merkel, C., and Alvarado, J. A., 1988, Tissue plasminogen activator in cultured human trabecular meshwork cells, *Invest. Ophthalmol. Vis. Sci.* **29:**401–405.

Smith, B. R., Gaster, R. N, Leopold, I. H., and Zeleznick, L. D., 1984, Forskolin, a potent adenylate cyclase activator, lowers rabbit intraocular pressure, *Arch. Ophthalmol.* **102:**146–148.

Southren, A. L., Dominguez, M. O., Gordon, G. G., Wenk, E. J., Hernandez, M. R., Dunn, M. W., and Weinstein, B. I., 1983a, Nuclear translocation of the cytoplasmic glucocorticoid receptor in the iris-ciliary body and adjacent corneoscleral tissue of the rabbit following topical administration of various glucocorticoids, *Invest. Ophthalmol. Vis. Sci.* **24:**147–152.

Southren, A. L., Gordon, G. G., Munnangi, P. R., Vitrek, J., Schwartz, J., Monder, C., Dunn, M. W., and Weinstein, B. I., 1983b, Altered cortisol metabolism in cells cultured from trabecular meshwork specimens obtained from patients with primary open-angle glaucoma, *Invest. Ophthalmol. Vis. Sci.* **24:**1413–1417.

Southren, A. L., Gordon, G. G., l'Hommedieu, D., Ravikumar, S., Dunn, M. W., and Weinstein, B. I., 1985, 5β-Dihydrocortisol: Possible mediator of the ocular hypertension in glaucoma, *Invest. Ophthalmol. Vis. Sci.* **26:**393–395.

Spector, A., and Garner, W. H., 1981, Hydrogen peroxide and human cataract, *Exp. Eye Res.* **33:**673–681.

Sramek, S. J., Wallow, I. H. L., Bindley, C., and Sterken, G., 1987, Fibronectin distribution in the rat eye. An immunohistochemical study, *Invest. Ophthalmol. Vis. Sci.* **28:**500–505.

Stern, F. A., and Bito, L. Z., 1982, Comparison of the hypotensive and other ocular effects of prostaglandins E_2 and $F_{2\alpha}$ on cat and rhesus monkey eyes, *Invest. Ophthalmol. Vis. Sci.* **22:**588–598.

Stone, R. A., and Glembotski, C. C., 1986, Immunoactive atrial natriuretic peptide in the rat eye: Molecular forms in anterior uvea and retina, *Biochem. Biophys. Res. Commun.* **134:**1022–1028.

Stur, M., Grabner, G., Dorda, W., and Zehetbauer, G., 1983, The effect of timolol on the concentrations of albumin and IgG in the aqueous humor of the human eye, *Am. J. Ophthalmol.* **96:**726–729.

Tripathi, B. J., and Tripathi, R. C., 1984, Effect of epinephrine *in vitro* on the morphology, phagocytosis, and mitotic activity of human trabecular endothelium, *Exp. Eye Res.* **39:**731–744.

Tripathi, B. J., Tripathi, R. C., Stefansson, K., Havran, W. L., and Fitch, F. W., 1986, Monoclonal antibodies to human trabecular endothelium. A preliminary report on production and characterization, *Exp. Eye Res.* **43:**863–866.

Tripathi, R. C., and Tripathi, B. J., 1982, Human trabecular endothelium, corneal endothelium, keratocytes, and scleral fibroblasts in primary cell culture. A comparative study of growth characteristics, morphology, and phagocytic activity by light and scanning electron microscopy, *Exp. Eye Res.* **35:**611–624.

Tripathi, R. C., Tripathi, B. J., and Spaeth, G. L., 1987, Localization of sialic acid moieties in the endothelial lining of Schlemm's canal in normal and glaucomatous eyes, *Exp. Eye Res.* **44:**293–306.

Tripathi, R. C., Millard, C. B., Tripathi, B. J., and Reddy, V., 1988, A molecule resembling fibroblast growth factor in aqueous humor, *Am. J. Ophthalmol.* **106:**230–231.

Tripathi, R. C., Millard, C. B., and Tripathi, B. J., 1989, Protein composition of human aqueous humor: SDS-PAGE analysis of surgical and post-mortem samples, *Exp. Eye Res.* **48:**117–130.

Trope, G. E., and Rumley, A. G., 1985, Catecholamines in human aqueous humor, *Invest. Ophthalmol. Vis. Sci.* **26:**399–401.

Usukura, J., Fain, G. L., and Bok, D., 1988, [^{3}H]Ouabain localization of Na-K ATPase in the epithelium of rabbit ciliary body pars plicata, *Invest. Ophthalmol. Vis. Sci.* **29:**606–614.

van Delft, J. L., van Haeringen, N. J., Glasius, E., Barthen, E. R., and Oosterhuis, J. A., 1987, Comparison of the effects of corticosteroids and indomethacin on the response of the blood-aqueous barrier to injury, *Curr. Eye Res.* **6:**419–425.

Waitzman, M. B., and Woods, W. D., 1971, Some characteristics of an adenyl cyclase preparation from rabbit ciliary process tissue, *Exp. Eye Res.* **12:**99–111.

Wax, M. B., and Coca-Prados, M., 1989, Receptor-mediated phosphoinositide hydrolysis in human ocular ciliary epithelial cells, *Invest. Ophthalmol. Vis. Sci.* **30:**1675–1679.

Wax, M. B., and Molinoff, P. B., 1987, Distribution and properties of β-adrenergic receptors in human iris–ciliary body, *Invest. Ophthalmol. Vis. Sci.* **28:**420–430.

Weinreb, R. N., Bloom, E., Baxter, J. D., Alvarado, J., Lan, N., O'Donnell, J., and Polansky, J. R., 1981, Detection of glucocorticoid receptors in cultured human trabecular cells, *Invest. Ophthalmol. Vis. Sci.* **21:**403–407.

Weinreb, R. N., Mitchell, M. D., and Polansky, J. R., 1983, Prostaglandin production by human trabecular cells: *In vitro* inhibition by dexamethasone, *Invest. Ophthalmol. Vis. Sci.* **24:**1541–1545.

Weinreb, R. N., Ryder, M. I., and Polansky, J. R., 1986, The cytoskeleton of the cynomolgus monkey trabecular cell. II. Influence of cytoskeleton-active drugs, *Invest. Ophthalmol. Vis. Sci.* **27:**1312–1317.

Weinreb, R. N., Polansky, J. R., Alvarado, J. A., and Mitchell, M. D., 1988, Arachidonic acid metabolism in human trabecular meshwork cells, *Invest. Ophthalmol. Vis. Sci.* **29:**1708–1712.

Weinstein, B. I., Munnangi, P., Gordon, G. G., and Southren, A. L., 1985, Defects in cortisol-metabolizing enzymes in primary open-angle glaucoma, *Invest. Ophthalmol. Vis. Sci.* **26:**890–893.

Williams, R. N., Bhattacherjee, P., and Eakins, K. E., 1983, Biosynthesis of lipoxygenase products by ocular tissues, *Exp. Eye Res.* **36:**397–402.

Wistrand, P. J., 1984, The use of carbonic anhydrase inhibitors in ophthalmology and clinical medicine, *Ann. N. Y. Acad. Sci.* **429:**609 -619.

Wistrand, P. J., Schenholm, M., and Lonnerholm, G., 1986, Carbonic anhydrase isoenzymes CA I and CA II in the human eye, *Invest. Ophthalmol. Vis. Sci.* **27:**419–428.

Worthen, D. M., and Cleveland, P. H., 1982, Fibronectin production by cultured human trabecular meshwork cells, *Invest. Ophthalmol. Vis. Sci.* **23:**265–269.

Yoshimura, N., Mittag, T. W., and Podos, S. M., 1987, Analysis of protein kinase activities in rabbit ciliary processes: Identification and characterization using exogenous substrates, *Exp. Eye Res.* **45:**45–56.

Yoshimura, N., Mittag, T. W., and Podos, S. M., 1989a, Calcium-dependent phosphorylation of proteins in rabbit ciliary processes, *Invest. Ophthalmol. Vis. Sci.* **30:**723–730.

Yoshimura, N., Mittag, T. W., and Podos, S. M., 1989b, Cyclic nucleotide-dependent phosphorylation of proteins in rabbit ciliary processes, *Invest. Ophthalmol. Vis. Sci.* **30:**875–881.

Yousufzai, S. Y. K., and Abdel-Latif, A. A., 1987, Alpha$_1$-adrenergic receptor induced subsensitivity and supersensitivity in rabbit iris–ciliary body, *Invest. Ophthalmol. Vis. Sci.* **28:**409–419.

Yousufzai, S. Y. K., Akhtar, R. A., and Abdel-Latif, A. A., 1986, Effects of substance P on inositol triphosphate accumulation, on contractile responses and on arachidonic acid release and prostaglandin biosynthesis in rabbit iris sphincter muscle, *Exp. Eye Res.* **43:**215–226.

Yousufzai, S. Y. K., Honkanen, R. E., and Abdel-Latif, A. A., 1987, Muscarinic cholinergic induced subsensitivity in rabbit iris–ciliary body, *Invest. Ophthalmol. Vis. Sci.* **28:**1630–1638.

Yue, B. Y. J. T., and Elvart, J. L., 1987, Biosynthesis of glycosaminoglycans by trabecular meshwork cells *in vitro*, *Curr. Eye Res.* **6:**959–967.

Yue, B. Y. J. T., Elner, V. M., Elner, S. G., and Davis, H. R., 1987, Lysosomal enzyme activities in cultured trabecular-meshwork cells, *Exp. Eye Res.* **44:**891–897.

Zimmerman, L. E., 1957, Demonstration of hyaluronidase-sensitive acid mucopolysaccharide in trabecula and iris in routine paraffin sections of adult human eyes, *Am. J. Ophthalmol.* **44:**1–4.

Zirm, M., 1982, Protein glaucoma. Overtaxing of flow mechanisms? *Ophthalmologica* **184:**155–161.

Lens 5

5.1. INTRODUCTION

The crystalline lens, with its unusually high protein content and unique arrangement of structural fibers, provides the refractive index necessary to focus images on the retina. To achieve this, the lens must be perfectly transparent; loss of transparency, or cataract, is found in all populations throughout the world and is a common cause of blindness if not treated surgically. Normal lens structure, metabolism, biochemistry, photochemistry, and development, as well as changes occurring during aging and/or in cataract formation, are under active investigation in hundreds of laboratories today. Studies on the soluble proteins, the crystallins, their three-dimensional structures, and their genes are generating an ever-growing body of information on the evolution and origins of the lens crystallins. Equally intensive efforts are being directed toward biochemical characterization of the lens fibers and elucidation of the role played by the gap junction-like protein, MP26 (main intrinsic polypeptide of M_r 26,000), in cell-to-cell communication.

Despite the abundant basic information currently available on the lens, space limitations imposed for a general text on eye biochemistry preclude extensive discussions of all aspects of lens biochemistry. Hence, only selected topics, mainly from recent literature, have been chosen for inclusion in this chapter; the reader is referred to several excellent reviews, surveys, symposia, and texts for detailed information as well as several hundred literature references that could not included in this chapter (Harding and Crabbe, 1984; Piatigorsky, 1984, 1987; Spector, 1984a,b, 1985; Maisel, 1985; Bloemendal, 1985; Lubsen *et al.*, 1988; Wistow and Piatigorsky, 1988; Piatigorsky *et al.*, 1988a).

5.1.1. Anatomic and Structural Features

The lens is an avascular transparent tissue enveloped in a basement membrane called the lens capsule. It is built up of two cell types: epithelial cells on the anterior surface underlying the capsule, and fiber cells that occupy the rest of the lens and comprise the major cellular component of the tissue (Fig. 5.1). Beginning in the embryo and continuing throughout life, the epithelial cells undergo mitosis at the equator and differentiate into elongated fiber cells. This process is characterized by massive accumulation of soluble crystallins and *de novo* synthesis of MP26, lipids, and other components for insertion into the rapidly elongating fiber cell plasma membrane. Electrical and metabolic communications between the fiber cells are established by the formation of junctional complexes that have some of the characteristics of gap junctions found in other tissues. With their differentiation into mature fiber cells, there is a concomitant loss of nuclei and other intracellular organelles. The mature fiber cell, surrounded by its lipid bilayer plasma membrane, contains two major components: the structural soluble (and insoluble) crys-

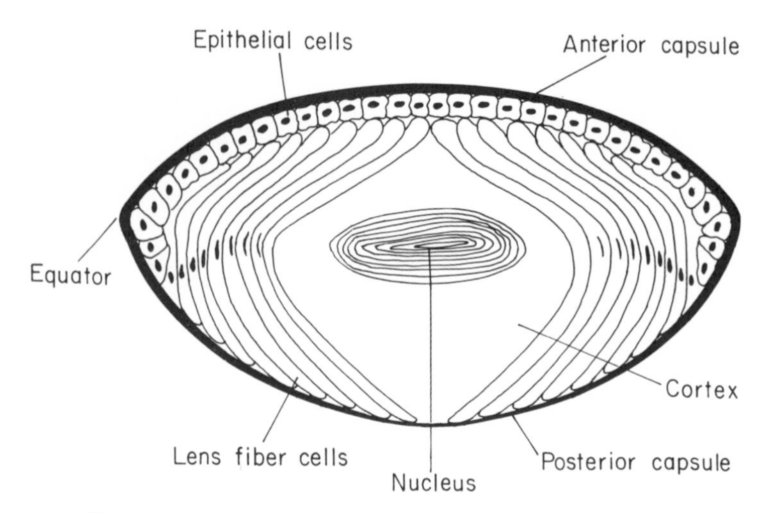

Figure 5.1. Schematic diagram of a mammalian lens in cross section.

tallins and cytoskeletal components. The latter are present only in certain regions of the lens and at certain periods of development. Some polyribosomes are also present in mature lens fiber cells.

New fiber cells are laid down as concentric layers on previously formed embryonic fibers. Thus, the nucleus contains the oldest cells, and the cortex the newest ones. There are no distinct morphological markers or barriers between these regions; rather, general physiological and biochemical gradients distinguish the transition from cortical to nuclear zones. There appears to be little or no turnover of protein in the lens nucleus; most if not all of the detectable protein synthesis occurs in the epithelium and in developing fiber cells in the peripheral cortex. Given the essential absence of new protein synthesis in the bulk of the lens, i.e., in the fiber cells, posttranslational modifications play a major role in modifying the protein composition of the mature lens.

5.1.2. Physical Basis of Transparency

The lens is transparent because it neither absorbs nor scatters significant amounts of light in the visible region of the spectrum. Lack of absorbance is due to the absence of molecular energy levels in the appropriate range for the absorption of photons (see review by Duncan and Jacob, 1984). Lack of light scatter has been more difficult to explain. The amount of light scattered by normal human lens is less than 5%, and most of this is associated with the fiber cell membranes (see review by Bettelheim, 1985). The volume fraction occupied by membranes in the lens is only about 0.05; hence, the cytoplasm is largely transparent. The lens cytoplasm contains about 35% solids, consisting almost entirely of soluble proteins, the crystallins. It is the close packing of these proteins, with essentially no fluctuations in refractive index between them, that forms the basis of lens transparency. Early theories to explain lens transparency based on the ordered structure of lens proteins in a paracrystalline state are no longer tenable. Instead, small-angle x-ray-scattering and light-scattering studies of lens extracts over a wide range of concentrations show convincingly that lens transparency is the result of short-range spatial order of the lens proteins (Delaye and Tardieu, 1983). In contrast to normal lenses, cataractous lenses scatter significant amounts of light. Either singly or in combination, loss of transparency in the cataractous lens is the result of many types of physical and/or biochemical change: formation of water-insoluble aggregates from soluble proteins, syneresis, membrane disintegration, and changes in the cytoskeletal organization of the lens fiber cell, to name but a few (see Section 5.6).

5.1.3. The State of Water

Approximately half of the total water of the lens is water of hydration, defined as bound water that is associated with the nonfreezable water content of the lens (Lahm *et al.*, 1985). Freezable water can be measured by differential scanning calorimetry, and total water by vacuum dehydration. There is a gradual decrease with age of the total water content in the nucleus and intermediate layers of human lens, although cortical layers remain essentially unchanged. The nonfreezable water content decreases with age throughout the lens. A collapse of the protein network causes a loss in the water of hydration and its appearance as bulk water (also called free water). This is considered the primary mechanism for syneresis, a molecular process that increases the refractive index difference between the scattering unit and its surroundings (see review by Bettelheim, 1985). Transparency is thus compromised, and a good correlation has been reported between high total water, low nonfreezable water, and turbidity in microsamples of human lenses removed for cataract surgery (Bettelheim *et al.*, 1986). In contrast, the nonfreezable water content of rat increases with age (Castoro and Bettelheim, 1986). Somewhat similarly in bovine lenses, there are "chronological" changes, or "internal age gradients," that are reflected by the lower nonfreezable water content of young (cortical) regions and the higher content in old (nuclear) regions (Castoro and Bettelheim, 1987).

5.2. CHEMICAL COMPOSITION

5.2.1. Water-Soluble Components

5.2.1.1. Electrolytes and Low-Molecular-Weight Solutes

The lens contains a broad range of electrolytes and low-molecular-weight solutes, some of which are listed in Table 5.1. More detailed information on the water-soluble nonprotein constituents of the lens, particularly their concentrations and distribution in a variety of species, have been summarized in the comprehensive review by Harding and Crabbe (1984). Only a few comments can be made about these substances in the present section. Glutathione is a key component of the lens redox system; its concentration varies with age and with species, being lowest in primates. Most of the glutathione in the lens, as in other cells, is in the reduced form, and it is thought to play an important role in protecting the tissue from oxidative damage (see Section 5.3.2.2). Ascorbic acid is also present at high levels in the lens and may function as an antioxidant. However, ascorbic acid also displays prooxidant properties under some conditions (Wolff *et al.*, 1987). The mechanism of this prooxidative activity appears to be through the formation of free radicals and other active oxygen species such as H_2O_2 in the presence of metal ions.

Table 5.1. Major Amino Acids and Low-Molecular-Weight Solutes in Lenses of Various Species[a]

Component[b]	Rat	Cattle	Rabbit	Human
Taurine	19.7	49.9	6.85	0.79
Serine	0.58	6.3	1.42	0.56
Proline	1.23	1.8	1.05	0.16
Glutamic acid	3.83	225.1	5.83	3.42
Glycine	0.31	93.3	1.79	0.79
Glutathione[c]	10.3	—	17.3	1.43
Alanine	2.21	26.1	1.27	1.34
Ammonia	2.0	0.5	1.01	—
Urea	2.7	3.8	5.98	4.7
Ascorbic acid	30.0	340.0	150.0	300.0
Pyridine nucleotides[d]				
NAD^+	134–460	163	310–988	208–570
NADH	29–324	26	337–487	51–210
$NADP^+$	10–56	33	9.2–18.8	15–20
NADPH	28–332	1.2	17.7–48	11–20

[a]Adapted from Harding and Crabbe, 1984.
[b]Quantities expressed as micrograms per gram wet weight except as indicated.
[c]Oxidized plus reduced (the major form) glutathione.
[d]Pyridine nucleotides are expressed as nanomol per gram wet weight.

Ascorbate is also known to induce cross-linking, cleavage, and insolubilization of crystallins (Garland *et al.*, 1986; Lohmann *et al.*, 1986; Ortwerth and Olesen, 1988a,b; Ortwerth *et al.*, 1988); there is compelling evidence that it plays an important role in the Maillard-type browning reaction found in aging lenses and in cataracts (see Sections 5.5.4 and 5.6.2.2, respectively).

Among other low-molecular-weight substances, pyridine nucleotides have been analyzed in nearly all species but display considerable variability, partly owing to losses during extraction and different methodologies employed (Harding and Crabbe, 1984). Several metabolic intermediates of glucose metabolism have also been analyzed by conventional techniques, but the development of nuclear magnetic resonance provides a more accurate assessment of both static and dynamic levels. Three inorganic ions, K^+, Na^+, and Ca^{2+}, show important metabolic fluxes in normal lens, and Ca^{2+} especially has been implicated in cataract and other lens pathology (see review by Hightower, 1985).

5.2.1.2. The Crystallins

The principal soluble structural components of lens fiber cells are the crystallins. Although considered here as soluble proteins, varying proportions are converted to water-insoluble proteins during aging and cataractogenesis. Three crystallins (α, β, and γ) have been recognized for nearly a century, and more recently, another class, δ crystallin, has been identified and found to be present exclusively in birds and reptiles. All mammalian lenses examined to date have only α, β, and γ crystallins with the exception of guinea pig lens, which contains an additional 38-kDa ζ-crystallin (Huang *et al.*, 1987). Many of the lens crystallins have been sequenced, and cDNA and genomic clones are being used to examine the relationships within and between the various families. From these findings it is now established that the lens crystallins of most species do not comprise three separate families; rather, they consist of two superfamilies, α and βγ (Driessen *et al.*, 1980; Piatigorsky, 1984, 1987; Quax-Jeuken *et al.*, 1985; Wistow and Piatigorsky, 1988; Lubsen *et al.*, 1988). The molecular architecture of most crystallins has been studied by x-ray crystallographic techniques, and their structural motifs established in a large number of cases. Several detailed and in-depth reviews are available on lens crystallins, their composition and structure, and most importantly the primary gene products and post-translational modifications (Bloemendal, 1982, 1985; Piatigorsky, 1984, 1987; Harding and Crabbe, 1984; Clayton, 1985; Lindley *et al.*, 1985; Slingsby, 1985; Wistow and Piatigorsky, 1988; Piatigorsky *et al.*, 1988a; Lubsen *et al.*, 1988). These reviews, some general and others more specific, include considerable information on the evolution and expression of the crystallins and their genes in a variety of species. A vast literature on all of these topics has been generated in recent years, but space limitations allow the inclusion of only a few highlights in this exciting field; original references for most of the discussion that follows are found in these reviews.

The crystallin nomenclature developed over the years in many laboratories is used throughout this chapter; however, it should be noted that nomenclature based on primary gene products has recently been recommended (Bloemendal *et al.*, 1989). This revised nomenclature will undoubtedly replace the many different systems currently in use. Four major families of immunologically distinct crystallins, α, β, γ, and δ, comprise about 90% of the soluble proteins of the lens (Table 5.2). Crystallins are isolated from the water-soluble and water-insoluble fractions of lens homogenates, and they are purified using a combination of physical methods such as ion-exchange chromatography, gel filtration, isoelectric focusing, SDS-PAGE, and chromatofocusing. With the exception of the six γ-crystallins and βs crystallin, which are monomeric, all of the crystallins are multimeric; i.e., they are aggregates containing various combinations of polypeptides. The crystallins are not evenly distibuted throughout the lens; rather, their total concentration increases as a relatively smooth gradient from the cortex to the nucleus. Their differential synthesis during lens development is genetically programmed both temporally

Table 5.2. A Brief Summary of Lens Crystallins[a]

		βγ superfamily		
Property	α	βH[b] βL[b]	γ	δ[c]
Molecular form	Multimeric	Multimeric	Monomeric	Multimeric
Molecular mass (kDa)	600–900	200 to 50	21[d]	200
Polypeptide composition	20 kDa αA2; αB2[e] αA1; αB1	20–30 kDa	21kDa[d]	48 (and 50) kDa
Thiol content	Low	High	High	Low

[a]Data adapted from Harding and Crabbe (1984) and Wistow and Piatigorsky (1988); see also recommendations for crystallin nomenclature by Bloemendal *et al.* (1989).
[b]The major β-crystallin is βB_p (βB2); βs-crystallin belongs to the γ-crystallin family (Bloemendal *et al.*, 1989).
[c]Found only in birds and reptiles, replacing γ-crystallins present in other species.
[d]Six monomeric types of γ-crystallin are known.
[e]αA2 and αB2 are the primary gene products.

and spatially, resulting in an uneven distribution of the individual crystallins throughout the lens fiber cells.

α-Crystallins. In the native state, the α-crystallins are the largest of the lens crystallins, with molecular masses ranging from about 600 to 900 kDa. They appear in the first peak eluted after gel filtration of lens extracts and are similar in all vertebrate species examined. The range of molecular weights of this heterogeneous population is a function not only of age but also of temperature (e.g., 4° or 37°) (Thomson and Augusteyn, 1988) as well as the ionic strength and pH of the extracting buffers (van den Oetelaar *et al.*, 1985). Denaturation of the native aggregates and SDS-PAGE electrophoresis reveals four major 20-kDa subunits, αA1, αA2, αB1, and αB2, and several minor ones in certain species. In the native state, the A and B chains are held together by noncovalent interactions (see review by Spector, 1985). Some rodents contain, in addition, another α-crystallin, αA^{ins}, that is identical to αA2 except that it contains an additional 23-residue peptide inserted between residues 63 and 64 (Cohen *et al.*, 1978; van den Heuvel *et al.*, 1985). The αA2 and αB2 polypeptides isolated from over 40 species show about 60% sequence homology and probably arose by a single duplication of an ancestral α-crystallin gene that gave rise to separate αA and αB genes. Quite unexpectedly, sequence data on small heat-shock proteins of *Drosophila* (Ingolia and Craig, 1982) and on p40 antigen of *Schistosoma mansoni* (Nene *et al.*, 1986) have revealed highly conserved regions similar to the C-terminal sequences of α crystallins. Although no function has yet been assigned to this conserved sequence, it may either have trypsin inhibitor activity or play a structural role.

It is now established that the products of the two α-crystallin genes are the αA2 and αB2 polypeptides; they are single-copy genes with extensive sequence homology. Recent work shows clearly that the αA1 and αB1 chains are polypeptides that have undergone posttranslational modifications. Although once thought to involve deamidation, there is strong evidence that cAMP-dependent phosphorylation of several serine residues in both the A and B chains is a major posttranslational event (Spector *et al.*, 1985). The phosphorylated polypeptides αA1 and αB1 arise by direct phosphorylation of the primary gene products, αA2 and αB2. Serine residue Ser-122 is the principal phosphorylation site in the α-crystallin A chain, but there are additional sites as well (Voorter *et al.*, 1986; Chiesa *et al.*, 1987). The B chain appears to be phosphorylated at two sites: Ser-59 and Ser-43 or Ser-45 (Chiesa *et al.*, 1988). In spite of extensive homology between the A and B chains, the phosphorylation sites are different, being located near the C and N terminals, respectively. The biological significance of phosphorylation of α-crystallins is not known but may be related to their different functional roles in the lens (Chiesa *et al.*, 1989). The A chains are synthesized and phosphorylated mainly, if not entirely, in differentiating fiber cells, where they may have lens-specific function(s); in contrast, synthesis and phosphorylation of the B chains is confined mainly to the epithelial cells, where they could play a more generalized, non-lens-specific role in epithelial physiology.

Spontaneous nonenzymatic cleavage of the primary gene products, leading to the production of truncated forms of α-crystallin, is another posttranslational modification (see review by Harding and Crabbe, 1984). The degradation appears to be stepwise and occurs at the C terminus of both the A2 and B2 chains. Other important transformations of α-crystallins include formation of high-molecular-weight (HMW) aggregates, known for many years to be associated with aging (see Section 5.5.1) and with cataractogenesis (see Section 5.6.1).

A 43-kDa crystallin aggregate has been detected and partially characterized in water-soluble and water-insoluble fractions of human lens. The polypeptide differs from actin (a 42-kDa globular protein) in electrophoretic behavior, immunologic reactivity, and amino acid composition (see reviews by Spector, 1984a, 1985). The 43-kDa aggregate isolated from human cataractous lenses was found to be composed of non-disulfide cross-linked crystallins (Roy *et al.*, 1984). Similar cross-linked polypeptides can be generated from bovine crystallins under oxidative conditions *in vitro* after 40 and 80 h incubation with either Fe^{3+} or Cu^{2+} (Garland *et al.*, 1986). Further studies using bovine lens α-crystallin show that a 43-kDa aggregate composed of both A and B chains can be generated from 20-kDa α-crystallin polypeptides by Fe^{2+}-catalyzed oxidation (McDermott *et al.*, 1988a). Sequence analyses of a 30-kDa fragment cleaved from the 43-kDa after limited proteolysis show that it is composed of three distinct components corresponding to A and B chain polypeptides in the ratio of 1:2. These investigations provided compelling evidence that the A and B chains in the parent 43-kDa aggregate are held together by covalently cross-linked nonreducible bonds.

βγ-Crystallins. The β- and γ-crystallins comprise the majority of adult vertebrate lens crystallins with the exception of birds and reptiles (see Table 5.2). Their nomenclature is operational, being derived from conventional gel filtration profiles; the heterogeneous βH ("heavy") and βL ("light") fractions are eluted in the second and third peaks, respectively, while the last peak contains the monomeric γ-crystallins. Further biochemical, physicochemical, and immunologic criteria are used to resolve and characterize the major aggregates and to elucidate their subunit structures.

A large number of β- and γ-crystallins have been sequenced, and in recent years many of their cDNAs and genes have been characterized. The gene structures and the protein structures deduced from x-ray crystallography show conclusively that an ancestral βγ gene was duplicated to produce both the family of β-crystallin subunits and a closely related family of γ-crystallin monomers (den Dunnen *et al.*, 1985b). More specifically, cDNA and amino acid sequence studies have established a remarkably close structural relationship between two members of these families, βB2 (formerly called βBp) and γII. The most strongly conserved regions in these two crystallins, aligned according to their four motifs, are shown in Fig. 5.2. Moreover, bovine γII, bovine βB2, and the major murine β-crystallin β23, all have two symmetrical domains, each folded into two similar "Greek key" motifs (for reviews see Piatigorsky, 1984; Harding and Crabbe, 1984; Wistow and Piatigorsky, 1988). The fourfold repeat in the tertiary structure can be seen in the primary structure (Fig. 5.2).

The β- and the γ-crystallin gene families contain eight and six primary gene products, respectively, and together they comprise the βγ-crystallin superfamily (for reviews see Harding and Crabbe, 1984; Piatigorsky, 1984, 1987; Quax-Jeuken *et al.*, 1985; Wistow and Piatigorsky, 1988). The highly homologous γ-crystallin genes are closely linked gene clusters: the six γ-crystallin genes of the rat are located in head-to-tail orientation on chromosome 9 (Moormann *et al.*, 1985; den Dunnen *et al.*, 1987), and those of the human genome are located on the long arm of chromosome 2, region q33–36 (den Dunnen *et al.*, 1985a; Shiloh *et al.*, 1986). In contrast, owing to the extensive heterogeneity of the β-crystallins, it has been predicted that their genes would be dispersed throughout the genome (Hogg *et al.*, 1987). There appear to be two linked βB2 genes, one of which has been mapped to chromosome 22,q11.2–q12.2; another human β-crystallin, βA3/A1, is located on chromosome 17.

The β-crystallins are extremely complex, consisting of a polydisperse group of

	3			6	7	11	13	34				37
γIIA	– Ile	– Thr	– Phe	– Tyr	– Glu ...	Phe ...	Gly ...	Ser	– Ile	– Arg	– Del	– Val
βBpA	– Ile	– Ile	– Ile	– Phe	– Glu ...	Phe ...	Gly ...	Ser	– Val	– Leu	– Del	– Val
βm A	– Ile	– Thr	– Ile	– Tyr	– Asp ...	Phe ...	Gly ...	Ser	– Leu	– Lys	– Del	– Val
γf A												
				46		51	53	73				
γIIB	– Trp	– Phe	– Val	– Tyr	– Glu ...	Tyr ...	Gly ...	Ser	Cys	– Arg	– Leu	– Ile
βBpB	– Trp	– Val	– Gly	– Tyr	– Glu ...	Cys ...	Gly ...	Ser	– Leu	– Arg	– Pro	– Ile
βm B	– Trp	– Ile	– Gly	– Tyr	– Glu ...	Phe ...	Gly ...	Ser	– Phe	– Arg	– Pro	– Ile
γf B	– Trp	– Met	– Leu	– Tyr	– Glu ...	Tyr ...	Gly ...	Ser	– Cys	– Arg	– Val	– Ile
				89		94	96	117				
γIIC	– Met	– Arg	– Ile	– Tyr	– Glu ...	Phe ...	Gly ...	Ser	– Val	– Arg	– Del	– Val
βBpC	– Ile	– Thr	– Leu	– Tyr	– Glu ...	Phe ...	Gly ...	Ser	– Val	– Arg	– Del	– Val
βm C	– Ile	– Thr	– Asn	– Phe	– Glu ...	Phe ...	Gly ...	Ser	Met	– Lys	– Del	– Ile
γf C	– Leu	– Arg	– Ile	– Tyr	– Glu ...	Phe ...	Gly ...	Ser	– Cys	– Lys	– Del	– Val
				125		130	132	157				
γIID	– Trp	– Val	– Ile	– Tyr	– Glu ...	Tyr ...	Gly ...	Ser	– Leu	– Arg	– Arg	– Val
βBpD	– Trp	– Val	– Gly	– Tyr	– Gln ...	Tyr ...	Gly ...	Ser	– Val	– Arg	– Arg	– Ile
βm D	– Trp	– Val	– Cys	– Tyr	– Gln ...	Tyr ...	Gly ...	Ser	– Ile	– Arg	– Arg	– Ile
γf D	– Trp	– Ile	– Leu	– Tyr	– Glu ...	Tyr ...	Gly ...	Ser	– Phe	– Arg	– Arg	– Val

Figure 5.2. Conserved residues in the known amino acid sequences of β and γ crystallins. The least variable amino acids are aligned according to the four motifs of the γII crystallin structure. The completely conserved glycine and serine residues are boxed. Note that the sequences corresponding to βB_p (βB2) and γII are the most highly conserved. (From Harding and Crabbe, 1984).

heteropolymers ranging in size from 50 to 200 kDa. Each of the six or seven polypeptides in this group is composed of varying numbers of 20- to 30-kDa subunits. In addition, a low-molecular-weight monomer, βs, has for many years been classified with the β-crystallins. However, cDNA sequencing shows that although βs has some characteristics of both members of the βγ superfamilies, it is more closely related to the γ- than to the β-crystallins (Wistow and Piatigorsky, 1988; Bloemendal *et al.*, 1989). Comparisons of βs-crystallins from various mammalian species with γ-crystallins show that the βs-crystallins are similar to one another in amino acid composition, size, charge, and secondary structures; there are, however, subtle but important differences between the βs- and γ- crystallins in amino acid composition and tertiary structure (Thomson *et al.*, 1989).

Eight β-crystallins have been characterized as primary gene products and classified into βA (acidic) and βB (basic) crystallins. The basic β-crystallins have both N- and C-terminal peptides extending from their four-motif core structures, together with a few small insertion sequences, whereas the acidic β-crystallins lack prominent C-terminal extensions but have many insertions. Three-dimensional models of β-crystallins have been constructed using interactive computer graphics techniques (Slingsby *et al.*, 1988). The sequences of the major β-crystallin polypeptide βB2, as well as a high-molecular-weight component βB1, and a representative of an acidic β-crystallin βA1/A3, were compared, and their common origin with γII-crystallin established. These three-dimensional models generate precise measurements of N- and C-terminal domains and suggest that the acidic and basic families have diverged in different ways from a common ancestor.

One of the β-crystallin polypeptides, βB2, is phosphorylated, as shown in recent studies on bovine lens (Kleiman *et al.*, 1988). Incubation of whole lenses with [^{32}P]orthophosphate results in the phosphorylation of a 27-kDa polypeptide that elutes in the βL-crystallin fraction isolated by gel filtration. The only β-crystallin that is phosphorylated is βB2, and the site of phosphorylation is Ser-203. Moreover, purified βB2 is a substrate for phosphorylation by cAMP-dependent protein kinase. These findings, as well as previous ones on α-crystallins, clearly establish that there are only three phosphorylated crystallin polypeptides in the lens: the A and B chains of α-crystallin and 27-kDa βB2-crystallin. The phosphorylation of specific crystallin species suggests that these polypeptides may be playing a hitherto unsuspected dynamic role in lens metabolism in addition to their structural role.

The γ-crystallins comprise a group of six homologous 21-kDa monomeric proteins. Although prominent in the lenses of most vertebrate species, they are absent in birds and possibly in some reptiles. The γ-crystallins are greatly enriched in the lens nucleus, but with aging they undergo oxidative, and other, changes, leading to their insolubilization (see Section 5.5.1). Of historical interest, the first lens crystallin to be sequenced was γII-crystallin (Croft, 1972); subsequently, complete primary sequences as well as the three-dimensional structures of other γ-crystallins have been determined (for reviews see Schoenmakers *et al.*, 1984; Wistow and Piatigorsky, 1988; Lubsen *et al.*, 1988). The secondary structures of the γ-crystallins are very similar, all being highly symmetrical and having β-pleated sheet structures. Their three-dimensional structures consist of four antiparallel β-helical motifs arranged as two symmetrical globular domains. Each in turn consists of two nearly identical "Greek key" motifs. There are, however, differences in core packing and in the number of exposed hydrophobic residues and ion pairs in mammalian γ-crystallins (Summers *et al.*, 1986). In addition, despite their structural similarities and high degree of sequence homology, there are inherent differences in their tertiary structures (Mandal *et al.*, 1987a). This has been demonstrated using bovine γ-crystallins II, III, and IV, which, under carefully controlled conditions, show striking differences in denaturation behavior and susceptibility to proteolysis.

The γ-crystallins of all species have a remarkably high thiol content, and this class of crystallins, perhaps more than others, is thought to play an important role in lens transparency (Blundell *et al.*, 1983). Amino acid analyses and x-ray crystallography show that γII, the major γ-crystallin in bovine lens and the most extensively studied, contains seven cysteine residues, six of them located in the N-terminal domain (Wistow *et al.*, 1983; Summers *et al.*, 1986). It is thought that exposed cysteine residues may interact with aromatic molecules such as tryptophan, provide pathways for electron transfer, and possibly even play a role in redox reactions. Studies on the reactivity of sulfhydryl groups of calf γII-, γIII-, and γIV-crystallins using a thiol-specific fluorescent probe demonstrate two classes of reactive cysteines with marked differences in spatial arrangements and microenvironments (Mandal *et al.*, 1987b). Other studies on amino acid compositions and N-terminal sequences in five γ-crystallins isolated from calf lens confirm the seven cysteine residues in γII. These studies also showed that γIIIa, γIIIb, and γIV have four to five cysteines, and βs has six (McDermott *et al.*, 1988b). Reduction of the γ-crystallin fraction with DTT causes an increase in sulfydryls, 81% of which can be accounted for in the γII fraction. The two residues forming the intramolecular disulfide bond of γII-crystallin are Cys-18 and Cys-22, a combination of cysteine residues that was not present in the other γ-crystallins examined.

δ-Crystallins. The δ-crystallins are the major soluble proteins of avian and reptile lenses, replacing γ-crystallins in these species (see reviews by Piatigorsky, 1984, 1987; Wistow and Piatigorsky, 1988). δ-Crystallin accounts for about 70% of the crystallins in developing chicken lens, but δ-crystallin mRNA disappears a few months after hatching, and the protein is no longer synthesized in mature chicken lens. The native molecule is a 200-kDa tetramer consisting of major 48-kDa and minor 50-kDa subunits. They can be synthesized *in vitro*, the relative proportions produced varying with the ionic strength and composition of the medium. Rather unexpectedly, mRNA produced from chicken δ1 cDNA synthesizes both subunits, yet two closely related δ-crystallin genes (δ1 and δ2) are present in the chicken. Although all reported cDNAs appear to be derived from the δ1 gene (see review by Piatigorsky, 1987), low levels of δ2 mRNA are present in embryonic chicken lens. Thus, both genes are transcriptionally active but appear to be differentially expressed. A possible explanation has been found by comparing the transcriptional ability of the core promoters for the two genes in a HeLa cell extract (Das and Piatigorsky, 1988). The results indicate that the δ1 promoter is more efficient than the δ2 promoter in initiating transcription.

Crystallins as Enzymes. Unexpected evolutionary relationships between the α-crystallins and nonlenticular proteins such as small heat-shock proteins and *Schistosoma* p40 antigen have been pointed out. Even more remarkable are several recent observations regarding the recruitment of enzymes as lens crystallins. For example, ε-crystallin, a major component of duck lens, expresses a high level of lactic dehydrogenase (LDH) activity that appears to be identical to that of duck heart LDH-B4 (Wistow *et al.*, 1987). The enzyme may have been recruited to play an extra role as a structural protein in the lens without gene duplication and divergence. Other "taxon-specific" crystallins closely related to well-known cellular enzymes have also been reported: δ2-crystallin of chicken has considerable sequence similarity to yeast and human argininosuccinate lyase, a urea cycle enzyme (see Fig. 7.20) (Wistow and Piatigorsky, 1987; Piatigorsky *et al.*, 1988b). In ducks and possibly in chickens, the same gene encodes these two different functions, δ2-crystallin and argininosuccinate lyase. Other examples include turtle τ-crystallin and squid S_{III}, which have sequence similarities and enzymatic properties of enolase and glutathione S-transferase, respectively (Wistow and Piatigorsky, 1987). In addition, ρ-crystallin of *Rana pipiens* shows approximately 40–50% sequence homologies to rat lens NADPH-dependent aldehyde reductase and to human liver aldose reductase (Carper *et al.*, 1987). Another recent report describes similarities between the 225 C-terminal amino acid sequences of the European commmon frog ε-crystallin and bovine lung prostaglandin F synthase; the finding of 77% identical and conservative substitutions without deletions or additions suggests that the proteins may be identical (Watanabe *et al.*, 1988). It is of interest to note that ε-crystallin of European common frog lens is different from the ε-crystallin of duck lens, the latter being identical to lactic dehydrogenase.

The long-held view that lens crystallins are "lens-specific" may have to be revised. A provocative hypothesis put forward recently implies that certain crystallins were recruited from nonlenticular tissues as lens proteins during evolution because of their thermodynamic stability; carried further, it has been predicted that nonlenticular functions may some day be found for all of the lens crystallins (Wistow and Piatigorsky, 1987; Piatigorsky *et al.*, 1988b). In this context, αB-crystallin (but not αA) has been detected in other tissues such as heart and lung. There is little doubt that "the proteins we know and love as crystallins are hand-me-downs well used in some earlier role" (Harding and Crabbe, 1984).

5.2.2. Structural Components

5.2.2.1. The Lens Fiber Cell Plasma Membrane

Effective methods have been developed for isolating highly purified fiber cell plasma membranes, thus opening the way for studying their important structural, biochemical, and physiological properties. Homogenization of decapsulated lens in dilute aqueous buffer followed by centrifugation yields a supernatant fraction containing the water-soluble crystallins and a pellet. The former comprises about 80–90% of the lens protein and is called the water-soluble (WS) fraction. The pellet, or insoluble fraction, contains, after extensive washing, three structural components: plasma membranes, some cytoskeletal proteins, and water-insoluble crystallins (see review by Alcala and Maisel, 1985). Most but not all of the water-insoluble crystallins and some of the cytoskeletal proteins can be solubilized in 7–8 M urea; this has been termed the urea-soluble (US) fraction. The residue remaining, termed the urea-insoluble (UI) fraction, consists mainly of lens fiber cell plasma membranes whose major components are phospholipids, integral (or intrinsic) proteins, and peripheral proteins. The major portion, 94%, of the crude lens plasma membrane fraction consists of fiber cell membranes (Alcala and Maisel, 1985).

MP26 and Gap Junction Protein(s). Gel electrophoresis under dissociating conditions (SDS-PAGE) of the crude plasma membrane fraction reveals several components, the most abundant being a 25- to 27-kDa polypeptide (Broekhuyse and Kuhlmann, 1974; Alcala *et al.*, 1975). This protein, shown to be present as a 26-kDa polypeptide in a wide variety of species, was named MP26 (26-kDa membrane protein) (Bloemendal *et al.*, 1977). The close sequence homology between frog, human, and other species suggests that MP26 has been highly conserved throughout evolution (Takemoto *et al.*, 1981). Using the bovine cDNA for MP26 isolated by Gorin *et al.* (1984) as a probe, the gene for human MP26 has been assigned to the cen-q14 region of the long arm of chromosome 12 (Sparkes *et al.*, 1986).

The importance of MP26 lies in its putative role as the main, if not only, gap junction protein of the lens fiber cell. This provocative notion is currently under dispute for several reasons: experimental findings are often in conflict, and lens fiber cell junctions appear to be atypical gap junctions. On the other hand, lens *epithelial* cells are believed to have typical gap junctions but do not contain MP26, a protein that is synthesized only by differentiating fiber cells (for reviews see Alcala and Maisel, 1985; Revel, 1985, 1987). Lens fiber cell plasma membranes are electrically and metabolically coupled through junctions that constitute 50–60% of the plasma membrane surfaces. MP26 is the principal intrinsic protein of fiber cell membranes and is localized in the junctional domains, as shown by studies using a monoclonal antibody specific for MP26 (Sas *et al.*, 1985).

There is additional circumstantial evidence that MP26 may be the principal gap junction-forming protein. A model of MP26 based on amino acid sequences deduced from cDNA clones suggests the presence of six membrane-spanning hydrophobic α-helices (characteristic of many intrinsic membrane proteins) and one amphiphilic transmembrane segment that could form an aqueous channel (Gorin *et al.*, 1984); moreover, reconstituted bovine lens junctions, consisting mainly (though not entirely) of MP26, form voltage-dependent channels (Zampighi *et al.*, 1985). More convincingly, freeze-fracture and immunogold labeling of purified MP26 in reconstituted liposomes show membrane-to-membrane close contacts and organized domains in the plane of the lipid bilayer similar to junctional complexes seen *in situ* (Dunia *et al.*, 1987). Nevertheless, arguing against MP26 as the gap junction protein of lens fiber cells is its presence in nonjunctional domains of the fiber cell membrane and its lack of sequence homology with gap junction proteins of either liver or heart (Revel *et al.*, 1987). It thus appears that MP26 may function in cell-to-cell channels in the lens, although its relationship to gap junction proteins of other tissues is unclear.

Given these uncertainties, another intrinsic membrane protein, MP70, has been implicated as a gap junction protein in lens fiber cell plasma membranes. Combined use of immunofluorescence microscopy and SDS-PAGE reveals concentrated domains of MP70 in the outer lens cortex that coincide with 16- to 17-nm fiber junctions in the same regions (Kistler *et al.*, 1985). However, the amounts of MP70 in the total lens plasma membranes are too small to account for the large number of junctions seen by electron microscopy. Moreover, MP70 appears to be confined to the cortical layer. Later work showed that the absence of MP70 in the inner cortex and nucleus is the result of endogenous proteolytic degradation of MP70 to MP38; this cleavage is effected by a Ca^{2+}-activated age-related endogenous lens protease (Kistler and Bullivant, 1987).

MP26 undergoes two major posttranslational modifications that are thought to modulate its functional activity: one is phosphorylation, and the other proteolytic degradation. Both MP26 and a smaller (19- to 20-kDa) intrinsic membrane protein are phosphorylated *in vitro* by an endogenous lens cAMP-dependent protein kinase (Johnson *et al.*, 1985; Garland and Russell, 1985; Louis *et al.*, 1985b). The major residue phosphorylated in MP26 isolated from decapsulated fragments of fetal calf lens is serine; in addition, some threonine is also phosphorylated, but tyrosine is not (Johnson *et al.*, 1986). This phosphorylation may be mediated by G proteins acting through a Ca^{2+}–calmodulin adenylate cyclase system (Louis *et al.*, 1987). There is, however, other strong evidence suggesting that these intrinsic plasma membrane proteins are also phosphorylated by an endogenous protein kinase C (Lampe *et al.*, 1986). This Ca^{2+}- and phospholipid-dependent enzyme appears to phosphorylate serine residues in the C-terminal region of MP26. Both phosphorylating systems are known to alter junctional permeability in other tissues, and they may be acting in concert in the lens to regulate the phosphorylation of MP26, which in turn could control the permeability of lens gap junctions.

The other posttranslational modification of MP26 is its slow age-related conversion to MP22 by limited proteolysis (Roy *et al.*, 1979; Broekhuyse and Kuhlmann, 1980). The two polypeptides are structurally and immunologically related (Zigler and Horwitz, 1981), and the physiological conversion is noticeable both in comparisons of old and young human lenses and in analyses of old (nuclear) and young (cortical) regions from a single lens. The conversion of MP26 to MP22 is accelerated in human and experimental cataract, and a protein similar in size to MP22 can be produced by a variety of exogenous proteases such as trypsin and V8 proteases. Several possible cleavage sites resulting in the formation of a low-molecular-weight peptide from intact MP26 have been suggested (Gorin *et al.*, 1984), and further details of its age-related cleavage are given in Section 5.5.2. Electron microscopic studies raise the possibility that two distinct classes of fiber cell junctions differing from each other in thickness are associated with the two different forms of MP26, intact and degraded (FitzGerald, 1987).

The gene for MP26 appears to be expressed only in the lens, specifically in its fiber cells and not in the epithelial cells. Recently, through the use of antisense RNA probes, polyclonal antibodies specific for MP26, and *in situ* hybridization, the expression of MP26 mRNA has been investigated in normal rat lens. One study showed that in the developing rat lens, the transcription of the gene for MP26 and the synthesis of MP26 are directly coupled in cells committed to terminal differentiation (Yancey *et al.*, 1988). The synthesis of MP26 occurs first in presumptive primary fibers and, after little or no time lag, in differentiating secondary fibers in the zone of elongation. The pattern of expression of the MP26 gene differs sharply from that of the βA1/A3 gene not only in cellular localization but also in the earlier appearance of the crystallin gene in differentiating cells. Similar findings have been reported in other studies using 3- to 4-week-old rat lenses (Bekhor, 1988). The MP26 mRNA transcripts appear first in elongating fibers at the equator; concentrations are reduced from the anterior fibers to the nucleus and then increase again in posterior fiber cells. In confirmation of previous studies, no gene transcription for MP26 mRNA could be detected in epithelial cells.

Other Structural Proteins. Nearly 20 proteins, in addition to MP26, that are associated with lens plasma membranes of various species have been described in a detailed and highly informative review (Alcala and Maisel, 1985). This section deals mainly with findings reported since that review appeared.

In addition to MP26 and MP70, at least one other protein (MP17) and several glycoproteins have also been characterized as intrinsic proteins of lens fiber cell plasma membranes. The phosphorylation of a 19- to 20-kDa protein by both cAMP-dependent protein kinase and protein kinase C was discussed above. A plasma membrane polypeptide of similar size, 17–19 kDa, has calmodulin-binding properties (Louis *et al.*, 1985a; Broekhuyse and van den Eijnden-van Raaij, 1986). These two peptides probably represent the same protein, which has now been designated MP17 (Mulders *et al.*, 1988). Use of hydrophobic photolabeling shows that it is the second most abundant lens membrane protein after MP26. The intrinsic membrane protein MP17 is present in several mammalian species but is absent in chickens. Moreover, similar to two other intrinsic proteins, MP26 and MP70, MP17 occurs only in fiber cells but not in the epithelium. Because of its calmodulin-binding activity and its ability to act as a substrate for cAMP-dependent protein kinase, MP17 may play an important regulatory role in lens metabolism.

The lens fiber cell membrane fraction contains about 3–4% carbohydrate consisting mainly of neutral sugars and amino sugars (Alcala and Maisel, 1985); all of them are components of membrane-bound glycolipids and glycoproteins. In recent years several of the glycoproteins have been identified using highly sensitive and specific lectins. A prominent 130-kDa Con A-binding glycoprotein is present in chicken and mammalian lens plasma membranes (Heslip *et al.*, 1986), and the use of multiple lectins has revealed glycoproteins with molecular weights ranging from 35,000 to 140,000 (Russell and Sato, 1986).

Another well-characterized component of lens fiber cell plasma membranes, termed EEP (EDTA-extractable protein), consists of two polypeptides of M_r 33,000 and 35,000. These proteins are bound to the phospholipid bilayer by Ca^{2+} and can be eluted from membranes with chelators of divalent cations. Hence, EEP is an extrinsic protein. It is present in a variety of species and is a substrate for phosphorylation by Ca^{2+}- and phospholipid-dependent protein kinase C *in vitro* (van den Eijnden-van Raaij *et al.*, 1987). Phosphorylation of lens plasma membrane proteins is not limited to EEP, since serine and, to a lesser extent, threonine residues of both MP26 and MP17 are also phosphorylated by protein kinase C and by cAMP-dependent adenylate cyclase. A recent study has shown that the 35-kDa polypeptide of EEP of bovine lens plasma membranes is immunologically related to calpactin I and contains phosphotyrosine residues (Russell *et al.*, 1987a). The 33-kDa protein from EEP may be the core protein of calpactin I. This protein is associated with actin, and its phosphorylation by specific tyrosine kinases implies a possible role in differentiation.

Development of monoclonal and polyclonal antibodies raised to a 115-kDa cytosolic lens antigen has led to the characterization of another extrinsic protein of lens plasma membranes (FitzGerald, 1988a). This protein is lens-specific and is localized in fiber cell plasma membranes. The 115-kDa polypeptide appears to be immunologically related to a previously described 95-kDa antigen (Ireland and Maisel, 1984a), a protein that was thought to be associated with beaded filaments, unusual structural components of chicken and other vertebrate lenses.

Several enzymes have been detected in lens plasma membranes using ultrastructural cytochemical and biochemical techniques (see review by Alcala and Maisel, 1985). The most extensively studied, Na^+,K^+-ATPase, is localized mainly in the epithelial cells and in outer cortical fiber cell membranes; as in most other cells, this enzyme is responsible for the energy-dependent inward transport of two K^+ and the outward transport of three Na^+ (see Chapter 1, Section 1.1.1.5). Another enzyme, a membrane-associated form of glyceraldehyde 3-phosphate dehydrogenase (of M_r 37,000), is loosely bound to lens membranes and therefore has been classified as an extrinsic protein. More recently, a Ca^{2+}-stimulated ATPase has been characterized in membrane-enriched fractions isolated from the epithelial and cortical regions of rabbit and bovine lenses (Borchman *et al.*, 1988). Its activity is only about one-fifth of that observed for Na^+,K^+-ATPase.

Lipids. The lens fiber cell plasma membranes, like those of most other cell types, are composed of approximately equal proportions of proteins and lipids (for reviews see Zelenka, 1984; Alcala and Maisel, 1985). Although the proteins have been extensively studied, relatively less attention has been directed toward the lipids apart from their overall characterization and a limited number of metabolic studies.

Cholesterol is the major sterol of lens plasma membranes, accounting for about 50–60% (or more in some species) of the total lipids. The ratios of cholesterol/phospholipid in chick lens fiber cell membranes and in gap junction membranes are 2.1 and 3.1, respectively, implying a localized accumulation of cholesterol in junctional areas (Alcala *et al.*, 1982/1983). Sphingomyelin is also relatively enriched in gap junction membranes. The presence of high concentrations of sphingomyelin and cholesterol and low levels of unsaturated fatty acids would contribute significantly to the rigidity of these membranes. The cholesterol content of lens plasma membranes can be modulated by certain drugs. For example, treatment of newborn rats with U18666A, an inhibitor of cholesterol synthesis

Table 5.3. Phospholipid Composition of Lens Fiber Cell Membranes in Various Species[a]

	Phospholipid[b] (% of total)				
Species	SPH	PC	PE	PS	PI
Human	47–56	2–5	9–18	6–15	1–4
Bovine	18–23	25–32	34	8–13	2
Rabbit	14–32	23–35	28–34	11–13	2
Rat	6–16	31–43	30–33	12–16	2–4
Monkey	46	7	20	9	1
Chicken	29	23–27	25–39	7	3

[a]Adapted from Zelenka (1984).
[b]The following abbreviations are used: SPH, sphingomyelin; PC, phosphatidylcholine; PE, phosphatidylethanolamine; PS, phosphatidylserine; PI, phosphatidylinositol.

and a potent hypocholesterolemic drug, causes a selective decrease in cholesterol content of both the total lens fiber cell plasma membranes and gap junction-enriched membranes (Fleschner and Cenedella, 1988). This restricted synthesis of gap junctional membrane is associated with a high incidence of cataract formation in these animals.

The phospholipid composition of lens plasma membranes is shown in Table 5.3. All species examined contain unusually high levels of sphingomyelin. Of the other phospholipids, phosphatidylethanolamine (PE) is present at somewhat higher levels than phosphatidylcholine (PC) in most species; human plasma membranes appear to have a strikingly low PC content. Lens plasma membranes contain mainly long-chain saturated fatty acids; the major species is palmitate (16:0), which accounts for one-third or more of the total. Two monounsaturated fatty acids, 18:1 and 24:1, are also prominent in lens plasma membranes (see review by Zelenka, 1984). The consequence of the high concentrations of sphingomyelin, cholesterol, and saturated fatty acids is a lipid bilayer of unusually low fluidity; in fact, fluorescence anisotropy measurements show that the lens plasma membrane is the least fluid of any eukaryotic membrane studied. In the native state, there is considerable evidence suggesting that protein–lipid interactions reduce and maintain the low fluidity of lens plasma membranes (see review by Zelenka, 1984).

Most of lens lipid biosynthesis occurs in the epithelium and outer cortex, with relatively little activity detectable in the nucleus. Incorporation of acetate into several of the lipid classes has been demonstrated in cultured bovine lens (Albers-Jackson and Do, 1987). Tritiated water is actively incorporated into digitonin-precipitable sterols in young rats, but there is a rapid decline in *de novo* biosynthesis of cholesterol in older animals (Cenedella, 1982). Cholesterol synthesis is regulated in part by an active hydroxymethylglutaryl (HMG) CoA reductase that has been demonstrated in cultured bovine lens epithelial cells (Hitchener and Cenedella, 1987). There is good correlation between rates of cholesterol synthesis, as measured by incorporation of tritiated water into digitonin-precipitable sterols, and enzyme activity. Both sterol synthesis and enzyme activity are increased in cells grown in lipoprotein-depleted medium, a situation that mimics the *in vivo* state, since the lens is not exposed to significant levels of lipoprotein.

Phosphatidylinositol (PI), although present at only low levels in the lens (Table 5.3), may, as in other tissues, play an important role in receptor-mediated membrane signal transduction (see Chapter 1, Section 1.3.2). van Heyningen (1957) reported unusually high concentrations of inositol in mammalian lenses, with values ranging from about 4 mM in the rat to about 40 mM in the human lens. This important observation remained dormant for many years, although it had been noted that inositol is actively transported into the lens. Later studies showed that ^{32}P is incorporated into PI of embryonic chick lens; the differentiation of embryonic chick lens epithelial cells into fiber cells is accompanied by increased synthesis and decreased turnover of PI (Zelenka, 1980). The half-lives of PI in epithelial and fiber cells are 5 and 63 h, respectively. More recently, the synthesis of phosphatidylinositol 4,5-bisphosphate (PIP_2) from labeled inositol and its hormone-stimulated breakdown to diacylglycerol (DG) and inositol trisphosphate (IP_3), has been demonstrated in rabbit lens epithelial cells (Vivekanandan and Lou, 1989). The initial product formed is phosphatidylinositol, which is rapidly phosphorylated by two phosphoinositide kinases to PIP_2. The breakdown of this key intermediate, presumably mediated by lens phospholipase C activity, is stimulated by a number of physiological agonists such as glucagon, epidermal growth factor, serotonin, and vitreous humor. Hormone-stimulated PI breakdown is correlated with the rate of epithelial cell division in the lens and may have other, as yet unknown, regulatory functions in this tissue.

5.2.2.2. Cytoskeleton

All three classes of cytoskeletal proteins found in nonocular tissues (see Chapter 1, Section 1.1.1.3) are also present in the lens. Techniques for their isolation and characterization have been improved in recent years, and the most commonly used method involves homogenization of the lens in Mg^{2+}-containing buffer of high ionic strength as the first step (see review by Alcala and Maisel, 1985). After centrifugation and repeated washings, the water-insoluble pellet is extracted in 6–8 M urea. Cytoskeletal components such as actin, vimentin, and spectrin present in this crude extract are purified by ion-exchange chromatography and/or SDS-PAGE. Beaded filament, a cytoskeletal component unique to the lens, is also isolated from the urea-soluble (US) fractions. Another cytoskeletal component, termed band 4.1 protein, a prominent component of the lattice network that lines the cytoplasmic face of erythrocyte plasma membranes, has also been detected in bovine (Aster *et al.*, 1984) and chicken (Granger and Lazarides, 1985) lens. At least six variants have been characterized, and their expression changes significantly during terminal differentiation of lens fiber cells.

Actin microfilaments have been detected by morphological as well as immunocytochemical methods in both epithelial and fiber cells in a wide variety of mammalian lenses (Alcala and Maisel, 1985). Lens actin has been identified biochemically by its isoelectric point of 5.5 and its amino acid composition, which is similar to that of muscle actin. Intermediate filaments (IFs) comprise a second class of lens cytoskeletal components. The major IF in lens, vimentin, is localized mainly in the cell cytoplasm and is highly conserved in a wide variety of species (Ellis *et al.*, 1984). Vimentin has been identified in epithelial and cortical fiber cells but is absent from nuclear fiber cells.

Microtubules, the third class of cytoskeletal proteins, have been detected in epithelial and cortical fiber cells and, like vimentin, are not found in nuclear fiber cells. The cytoskeleton is thought to play a major role in lens fiber cell differentiation, particularly in the elongation of the cell bodies (see review by Alcala and Maisel, 1985).

Beaded-chain filaments, cytoskeletal components that appear to be unique to the lens, have been identified in the fiber cells of chicken and several other vertebrates lenses. The name is derived from their characteristic appearance: globular protein particles about 12 nm in diameter attached to a 5-nm-diameter filament backbone. Two major urea-soluble (US) proteins of M_r 86,000–95,000 and 47,000–49,000 are associated with these filaments in chick lens (Ireland and Maisel, 1984a). The 49-kDa polypeptide of chick lens is distinguishable from actin not only in size and amino acid composition but also in lack of immunologic cross-reactivity. The beaded-chain filaments are thought to act as attachment sites for crystallins.

Phosphorylation of cytoskeletal proteins plays an important role in their interactions with one another and in their associations with cellular membranes in a variety of nonocular tissues. The functional components for at least two phosphorylating systems have been detected in lens: cAMP-dependent protein kinase (type I) and Ca^{2+}- and phospholipid-dependent protein kinase C. Endogenous substrates for protein kinases are MP26 and calpactin I, the latter being phosphorylated by a specific tyrosine kinase. Cytoskeletal components of the chick lens are also phosphorylated after incubation with [^{32}P]orthophosphate (Ireland and Maisel, 1984b). The phosphorylation of both vimentin and 47-kDa beaded filament protein is stimulated by β-adrenergic drugs such as isoproterenol and epinephrine; moreover, phosphorylation is increased by forskolin, an agent that directly activates the catalytic site of adenylate cyclase (Ireland and Maisel, 1987). These responses appear to be mediated by β-adrenergic receptors coupled to the adenylate cyclase system, and the phosphorylation of lens cytoskeletal proteins could play an important role in fiber elongation and differentiation, as well as in accommodation.

5.2.2.3. Lens Capsule

The lens capsule has been recognized as a specialized type of collagen since the pioneering work of Pirie (1951); later investigations by Dische and others have been reviewed by Harding and Crabbe (1984). Now classified as a typical basement membrane secreted by, and tightly attached to, the epithelium, the lens capsule acts as an important permeability barrier for substances entering and leaving the lens. Its principal structural component is type IV collagen, and lens capsule is often used as a model for studying the chemistry and immunology of type IV collagens in other tissues (Brinker *et al.*, 1985). The chemical composition of basement membrane collagen from bovine lens capsule is typical for type IV collagen; glycine comprises approximately one-third of its residues, and it contains about 8% carbohydrate (Table 5.4). The structure of lens capsule type IV collagen, as determined by rotary shadowing electron microscopy, is predominantly that of a triple helix; it also contains nonhelical globular domains (see Chapter 1, Section 1.2.1.1). Type IV collagen does not show typical 67-nm banding; rather, it forms an open nonfibrillar network composed of tetrameric units joined at their N termini.

5.3. METABOLISM

Detailed reviews of lens metabolism by Harding and Crabbe (1984) and Cheng and Chylack (1985) include all aspects of this extensively studied field up to about 1982. Their chapters have provided the background for the abbreviated and updated discussion of lens metabolism that follows.

Table 5.4. Chemical Composition of Bovine Anterior Lens Capsule Procollagen[a]

Component	Abundance[b]
Amino acids[c]	
3-Hydroxyproline	12.10
4-Hydroxyproline	82.50
Aspartic acid	40.50
Threonine	31.80
Serine	51.85
Glutamic acid	105.40
Proline	67.30
Glycine	264.00
Alanine	44.40
Valine	30.72
Isoleucine	28.90
Leucine	56.15
Phenylalanine	33.90
Hydroxylysine	35.30
Lysine	21.20
Arginine	38.00
Sugars	
Galactose	3.0
Glucose	3.7
Mannose	0.7
Glucosamine	0.4
Galactosamine	0.1

[a]Adapted from Brinker *et al.* (1985).
[b]Abundance of amino acids is expressed as residues per 1000 amino acid residues; that of sugars in percentage of dry weight.
[c]Includes only the major amino acids.

5.3.1. Carbohydrate Metabolism

The avascular lens relies on the aqueous humor as the major source of oxygen, glucose, and other nutrients needed to support its normal metabolic activity. Metabolism of glucose via glycolysis provides at least two-thirds of the lens ATP. The remainder is generated by oxidative metabolism through the tricarboxylic acid cycle; this occurs mainly, if not entirely, in the epithelium, since fiber cells are devoid of mitochondria. It is the single layer of epithelial cells that contains the majority of enzymes involved in carbohydrate metabolism (Harding and Crabbe, 1984).

Glucose enters the lens by a passive, facilitated, insulin-independent transport mechanism that is stereospecific for D-glucose; there is no significant difference in uptake between the α and β anomers of D-glucose, although rat lens shows a preference for the β anomer of D-galactose over the α anomer (Okuda *et al.*, 1987). The translocation of glucose is mediated by a transporter (or carrier) that is reversibly inhibited by cytochalasin B; in monkey lens, the transporter comprises only about 0.5% of the total membrane protein (Lucas and Zigler, 1987). Cytochalasin B can be bound irreversibly to the glucose transporter by photoaffinity labeling, a property that has been exploited to show that the lens glucose transporter is unexpectedly located primarily in membranes derived from the cortex and nucleus, with very little being detectable in capsule-epithelium preparations (Lucas and Zigler, 1988). The lens glucose transporter is similar to that of erythrocytes; both are intrinsic membrane glycoproteins that migrate on SDS-PAGE as broad 53-kDa bands.

5.3.1.1. Glucose Metabolism and Energy Production

The major pathways of glucose metabolism in the lens are shown in Fig. 5.3. All of the enzymes in this highly coordinated system are present in the lens, and many have been isolated and purified. Detailed studies of their kinetics and, most importantly, changes occurring in aging and/or in cataract formation have been summarized in two review articles (Harding and Crabbe, 1984; Cheng and Chylack, 1985). Owing to space limitations, only brief mention can be made of a few salient points. Three kinases play key roles in the control of glycolysis and in the generation of ATP in the lens. Initial phosphorylation of free glucose is regulated by hexokinase in a reaction considered to be a rate-limiting step in glycolysis. Two forms of lens hexokinase, types I and II, are present in most species and differ from one another in kinetic properties and sensitivities to heat inactivation. As in other tissues, the two ATP-generating steps in lens glycolysis are mediated by phosphoglycerate kinase and pyruvate kinase (Fig. 5.3).

Most of the pyruvate produced by glycolysis is converted to lactate by lactic dehydrogenase. All five known isoenzymic forms of this enzyme are present in lens. The activity of lactic dehydrogenase, which uses NADH as its cofactor, is regulated in part by the activity of glyceraldehyde 3-phosphate (G3P) dehydrogenase, an enzyme that generates NADH during the glycolytic breakdown of glucose. In the presence of KCN or excess glucose, the concentration of G3P increases and reaches a new steady-state level in response to changes in the $NADH/NAD^+$ ratio under a particular experimental condition (Cheng *et al.*, 1988). These findings support the view that measurements of pyridine nucleotides by *in vivo* redox fluorometry is an accurate reflection of the metabolic status of the lens (Tsubota *et al.*, 1987). Apart from the availability of NADH, other factors may also control lactate production in the intact lens. For example, increased intracellular concentrations of calcium have been shown to inhibit lactate production in cultured rabbit lens (Hightower and Harrison, 1987). The effect appears to be a specific one and is not caused by chelation of ATP by excess calcium.

Mitochondria-containing epithelial cells comprise only a small fraction of the total lens mass, yet the tricarboxylic acid cycle operating in this cell layer is thought to supply at least one-third of the energy requirements of the lens. All of the enzymes necessary for the metabolism of acetyl CoA, as well as components of electron transport, are present in the lens. They are localized mainly in the epithelium and outer cortex.

The pentose phosphate pathway (Fig. 5.3) accounts for about 10–20% of the glucose metabolized by the lens. It is not only the major source of CO_2 produced in the tissue, but is also important in generating 2 reducing equivalents of NADPH per mol of glucose oxidized. The NADPH is used in several key metabolic systems in the lens such as reductive synthesis of fatty acids, conversion of glucose to sorbitol by aldose reductase, maintenance of glutathione (GSH) in a reduced state, and supply of pentoses for nucleic acid synthesis. The first two enzymes in the pentose phosphate pathway, G6PD and 6PGD, have been extensively studied in the lens. The pentose phosphate pathway (or

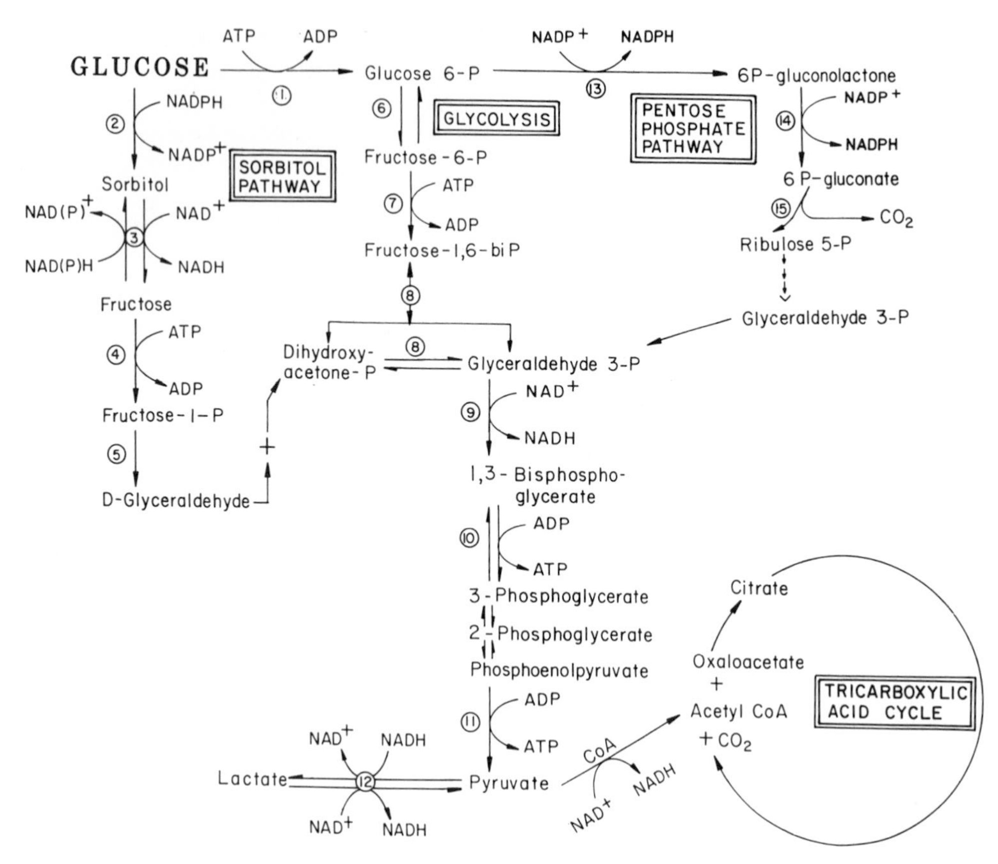

Figure 5.3. An overview of glucose metabolism in the lens. The major pathways shown are glycolysis, pentose phosphate pathway, sorbitol pathway, and tricarboxylic acid cycle. The enzymes shown in encircled numbers are: 1, hexokinase; 2, aldose reductase; 3, polyol dehydrogenase; 4, ketohexokinase; 5, aldolase; 6, phosphoglucose isomerase; 7, 6-phosphofructokinase; 8, aldolase and triose phosphate isomerase; 9, glyceraldehyde 3-phosphate dehydrogenase; 10, phosphoglycerate kinase; 11, pyruvate kinase; 12, lactic dehydrogenase; 13, glucose 6-phosphate dehydrogenase (G6PD); 14, lactonase; 15, 6-phosphogluconate dehydrogenase (6PGD). The pentose phosphate pathway and glycolysis are linked by two enzymes (not shown): transketolase and transaldolase.

hexose monophosphate shunt, as it is sometimes called) can be activated when there is a metabolic demand for NADPH, for example, in the reductive conversion of excess glucose or galactose to their respective sugar alcohols or in conditions of oxidative stress when increased intracellular levels of GSH are required. This pathway is also stimulated in the presence of pyrroline 5-carboxylate (P5C), which is reduced to proline by P5C reductase in the presence of either NADH or NADPH (Shiono *et al.*, 1985). This enzyme, an intermediate in glutamate and ornithine metabolism (see Fig. 7.20), has been detected in the lens and may play a role in modifying cellular redox potentials through the generation of $NADP^+$.

Other potential sources of NADPH in addition to the pentose phosphate pathway have recently been detected in capsule-epithelium preparations of rabbit lens (Winkler and Solomon, 1988). The combined activities of two $NADP^+$-dependent cytosolic enzymes, isocitrate dehydrogenase and malic enzyme, are comparable in NADPH-generating potential to glucose 6-phosphate dehydrogenase, suggesting that mitochondria-derived enzymes may be involved in protecting the lens from oxidative stress.

Although the physical basis of lens transparency is well understood (see Section 5.1.2), the biochemical basis is still uncertain. There is evidence supporting the view that lens transparency is directly related to the unique and complex nature of its energy metabolism. An overview of the interacting pathways of carbohydrate metabolism in the lens has been given in Fig. 5.3. Glycolysis is regulated by two essentially irreversible steps catalyzed by hexokinase and phosphofructokinase; ATP is generated in two steps catalyzed by phosphoglycerate kinase and pyruvate kinase. The activities of competing pathways in the metabolism of glucose have been extensively investigated using conventional methods under basal conditions and under a variety of stress-induced conditions, but the findings are often inconsistent. In recent years, with the development of ^{31}P-NMR spectroscopy and other noninvasive techniques, questions regarding dynamic changes in metabolites and the energy status of the lens are now being addressed (see reviews by Greiner *et al.*, 1985; Cheng and Chylack, 1985; Farnsworth and Schleich, 1985). Inorganic phosphate (Pi) as well as intracellular organophosphates can be accurately detected and quantitated under a variety of experimental conditions. For example, there are significant differences in phosphorus metabolite profiles as well as marked shifts in glucose metabolism from glycolysis to pentose phosphate pathway in adult rabbit lenses compared to those of juvenile animals (Williams *et al.*, 1988). These lenses remained transparent under standard organ-culture conditions but, when incubated in NaCl-deficient media, they developed opacities. However, no correlation could be found between transparency and changes in NMR-visible phosphorus metabolite profiles. Other studies on the energy status of human lenses show that the ATP/Pi ratio does not vary with age provided that the lenses examined had been preserved in moist chambers on ice (Cheng and Aguayo, 1988). Moreover, the ATP content of cataractous lenses before incubation is similar to that of clear lenses. Thus, NMR spectroscopy has produced expected results in some cases and unexpected ones in others. Further studies are needed to exploit the maximum potential of these new techniques.

5.3.1.2. Sorbitol Pathway

The existence of this pathway in the lens derives from the discovery by van Heyningen (1959) of an association between lenticular accumulation of polyols and dietary or experimentally induced excesses of xylose, galactose, or glucose. All animals developed cataracts, an observation that has stimulated 30 years of biochemical, morphological, and molecular biological research (see Section 5.6.2.1). The enzymatic basis of the sorbitol pathway in rat lens was published shortly afterward (van Heyningen, 1962).

Despite extensive investigations over the years, the function of the sorbitol pathway in the normal lens remains obscure. The pathway contains two enzymes: aldose reductase (AR), which uses NADPH as its cofactor, and polyol dehydrogenase (PD), an NAD^+-dependent enzyme (Fig. 5.3). Many of the details of this pathway, as well as relevant literature references, have appeared in two excellent reviews (Harding and Crabbe, 1984; Cheng and Chylack, 1985). The first enzyme, AR, is known in other tissues such as placenta and seminal vesicles; it is also present in the retinal vasculature. The enzyme is widely distibuted in lenses of a variety of animal species and in humans, although the activity in human lens is only 16–30% of that present in rat or rabbit lens. In humans AR is localized mainly in the epithelium (Jedziniak *et al.*, 1981). The K_m for glucose is 200 mM, implying that although glucose could be metabolized by this pathway in poorly controlled diabetics, little glucose would be expected to be reduced by AR under normal physiological conditions in nondiabetic individuals. The purified enzyme from bovine lens is a monomeric 37-kDa protein that catalyzes the reduction of glucose and galactose as well as a variety of other aldehydes (see Table XXXIV in Harding and Crabbe, 1984). The finding that glyceraldehyde 3-phosphate is a very effective substrate for AR suggests a role for this enzyme in the synthesis of membrane phosphoglycerides through the formation of glycerol 3-phosphate.

The possibility that inhibition of AR could prevent or ameliorate the development of sugar cataracts (e.g., in diabetes or in galactosemia) has stimulated an intensive search for AR inhibitors. During the past decades, a large number of structurally diverse AR inhibitors have been developed and tested in animals (Harding and Crabbe, 1984; Kinoshita, 1986); some of their effects on experimentally induced cataracts are discussed in Section 5.6.2.1. Direct measurements of AR activity in organ-cultured rabbit lenses in the presence of AR inhibitors have been reported using ^{13}C-NMR spectroscopy (Williams and Odom, 1987). Using sorbitol production as the main parameter, these studies suggest that the most effective inhibitors are tolrestat, sorbinil, and sulindac.

The second enzyme in the sorbitol pathway, polyol dehydrogenase (PD), is present in lenses of most species, although it has not been studied as extensively as AR. It catalyzes the oxidation of sorbitol and xylitol but not dulcitol (galactitol) to the corresponding ketosugar. Of interest are the observations that although AR activity in human lens is low in comparison to animal lenses, the opposite is true of PD (Jedziniak *et al.*, 1981). The enzyme has an unusually high capacity to catalyze the interconversion of sorbitol and fructose. The activity in the direction from fructose to sorbitol is ten times greater than the reverse reaction; under equilibrium conditions, both sorbitol and fructose could be present in human lens. Fructose may be further metabolized in some species; for example, in cattle lens this sugar is phosphorylated by ketohexokinase and converted to glycolytic intermediates with the potential of generating 1 mol of ATP per mol of fructose (Ohrloff *et al.*, 1982). Moreover, in rat lenses incubated with labeled exogenous fructose, measurements of ^{31}P-NMR spectra show that it is a suboptimal but usable substrate for glycolysis (Cheng *et al.*, 1985). Given the limited permeability of fructose, these findings imply a role for the sorbitol pathway as a secondary energy source in the lens of some species.

5.3.2. Glutathione Metabolism and Function

Glutathione (GSH), present at relatively high concentration in the lens, especially in the epithelium, is a γ-glutamyl-cysteinyl-glycine tripeptide (Fig. 5.4). Among its many important functions are (1) maintenance of protein sulfhydryl groups in the reduced form, (2) protection from oxidative damage by detoxification of H_2O_2, (3) removal of xenobiotics by conjugation with hydrophobic compounds having an electrophilic center, a reaction catalyzed by glutathione S-transferase, and (4) participation in amino acid transport as a γ-glutamyl donor to the α-amino groups of acceptor amino acids such as cysteine or glutamine. This reaction is catalyzed by the membrane-bound enzyme γ-glutamyl transpeptidase. Glutathione may also play an indirect role in ion transport, particularly of Na^+ and K^+, by protecting the sulfhydryl groups of Na^+,K^+-ATPase from oxidation (Cheng and Chylack, 1985).

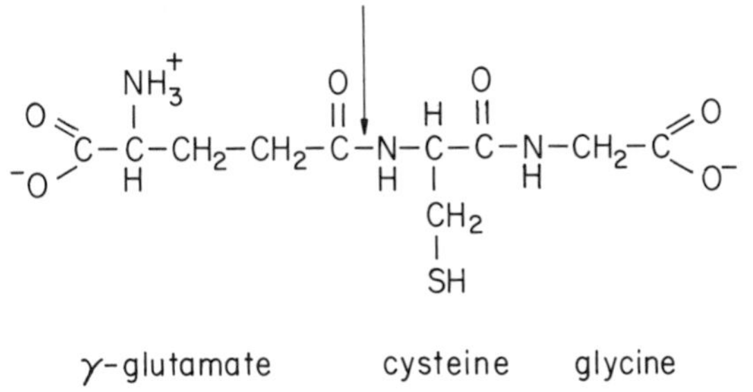

Figure 5.4. Structure of glutathione, a tripeptide of γ-glutamate, cysteine, and glycine. The peptide bond indicated by the arrow is atypical because it is between the nitrogen atom of cysteine and the carboxyl group of the δ-carbon of glutamate rather than with the usual α-carbon carboxyl found in most peptides.

γ-glutamate cysteine glycine

5.3.2.1. Biosynthesis and Degradation

All of the enzymes required for the synthesis and degradation of GSH have been found in the lens (see reviews by Harding and Crabbe, 1984; Cheng and Chylack, 1985). Under normal conditions, a steady-state concentration of glutathione is maintained through operation of the γ-glutamyl cycle (Fig. 5.5). Glutathione is synthesized sequentially by two ATP- and Mg^{2+}-requiring enzymes, γ-glutamylcysteine synthetase and glutathione synthetase. Approximately 12% of the total ATP formed by the lens is used for GSH synthesis. Synthetic activity declines with age in humans and in several animal species; moreover, the levels of GSH are low in most forms of cataract. The decline in synthetic activity in aging lenses appears to be caused by a reduction in γ-glutamylcysteine synthetase activity (Rathbun, 1984). Two enzymes involved in the degradation of GSH in the lens have also been characterized: γ-glutamyltransferase and 5-oxoprolinase.

Other enzymes of glutathione metabolism in the lens play important roles in detoxification. As shown in Fig. 1.9, glutathione peroxidase in the presence of GSH catalyzes the decomposition of H_2O_2, a reaction that is coupled to glutathione reductase. The

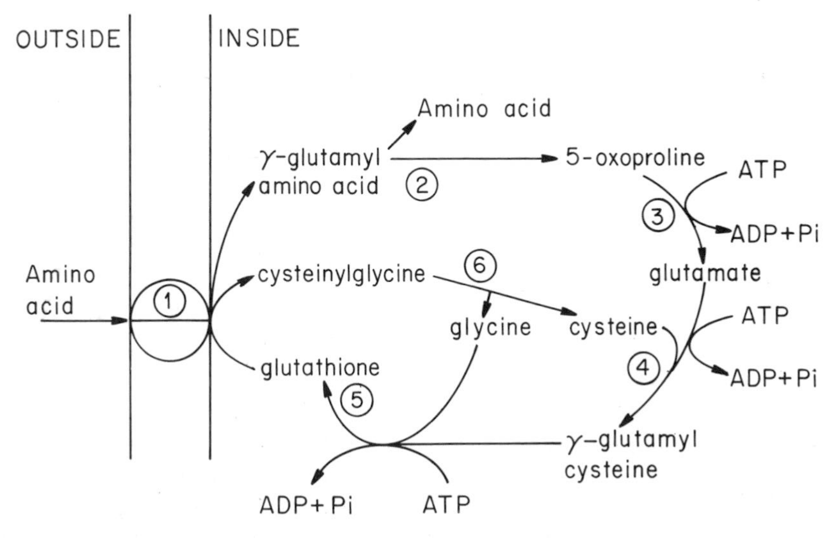

Figure 5.5. The γ-glutamyl cycle. The enzymes shown in encircled numbers are: 1, membrane-bound α-glutamyl transferase; 2, γ-glutamyl cyclotransferase; 3, 5-oxoprolinase; 4, γ-glutamylcysteine synthetase; 5, glutathione synthetase; 6, dipeptidase. (From Harding and Crabbe, 1984.)

oxidized form of glutathione (GSSG) is produced, and GSH is resynthesized by NADPH generated mainly from the pentose phosphate pathway. There is considerable species variation in the activities of the individual enzymes in this redox cycle, and there are variations with age as well (Sethna *et al.*, 1982/1983). In the monkey lens, glutathione peroxidase activity is low at birth, increases until adulthood, and then declines; glutathione reductase activity decreases from birth until juvenile life and afterward remains at a rather constant level (Rathbun *et al.*, 1986). This enzyme appears to be rate-limiting in the glutathione redox cycle. Many lines of evidence suggest that this redox cycle plays an important role in H_2O_2 detoxification in the lens, but in addition, catalase contributes a certain amount of H_2O_2 detoxifying activity (Bhuyan and Bhuyan, 1977). The other important enzyme in glutathione metabolism, glutathione S-transferase, is discussed in Section 5.3.2.3.

5.3.2.2. Protection from Oxidative Damage

In the majority of aerobic cells, univalent reduction of oxygen generates superoxide anion, a free radical that undergoes dismutation by superoxide dismutase, resulting in the formation of H_2O_2 (see Fig. 1.9). This potentially toxic substance is present in human aqueous humor at a concentration of about 30 μM and in rabbits in the range of 50–70 μM. It was suggested by Pirie (1965) that aqueous humor H_2O_2 is formed from oxygen during the oxidation of ascorbic acid in a reaction catalyzed by light and riboflavin; Pirie (1965) also proposed that H_2O_2 entering the lens is detoxified enzymatically by the glutathione redox cycle described above, coupled to the pentose phosphate pathway. A possible association between human cataract and elevated levels of aqueous humor H_2O_2 (Spector and Garner, 1981) has generated considerable interest in the role of the glutathione redox cycle and the pentose phosphate pathway in protecting the lens from H_2O_2 toxicity.

Both whole rabbit lenses (Giblin and McCready, 1983) and cultured epithelial cells (Giblin *et al.*, 1985) are readily able to detoxify exogenous H_2O_2. There are no overt signs of tissue damage, nor is there any decrease in cellular GSH after exposure to levels of H_2O_2 as high as 0.05 mM. This absence of H_2O_2 toxicity is strongly correlated with a dramatic stimulation of the pentose phosphate pathway, which appears to be a major factor in maintaining adequate supplies of GSH to protect the lens from H_2O_2-induced toxicity. A similar response is demonstrable in lenses of young (4-day-old) rabbits, but cultured lenses from old (8-year-old) animals have considerably less capacity to detoxify H_2O_2 possibly because of their lower glutathione reductase activity and diminished pentose phosphate pathway responses (Reddan *et al.*, 1988). Treatment of rabbit lenses with 1,3-bis(chloroethyl)-1-nitrosourea (BCNU), an inhibitor of glutathione reductase, prior to exposure to normally well-tolerated levels of H_2O_2 results in a significant accumulation of GSSG and marked disturbances in cation transport (Giblin *et al.*, 1987). Nevertheless, the effects of combined BCNU–H_2O_2 treatment are partially reversible, as shown by the restoration of normal intracellular levels of K^+ and reversal of $^{86}Rb^+$ efflux after incubation of treated lenses in normal medium.

Other studies using cultured rabbit lens epithelial cells show that the specific targets of H_2O_2 toxicity are the sulfhydryl groups of the epithelial cell membranes (Hightower *et al.*, 1989). Brief exposure of the cells to relatively high (1 mM) levels of H_2O_2 induces loss of membrane thiols as well as alterations in ion homeostasis; these changes occur only when cellular levels of GSH are severely reduced and are not obvious with lower concentrations of H_2O_2. Whether physiological levels of H_2O_2 in the aqueous humor have any damaging effects on lens metabolism or cellular function still remains to be established. In still another approach to the problem of oxidative damage, it has been shown that exposure of cultured rabbit lenses to hyperbaric O_2 causes a depletion of GSH and a corresponding increase in GSSG (Giblin *et al.*, 1988). The nucleus, which has a low endogenous level of GSH, appears to be more susceptible to oxygen-induced damage than the capsule-epithelium. Rather surprisingly, however, dense opacities did not develop under these conditions despite the sharp drop in GSH levels.

Studies using bovine lenses give a somewhat different impression of H_2O_2-induced injury and the protective role of GSH. Exposure of bovine lenses to physiological concentrations of H_2O_2 inhibits the influx of $^{86}Rb^+$ by direct modification of membrane-bound Na^+,K^+-ATPase (Garner *et al.*, 1983). The primary effects of H_2O_2 on the enzyme are loss of external Na^+ stimulation and alteration of K^+ stimulation, the overall result being an inhibition of pump-dependent K^+ influx (Garner *et al.*, 1986). The question of a protective role for GSH has been addressed directly by exposing cultured bovine epithelial cells to glutathione monoethylester; this GSH derivative is readily transported into the cell and hydrolyzed, resulting in an increase in cellular levels of free GSH (Spector *et al.*, 1987). By using $^{86}Rb^+$ accumulation, ATP concentration, and glyceraldehyde 3-phosphate dehydrogenase (G3PD) activity as parameters of oxidative damage induced by exposure of the cells to H_2O_2, it could be shown that, except for G3PD activity, elevated levels of GSH do not protect the cell from oxidative insult; they may in fact be somewhat harmful. A far more effective agent in mediating the recovery of cultured lens epithelial cells from oxidative damage is thioredoxin, a 12-kDa dithiol polypeptide that, even at very low concentrations, is able to reduce protein disulfides (Spector *et al.*, 1988). Incubation of *E. coli* thioredoxin with cultured bovine epithelial cells after they had been exposed to high concentrations (0.2–0.3 mM) of H_2O_2 significantly enhances the ability of the cells to recover from peroxide-induced oxidation.

5.3.2.3. Detoxification of Xenobiotics

Drugs, atmospheric pollutants, and other potentially toxic substances are detoxified in the lens by glutathione S-transferases (GST), a group of enzymes that catalyze the conjugation of a diverse group of electrophilic xenobiotics to GSH (Awasthi *et al.*, 1980). A diagrammatic scheme has been given in Fig. 4.2 showing the initial conjugation and subsequent enzymatic steps leading to the formation of mercapturic acid. Some forms of glutathione S-transferase express selenium-independent glutathione peroxidase II activity toward lipid hydroperoxides. Multiple forms of glutathione S-transferase are found in most tissues, the three major ones being designated as α, μ, and π. Their expression is tissue specific. Two isoenzymes of glutathione S-transferase in bovine lens, designated GST 7.4 and GST 5.6, are homodimers composed of 23.5-kDa subunits (Ahmad *et al.*, 1988). Although these two enzymatic forms of glutathione S-transferase are related chemically and immunologically, they differ in pI values and in peptide fingerprints. Both belong to the μ class of glutathione S-transferases.

Relatively little is known of specific xenobiotics metabolized by the mercapturic acid pathway in lens, but a recent study has shown that the pathway is operative in rabbits (Iwata and Maesato, 1988). Oral administration of naphthalene to rabbits results in cataract formation, and oxidized derivatives of naphthalene conjugated with GSH were identified in lens extracts. One of the major metabolites, N-acetyl-S-(1,2-dihydro-2-hydroxynaphthyl) cysteine was shown to be an intermediate in the mercapturic acid pathway.

5.3.3. Protein Metabolism

Some general aspects of protein synthesis are discussed briefly in Chapter 1, Section 1.1.2, and a diagram showing the major steps in the synthesis of mRNA transcripts from nuclear DNA, and the translation of these transcripts in the cytosol, is given in Fig. 5.6. The many important basic studies on protein synthesis in the lens carried out during the 1970s and early 1980s have been comprehensively reviewed by Harding and Crabbe (1984); hence, this section includes only a brief outline and update of this broad topic.

5.3.3.1. Synthesis

Active protein synthesis in the lens occurs principally in the epithelium and peripheral cortex. It is generally thought that little synthesis or turnover occurs in the nucleus, since the mature fiber cells in this region are devoid of nuclei and other cytoplasmic organelles. Nevertheless, these fiber cells do contain DNA segments that presumably serve as templates for DNA repair enzyme, an activity that disappears with terminal cell differentiation (see review by Harding and Crabbe, 1984).

Polyribosomes and messengers from lenses of calf, chicken, and other species can be translated in a number of cell-free systems and in oocytes from *Xenopus laevis*. Synthesis of the major classes of crytallins, as well as lens membrane-specific proteins, has been demonstrated using calf lens polyribosomes in a reticulocyte lysate. Not only calf but also fetal human lens mRNAs encoding both crystallin and noncrystallin proteins have been isolated and translated in a heterologous cell-free system (Ringens *et al.*, 1982). In these and other studies, polyribosomes were derived from whole lenses, but the generally prevailing view is that the polyribosomes involved in protein synthesis in the lens are present mainly, if not entirely, in the epithelium and peripheral cortex, where differentiation is occurring. It is therefore of interest that polyribosomes isolated from the nuclear region of bovine and human lenses are active in directing the *de novo* synthesis of proteins both *in vitro* and in a reticulocyte cell-free system (Ozaki *et al.*, 1985). Water-soluble (α- and β-crystallins) as well as water-insoluble (actin and vimentin) proteins were identified as translation products of nucleus-derived polyribosomes.

At least three mRNAs have been isolated from lenses of various species and shown to direct the synthesis of specific crystallins (see review by Harding and Crabbe, 1984). A 14 S calf lens poly(A)$^+$mRNA has been identified that codes αA2-crystallin synthesis as well as an additional 24-kDa crystallin, and a 10 S mRNA isolated from calf lens synthesizes either αB2- or β-crystallins. More recently, it has been shown that the 14 S mRNA fraction of rat lens comprises two αA2 mRNAs containing about 1250 and 1350 nucleotides as well as the αA^{ins} mRNA, which is similar in size to the largest of the αA2 mRNAs (Dodemont *et al.*, 1985). The 14 S mRNA fraction from calf lens isolated in these studies harbors a heterogeneous population of αA2 mRNAs. In embryonic chick lenses, both 21 S and 17 S mRNAs direct the synthesis of the δ-crystallin subunit, and 10 S and 14 S mRNAs code for α-crystallin subunits. There are at least four similar γ-crystallin mRNAs in the mouse lens.

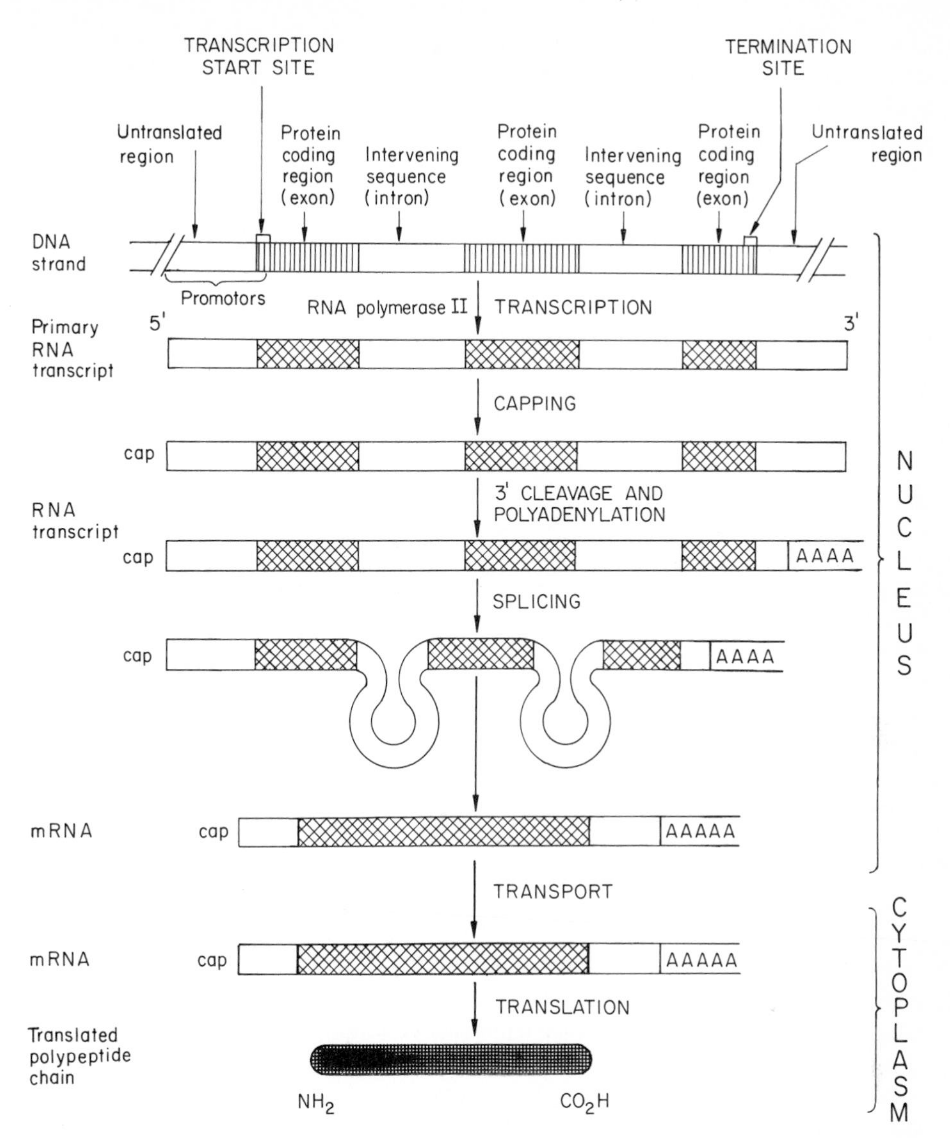

Figure 5.6. General pathway for polypeptide synthesis from DNA (see discussion in Chapter 1, Section 1.1.2). Coding sequences (exons) are separated by intervening sequences (introns), which are removed in the conversion of the primary transcript into functional mRNA. Transcription is initiated by RNA polymerase II, and the next sequential steps are (1) addition of a methylated cap structure at the 5′ end of each primary transcript and (2) addition of poly-A at the 3′-hydroxyl end. The splicing of primary transcripts in the nucleus results in the production of functional mRNA. The introns that are deleted do not accumulate; rather, they are probably destroyed. The completed mRNA is then transported to the cytoplasm for translation.

Possible abnormalities in functional mRNA in inherited or induced cataract have been noted: an mRNA that directs the synthesis of 27-kDa β-crystallin polypeptide appears to be absent (or nondetectable) in the Philly mouse lens (Carper *et al.*, 1982). In galactosemic rat lenses, there may be decreases in crystallin mRNAs (Shinohara *et al.*, 1982) and/or fluctuations in noncrystallin mRNAs (Hsu *et al.*, 1987a; Bekhor, 1988).

Several studies on crystallin ontogeny in the lens have been undertaken in an effort to address the complex question of developmental regulation of these proteins. It has long been recognized that synthesis of the individual polypeptides comprising each class of lens crystallin is temporally and spatially regulated. Crystallin synthesis is regulated by differential gene expression, and specific crystallins appear at different times and in different regions throughout development. The primary gene products undergo numerous posttranslational changes whose precise nature is under active investigation. In the rat, protein accumulation, synthesis, and mRNA translation in a reticulocyte lysate show transitions from embryonic to adult crystallin expression that begin 1 week after birth (Carper *et al.*, 1985). Three β-crystallins are not synthesized in embryonic rat lens, and for at least one of them, a functional mRNA could not be detected until the first postnatal day. The 27-kDa polypeptide characteristic of adult β-crystallin is detectable only after about the fifth postnatal day.

Differential regulation of crystallin synthesis with (what appears to be) random posttranslational modifications during development has been reported in human lens (Thomson and Augusteyn, 1985). The major protein in neonatal lens is α-crystallin, and β-crystallin is the next most abundant polypeptide; γ-crystallin does not appear to be synthesized in neonatal lenses. Modifications of existing polypeptides such as deamidation and degradation occur within each crystallin class, and by birth, at least 20 different modified derivatives are already distinguishable. In another study, HPLC, SDS-PAGE, and immunoblotting were used to analyze the crystallins in cortical and nuclear regions of 33 normal human lenses from donors ranging in age from newborn to 70 years old (McFall-Ngai *et al.*, 1985). In both regions, a gradual age-related decrease of 19- and 21-kDa proteins (γ-crystallins) was found, data somewhat in conflict with those of Thomson and Augusteyn (1985) who reported that γ-crystallin synthesis ceases shortly after birth. McFall-Ngai *et al.* (1985) also noted increased insolubilization in high-molecular-weight aggregates in nuclear samples in 35- to 45-year-old individuals. The aggregates consist partly of α-crystallins but with increasing age have higher proportions of β- and γ-crystallins.

Owing to uncertainties in obtaining normal human donor lenses and the scarcity of cataract lenses following intracapsular extraction, many efforts are being made to culture human lens epithelium. A recent study has shown that primary cultures of lens epithelium can be established from young donors, and after reaching confluency, the cells are able to be successfully subcultured through at least three passages (Reddy *et al.*, 1988). These cells are active in synthesizing crystallins, with α-crystallins accounting for about 13% of the total protein synthesized during the first two passages. Crystallin synthesis declines rapidly, however, during the third passage.

5.3.3.2. Posttranslational Modifications

The structural proteins of lens, both the soluble crystallins and the fiber cell plasma membrane components, undergo major and complex enzymatic and nonenzymatic post-synthetic modifications. One of the most important modifications of primary gene products is the phosphorylation of soluble αA2-, αB2-, and βB2-crystallins and plasma membrane proteins (MP26 and others). These topics have been discussed in Sections 5.2.1.2 and 5.2.2.2. In addition, oxidative changes occur to varying degrees in all lenses, and there is evidence supporting the view that oxidative insult to plasma cell membranes may be an initiating event in cataract formation (Spector, 1984a,b, 1985). The present discussion is limited to two extensively studied postsynthetic modifications of lens proteins: (1) formation of high-molecular-weight (HMW) aggregates and (2) proteolytic degradation.

Formation of High-Molecular-Weight Aggregates. Four types of HMW aggregates have been defined by Harding and Crabbe (1984): HMW1, isolated from bovine lens α-crystallin fractions; HMW2, isolated directly from total soluble lens homogenates by gel filtration; HMW3, a disulfide-linked aggregate; and HMW4, a covalently linked aggregate having no disulfide linkages. The latter two are found in aging and/or cataractous lenses together with water-insoluble HMW aggregates. The first type, HMW1, has been isolated from bovine lens and has a M_r greater than 10^6; its amino acid composition and subunit size are indistinguishable from α-crystallin (Spector *et al.*, 1971). The HMW2 fraction isolated from bovine lens is similar to HMW1, but in rabbit, human, and other species, it consists of a mixture of crystallins. The HMW aggregates of most species are localized mainly, if not entirely, in the nucleus.

Many studies have been carried out on human HMW2, but there are conflicting and discrepant data (Harding and Crabbe, 1984). The amino acid analyses of human HMW2 do not correspond to any known crystallin, and the subunit composition shows it to be a mixture of the three major crystallins. Moreover, these aggregates contain low-molecular-weight peptides possibly resulting from increased endopeptidase activity and/or diminished activities of exopeptidases. Degraded polypeptides of M_r <18,000 are present in increasing quantities in HMW proteins of aging lens and may play a role in the formation of these aggregates (Srivastava, 1988a). Highly variable amounts of HMW aggregates are present in human lenses, with different isolation methods giving yields ranging from 2–30% of the total protein; moreover, only about 50% of HMW2 is protein. The possibility that HMW2 is an artifact arising during its extraction and purification has been raised by Harding and Crabbe (1984).

There may be uncertainties regarding the precise nature of HMW aggregates in human lenses, but studies on rat lenses show a rather clearly defined age-related increase in HMW content (Swamy and Abraham, 1987). With advancing age, there appear to be declines in both γ-crystallins and reactive sulfhydryls, as well as increased glycation in HMW aggregates.

Many substances are capable of inducing the formation of HMW aggregates in animal lenses. An accumulation of calcium in rabbit lenses leads to the slow formation of HMW aggregates and concomitant appearance of opacities (see review by Hightower, 1985). The precise role of Ca^{2+} in inducing HMW aggregation has not been established, but an early product of protein cross-linking in Ca^{2+}-treated rabbit lenses is a 55-kDa dimer of β-crystallin (Lorand *et al.*, 1985). Formation of the dimer is mediated by a Ca^{2+}-activated lens transglutaminase acting on endogenous endo-γ-glutaminyl acceptors present in βH- and βL-crystallins (Velasco and Lorand, 1987). Dimer formation is inhibited by 75 mM histamine and by other amines at much lower concentrations (Lorand *et al.*, 1987).

Other studies also implicate β-crystallin in aggregate formation. Incubation of bovine lens crystallins with potassium cyanate results in the carbamylation of the proteins and the formation of disulfide-bonded protein aggregates over a period of several days (Beswick and Harding, 1987a). Carbamylated βL-crystallin alone is capable of acting as an aggregation center for other unmodified proteins. Carbamylation may cause unfolding of the protein chains and, under oxidative conditions, could generate interprotein disulfide linkages; the formation of aggregates of the HMW3 type similar to those found in cataractous lenses could be envisaged under these conditions.

Proteolytic Degradation and Ubiquitin Conjugation. It is now recognized that lens contains a rather sizable number of proteolytic enzymes as well as a ubiquitin conjugation system. Researches during the 1960s and 1970s on two of the proteolytic enzymes, a neutral proteinase (or endopeptidase) and the major exopeptidase of the lens, leucine aminopeptidase, have been reviewed in detail by Harding and Crabbe (1984). Hence, only a brief discussion of this early work is included in the present section; the major emphasis is on more recent findings, and for this purpose, proteolytic degradation is considered under the following headings: (1) endopeptidases (neutral proteinases and calpains I and II, (2) exopeptidases, and (3) membrane-associated protease. The ubiquitin ATP-dependent conjugation system is also discussed.

Endopeptidases (neutral proteinases and calpains I and II). A neutral endopeptidase in bovine and human lenses has been known for many years (see review by Harding and Crabbe, 1984). The purified enzyme from bovine lens has a M_r of 700,000 in the native state and on SDS-PAGE dissociates into at least eight subunits ranging in size from 24 to 32 kDa (Ray and Harris, 1985). The enzyme is not unique to lens, since a similar, if not identical, enzyme has been isolated from bovine pituitaries. Other studies show that neutral proteinase preparations from bovine lens comprise a family of at least three activities that are distinguishable by their specificities toward synthetic peptide substrates (Wagner *et al.*, 1986a,b). The partially purified enzyme hydrolyzes peptides with nonpolar as well as negatively charged side chains; it also displays trypsin-like activity and hydrolyzes modified proteins at physiological temperatures. For example, glutamine synthetase oxidized in the presence of ascorbate and $FeCl_3$ is hydrolyzed at a rate of 1–2% per hour, whereas the native enzyme is only marginally degraded.

Another endopeptidase, 25-kDa serine proteinase, is found in bovine lens homogenates associated with the α-crystallin fraction (Srivastava and Ortwerth, 1983). Similar to the neutral proteinase described above, it also shows trypsin-like activity. The purified enzyme has a pH optimum between 7.2 and 8.2, and hydrolyzes synthetic substrates as well as the B chain of α-crystallin. This activity is not detectable in crude extracts owing to the presence of endogenous trypsin inhibitors.

Calpains are sulfhydryl (cysteine) Ca^{2+}-dependent neutral endopeptidases ubiquitously distributed in mammalian and avian tissues. Two distinct forms of the enzyme have been characterized: calpain I, activated at a low (approximately 10 μM) concentration of Ca^{2+}, and calpain II, requiring high (1 mM) Ca^{2+} for activity. Endogenous inhibitors of both types of calpain, designated calpastatins, are also found in most tissues. Induction of proteolysis by Ca^{2+} in lens cytoskeletal preparations (Roy *et al.*, 1983; Ireland and Maisel, 1984c) and in lens homogenates (Russell, 1984) was the first indication that calpain enzymes may be present in the lens. Simultaneous with these observations, Yoshida *et al.* (1984) isolated a highly purified calpain II from the cytosolic fraction of bovine lens; this preparation was active in degrading both the A and B chains of α-crystallin, producing 18-kDa and 19.5-kDa polypeptides, respectively. The chains are cleaved at their C terminals, the A chain mainly at the Arg^{163}-Glu^{164} linkage, and the B chain principally at the Thr^{170}-Ala^{171} linkage (Yoshida *et al.*, 1986).

Calpain I activity has also been detected in bovine lens. Both calpains, as well inhibitor calpastatin, are found mainly in the epithelium and cortex of the lens; no activity is detectable in the nucleus (Yoshida *et al.*, 1985). Purified calpain II from rat lens is a protein of M_r approximately 120,000 composed of 80- and 28-kDa subunits (David and Shearer, 1986). Several endogenous substrates for calpain have been characterized in rat (David and Shearer, 1986) and bovine (Ireland and Maisel, 1984c) lens; these include vimentin, intrinsic membrane proteins, and crystallins. The calpain II enzymes from bovine and rat lenses appear to have different substrate specificities, since the rat enzyme degrades β-crystallin but the bovine enzyme does not. Calpain II activity has also been detected in human lens, but the specific activity is only about 3% of that present in rat lens (David *et al.*, 1989). High levels of inhibitor calpastatin are also present in human lens, especially in the nucleus.

Exopeptidases. Proteolysis in most tissues is believed to be initiated by endopeptidases, and the final cleavage of peptides to amino acids is carried out by aminopeptidases (exopeptidases). In the lens, the major exopeptidase is leucine aminopeptidase, which catalyzes the hydrolysis of N-terminal amino acids from a wide range of synthetic peptides. This enzyme has been extensively studied in the lens; its structural, kinetic, and catalytic properties are now well understood from studies using highly purified preparations from bovine lens (see review by Harding and Crabbe, 1984). Substrate specificities have been established for both the Mn^{2+}- and Mg^{2+}-activated forms of the enzyme whose pH optima are about 8.5–9.0. Leucine aminopeptidase is a 320-kDa protein composed of six identical 54-kDa subunits, each of which has two metal-binding sites. The amino acid sequence of 472 out of the 478 residues of this enzyme has been published (see Fig. 71 in Harding and Crabbe, 1984). This review also summarizes nearly two decades of research on lens leucine aminopeptidase.

Although readily detected in lenses of various animal species, there have been uncertainties about the presence of leucine aminopeptidase in human lens until rather recently. It appears that young animals had been used in most of the early studies, whereas for humans, the lenses analyzed were from elderly donors. Leucine aminopeptidase activity can, however, be demonstrated in young human lenses by immunologic methods. Moreover, the human enzyme has been purified, and comparison of its amino acid sequences with purified enzymes from cattle and hog tissues show that they share considerable homology (Taylor *et al.*, 1984; Taylor and Davies, 1987). The specific activity of leucine aminopeptidase in young human lenses is in fact comparable to that of lenses from young animals, but it is considerably attenuated in older lenses of all species.

Another aminopeptidase has been isolated and characterized in bovine lens (Sharma and Ortwerth, 1986). Unlike leucine aminopeptidase, this enzyme does not require Mg^{2+} or Mn^{2+} for activation, and its pH optimum is at 6.0. Based on substrate specificities and other characteristics, the enzyme has been classified as aminopeptidase III. This exopeptidase would be more active than leucine aminopeptidase at the physiological pH of the lens (between 7.0 and 7.2) and therefore may have an important functional role *in vivo*.

Membrane-associated protease. A 68-kDa proteinase isolated from bovine lens fiber cell membranes has been purified to homogeneity; it is a tetramer composed of 17-kDa subunits (Srivastava, 1988b). The enzyme shows maximum activity at pH 7.8 and is inhibited by all serine protease inhibitors tested. Its substrate specificity toward synthetic peptides and its catalytic properties distinguish it from the other major endopeptidases described above (neutral proteinases, calpains I and II, and 25-kDa trypsin-like proteinase). It is tempting to speculate that this membrane-associated protease could play a role in the age-related conversion of MP26 to MP22.

Ubiquitin ATP-dependent conjugation system. Ubiquitin is a small, 8.5-kDa, protein present in the cytosol of many types of mammalian cells. It is an unusual protein in that its carboxy-terminal glycine residues become covalently linked to the ϵ-amino groups of lysine residues of proteins destined for degradation. The energy for formation of these bonds comes from ATP. At least three enzymes have been characterized that mediate the ATP-dependent conjugation of "damaged" proteins to ubiquitin. The most commonly used source of these conjugating enzymes is reticulocyte lysate (fraction II). The presence of protein linked to ubiquitin induces hydrolysis through an as yet unknown mechanism. This system is often considered an alternative, or a supplement, to the lysosomal degradation of proteins.

The ubiquitin conjugation system has been detected in rabbit, bovine, and human lenses; the highest activity is present in the epithelium, and the lowest in the nucleus (Jahngen *et al.*, 1986a,b). Ubiquitin conjugation in the lens is ATP-dependent, and the major conjugates formed are with endogenous lens proteins of M_r greater than 150 kDa. There is a striking difference in the relative distribution of free and conjugated ubiquitin in young (epithelial) and old (nuclear) tissue, with a strong tendency toward diminished proteolytic capabilities in the older tissue.

Glutathione–Protein Mixed Disulfides. Glutathione is found in unusually high concentrations in the rabbit and rat lens, but the levels are considerably lower in human lens (Table 5.1). Approximately 95% is in the reduced (GSH) state, and the remainder is present as oxidized glutathione (GSSG). The oxidized form is thought to participate in the formation of glutathione–protein mixed disulfides, possibly through the following reaction, which may proceed in either one (Harding and Crabbe, 1984) or two (Spector, 1985) steps:

$$\text{Protein-SH} + \text{GSSG} \rightleftharpoons \text{Protein-SSG} + \text{GSH}$$

Until rather recently, it was not possible to quantitate the levels of protein-SSG accurately in the lens; however, new and sensitive chromatographic methods are now available for obtaining reliable estimations. Low endogenous levels of glutathione–protein mixed disulfides have been detected in γ-crystallins isolated from cultured calf capsule-epithelial cells (Spector *et al.*, 1986) and in TCA-insoluble protein fractions from organ-cultured rat lenses (Lou *et al.*, 1986). In both preparations, as well as in cultured monkey lens, the formation of mixed disulfides is dramatically increased after exposure to H_2O_2. Other studies have shown that the attachment of glutathione to lens proteins, specifically α- and γII-crystallins, causes an unfolding of the molecules and leads to conformational destabilization of the proteins (Liang and Pelletier, 1988). The destabilization of α-crystallin, but not of γII crystallin, appears to increase its susceptibility to trypsin digestion.

Although not directly related to glutathione–protein mixed disulfide formation, the detection of glutathione–insulin transhydrogenase activity in lens epithelium suggests a possible enzymatic mechanism for sulfhydryl–disulfide interchange (Darrow *et al.*, 1988). The enzyme cleaves the A and B chains of insulin, the preferred substrate, and also mediates the formation of disulfide-linked high-molecular-weight protein aggregates. Of interest was the finding that an immunoreactive protein sharing at least one common epitope with the enzyme is associated with soluble β-crystallin of bovine lens. This was thought to be another example of an enzyme that has been recruited to serve as a structural protein in the lens.

5.4. PHOTOCHEMICAL REACTIONS

The continuous entry of optical radiation into the lens makes this tissue especially prone to photochemical stress. Potentially damaging photosensitization and photooxidative changes occur in the human lens throughout life and may be among the principal factors involved in cataractogenesis. Two excellent detailed review articles on the photochemistry (Dillon, 1985) and photobiology (Zigman, 1985) of the lens form the basis of the brief summary presented in this section. Most of the original literature references in this field until 1983 are to be found in these review articles; more recent findings are cited in the overview presented below.

Ultraviolet (UV) wavelengths of <400 nm reaching the eye are filtered out by the cornea and the lens; hence, under normal circumstances little or no UV light reaches the retina. All light below 295 nm is cut off by the cornea, and the lens absorbs most of the 295- to 400-nm ambient radiation. This is especially true after the development of yellow pigmentation in the adult lens. The interactions of endogenous as well as exogenous biomolecules with UV light has been extensively studied in the lens, and the present section is limited to only certain aspects of this broad and complex subject.

The major lens biomolecules that absorb near-UV light are free or bound aromatic amino acids (tryptophan, tyrosine, and phenylalanine) as well as numerous pigments and fluorescent chromophores. Among the amino acids, tryptophan absorbs 95% or more of the radiant energy (Dillon, 1985). The other major substance that absorbs light is 3-hydroxykynurenine (3-OH Kyn); the fluorescent glucoside of 3-OH Kyn was first identified in human lens by van Heyningen (1971). Other classic studies by Pirie (1971), using human lens homogenates, demonstrated that sunlight in the presence of air causes the oxidative cleavage of tryptophan to N-formylkynurenine (N-FKyn). Moreover, a direct relationship between coloration in human lenses and the photooxidation of tryptophan by near-UV radiation was established at about the same time (Grover and Zigman, 1972). Thus, photochemical and photooxidative reactions play important roles in normal lens biochemistry and are thought to be involved in both pigment formation and cataractogenesis.

As discussed by Dillon (1985), there are several essential parameters in photochemical processes in general and in the lens in particular. The first is the presence of endogenous biomolecules that absorb near-UV light; apart from tryptophan and 3-OH Kyn glucoside, other intrinsic photosensitizers in the lens are riboflavin and N-FKyn. Given the presence of photosensitizers, two other factors determine the rate or extent of photochemical reactions: O_2 can increase the rate of photolysis, and substances such as glutathione, ascorbic acid, and vitamin E can under certain conditions act as quenchers. Finally, the actual site of photolytic attack, whether it is directed toward membranes, soluble proteins, or both, is of prime importance.

Photooxidation of lens proteins causes brown colorations, and the question of whether it is free or protein-bound tryptophan that is the target for near-UV-induced changes leading to increased pigmentation has been discussed in detail by Zigman (1985). There is indirect, though compelling, evidence that this pigmentation results from the generation of free radicals and other reactive species produced after exposure of endogenous (or exogenous) photosensitizers to near-UV radiation. For example, singlet oxygen is generated during the dye-sensitized photooxidation of lens crystallins, and most importantly this photooxidation is associated with the production of non-disulfide cross-linked protein and the appearance of blue nontryptophan (NT) fluorescence (Goosey *et al.*, 1980). Other studies on photodynamic processes in human lens show that oxidation products of tryptophan, identified as 3-OH Kyn and N-FKyn, can act as photosensitizers in the production of singlet oxygen (Zigler and Goosey, 1981). Photopolymerization of proteins from calf lens has also been demonstrated in both soluble and membranous fractions (Dillon, 1984; Dillon *et al.*, 1985). These photosensitized reactions are mediated by 3-OH Kyn glucoside, a highly efficient generator of singlet oxygen (Andley and Clark, 1989). Similar changes are seen in the outer layer of gray squirrel lenses after exposure to 365-nm radiation (Zigman *et al.*, 1988). Cross-linking of proteins and increase in a 20-kDa peptide, the most prominent consequences of near-UV radiation, are probably induced by endogenous photosensitizers through the generation of singlet oxygen.

Excited-state and stable free radicals of photooxidized tryptophan present in human lens form active species that can bind and alter structural lens proteins. Several lens enzymes have been shown to be inactivated by near-UV light in the presence of tryptophan or some of its oxidation products (see review by Zigman, 1985). These include catalase, glutathione reductase, Na^+,K^+-ATPase, Mg^{2+}-ATPase, xanthine oxidase, and cytochrome oxidase. Superoxide dismutase appears to be insensitive to near-UV inactivation. In the case of Na^+,K^+-ATPase of rat lens, the decreased activity of the enzyme by near-UV radiation is correlated with lens swelling and development of opacities (Torriglia and Zigman, 1988). Whether tryptophan residues in the enzyme or photoproducts arising from tryptophan are the primary targets of UV radiation in the rat lens remains to be established.

The activities of several human lens enzymes involved in glucose metabolism, many of which are in close association with the β-crystallins, are greatly diminished by active oxygen species generated by the reaction of photosensitizers such as methylene blue or riboflavin with light (Jedziniak *et al.*, 1987). Of the enzymes studied, glutathione reductase and glyceraldehyde 3-phosphate dehydrogenase are inactivated most rapidly, and aldehyde dehydrogenase and polyol dehydrogenase somewhat slower. Exposure of rat lenses to near-UV radiation at 300 nm causes slight but statistically significant deactivation of hexokinase in rat, rabbit, and calf lenses (Tung *et al.*, 1988). Glucose, catalase, and ascorbic acid, as well as attached vitreous, exert a protective effect on the deactivation. These and other findings raise the possibility that H_2O_2 generated by near-UV radiation is mainly responsible for the hexokinase inactivation. This view is supported by experiments suggesting that H_2O_2 generated in the lens or in surrounding media poses a greater oxidative stress than free radicals such as superoxide anion or hydroxyl radicals (Zigler *et al.*, 1985a; Jernigan, 1985; Kletzky *et al.*, 1986).

The foregoing discussion has provided a few examples of the damaging effects of near-UV radiation on lens proteins and enzymes. They are caused by the interaction of light with endogenous photosensitizers, resulting in the generation of singlet oxygen, free radicals, and other active species such as H_2O_2. However, active oxygen species such as H_2O_2 could also accumulate in the lens by light-independent mechanisms, i.e., by uptake from ocular fluids surrounding the tissue *in vivo* or through the metabolically generated univalent reduction of oxygen intracellularly (see Chapter 1, Section 1.4.1). Given the low rate of oxidative metabolism in the lens, photochemically generated H_2O_2 would seem to represent the major source, and under normal conditions H_2O_2 will be detoxified by protective mechanisms known to be present in the lens.

Two types of photooxidation mechanisms, termed types I and II, are now recognized (see review by Dillon, 1985). In the type I reaction, a sensitizer in the triplet state reacts directly with a substrate by abstracting a proton and initiating a free radical process; the major active species generated are superoxide anion and H_2O_2. In the type II reaction, the sensitizer reacts directly with oxygen, forming singlet oxygen. Among many photosensitizers under investigation, riboflavin has attracted considerable attention in recent years because it is intrinsic to the lens. When added exogenously to cultured rat lenses as a photosensitizer, carrier-mediated transport systems are damaged, an effect that can be blocked by catalase (Jernigan *et al.*, 1981). Riboflavin-sensitized production of H_2O_2 in the presence of light and oxygen was shown in later experiments (Jernigan, 1985), suggesting that the photooxidation is through a type I mechanism. However, other investigations on conformational changes resulting from photolysis induced in calf lens α-crystallin by photosensitized riboflavin suggest that both reaction types may be involved (Andley, 1988). The mechanism of β-crystallin photolysis, characterized by conformational changes leading to the formation of supra-aggregated proteins, is different from that of α-crystallins, since type I photosensitization prevails (Andley and Clark, 1988a). Subtle differences in the microenvironments of cysteine and tryptophan residues in the crystallins provides the physical basis for their different reactivities to riboflavin-sensitized photolysis. The photooxidation of calf γ-crystallin, studied by either a riboflavin-sensitized reaction or by direct photolysis after 300-nm irradiation, results in changes in the tertiary structure of the protein and subsequent insolubilization (Andley and Clark, 1988b). The mechanism of photooxidation involves the generation of H_2O_2; therefore, this is a type I riboflavin-sensitized reaction similar to that observed for β-crystallin.

Despite extensive investigations, some of which have been discussed above, the phototoxicity of riboflavin is still not completely understood because of the complexity of riboflavin photochemistry. In cultured bovine lens epithelial cells, it has been shown that photosensitized riboflavin stimulates both the formation of ascorbyl semiquinone radicals and the consumption of oxygen by ascorbate in a concentration-dependent manner (Wolff *et al.*, 1987). Photoexcited riboflavin mediates a one-electron oxidation of ascorbate that generates riboflavin radical; the latter may then autooxidize, producing superoxide anion. Impaired transport in the cultured epithelial cells, measured as $^{86}Rb^+$ uptake, depends on the presence of ascorbate. These studies also showed that both Cu^{2+} and Fe^{2+} catalyze ascorbate oxidation. Their toxic effects appear to be caused by the generation of reactive species such as H_2O_2.

To summarize briefly the forgoing discussion, the prevailing view is that tryptophan photoproducts act as photosensitizers and appear to be involved (through singlet oxygen and/or H_2O_2) in the protein aggregation and pigmentation resulting from near-UV radiation. Nevertheless, studies in model (nonlens) systems show that tryptophan is photolyzed mainly at wavelengths <295 nm; these wavelengths are filtered out by the cornea and do not reach the lens. Some photolysis of tryptophan is thought to occur at wavelengths >300 nm and apparently requires oxygen. Photolysis has been demonstrated in aqueous solutions of tryptophan exposed to 337.1-nm radiation, and singlet oxygen appears to play an important role (Borkman *et al.*, 1986). However, later sudies using Raman spectroscopy show that exposure of intact guinea pig lens to UV radiation of long wavelength (325 nm *in vitro* or 353 nm *in vivo*) causes no tryptophan photolysis, possibly because of the low oxygen levels in intact lenses (Barron *et al.*, 1987). There is, however, a dramatic production of a 457.9-nm-excited fluorophor resulting from irradiation that does not accumulate during aging. Although there is no photolysis of tryptophan after long-wave UV exposure in the guinea pig lens, there is a striking loss of sulfhydryl and increase in disulfide content across the visual axis (Barron *et al.*, 1988a). The photosensitizers responsible for this conversion were not identified.

A newly developed microbeam fluorescence/Raman imaging system has been used to demonstrate a 441.6-nm-excited green fluorophor in human lenses (Yu *et al.*, 1988). This fluorophor is not produced by photocatalyzed reaction(s); rather, it is a metabolic product resulting from normal lens aging. The distribution of another metabolically generated 488.0-nm-excited fluorophor in the equatorial plane of human lens has also been described using the same technique (Barron *et al.*, 1988b). The three-dimensional grid maps, topographic contour maps, and six-color intensity interval maps of the distribution of the fluorophor are especially striking. Preliminary studies show that the known fluorophor 3-OH Kyn glucoside is the precursor of the 488.0-nm-excited green fluorophor that is detectable in human lenses older than 10 yr. Since these fluorophores are present at highest concentrations in the nucleus, they are probably metabolically generated and not formed from UV radiation.

5.5. AGING

A broad spectrum of molecular, biochemical, and structural changes occur in the lens during normal development, maturation, and aging. Despite the large body of information that has become available during the past two decades on these changes, only a brief summary, mainly of recent findings, can be included in this section. Aging is intimately associated with lens opacification (cataract), and the latter is in fact often considered to be an accelerated form of aging. For clarity of presentation in a general chapter on lens biochemistry, the present overview of aging is limited to changes found in normal noncataractous lenses; the biochemistry of human and experimentally induced cataract is discussed in Section 5.6.

Lens fiber cells are not shed, nor do they degenerate; rather, they are continually displaced inwardly toward the center as they mature. Hence, the proteins in old (nuclear) lens fiber cells are as old as the organism itself. Because of the loss of intracellular organelles, there is little or no detectable protein synthesis in the nuclear region; metabolic activity in the lens appears to be confined to the superficial cortical and epithelial layers.

Aging of the lens is a normal, though complex, physiological process. Age-related changes in the lens are studied using three types of experimental approach: (1) comparisons of nuclear (old) and cortical (new) regions of single lenses; (2) analyses of concentric microdissected layers of individual lenses; or (3) comparisons of specific components in individual lenses of various ages. The last two approaches have recently been integrated into a single technique by several laboratories (Li *et al.*, 1985; McFall-Ngai *et al.*, 1985; Pierscionek-Balcerzak and Augusteyn, 1985). In one study, the distribution of water-soluble (WS), urea-soluble (US), and urea-insoluble (UI) proteins was examined in human noncataractous lenses ranging from 1.8 to 65 years of age (Li *et al.*,

1986). With the dry weight of lens tissue used as a reference, the amounts of WS, US, and UI proteins were found to be similar in cortical and nuclear regions until the age of 30. However, from age 40 onward there is a continuous cortex-to-nucleus increase in insoluble protein; the increase in nuclear protein insolubilization in lenses of fifth- to sixth-decade individuals involves mainly the UI fraction, whereas in younger lenses this increase is due mainly to US proteins. Also noted was a disappearance of 20- to 22-kDa soluble proteins in the deep cortex and nucleus of all lenses except those of a 1.8-year-old infant. This striking loss of 20- to 22-kDa polypeptide has also been observed in other laboratories as well (Bessems *et al.*, 1983; McFall-Ngai *et al.*, 1985). Apart from nuclear protein insolubilization and loss of 20- to 22-kDa crystallins, other major age-related biochemical changes in the lens include formation of yellow pigmentation and nontryptophan (NT) fluorescence, glycosylation of crystallins, changes in lipid composition, and cleavage of fiber cell membrane protein MP26.

5.5.1. Crystallins

Age-related changes in the lens crystallins occur as a result of both differential synthesis during development and postsynthetic modifications throughout life (see reviews by Harding and Dilley, 1976; Hoenders and Bloemendal, 1981, 1983; Harding and Crabbe, 1984). Although the postsynthetic changes occur predominantly in the nuclear region and changes resulting from differentiation are expressed mainly in outer cortical crystallins, there is nevertheless a considerable overlap. The documented age-related changes occurring in the crystallins include loss of sulfhydryl and formation of disulfide bridges, cross-linking, accumulation of high-molecular-weight (HMW) aggregates, increase in insoluble protein associated with browning and nontryptophan (NT) fluorescence, photooxidation of tryptophan, production of photosensitizers, generation of free radicals, proteolysis of polypeptide chains, deamidation of glutamine and asparagine residues, racemization of aspartic acid, and nonenzymatic glycation. In recent years, improved resolution of the crystallins has been achieved by the use of high-pressure gel-permeation and high-pressure ion-exchange chromatography. These techniques, together with conventional ones such as SDS-PAGE and isoelectric focusing, provide more precise characterizations of the lens crystallins than was possible before.

The soluble crystallins in human lenses from donors ranging in age from newborn to 70 years old have been analyzed in serial sections through the optical axis of the lens using HPLC gel filtration and SDS-PAGE (McFall-Ngai *et al.*, 1985). There are both gradual and abrupt changes in crystallin concentration in cortical and nuclear regions with aging. One of the most striking changes is the gradual age-related loss of 19- to 21-kDa γ-crystallins from both cortical and nuclear regions. These proteins have anomalous electrophoretic behavior on SDS gels if not subjected to prior heat denaturation. This may have resulted from age-related modifications in their charge characteristics (Zigler *et al.*, 1985b). Three populations of γ-crystallins are readily resolved and identified in young lenses, but with increasing age, two of the species are no longer separable. There is an age-related increase in heterogeneity of the γ-crystallins; moreover, they become more acidic with age.

Similar changes can be induced in unmodified γ-crystallin (obtained from the expression of a gene construct integrated into mouse L cells) after exposure to ascorbic acid in the presence of iron and oxygen (Russell *et al.*, 1987b). The alterations of γ-crystallin in this system indicate an ascorbic acid-induced oxidation and imply that a similar mechanism may explain the changes occurring *in vivo* in aging γ-crystallins. Other studies using antisera to synthetic peptides corresponding to the N- and C-terminal regions of the γ_{1-2} crystallin gene prepared from human lens (Schoenmakers *et al.*, 1984) demonstrate extensive covalent modifications during aging in normal human lenses (Takemoto *et al.*, 1987b). Whether these apparent covalent modifications of the N- and C-terminal regions of γ-crystallins are mediated by endogenous lens proteases or by oxidative changes or both remains to be established.

The γ-crystallins contain six or seven cysteine sulfhydryl residues per monomer, and although a marked loss in protein thiol occurs in the nucleus of older rodent lenses, this change is not found in aging human lenses (Kuck *et al.*, 1982). In rodents, γ-crystallins undergo increasing insolubilization with age, thought to result from oxidation of thiol groups and formation of disulfides. Support for this concept has come from Raman spectroscopy, an important tool not only for monitoring the conversion of sulfhydryl to disulfide but also for measuring the degree of hydration or dehydration in the lens nucleus. There is now direct evidence for age-related conversion of -SH to -SS- in the nucleus of rodent lenses (Yu *et al.*, 1985; Ozaki *et al.*, 1983, 1987). Not only exposed but also buried protein sulfhydryl groups are converted to disulfides; this transformation is correlated with the hard, relatively dehydrated nucleus found in the aging rodent lens. Biochemical studies show that the -SS- bonds formed in aging rat lenses are intramolecular and not intermolecular (Hum and Augusteyn, 1987b). There are no substantial changes in the molecular weight distribution of the proteins, and it has been suggested that the γ-crystallin molecules assume a more compact spatial arrangement in the nucleus as a result of intramolecular sulfhydryl bond formation. This is in contrast to human and guinea pig lenses, where age- related loss of -SH is not accompanied by the formation of -SS- bonds (Yu *et al.*, 1985). The loss of sulfhydryl in this case may be the result of increased oxidation of glutathione (GSH), formation of protein sulfhydryl, and reduction of the mixed disulfide by glutathione reductase and NADPH to yield the original protein sulfhydryl and oxidized glutathione (GSSG). The latter is then thought to be extruded from the lens.

Although β-crystallins are the most abundant polymeric soluble proteins of human and other lenses, relatively little is known regarding their age-related synthesis. β Basic principal polypeptide, βB_p (βB2), is the major β-crystallin in bovine and human lens. It bears extensive sequence homology with γII crystallin (see Section 5.2.1.2); moreover, bovine βB2 displays unusual heat stability (Mostafapour and Schwartz, 1981/1982). Human βB2-crystallin is also heat stable, which provides a convenient method for its analysis (McFall-Ngai *et al.*, 1986). This polypeptide is synthesized in new cortical fiber cells throughout life, but its concentration in old nuclear fiber cells decreases significantly with age. Posttranslational changes result in a decrease in the molecular mass of βB2 from 26 to 22–23 kDa, a consequence of *in vivo* proteolysis of the N-terminal region of the molecule (Takemoto *et al.*, 1987c). The cleavage occurs mainly in young lenses (0–10 years), with very little being detectable in older ones (>40 years). Other studies utilizing a monoclonal antibody have shown an age-related increase in human lens 27-kDa β-crystallin (Russell *et al.*, 1985). Although little immunoreactivity is present in fetal lenses, appreciable amounts are found at all postnatal ages, particularly in the cortical region. Other changes in human β-crystallins include an age-related decrease in a 29-kDa polypeptide and an increase in the 27-kDa polypeptide, but only until about age 5; afterward there is a slow decrease until age 86 (Alcala *et al.*, 1988). At all ages, the 29-kDa β-crystallin is detectable in the superficial fiber cells. However, the 27-kDa polypeptide shows a changing distribution pattern during aging, persisting in the superficial cortex but showing marked losses in the deep cortical and nuclear regions in older lenses. The 27-kDa β-crystallin is believed to play an important role in secondary fiber cell differentiation; its absence in a mouse model of hereditary cataract that is associated with a defect in fiber cell differentiation (Carper *et al.*, 1986) supports this view.

The foregoing discussion has centered on age-related changes in human and rodent crystallins. Bovine lens crystallins have also been extensively investigated (see reviews by Harding and Dilley, 1976; Hoenders and Bloemendal, 1981, 1983; Harding and Crabbe, 1984). With increasing age, bovine α-crystallin, but not β- or γ-crystallins, undergoes changes in tertiary structure involving tryptophan, tyrosine, and cysteine residues (Liang *et al.*, 1985). These conformational changes, detected by circular dichroism (CD), absorption, and UV-fluorescence measurements, show a decrease in tryptophan fluorescence and increase in nontryptophan (NT) fluorescence as well as decreases in accessible sulfhydryl groups. Other studies on bovine lens crystallins show a gradual decrease in the proportion of soluble α-crystallin in the nucleus (Bessems *et al.*, 1986). In addition, and similar to the human lens study described above, there is an increase in both charge and size heterogeneity of monomeric bovine lens γ-crystallins with age. Analyses of concentric layers of lenses from prenatal to 15-year-old cattle show that whereas γ-crystallins account for about 22% of the proteins synthesized prenatally, this value drops to about 4% at birth; the major monomeric species at that time is βs-crystallin (Pierscionek and Augusteyn, 1988). The β-crystallins increase from about 30% to 40% in prenatal and postnatal lenses, respectively; in older tissue, some of the βH-crystallin synthesized is converted to a higher-molecular-weight form (HMWβ), possibly in association with βB1 polypeptides. In bovine lens, as in other species, α-crystallin is formed at a constant rate throughout life, but because of major structural modifications it becomes progessively less soluble with increasing age.

Studies on interactions between purified A and B subunits of bovine α-crystallin *in vitro* show that at a ratio lower than two A chains to one B chain, an abnormal molecule termed α-neoprotein is formed (Manski and Malinowski, 1980). This protein differs from native α-crystallin in quaternary structure, a property that led to its identification in lenses *in vivo* (Manski and Malinowski, 1985). Although it is not detectable in lenses from young rats or calf, substantial amounts are present in the water-insoluble fraction of 2-year-old bovine lens nucleus and in adult rat lens. Exposure of α-crystallin to UV radiation causes the formation of α-neoprotein from aged nuclear bovine α-crystallin but not from calf cortical α-crystallin (Hibino *et al.*, 1988). It was also noted that older α-crystallin is photolyzed at a faster rate than young α-crystallin, possibly the consequence of an age-related formation of protein-bound photosensitizer *in vivo*.

A high-molecular-weight (HMW) class of crystallin, composed mainly of α-crystallin polypeptides, was first identified in bovine lens by Spector *et al.* (1971). It accumulates in an age-dependent manner in most animal lenses as well as in the human lens (Jedziniak *et al.*, 1978). Of the four types of HMW aggregates classified by Harding and Crabbe (1984), the first two (HMW1 and HMW2) are probably formed in normal lenses as posttranslational modifications, whereas the last two appear to be present mainly, if not entirely, in aging and/or cataractous lenses. The age-related increase in the concentration of HMW aggregates is primarily in the nuclear region of normal lenses (Spector, 1984a,

1985). The bonds holding the aggregates are weak, noncovalent linkages that can be disrupted by detergents and other agents. It is thought that HMW aggregates are intermediates in the formation of water-insoluble (urea-soluble) protein, but this has not been proven. Thus, despite extensive investigation, neither the precise chemical nature of HMW aggregates nor the mechanism of their formation is known with certainty. A recent study on HMW fractions isolated from the nucleus of calf and adult bovine lenses shows age-related differences in tertiary structure as manifested by their fluorescence and circular dichroism (CD) properties (Messmer and Chakrabarti, 1988). The HMW aggregates in bovine lens are polydisperse populations of different conformation; because of the apolar (hydrophobic) nature of the predominant α-crystallin polypeptides, their association in HMW aggregates is mainly through noncovalent interactions.

The HMW aggregates of human lens appear to have a more complex composition; the presence of 43-kDa polypeptide in disulfide-linked aggregates has been described by Spector (1984a,b, 1985). Several degraded proteins are also found in HMW aggregates from human lens; one of them, a 9.6-kDa polypeptide, is a cleavage product of the α-crystallin A chain (Roy and Spector, 1978). It increases during aging from 7% of the water-soluble (WS) proteins to 36% of the water-insoluble (WI) proteins. In further studies, the proteins of human lenses of various ages have been separated into three fractions: water soluble (WS), urea soluble (US) and urea insoluble (UI) (Srivastava, 1988a). The amounts of degraded polypeptides smaller than 18 kDa increase with age; moreover, the polypeptides isolated from either WS or US fractions self-aggregate during storage, resulting in the formation of polymers ranging in size from 18 to 1500 kDa. These findings, as well as others showing that the degraded polypeptides are immunoreactive toward both α- and γ-crystallins, suggest that an age-related polymerization of degraded polypeptides into HMW aggregates leads to their insolubilization. The origin of the degraded polypeptides has not been established, but they may be derived through the activation of endogenous proteases. Among the endogenous substances that may control protease activity in human lens is a trypsin inhibitor activity (Sharma *et al.*, 1987; Srivastava and Ortwerth, 1989). This activity is due mainly to the α-crystallins in young lenses, but with age, it is increasingly associated with the HMW aggregates. Beyond the age of 60, the inhibitor activity is found mainly in the WI fraction.

5.5.2. Fiber Cell Plasma Membranes and Cytoskeleton

Major intrinsic protein MP26 is the most abundant component of lens fiber plasma membranes, but whether it functions as a gap junction-forming protein is still controversial (see Section 5.2.2.1). Although an age-related conversion of MP26 to MP22 in human and bovine fiber cell plasma membranes has been recognized for many years (Roy *et al.*, 1979; Horwitz *et al.*, 1979,; Alcala *et al.*, 1980; Bouman *et al.*, 1980), the site(s) of cleavage could not be established with certainty in these early studies. More recently, antisera made against the C- and N-terminal peptides of bovine MP26 have been used as probes to identify subtle covalent changes in the molecule not recognizable by conventional polyclonal antisera (Takemoto *et al*, 1985; Takemoto and Takehana, 1986a). These studies demonstrate that an age-dependent production of 15- and 20-kDa polypeptides occurs at the N terminus of the molecule, while cleavage of MP26 to form MP22 was shown to occur at the C terminus. Thus, both ends of the MP26 molecule are degraded in an age-dependent manner, possibly by proteolytic cleavage, *in vivo*.

There is also evidence for proteolytic breakdown of other proteins of higher molecular weight in older plasma membrane fibers. Radioimmunoassays for 57-, 70-, 82-, and 100-kDa proteins in cortical and nuclear regions of sheep lens suggest that each of these proteins follows a specific and characteristic pattern of degradation (Kistler *et al.*, 1986). A recently described 115-kDa extrinsic membrane protein, possibly a component of the beaded filament, also appears to be degraded in bovine lens in an age-related manner (FitzGerald, 1988b). Immunoreactive fragments not arising as artifacts during the isolation procedure were more abundant in the old (nuclear) region than in the cortex.

Lens fiber cell membranes have an unusually high content of sphingomyelin, cholesterol, and saturated fatty acid, resulting in a lipid bilayer of low fluidity and high rigidity. It has long been recognized that the cholesterol/phospholipid (C/PL) ratio is higher in the nucleus than in cortical regions, a difference that is associated with increased rigidity of nuclear fiber membranes (see review by Alcala and Maisel, 1985). Precise changes in the absolute amounts of lipids throughout the lens have been obtained by analyzing consecutive layers of single human lenses (Li *et al.*, 1985). The phospholipid content is relatively constant throughout the lens except in the inner nucleus, where it drops precipitously. However, the distribution of cholesterol is more complex; it is low in the outer cortical region, rises significantly in the inner cortex, and then falls again in the nucleus. Thus, the general increase in C/PL ratio from the epithelium (C/PL of 0.8) to the nucleus (C/PL of 3.5) results from both the low cholesterol content in the outer cortex and the large decrease in phospholipids in the nucleus. Further studies on concentric layers of human lens show a complex age-related disappearance of phosphatidylethanolamine (PE) and decrease in phosphatidylserine (PS) in the nucleus (Li *et al.*, 1987). There appears to be a continuous regional and temporal increase in C/PL ratio with age; the rigidity of nuclear fibers begins to increase from about the second decade of life and continues until the age of 63, the oldest lens analyzed in this series. Bovine lenses have a lower cholesterol and phospholipid content than human lenses, but analyses of concentric layers show a similar cortex-to-nucleus increase in C/PL ratio from 0.5 to about 2.0 (Li and So, 1987). As in human lenses, this is due mainly to a relative decrease in phospholipid from the cortex to the nucleus.

The foregoing studies suggest that in both human and bovine lens, the high C/PL ratio in the nucleus reflects a high membrane rigidity in this older region of the lens. Further insight into age-related changes in bovine and human membranes has been obtained by examining changes in intrinsic and extrinsic fluorescence intensity using specific probes for surface proteins and the interior of membranes (Liang *et al.*, 1989). The tryptophan residues in plasma membrane proteins are in a hydrophobic environment and show no changes with aging. There are however, decreases in fluorescence intensity in older membranes and increases in anisotropy with aging. A higher degree of structural order, indicating an increased rigidity, appears to be the consequence of the increased C/PL ratios in aging lenses.

Cytoskeletal components of the lens undergo numerous morphological and biochemical changes with aging, although many of the details are still unclear (for reviews see Maisel, 1984; Alcala and Maisel, 1985). In general, it appears that degradation of cytoskeletal proteins commences with the formation of mature secondary fibers. There is on the whole a gradual loss of spectrin, vimentin, and actin from the deep cortical and nuclear fiber cells, although these proteins persist in the superficial cortical cells. The intermediate filament of the lens, vimentin, has been detected morphologically in an 80-

year-old human lens, but only in the epithelial cells. Biochemical analyses of sequential fiber cell layers of human lens show a marked decrease in vimentin in the deep layers of the cortex and the complete absence of this and other cytoskeletal components from adult human nucleus; they are, however, detectable in newborn lens (Maisel and Ellis, 1984). There is some evidence suggesting that vimentin may be degraded by Ca^{2+}-activated calpain II.

5.5.3. Enzyme Activities

It has long been recognized that the activities of most lens enzymes decline with age (Table 5.5). The extent depends in part on how the enzyme activity is expressed, i.e., as wet weight of tissue, as milligrams of protein, or as total activity per lens. Moreover, with any of these parameters used as reference for calculation, the enzyme activity will vary considerably according to whether young (cortical) or old (nuclear) regions are chosen for analysis.

The catalytic activities of glycolytic and oxidative enzymes in the rat lens are higher in the equatorial (young) region than in the nuclear (old) region (Ohrloff and Hockwin, 1983). Many enzymes also display altered electrophoretic mobility with aging; moreover, certain enzymes lose their catalytic activity, even though they are immunologically detectable. The specific activities of most glycolytic enzymes in whole rat lens decline with age when calculated on the basis of wet weight of tissue and plotted on a semilogarithmic scale (Bours *et al.*, 1988). However, the specific activities of the same group of enzymes, calculated per milligram protein, remains fairly constant with age except for aldolase. This enzyme displays a marked decline in specific activity in the aging rat lens, possibly the result of posttranslational modification such as denaturation that leads to its inactivation (Dovrat and Gershon, 1983). The same may be true for G6PD, a key enzyme in the pentose phosphate pathway (see Fig. 5.3). By using an antibody prepared against denatured G6PD that does not recognize "native" G6PD, it was shown that antigenically cross-reactive but catalytically inactive G6PD molecules accumulate in rat lens nucleus (Dovrat *et al.*, 1986).

Table 5.5. Lens Enzymes Showing Decreased Activity in Aging

Enzyme(s)	Species	Reference
Glycolytic or oxidative	Rat	Ohrloff *et al.* (1984); Dovrat *et al.* (1986); Bours *et al.* (1988)
Methionine adenosyltransferase	Rat	Geller *et al.* (1988)
Superoxide dismutase	Human	Ohrloff *et al.* (1984); Scharf *et al.* (1987)
Endopeptidases	Rat Human	Fleshman and Wagner (1984); Taylor and Davies (1987)
Exopeptidases	Human	Taylor and Davies (1987)
Glutathione peroxidase	Human Monkey[a]	Rathbun and Bovis (1986); Rathbun *et al.* (1986)
Glutathione reductase	Monkey Rabbit Human	Rathbun *et al.* (1986); Reddan *et al.* (1988); Ohrloff *et al.* (1984)

[a]Glutathione peroxidase activity in monkey lens increases until adulthood and then declines.

Superoxide dismutase (SOD) is present at low but detectable levels in the lens (Bhuyan and Bhuyan, 1978), and in normal whole human lens its activity does not decline with age (Ohrloff *et al.*, 1984). However, comparison of nuclear and equatorial samples shows a large decline in SOD activity with age in both regions. Use of an antibody raised to denatured SOD shows an accumulation of catalytically inactive but antigenically reactive SOD molecules in aging human lens (Scharf *et al.*, 1987).

The activities of enzymes of glutathione synthesis—glutathione synthetase and γ-glutamylcysteine synthetase—decline sharply with age in human, bovine, rabbit, and dog lenses (Sethna *et al.*, 1982/1983). However, other enzymes in glutathione metabolism display species variation in age-related changes in activity. For example, in both bovine (Ohrloff *et al.*, 1984) and monkey (Rathbun *et al.*, 1986) lens, glutathione reductase activity is high at birth, drops rather steeply during the next few years, and then declines gradually throughout life. In the human lens, there is only a slow gradual loss of activity throughout life, regardless of how the enzyme activity is calculated (Rathbun and Bovis, 1986). Glutathione reductase activity in cultured epithelial cells from 8-year-old rabbits is 50% lower than in cells cultured from 4-day-old rabbits (Reddan *et al.*, 1988). However, the levels of glutathione peroxidase, hexokinase, and G6PD are comparable in young and old cells. Nevertheless, cells from younger donors are better able to sustain oxidative damage from H_2O_2 than those from older donors, possibly because of more effective stimulation of the pentose phosphate pathway in young cells. Glutathione peroxidase is low in fetal and neonatal human (Rathbun and Bovis, 1986) and monkey (Rathbun *et al.*, 1986) lens; however, it increases with age until adulthood and then declines slowly. In bovine lens, this enzyme displays high activity in young animals and afterward shows a gradual decline (Ohrloff *et al.*, 1984).

5.5.4. Glycation

The nonenzymatic glycation (or glycosylation) of a protein occurs by condensation of a sugar molecule with a protein amino group to form a covalent adduct. In hemoglobin, the major sites of glycation are valine and lysine residues (see review by Harding, 1985). The initial product formed is a labile Schiff base, which undergoes Amadori rearrangement to a ketoamine. This is the first step in the nonenzymatic browning of food (the Maillard reaction) and may be involved in aging and in cataract formation in the lens. Additional sugar molecules can condense with the ε-amino as well as the α-amino groups of the protein, forming cross-links that decrease the solubility of the protein. The extent to which this reaction proceeds, either *in vivo* or *in vitro*, depends on the concentration of glucose present, the number of potentially reactive amino groups on the protein, and other less well-defined factors.

The ε-amino groups of lysine in bovine lens crystallins, particularly the high-molecular-weight (HMW) aggregates of α-crystallin, are glycated nonenzymatically *in vivo* in an age-related manner (Chiou *et al.*, 1981). Moreover, incubation of human and bovine lens proteins with reducing sugars in the absence of oxygen leads to the formation of fluorescent yellow pigments and nondisulfide protein cross-links (Monnier and Cerami, 1981). The pigment formed is strikingly similar to that found in diabetic lenses and in cataract (see Section 5.6). Both of these observations imply the existence of a Maillard reaction in the lens. Age-related glycation of rat lens crystallins has also been observed (Swamy and Abraham, 1987). The glycated proteins in older rat lenses are found mainly in the urea-soluble (US) fraction, and the increase in glycated proteins coincides with an increase in insoluble HMW aggregates.

Human lens crystallins, soluble and insoluble, contain an average of 0.028 nmol of glycated lysine per nmol of crystallin monomer (Garlick *et al.*, 1984). Unlike the rat lens, similar levels of glycation are found in the soluble and the insoluble crystallins. Glycation of human lens crystallins increases with age in a linear manner: 1.3% of lysine residues are glycated in the infant lens, 2.7% in the 50-year-old lens, and, by extrapolation of the data, approximately 4.2% of human lens crystallins would be glycated in older age groups. These are surprisingly low values; in spite of their longevity, lens proteins undergo far less glycation than rapidly turning over proteins such as hemoglobin or serum albumin. The relatively slow rate of glycation of lens crystallin may be the consequence of several factors: (1) the concentration of glucose in the lens is relatively low (1 mM in normal humans); (2) the predominantly β-pleated structures of the crystallins provide only a limited number of exposed lysine residues for glycation; and (3) lens crystallins have a smaller percentage of lysine residues than highly glycated proteins such as hemoglobin or albumin, and some of their α-amino groups may be blocked.

Apart from glucose, ascorbic acid (Fig. 5.7) is also capable of forming adducts with bovine lens crystallins (Bensch *et al.*, 1985). The reaction with ascorbate is faster than that occurring with glucose, although the fluorescence spectra are similar. In both cases the brown coloration formed appears to represent a nonenzymatic Maillard-type reaction. Ascorbic acid is present at much higher concentrations than glucose in normal lens. Hence, it may play an important role in the glycation of lens crystallins, and recent findings support this view. The browning and aggregation of bovine β- and γ-crystallins, but not α-crystallin, can be produced *in vitro* in the presence of either ascorbic acid or dehydroascorbic acid; under similar experimental conditions, glucose, even at concentrations as high as 100 mM over a 6-week period, is ineffective in generating cross-linked proteins (Ortwerth *et al.*, 1988). The physiological importance of ascorbate glycation is that it occurs under low oxygen tension with as little as 2 mM ascorbic acid, the average concentration in normal lens. The glycating species is the oxidized form of ascorbic acid, since the formation of covalent adducts, as well as cross-linked proteins, is inhibited by glutathione (GSH) and other reducing agents (Ortwerth and Olesen, 1988a). These interesting findings suggest that *in vitro*, and possibly *in vivo*, GSH may inhibit glycation by maintaing ascorbic acid in the reduced state, a form that is not active in glycation. In aging and in cataractous lenses, where the concentrations of oxidized ascorbic acid are presum-

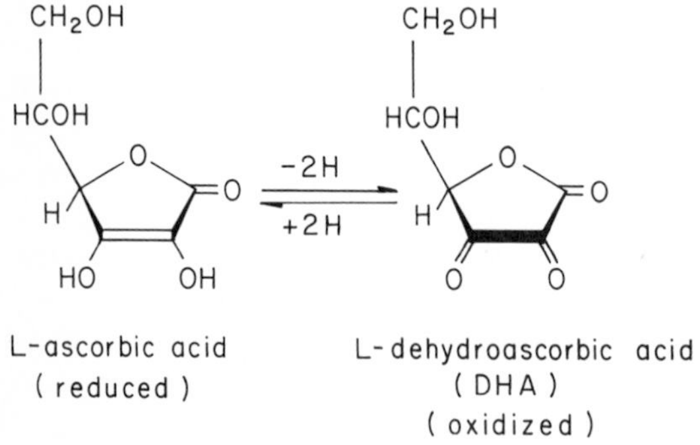

CH_2OH HCOH ... L-ascorbic acid (reduced) $\underset{+2H}{\overset{-2H}{\rightleftharpoons}}$ CH_2OH HCOH ... L-dehydroascorbic acid (DHA) (oxidized)

Figure 5.7. Structures of L-ascorbic acid and L-dehydroascorbic acid. The reduced form (ASA) with an enediol structure at carbon atoms 2 and 3 is unstable and in solution at physiological pH readily undergoes oxidation to dehydroascorbic acid (DHA), possibly through an ascorbyl radical intermediate. The equilibrium concentrations of each species depends on the levels of oxygen present; in the lens, where oxygen levels are low, the reduced species (ASA) is probably the major one.

ably normal but the levels of GSH may be low or depleted, increased glycation would be expected. Other studies show that incubation of calf lens extracts with 20 mM ascorbic acid induces the formation of HMW aggregates; moreover, the cross-linking by ascorbate appears to be between the βH- and α-crystallins (Ortwerth and Olesen, 1988b). The principal ascorbate-modified amino acid is lysine; in addition, a small amount of glycated arginine was detected. The mechanism of glycation by ascorbate is still unknown, but studies using a model system have shown that ascorbic acid autooxidation by GSH is enhanced in the presence of metal ions, suggesting involvement of oxygen free radicals (Wolff *et al.*, 1987; Winkler, 1987). Whether glycation *in vivo* proceeds through the formation of an ascorbyl free radical generated by GSH in the presence of endogenous metal ions remains to be established.

5.5.5. Pigments and Fluorescent Substances

The increase of insoluble protein in aging human lens is accompanied by increased yellow coloration and nontryptophan (NT) fluorescence. One of the factors inducing these changes is the UV radiation absorbed by the lens throughout the lifetime of the individual (Lerman, 1980, 1983). Two mechanisms have been proposed to account for these changes: one is a direct process in which radiation is absorbed by endogenous chromophores such as tryptophan and other aromatic amino acids, and the other is an indirect process, i.e., absorption by extraneous photosensitizing compounds such as drugs. Three aging parameters in the lens have been described by Lerman (1983): generation of a series of fluorescent chromophores absorbing at increasingly longer wavelengths than tryptophan, a deeping of yellow coloration in the nucleus, and a progressive cross-linking and insolubilization of lens crystallins by fluorescent pigments. The level of fluorogens (UV-absorbing chromophores) is relatively low in normal lenses below 10 years of age, but there is a steady increase in these substances, especially in the nucleus, as the lens ages.

As discussed in Section 5.4, tryptophan residues in the lens are thought to be the major photochemically active species in UV-induced changes in the aging lens; pigment formation and protein aggregation involve singlet oxygen and other reactive free radicals. Brown nuclear cataracts are an accelerated form of the normal aging process in which photochemically induced fluorescent pigments accumulate mainly in the nuclear region (Lerman, 1983). The age-related yellowing of the human lens is in fact beneficial to vision, since it allows the lens to act as a filter in protecting the aging retina from cumulative photochemical damage (Lerman, 1988). However, the increasing amount of UV radiation absorbed by the lens during the lifetime of the individual may be a risk factor in the development of cortical cataracts (see Section 5.6.9.2).

Whether tryptophan plays a direct or indirect role in chromophore production during aging has not been established with certainty. Kynurenine derivatives, products of tryptophan oxidation, can function as photosensitizers and have been detected in clear aging lenses and in cataract. In a recent study of human lenses showing increased nuclear pigmentation with age, there was no corresponding increase in tryptophan in water-soluble extracts from either cortical or nuclear regions (Bessems and Hoenders, 1987). Instead, there was a significant increase in a particular fluorophore (excitation and emission maxima of 345 and 425 nm, respectively) in old clear lenses and in those with nuclear cataracts. This fluorophor is not an oxidation product of free tryptophan; it appears to be associated with the γ-crystallins and may represent an oxidized fragment of this crystallin.

A yellow pigment that fluoresces at 440 nm with excitation at 370 nm is thought to arise from either UV-oxidized tryptophan residues (see review by Zigman, 1985) or a Maillard reaction following glycation of lens crystallins. This nontryptophan (NT) fluorescence has been examined further using a new fluorometric technique developed by Liang *et al.* (1988). The method utilizes front-surface illumination of samples oriented 60° relative to the incident beam; this minimizes reflected light and therefore can be used with solid or turbid samples. Studies on insoluble crystallins from old and young bovine lenses show a red-shift of tryptophan fluorescence, indicating a less hydrophobic environment with aging. Moreover, tryptophan residues are more exposed in older samples as a result of unfolding of crystallins (Liang *et al.*, 1985). Another age-related difference is the appearance of a 370/440-nm nontryptophan (NT) fluorescence peak in insoluble powdered samples from old bovine lenses. This peak, which is also present in human lenses, shows a shift to longer wavelengths in 60- to 75-year-old lenses.

Two metabolically generated fluorophores have been detected in human lens using a newly developed Raman imaging system. One is a 441.6-nm-excited green fluorophor (Yu *et al.*, 1988), and the other a 488.0-nm-excited fluorophor (Barron *et al.*, 1988b). The latter is detectable in human lenses over 10 years of age.

5.6. CATARACT

Cataract in its broadest definition is an opacity, either partial or total, of the lens. Although cataracts can occur as a result of disease, injury, exposure to toxic substances, or congenital anomalies, the majority of cataracts in humans are associated with complex aging processes in the lens. Three major types of senile cataracts are now recognized clinically: cortical (including anterior subcapsular), posterior subcapsular, and nuclear. In recent years precise and objective systems have been developed for both *in vivo* and *in vitro* classification of cataracts (see Section 5.6.9.1). These include the Lens Opacities Case-Control Classification System (LOCS) developed by the Cooperative Cataract Research Group (CCRG) in America and the Oxford Clinical Cataract Classification and Grading System used extensively in England and Europe.

Biochemical changes in human cataracts have been studied for several decades, and parallel research on experimentally induced cataracts in laboratory animals leave no doubt that cataractogenesis is a highly complex, multifactoral process: it is intimately associated with aging and, in all probability, is initiated by oxidative changes. The great variety of metabolic insults that can cause cataracts has been described in great detail by Harding and Crabbe (1984); this review is highly recommended for in-depth descriptions of more than 30 types of cataract, covering the literature until about 1982. In addition, the articles by Spector (1984a,b, 1985) and by Chylack (1984) provide extensive background material on the biochemical basis of human senile cataract. Since a complete discussion of this vast topic is far beyond the scope of a general text on eye biochemistry, only certain topics have been selected for inclusion. The first sections (5.6.1 to 5.6.7) cover model (*in vitro*) systems, experimentally induced cataracts in animals, and hereditary cataracts in mice; the last two sections (5.6.8 and 5.6.9) describe biochemical changes in human cataracts and epidemiologic studies.

5.6.1. Lens Proteins

The modifications in lens proteins during cataract formation are remarkably similar to those found in aging (Harding and Dilley, 1976; Harding, 1981; Harding and Crabbe, 1984; Spector, 1984a,b, 1985). These changes, mediated by enzymatic as well as nonenzymatic mechanisms, include tryptophan oxidation, deamidation, increase in yellow pigmentation, increase in nontryptophan (NT) fluorescence, mixed disulfide and protein–protein disulfide bond formation, cross-linking, increase in insoluble protein, glycation, and formation of high-molecular-weight (HMW) aggregates.

Many of the changes described above occur before the appearance of opacities, and some of them can be produced experimentally when lens proteins are subjected to photooxidative damage by UV light or by reaction with exogenous sensitizers. The modifications induced under these conditions appear to be mediated by singlet oxygen generated photodynamically (Goosey *et al.*, 1980; Zigler and Goosey, 1981, 1984). Ultraviolet radiation of bovine lens crystallins results in cross-linking of proteins, formation of disulfide and nondisulfide covalent bonds, development of blue fluorescence, pigmentation, and aggregation to high-molecular-weight species some of which are insoluble. The chromophores responsible for absorbance of UV light, N-formylkynurenine and others, are present in the lens (van Heyningen, 1971; Pirie, 1971; Grover and Zigman, 1972). The site of UV absorption is presumably in the anterior cortex, but the oxidative damage caused by the generation of singlet oxygen appears to be mainly in the nucleus, possibly because of the reduced antioxidant capacity in this region (Zigler and Goosey, 1984).

Ultraviolet laser irradiation (292 and 298 nm) of whole rat lenses causes changes in lens proteins similar to those found in human cataract (Borkman, 1984). Loss of tryptophan fluorescence (which appears to be O_2-dependent), increased NT fluorescence, and cross-linking of crystallin polypeptides are the most prominent changes. Protein modifications can also be induced by 337-nm radiation, a wavelength that is not absorbed by tryptophan. Since exogenous sensitizers were not added, the protein modifications were probably mediated by endogenous sensitizers such as kynurenine derivatives.

Nonenzymatic conformational changes can be induced in bovine α- and γ-crystallins by incubation with glucose 6-phosphate (Beswick and Harding, 1987b). Glycation by glucose 6-phosphate results in alterations of isoelectric points and changes in the tertiary/quaternary structures of these proteins; there is, however, no detectable increase in interprotein disulfide binding. These experiments were performed under N_2 and hence cannot be attributed to oxidative changes. Modification of γ-crystallin by glucose 6-phosphate results in a net gain of three negative charges, causing extensive surface changes that could facilitate interactions between γ-crystallins localized mainly in the lens nucleus. The hydration of γ-crystallins is also affected and, together with alterations in surface charge, could lead to changes in refractive index. This would result in light scattering and decreased light transmission similar to that found in opaque areas in cataractous lenses.

A long-recognized characteristic of aging and cataract is the marked increase in water-insoluble (WI), urea-soluble (US) protein. In the bovine lens this fraction consists mainly of α-crystallin, but in human lens it contains a mixture of crystallins (Harding, 1981; Harding and Crabbe, 1984). Until recently, physicochemical measurements of water-insoluble crystallins have been difficult and somewhat unreliable; however, front-surface illumination of samples oriented 60° relative to the incident beam allows accurate fluorescence measurements of solid or powdered protein samples (Liang *et al.*, 1988). These studies show that insolubilization involves a partial unfolding and exposure of hydrophobic groups resulting in a greater exposure of tryptophan residues. Proteins undergoing such conformational changes have a higher index of refraction than soluble proteins; therefore, the amount of light scattered would be increased. In addition, the partially unfolded state may facilitate the formation of HMW aggregates and enhance their association with the fiber cell plasma membranes.

Cold cataract is a temperature-dependent reversible opacity that can be induced in the nucleus of young rat, rabbit, and bovine lenses. Cryoprecipitation of lens proteins has been recognized for many years (Lerman and Zigman, 1965); the phenomenon is associated with protein–water interactions and is specifically related to the γ-crystallins, which are especially concentrated in the innermost region of the nucleus (Lerman *et al.*, 1966, 1983; Blundell *et al.*, 1983). Precipitation in the cold results from phase separation of the reactants, which causes the production of protein-rich and protein-poor domains. The resulting fluctuations in refractive index in these regions cause scattering of incident light and opacification. Certain "large scattering elements" enriched in low-molecular-weight species (probably γ-crystallins) have been visualized in calf lens nuclear cytoplasm by electron microscopy; these domains are thought to be responsible for cold cataract opacification (Gulik-Krzywicki *et al.*, 1984). Studies using individual fractions of γ-crystallins of calf lens in solution show that the concentration of γIV-crystallin is the major determinant of opacification temperature (Siezen *et al.*, 1985). It is concentrated mainly in the nucleus of young lenses as a result of its high rate of synthesis during embryonic development. Aged nucleus contains very little γIV- (or γIII) crystallin (Slingsby and Miller, 1983), possibly because of its insolubilization and association with α- and β-crystallins during the formation of HMW aggregates. A definite relationship between cold cataract and human senile cataract has not been established. However, γIV-crystallin plays a central role in the model system of nuclear opacification studied by Siezen *et al.* (1985) and, despite its low concentration in adult lens, it may nevertheless play a role in the development of human nuclear cataracts.

5.6.2. Sugar Cataracts

Clinicians have long recognized an association between diabetes and cataract, and epidemiologic studies suggest that diabetics are at high risk for developing cataract (see Section 5.6.9.2). Infants with galactosemia develop cataracts early in life, and cataracts can be induced in young rats by feeding diets containing 50% galactose. It was van Heyningen (1959) who first demonstrated that sugar alcohols (polyols) accumulate in the lenses of alloxan-induced diabetic rats and in young rats maintained on diets high in either galactose or xylose. The observation that lens extracts can reduce xylose, galactose, and glucose to their corresponding sugar alcohols in the presence of NADPH in an enzymatic reaction mediated by aldose reductase (van Heyningen, 1962) opened decades of research on the mechanism of sugar cataracts and its prevention by aldose reductase inhibitors.

5.6.2.1. Experimentally Induced Sugar Cataracts

The major pathways of glucose metabolism in the lens are shown diagrammatically in Fig. 5.3. Hexokinase has a very high affinity for glucose, and most of the glucose

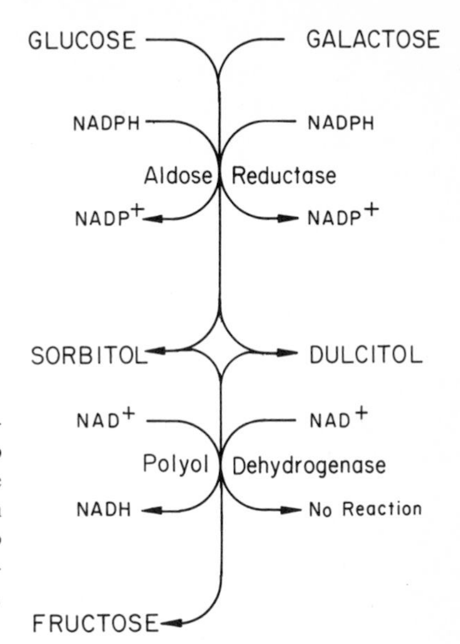

Figure 5.8. Sorbitol (or polyol) pathway of glucose and galactose metabolism. The reduction of glucose and galactose to corresponding polyols (sugar alcohols) is catalyzed by aldose reductase (AR) using NADPH as cofactor. The second step in this pathway is the oxidation of sorbitol (but not dulcitol) to fructose by polyol dehydrogenase (PD) using NAD^+ as cofactor. This step is reversible in human lens (Jedziniak *et al.*, 1981).

entering the lens is phosphorylated and metabolized through either the Embden–Meyerhof glycolytic pathway or the pentose phosphate pathway. However, in the diabetic lens, or in patients with inborn errors of galactose metabolism (galactosemia and galactokinase deficiency), or in animals fed high levels of galactose or xylose, substantial amounts of sugar are reduced by aldose reductase (AR), an NADPH-dependent enzyme that catalyzes the first step in the sorbitol pathway (Fig. 5.8). The polyols produced from glucose and galactose are sorbitol and galactitol (dulcitol), respectively. The second enzyme in the sorbitol pathway, polyol dehydrogenase (PD), uses NAD^+ as a cofactor and catalyzes the oxidation of sorbitol to fructose; it is not active on dulcitol. These two enzymes constitute the sorbitol pathway (van Heyningen, 1962). The direct association between accumulation of sugar alcohols (polyols) in the lens and cataract formation, first described in the rat lens, has also been observed in other experimental animals. The brief discussion that follows is based on several excellent reviews summarizing biochemical as well as morphological aspects of sugar cataracts and, most importantly, their prevention (or delay) by aldose reductase (AR) inhibitors (Harding and Crabbe, 1984; Kador and Kinoshita, 1984; Kador *et al.*, 1985, 1986; Kinoshita, 1986). Because of space limitations, only a few original literature references before 1982 are included in the overview that follows; they are, however, available in the review articles cited above.

Sugar cataracts have been studied mainly in galactose-fed animals. According to the osmotic hypothesis of galactose-induced cataract (Kinoshita *et al.*, 1962), the lens begins to swell in response to the AR-initiated accumulation of galactitol. The hyperosmotic effects of galactose are greater than those of glucose for two reasons: (1) owing to the greater affinity of AR for galactose than for glucose, galactitol is formed more rapidly than sorbitol, and (2) whereas sorbitol can be further metabolized, galactitol cannot. Hence, substantial amounts of polyols accumulate in galactose-induced cataracts, and considerably less in experimentally induced diabetic cataracts. A recent study on galactose cataract shows that there is a considerable influx of water, the excess being present mainly as free water in lakes and pools rather than as bound water of hydration of proteins (Wang and Bettelheim, 1988). This influx of water in experimentally induced sugar cataracts causes changes in membrane permeability, leading to electrolyte imbalance (see reviews by Kador *et al.*, 1985, 1986; Kinoshita, 1986). The normally high K^+/Na^+ ratio in the lens decreases as a result of influx of Na^+. This is accompanied by a partial depletion of NADPH and dramatic decreases in reduced glutathione (GSH), ATP, inositol, and other essential metabolites; in addition, there is impaired uptake of amino acids and decreased protein synthesis. The lens fibers become swollen and eventually form visible vacuoles. Nuclear opacification is observed in the final stages.

A complete summary of changes occurring in galactose cataract has been compiled (see Table XLVI in Harding and Crabbe, 1984), and a speculative scheme showing major biochemical changes and their interrelationships is given in Fig. 5.9. Restoration of a normal diet in the early stages leads to a reversal of galactose-induced cataracts in the rat. Further evidence supporting the polyol-osmotic theory of sugar cataract formation has come from studies showing a substantial drop in inositol accumulation in rat lenses incubated overnight in 30 mM galactose (Kawaba *et al.*, 1986). This effect could be

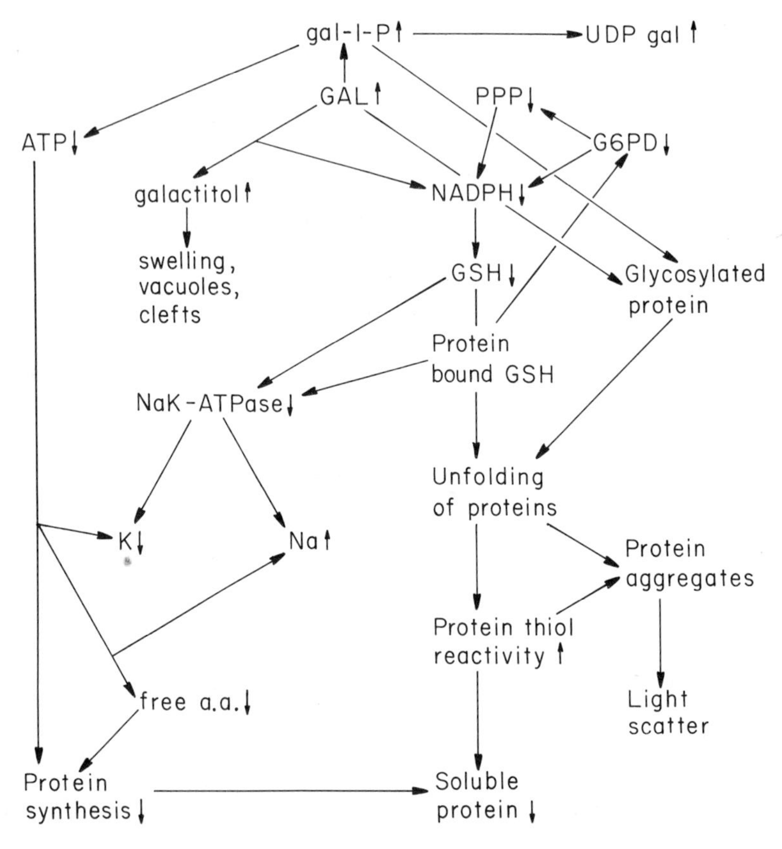

Figure 5.9. Speculative scheme of changes occurring in galactose cataract. The following abbreviations are used: gal-1-P, galactose 1-phosphate; G6PD, glucose 6-phosphate dehydrogenase; ppp, pentose phosphate pathway; GSH, reduced glutathione. (From Harding and Crabbe, 1984.)

abolished either by the AR inhibitor sorbinil or by rendering the galactose medium hypertonic in order to offset the osmotic effect of polyol accumulation in the lens. Other studies on the mechanism of GSH depletion in galactosemic and diabetic rats also suggest that osmotic stress is an important component of sugar cataract (Lou *et al.*, 1988). One of the earliest changes is an impaired ability of the lens to accumulate amino acids, and the depletion of GSH is considered in terms of its reduced biosynthesis and faster efflux resulting from increased permeability of the lens membranes.

The most compelling evidence for an involvement of AR in the development of sugar cataracts is the ability of AR inhibitors to either prevent or delay their onset. This has been clearly demonstrated in galactose-induced cataract in the rat. Aldose reductase inhibitors not only prevent or delay the appearance of cataract, but they may even reverse the process if administered prior to the formation of nuclear opacities (Kador *et al.*, 1986). These inhibitors also prevent cataract formation in some forms of experimentally induced diabetes. The names and structures of many AR inhibitors have been published (Kador and Kinoshita, 1984; Kador *et al.*, 1985; Kinoshita, 1986), and of those listed, the most extensively investigated include sorbinil, tolrestat, AL 1576, statil, and ONO 2235. Sorbinil is able to reverse the pathological process in rats maintained on a 50% galactose diet for 5 days and is as effective as withdrawal of galactose from the diet (Hu *et al.*, 1983). Topical sorbinil can also reverse galactose cataracts and, rather unexpectedly, appeared to be acting systemically rather than locally (Hu *et al.*, 1984).

Substantial decreases in phospholipid precursors such as phosphorylcholine, glycerophosphorylcholine, and glycerophosphorylethanolamine have been observed in lenses of streptozotocin-induced diabetic rats (Cheng and Gonzalez, 1986). More recent studies on hyperglycemic rat lens show not only a depletion of phosphorylcholine and ATP but also a drastic reduction in choline influx (Lou *et al.*, 1989). These changes were thought to be the result of membrane defects caused by osmotic damage induced under hyperglycemic stress. The depletion of phosphorylcholine and ATP, as well as the impaired uptake of choline, could be prevented by aldose reductase inhibitors AL 1576, AL 1567, and AL 1750.

There is little doubt that AR plays an important role in the production of sugar cataracts, but the osmotic mechanism may not be the only explanation for their development. Alternative hypotheses have been proposed, especially for diabetic cataract (Harding and Crabbe, 1984). As noted above, glucose is a relatively poor substrate for AR, and the amount of sorbitol that accumulates in experimental diabetic cataracts may be too little to generate the osmotic effects found for galactose. In human lens, the activity of AR is at least an order of magnitude lower than in the rat, and the activity of polyol dehydrogenase is considerably higher. This implies that sorbitol could be removed as rapidly as it is formed, and little accumulation would be expected in human lens. Glucose, or glucose 6-phosphate, or fructose may, however, be involved in an entirely different way in diabetic cataracts, namely, through the formation of covalent adducts with ϵ-lysine groups of lens proteins, causing conformational changes in lens proteins and Maillard browning reactions (see Section 5.6.2.2).

Several studies have shown striking changes in protein synthesis and in crystallin and MP26 mRNAs in the galactosemic rat lens. There is a marked reduction, followed by total cessation, of crystallin synthesis in the fiber cells of rats maintained on a 50% galactose diet over a period of 18 days (Shinohara *et al.*, 1982). Protein synthesis in the epithelial cells is normal or even somewhat enhanced. Studies of *in vitro* translation in a rabbit reticulocyte lysate show that after 12 days on a galactose diet, the lenses had lost half of the cortical crystallin mRNAs and all of the crystallin mRNAs from nuclear fiber cells. These findings were interpreted in terms of impaired utilization and subsequent degradation of crystallin mRNAs, possibly resulting from the ionic imbalance in lenses of galactose-fed rats. Similar observations on greatly reduced protein synthesis were reported in rats maintained on 50% galactose for periods up to 21 days (Williams *et al.*, 1985).

A more recent study has shown that continuous exposure of rat lenses to dietary galactose for 45 days leads to significant fluctuations in mRNA synthesis and survival (Hsu *et al.*, 1987a). Although noncrystallin gene activity is maintained, crystallin activity is significantly altered. Several minor mRNA translation products not present in normal lenses were detected in lenses of galactose-fed rats. The use of soft laser densitometry for quantitation of *in vitro* translation products of mRNAs showed that although the total levels are similar in controls and in galactose-fed rats, with or without sorbinil, the relative proportions of α- and γ-crystallin mRNA translation products are highly variable (Hsu *et al.*, 1987b). An MP26 probe isolated by cDNA cloning and transcribed into antisense [^{35}S]-UTP-labeled RNA has been used to localize and compare the MP26 mRNA in normal and galactosemic lenses by *in situ* hybridization (Bekhor, 1988). In galactose-induced cataractous lenses, MP26 mRNA is found at lower concentrations at the bow than in normal lenses. Moreover, it not only persists, but its level is even enhanced in viable areas of cortical fiber cells. In contrast, and not unexpectedly, MP26 mRNA is only barely detectable in damaged regions of the cataractous lens.

5.6.2.2. Glycation

The nonenzymatic glycation of lens crystallins as a normal physiological process in aging has been discussed in Section 5.5.4. It may also be a major factor in the predisposition of diabetics to cataract (Harding, 1985), a view supported by the observation that diabetic crystallin samples contain about twice as much glycated crystallin as age-matched controls (Garlick *et al.*, 1984). This finding has been confirmed in other laboratories. In model systems studied by Stevens *et al.* (1978), incubation of bovine lens crystallins with glucose or glucose 6-phosphate leads to the glycation of ϵ-amino groups of lysine residues, formation of HMW aggregates, and production of opalescence.

Crystallins from diabetic rats show enhanced glycation compared to normal controls (Stevens *et al.*, 1978), and HMW aggregates linked in part by disulfide bonds accumulate in cataractous lenses of both diabetic and galactosemic rats (Monnier *et al.*, 1979). Further studies using molecular seive HPLC techniques to separate water-soluble (WS) and urea-soluble (US) crystallins in diabetic cataractous rat lenses show that the US crystallins in particular become increasingly glycated as hyperglycemia develops (Perry *et al.*, 1987). There is a decrease in reactive thiols associated with the formation of disulfide-linked HMW aggregates as well as a decrease in γ-crystallins in the WS fractions. In all, glycation causes conformational changes in lens crystallins, resulting in unfolding of their tertiary structures (Harding and Crabbe, 1984; Liang and Chylack, 1987). Increased AR activity in hyperglycemia may contribute indirectly to diabetic cataract by diverting NADPH from the pentose phosphate pathway to the sorbitol pathway, resulting in a relative depletion of NADPH and hence a decrease in GSH levels and loss of the normal reducing environment in the lens.

Glycation of bovine lens crystallins by glucose 6-phosphate causes changes in the tertiary/quaternary structure of bovine α-and γ-crystallins (Beswick and Harding, 1987b). These and other findings suggest that conformational changes in lens crystallins induced by glycation may be involved in the etiology of human cataracts, especially among diabetics. Galactose also forms adducts with lens proteins, in fact, at a rate somewhat faster than glucose (Huby and Harding, 1988). The glycation can be inhibited by aspirin, perhaps through a covalent type of acetylation of ϵ-amino groups (Rao *et al.*, 1985), but more probably by either noncovalent association with the proteins, or through its action as a free radical quencher. Glutathione (GSH) at physiological concentrations also inhibits galactosylation of lens proteins, possibly by blocking its initial binding (Huby and Harding, 1988). Although the mechanisms appear to be different, it is interesting to note that GSH and other reducing agents also inhibit the glycation of lens crystallins by ascorbic acid (Ortwerth and Olesen, 1988a). The rate of adduct formation by galactose and glucose 6-phosphate can be substantially decreased by 5-hydroxybendazac, a metabolite of the experimental anticataract drug bendazac (Lewis and Harding, 1988).

The findings summarized in the foregoing discussion suggest that glycation of lens crystallins leads to major changes in their conformation; this results in the formation of disulfide-linked HMW aggregates, which are believed to play a role in light scattering and opacification. However, other lens proteins can also be glycated; for example, incubation of bovine lens epithelial cells in a high-glucose-containing medium (30.5 mM) results in glycation of Na^+,K^+-ATPase (Garner and Spector, 1986; Garner *et al.*, 1987). The Na^+,K^+-ATPase-dependent transport of K^+ is inhibited, and studies using isolated semi-purified enzyme show in addition a decreased rate of ATP hydrolysis at near-saturating substrate concentrations. The aldose reductase inhibitor AL 1576 has a small but significant stimulatory effect on the rate of ATP hydrolysis at near-saturating concentrations by the unmodified enzyme; it also causes an increase in pump-dependent K^+ influx. However, the drug has an even more striking effect on the glucosylated enzyme, restoring both K^+ transport and hydrolytic activity to near-normal levels. These findings suggest a second possible role for AR inhibitors, namely, direct stimulation of glucosylated Na^+,K^+-ATPase of the lens epithelium.

5.6.3. Radiation Cataracts and Oxidative Stress

The many types of radiation that can produce cataracts in experimental animals and in model systems *in vitro* have been reviewed in detail by Harding and Crabbe (1984). The radiation sources described in this extensive and informative article include x rays, β and γ radiation, neutron radiation, protons and heavy ions, atomic bomb, infrared, microwave, ultraviolet (UV), sunlight, and visible radiation. Evidence for a cause-and-effect relationship between radiation and cataractogenesis in a variety of experimental animals is convincing, but the biochemical basis remains elusive. In this context, recent flow charts on possible biochemical changes resulting from microwave and ionizing radiation offer some provocative views on the etiology of these forms of radiation cataract (Lipman *et al.*, 1988). The most extensively studied among the cataractogenic agents are x ray and UV radiation. Photooxidative changes, especially from UV radiation, that occur during aging and in cataract formation have attracted considerable attention in recent years. A prevailing view today is that the generation of singlet oxygen and/or free radicals, either directly or indirectly, by UV radiation is a major factor in cataractogenesis (for reviews see Wiegand *et al.*, 1984; Harding and Crabbe, 1984; Zigman, 1985; Lipman *et al.*, 1988).

Three potential targets of radiation-induced free radical and oxidative damage in the lens are proteins, lipids, and DNA. Lens fiber cell plasma membranes and epithelial cell membranes, which comprise about 3–5% of the wet weight of the lens, contain only about 2% polyunsaturated fatty acids (Wiegand *et al.*, 1984). Malondialdehyde (MDA), a thiobarbituric acid (TBA)-reactive breakdown product of peroxidized lipids, is produced after illumination of lenses by fluorescent light and also appears to be elevated in cataractous lenses. However, MDA production is only an indirect measure of lipid peroxidation, and its presence in human cataract may be unrelated to sunlight (see Harding and Crabbe, 1984). Ultraviolet radiation in combination with photosensitizers induces DNA repair synthesis in lens epithelium, but a clear relationship to cataract formation remains to be established (Wiegand *et al.*, 1984).

The lens crystallins are undoubtedly the major targets of UV radiation, and photooxidation of lens proteins by direct action on their tryptophan residues causes yellowing, increased fluorescence, and cross-linking. Singlet oxygen, but not H_2O_2, is generated after exposure of α-crystallin to UV radiation in the presence of a photosensitizer and oxygen (Goosey *et al.*, 1980); moreover, other oxidation products of tryptophan, such as N-formylkynurenine and its hydroxylated glucoside, can act as photosensitizers in the production of singlet oxygen (Zigler and Goosey, 1981). It has often been proposed that sunlight is a major causative factor in brown nuclear cataract, and many arguments have been given both supporting and refuting this hypothesis. Although model systems *in vitro* do not necessarily mimic the *in vivo* situation, biochemical analyses of cataractous lenses and epidemiologic studies nevertheless suggest a possible link between exposure to UV radiation and development of cortical opacities (Taylor *et al.*, 1988).

Apart from the UV-induced production of singlet oxygen, other reactive oxygen species such as superoxide anion (O_2^-), hydroxyl radical (OH·), and H_2O_2 also appear to be produced metabolically in the lens. The mechanism of their generation in the lens is under active investigation and has not yet been completely clarified. In the normal lens, the levels of these oxidants are probably very low owing to the presence of well-known protective enzymes such as glutathione peroxidase, superoxide dismutase, and catalase (Bhuyan and Bhuyan, 1977, 1978). Ascorbic acid, found at high concentrations in the lens, plays a pivotal role since it can act as both an antioxidant and a prooxidant. In model systems, ascorbate is able to scavenge various reactive oxygen species, but on the other hand, it can also reduce transition metals such as copper and iron, which may generate free radicals through Haber–Weiss or Fenton-type reactions (see Fig. 1.9). Incubation of rat lenses with a xanthine–xanthine oxidase free radical-generating system causes a marked decrease in the ability of the lens to accumulate $^{86}Rb^+$ or α-aminoisobutyric acid against a concentration gradient (Varma *et al.*, 1986). Inclusion of ascorbate in the incubation medium provides significant protection against the toxic effects of this free radical-generating system. However, other *in vitro* studies on human lens γ-crystallin (Russell *et al.*, 1987b) and on cultured bovine epithelial cells (Wolff *et al.*, 1987) support a prooxidative role for ascorbic acid. The autooxidation of ascorbate, for example, in the presence of GSH would produce H_2O_2, whose potentially toxic effects are discussed below. In addition, ascorbate may play a role in cataractogenesis through the formation of adducts with lens crystallins (Bensch *et al.*, 1985; Ortwerth *et al.*, 1988). The nonenzymatic browning and aggregation of lens crystallins is thought to occur through a Maillard-type reaction and may be an important feature of the aging lens. There is considerable speculation that it is also involved in the nuclear browning found in human cataracts.

Oxidative stress is now considered one of the most important components of senile cataract formation (Spector, 1984a,b, 1985; Chylack, 1984). It is generally thought that lens opacification may be the last step in a complex process initiated by oxidative modifications to lens proteins through either photooxidation, UV radiation, or endogenous univalent reduction of oxygen (Varma *et al*, 1984). The oxygen requirements of the lens are very small, since most of the ATP is produced through glycolysis. Low levels of reactive species of oxygen are probably present in the lens, and there is convincing evidence that free radicals (O_2^- and OH·) as well as H_2O_2 play a role in cataractogenesis. The levels of aqueous humor H_2O_2 are reported to be somewhat elevated in certain patients undergoing cataract surgery (Spector and Garner, 1981); they may also be elevated in some forms of experimentally induced cataract (Bhuyan and Bhuyan, 1984). Exposure of rat lenses to high (1 mM) concentrations of H_2O_2 causes an inhibition of $^{86}RB^+$ influx and also leads to disturbances in the active and passive fluxes of Na^+ and K^+. One of the major effects of H_2O_2 is the direct modification of Na^+,K^+-ATPase (Garner *et al.*, 1983), which alters the Na^+ stimulation of ATP hydrolysis and the Na^+ inhibition of *p* -nitrophenylphosphate hydrolysis but has little effect on the K^+ control of hydrolysis of these substrates (Garner *et al.*, 1984). The overall result is an inhibition of pump-dependent K^+ influx (Garner *et al.*, 1986).

The major mechanism for detoxification of H_2O_2 in the lens is through stimulation of the pentose phosphate pathway, which appears to provide extra reducing potential in situations where the levels of H_2O_2 are slightly higher than the physiological limits of approximately 40–50 μM (see Section 5.3.2.2). However, exposing rabbit lens epithelial cells to high (1 mM) concentrations of H_2O_2 causes loss of epithelial membrane sulhydryl groups in addition to altering ion homeostasis (Hightower *et al.*, 1989). Long-term incubation of human lens proteins with relatively low levels of H_2O_2 (<1 mM) results in a rapid oxidation of cysteine and methionine residues but does not affect any other amino acids (McNamara and Augusteyn, 1984). These changes are accompanied by a threefold increase in nontryptophan (NT) fluorescence, the latter reflecting conformational changes in the lens proteins. Up to 75% of the proteins become insolubilized and consist of covalently cross-linked polymers that appear to be similar to those found in cataractous lenses.

Incubation of calf lenses with somewhat higher concentrations of H_2O_2 for short periods of time suggests that one of earliest changes is a depletion of reduced glutathione (GSH) with resultant formation of oxidized glutathione (GSSG), the latter reacting with protein sulfhydryl to form mixed protein–glutathione disulfide (Siezen *et al.*, 1989). Further depletion of GSH results in the formation of inter- and intramolecular protein–protein disulfides. The βH-crystallins, with their high sulfhydryl content, are the most susceptible to oxidative modifications; moreover, both β- and γ-crystallins become progressively more acidic with increasing concentrations of H_2O_2. These changes have been observed in aging human lenses (Thomson and Augusteyn, 1985; Zigler *et al.*, 1985b) and are also known to occur in cataractous lenses. The γ-crystallins undergo reversible and irreversible charge modifications in the presence of H_2O_2. These studies show clearly that protein oxidation induced by H_2O_2 precedes aggregation.

5.6.4. Selenite Cataract

Selenium (Se) is an integral component of glutathione peroxidase type I, an enzyme that reduces both H_2O_2 and lipid hydroperoxides in the presence of glutathione (GSH). It is an essential trace mineral; tissue levels drop precipitously in animals maintained on Se-poor diets, and cataracts may develop after long-term deficiencies. An excess of Se following subcutaneous injection into young rats also causes cataracts (see review by Shearer *et al.*, 1987). Studies with ^{75}Se show that it is rapidly taken up in lenses of young (14-day-old) rats, whereas adult animals accumulate relatively little. Administration of selenium results in a rapidly developing nuclear cataract and a more slowly forming cortical opacification; the former is permanent, but the latter eventually clears (Anderson *et al.*, 1986). The nuclear cataracts, characterized by deposits of opaque particles, develop about 4 days after the injection of Se (Shearer *et al.*, 1987).

Early histological changes in the epithelium include suppression of mitosis and nuclear fragmentation (Anderson *et al.*, 1986; Shearer *et al.*, 1987). The percentage of germinative epithelium in prophase decreases, and there is a 60–70% reduction in DNA synthesis a few hours after Se injection. These observations suggest that one of the primary effects of Se is to block the cell cycle in the germinative zone of lens epithelium in S and/or G_2 phases and to interfere with normal fiber cell formation. A more precise understanding of selenium damage has been obtained using a newly developed direct chemical method for estimating DNA synthesis and net rates of lens epithelial cell differentiation *in vivo* (Cenedella, 1987). The method utilizes incorporation of systemically injected [^{3}H]thymidine into lens DNA to measure the time course of DNA synthesis in the epithelium and the net movement of epithelial cells via differentiation to fiber cells. If Se is present during DNA synthesis, there is a marked decrease in DNA labeling, and the net migration time is considerably prolonged (Cenedella, 1989). Selenium has little effect on these parameters if administered after the S phase. Therefore, Se damage appears to be confined to germinative epithelial cells in S or pre-S phases of the cell cycle.

Even though glutathione and NADPH levels are depressed in selenite cataract, there appears to be a reserve of pentose phosphate pathway activity (Shearer *et al.*, 1987). There is no increase in either oxidized glutathione (GSSG) or in protein-bound glutathione; hence, selenite cataract is not caused by disulfide-linked aggregates (David and Shearer, 1984a). Another possible effect of selenite is the inhibition of Na^+,K^+-ATPase, which may be responsible for some of the early changes in selenite-induced cataract (Bergad and Rathbun, 1986). This could contribute indirectly to increased calcium uptake, as discussed below.

The major physiological change in selenite cataract appears to be related to calcium homeostasis. The calcium concentration in whole lens increases about fivefold after selenite injection (Shearer and David, 1982/1983), an increase that is especially marked in the nucleus and appears before the development of cataracts (David and Shearer, 1984b). Two mechanisms for the increase in lens calcium have been proposed: (1) oxidation of functional sulfhydryl groups on Ca^{2+}-ATPase, which could impair Ca^{2+} export, and (2) oxidation of membrane sulfhydryl groups, which could enhance the influx of Ca^{2+} through the ion channels (Shearer *et al.*, 1987). Recently, an *in vitro* model for selenite cataract has been developed using cultured rabbit lenses; exposure to 0.1 mM Se causes a threefold increase in lenticular Na^+ levels 48 h after removal of Se from the medium (Hightower and McCready, 1989). The levels of Ca^{2+} remain normal initially and only begin to rise after the lenses are swollen and opaque. These findings suggest that the initial cytotoxic effect of Se is inhibition of the Na^+,K^+-ATPase pump with ensuing Na^+ accumulation. Selenium does not appear to inhibit the Ca^{2+} pump, since this cation accumulates at a later stage, possibly because of the osmotic stress created by inhibition of the Na^+ pump and concomitant hydration of the lens.

Regardless of the mechanism for Ca^{2+} accumulation, a major consequence of the raised levels is a nearly fivefold increase in urea-soluble (US) proteins and a striking accumulation of proteolyzed crystallin polypeptides and proteolyzed membrane proteins before the appearance of opacities (David and Shearer, 1984b; David *et al.*, 1987, 1988). This proteolysis probably resulted from activation of calpain II, a Ca^{2+}-dependent neutral protease (see Section 5.3.3.2). Calpain II is activated by calcium levels >50 μM, and the activity can be demonstrated in the nucleus of rats before the onset of selenium cataract (Hightower *et al.*, 1987). An unusually large fraction (7%) of the total calcium is present in the nucleus in the free, physiologically active form by the fourth day after Se injection. At later stages, calapin II activity decreases, possibly as a result of its autolysis at high concentrations of calcium. The urea-soluble protein that accumulates in selenite cataract consists primarily of insolubilized β- and γ-crystallins, some of which are produced by limited proteolysis of 26.5-kDa βL-crystallin. Proteolyzed β-crystallin and intact γ-crystallins are thought to associate noncovalently, resulting in the formation of light-scattering aggregates. Calpain II may also be involved in the cleavage of putative gap junction protein MP26, which undergoes limited proteolysis at the C terminus to 24- and 22-kDa fragments in selenite cataract (David *et al.*, 1988).

5.6.5. Cyanate and Carbamylation

The importance of conformational changes in lens crystallins in the etiology of cataracts has been stressed in several reviews (Harding and Dilley, 1976; Harding, 1981; Harding and Crabbe, 1984). These changes could interfere with the short-range spatial order of the lens proteins essential for lens transparency. A wide variety of substances are now known to be capable of initiating conformational changes, and among them the chemical modification of crystallins by cyanate is a particularly useful model.

Cyanate, which is in equilibrium with urea in the blood and in other body fluids including aqueous humor, has been shown to induce conformational changes in bovine α-crystallin *in vitro* (Beswick and Harding, 1984). The major reaction is the carbamylation of ε-amino groups, particularly lysine; the blocking of positive charges on the protein surface was thought to destabilize the protein. This alters the secondary and tertiary structure of the protein and promotes intermolecular disulfide bonding. Rat lenses incubated with cyanate develop opacities that, in the early stages, may be caused by phase separation resulting from altered interactions between modified proteins and the surrounding water as well as with unmodified neighboring proteins (Crompton *et al.*, 1985). This was considered as part of a general phenomenon in which any agent that causes a decrease in positive surface charges or an increase in negative charges (e.g., as in deamidation or reaction with GSSG) induces conformational changes leading to cataract.

Preincubation of lenses with aspirin prevents the cyanate-induced increase in the phase separation temperature (Crompton *et al.*, 1985). Aspirin, which appears to be acting by transfer of acetyl groups to protein amino groups (see also Rao *et al.*, 1985), prevents the carbamylation and the development of opacities in this model system. In addition to aspirin, bendazac, a putative anticataract agent, also prevents the cyanate-induced increase in phase separation temperature in incubated rat lenses (Lewis *et al.*, 1986). Both bendazac and its metabolite, 5-hydroxybendazac, effectively inhibit the carbamylation of lens proteins by cyanate, and the mechanism of their preventive action appears to be similar to that of aspirin.

Nonenzymatic carbamylation of lens proteins may be a risk factor in human cataract formation in cases of raised blood levels of urea, as in severe diarrhea or in renal failure (Beswick and Harding, 1987a). Carbamylation may also occur as part of the normal aging process and could be a factor in cataractogenesis in otherwise healthy individuals (Harding and Crabbe, 1984; Harding, 1985). This view is supported by studies using model systems *in vitro*. For example, incubation of bovine lens crystallins with cyanate results in the formation of substantial amounts of disulfide-bonded aggregates. They are similar to the HMW3 type of aggregate found in aging and/or cataractous lenses described by Harding and Crabbe (1984). Carbamylated βL-crystallin itself does not aggregate, but when it is added to mixed lens proteins, it induces aggregation even though the other crystallins present had not been chemically modified. These findings suggest that carbamylated βL-crystallin may act as a nucleus for aggregation; its modification induces an unfolding and possible exposure of sulfhydryl groups that could become oxidized, thus leading to the formation of disulfide-bonded aggregates.

5.6.6. U18666A Cataract

Treatment of newborn rats with U18666A, a potent inhibitor of cholesterol biosynthesis, causes marked alterations in sterol/phospholipid ratios in the lens and development of irreversible nuclear cataracts in the majority of treated animals (Cenedella, 1983). Lenses of treated rats synthesize sterol at approximately one-half the rate of controls, and desmosterol accounts for 50–75% of total sterol. The major loss of sterol occurs in the lens cortex; synthesis in the nuclear region is only marginally affected. Further studies showed that the drug does not cause any substantial changes in either the metabolism or composition of lens phospholipids (Cenedella, 1985). The decrease in sterol content of the lens cortex of U18666A-treated rats affects mainly the gap junctional membranes (Fleschner and Cenedella, 1988). This results in a significant 50% decrease in the sterol/phospholipid ratio in these specific membranes but only marginal decreases in membranes isolated from other regions of the lens.

An increase in the limited proteolysis of the putative gap junction protein MP26 has also been observed in cortical and nuclear fibers of U18666A-treated rats (Alcala *et al.*, 1985). Not only is MP26 largely replaced by MP23-24, but ultrastructural studies show a complete disappearance of identifiable intercellular junctions in the opaque fiber cells (Kuszak *et al.*, 1988). Steady-state fluorescence anisotropy measurements of parinaric acid probes indicate an alteration in lipid–protein interactions in cortical membranes of cataractous lenses from U18666A-treated rats; this results in more highly ordered membranes than those in control or in treated but clear lenses (Rintoul *et al.*, 1987).

A working hypothesis based on experimental findings reported to date suggests that the inhibition of cholesterol synthesis in the plasma membranes of the lens cortex leads to disruption of the gap junctional complexes and loss of permeability control and normal ionic homeostasis in U18666A-treated rats. Increased proteolytic cleavage of plasma membrane protein and ultimate death of nuclear fiber cells are thought to be the consequence of metabolic and structural insults that eventually lead to cataract formation.

5.6.7. Animal Models of Hereditary Cataract

Several animal models with congenital cataract are now recognized, and three in particular have been extensively investigated. Each displays unique biochemical and ultrastructural changes, in some cases resembling those found in human cataract.

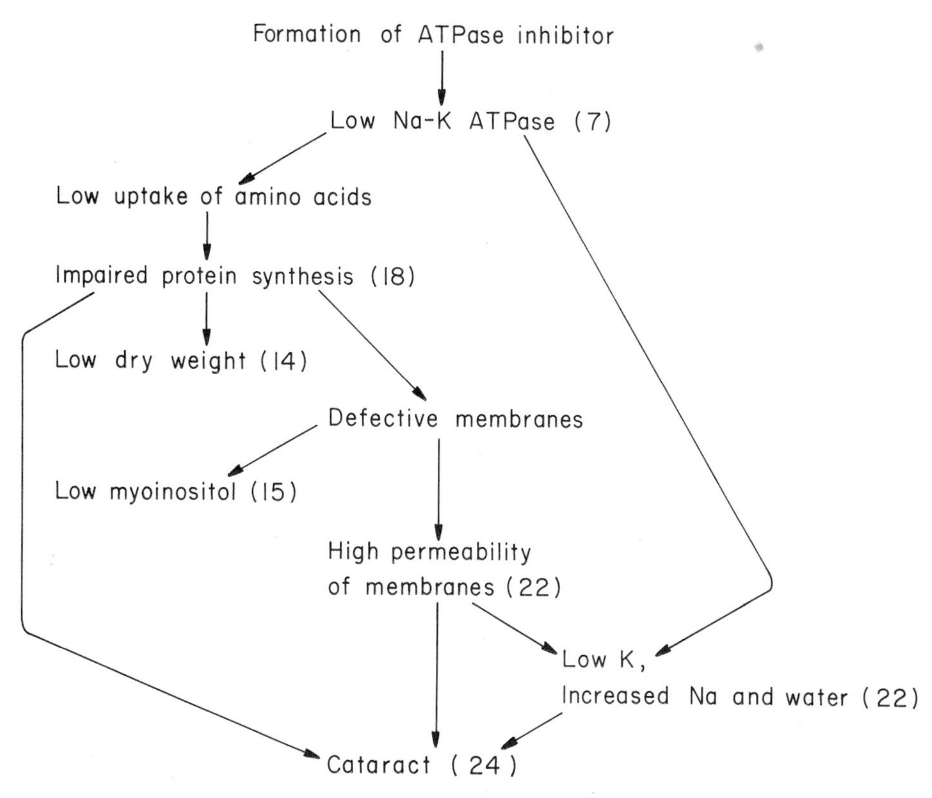

Figure 5.10. Tentative scheme of major early changes in the Nakano cataract. The primary defect is assumed to be the formation of a Na^+,K^+-ATPase inhibitor. The numbers in parentheses are the days after birth when the events occur. (From Harding and Crabbe, 1984.)

5.6.7.1. Nakano Mouse Cataract

Nuclear opacities appear in the Nakano mouse lens about 24 days after birth, but significant biochemical changes are detectable by the seventh postnatal day. One of the earliest is a decrease in Na^+,K^+-ATPase activity, which, by the 13th postnatal day, is 50% lower in the Nakano mouse lens than in age-matched controls. The decline in Na^+,K^+-ATPase activity is followed by increased permeability to Na^+ and water and decreased influx of K^+ and/or $^{86}Rb^+$. The inactivation of this specific enzyme was thought to caused by an endogenous low-molecular-weight heat-stable Na^+,K^+-ATPase inhibitor that has been detected in cultured Nakano epithelial cells (Russell *et al.*, 1977) as well as in the native lens (Fukui *et al.*, 1978). The osmotic changes probably limit the uptake of amino acids, which could later lead to impaired protein synthesis. A speculative flow chart showing some of the early changes in the Nakano mouse cataract is given in Fig. 5.10.

There are diminished amounts of γ-crystallin in soluble HMW aggregates isolated from homogenates of Nakano mouse lens (Russell *et al.*, 1979), possibly resulting from conversion of this specific crystallin to water-insoluble protein (Roy *et al.*, 1982). Membrane-associated disulfide-linked cytosolic polypeptides are present at higher levels in the Nakano cataractous lens than in normal controls; moreover, these fractions appear to be enriched in γ-crystallin. Many of the changes in protein composition and/or structure in the final stages of the Nakano cataract bear a close resemblance to those found in human cataract, even though the initiating events may differ.

Incubation of cultured Nakano mouse lens epithelial cells with orthovanadate stimulates the incorporation of [^{3}H]thymidine into DNA in a dose-dependent manner (Jones and Reid, 1984). Insulin acts synergistically with vanadate in stimulating DNA synthesis but not in increasing cell number beyond that resulting from insulin alone. Although vanadate is a known inhibitor of Na^+,K^+-ATPase, the concentration required to induce 50% inhibition of the enzyme is two orders of magnitude greater than that needed to stimulate thymidine uptake. Hence, inhibition of Na^+,K^+-ATPase does not appear to be the primary mechanism of vanadate-stimulated DNA synthesis in Nakano lens epithelial cells. Further studies show a strong similarity in the vanadate concentrations and the time course required for stimulation of both DNA synthesis and protein tyrosine phosphorylation in cultured Nakano lens epithelial cells (Gentleman *et al.*, 1987). However, the effect is not through direct activation of protein tyrosine kinase(s); rather, vanadate inhibits a membrane-associated phosphotyrosine-specific phosphatase activity. The β-subunit of insulin receptor is a tyrosine protein kinase, and vanadate inhibition of phosphatase activity could lead to persistent phosphorylation and activation of this receptor. Thus, phosphotyrosine proteins may play a role in the regulation of DNA synthesis and of cell division in Nakano mouse lens epithelial cells; however, a relationship to cataract formation has yet to be established.

5.6.7.2. Philly Mouse Cataract

In this animal model of congenital cataract derived from Swiss–Webster mice, the lens epithelial cells fail to differentiate into fiber cells, and normal lens function appears to be lost as early as the seventh postnatal day (Uga *et al.*, 1980). Osmotic changes measurable from about the 20th postnatal day precede the formation of visible cataracts, which occurs about 5–6 weeks postnatally (Kador *et al.*, 1980). Defects in lens permeability possibly related to abnormalities in membrane glycoprotein biosynthesis (Garadi *et al.*, 1983) or to increased turnover of membrane phospholipids (Andrews *et al.*, 1984) have been observed. Crystallin metabolism is disturbed (Piatigorsky *et al.*, 1980), and more specifically, there is a selective absence of 27-kDa β-crystallin in Philly mouse cataract lenses (Zigler *et al.*, 1981). This results in major changes in the physicochemical state of the lens cytoplasm, abnormalities in protein–protein interactions, and dramatically different behavior of the phase separation temperature, T_c, compared to normal controls and to Philly mouse hybrids (Clark and Carper, 1987). Marked differences are apparent just 1 day after birth, when the lenses are still perfectly clear, and measurements of T_c during early postnatal development can be used to predict the development of opacities in mature lenses. The changes in T_c precede biochemical abnormalities in crystallin synthesis, suggesting that factors affecting the T_c of the cytoplasm may influence gene expression.

A major advance in understanding the etiology of the Philly mouse cataract came with the observation that, whereas the mRNA for 27-kDa β-crystallin is detectable in the normal mouse lens 5–10 days postnatally, no functional 27-kDa β-crystallin mRNA could be detected in homozygous Philly mouse lenses (Carper *et al.*, 1982). Use of a monoclonal antibody to 27-kDa polypeptide has shown that in normal mouse lens, this crystallin is localized to elongating and differentiating fiber cells of the equatorial region by about the second postnatal day, and it is a major polypeptide component in 16-day-old lenses (Carper *et al.*, 1986). In contrast, 27-kDa polypeptide is completely absent from Philly mouse lens proteins; the monoclonal antibody did not recognize any antigen, suggesting that either it is not present or the protein has been altered. This finding is of interest in view of a possible relationship between 27-kDa crystallin and the major β-crystallin of mammalian lens, βB2 (βB_P), a polypeptide that is synthesized in the cortical layers of the lens and in aging undergoes several posttranslational modifications (see Section 5.5.1). Moreover, βB2-crystallin displays unusual heat stability properties (Mostafapour and Schwartz, 1981/1982; McFall-Ngai *et al.*, 1986), and recent investigations show that the Philly mouse lens lacks a heat-stable protein (Nakamura *et al.*, 1988). Although an mRNA coding for βB2-crystallin is present in the Philly mouse lens, normal βB2 cannot be detected. Instead, it is replaced by a protein that is somewhat smaller in size and more acidic than normal βB2. The N-terminal amino acid residues of the two proteins are similar, but the Philly mouse lens protein lacks part of the C-terminal half of normal βB2. Thus, synthesis of an altered βB2-crystallin appears to be the principal defect in this hereditary cataract.

5.6.7.3. Emory Mouse Cataract

Emory mice develop a late-onset cataract 6–8 months postnatally; hence, this is an attractive model for studying human senile cataract (Kuck *et al.*, 1981/1982). Moreover, certain morphological and histological changes resemble those found in human senile cataract. At the ultrastructural level, an extensive formation of ridges and other altered structures on superficial cortical plasma membranes have been described (Lo and Kuck, 1987).

The biochemical basis of cataract formation in this model has not been clarified. Lipid peroxidation, estimated by malondialdehyde production, coupled with decreased activities of defensive enzymes may be among the factors leading to altered membrane function in the Emory mouse lens (Bhuyan *et al.*, 1982/1983). Certain proteases display increased activity (Swanson *et al.*, 1985), and it has been suggested that enhanced conversion of MP26 to MP22 could account for the appearance of ridges and the development of other morphological changes in this animal model (Lo and Kuck, 1987). Other studies utilizing antisera prepared against synthetic peptides corresponding to various regions of MP26 demonstrate covalent changes in the MP26 molecule of the Emory mouse lens (Takemoto *et al.*, 1988). Of the antisera tested, it was found that, similar to human cataract, the anti-$MP26_{229-237}$ serum binds significantly more to MP26 of Emory mouse cataracts than it does to MP26 of age-matched control mice. This increase in binding during the opacification may reflect a covalent change within the 229–237 sequence; alternatively, it may reflect a covalent change elsewhere in the MP26 molecule, resulting in increased availability of the epitope present in the 229–237 sequence.

Enhanced conversion of soluble to insoluble protein and some changes in total disulfide have been reported in late-stage opaque Emory mouse lenses compared to normal age-matched controls (Kuck and Kuck, 1983). However, laser Raman spectroscopy failed to reveal any differences in either sulfhydryl levels or disulfide bond formation in the early (grade 1) cataracts of Emory mouse lenses and control mice (DeNagel *et al.*, 1988).

5.6.8. Human Cataract

Major advances have been made, especially in recent years, in our understanding of normal noncataractous human and animal lens biochemistry. An important component of lens biochemistry involves metabolic, physical, and structural changes that occur during aging, many of which are intimately associated with cataract formation; often there is no clear distinction between the two processes. The question of why some aging lenses develop cataracts and others do not still remains unanswered. Using other approaches, considerable insight has been gained on mechanisms of opacification through studies on experimentally induced animal cataracts and in genetic animal models. Many of the findings reported may be relevant to human cataract, although the relationships are not always clear.

The study of biochemical changes associated with human senile cataract is exceedingly complex owing to many factors: (1) normal changes that occur with aging are superimposed on cataractous changes, and events that lead to the formation of opacities cannot always be distinguished from the aging process; (2) lenses obtained from cataract surgery are usually not completely opaque; if the whole lens is analyzed, changes in the chemistry of small opaque regions would be masked by the essentially normal chemistry of the major part of the lens; (3) a cataract is the end stage of a series of metabolic, physicochemical, and structural changes, and analyses of cataractous lenses do not necessarily reveal the initiating factor(s); and (4) until rather recently, cataracts were not classified either by location of the opacities or by the extent of yellow pigmentation, so biochemical correlates were (and still are) sometimes difficult to interpret.

In addition to the foregoing problems, the increased use of extracapsular surgery means that fewer lenses suitable for biochemical or biophysical analyses are available. For example, whereas a statistically significant decrease in nonfreezable water in opaque regions versus clear areas can be demonstrated in cataractous lenses removed by intracapsular surgery, these differences are not significant in extracapsular specimens (Bettelheim *et al.*, 1986). Many current studies utilize eye bank lenses, and other alternatives to fresh intracapsularly extracted lenses are being developed. Improved methods are now available for culturing human lens epithelium, and noninvasive techniques such as NMR and fluorescence spectroscopy provide valuable information on metabolic and structural changes in the intact lens (Garner, 1984; Cheng and Chylack, 1985; Farnsworth and Schleich, 1985; Greiner *et al.*, 1985).

A unified theory that can account for the heterogeneous changes leading to the most common type of cataract, human senile cataract, remains to be established. However, one of the major initiating events is thought to be oxidative damage, and one such damaging agent could be H_2O_2 (Spector 1984a,b, 1985). Whether it is produced metabolically within the lens or enters from the aqueous humor, if H_2O_2 is present at a high enough concentration it is able to uncouple membrane-associated Na^+,K^+-ATPase and inhibit pump-dependent K^+ influx (Garner *et al.*, 1983, 1984, 1986). Another major source of oxidative damage is the UV-induced generation of singlet oxygen mediated by photosensitizers such as tryptophan or its derivatives (Goosey *et al.*, 1980; Zigler and Goosey, 1981). Excited-state and stable free radicals of photooxidized tryptophan form active species that can bind and alter structural lens proteins (see Section 5.4). Other initiating factors include glucose, cyanate, and steroids, which may be acting through chemical modifications of lens crystallins (Harding and Crabbe, 1984). Once modified, e.g., unfolded, these proteins could be more susceptible to oxidative changes.

Whatever the initiating event(s), the sequelae (but not necessarily the sequence) of secondary changes leading to cataract formation are numerous and include decrease in water-soluble crystallins, protein unfolding, increase in water-insoluble protein, oxidation of cysteine and methionine residues, formation of high-molecular-weight (HMW) disulfide-linked aggregates, changes in membrane permeability leading to osmotic imbalance, loss of glutathione (GSH), phase separation of proteins in the cytoplasm, and others (see reviews by Chylack, 1984; Harding and Crabbe, 1984; Spector, 1984a,b, 1985). Several provocative schemes depicting cataract formation have been published in these reviews.

A major change that emerges from all studies is the formation of S-S linkages, but whether this is a causative or stabilizing factor in aging or in nuclear cataracts is still a debatable question (Barron *et al.*, 1988a). Another major change in nuclear cataracts known for more than two decades (Pirie, 1968) is the appearance of fluorescent yellow and brown colorations and the accumulation of water-insoluble and urea-insoluble protein. Yet in spite of the great diversity of metabolic and structural changes found in human nuclear cataract, glucose metabolism does not seem to be impaired (Wolfe and Chylack, 1986). Cataractous lenses obtained by intracapsular extraction and classified according to CCRG guidlines have normal glycolytic function, and only in some cases do they produce small amounts of sorbitol when incubated in media containing as much as 16.5 mM glucose.

Many lines of investigation on the etiology of human senile cataract are being pursued, but space limitations permit the selection of only three general topics currently under investigation.

5.6.8.1. Proteins: Conformational Changes, Oxidation, and Aggregation

Conformational changes that make protein thiols more susceptible to oxidation and disulfide bond formation—and lead to the formation of water-insoluble HMW aggregates in human nuclear cataracts—have been recognized for many years (Harding, 1972, 1973; Spector and Roy, 1978). These changes have been discussed in detail in several reviews (Spector, 1984a,b; Harding and Crabbe, 1984). They are thought to alter the surface charge distributions of the lens crystallins and may be caused by deamidation, oxidation of cysteine and methionine, glycosylation, carbamylation, addition of glutathione, racemization of aspartyl residues, degradation of proteins, and sunlight. A schematic diagram showing how such conformational changes in lens proteins could lead to the formation of HMW3 and HMW4 aggregates found in human nuclear cataracts is shown in Fig. 5.11.

Oxidation of both methionine and cysteine in soluble as well as membranous fractions is found in cataractous lenses (Garner and Spector, 1980). More than 50% of the methionine and 75–100% of the cysteine are in the oxidized state, and extensive formation of disulfide bonds occurs, particularly in the membrane fractions. Further studies have been reported on the state of sulfhydryl groups in the crystallins of cortical and nuclear regions of cataractous lenses classified according to nuclear color (Hum and Augusteyn, 1987a). The most prominent changes occurring with increasing development of color are (1) progressive decreases in sulfhydryl content of the crystallins in the nucleus

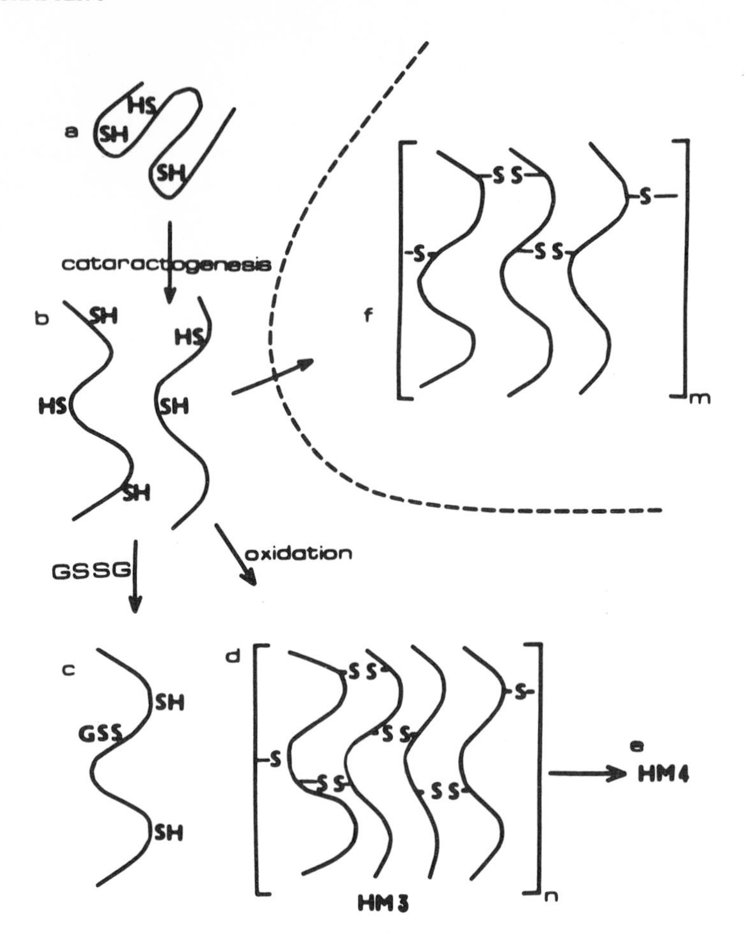

Figure 5.11. Schematic diagram showing changes in lens proteins during the development of human senile cataract: (a) native protein; (b) unfolded protein of nuclear cataract; (c) formation of mixed disulfide; (d) HMW3; and (f) disulfide-linked products formed during aerobic extraction of lens proteins. (From Harding and Crabbe, 1984.)

and (2) increases in both disulfide bond formation and in urea-insoluble proteins, the latter accounting for as much as 70% of the total cysteine pool in the nuclear region. Although there was no evidence that a specific crystallin was responsible for the changes, other studies described below suggest that the γ-crystallins (and, to a lesser extent, β-crystallins) play important roles in human cataractogenesis.

Analyses of crystallins from cortical and nuclear regions of human cataractous lenses fractionated by high-pressure gel-permeation chromatography (HPGPC) reveal a sharp decrease in soluble γ-crystallin in the nuclear region of advanced nuclear cataracts (Bessems *et al.*, 1983). In addition, a 355/420-nm fluorophore has been detected that appears to be loosely bound to γ1-crystallin in nuclear cataracts (Bessems *et al.*, 1987a). The association of this specific γ-crystallin species with a nontryptophan (NT) fluorophore was thought to alter its solubility properties and induce conformational changes leading to its increased susceptibility to aggregation and insolubilization. Microdissection of human cataractous lenses is now feasible (Horwitz *et al.*, 1981), and polyclonal antisera against human β-crystallin and against synthetic peptides corresponding to the N- and C-terminal sequences of bovine βB2-crystallin have been used to quantitate binding in clear and opaque areas of individual CCRG-classified lenses (Takemoto *et al.*, 1987d). Some but not all of the opacified regions showed changes in the β-crystallin class of polypeptides, mainly in sections taken from the anterior cortex.

The formation of HMW aggregates is part of the natural aging process in normal clear lenses; in the bovine lens these aggregates consist mainly of α-crystallin, but in human lenses many crystallins and other polypeptides appear to be involved in their formation (see reviews by Spector, 1984a,b, 1985). These aggregates have been classified into four types, two of which (HMW3 and HMW4) are found only in cataractous lenses (Harding and Crabbe, 1984). Their apparent M_r is 5×10^6 or greater; depending on the methods and conditions of isolation, HMW aggregates are found in both the water-soluble and water-insoluble fractions and accumulate mainly, though not entirely, in the nucleus of cataractous lenses. Detailed reviews of their composition, relationship and/or association with the fiber cell membranes, and possible origin have appeared (Spector, 1984a,b, 1985). These aggregates are stabilized by covalent disulfide bonds and contain both 10- and 43-kDa polypeptides (Spector *et al.*, 1979; Garner *et al.*, 1979) as well as 20-kDa components, the latter probably representing crystallin polypeptides. The 10-kDa polypeptides are heterogeneous in size and composition, and recent investigations using antisera raised against lens crystallins and against synthetic peptides corresponding to α- and βB2-crystallins suggest that the 10-kDa polypeptides arise from *in vivo* proteolysis of γ-crystallins; they do not appear to originate from either α- or β-crystallins (Takemoto *et al.*, 1989). The γ-crystallin polypeptide may be particularly susceptible to proteolytic cleavage, a possible point of attack being the peptide linkage between the two symmetrical domains. This would yield 10-kDa fragments. It is of interest that several active peptidases are present in the lens (see Section 5.3.3.2), and alterations in their activity and/or of endogenous protease inhibitors (Srivastava and Ortwerth, 1989) could give rise to 10-kDa polypeptides such as those found in cataractous lenses. The 43-kDa polypeptide is present in all disulfide-linked aggregates and appears to be an intrinsic membrane protein (Spector, 1984a,b, 1985).

Membrane–crystallin interactions occurring during cataractogenesis have been studied by a number of techniques. The use of two-dimensional diagonal electrophoresis shows that much of the known intermolecular disulfide bonding found in cataractous lenses is localized to the fiber cell membrane (Takemoto and Hansen, 1982). Moreover, the amounts of membrane-associated crystallins are greatly increased in fractions of high density isolated on discontinuous sucrose gradients (Takehana and Takemoto, 1987). Oxidation commences at the membrane and in nuclear cataracts appears to involve the γ-crystallins (Spector, 1985). This has been convincingly demonstrated using monospecific antisera raised against lens crystallins (Kodama and Takemoto, 1988). In 10 out of 13 cataractous lenses examined, γ-crystallin was shown to be in close association with highly purified plasma membranes, presumably linked through intermolecular disulfide bonds. Four of the lenses showed disulfide-bonded β-crystallins and none showed α-crystallin association. Significantly, no disulfide bonding was found in normal lenses. These findings raise the possibility of a specific role for γ-crystallins in cataractogenesis; although comprising only a small percentage of the total crystallins, they have a high thiol content, and some of the sulfhydryls are localized on the surface. Oxidation of these groups could cause destabilization and unfolding of the molecule; in the case of nuclear cataracts, this could lead to selective intermolecular disulfide interactions with the membranes. The mechanism of these putative changes is not entirely clear. In cortical cataracts, it is thought that the HMW disulfide-linked aggregates are "cytosolic" rather than membrane associated (Spector, 1985; Kodama and Takemoto, 1988).

The investigations cited above, as well as others, lend strong support to the view that oxidation of protein thiols precedes protein modifications such as accumulation of urea-insoluble material, polypeptide cross-linking, and formation of HMW aggregates. One possible oxidizing agent that could be involved in cataract formation is H_2O_2. The levels of H_2O_2 in the aqueous humor were found to be elevated in some patients undergoing cataract surgery (Spector and Garner, 1981), an observation that could not be confirmed in a later study on 31 cataract patients (Tomba *et al.*, 1984/1985). This important and controversial point requires further clarification. Whether the putative initial insult, e.g., H_2O_2, is derived from adjacent ocular tissues or whether it originates from within the lens itself has not yet been established. Whatever the case may be, if present at high enough concentration, H_2O_2 could exert toxic effects on essential membrane components such as Na^+,K^+-ATPase (Garner *et al.*, 1986; Spector 1984a,b, 1985). However, the lens ordinarily has efficient defense mechanisms against oxidative stress (see Section 5.3.2.2); in model systems there is an increase in pentose phosphate pathway activity after exposure of incubated lenses to H_2O_2, suggesting that the lens has a high capacity to generate reducing potential through this pathway in the form of NADPH. Moreover, key enzymes that provide protection from oxidative damage have been detected in the lens; these include superoxide dismutase, catalase, and glutathione peroxidase. In a study of 75 senile nuclear cataracts, no changes were found in catalase activity, but the activities of the other two enzymes were severely depressed, especially in the nuclear region (Fecondo and Augusteyn, 1983). The decreased activities of these enzymes do not appear to be age-related, since all of the groups examined fell into a mean age range of 67–80. The loss of superoxide dismutase activity may be explained in part by the finding that catalytically inactive but immunologically reactive superoxide dismutase molecules accumulate in many types of cataractous lenses (Scharf and Dovrat, 1986). The activity in mature dense cataracts is particularly low or even nondetectable, yet the inactive enzyme molecule can still be detected immunologically.

The possibility that other active oxygen species, in addition to H_2O_2, may initiate oxidative changes in lens proteins has long been suspected. Photooxidation of lens proteins has been demonstrated *in vitro* by a variety of techniques; the UV-induced conformational changes in lens crystallins in the presence of photosensitizers and O_2 is directly related to the production of singlet oxygen (Goosey *et al.*, 1980).

Endogenous photosensitizers in human lens include N-formylkynurenine (N-FKyn) and its derivatives (Pirie, 1971; van Heyningen, 1971). Near-UV photooxidation of tryptophan residues of lens crystallins is the most probable source of N-FKyn and related photosensitizers (Zigler and Goosey, 1981), and this may be one of the ways in which intracellular oxidants are generated in the lens (Andley and Clark, 1989). Although not directly demonstrated *in vivo* in human lens, the *in vitro* photolysis of human lens α-crystallin by 300-nm radiation is accompanied by the generation of superoxide anion (O_2^-) and H_2O_2. The UV radiation causes an alteration of protein tertiary structure, loss of tryptophan fluorescence, and increase in nontryptophan (NT) fluorescence, probably resulting from the formation of N-FKyn or its derivatives. It has been postulated that N-FKyn can react via its triplet state with O_2 to generate singlet oxygen, or with reducing substances to produce free radicals. The latter, in the presence of O_2, could generate both O_2^- and H_2O_2. Moreover, these substances could interact, giving rise to hydroxyl radical, the most reactive of all free radicals. Of interest was the observation that addition of superoxide dismutase to the incubation solution prior to photolysis causes a three- to fourfold increase in H_2O_2. These and other findings imply that O_2^- is one of the major active oxygen species generated by UV irradiation and that only a fraction of it is converted to H_2O_2 in the absence of superoxide dismutase. Although highly speculative, the reports of decreased activity of this enzyme in cataractous lenses (Fecondo and Augusteyn, 1983; Scharf and Dovrat, 1986) suggests a possible mechanism for UV-induced cataractogenesis.

Cross-linking and insolubilization of lens crystallins in aging and in cataractogenesis is thought to involve mainly the formation of disulfide linkages. In addition, small but significant amounts of covalent nondisulfide bonds are also present in nuclear cataracts (Harding and Crabbe, 1984; Spector, 1985). Dye-sensitized photooxidation of lens proteins mediated by singlet oxygen is associated with the production of nondisulfide cross-linked protein and the appearance of blue nontryptophan fluorescence (Goosey *et al.*, 1980). The chemical nature of nondisulfide linkages formed during photodynamic processes or in cataractogenesis has not been clarified, but oxidation products of cysteine (cystine and cysteic acid) and methionine (the sulfoxide and sulfone) are thought to be involved. Another oxidative degradation product of cystine found in nuclear cataracts is the symmetrical thioether lanthionine, formed by photooxidation of cystine possibly via an alanyl radical intermediate (Bessems *et al.*, 1987b). Interaction with the ϵ-amino group of lysine or the imidazol nitrogen of histidine would produce lysinoalanine and histidinoalanine, respectively. The latter has in fact been reported in cataractous lenses; it accumulates in increasing amounts with deepening nuclear color and may be related to hardening of the nucleus in deep brown, grade IV, cataracts (Kanayama *et al.*, 1987).

5.6.8.2. Glycation and Diabetes

Glycation is one of many posttranslational modifications of lens proteins that accompanies aging (see Section 5.5.4). This nonenzymatic process involves the condensation of a sugar molecule with a protein amino group, usually the ϵ-amino group of lysine, to form a covalent adduct. Amadori rearrangement to a ketoamine is the first step in nonenzymatic browning (the Maillard reaction). In later stages, the formation of pigmented condensation products is accompanied by cross-linking and insolubilization of the glycated proteins. Recently these advanced-stage glycosylation end products have been isolated by HPLC and measured fluorometrically, as described below (Oimomi *et al.*, 1988).

Glycation is increased in experimentally induced hyperglycemia in rats, and studies using model systems show that glycated crystallins undergo major conformational changes that may be related to lens opacification. The finding that nonenzymatic glycation of rat lens proteins by glucose 6-phosphate *in vitro* causes increased opalescence suggested that glycation could play a role in the etiology of human diabetic cataracts (Stevens *et al.*, 1978). To what extent this may be true is still controversial. Direct analyses of human lenses revealed only minimal amounts of glycated proteins, and no detectable differences could be found between cataractous, diabetic, or normal lenses (Pande *et al.*, 1979). Nevertheless, Maillard reaction products have been detected in pooled human cataractous lenses, particularly those that are heavily pigmented (Monnier and Cerami,

1983). Two other independent investigations suggest an approximately twofold increase in glycated proteins in human diabetic lenses compared to age-matched controls (Garlick *et al.*, 1984; Rao and Coltier, 1986). That glycation may play a role in the development of both diabetic and nondiabetic cataracts has come from analyses of early and advanced-stage products of the Maillard reaction (Oimomi *et al.*, 1988). The latter, measured as 370/440-nm fluorescence of fractions isolated by HPLC, are especially prominent in the nucleus and cortex of both diabetic and senile cataracts.

The role of glycation in diabetic cataract formation has also been investigated by fluorescence measurements (emission spectra, quantum yield, and polarization) of α-crystallins isolated from fetal, young, senile nondiabetic, and diabetic lenses (Liang, 1987). The results indicate that diabetic effects such as glycation cause greater fluorescence changes than those associated with aging alone. Tryptophan oxidation, increase in nontryptophan fluorescence, and mixed disulfide formation resulting from glycation cause major conformational changes that involve partial unfolding of the protein and increased susceptibility to aggregation. Further comparisons of glycated and nonglycated α-crystallins isolated from human diabetic cataractous lenses show that glycation causes significant changes in the tertiary structure of the protein (Liang and Chylack, 1987). A marked decrease in sulfhydryl groups with concomitant increase in mixed disulfides was found in glycated versus nonglycated α-crystallin, implying that glycation-induced conformational changes in α-crystallin lead to increased disulfide formation.

Apart from, or in addition to, adducts formed with glucose, nonenzymatic condensation of ascorbate with lens proteins and subsequent Maillard-type browning reactions may be involved in cataractogenesis. Ascorbic acid is present at high concentration in the lens, mainly in the reduced form (Fig. 5.7). It is, however, either the oxidized form (dehydroascorbic acid) or more probably the lactone (diketogulonic acid) that condenses with the lysine residues of lens proteins to form covalent adducts (B. Ortwerth, personal communication); their formation is significantly faster than that observed for sugars (Bensch *et al.*, 1985; Ortwerth *et al.*, 1988). The relative proportion of dehydroascorbic acid in the lens increases during the development of nuclear cataracts from about 11% of the total ascorbate in control lenses to 26% in hypermature brunescent cataracts (Lohmann *et al.*, 1986). Moreover, electron spin resonance (ESR) measurements show that this increase is associated with the production of ascorbyl free radical, thought to be the first oxidation product of ascorbic acid. Thus, increased availability of an oxidized or free radical form of ascorbic acid that can react with lens proteins and produce Maillard-type browning products could be responsible for some of the colorations in brunescent nuclear cataracts.

5.6.8.3. Membrane Changes and Ion Fluxes

Ionic imbalance is found in most types of cataracts, suggesting that impaired transport function and/or defects in membrane permeability are involved in lens opacification. Increased levels of Ca^{2+} have been reported in experimentally induced cataracts (see review by Hightower, 1985), and in these cases, the opacifications may have been caused by activation of Ca^{2+}-dependent calpain with subsequent proteolysis of lens proteins. The increased Na^+ and decreased K^+ found in some cataractous lenses does not appear to be caused by impaired cation pump function, since $^{86}Rb^+$ transport in these lenses is similar to that found in normal controls (Pasino and Maraini, 1982; Maraini and Pasino, 1983).

Instead, there is evidence from studies on $^{86}Rb^+$ efflux in the presence and absence of $BaCl_2$ suggesting that the ion imbalance in some cataractous lenses is the result of membrane damage (Gandolfi *et al.*, 1985). This may apply particularly to cataractous lenses with high Na^+ and Ca^{2+} contents, since they appear to have suffered general but graded increases in membrane permeability (Lucas *et al.*, 1986). The movement of K^+ (as measured by $^{86}Rb^+$ efflux) is not compromised in cataractous lenses with low Na^+ content. Other studies show that cataractous lenses with higher than normal levels of Na^+ and Ca^{2+} have a markedly reduced ability to accumulate tyrosine against a concentration gradient (Marcantonio and Duncan, 1987). These lenses incorporate a higher proportion of amino acid into HMW proteins than cataractous lenses with normal cation content. The foregoing studies, and others, leave the general impression that cortical cataracts may be caused by membrane damage leading to permeability changes and ionic imbalance, whereas senile nuclear cataracts are the result of conformational changes in lens proteins initiated by oxidative insult and resulting in unfolding, cross-linking, and insolubilization of the proteins.

Early investigations on changes in Na^+,K^+-ATPase activity in cataractous lenses were often inconclusive. However, recent studies on the kinetics of unmodified and H_2O_2-modified enzyme provide a basis for interpreting changes in Na^+,K^+-ATPase activity found in some cataracts. This enzyme transports three Na^+ outwardly and two K^+ inwardly for each mol of ATP hydrolyzed. It exists in two reversible conformational states: in one of them, E1, Na^+ is bound more efficiently, and the catalytic subunit has a high-affinity binding site for ATP; in the other, E2, K^+ is bound more efficiently, and the catalytic subunit has a lower affinity for ATP. The hydrolysis of $Mg^{2+}ATP$ by Na^+,K^+-ATPase cannot be described using the Henri/Michaelis–Menten relationship; instead, it is best described as substrate activation or negative kinetic cooperativity (Garner *et al.*, 1984). This is the kinetic behavior for the unmodified enzyme, whereas the H_2O_2-modified enzyme displays classical positive kinetic cooperativity. The Na^+ control of hydrolysis is altered after exposure to H_2O_2, but the K^+ control is unaffected. By using substrate-hydrolysis kinetics to measure Na^+,K^+-ATPase activity in 26 cataractous lenses classified according to CCRG guidelines, four distinct kinetic types have been identified (Garner and Spector, 1986). Negative kinetic cooperativity, Michaelis–Menten, and substrate-inhibition kinetics were found in both the cataractous population and among unclassified lenses; however, lenses showing positive kinetic cooperativity or no activity at all were found only in the cataractous lenses. It may not be a coincidence that the relative fraction of cataractous lenses showing positive kinetic cooperativity is similar to that of cataract patients with elevated levels of aqueous humor H_2O_2 (Spector and Garner, 1981). Moreover, substrate-inhibition kinetics is prevalent among diabetic individuals and was also observed in lenses incubated with high levels of glucose. These findings imply that either H_2O_2 or glycation (or both) may alter the steady-state kinetics of $Mg^{2+}ATP$ hydrolysis in cataractous and/or diabetic individuals by direct modification(s) of Na^+,K^+-ATPase. One of the consequences of enzyme modification in these cases is an impairment of K^+ influx.

The major polypeptide of lens fiber cell plasma membranes is MP26, an intrinsic channel-forming protein that plays a key role in regulating ion fluxes and intercellular communication in the lens. Whether it is the principal or the only gap junction protein of the lens is still controversial (see Section 5.2.2.1). Comparisons of MP26 from normal and cataractous lenses reconstructed into liposomes show significant differences in permeability as a function of pH (Gooden *et al.*, 1985). The channel-forming properties of MP26 from cataractous lenses appear to be altered, possibly because of covalent changes in the N-terminal region of MP26 in cataractous lenses (Takemoto *et al.*, 1986; Takemoto and Takehana, 1986b). These studies were carried out on membranes isolated from whole decapsulated cataractous lenses and demonstrated, using anti-229 probe, that covalent changes had occurred in the internal sequence of residues 229–237 of the MP26 molecule. Further studies using antisera to synthetic peptides of MP26 have been carried out on opaque and clear areas of microdissected CCRG-classified lenses obtained after intracapsular extraction (Takemoto *et al.*, 1987a). In this case, increased binding of anti-229 serum was found in opaque areas in two out of the four lenses examined, and the changes corresponded to anterior cortical opacifications. Changes in structure and/or sequences of MP26 were not detectable in opaque regions located in the equatorial or posterior regions of the same lens. These findings suggest that opacifications of the anterior cortex differ from those in other regions because of either biochemical changes in the lens epithelium or a particular susceptibility of the anterior lens to incoming light.

5.6.9. Classification, Epidemiology, Risk Factors, and Therapy

5.6.9.1. Classification

Considerable effort is being invested to develop reliable and objective methods to document the type and extent of cataract *in vivo* as well as *in vitro*. This is essential for many purposes: epidemiologic studies, assessment of risk factors, efficacy of therapy, and correlation with biochemical changes measured after cataract extraction. Among the newly developed techniques, Scheimpflug photography using the Topcon SL 45 slit-lamp camera has been extensively investigated (Hockwin *et al.*, 1983, 1984; Datiles *et al.*, 1987). The photographs are analyzed by quantitative linear microdensitometry and show high reproducibility. However, this method, although automated, objective, and precise, is applicable mainly to early or moderately advanced nuclear cataracts; it is not considered reliable for either anterior or posterior cataracts or for mixed cortical and nuclear cataracts (see review by Chylack and Cheng, 1985). Cortical cataracts are asymmetric and hence are difficult to assess except by retroillumination. Currently the most widely used method for precise documentation of cortical cataracts involves retroillumination and photography with the Neitz CTR Cataract Camera. In a detailed study of 11 selected volunteers ranging in age from 24 to 87 years, nuclear cataracts were assessed by Topcon SL 45 slit-lamp photographaphy while cortical cataracts were evaluated from Neitz CTR photographs (Chylack *et al.*, 1987). Although both are objective techniques, these studies showed how sources of variance affect the results. It is generally acknowledged that inter- and intraobserver evaluations can achieve a high rate of objectivity, although equivocal results are often obtained regardless of precautions taken. Recent studies have shown the feasibility of grading nuclear cataracts using two different systems: the regular Topcon SL 5D slit lamp and the Topcon SL 45 Scheimpflug camera (West *et al.*, 1988). Despite high interobserver reliability in grading the two sets of photographs, only fair association was found with evaluation by clinical examination. Of the two types of photographs, slit-lamp photographs with the regular Topcon SL 5D camera showed better association with clinical grading than Scheimpflug photographs, in part because of intereference from nuclear color in the latter technique. Slit-lamp rather than Scheimpflug photography thus appears to be the method of choice for epidemiologic surveys of nuclear cataracts.

Objective methods for assessing cataract along similar lines to those described above have been developed by the Oxford group in England (Brown *et al.*, 1987). The Oxford Clinical Cataract Classification and Grading System is based on three components: resolution target projection ophthalmoscopy (acuity scope), clinical definition of the cataract visible with the slit-lamp microscope, and two photographic methods (slit lamp and retroillumination). Another system is the Lens Opacities Case-Control Classification System (LOCS), developed jointly at several eye centers in the United States; LOCS is considered a simple system that includes both clinical grading with the slit lamp and evaluation of two types of standard photographs: black-and-white Neitz CTR retroilluminated photographs for grading of cortical and posterior subcapsular cataracts and a single color slit-lamp photograph for classifying nuclear color and opalescence (Leske *et al.*, 1988; Chylack *et al.*, 1988). The system is highly reproducible and easy to implement; moreover, there is good agreement between clinical and photograph-derived gradings, especially for nuclear cataracts. In an entirely different approach, a psychometric technique has been developed for identifying visual features of nuclear cataract using multidimensional scaling (Getty *et al.*, 1989). These studies revealed that there are many more systematic visual distinctions that can be made of nuclear cataracts than are now recognized by clinical examination.

The American Cooperative Cataract Research Group (CCRG) classification system is now used extensively for localization and semiquantitative evaluation of cataracts in extracted human lenses (Chylack *et al.*, 1983, 1985; Chylack and Tung, 1986). The statistical and anatomic validity of the system is supported by the high inter- and intraobserver reproducibility reported. Although obviously applicable only to lenses removed by intracapsular extraction, the classification system has the potential for *in vivo* modification.

An ultimate goal in cataract research is to correlate and localize the opacifications observed clinically or photographically (prior to extraction) with biochemical changes. To this end, the microdissection technique of Horwitz *et al.* (1981) has been successfully applied to examining a variety of biochemical changes in CCRG-classified lenses. Using a different technique, protein profiles have been examined in serial frozen sections from normal and cataractous lenses that had been classified by Scheimpflug photography before extraction (Hockwin *et al.*, 1986). The technique has the potential of detecting subtle regional differences in protein composition that could be correlated with photographs taken prior to extraction. Other studies have been reported on the use of quasielastic light scattering (QLS) spectroscopy to measure molecular changes associated with early human cataract development (Benedek *et al.*, 1987). The Brownian movement of lens proteins causes fluctuations in the intensity of light scattered from a small region illuminated by a weak incident laser beam. The fluctuations are produced by two molecular species: α-crystallins and large protein aggregates about 100 nm in size. The early development of cataract is perceived as a redistribution of protein between the two molecular species and is most accurately demonstrated in the nucleus.

5.6.9.2. Epidemiology: Risk Factors

Ultraviolet radiation has long been suspected as a risk factor in human cataract formation. A summary of early studies shows a striking correlation between a number of parameters associated with UV exposure in experimental animals and similar changes in human (unclassified) cataracts (Borkman, 1984). Tryptophan residues in lens proteins are the main absorbers of UV-B radiation (290–320 nm); this exposure leads to the production of photosensitizers such as N-formylkynurenine and its derivatives. In the presence of O_2 these substances give rise to fluorescent products, and further photolytic reactions lead to the cross-linking and breakdown of lens proteins (see Section 5.4). Early epidemiologic studies were not able to establish clearly the major site(s) of UV damage in cataractous lenses (see review by Zigman, 1985). Moreover, the amounts of UV exposure in the cataract populations examined were not accurately estimated.

Recently, however, a population-based, case-control study of 838 Chesapeake Bay watermen has revealed a statistically significant association between total cumulative UV-B exposure and risk of developing cortical cataracts (Taylor *et al.*, 1988). These watermen were selected for study because they formed a stable occupational group and their annual exposure to sunlight from age 16 could be accurately calculated. This study showed that UV-B exposure was approximately 21% higher in watermen with cortical opacities than in those without opacities. No association was found between nuclear cataracts and UV-B exposure. Another case-control study on a rural population in eastern Maryland has revealed a significant association between UV-B exposure and increased risk of posterior subcapsular cataracts (Bochow *et al.*, 1989).

Nutritional, metabolic, and antioxidant status, as well as diabetes, are being investigated as possible risk factors in cataract. A large number of parameters has been examined in a population residing in southeast Scotland by Clayton *et al.* (1984), and several factors, either singly or combined, appear to be associated with cataract. Among them may be certain plasma components such as calcium, urea, total protein, cholesterol, and glucose. Studies on 463 cataract patients in Germany point to an association between posterior subcapsular cataract and elevated blood levels of triglycerides and glucose (Jahn *et al.*, 1986). Further studies in Oxfordshire, England concur that individuals with high or moderately high levels of glucose, or those with abnormal glucose tolerance curves, are predisposed to cataract (van Heyningen and Harding, 1986). Diabetics are at greater risk than nondiabetics to develop cataracts (van Heyningen and Harding, 1988), and females have higher rates than males (Klein *et al.*, 1985; Harding and van Heyningen, 1987). A prospective study of 134 diabetic patients showed a significant and progressive increase (4–7% every 6 months) in light scattering in the anterior superficial lens cortex over 24 months, as measured by Scheimpflug (Topcon SL 45) photography (Dobbs *et al.*, 1987). However, use of this system to assess cortical changes in the lens has been challenged (see discussion following Dobbs *et al.*, 1987). Whether moderate elevations of blood glucose play a role in human diabetic cataracts through the osmotic effects of accumulated sorbitol has not been clearly established. However, the increased immunoreactive staining for aldose reductase observed in anterior and posterior superficial layers of diabetic lenses compared to normal ones supports the view that enhanced production of sorbitol in localized areas of diabetic lenses could exert significant osmotic stress (Akagi *et al.*, 1987). More than likely, however, it appears that in diabetic individuals with even moderately raised levels of glucose, glycation of lens proteins with subsequent cross-linking and insolubilization plays an important role in cataractogenesis (van Heyningen and Harding, 1986). Steroids are also associated with an 80% increase in the risk of developing cataract (Harding and van Heyningen, 1988).

A deficiency in selenium, as well as an excess, causes cataracts in rats (see Section 5.6.4), yet no relationship between lens opacities and plasma selenium levels were found in a large male population examined in Malmo, Sweden (Akesson *et al.*, 1987). Other nutritional factors have been linked to cataract in animal studies, and analyses of blood levels of vitamins and micronutrients in a group of 112 human subjects suggest an association between reduced risk of cortical cataract and elevated levels of vitamin D and carotenoids (Jacques *et al.*, 1988a). The finding of an "inverse" relationship between vitamin D and risk of cortical cataracts was unexpected. Optimum blood levels of vitamin D are dependent on sufficient exposure to sunlight; in the study of 838 watermen reported above, increased exposure to sunlight, implying maximum blood vitamin D levels, showed a positive correlation with development of cortical cataracts. One explanation of this discrepancy is that persons with cataract may spend less time outdoors and hence would have lower blood levels of vitamin D.

Oxidative stress is thought to play an important role in senile cataract, and measurements of the antioxidant status of 112 subjects aged 40 to 70 years, with and without cataract, tend to support this view (Jacques *et al.*, 1988b). Subjects with elevated plasma levels of at least two out of three antioxidant vitamins (vitamin E, vitamin C, and carotenoids) are at reduced risk of cataracts compared to individuals with low levels of one or more of these vitamins. Nevertheless, although antioxidant defense may play a role in cataractogenesis, there are no data showing that blood levels of these substances reflect the levels found in the lens. Another vitamin, riboflavin, may also be implicated in cataractogenesis, judging from a study made on human lens epithelium following cataract surgery (Horwitz *et al.*, 1987). Flavin adenine dinucleotide (FAD) is an essential cofactor of glutathione reductase, an enzyme that maintains glutathione in the reduced form. Of 32 epithelial preparations examined from cataractous lenses, 14 had no measurable glutathione reductase activity. The activity in 8 of these lens epithelial preparations could, however, be restored by addition of FAD, suggesting that some cataract patients may have a localized deficiency of riboflavin or possibly faulty cellular synthesis of FAD.

General health status and hereditary factors have also been investigated. For example, individuals with renal failure and severe diarrhea, as well as diabetes (discussed above), are at risk for cataracts according to a case-control study of 300 cataract patients and 609 age-matched controls in Oxfordshire, England (Harding and van Heyningen, 1987). It has long been recognized that hereditary defects in galactose metabolism (galactose 1-phosphate:uridyltransferase and galactokinase deficiencies) are associated with cataracts. The former is manifested from birth and can be alleviated by milk-free diets, whereas the latter is milder and of late onset. Although homozygotes for galactokinase deficiency are at risk for developing presenile cataracts, the case of heterozygotes is still controversial. In one survey of 39 patients who had developed cataracts before the age of 55, three were found to be heterozygous for galactokinase deficiency, and two of them had consumed large amounts of dairy products in adulthood (Stambolian *et al.*, 1986). This finding implies but does not prove that restriction of dietary dairy products could be beneficial in preventing cataract in individuals with family histories of galactokinase deficiency.

5.6.9.3. Therapy

Current understanding of the biochemical, functional, and structural changes in the lens that are associated with human as well as experimental cataract provides a broad theoretical basis for therapeutic approaches. The rationale for development and testing of anticataract drugs has been discussed, and at least 10 different types of therapy can be matched to corresponding biochemical and/or metabolic damage in the lens (Brown and Bron, 1985; Bron *et al.*, 1987). The majority of these drugs are being tested on experimental animals, but many of them should soon be available for human clinical trials. The most extensively investigated anticataract drugs are aldose reductase inhibitors. A large number of these inhibitors has been developed and are now commercially available (Kador *et al.*, 1985, 1986; Kinoshita, 1986). There is extensive proof of their efficacy in the prevention and/or reversal of sugar cataracts in experimental models, and the major usefulness of aldose reductase inhibitors in humans appears to be for individuals with inborn errors of galactose metabolism and in juvenile diabetics.

Aspirin is also an aldose reductase inhibitor; moreover, it is able to acetylate lens proteins under certain experimental conditions *in vitro*. Assuming that this may also be true *in vivo*, acetylation could prevent nonenzymatic carbamylation, glycosylation, and steroid-binding of lens proteins (van Heyningen and Harding, 1986). Aspirin-like analgesics such as paracetamol and ibuprofen (Harding and van Heyningen, 1987, 1988) as well as aspirin appear to protect against cataract, the latter shown in a recent detailed study by Eckerskorn and co-workers (1987) using Scheimpflug photography for the classification of cataracts. Nevertheless, results of another study cast some doubt on the protective effect of aspirin. A population-based survey of 838 watermen showed that neither large doses nor frequent use of aspirin either protects against or retards lens opacities (West *et al.*, 1987). Clearly, further studies are needed to resolve this controversial question.

5.7. REFERENCES

Ahmad, H., Singh, S. V., Medh, R. D., Ansari, G. A. S., Kurosky, A., and Awasthi, Y. C., 1988, Differential expression of α, μ and π classes of isozymes of glutathione S-transferase in bovine lens, cornea, and retina, *Arch. Biochem. Biophys.* **266:**416–426.

Akagi, Y., Kador, P. F., and Kinoshita, J. H., 1987, Immunohistochemical localization for aldose reductase in diabetic lenses, *Invest. Ophthalmol. Vis. Sci.* **28:**163–167.

Akesson, B., Bengtsson, B., and Steen, B., 1987, Are lens opacities related to plasma selenium and glutathione peroxidase in man? *Exp. Eye Res.* **44:**595–596.

Albers-Jackson, B., and Do, A. N., 1987, Lipid synthesis in the bovine lens, *Curr. Eye Res.* **6:**1161–1164.

Alcala, J., and Maisel, H., 1985, Biochemistry of lens plasma membranes and cytoskeleton, in: *The Ocular Lens. Structure, Function, and Pathology* (H. Maisel, ed.), Marcel Dekker, New York, pp. 169–222.

Alcala, J., Lieska, N., and Maisel, H., 1975, Protein composition of bovine lens cortical fiber cell membranes, *Exp. Eye Res.* **21:**581-595.

Alcala, J., Valentine, J., and Maisel, H., 1980, Human lens fiber cell plasma membranes. I. Isolation, polypeptide composition and changes associated with ageing, *Exp. Eye Res.* **30:**659–677.

Alcala, J., Katar, M., and Maisel, H., 1982/1983, Lipid composition of chick lens fiber cell gap junctions, *Curr. Eye Res.* **2:**569–578.

Alcala, J., Cenedella, R. J., and Katar, M., 1985, Limited proteolysis of MP26 in lens fiber plasma membranes of the U18666A-induced cataracts in rats, *Curr. Eye Res.* **4:**1001–1005.

Alcala, J., Katar, M., Rudner, G., and Maisel, H., 1988, Human beta crystallins: Regional and age related changes, *Curr. Eye Res.* **7:**353–359.

Anderson, R. S., Shearer, T. R., and Claycomb, C. K., 1986, Selenite-induced epithelial damage and cortical cataract, *Curr. Eye Res.* **5:**53–61.

Andley, U. P., 1988, Spectroscopic studies on the riboflavin-sensitized conformational changes of calf lens α-crystallin, *Exp. Eye Res.* **46:**531–544.

Andley, U. P., and Clark, B. A., 1988a, Conformational changes of β_H-crystallin in riboflavin-sensitized photooxidation, *Exp. Eye Res.* **47:**1–15.

Andley, U. P., and Clark, B. A., 1988b, Spectroscopic studies on the photooxidation of calf-lens γ-crystallin, *Curr. Eye Res.* **7:**571–579.

Andley, U. P., and Clark, B. A., 1989, Generation of oxidants in the near-UV photooxidation of human lens α-crystallin, *Invest. Ophthalmol. Vis. Sci.* **30:**706–713.

Andrews, J. S., Leonard-Martin, T., and Kador, P. F., 1984, Membrane lipid biosynthesis in the Philly mouse lens. I. The major phospholipid classes, *Curr. Eye Res.* **3:**279–285.

Aster, J. C., Brewer, G. J., Hanash, S. M., and Maisel, H., 1984, Band 4.1-like proteins of the bovine lens, *Biochem J.* **224:** 609–616.

Awasthi, Y. C., Saneto, R. P., and Srivastava, S. K., 1980, Purification and properties of bovine lens glutathione S-transferase, *Exp. Eye Res.* **30:**29–39.

Barron, B. C., Yu, N.-T., and Kuck, J. F. R., Jr., 1987, Tryptophan Raman/457.9-nm-excited fluorescence of intact guinea pig lenses in aging and ultraviolet light, *Invest. Ophthalmol. Vis. Sci.* **28:**815–821.

Barron, B. C., Yu, N.-T., and Kuck, J. F. R., Jr., 1988a, Raman spectroscopic evaluation of aging and long-wave UV exposure in the guinea pig lens: A possible model for human aging, *Exp. Eye Res.* **46:**249–258.

Barron, B. C., Yu, N.-T., and Kuck, J. F. R., Jr., 1988b, Distribution of a 488.0-nm-excited fluorophor in the equatorial plane of the human lens by a laser Raman microprobe: A new concept in fluorescence studies, *Exp. Eye Res.* **47:**901–904.

Bekhor, I., 1988, MP26 messenger RNA sequences in normal and cataractous lens, *Invest. Ophthalmol. Vis. Sci.* **29:**802–813.

Benedek, G. B., Chylak, L. T., Jr., Libondi, T., Magnante, P., and Pennett, M., 1987, Quantitative detection of the molecular changes associated with early cataractogenesis in the living human lens using quasielastic light scattering, *Curr. Eye Res.* **6:**1421–1432.

Bensch, K. G., Fleming, J. E., and Lohmann, W., 1985, The role of ascorbic acid in senile cataract, *Proc. Natl. Acad. Sci. USA* **82:**7193–7196.

Bergad, P. L., and Rathbun, W. B., 1986, Inhibition of Na,K-ATPase by sodium selenite and reversal by glutathione, *Curr. Eye Res.* **5:**919–923.

Bessems, G. J. H., and Hoenders, H. J., 1987, Distribution of aromatic and fluorescent compounds within single human lenses, *Exp. Eye Res.* **44:**817–824.

Bessems, G. J. H., Hoenders, H. J., and Wollensak, J., 1983, Variation in proportion and molecular weight of native crystallins from single human lenses upon aging and formation of nuclear cataract, *Exp. Eye Res.* **37:**627–637.

Bessems, G. J. H., de Man, B. M., Bours, J., and Hoenders, H. J., 1986, Age-related variations in the distribution of crystallins within the bovine lens, *Exp. Eye Res.* **43:**1019–1030.

Bessems, G. J. H., Keizer, E., Wollensak, J., and Hoenders, H. J., 1987a, Non-tryptophan fluorescence of crystallins from normal and cataractous human lenses, *Invest. Ophthalmol. Vis. Sci.* **28:**1157–1163.

Bessems, G. J. H., Rennen, H. J. J. M., and Hoenders, H. J., 1987b, Lanthionine, a protein cross-link in cataractous human lenses, *Exp. Eye Res.* **44:**691–695.

Beswick, H. T., and Harding, J. J., 1984, Conformational changes induced in bovine lens α-crystallin by carbamylation, *Biochem. J.* **223:**221–227.

Beswick, H. T., and Harding, J. J., 1987a, High-molecular-weight crystallin aggregate formation resulting from non-enzymic carbamylation of lens crystallins: Relevance to cataract formation *Exp. Eye Res.* **45:**569–578.

Beswick, H. T., and Harding, J. J., 1987b, Conformational changes induced in lens α- and γ-crystallins by modification with glucose 6-phosphate, *Biochem. J.* **246:**761–769.

Bettelheim, F. A., 1985, Physical basis of lens transparency, in: *The Ocular Lens. Structure, Function, and Pathology* (H. Maisel, ed.), Marcel Dekker, New York, pp. 265–300.

Bettelheim, F. A., Castoro, J. A., White, O., and Chylack, L. T., Jr., 1986, Topographic correspondence between total and non-freezable water content and the appearance of cataract in human lenses, *Curr. Eye Res.* **5:**925–932.

Bhuyan, K. C., and Bhuyan, D. K., 1977, Regulation of hydrogen peroxide in eye humors. Effect of 3-amino-1H-1,2,4-triazole on catalase and glutathione peroxidase of rabbit eye, *Biochim. Biophys. Acta* **497:**641–651.

Bhuyan, K. C., and Bhuyan, D. K., 1978, Superoxide dismutase of the eye. Relative functions of superoxide dismutase and catalase in protecting the ocular lens from oxidative damage, *Biochim. Biophys. Acta* **542:**28–38.

Bhuyan, K. C., and Bhuyan, D. K., 1984, Molecular mechanism of cataractogenesis: III. Toxic metabolites of oxygen as initiators of lipid peroxidation and cataract, *Curr. Eye Res.* **3:**67–81.

Bhuyan, K. C., Bhuyan, D. K., Kuck, J. F. R., Jr., Kuck, K. D., and Kern, H. L., 1982/1983, Increased lipid peroxidation and altered membrane functions in Emory mouse cataract, *Curr. Eye Res.* **2:**597–606.

Bloemendal, H., 1982, Lens proteins, *CRC Crit. Rev. Biochem.* **12:**1–38.

Bloemendal, H., 1985, Lens research: From protein to gene, *Exp. Eye Res.* **41:**429–448.

Bloemendal, H., Vermorken, A. J. M., Kibbelaar, M., Dunia, I., and Benedetti, E. L., 1977, Nomenclature for the polypeptide chains of lens plasma membranes, *Exp. Eye Res.* **24:**413–415.

Bloemendal, H., Piatigorsky, J., and Spector, A., 1989, Recommendations for crystallin nomenclature, *Exp. Eye Res.* **48:**465–466.

Blundell, T. L., Lindley, P. F., Miller, L. R., Moss, D. S., Slingsby, C., Turnell, W. G., and Wistow, G., 1983, Interactions of γ crystallin in relation to eye-lens transparency, *Lens Res.* **1:**109–131.

Bochow, T. W., West, S. K., Azar, A., Munoz, B., Sommer, A., and Taylor, H. R., 1989, Ultraviolet light exposure and risk of posterior subcapsular cataracts, *Arch. Ophthalmol.* **107:**369–372.

Borchman, D., Delamere, N. A., and Paterson, C. A., 1988, Ca-ATPase activity in the rabbit and bovine lens, *Invest. Ophthalmol. Vis. Sci.* **29:**982–987.

Borkman, R. F., 1984, Cataracts and photochemical damage in the lens, in: *Human Cataract Formation, Ciba Foundation Symposium 106* (J. Nugent and J. Whelan, eds.), Pitman, London, pp. 88–109.

Borkman, R. F., Hibbard, L. B., and Dillon, J., 1986, The photolysis of tryptophan with 337.1 nm laser radiation, *Photochem. Photobiol.* **43:**13–19.

Bouman, A. A., de Leeuw, A. L. M., and Broekhuyse, R. M., 1980, Lens membranes. XII. Age-related changes in polypeptide compositon of bovine lens fiber membranes, *Exp. Eye Res.* **31:**495–503.

Bours, J. Fink, H., and Hockwin, O., 1988, The quantification of eight enzymes from the ageing rat lens, with respect to sex differences and special reference to aldolase, *Curr. Eye Res.* **7:**449–455.

Brinker, J. M., Pegg, M. T., Howard, P. S., and Kefalides, N. A., 1985, Immunochemical characterization of type IV procollagen from anterior lens capsule, *Collagen Rel. Res.* **5:**233–244.

Broekhuyse, R. M., and Kuhlmann, E. D., 1974, Lens membranes 1. Composition of urea-treated plasma membranes from calf lens, *Exp. Eye Res.* **19:**297–302.

Broekhuyse, R. M., and Kuhlmann, E. D., 1980, Lens membranes XI. Some properties of human lens main intrinsic protein (MIP) and its enzymatic conversion into a 22000 dalton polypeptide, *Exp. Eye Res.* **30:**305–310.

Broekhuyse, R. M., and van den Eijnden-van Raaij, A. J. M., 1986, Calcium binding proteins in lens and other tissues, *Lens Res.* **3:**17–33.

Bron, A. J., Brown, N. A. P, Sparrow, J. M., and Shun-Shin, G. A., 1987, Medical treatment of cataract, *Eye* **1:**542–550.

Brown, N. A. P., and Bron, A. J., 1985, Medical therapy in the prevention of cataract, *Trans. Ophthalmol. Soc. U.K.* **104:**748–754.

Brown, N. A. P., Bron, A. J., Ayliffe, W., Sparrow, J., and Hill, A. R., 1987, The objective assessment of cataract, *Eye* **1:**234–246.

Carper, D., Shinohara, T., Piatigorsky, J., and Kinoshita, J. H., 1982, Deficiency of functional messenger RNA for a developmentally regulated β-crystallin polypeptide in a hereditary cataract, *Science* **217:**463–464.

Carper, D., Russell, P., Shinohara, T., and Kinoshita, J. H., 1985, Differential synthesis of rat lens proteins during development, *Exp. Eye Res.* **40:**85–94.

Carper, D., Smith-Gill, S. J., and Kinoshita, J. H., 1986, Immunocytochemical localization of the 27K β-crystallin polypeptide in the mouse lens during development using a specific monoclonal antibody: Implications for cataract formation in the Philly mouse, *Dev. Biol.* **113:**104–109.

Carper, D., Nishimura, C., Shinohara, T., Dietzchold, B., Wistow, G., Craft, C., Kador, P., and Kinoshita, J. H., 1987, Aldose reductase and p-crystallin belong to the same protein superfamily as aldehyde reductase, *FEBS Lett.* **220:**209–213.

Castoro, J. A., and Bettelheim, F. A., 1986, Distribution of the total and non-freezable water in rat lenses, *Exp. Eye Res.* **43:**185–191.

Castoro, J. A., and Bettelheim, F. A., 1987, Water gradients across bovine lenses, *Exp. Eye Res.* **45:**191–195.

Cenedella, R. J., 1982, Sterol synthesis by the ocular lens of the rat during postnatal development, *J. Lipid Res.* **23:**619–626.

Cenedella, R. J., 1983, Source of cholesterol for the ocular lens, studied with U18666A: A cataract-producing inhibitor of lipid metabolism, *Exp. Eye Res.* **37:**33–43.

Cenedella, R. J., 1985, Regional distribution of lipids and phospholipase A_2 activity in normal and cataractous rat lens, *Curr. Eye Res.* **4:**113–120.

Cenedella, R. J., 1987, Direct chemical measurement of DNA synthesis and net rates of differentiation of rat lens epithelial cells *in vivo*: Applied to the selenium cataract, *Exp. Eye Res.* **44:**677–690.

Cenedella, R. J., 1989, Cell cycle specific effects of selenium on the lens epithelium studied *in vivo* by the direct chemical approach, *Curr. Eye Res.* **8:**429–433.

Cheng, H.-M., and Aguayo, J. B., 1988, The energy status of lenses from human donor eyes, *Exp. Eye Res.* **46:**451–456.

Cheng, H.-M., and Chylack, L. T., Jr., 1985, Lens metabolism, in: *The Ocular Lens. Structure, Function, and Pathology* (H. Maisel, ed.), Marcel Dekker, New York, pp. 223–264.

Cheng, H.-M., and Gonzalez, R. G., 1986, The effect of high glucose and oxidative stress on lens metabolism, aldose reductase, and senile cataractogenesis, *Metabolism* **35:**10–14.

Cheng, H.-M., Gonzalez, R. G., Barnett, P. A., Aguayo, J. B., Wolfe, J., and Chylack, L. T., Jr., 1985, Sorbitol/fructose metabolism in the lens, *Exp. Eye Res.* **40:**223–229.

Cheng, H.-M, Gonzalez, R. G., von Saltza, I., Chylack, L. T., Jr., and Hutson, N. J., 1988, Glucose flux and the redox state of pyridine dinucleotides in the rat lens, *Exp. Eye Res.* **46:**47–952.

Chiesa, R., Gawinowicz-Kolks, M. A., and Spector, A., 1987, The phosphorylation of the primary gene products of α-crystallin, *J. Biol. Chem.* **262:**1438–1441.

Chiesa, R., Gawinowicz-Kolks, M. A., Kleiman, N. J., and Spector, A., 1988, Definition and comparison of the phosphorylation sites of the A and B chains of bovine α-crystallin, *Exp. Eye Res.* **46:**199–208.

Chiesa, R., McDermott, M. J., and Spector, A., 1989, Differential synthesis and phosphorylation of the α-crystallin A and B chains during bovine lens fiber cell differentiation, *Curr. Eye Res.* **8:**151–158.

Chiou, S.-H., Chylack, L. T., Jr., Tung, W. H., and Bunn, H. F., 1981, Nonenzymatic glycosylation of bovine lens crystallins; Effect of aging, *J. Biol. Chem.* **256:**5176–5180.

Chylack, L. T., Jr., 1984, Mechanisms of senile cataract formation, *Ophthalmology* **91:**596–602.

Chylack, L. T., Jr., and Cheng, H.-M., 1985, Clinical implications of research on lens and cataract, in: *The Ocular Lens. Structure, Function, and Pathology* (H. Maisel, ed.), Marcel Dekker, New York, pp. 439–466.

Chylack, L., and Tung, W., 1986, Inexpensive stereoscopic CCRG camera for lens/cataract photography in vitro, *Invest. Ophthalmol. Vis. Sci.* **27:**118–122.

Chylack, L., Lee, M., Tung, W., and Cheng, H.-M., 1983, Classification of human senile cataractous change by the American Cooperative Cataract Research Group (CCRG) method, *Invest. Ophthalmol. Vis. Sci.* **24:**424–431.

Chylack, L. T., Jr., Rosner, B., Garner, W., Giblin, F., Waldron, W., Wolfe, J., Leske, M. C., and White, O., 1985, Validity and reproducibility of the cooperative cataract research group (CCRG) cataract classification system, *Exp. Eye Res.* **40:**135–147.

Chylack, L. T., Jr., Rosner, B., Cheng, H.-M., McCarthy, D., and Pennett, M., 1987, Sources of variance in the objective documentation of human cataractous change with topcon SL-45 and Neitz-CTR retroillumination photography and computerized image analysis, *Curr. Eye Res.* **6:**1381–1390.

Chylack, L. T, Jr., Leske, M. C., Sperduto, R., Khu, P., and McCarthy, D., 1988, Lens opacities classification system, *Arch. Ophthalmol.* **106:**330–334.

Clark, J. I., and Carper, D., 1987, Phase separation in lens cytoplasm is genetically linked to cataract formation in the Philly mouse, *Proc. Natl. Acad. Sci. USA* **84:**122–125.

Clayton, R. M., 1985, Developmental genetics of the lens, in: *The Ocular Lens. Structure, Function, and Pathology* (H. Maisel, ed.), Marcel Dekker, New York, pp. 61–92.

Clayton, R. M., Cuthbert, J., Seth, J., Phillips, C. I., Bartholomew, R. S., and Reid, J. M., 1984, Epidemiological and other studies in the assessment of factors contributing to cataractogenesis, in: *Human Cataract Formation, Ciba Foundation Symposium 106* (J. Nugent and J. Whelan, eds.), Pitman, London, pp. 25–47.

Cohen, L. H., Westerhuis, L. W., de Jong, W. W., and Bloemendal, H., 1978, Rat α-crystallin A chain with an insertion of 22 residues, *Eur. J. Biochem.* **89:**259–266.

Croft, L. R., 1972, The amino acid sequence of γ-crystallin (fraction II) from calf lens, *Biochem. J.* **128:**961–970.

Crompton, M., Rixon, K. C., and Harding, J. J., 1985, Aspirin prevents carbamylation of soluble lens proteins and prevents cyanate-induced phase separation opacities *in vitro*: A possible mechanism by which aspirin could prevent cataract, *Exp. Eye Res.* **40:**297–311.

Darrow, R. M., Morris, J. I., Organisciak, D. T., and Varandani, P. T., 1988, The occurrence of glutathione-insulin transhydrogenase (protein-disulfide interchange enzyme) in the lens, *Curr. Eye Res.* **7:**861–869.

Das, G. C., and Piatigorsky, J., 1988, Promoter activity of the two chicken δ-crystallin genes in a Hela cell extract, *Curr. Eye Res.* **7:**331–340.

Datiles, M. B., Edwards, P. A., Trus, B. L., and Green, S. B., 1987, *In vivo* studies on cataracts using the Scheimpflug slit lamp camera, *Invest. Ophthalmol. Vis. Sci.* **28:**1707–1710.

David, L. L., and Shearer, T. R., 1984a, State of sulfhydryl in selenite cataract, *Toxicol. Appl. Pharmacol.* **74:**109–115.

David, L. L., and Shearer, T. R., 1984b, Calcium-activated proteolysis in the lens nucleus during selenite cataractogenesis, *Invest. Ophthamol. Vis. Sci.* **25:**1275–1283.

David, L. L., and Shearer, T. R., 1986, Purification of calpain II from rat lens and determination of endogenous substrates, *Exp. Eye Res.* **42:**227–238.

David, L. L., Dickey, B. M., and Shearer, T. R., 1987, Origin of urea-soluble protein in the selenite cataract. Role of β-crystallin proteolysis and calpain II, *Invest. Ophthalmol. Vis. Sci.* **28:**1148–1156.

David, L. L., Takemoto, L. J., Anderson, R. S., and Shearer, T. R., 1988, Proteolytic changes in main intrinsic polypeptide (MIP26) from membranes in selenite cataract, *Curr. Eye Res.* **7:**411–417.

David, L. L., Varnum, M. D., Lampi, K. J., and Shearer, T. R., 1989, Calpain II in human lens, *Invest. Ophthalmol. Vis. Sci.* **30:**269–275.

Delaye, M., and Tardieu, A., 1983, Short-range order of crystallin proteins accounts for eye lens transparency, *Nature* **302:**415–417.

DeNagel, D. C., Bando, M., Yu, N.-T., and Kuck, J. F. R., Jr., 1988, A Raman study of disulfide and sulfhydryl in the Emory mouse cataract, *Invest. Ophthalmol. Vis. Sci.* **29:**823–826.

den Dunnen, J. T., Jongbloed, R. J. E., Geurts van Kessel, A. H. M., and Schoenmakers, J. G. G., 1985a, Human lens γ-crystallin sequences are located the in p12-qter region of chromosome 2, *Hum. Genet.* **70:**217–221.

den Dunnen, J. T., Moormann, R. J. M., and Schoenmakers, J. G. G., 1985b, Rat lens β-crystallins are internally duplicated and homologous to γ-crystallins, *Biochim. Biophys. Acta* **824:**295–303.

den Dunnen, J. T., Szpirer, J., Levan, G., Islam, Q., and Schoenmakers, J. G. G., 1987, All six rat γ-crystallin genes are located on chromosome 9, *Exp. Eye Res.* **45:**747–750.

Dillon, J., 1984, Photolytic changes in lens proteins, *Curr. Eye Res.* **3:**145–150.

Dillon, J., 1985, Photochemical mechanisms in the lens, in: *The Ocular Lens. Structure, Function, and Pathology* (H. Maisel, ed.), Marcel Dekker, New York, pp. 349–366.

Dillon, J., Roy, D., and Spector, A., 1985, The photolysis of lens fiber membranes, *Exp. Eye Res.* **40:**53–60.

Dobbs, R. E., Smith, J. P., Chen, T., Knowles, W., and Hockwin, O., 1987, Long-term follow-up of lens changes with Scheimpflug photography in diabetics, *Ophthalmology* **94:**881–890.

Dodemont, H., Groenen, M., Jansen, L., Schoenmakers, J., and Bloemendal, H., 1985, Comparison of the crystallin mRNA populations from rat, calf and duck lens. Evidence for a longer αA_2-mRNA and two distinct αB_2-mRNAs in the birds, *Biochim. Biophys. Acta* **824:**284–294.

Dovrat, A., and Gershon, D., 1983, Studies on the fate of aldolase molecules in the aging rat lens, *Biochim. Biophys. Acta* **757:**164–167.

Dovrat, A., Scharf, J., Eisenbach, L., and Gershon, D., 1986, G6PD molecules devoid of catalytic activity are present in the nucleus of the rat lens, *Exp. Eye Res.* **42:**489–496.

Driessen, H. P. C., Herbrink, P., Bloemendal, H., and de Jong, W. W., 1980, The β-crystallin Bp chain is internally duplicated and homologous with γ-crystallin, *Exp. Eye Res.* **31:**243–246.

Duncan, G., and Jacob, T. J. C., 1984, The lens as a physicochemical physicochemical system, in: *The Eye*, 3rd ed. Vol. 1b (H. Davson, ed.), Academic Press, Orlando, FL, pp. 169–206.

Dunia, I., Manenti, S., Rousselet, A., and Benedetti, E. L., 1987, Electron microscopic observations of reconstituted proteoliposomes with the purified major intrinsic membrane protein of eye lens fibers, *J. Cell Biol.* **105:**1679–1689.

Eckerskorn, U., Hockwin, O., Muller-Breitenkamp, R., Chen, T. T., Knowles, W., and Dobbs, R. E., 1987, Evaluation of cataract-related risk factors using detailed classification systems and multivariate statistical methods, *Dev. Ophthalmol.* **15:**82–91.

Ellis, M., Alousi, S., Lawniczak, J., Maisel, H., and Welsh, M., 1984, Studies on lens vimentin, *Exp. Eye Res.* **38:**195–202.

Farnsworth, P. N., and Schleich, T., 1985, Progress toward the establishment of nuclear magnetic resonance measurements as an index of *in vivo* lens functional integrity, *Curr. Eye Res.* **4:**291–297.

Fecondo, J. V., and Augusteyn, R. C., 1983, Superoxide dismutase, catalase and glutathione peroxidase in the human cataractous lens, *Exp. Eye Res.* **36:**15–23.

FitzGerald, P. G., 1987, Main intrinsic polypeptide proteolysis and fiber cell membrane domains, *Invest. Ophthalmol. Vis. Sci.* **28:**795–805.

FitzGerald, P. G., 1988a, Immunochemical characterization of a M_r 115 lens fiber cell-specific extrinsic membrane protein, *Curr. Eye Res.* **7**:1243–1253.

FitzGerald, P. G., 1988b, Age-related changes in a fiber cell-specific extrinsic membrane protein, *Curr. Eye Res.* **7**:1255–1262.

Fleschner, C. R., and Cenedella, R. J., 1988, Specific restriction of cholesterol from cortical lens gap junctional membrane in the U18666A cataract, *Curr. Eye Res.* **7**:1029–1034.

Fleshman, K. R., and Wagner, B. J., 1984, Changes during aging in rat lens endopeptidase activity, *Exp. Eye Res.* **39**:543–551.

Fukui, H. N., Merola, L. O., and Kinoshita, J. H., 1978, A possible cataractogenic factor in the Nakamo mouse lens, *Exp. Eye Res.* **26**:477–485.

Gandolfi, S. A., Tomba, M. C., and Maraini, G., 1985, 86-Rb efflux in normal and cataractous human lenses, *Curr. Eye Res.* **4**:753–758.

Garadi, R., Reddy, V. N., Kador, P. F., and Kinoshita, J. H., 1983, Membrane glycoproteins of Philly mouse lens, *Invest. Ophtalmol. Vis. Sci.* **24**:1321–1324.

Garland, D., and Russell, P., 1985, Phosphorylation of lens fiber cell membrane proteins, *Proc. Natl. Acad. Sci. USA* **82**:653–657.

Garland, D., Zigler, J. S., Jr., and Kinoshita, J., 1986, Structural changes in bovine lens crystallins induced by ascorbate, metal and oxygen, *Arch. Biochem. Biophys.* **251**:771–776.

Garlick, R. L., Mazer, J. S., Chylack, L. T., Jr., Tung, W. H., and Bunn, H. F., 1984, Nonenzymatic glycation of human lens crystallin. Effect of aging and diabetes mellitus, *J. Clin. Invest.* **74**:1742–1749.

Garner, M. H., and Spector, A., 1980, Selective oxidation of cysteine and methionine in normal and senile cataractous lenses, *Proc. Natl. Acad. Sci. USA* **77**:1274–1277.

Garner, M. H., and Spector, A., 1986, ATP hydrolysis kinetics by Na,K-ATPase in cataract, *Exp. Eye Res.* **42**:339–348.

Garner, M. H., Garner, W. H., and Spector, A., 1984, Kinetic cooperativity change after H_2O_2 modification of (Na,K)-ATPase, *J. Biol. Chem.* **259**:7712–7718.

Garner, M. H., Garner, W. H., and Spector, A., 1986, H_2O_2-modification of Na,K-ATPase, *Invest. Ophthalmol. Vis. Sci.* **27**:103–107.

Garner, M. H., Wang, G.-M., and Spector, A., 1987, Stimulation of glycosylated lens epithelial Na,K-ATPase by an aldose reductase inhibitor, *Exp. Eye Res.* **44**:339–345.

Garner, W., 1984, The application of non-invasive techniques to the study of cataract development on the metabolic and the protein molecular level, in: *Human Cataract Formation, Ciba Foundation Symposium 106* (J. Nugent and J. Whelan, eds.), Pitman, London, pp. 248–265.

Garner, W. H., Garner, M. H., and Spector, A., 1979, Comparison of the 10,000 and 43,000 dalton polypeptide populations isolated from the water soluble and insoluble fractions of human cataractous lenses, *Exp. Eye Res.* **29**:257–276.

Garner, W. H., Garner, M. H., and Spector, A., 1983, H_2O_2-induced uncoupling of bovine lens Na^+, K^+-ATPase, *Proc. Natl. Acad. Sci. USA* **80**:2044–2048.

Geller, A. M., Kotb, M. Y. S., Jernigan, H. M., Jr., and Kredich, N. M., 1988, Methionine adenosyltransferase and S-adenosylmethionine in the developing rat lens, *Exp. Eye Res.* **47**:197–204.

Gentleman, S., Reid, T. W., and Martensen, T. M., 1987, Vanadate stimulation of phosphotyrosine protein levels in quiscent Nakano mouse lens cells, *Exp. Eye Res.* **44**:587–594.

Getty, D. J., Pickett, R. M., Chylack, L. T., Jr., McCarthy, D. F., and Huggins, A. W. F., 1989, An enriched set of features of nuclear cataract identified by multidimensional scaling, *Curr. Eye Res.* **8**:1–7.

Giblin, F. J., and McCready, J. P., 1983, The effect of inhibition of glutathione reductase on the detoxification of H_2O_2 by rabbit lens, *Invest. Ophthalmol. Vis. Sci.* **24**:113–118.

Giblin, F. J., McCready, J. P., Reddan, J. R., Dziedzic, D. C., and Reddy, V. N., 1985, Detoxification of H_2O_2 by cultured rabbit lens epithelial cells: Participation of the glutathione redox cycle, *Exp. Eye Res.* **40**:827–840.

Giblin, F. J., McCready, J. P., Schrimscher, L., and Reddy, V. N., 1987, Peroxide-induced effects on lens cation transport following inhibition of glutathione reductase activity *in vitro*, *Exp. Eye Res.* **45**:77–91.

Giblin, F. J., Schrimscher, L., Chakrapani, B., and Reddy, V. N., 1988, Exposure of rabbit lens to hyperbaric oxygen *in vitro*: Regional effects on GSH level, *Invest. Ophthalmol. Vis. Sci.* **29**:1312–1319.

Gooden, M. M., Takemoto, L. J., and Rintoul, D. A., 1985, Reconstitution of MIP26 from single human lenses into artificial membranes. I. Differences in pH sensitivity of cataractous vs. normal human lens fiber cell proteins, *Curr. Eye Res.* **4**:1107–1115.

Goosey, J. D., Zigler, J. S., Jr., and Kinoshita, J. H., 1980, Cross-linking of lens crystallins in a photodynamic system: A process mediated by singlet oxygen, *Science* **208:**1278–1280.

Gorin, M. B., Yancey, S. B., Cline, J., Revel, J.-P., and Horwitz, J., 1984, The major intrinsic protein (MIP) of the bovine lens fiber membrane: Characterization and structure based on cDNA cloning, *Cell* **39:**49–59.

Granger, B. L., and Lazarides, E., 1985, Appearance of new variants of membrane skeletal protein 4.1 during terminal differentiation of avian erythroid and lenticular cells, *Nature* **313:**238–241.

Greiner, J. V., Kopp, S. J., and Glonek, T., 1985, Phosphorous nuclear magnetic resonance and ocular metabolism, *Surv. Ophthalmol.* **30:**189–202.

Grover, D., and Zigman, S., 1972, Coloration of human lenses by near ultraviolet photo-oxidized tryptophan, *Exp. Eye Res.* **13:**70–76.

Gulik-Krzywicki, T., Tardieu, A., and Delaye, M., 1984, Spatial reorganization of low-molecular-weight proteins during cold cataract opacification, *Biochim. Biophys. Acta* **800:**28–32.

Harding, J. J., 1972, Conformational changes in human lens proteins in cataract, *Biochem. J.* **129:**97–100.

Harding, J. J., 1973, Disulphide cross-linked protein of high molecular weight in human cataractous lens, *Exp. Eye Res.* **17:**377–383.

Harding, J. J., 1981, Changes in lens proteins in cataract, in: *Molecular and Cellular Biology of the Eye Lens* (H. Bloemendal, ed.), John Wiley & Sons, New York, pp. 327–366.

Harding, J. J., 1985, Nonenzymatic covalent posttranslational modification of proteins *in vivo*, *Adv. Protein Chem.* **37:**247–334.

Harding, J. J., and Crabbe, M. J. C., 1984, The lens: Development, proteins, metabolism and cataract, in: *The Eye*, 3rd ed., Vol. 1b (H. Davson, ed.), Academic Press, Orlando, FL, pp. 207–492.

Harding, J. J., and Dilley, K. J., 1976, Structural proteins of the mammalian lens: A review with emphasis on changes in development, aging and cataract, *Exp. Eye Res.* **22:**1–73.

Harding, J. J., and van Heyningen, R., 1987, Epidemiology and risk factors for cataract, *Eye* **1:**537–541.

Harding, J. J., and van Heyningen, R., 1988, Drugs, including alcohol, that act as risk factors for cataract, and possible protection against cataract by aspirin-like analgesics and cyclopenthiazide, *Br. J. Ophthalmol.* **72:**809–814.

Heslip, J., Bagchi, M., Zhang, S., Alousi, S., and Maisel, H., 1986, An intrinsic membrane glycoprotein of the lens, *Curr. Eye Res.* **5:**949–958.

Hibino, K., Du, J., Dillon, J., and Malinowski, K., 1988, Age-dependent effect of UV light in abnormal alpha neoprotein formation in the lens, *Curr. Eye Res.* **7:**1113–1124.

Hightower, K. R., 1985, Cytotoxic effects of internal calcium on lens physiology: A review, *Curr. Eye Res.* **4:**453–459.

Hightower, K. R., and Harrison, S. E., 1987, The influence of calcium on glucose metabolism in the rabbit lens, *Invest. Ophthalmol. Vis. Sci.* **28:**1433–1436.

Hightower, K. R., and McCready, J. P., 1989, Effects of selenium on ion homeostasis and transparency in cultured lenses, *Invest. Ophthalmol. Vis. Sci.* **30:**171–175.

Hightower, K. R., David, L. L., and Shearer, T. R., 1987, Regional distribution of free calcium in selenite cataract: Relation to calpain II, *Invest. Ophthalmol. Vis. Sci.* **28:**1702–1706.

Hightower, K. R., Reddan, J. R., and Dziedzic, D. C., 1989, Susceptibility of lens epithelial membrane SH groups to hydrogen peroxide, *Invest. Ophthalmol. Vis. Sci.* **30:**569–574.

Hitchener, W. R., and Cenedella, R. J., 1987, HMG CoA reductase activity of lens epithelial cells: Compared with true rates of sterol synthesis, *Curr. Eye Res.* **6:**1045–1049.

Hockwin, O., Weigelin, E., Laser, H., and Dragomirescu, V., 1983, Biometry of the anterior eye segment by Scheimpflug photography, *Ophthalmic Res.* **15:**102–108.

Hockwin, O., Lerman, S., and Ohrloff, C., 1984, Investigation on lens transparency and its disturbances by microdensitometric analyses of Scheimpflug photographs, *Curr. Eye Res.* **3:**15–22.

Hockwin, O., Ahrend, M. H. J., and Bours, J., 1986, Correlation of Scheimpflug photography of the anterior eye segment with biochemical analysis of the lens, *Albrecht von Graefes Arch. Klin. Exp. Ophthalmol.* **224:**265–270.

Hoenders, H. J., and Bloemendal, H., 1981, Aging of lens proteins, in: *Molecular and Cellular Biology of the Eye Lens* (H. Bloemendal, ed.), John Wiley & Sons, New York, pp. 327–365.

Hoenders, H. J., and Bloemendal, H., 1983, Lens proteins and aging, *J. Gerontol.* **38:**278–286.

Hogg, D., Gorin, M. B., Heinzmann, C., Zollman, S., Mohandas, T., Klisak, I., Sparkes, R. S., Breitman, M., Tsui, L.-C., and Horwitz, J., 1987, Nucleotide sequence for the cDNA of the bovine βB2 crystallin and assignment of the orthologous human locus to chromosome 22, *Curr. Eye Res.* **6:**1335–1342.

Horwitz, J., Robertson, N. P., Wong, M. M., Zigler, J. S., and Kinoshita, J. H., 1979, Some properties of lens plasma membrane polypeptides isolated from normal human lenses, *Exp. Eye Res.* **28:**359–365.

Horwitz, J., Neuhaus, R., and Dockstader, J., 1981, Analysis of microdissected cataractous human lenses, *Invest. Ophthalmol. Vis. Sci.* **21:**616–619.

Horwitz, J., Dovrat, A., Straatsma, B. R., Revilla, P. J., and Lightfoot, D. O., 1987, Glutathione reductase in human lens epithelium: FAD-induced *in vitro* activation, *Curr. Eye Res.* **6:**1249–1256.

Hsu, M.-Y., Jaskoll, T. F., Unakar, N. J., and Bekhor, I., 1987a, Survival of fiber cells and fiber-cell messenger RNA in lens of rats maintained on a 50% galactose diet for 45 days, *Exp. Eye Res.* **44:**577–586.

Hsu, M.-Y., Davis, C., Jaskoll, T. F., Zeineh, R. A., Unakar, N. J., and Bekhor, I., 1987b, Crystallin mRNA product levels in lens undergoing reversal and inhibition of galactose cataracts, *Invest. Ophthalmol. Vis. Sci.* **28:**1413–1421.

Hu, T.-S., Datiles, M., and Kinoshita, J. H., 1983, Reversal of galactose cataract with sorbinil in rats, *Invest. Ophthalmol. Vis. Sci.* **24:**640–644.

Hu, T.-S., Merola, L. O., Kuwabara, T., and Kinoshita, J. H., 1984, Prevention and reversal of galactose cataract in rats with topical sorbinil, *Invest. Ophthalmol. Vis. Sci.* **25:**603–605.

Huang, Q.-L., Russell, P., Stone, S. H., and Zigler, J. S., Jr., 1987, Zeta-crystallin, a novel lens protein from the guinea pig, *Curr. Eye Res.* **6:**725–732.

Huby, R., and Harding, J. J., 1988, Non-enzymic glycosylation (glycation) of lens proteins by galactose and protection by aspirin and reduced glutathione, *Exp. Eye Res.* **47:**53–59.

Hum, T. P., and Augusteyn, R. C., 1987a, The state of sulphydryl groups in proteins isolated from normal and cataractous human lenses, *Curr. Eye Res.* **6:**1091–1101.

Hum, T. P., and Augusteyn, R. C., 1987b, The nature of disulphide bonds in rat lens proteins, *Curr. Eye Res.* **6:**1103–1108.

Ingolia, T. D., and Craig, E. A., 1982, Four small *Drosophila* heat shock proteins are related to each other and to mammalian α-crystallin, *Proc. Natl. Acad. Sci. USA* **79:**2360–2364.

Ireland, M., and Maisel, H., 1984a, A cytoskeletal protein unique to lens fiber cell differentiation, *Exp. Eye Res.* **38:**637–645.

Ireland, M., and Maisel, H., 1984b, Phosphorylation of chick lens proteins, *Curr. Eye Res.* 3:961–968.

Ireland, M., and Maisel, H., 1984c, Evidence for a calcium activated protease specific for lens intermediate filaments, *Curr. Eye Res.* **3:**423–429.

Ireland, M. E., and Maisel, H., 1987, Adrenergic stimulation of lens cytoskeletal phosphorylation, *Curr. Eye Res.* **6:**489–496.

Iwata, S., and Maesato, T., 1988, Studies on the mercapturic acid pathway in the rabbit lens, *Exp. Eye Res.* **47:**479–488.

Jacques, P. F., Hartz, S. C., Chylack, L. T., McGandy, R. B., and Sadowski, J. A., 1988a, Nutritional status in persons with and without senile cataract: Blood vitamin and mineral levels, *Am. J. Clin. Nutr.* **48:**152–158.

Jacques, P. F., Chylack, L. T., McGandy, R. B., and Hartz, S. C., 1988b, Antioxidant status in persons with and without senile cataract, *Arch. Ophthalmol.* **106:**337–340.

Jahn, C. E., Janke, M., Winowski, H., Bergmann, K. V., Leiss, O., and Hockwin, O., 1986, Identification of metabolic risk factors for posterior subcapsular cataract, *Ophthalmic Res.* **18:**112–116.

Jahngen, J. H., Eisenhauer, D., and Taylor, A., 1986a, Lens proteins are substrates for the reticulocyte ubiquitin conjugation system, *Curr. Eye Res.* **5:**725–733.

Jahngen, J. H., Haas, A. L., Ciechanover, A., Blondin, J., Eisenhauer, D., and Taylor, A., 1986b, The eye lens has an active ubiquitin–protein conjugation system, *J. Biol. Chem.* **261:**13760–13767.

Jedziniak, J. A., Nicoli, D. F., Baram, H., and Benedek, G. B., 1978, Quantitative verification of the existence of high molecular weight protein aggregates in the intact normal human lens by light-scattering spectroscopy, *Invest. Ophthalmol. Vis. Sci.* **17:**51–57.

Jedziniak, J. A., Chylack, L. T., Jr., Cheng, H.-M., Gillis, M. K., Kalustian, A. A., and Tung, W. H., 1981, The sorbitol pathway in the human lens: Aldose reductase and polyol dehydrogenase, *Invest. Ophthalmol. Vis. Sci.* **20:**314–326.

Jedziniak, J., Arredondo, M., and Andley, U., 1987, Oxidative damage to human lens enzymes, *Curr. Eye Res.* **6:**345–350.

Jernigan, H. M., Jr., 1985, Role of hydrogen peroxide in riboflavin-sensitized photodynamic damage to cultured rat lenses, *Exp. Eye Res.* **41:**121–129.

Jernigan, H. M., Jr., Fukui, H. N., Goosey, J. D., and Kinoshita, J. H., 1981, Photodynamic effects of rose bengal or riboflavin on carrier-mediated transport systems in rat lens, *Exp. Eye Res.* **32:**461–466.

Johnson, K. R., Panter, S. S., and Johnson, R. G., 1985, Phosphorylation of lens membranes with a cyclic AMP-dependent protein kinase purified from the bovine lens, *Biochim. Biophys. Acta* **844:**367–376.

Johnson, K. R., Lampe, P. D., Hur, K. C., Louis, C. F., and Johnson, R. G., 1986, A lens intracellular junction protein, MP26, is a phosphoprotein, *J. Cell Biol.* **102:**1334–1343.

Jones, T. R., and Reid, T. W., 1984, Sodium orthovanadate stimulation of DNA synthesis in Nakano mouse lens epithelial cells in serum-free medium, *J. Cell. Physiol.* **121:**199–205.

Kador, P. F., and Kinoshita, J. H., 1984, Diabetic and galactosaemic cataracts, in: *Human Cataract Formation, Ciba Foundation Symposium 106* (J. Nugent and J. Whelan, eds.), Pitman, London, pp. 110–131.

Kador, P. F., Fukui, H. N., Fukushi, S., Jernigan, H. M., Jr., and Kinoshita, J. H., 1980, Philly mouse: A new model of hereditary cataract, *Exp. Eye Res.* **30:**59–68.

Kador, P. F., Robison, W. G., Jr., and Kinoshita, J. H., 1985, The pharmacology of aldose reductase inhibitors, *Annu. Rev. Pharmacol. Toxicol.* **25:**691–714.

Kador, P. F., Akagi, Y., and Kinoshita, J. H., 1986, The effect of aldose reductase and its inhibition on sugar cataract formation, *Metabolism (Suppl.)* **35:**15–19.

Kanayama, T., Miyanaga, Y., Horiuchi, K., and Fujimoto, D., 1987, Detection of the cross-linking amino acid, histidinoalanine, in human brown cataractous lens protein, *Exp. Eye Res.* **44:**165–169.

Kawaba, T., Cheng, H.-M., and Kinoshita, J. H., 1986, The accumulation of myoinositol and rubidium ions in galactose- exposed rat lens, *Invest. Ophthalmol. Vis. Sci.* **27:**1522–1526.

Kinoshita, J. H., 1986, Aldose reductase in the diabetic eye, *Am. J. Ophthalmol.* **102**:685–692.

Kinoshita, J. H., Merola, L. O., and Dikmak, E., 1962, Osmotic changes in experimental galactose cataracts, *Exp. Eye Res.* **1:**405–410.

Kistler, J., and Bullivant, S., 1987, Protein processing in lens intercellular junctions: Cleavage of MP70 to MP38, *Invest. Ophthalmol. Vis. Sci.* **28:**1687–1692.

Kistler, J., Kirkland, B., and Bullivant, S., 1985, Identification of a 70,000-D protein in lens membrane junctional domains, *J. Cell Biol.* **101:**28–35.

Kistler, J., Kirkland, B., Gilbert, K., and Bullivant, S., 1986, Aging of lens fibers. Mapping membrane proteins with monoclonal antibodies, *Invest. Ophthalmol. Vis. Sci.* **27:**772–780.

Kleiman, N. J., Chiesa, R., Gawinowicz-Kolks, M. A., and Spector, A., 1988, Phosphorylation of β-crystallin B2 (βBp) in the bovine lens, *J. Biol. Chem.* **263:**14978–14983.

Klein, B. E. K., Klein, R., and Moss, S. E., 1985, Prevalence of cataracts in a population-based study of persons with diabetes mellitus, *Ophthalmology* **92:**1191–1196.

Kletzky, D. L., Tung, W. H., and Chylack, L. T., Jr., 1986, The protective effect of glucose on soluble rat lens hexokinase in the presence of oxidative stress, *Curr. Eye Res.* **5:**433–439.

Kodama, T., and Takemoto, L., 1988, Characterization of disulfide-linked crystallins associated with human cataractous lens membranes, *Invest. Ophthalmol. Vis. Sci.* **29:**145–149.

Kuck, J. F. R., and Kuck, K. D., 1983, The Emory mouse cataract: Loss of soluble protein, glutathione, protein sulfhydryl and other changes, *Exp. Eye Res.* **36:**351–362.

Kuck, J. F. R., Kuwabara, T., and Kuck, K. D., 1981/1982, The Emory mouse cataract: An animal model for human senile cataract, *Curr. Eye Res.* **1:**643–649.

Kuck, J. F. R., Yu, N.-T., and Askren, C. C., 1982, Total sulfhydryl by Raman spectroscopy in the intact lens of several species: Variations in the nucleus and along the optical axis during aging, *Exp. Eye Res.* **34:** 23–37.

Kuszak, J. R., Khan, A. R., and Cenedella, R. J., 1988, An ultrastructural analysis of plasma membrane in the U18666A cataract, *Invest. Ophthalmol. Vis. Sci.* **29:**261–267.

Lahm, D., Lee, L. K., and Bettelheim, F. A., 1985, Age dependence of freezable and nonfreezable water content of normal human lenses, *Invest. Ophthalmol. Vis. Sci.* **26:**1162–1165.

Lampe, P. D., Bazzi, M. D., Nelsestuen, G. L., and Johnson, R. G., 1986, Phosphorylation of lens intrinsic membrane proteins by protein kinase C, *Eur. J. Biochem.* **156:**351–357.

Lerman, S., 1980, *Radiant Energy and the Eye*, Macmillan, New York. Lerman, S., 1983, An experimental and clinical evaluation of lens transparency and aging, *J. Gerontol.* **38:**293–301.

Lerman, S., 1988, Ocular phototoxicity, *N. Engl. J. Med.* **319:**1475–1477.

Lerman, S., and Zigman, S., 1965, The metabolism of the lens as related to aging and experimental cataractogenesis, *Invest. Ophthalmol.* **4:**643–660.

Lerman, S., Zigman, S., and Forbes, W. F., 1966, Properties of a cryoprotein in the ocular lens, *Biochem. Biophys. Res. Commun.* **22:**57–61.

Lerman, S., Megaw, J. M., Gardner, K., Ashley, D., Long, R. C., Jr., and Goldstein, J. H., 1983, NMR analyses of the cold cataract. II. Studies on protein solutions, *Invest. Ophthalmol. Vis. Sci.* **24:**99–105.

Leske, M. C., Chylack, L. T., Jr., Sperduto, R., Khu, P., Wu, S.-Y., and McCarthy, D., 1988, Evaluation of a lens opacities classification system, *Arch. Ophthalmol.* **106:**327–329.

Lewis, B. S., and Harding, J. J., 1988, The major metabolite of bendazac inhibits the glycosylation of soluble lens proteins: A possible mechanism for a delay in cataractogenesis, *Exp. Eye Res.* **47:**217–225.

Lewis, B. S., Rixon, K. C., and Harding, J. J., 1986, Bendazac prevents cyanate binding to soluble lens proteins and cyanate-induced phase-separation opacities *in vitro*: A possible mechanism by which bendazac could delay cataract, *Exp. Eye Res.* **43:**973–979.

Li, L.-K., and So, L., 1987, Age dependent lipid and protein changes in individual bovine lenses, *Curr. Eye Res.* **6:**599–605.

Li, L.-K., So, L., and Spector, A., 1985, Membrane cholesterol and phospholipid in consecutive concentric sections of human lenses, *J. Lipid Res.* **26:**600–609.

Li, L.-K., Roy, D., and Spector, A., 1986, Changes in lens protein in concentric fractions from individual normal human lenses, *Curr. Eye Res.* **5:**127–135.

Li, L.-K., So, L., and Spector, A., 1987, Age-dependent changes in the distribution and concentration of human lens cholesterol and phospholipids, *Biochim. Biophys. Acta* **917:**112–120.

Liang, J. N., 1987, Fluorescence study of the effects of aging and diabetes mellitus on human lens α-crystallin, *Curr. Eye Res.* **6:**351–355.

Liang, J. N., and Chylack, L. T., Jr., 1987, Spectroscopic study on the effects of nonenzymatic glycation in human α-crystallin, *Invest. Ophthalmol. Vis. Sci.* **28:**790–794.

Liang, J. N., and Pelletier, M. R., 1988, Destabilization of lens protein conformation by glutathione mixed disulfide, *Exp. Eye Res.* **47:**17–25.

Liang, J. N., Bose, S. K., and Chakrabarti, B., 1985, Age-related changes in protein conformation in bovine lens crystallins, *Exp. Eye Res.* **40:**461–469.

Liang, J. N., Pelletier, M. R., and Chylack, L. T., Jr., 1988, Front surface fluorometric study of lens insoluble proteins, *Curr. Eye Res.* **7:**61–67.

Liang, J. N., Rossi, M. R., and Andley, U. P., 1989, Fluorescence studies on the age related changes in bovine and human lens membrane structure, *Curr. Eye Res.* **8:**293–298.

Lindley, P. F., Narebor, M. E., Summers, L. J., and Wistow, G. J., 1985, The structure of lens proteins, in: *The Ocular Lens. Structure, Function, and Pathology* (H. Maisel, ed.), Marcel Dekker, New York, pp. 123–167.

Lipman, R. M., Tripathi, B. J., and Tripathi, R. C., 1988, Cataracts induced by microwave and ionizing radiation, *Surv. Ophthalmol.* **33:**200–210.

Lo, W.-K., and Kuck, J. F. R., 1987, Alterations in fiber cell membranes of Emory mouse cataract: A morphologic study, *Curr. Eye Res.* **6:**433–444.

Lohmann, W., Schmehl, W., and Strobel, J., 1986, Nuclear cataract: Oxidative damage to the lens, *Exp. Eye Res.* **43:**859–862.

Lorand, L., Conrad, S. M., and Velasco, P. T., 1985, Formation of a 55000-weight cross-linked β crystallin dimer in the Ca^{2+}-treated lens. A model for cataract, *Biochemistry* **24:**1525–1531.

Lorand, L., Conrad, S. M., and Velasco, P. T., 1987, Inhibition of β-crystallin cross-linking in the Ca^{2+}-treated lens, *Invest. Ophthalmol. Vis. Sci.* **28:**1218–1222.

Lou, M. F., McKellar, R., and Chyan, O., 1986, Quantitation of lens protein mixed disulfides by ion-exchange chromatography, *Exp. Eye Res.* **42:**607–616.

Lou, M. F., Dickerson, J. E., Jr., Garadi, R., and York, B. M., Jr., 1988, Glutathione depletion in the lens of galactosemic and diabetic rats, *Exp. Eye Res.* **46:**517–530.

Lou, M. F., Garadi, R., Thomas, D. M., Mahendroo, P. P., York, B. M., Jr., and Jernigan, H. M., Jr., 1989, The effect of an aldose reductase inhibitor on lens phosphorylcholine under hyperglycemic conditions: Biochemical and NMR studies, *Exp. Eye Res.* **48:**11–24.

Louis, C. F., Johnson, R., and Turnquist, J., 1985a, Identification of the calmodulin-binding components in bovine lens plasma membranes, *Eur. J. Biochem.* **150:**271–278.

Louis, C. F., Johnson, R., Johnson, K., and Turnquist, J., 1985b, Characterization of the bovine lens plasma membrane substrates for cAMP-dependent protein kinase, *Eur. J. Biochem.* **150:**279–286.

Louis, C. F., Mickelson, J. R., Turnquist, J., Hur, K. C., and Johnson, R., 1987, Regulation of lens cyclic nucleotide metabolism by Ca^{2+} plus calmodulin, *Invest. Ophthalmol. Vis. Sci.* **28:**806–814.

Lubsen, N. H., Aarts, H. J. M., and Schoenmakers, J. G. G., 1988, The evolution of lenticular proteins: The β- and α-crystallin super gene family, *Prog. Biophys. Mol. Biol.* **51:**47–76.

Lucas, V. A., and Zigler, J. S., Jr., 1987, Transmembrane glucose carriers in the monkey lens, *Invest. Ophthalmol. Vis. Sci.* **28:**1404–1412.

Lucas, V. A., and Zigler, J. S., Jr., 1988, Identification of the monkey lens glucose transporter by photoaffinity labelling with cytochalasin B, *Invest. Ophthalmol. Vis. Sci.* **29:**630–635.

Lucas, V. A., Duncan, G., and Davies, P., 1986, Membrane permeability characteristics of perfused human senile cataractous lenses, *Exp. Eye Res.* **42:**151–165.

Maisel, H., 1984, Cytoskeletal proteins of the ageing human lens, in: *Human Cataract Formation, Ciba Foundation Symposium 106* (J. Nugent and J. Whelan, eds.), Pitman, London, pp. 163–176.

Maisel, H., (ed.), 1985, *The Ocular Lens: Structure, Function and Pathology*, Marcel Dekker, New York.

Maisel, H., and Ellis, M., 1984, Cytoskeletal proteins of the aging human lens, *Curr. Eye Res.* **3:**369–381.

Mandal, K., Chakrabarti, B., Thomson, J., and Siezen, R. J., 1987a, Structure and stability of γ-crystallins. Denaturation and proteolysis behavior, *J. Biol. Chem.* **262:**8096–8102.

Mandal, K., Bose, S. K., Chakrabarti, B., and Siezen, R. J., 1987b, Structure and stability of γ-crystallins. II. Differences in microenvironments and spatial arrangements of cysteine residues, *Biochim. Biophys. Acta* **911:**277–284.

Manski, W. J., and Malinowski, K. C., 1980, Regulatory effect of A to B cattle lens α-crystallin subunit ratios on the quaternary structure of macromolecules formed by their assembly, *J. Biol. Chem.* **255:**1572–1576.

Manski, W., and Malinowski, K., 1985, Alpha neoprotein molecules in normal lenses from animals of different ages and in cataractous lenses, *Exp. Eye Res.* **40:**179–190.

Maraini, G., and Pasino, M., 1983, Active and passive rubidium influx in normal human lenses and in senile cataracts, *Exp. Eye Res.* **33:**543–550.

Marcantonio, J., and Duncan, G., 1987, Amino acid transport and protein synthesis in human normal and cataractous lenses, *Curr. Eye Res.* **6:**1299–1308.

McDermott, M. J., Chiesa, R., and Spector, A., 1988a, Fe^{2+} oxidation of α-crystallin produces a 43,000 Da aggregate composed of A and B chains cross-linked by non-reducible covalent bonds, *Biochem. Biophys. Res. Commun.* **157:**626–631.

McDermott, M. J., Gawinowicz-Kolks, M. A., Chiesa, R., and Spector, A., 1988b, The disulfide content of calf γ-crystallin, *Arch. Biochem. Biophys.* **262:**609–619.

McFall-Ngai, M. J., Ding, L.-L., Takemoto, L. J., and Horwitz, J., 1985, Spatial and temporal mapping of the age-related changes in human lens crystallins, *Exp. Eye Res.* **41:**745–758.

McFall-Ngai, M., Horwitz, J., Ding, L.-L., and Lacey, L., 1986, Age-dependent changes in the heat-stable crystallin, βBp, of the human lens, *Curr. Eye Res.* **5:**387–394.

McNamara, M., and Augusteyn, R. C., 1984, The effects of hydrogen peroxide on lens proteins: A possible model for nuclear cataract, *Exp. Eye Res.* **38:**45–56.

Messmer, M., and Chakrabarti, B., 1988, High-molecular-weight protein aggregates of calf and cow lens: Spectroscopic evaluation, *Exp. Eye Res.* **47:**173–183.

Monnier, V. M., and Cerami, A., 1981, Nonenzymatic browning *in vivo*: Possible process for aging of long-lived proteins, *Science* **211:**491–493.

Monnier, V. M., and Cerami, A., 1983, Detection of nonenzymatic browning products in the human lens, *Biochim. Biophys. Acta* **760:**97–760.

Monnier, V. M., Stevens, V. J., and Cerami, A., 1979, Nonenzymatic glycosylation, sulfhydryl oxidation, and aggregation of lens proteins in experimental sugar cataracts, *J. Exp. Med.* **150:**1098–1107.

Moormann, R. J. M., den Dunnen, J. T., Heuyerjans, J., Jongbloed, R. J. E., van Leen, R. W., Lubsen, N. H., and Schoenmakers, J. G. G., 1985, Characterization of the rat γ-crystallin gene family and its expression in the eye lens, *J. Mol. Biol.* **182**:419–430.

Mostafapour, M. K., and Schwartz, C. A., 1981/1982, Purification of a heat-stable beta-crystallin polypeptide of the bovine lens, *Curr. Eye Res.* **1:**517–522.

Mulders, J. W. M., Voorter, C. E. M., Lamers, C., de Haard-Hoeckman, W. A., Montecucco, C., van de Ven, W. J. M., Bloemendal, H., and de Jong, W. W., 1988, MP17, a fiber-specific intrinsic membrane protein from mammalian eye lens, *Curr. Eye Res.* **7:**207–219.

Nakamura, M., Russell, P., Carper, D. A., Inana, G. and Kinoshita, J. H., 1988, Alteration of a developmentally regulated, heat-stable polypeptide in the lens of the Philly mouse, *J. Biol. Chem.* **263:**19218–19221.

Nene, V., Dunne, D. W., Johnson, K. S., Taylor, D. W., and Cordingley, J. S., 1986, Sequence and expression of a major egg antigen from *Schistosoma mansoni*. Homologies to heat shock proteins and alpha-crystallins, *Mol. Biochem. Parasitol.* **21:**179–188.

Ohrloff, C., and Hockwin, O., 1983, Lens metabolism and aging: Enzyme activities and enzyme alterations in lenses of different species during the process of aging, *J. Gerontol.* **38:**271–277.

Ohrloff, C., Zierz, S., and Hockwin, O., 1982, Investigations of the enzymes involved in the fructose breakdown in the cattle lens, *Ophthalmic Res.* **14:**221–229.

Ohrloff, C., Hockwin, O., Olson, R., and Dickman, S., 1984, Glutathione peroxidase, glutathione reductase and superoxide dismutase in the aging lens, *Curr. Eye Res.* **3:**109–115.

Oimomi, M., Maeda, Y., Hata, F., Kitamura, Y., Matsumoto, S., Baba, S., Iga, T., and Yamamoto, M., 1988, Glycation of cataractous lens in non-diabetic senile subjects and in diabetic patients, *Exp. Eye Res.* **46:**415–420.

Okuda, J., Kawamura, M., and Didelot, S., 1987, Anomeric preference in uptake of D-glucose and of D-galactose by rat lenses, *Curr. Eye Res.* **6:**1223–1226.

Ortwerth, B. J., and Olesen, P. R., 1988a, Glutathione inhibits the glycation and crosslinking of lens proteins by ascorbic acid, *Exp. Eye Res.* **47:**737–750.

Ortwerth, B. J., and Olesen, P. R., 1988b, Ascorbic acid-induced crosslinking of lens proteins: Evidence supporting a Maillard reaction, *Biochim. Biophys. Acta* **956:**10–22.

Ortwerth, B. J., Feather, M. S., and Olesen, P. R., 1988, The precipitation and cross-linking of lens crystallins by ascorbic acid, *Exp. Eye Res.* **47:**155–168.

Ozaki, L., Jap, P., and Bloemendal, H., 1985, Protein synthesis in bovine and human nuclear fiber cells, *Exp. Eye Res.* **41:**569–575.

Ozaki, Y., Mizuno, A., Itoh, K., Yoshiura, M. Iwamoto, T., and Iriyama, K., 1983, Raman spectroscopic study of age-related structural changes in the lens proteins of an intact mouse lens, *Biochemistry* **22:**6254–6259.

Ozaki, Y., Mizuno, A. Itoh, K., and Iriyama, K., 1987, Inter- and intramolecular disulfide bond formation and related structural changes in the lens proteins. A Raman spectroscopic study *in vivo* of lens aging, *J. Biol. Chem.* **262:**15545–15551.

Pande, A., Garner, W. H., and Spector, A., 1979, Glucosylation of human lens protein and cataractogenesis, *Biochem. Biophys. Res. Commun.* **89:**1260–1266.

Pasino, M., and Maraini, G., 1982, Cation pump activity and membrane permeability in human senile cataractous lenses, *Exp. Eye Res.* **34:**887–893.

Perry, R. E., Swamy, M. S., and Abraham, E. C., 1987, Progressive changes in lens crystallin glycation and high-molecular-weight aggregate formation leading to cataract development in streptozotocin-diabetic rats, *Exp. Eye Res.* **44:**269–282.

Piatigorsky, J., 1984, Lens crystallins and their gene families, *Cell* **38:**620–621.

Piatigorsky, J., 1987, Gene expression and genetic engineering in the lens, *Invest. Ophthalmol. Vis. Sci.* 28:9–28.

Piatigorsky, J., Kador, P. F., and Kinoshita, J. H., 1980, Differential synthesis and degradation of protein in the hereditary Philly mouse cataract, *Exp. Eye Res.* **30:**69–78.

Piatigorsky, J., Shinohara, T., and Zelenka, P. S. (eds.), 1988a, *Molecular Biology of the Eye: Genes, Vision and Ocular Disease*, Alan R. Liss, New York.

Piatigorsky, J., O'Brien, W. E., Norman, B. L., Kalumuck, K., Wistow, G. J., Borras, T., Nickerson, J. M., and Wawrousek, E. F., 1988b, Gene sharing by δ-crystallin and argininosuccinate lyase, *Proc. Natl. Acad. Sci. USA* **85:**3479–3483.

Pierscionek-Balcerzak, B., and Augusteyn, R. C., 1985, A new method for studying protein changes in the human lens during ageing and cataract formation, *Australian J. Optom.* **68:**49–53.

Pierscionek, B., and Augusteyn, R. C., 1988, Protein distribution patterns in concentric layers from single bovine lenses: Changes with development and ageing, *Curr. Eye Res.* **7:**11–23.

Pirie, A., 1951, Composition of ox lens capsule, *Biochem. J.* **48:**368–371.

Pirie, A., 1965, Glutathione peroxidase in lens and a source of hydrogen peroxide in aqueous humour, *Biochem. J.* **96:**244–253.

Pirie, A., 1968, Color and solubility of the proteins of human cataracts, *Invest. Ophthalmol.* **7:**634–650.

Pirie, A., 1971, Formation of N'-formylkynurenine in proteins from lens and other sources by exposure to sunlight, *Biochem. J.* **125:**203–208.

Quax-Jeuken, Y., Driessen, H., Leunissen, J., Quax, W., de Jong, W., and Bloemendal, H., 1985, βs-Crystallin: Structure and evolution of a distinct member of the βγ-superfamily, *EMBO J.* **4:**2597–2602.

Rao, G. N., and Cotlier, E., 1986, Free epsilon amino groups and 5-hydroxymethylfurfural contents in clear and cataractous human lenses, *Invest. Ophthalmol. Vis. Sci.* **27:**98–102.

Rao, G. N., Lardis, M. P., and Cotlier, E., 1985, Acetylation of lens crystallins: A possible mechanism by which aspirin could prevent cataract formation, *Biochem. Biophys. Res. Commun.* **128:**1125–1132.

Rathbun, W. B., 1984, Lenticular glutathione synthesis: Rate-limiting factors in its regulation and decline, *Curr. Eye Res.* **3**:101–108.

Rathbun, W. B., and Bovis, M. G., 1986, Activity of glutathione peroxidase and glutathione reductase in the human lens related to age, *Curr. Eye Res.* **5**:381–385.

Rathbun, W. B., Bovis, M. G., and Holleschau, A. M., 1986, Glutathione peroxidase, glutathione reductase and glutathione- S-transferase activities in the rhesus monkey lens as a function of age, *Curr. Eye Res.* **5**:195–199.

Ray, K., and Harris, H., 1985, Purification of neutral lens endopeptidase: Close similarity to a neutral proteinase in pituitary, *Proc. Natl. Acad. Sci. USA* **82**:7545–7549.

Reddan, J. R., Giblin, F. J., Dziedzic, D. C., McCready, J. P., Schrimscher, L., and Reddy, V. N., 1988, Influence of the activity of glutathione reductase on the response of cultured lens epithelial cells from young and old rabbits to hydrogen peroxide, *Exp. Eye Res.* **46**:209–221.

Reddy, V. N., Lin, L.-R., Arita, T., Zigler, J. S., Jr., and Huang, Q. L., 1988, Crystallins and their synthesis in human lens epithelial cells in tissue culture, *Exp. Eye Res.* **47**:465–478.

Revel, J. P., Nicholson, B., and Yancey, S. B., 1985, Chemistry of gap junctions, *Annu. Rev. Physiol.* **47**:263–279.

Revel, J. P., Yancey, S. B., Nicholson, B., and Hoh, J., 1987, Sequence diversity of gap junction proteins, in: *Junctional Complexes of Epithelial Cells, Ciba Foundation Symposium 125* (G. Bok and S. Clark, eds.), John Wiley & Sons, Chichester, pp. 108–127.

Ringens, P. J., Hoenders, H. J., and Bloemendal, H., 1982, Cell-free translation of human lens polyribosomes, *Exp. Eye Res.* **34**:831–834.

Rintoul, D. A., Cundy, K. V., and Cenedella, R. J., 1987, Physical properties of membranes and membrane lipids from the fiber cell of the U18666A-cataractous rat, *Curr. Eye Res.* **6**:1343–1348.

Roy, D., and Spector, A., 1978, Human insoluble lens protein. II. Isolation and characterization of a 9600 dalton polypeptide, *Exp. Eye Res.* **26**:445–459.

Roy, D., Spector, A., and Farnsworth, P. N., 1979, Human lens membrane: Comparison of major intrinsic polypeptides from young and old lenses isolated by a new methodology, *Exp. Eye Res.* **28**:353–358.

Roy, D., Garner, M. H., Spector, A., Carper, D., and Russell, P., 1982, Investigation of Nakano lens proteins, *Exp. Eye Res.* **34**:989–920.

Roy, D., Chiesa, R., and Spector, A., 1983, Lens calcium activated proteinase: Degradation of vimentin, *Biochem. Biophys. Res. Commun.* **116**:204–209.

Roy, D., Dillon, J., Wada, E., Chaney, W., and Spector, A., 1984, Nondisulfide polymerization of γ- and β-crystallins in the human lens, *Proc. Natl. Acad. Sci. USA* **81**:2878–2881.

Russell, P., 1984, *In vitro* alterations similar to posttranslational modification of lens proteins, *Invest. Ophthalmol. Vis. Sci.* **25**:209–212.

Russell, P., and Sato, S., 1986, A study of lectin-binding to the water-insoluble proteins of the lens, *Exp. Eye Res.* **42**:95–106.

Russell, P., Fukui, H. N., Tsunematsu, Y., Huang, F. L., and Kinoshita, J. H., 1977, Tissue culture of lens epithelial cells from normal and Nakano mice, *Invest. Ophthalmol. Vis. Sci.* **16**:243–246.

Russell, P., Smith, S. G., Carper, D. A., and Kinoshita, J. H., 1979, Age and cataract-related changes in the heavy molecular weight proteins and gamma crystallin composition of the mouse lens, *Exp. Eye Res.* **29**:245–255.

Russell, P., Carper, D. A., Chiogioji, A., and Reddy, V., 1985, The comparison of human lens crystallins using three monoclonal antibodies, *Invest. Ophthamol. Vis. Sci.* **26**:1028–1031.

Russell, P., Zelenka, P., Martensen, T., and Reid, T. W., 1987a, Identification of the EDTA-extractable protein in lens as calpactin I, *Curr. Eye Res.* **6**:533–538.

Russell, P., Garland, D., Zigler, J. S., Jr., Meakin, S. O., Tsui, L.-C., and Breitman, M.L., 1987b, Aging effects of vitamin C on a human lens protein produced in vitro, *FASEB J.* **1**:32–35.

Sas, D. F., Sas, M. J., Johnson, K. R., Menko, A. S., and Johnson, R. G., 1985, Junctions between lens fiber cells are labeled with a monoclonal antibody shown to be specific for MP26, *J. Cell Biol.* **100**:216–225.

Scharf, J., and Dovrat, A., 1986, Superoxide dismutase molecules in human cataractous lenses, *Ophthalmic Res.* **18**:332–337.

Scharf, J., Dovrat, A., and Gershon, D., 1987, Defective superoxide dismutase molecules accumulate with age in human lenses, *Albrecht von Graefes Arch. Klin. Exp. Ophthalmol.* **225**:133–136.

Schoenmakers, J. G. G., Den Dunnen, J. T., Moormann, R. J. M., Jongbloed, R., van Leen, R. W., and Lubsen, N. H., 1984, The crystallin gene families, in: *Human Cataract Formation, Ciba Foundation Symposium 106* (J. Nugent and J. Whelan, eds.), Pitman, London, pp. 208–218.

Sethna, S. S., Holleschau, A. M., and Rathbun, W. B., 1982/1983, Activity of glutathione synthesis enzymes in human lens related to age, *Curr. Eye Res.* **2**:735–742.

Sharma, K. K., and Ortwerth, B. J., 1986, Isolation and characterization of a new aminopeptidase from bovine lens, *J. Biol. Chem.* **261**:4295–4301.

Sharma, K. K., Olesen, P. R., and Ortwerth, B. J., 1987, The binding and inhibition of trypsin by α-crystallin, *Biochim. Biophys. Acta* **915**:284–291.

Shearer, T. R., and David, L. L., 1982/1983, Role of calcium in selenium cataract, *Curr. Eye Res.* **2**:777–784.

Shearer, T. R., David, L. L., and Anderson, R. S., 1987, Selenite cataract: A review, *Curr. Eye Res.* **6**:289–300.

Shiloh, Y., Donlon, T., Bruns, G., Breitman, M. L., and Tsui, L.-C., 1986, Assignment of the human γ-crystallin gene cluster (CRYG) to the long arm of chromosome 2, region q33-36, *Hum. Genet.* **73**:17–19.

Shinohara, T., Piatigorsky, J., Carper, D. A., and Kinoshita, J. H., 1982, Crystallin synthesis and crystallin mRNAs in galactosemic rat lenses, *Exp. Eye Res.* **34**:39–48.

Shiono, T., Kador, P. F., and Kinoshita, J. H., 1985, Stimulation of the hexose monophosphate pathway by pyrroline-5-carboxylate reductase in the lens, *Exp. Eye Res.* **41**:767–775.

Siezen, R. J., Fisch, M. R., Slingsby, C., and Benedek, G. B., 1985, Opacification of γ-crystallin solutions from calf lens in relation to cold cataract formation, *Proc. Natl. Acad. Sci. USA* **82**:1701–1705.

Siezen, R. J., Coppin, C. M., Kaplan, E. D., Dwyer, D., and Thomson, J. A., 1989, Oxidative modifications to crystallins induced in calf lenses *in vitro* by hydrogen peroxide, *Exp. Eye Res.* **48**:225–235.

Slingsby, C., 1985, Structural variation in lens crystallins, *Trends Biochem. Sci.* **10**:281–284.

Slingsby, C., and Miller, L. R., 1983, Purification and crystallization of mammalian lens γ-crystallins, *Exp. Eye Res.* **37**:517–530.

Slingsby, C., Driessen, H. P. C., Mahadevan, D., Bax, B., and Blundell, T. L., 1988, Evolutionary and functional relationships between the basic and acidic β-crystallins, *Exp. Eye Res.* **46**:375–403.

Sparkes, R. S., Mohandas, T., Heinzmann, C., Gorin, M. B., Horwitz, J., Law, M. L., Jones, C. A., and Bateman, J. B., 1986, The gene for the major intrinsic protein (MIP) of the ocular lens is assigned to human chromosome 12cen-q14, *Invest. Ophthalmol. Vis. Sci.* **27**:1351–1354.

Spector, A., 1984a, The search for a solution to senile cataracts, *Invest. Ophthalmol. Vis. Sci.* **25**:130–146.

Spector, A., 1984b, Oxidation and cataract, in: *Human Cataract Formation, Ciba Foundation Symposium 106* (J. Nugent and J. Whelan, eds.), Pitman, London, pp. 48–64.

Spector, A., 1985, Aspects of the biochemistry of cataract, in: *The Ocular Lens. Structure, Function, and Pathology* (H. Maisel, ed.), Marcel Dekker, New York, pp. 405–438.

Spector, A., and Garner, W. H., 1981, Hydrogen peroxide and human cataract, *Exp. Eye Res.* **33**:673–681.

Spector, A., and Roy, D., 1978, Disulfide-linked high molecular weight protein associated with human cataract, *Proc. Natl. Acad. Sci. USA* **75**:3244–3248.

Spector, A., Freund, T., Li, L.-K., and Augusteyn, R. C., 1971, Age-dependent changes in the structure of alpha crystallin, *Invest. Ophthalmol.* **10**:677–686.

Spector, A., Garner, M. H., Garner, W. H., Roy, D., Farnsworth, P., and Shyne, S., 1979, An extrinsic membrane polypeptide associated with high-molecular-weight protein aggregates in human cataract, *Science* **204**:1323–1326.

Spector, A., Chiesa, R., Sredy, J., and Garner, W., 1985, cAMP-dependent phosphorylation of bovine lens α-crystallin, *Proc. Natl. Acad. Sci. USA* **82**:4712–4716.

Spector, A., Wang, G.-M., and Huang, R.-R. C., 1986, A new HPLC method to determine glutathione-protein mixed disulfide, *Curr. Eye Res.* **5**:47–51.

Spector, A., Huang, R.-R. C., Wang, G.-M., Schmidt, C., Yan, G.-Z., and Chifflet, S., 1987, Does elevated glutathione protect the cell from H_2O_2 insult? *Exp. Eye Res.* **45**:453–465.

Spector, A., Yan, G.-Z., Huang, R.-R. C., McDermott, M. J., Gascoyne, P. R. C., and Pigiet, V., 1988, The effect of H_2O_2 upon thioredoxin-enriched lens epithelial cells, *J. Biol. Chem.* **263**:4984–4990.

Srivastava, O. P., 1988a, Age-related increase in concentration and aggregation of degraded polypeptides in human lenses, *Exp. Eye Res.* **47**:525–543.

Srivastava, O. P., 1988b, Characterization of a highly purified membrane proteinase from bovine lens, *Exp. Eye Res.* **46**:269–283.

Srivastava, O. P., and Ortwerth, B. J., 1983, Isolation and characterization of a 25K serine proteinase from bovine lens cortex, *Exp. Eye Res.* **37**:597–612.

Srivastava, O. P., and Ortwerth, B. J., 1989, The effects of aging and cataract formation on the trypsin inhibitor activity of human lens, *Exp. Eye Res.* **48**:25–36.

Stambolian, D., Scarpino-Myers, V., Eagle, R. C., Jr., Hodes, B., and Harris, H., 1986, Cataracts in patients heterozygous for galactokinase deficiency, *Invest. Ophthalmol. Vis. Sci.* **27:** 429–433.

Stevens, V. J., Rouzer, C. A., Monnier, V. M., and Cerami, A., 1978, Diabetic cataract formation: Potential role of glycosylation of lens crystallins, *Proc. Natl. Acad. Sci. USA* **75:**2918–2922.

Summers, L. J., Slingsby, C., Blundell, T. L., den Dunnen, J. T., Moormann, R. J. M., and Schoenmakers, J. G. G., 1986, Structural variation in mammalian γ-crystallins based on computer graphics analyses of human, rat and calf sequences. 1. Core packing and surface properties, *Exp. Eye Res.* **43:**77–92.

Swamy, M. S., and Abraham, E. C., 1987, Lens protein composition, glycation and high molecular weight aggregation in aging rats, *Invest. Ophthalmol. Vis. Sci.* **28:**1693–1701.

Swanson, A. A., Davis, R. M., Meinhardt, N. C., Kuck, K. D., and Kuck, J. F. R., Jr., 1985, Proteases in the Emory mouse cataract, *Invest. Ophthalmol. Vis. Sci.* **26:**1035–1037.

Takehana, M., and Takemoto, L., 1987, Quantitation of membrane- associated crystallins from aging and cataractous human lenses, *Invest. Ophthalmol. Vis. Sci.* **28:**780–784.

Takemoto, L. J., and Hansen, J. S., 1982, Intermolecular disulfide bonding of lens membrane proteins during human cataractogenesis, *Invest. Ophthalmol. Vis. Sci.* **22:**336–342.

Takemoto, L., and Takehana, M., 1986a, Major intrinsic polypeptide (MIP26K) from human lens membrane: Characterization of low-molecular-weight forms in the aging human lens, *Exp. Eye Res.* **43:**661–667.

Takemoto, L., and Takehana, M., 1986b, Covalent change of major intrinsic polypeptide (MIP26K) of lens membrane during human senile cataractogenesis, *Biochem. Biophys. Res. Commun.* **135:**965–971.

Takemoto, L. J., Hansen, J. S., and Horwitz, J., 1981, Interspecies conservation of the main intrinsic polypeptide (MIP) of the lens membrane, *Comp. Biochem. Physiol.* **68B:**101–106.

Takemoto, L. J., Hansen, J. S., and Horwitz, J., 1985, Antisera to synthetic peptides of lens MIP26K (major intrinsic polypeptide): Characterization and use as site-specific probes of membrane changes in the aging human lens, *Exp. Eye Res.* **41:**415–422.

Takemoto, L., Takehana, M., and Horwitz, J., 1986, Antisera to synthetic peptides of MIP26K as probes of membrane changes during human cataractogenesis, *Exp. Eye Res.* **42:**497–501.

Takemoto, L., Kodama, T., and Takemoto, D., 1987a, Antisera to synthetic peptides of MIP26K as probes of changes in opaque vs. transparent regions within the same human cataractous lens, *Exp. Eye Res.* **45:**179–183.

Takemoto, L., Kodama, T., and Takemoto, D., 1987b, Covalent changes at the N- and C-terminal regions of gamma crystallin during aging of the normal human lens, *Exp. Eye Res.* **45:**207–214.

Takemoto, L., Takemoto, D., Brown, G., Takehana, M., Smith, J., and Horwitz, J., 1987c, Cleavage from the N-terminal region of βBp crystallin during aging of the human lens, *Exp. Eye Res.* **45:**385–392.

Takemoto, L., Kodama, T., Wolfe, J., and Chylack, L., 1987d, Comparison of microdissected sections from the human cataractous lens by antisera to synthetic peptides, *Invest. Ophthalmol. Vis. Sci.* **28:**1210–1213.

Takemoto, L., Kuck, J., and Kuck, K., 1988, Changes in the major intrinsic polypeptide (MIP26K) during opacification of the Emory mouse lens, *Exp. Eye Res.* **47:**329–336.

Takemoto, L., Straatsma, B., and Horwitz, J., 1989, Immunochemical characterization of the major low molecular weight polypeptide (10K) from human cataractous lenses, *Exp. Eye Res.* **48:**261–270.

Taylor, A., and Davies, K. J. A., 1987, Protein oxidation and loss of protease activity may lead to cataract formation in the aged lens, *Free Radical Biol. Med.* **3:**371–377.

Taylor, A., Surgenor, T., Thomson, D. K. R., Graham, R. J., and Oettgen, H., 1984, Comparison of leucine aminopeptidase from human lens, beef lens and kidney, and hog lens in kidney, *Exp. Eye Res.* **38:**217–329.

Taylor, H. R., West, S. K., Rosenthal, F. S., Munoz, B., Newland, H. S., Abbey, H., and Emmett, E. A., 1988, Effect of ultraviolet radiation on cataract formation, *N. Engl. J. Med.* **319:**1429–1433.

Thomson, J. A., and Augusteyn, R. C., 1985, Ontogeny of human lens crystallins, *Exp. Eye Res.* **40:**393–410.

Thomson, J. A., and Augusteyn, R. C., 1988, On the structure of α-crystallin: The minimum molecular weight, *Curr. Eye Res.* **7:**563–569.

Thomson, J. A., Siezen, R. J., Kaplan, E. D., Messmer, M., and Chakrabarti, B., 1989, Comparative studies of βs-crystallins from human, bovine, rat and rabbit lenses, *Curr. Eye Res.* **8:** 139–149.

Tomba, M. C., Gandolfi, S. A., and Maraini, G., 1984/1985, Search for an oxidative stress in human senile cataract. Hydrogen peroxide and ascorbic acid in the aqueous humour and malondialdehyde in the lens, *Lens Res.* **2:**263–276.

Torriglia, A., and Zigman, S., 1988, The effect of near-UV light on Na-K-ATPase of the rat lens, *Curr. Eye Res.* **7:**539–548.

Tsubota, K., Laing, R. A., and Kenyon, K. R., 1987, Noninvasive measurements of pyridine nucleotide and flavoprotein in the lens, *Invest. Ophthalmol. Vis. Sci.* **28:**785–789.

Tung, W. H., Chylack, L. T., Jr., and Andley, U. P., 1988, Lens hexokinase deactivation by near-UV irradiation, *Curr. Eye Res.* **7:**257–263.

Uga, S., Kador, P. F., and Kuwabara, T., 1980, Cytological study of Philly mouse cataract, *Exp. Eye Res.* **30:**79–92.

van den Eijnden-van Raaij, A. J. M., Feijen, A., and Snoek, G. T., 1987, EDTA-extractable proteins from calf lens fiber membranes are phosphorylated by Ca^{2+}-phospholipid-dependent protein kinase, *Exp. Eye Res.* **45:**215–225.

van den Heuvel, R., Hendriks, W., Quax, W., and Bloemendal, H., 1985, Complete structure of the hamster αA crystallin gene, *J. Mol. Biol.* **185:**273–284.

van den Oetelaar, P. J. M., Clauwaert, J., van Laethem, M., and Hoenders, H. J., 1985, The influence of isolation conditions on the molecular weight of bovine α-crystallin, *J. Biol. Chem.* **260:**14030–14034.

van Heyningen, R., 1957, *meso*Inositol in the lens of mammalian eyes, *Biochem. J.* **65:**24–28.

van Heyningen, R., 1959, Formation of polyols by the lens of the rat with "sugar" cataract, *Nature* **184:**194–195.

van Heyningen, R., 1962, The sorbitol pathway in the lens, *Exp. Eye Res.* **1:**396–404.

van Heyningen, R., 1971, Fluorescent glucoside in the human lens, *Nature* **230:**393–394.

van Heyningen, R., and Harding, J. J., 1986, Do aspirin-like analgesics protect against cataract?, *Lancet* **1:**1111–1113.

van Heyningen, R., and Harding, J. J., 1988, A case-control study of cataract in Oxfordshire: Some risk factors, *Br. J. Ophthalmol.* **72:**804–808.

Varma, S. D., Chand, D., Sharma, Y. R., Kuck, J. F., Jr., and Richards, R. D., 1984, Oxidative stress on lens and cataract formation: Role of light and oxygen, *Curr. Eye Res.* **3:**35–57.

Varma, S. D., Morris, S. M., Bauer, S. A., and Koppenol, W. H., 1986, *In vitro* damage to rat lens by xanthine–xanthine oxidase: Protection by ascorbate, *Exp. Eye Res.* **43:**1067–1076.

Velasco, P. T., and Lorand, L., 1987, Acceptor-donor relationships in the transglutaminase-mediated cross-linking of lens β-crystallin subunits, *Biochemistry* **26:**4629–4634.

Vivekanandan, S., and Lou, M. F., 1989, Evidence for the presence of phosphoinositide cycle and its involvement in cellular signal transduction in the rabbit lens, *Curr. Eye Res.* **8:**101–111.

Voorter, C. E. M., Mulders, J. W. M., Bloemendal, H., and de Jong, W. W., 1986, Some aspects of the phosphorylation of α-crystallin A, *Eur. J. Biochem.* **160:**203–210.

Wagner, B. J., Margolis, J. W., and Abramovitz, A. S., 1986a, The bovine lens neutral proteinase comprises a family of cysteine-dependent proteolytic activities, *Curr. Eye Res.* **5:** 863–868.

Wagner, B. J., Margolis, J. W., Garland, D., and Roseman, J. E., 1986b, Bovine lens neutral proteinase preferentially hydrolyses oxidatively modified glutamine synthetase, *Exp. Eye Res.* **43:**1141–1143.

Wang, X., and Bettelheim, F. A., 1988, Distribution of total and non-freezable water contents of galactosemic rat lenses, *Curr. Eye Res.* **7:**771–776.

Watanabe, K., Fujii, Y., Nakayama, K., Ohkubo, H., Kuramitsu, S., Kagamiyama, H., Nakanishi, S., and Hayaishi, O., 1988, Structural similarity of bovine lung prostaglandin F synthase to lens ϵ-crystallin of the European common frog, *Proc. Natl. Acad. Sci. USA* **85:**11–15.

West, S. K., Munoz, B. E., Newland, H. S., Emmett, E. A., and Taylor, H. R., 1987, Lack of evidence for aspirin use and prevention of cataracts, *Arch. Ophthalmol.* **105:**1229–1231.

West, S. K., Rosenthal, F., Newland, H. S., and Taylor, H. R., 1988, Use of photographic techniques to grade nuclear cataracts, *Invest. Ophthalmol. Vis. Sci.* **29:**73–77.

Wiegand, R. D., Jose, J. G., Rapp, L. M., and Anderson, R. E., 1984, Free radicals and damage to ocular tissues, in: *Free Radicals in Molecular Biology, Aging, and Disease* (D. Armstrong, R. S. Sohol, R. G. Cutler, and T. F. Slater, eds.), Raven Press, New York, pp. 317–353.

Williams, E. H., Chaplain, T. L., and Meakem, T., 1985, A temporal and spatial study of the synthesis and degradation of water-soluble and insoluble proteins in galactosemic rat lenses, *Exp. Eye Res.* **41:**475–486.

Williams, W. F., and Odom, J. D., 1987, The utilization of ^{13}C and ^{31}P nuclear magnetic resonance spectroscopy in the study of the sorbitol pathway and aldose reductase inhibition in intact rabbit lenses, *Exp. Eye Res.* **44:**717–730.

Williams, W. F., Austin, C. D., Farnsworth, P. N., Groth-Vasselli, B., Willis, J. A., and Schleich, T., 1988, Phosphorus and proton magnetic resonance spectroscopic studies on the relationship between transparency and glucose metabolism in the rabbit lens, *Exp. Eye Res.* **47:**97–112.

Winkler, B. S., 1987, *In vitro* oxidation of ascorbic acid and its prevention by GSH, *Biochim. Biophys. Acta* **925:**258–264.

Winkler, B. S., and Solomon, F., 1988, High activities of NADP$^+$-dependent isocitrate dehydrogenase and malic enzyme in rabbit lens epithelial cells, *Invest. Ophthalmol. Vis. Sci.* **29:**821–823.

Wistow, G., and Piatigorsky, J., 1987, Recruitment of enzymes as lens structural proteins, *Science* **236:**1554–1556.

Wistow, G. J., and Piatigorsky, J., 1988, Lens crystallins: The evolution and expression of proteins for a highly specialized tissue, *Annu. Rev. Biochem.* **57:**479–504.

Wistow, G., Turnell, B., Summers, L., Slingsby, C., Moss, D., Miller, L., Lindley, P., and Blundell, T., 1983, X-ray analysis of the eye lens protein γ-II crystallin at 1.9 A resolution, *J. Mol. Biol.* **170:**175–202.

Wistow, G. J., Mulders, J. W. M., and de Jong, W. W., 1987, The enzyme lactate dehydrogenase as a structural protein in avian and crocodilian lenses, *Nature* **326:**622–624.

Wolfe, J. K., and Chylack, L. T., Jr., 1986, Glucose metabolism by human cataracts in culture, *Exp. Eye Res.* **43:**243–249.

Wolff, S. P., Wang, G.-M., and Spector, A., 1987, Pro-oxidant activation of ocular reductants. 1. Copper and riboflavin stimulate ascorbate oxidation causing lens epithelial cytotoxicity in vitro, *Exp. Eye Res.* **45:**777–789.

Yancey, S. B., Koh, K., Chung, J., and Revel, J.-P., 1988, Expression of the gene for main intrinsic polypeptide (MIP): Separate spatial distributions of MIP and β-crystallin gene transcripts in rat lens development, *J. Cell Biol.* **106:**705–714.

Yoshida, H., Murachi, T., and Tsukahara, I., 1984, Limited proteolysis of bovine lens α-crystallin by calpain, A Ca^{2+}- dependent cysteine proteinase, isolated from the same tissue, *Biochim. Biophys. Acta* **798:**252–259.

Yoshida, H., Murachi, T., and Tsukahara, I., 1985, Distribution of calpain I, calpain II, and calpastatin in bovine lens, *Invest. Ophthalmol. Vis. Sci.* **26:**953–956.

Yoshida, H., Yumoto, N., Tsukahara, I., and Murachi, T., 1986, The degradation of α-crystallin at its carboxyl-terminal portion by calpain in bovine lens, *Invest. Ophthalmol. Vis. Sci.* **27:**1269–1273.

Yu, N.-T., DeNagel, D. C., Pruett, P. L., and Kuck, J. F. R., Jr., 1985, Disulfide bond formation in the eye lens, *Proc. Natl. Acad. Sci. USA* **82:**7965–7968.

Yu, N.-T., Cai, M.-Z., Ho, D. J.-Y., and Kuck, J. F. R., Jr., 1988, Automated laser-scanning-microbeam fluorescence/Raman image analysis of human lens with multichannel detection: Evidence for metabolic production of a green fluorophor, *Proc. Natl. Acad. Sci. USA* **85:**103–106.

Zampighi, G. A., Hall, J. E., and Kreman, M., 1985, Purified lens junctional protein forms channels in planar lipid films, *Proc. Natl. Acad. Sci. USA* **82:**8468–8472.

Zelenka, P. S., 1980, Changes in phosphatidylinositol metabolism during differentiation of lens epithelial cells into lens fiber cells in the embryonic chick, *J. Biol. Chem.* **255:**1296–1300.

Zelenka, P. S., 1984, Lens lipids, *Curr. Eye Res.* **3:**1337–1359.

Zigler, J. S., Jr., and Goosey, J. D., 1981, Photosensitized oxidation in the ocular lens: Evidence for photosensitizers endogenous to the human lens, *Photochem. Photobiol.* **33:**869–874.

Zigler, J. S., Jr., and Goosey, J. D., 1984, Singlet oxygen as a possible factor in human senile nuclear cataract development, *Curr. Eye Res.* **3:**59–65.

Zigler, J. S., Jr., and Horwitz, J., 1981, Immunochemical studies on the major intrinsic polypeptides from human lens membrane, *Invest. Ophthalmol. Vis. Sci.* **21:**46–51.

Zigler, J. S., Jr., Carper, D. A., and Kinoshita, J. H., 1981, Changes in lens crystallins during cataract development in the Philly mouse, *Ophthalmic Res.* **13:**237–251.

Zigler, J. S., Jr., Jernigan, H. M., Jr., Garland, D., and Reddy, V. N., 1985a, The effects of "Oxygen radicals" generated in the medium on lenses in organ culture: Inhibition of damage by chelated iron, *Arch. Biochem. Biophys.* **241:**163–172.

Zigler, J. S., Jr., Russell, P., Takemoto, L. J., Schwab, S. J., Hansen, J. S., Horwitz, J., and Kinoshita, J. H., 1985b, Partial characterization of three distinct populations of human γ-crystallins, *Invest. Ophthalmol. Vis. Sci.* **26:**525–531.

Zigman, S., 1985, Photobiology of the lens, in: *The Ocular Lens. Structure, Function, and Pathology* (H. Maisel, ed.), Marcel Dekker, New York, pp. 301–347.

Zigman, S., Paxhia, T., and Waldron, W., 1988, Effects of near-UV radiation on the protein of the grey squirrel lens, *Curr. Eye Res.* **7:**531–537.

Vitreous 6

6.1. INTRODUCTION AND GENERAL FEATURES

The space between the lens, ciliary body, and retina is filled with a transparent gel or liquid that is variously termed the vitreous body, vitreous humor, vitreus (Denlinger *et al.*, 1980; Balazs and Denlinger, 1984), or vitreous. The latter term seems most appropriate and is used throughout this chapter. The vitreous is an extracellular matrix that occupies the major part of the globe in most if not all vertebrate species. A large body of experimental evidence supports the view that the vitreous is an extended extracellular matrix of the retina and could be considered embryologically as the basement membrane of the retina (Swann, 1980). Other concepts have also been developed in which the vitreous is considered as a "typical connective tissue compartment surrounded by epithelial (lens, ciliary body) and neuroglial (retinal Müller cells) cells which form basal laminae around it" (Balazs and Denlinger, 1984). Further studies at the molecular level are clearly needed to reconcile these differing points of view.

Apart from its space-filling function, the vitreous also provides mechanical support to the surrounding ocular tissues, and because of its viscoelastic properties, it serves as a shock absorber against mechanical impact (see review by Balazs and Denlinger, 1984). It has an organized structure, as described in Section 6.2.1, and is free of cellular components except for a small number of pleiomorphic macrophage-like cells localized on the peripheral surface, mainly in the ora serrata region (Szirmai and Balazs, 1958; Hamburg, 1959; Bloom and Balazs, 1965).

Several reviews and monographs on the structure, biochemistry, and physiology of normal vitreous are available (Balazs, 1961, 1965; Pirie, 1969; Berman and Voaden, 1970; Swann, 1980, 1987; Gloor, 1981; Moorhead, 1983; Balazs and Denlinger, 1984); in addition, pathological conditions of the vitreous—hemorrhage, vitreoretinal degenerations, and other disorders—have been summarized by Streeten (1982). Of these, proliferative vitreoretinopathy (PVR) has received the most attention in recent years, and some aspects of its pathogenesis at the cellular level are described in a research update of retinal detachment (Steinberg, 1986). This complication of retinal detachment occurs in untreated cases as well as in about 10% of retinas that had been surgically reattached. The cellular events that characterize PVR are (1) migration of glial cells, pigment epithelial cells, and fibrocytes into the vitreous cavity, where they proliferate and ultimately undergo extensive transformation and dedifferentiation, (2) interaction of these cells with endogenous membranous components of the vitreous (probably collagen), leading to the formation of vitreal, epiretinal, and subretinal contratactile membranes, and (3) traction retinal detachment. Many experimental models of PVR are available, and its pathogenesis as well as treatment by vitrectomy and other surgical and pharmacological therapies have been reviewed in detail by Machemer (1988). The causes and prevention of PVR are important clinical problems whose ultimate solution lies in an understanding of the biochemistry of the vitreous and its role as a "tissue culture medium" in which these, and other, pathological transformations take place. The present chapter is limited to a discussion of the normal vitreous only.

6.2. CHEMICAL COMPOSITION

The vitreous contains about 98% water and 0.1 % colloids; the rest of the solid matter consists of ions and low-molecular-weight solutes. The two major structural components are collagen and hyaluronic acid (HA). A number of noncollagenous proteins and glycoproteins have also been isolated and partially characterized. The vitreous is not fully developed at birth, and postnatal changes in both chemical composition and volume occur in all species examined (Balazs *et al.*, 1959; Berman and Michaelson, 1964; Balazs and Denlinger, 1984). The collagen content increases, but its concentration appears to remain constant as the vitreous volume expands during development; in contrast, the absolute concentration of HA increases about two- to fourfold from birth to adulthood, depending on the species. All studies show that the vitreous is not fully developed biochemically until the eye reaches adult size (for reviews see Gloor, 1981; Balazs and Denlinger, 1984; Swann, 1987). There are also important changes in aging, commencing at about 40 years of age in humans, as discussed in Section 6.4.

6.2.1. Molecular Organization

The vitreous has a unique molecular structure best described in terms of an insoluble primary network of randomly oriented collagen fibers and a secondary network of polyanionic hyaluronic acid (HA) molecules. The former confers gel-like properties to the tissue, whereas the latter provides a viscoelastic consistency that resists compression and also stabilizes the collagen network. Soluble proteins and glycoproteins may contribute to the molecular organization of the vitreous, although this has not yet been established. A schematic diagram depicting the organization of the major structural components of the gel vitreous, collagen and HA, is given in Fig. 6.1.

Topographic variations in both collagen and HA have been noted in most species (for reviews see Balazs, 1961, 1965; Balazs and Denlinger, 1984). The highest density of collagen fibrils is in the cortical gel bordering the ciliary epithelium and the retina at the ora serrata, and the lowest is in the center of the vitreous and in the anterior region. Chemical analyses of HA and soluble proteins in frozen-and-thawed vitreous samples have shown regional distributions similar to those of the collagen fibrils. However, other data suggest that use of this technique may introduce artifacts, since manual dissection of fresh calf vitreous shows no correlation in the distribution of HA and soluble protein to collagen (Swann and Constable, 1972a). Although HA is present at highest concentration in the peripheral vitreous, soluble proteins appear to be evenly distributed throughout the gel of freshly dissected vitreous. Another major aspect of vitreous structure relates to the large variations in the proportion of gel (collagen-containing) vitreous to liquid vitreous as functions of both species and age (see Sections 6.2.1.1 and 6.4, respectively).

A basal lamina mediates the attachment of the vitreous to three surrounding tissues: the lens, the ciliary epithelium, and glial cells of the retina. The dense structure of this

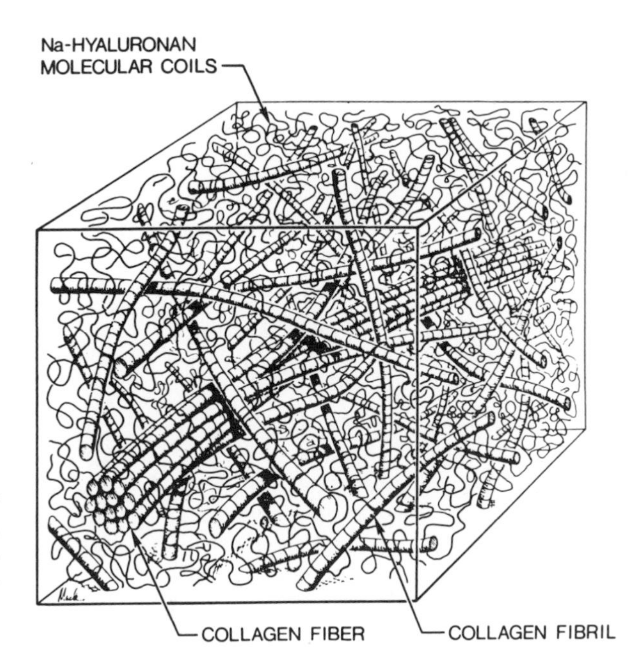

Figure 6.1. Schematic diagram of the molecular organization of gel vitreous showing the random network of collagen fibrils and the hydrated coils of hyaluronic acid. (From Balazs and Denlinger, 1984.)

basal lamina is thought to impede the movement of macromolecules and, in addition, may act as a barrier to the entrance of molecules larger than 15–20 nm in diameter (Balazs and Denlinger, 1984). However, the primary protection against the entry of foreign substances into the vitreous appears to be the blood–retinal barrier combined with posterior transport of dissolved solutes, gases, and water across the retina into the choroidal circulation (Streeten, 1982; Moseley and Foulds, 1982; Moseley *et al.*, 1984). The cortical gel vitreous is not attached to a basal lamina at the border of the posterior chamber; hence, the cortical gel is directly exposed to the aqueous, and it is through this surface that large molecules such as hyaluronic acid could diffuse out of the vitreous after implantation (Balazs and Denlinger, 1984). However, in the untraumatized normal eye, very little hyaluronic acid appears to enter the aqueous from the vitreous (see Chapter 4, Section 4.1.1.4).

6.2.1.1. Collagen

As discussed above, the gel-forming component of the vitreous is collagen, but there are marked species variations in the proportions of gel to liquid vitreous (Table 6.1). There are also changes in these proportions with age. In cattle, sheep, rabbit, guinea pig, and frog, all of the vitreous is in the gel state throughout life. In humans and in rhesus monkey, the volume of liquid vitreous as a fraction of total vitreous space increases throughout life (see Fig. 1 in Balazs and Denlinger, 1984). At the other extreme, the vitreous of certain species such as owl monkey and squid is predominantly liquid.

Despite its apparent simplicity and abundance, it has taken many years to define the chemical nature of vitreous collagen. A "residual protein" of the vitreous recognized since the turn of the century was first tentatively identified by x-ray and amino acid analyses as

Table 6.1. Rheological State and Concentrations of Collagen and Protein in Vitreous of Various Species[a]

Species	Rheological state (% total)		Collagen (μg/ml)	Protein (μg/ml)
	Gel	Liquid		
Cattle	100	0	52–112	380–970
Sheep	100	0	67–82	630–700
Rabbit	100	0	75–900	44–81
Guinea pig	100	0	10–140	800–900
Frog	100	0	180–190	
Human	40–80	20–60	40–120	450–1100
Monkey				
Rhesus	60	40		80–140
Owl	2	98	1250–2580	33–110
Chicken	37	63	100–200	33–56
Carp	40	60	600–900	
Squid	2	98		

[a]Adapted from Balazs and Denlinger (1984).

a collagen-like protein (Pirie *et al.*, 1948). Early electron microscopic studies of the vitreous revealed a random network of thin collagen-like fibrils 10–20 nm in width but lacking any recognizable axial periodicity (see review by Swann, 1980). The average diameters of human and bovine collagen fibrils are 15 and 20 nm, respectively. In spite of their unusual morphology, these fibrils were considered as belonging to the family of collagens, and the principal collagen of bovine vitreous was identified some years later as type II collagen (Swann *et al.*, 1972; Newsome *et al.*, 1976; Swann and Sotman, 1980; Burke, 1980). It is composed of three α1(II) polypeptide chains similar but not identical to the type II collagen isolated from pepsin digests of cartilage, the differences being mainly in amino acid composition and distribution of CNBr (cyanogen bromide) cleavage peptides. Vitreous type II collagen has a lower content of alanine and higher amounts of aspartic acid, valine, and leucine than cartilage type II collagen. Moreover, the α chain from vitreous collagen has a significantly higher content of galactosylglucose side chains than the corresponding fraction from cartilage.

Minor components with slower electrophoretic mobilities were also detected (but not identified) in the investigations described above, and more recent studies show clearly that vitreous contains an additional genetically distinct form of collagen. By different extraction and fractionation techniques, type II collagen was again shown to be the major collagen in bovine vitreous, but other "cartilage–phosphate-soluble" collagens (C-PS1 and C-PS2) were also characterized by SDS-PAGE and amino acid analyses (Ayad and Weiss, 1984). The atypical collagen type II with covalently linked terminal peptides observed by Swann and Sotman (1980) is probably the same as C-PS1 collagen and its higher-M_r forms. Moreover, as described below, this component in all probability represented type IX collagen, which had not been characterized at the time these investigations were carried out.

Developmental studies of vitreous collagen have shed considerable light not only on the identity of the genetically distinct polypeptide chains but also on their origin. In the chicken embryo, two different cell types synthesize vitreous collagen at different stages of development (Newsome *et al.*, 1976). In early embryonic stages, neural retina is the major, if not only, source of vitreous type II collagen, whereas later in development the synthetic activity appears to reside in the vitreous. Independent studies showed that chicken neural retinal cells in culture synthesize at least two genetically distinct types of collagen, one clearly identified as type II and the other composed of multiple, very high-molecular-weight pepsin-resistant forms of collagen with covalent nonhelical regions (Linsenmayer and Little, 1978).

More recently, the use of type-specific monoclonal antibodies that recognize different epitopes in the helical domains of collagen have been used histochemically to demonstrate the coordinate deposition by the neural retina of collagen types II and IX in developing chicken vitreous (Fitch *et al.*, 1988). Type IX collagen is a non-fibril-forming molecule composed of three helical and four nonhelical domains; it also serves as a core protein for a single glycosaminoglycan side chain covalently bound to its α2(IX) chain (van der Rest and Mayne, 1988). In cartilage, type IX collagen is linked to type II collagen by a hydroxypyridinium cross-link that is very close to the glycosaminoglycan attachment site in the nonhelical NC3 domain. These findings led to a model in which type IX collagen, oriented along the surface of type II collagen fibrils, may function as a bridge, linking these fibrils to each other and to nonfibrillar matrix components.

Whether this is also true for the vitreous remains to be established, but further insight into the structure of adult chicken vitreous has been obtained by immunoelectron microscopy using rotary shadowing with platinum (Wright and Mayne, 1988). Fine details of two independent fibrillar systems could be distinguished by these sensitive and specific techniques. One is the major type II collagen measuring 27.9 nm in diameter; this fibril is completely coated by a chondroitinase ABC-susceptible proteoglycan. These studies also showed that about 20–30% of the free collagen fibrils in chicken vitreous have the characteristic dimensions and morphology of type IX collagen; its identity was further established by the isolation of pepsin-resistant type IX collagen fragments. Another observation was the close association of tenascin molecules with the chondroitin sulfate of some but not all of the collagen fibrils (see discussion by Wright and Mayne, 1988).

It is interesting to point out that a chondroitinase ABC-sensitive chondroitin sulfate proteoglycan was detected many years ago in the vitreous (Allen *et al.*, 1977) and in retrospect probably represented the proteoglycan located in the nonhelical domain of type IX collagen. The second, less abundant, fibril in chicken vitreous has been termed a "beaded fibril"; its average diameter is 22.6 nm, and it exists in open and closed forms with axial periodicities of 35 to 50 nm and 49.8 nm, respectively. These structures may be closely related, if not identical, to the zonular fibrils. The latter have been characterized as elastin microfilaments, as discussed in Section 6.2.2.

6.2.1.2. Hyaluronic Acid

Hyaluronic acid (HA) is a linear unbranched nonsulfated polyanionic glycosaminoglycan composed entirely of one disaccharide repeating unit: D-glucuronic acid and N-acetyl-D-glucosamine (Fig. 6.2). It is the only glycosaminoglycan that contains no covalently linked core protein. The size and shape of HA in solution in a variety of tissues including the vitreous have been extensively investigated (for reviews see Balazs, 1965; Berman and Voaden, 1970; Swann, 1980; Balazs and Denlinger, 1984). Physicochemical

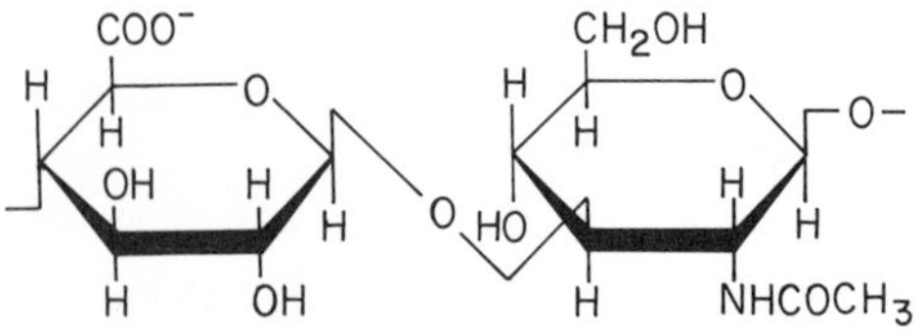

Figure 6.2. Repeating disaccharide unit of hyaluronic acid. The glucuronic acid and N-acetylglucosamine residues are joined by β1,3 glycosidic bonds; the linkage between the disaccharides is a β1,4 glycosidic bond. Hyaluronic acid is a polyanion at physiological pH, the net negative charge resulting from ionization of the carboxyl groups of the glucuronic acid residues.

studies on the dimensions of vitreous HA show that it has a large spheroidal molecular domain with an axial ratio of about 2–3:1 and a diameter of 0.2 to 0.5 μm. Its configuration is that of a somewhat stiff open coil that occupies a large domain in solution. Molecules of this size would overlap and become entangled at concentrations greater than 300–500 μg/ml. As shown in Table 6.2, concentrations of HA in this range are found in many, but not all, species. Such interactions may be especially prominent in the cortical layer, where HA concentrations usually reach considerably high levels.

The M_r range of HA in human, rhesus monkey, and owl monkey vitreous is from 3.0 to 4.5×10^6, and in these species, HA does not seem to be highly polydisperse (Balazs and Denlinger, 1984). Bovine vitreous, on the other hand, is of lower M_r, with a range of about 3.0×10^5 to 2.0×10^6. Polydispersity is suggested not only from light-scattering studies but also from the finding that five to seven discrete species of HA can be isolated by DEAE-Sephadex chromatography; size and/or M_r polydispersity was inferred from the variations in limiting viscosity numbers found among the individual fractions (Berman, 1963). Some very high-M_r species appear to be present in the central region of bovine

Table 6.2. Hyaluronic Acid Content in the Vitreous of Various Species[a]

Species	Hyaluronic acid (μg/ml)
Cattle	110–1070
Sheep	100–400
Dog	40–60
Rabbit	20–50
Guinea pig	10–20
Chicken	15–30
Carp	600–700[b]
Human	100–400
Squid	260–360
Monkey	
Rhesus	100–180
Owl	300–600[c]
	800–900[b]

[a]Adapted from Balazs and Denlinger (1984).
[b]The HA concentration of the thin cortical layer is 800–900 μg/ml.
[c]In the liquid vitreous.

vitreous. However, the possibility has been raised that HA polydispersity, especially in bovine vitreous, arises from artifactuous degradation prior to fractionation (Balazs and Denlinger, 1984). In this context, the question of HA polydispersity has recently been reexamined using DEAE-Sephacel as the fractionating agent; chromatography of HA on this anion-exchange resin yields 10–30 fractions from sources previously considered to be homogeneous, e.g., rooster comb and umbilical cord (Armand and Reyes, 1983; Armand and Chakrabarti, 1987).

Fractionation of HA collected from the central portion of human vitreous (liquid vitreous) and from the cortical region (gel vitreous) on DEAE-Sephacel yields about 15–20 discrete fractions, with little variation in either chemical composition or limiting viscosity number. This supports the view of polydispersity but suggests that it is not related to either chemical composition or M_r of the HA molecules. The physical basis for the appearance of multiple fractions of HA after DEAE-Sephacel (or DEAE-Sephadex) chromatography may be different binding affinities of individual species for the resin, which allows them to be selectively eluted at different salt concentrations and/or ionic strengths. This would be consistent with differing charge densities and/or conformations among the various molecular species, and circular dichroism (CD) measurements of these samples suggests such possibilities (Armand and Chakrabarti, 1987). In these studies several HA fractions were found to display CD minima at 210 nm, with molar ellipticity values ranging from about 13 to 17 $\times$ 10^3 deg·cm^2/dmol. However, HA fractions from liquid vitreous showed lower ellipticity values than those from gel vitreous and in addition displayed a weak positive signal above 240 nm. Thus, there appear to be subtle but measurable conformational differences between gel (cortical) and liquid (central) vitreous hyaluronic acid.

Hyaluronic acid is thought to play an important role in maintaining the integrity of the gel vitreous. The molecular organization of human, monkey, cattle, sheep, and fish gel vitreous is perceived as a collagen network suspended in a relatively concentrated, viscous HA solution (Balazs and Denlinger, 1984). The space between the collagen fibrils is entirely filled with large hydrated HA molecules (see Fig. 6.1), and there is considerable frictional interaction between the two structural components. In these species, HA stabilizes the vitreous gel, but in others such as rabbit, dog, cat, rodents, and chicken, the HA concentration appears to be too low to serve a similar function. Instead, the relatively high concentration of collagen may be sufficient to maintain the gel state without secondary interaction with HA.

Vitreous liquefaction, or synchysis, is known to occur in aging in the normal human eye and is also associated with pathological processes such as vitreous detachment and rhegmatogenous retinal detachment. The latter are not necessarily age-related disorders. Synchysis is thought to be caused by changes in the chemical state and/or conformation of HA; additionally, there may be alterations in its interaction with collagen fibrils in those species in which HA contributes to the stabilization of the gel vitreous (human, monkey, cattle, sheep, and fish). Although this question has been addressed many times for several decades using a variety of experimental approaches (see review by Balazs and Denlinger, 1984), a recent study has reinvestigated the possible role of light in generating active species of oxygen that could effect gel-to-liquid transformations in the vitreous (Ueno *et al.*, 1987). For this purpose, rabbits were injected *in vivo* with the photosensitizer riboflavin, and calf vitreous was treated *in vitro* with either riboflavin or methylene blue. After irradiation for periods of up to 6 h, substantial liquefaction was observed in calf vitreous but almost none in rabbit vitreous. Hyaluronidase (without irradiation) was far more effective in causing vitreous liquefaction in calf vitreous than either of the photosensitizers. These findings imply that loss of gel vitreous structure can result from extensive depolymerization of HA by hyaluronidase, whereas the minimal chemical and/or conformational changes induced by photosensitizing reactions are less dramatic although they may be more physiologically relevant. The failure to induce liquefaction in rabbit vitreous *in vivo* may reflect its low concentration in the vitreous and hence its questionable role in maintaining the gel structure. Alternatively, free radical scavengers may be present in the living eye that could prevent a buildup of photochemically induced active oxygen species.

6.2.1.3. Soluble Proteins

The vitreous contains about 12 or more noncollagenous proteins and glycoproteins, some originating from the serum and some tissue-specific. Descriptions of early immunochemical studies have been given in several reviews (Swann, 1980, 1987; Balazs and Denlinger, 1984). The concentration of soluble protein in a number of species (e.g., dog, lamb, calf, and pig) appears to be remarkably constant, ranging from about 1.0 to 1.2 mg/ml (Chen and Chen, 1981). These levels are only about 1–3% of those present in serum or plasma. Regardless of the analytical methods used, serum albumin is the major single vitreous protein found in most if not all species examined to date.

Analyses of soluble proteins from calf, sheep, human, bovine, and rabbit vitreous by SDS-PAGE reveal rather strikingly different protein patterns from one species to the next, but in all cases the electrophoretic profiles are readily distinguishable from those of serum (Swann, 1980). Fractionation of vitreous supernatant on a cesium chloride gradient (to remove hyaluronic acid) followed by gel chromatography and SDS-PAGE reveals a total of 15 discrete protein bands in bovine, rabbit and sheep vitreous. However, only five of the polypeptides were found in all three species; the other ten proteins were present at varying levels or were not detectable at all in some of the species. Other studies using electrofocusing to resolve the proteins in fetal calf, dog, and adult bovine vitreous showed some similarities between vitreous and serum proteins of pI <6.0 (Chen and Chen, 1981). Here too, no clear-cut conclusions could be drawn on either the identity or the origin of vitreous proteins.

Analyses of rat vitreous using disk electrophoresis to resolve the protein mixtures showed marked differences between vitreous and serum (Hawkins, 1986). Immunoelectrophoresis revealed that of all the proteins present, only albumin and transferrin were immunologically identical in the two fluids. Transferrin is also thought to be present in other species, and the question of iron-binding proteins was further examined in monkey vitreous using isoelectric focusing and immunoelectrophoretic techniques (van Bockxmeer *et al.*, 1983). The concentration of proteins in rhesus monkey vitreous is about 217 μg/ml, considerably lower than that in most other species. Of this, about 40% is serum albumin, and 30% appears to consist of iron-binding proteins: transferrin and/or lactoferrin. Both migrate as 78-kDa proteins on SDS-PAGE, but lactoferrin exhibits anomalous electrophoretic behavior. The iron-binding proteins can, however, be distinguished from one another by immunoelectrophoretic techniques, and the conclusion drawn from these studies was that lactoferrin is a major iron-binding protein of monkey vitreous. The latent iron-binding capacity of the monkey vitreous lactoferrin–transferrin system was calculated to be 60 ng Fe per eye, equivalent to the hemoglobin iron content of

570,000 erythrocytes or 0.1 μl of blood. Thus, iron-binding proteins appear to be major components of the vitreous and serve important protective roles in cases of vitreous hemorrhage or iron foreign body toxicity.

The vitreous has a relatively high content of soluble glycoproteins, estimated as about 20% of the total protein (Balazs and Denlinger, 1984). Glycoproteins are thought to originate from ocular tissues surrounding the vitreous and not from the blood. Some have been partially characterized in bovine vitreous. Interphotoreceptor retinol-binding protein (IRBP) (see Chapter 7, Section 7.3.3), a 140-kDa glycolipoprotein, has been detected in monkey vitreous using an ELISA for its quantitation (Wiggert *et al.*, 1986). It was further identified by Western blot and [^{3}H]retinol-binding studies. This retinol-binding protein is synthesized in the photoreceptor cells of the retina and in the pineal, and its presence in the vitreous (and aqueous), although somewhat unexpected, suggests that it may have functional roles outside of the retina and the pineal. Another glycoprotein isolated from bovine vitreous is a 5.7-kDa peptide that inhibits both angiogenesis and collagenase activity (Taylor and Weiss, 1985). The exact contribution of these glycoproteins, as well as others, to the total pool of vitreous glycoproteins cannot be calculated from available data.

Two other proteins that may have important functions in the vitreous have recently been characterized using specifically developed methods for their identification. One is a neutrophil elastase inhibitor, distinguishable from serum inhibitors by several criteria and relatively specific for human leukocyte elastase (Arsenis *et al.*, 1988). Inhibitors of this type may play a role in resisting and/or modulating neovascularization in the vitreous. Another protein, tissue plasminogen activator (t-PA), is a potent fibrinolytic agent. Although detectable in only trace amounts using a [^{125}]fibrin-coated well assay, development of an ELISA led to its quantitation in dog and calf vitreous; very little is present in monkey vitreous (Geanon *et al.*, 1987). The origin of t-PA in the vitreous is unknown, but its presence in the vitreous, albeit at very low levels, suggests a fibrolytic role in the event of vitreous hemorrhage, or other function(s) such as destructive remodeling or turnover of vitreous matrix components.

New approaches to an understanding of the structural and functional organization of the vitreous are being developed, and one of them, *in situ* circular dichroism (CD) spectroscopy, may be applicable to studies of vitreous proteins (Ueno and Chakrabarti, 1988). The CD spectrum of intact vitreous shows a minimum at 206 nm with a shoulder at 220 nm and one small positive peak at about 252 nm. Analysis of the individual components in the spectrum shows that noncollagen protein makes the major contribution to the overall dichroic strength of the vitreous; contributions from HA and collagen are small but measurable. Further development of this technique may provide a tool for evaluating possible structural changes in vitreous proteins in aging or in pathological conditions.

6.2.2. Zonules

The zonules, or zonular fibers, in the anterior vitreous are distinctly different chemically, structurally, and immunologically from the collagenous fibrils. Early studies showed that they are resistant to collagenase digestion, contain no hydroxyproline, and have a much higher cystine content than vitreous collagen; morphologically the zonular fibers bear a strong resemblance to microfibrils associated with elastin (see reviews by Pirie, 1969; Swann, 1980).

The finding that ocular zonular fibrils and elastic tissue microfibrils are related antigenically established a close similarity, if not identity, between the two structures (Streeten *et al.*, 1981). Antibody to bovine zonule raised in rabbits shows not only heavy binding but also a distinct 35- to 45-nm periodicity in microfibrils of elastic tissue in human and bovine ciliary body, calf ligamentum nuchae, and chick aorta. Biochemical analyses of human and bovine zonules have corroborated immunologic and morphological studies (Streeten *et al.*, 1983). The zonules were shown to be composed of noncollagenous acidic glycoprotein with an unusually high cystine content; other peptides with electrophoretic mobilities similar to those found in elastic microfibrils in other tissues were also detected.

In later studies of the extractable proteins of bovine zonular fibers using immunoblotting and SDS-PAGE, two PAS-positive 32-kDa and 250-kDa glycoproteins were identified as the major nonserum components (Streeten and Gibson, 1988). A 67-kDa band was characterized by immunologic reaction and electrophoretic mobility as bovine serum albumin; in addition, a 65-kDa band was detected that may represent vitronectin, a common adhesion protein found in serum in addition to the major one, fibronectin. The latter does not appear to be present in the zonular extracts examined in these studies. To date, two distinctive proteins have been characterized in microfibrils in other tissues; one of them is 31-kDa microfibrillar-associated glycoprotein (MAGP), and the other is 350-kDa fibrillin (Sakai *et al.*, 1986). Other components thought to be present in the microfibrils surrounding elastic tissues are lysyl oxidase and a 35-kDa protein with amine oxidase activity (see Mecham *et al.*, 1988). However, some tissues (e.g., ocular zonules, ligaments, and tendons) contain microfibrillar structures that are not associated with elastin but share common antigenic determinants with elastin-containing microfibrils. The development of a library of monoclonal antibodies to antigens in bovine ocular zonule should clarify some of these questions (Mecham *et al.*, 1988). In the studies of Streeten and Gibson (1988), no firm evidence could be obtained for the presence of fibronectin in bovine zonular extracts; moreover, neither fibrillin nor MAGP could be clearly identified.

The morphological studies of Wright and Mayne (1988) suggest that the "beaded fibrils" observed in electron micrographs of zonular fibers from chicken vitreous prepared after rotary shadowing with platinum may be closely related, if not identical, to elastin microfibrils in bovine zonules and other tissues. These fibrils are 22.6 nm in diameter and have an axial periodicity of 49.8 nm in the closed form and and 35 to 50 nm in the open form. Further morphological studies combined with biochemical and immunologic investigations should in the future provide the additional data needed for a precise characterization of zonular fibers.

6.2.3. Lipids

The pellet obtained after centrifugation of rabbit vitreous has been reported to contain significant amounts of saline- and guanidine-extractable palmitic and stearic acid; these fatty acids accounted for about 7% (w/w) of the total residual fraction (Swann *et al.*, 1975). The chloroform:methanol-extractable lipid content of insoluble residues of bovine, sheep, and dog vitreous varies from about 1–14% of the total when expressed as micrograms per milliliter of native vitreous (Swann, 1980). Fatty acids have also been identified in human vitreous (Reddy *et al.*, 1986). As in the rabbit, the major saturated fatty acids are palmitate (24–27 mol %) and stearate (16–20 mol %), and the major unsaturated fatty acids are oleate (22–24 mol %) and arachidonate (14–20 mol %). There appears to be little variation with age. Evidence for active lipid metabolism was also found in these studies (see Section 6.3).

6.2.4. Low-Molecular-Weight Solutes

Ions and organic solutes present in the vitreous are thought to originate from two principal sources: adjacent ocular tissues (lens, ciliary epithelium, and retina) and blood plasma (for reviews see Gloor, 1981; Balazs and Denlinger, 1984). The barriers that control the entry of substances into the vitreous are located in the vascular endothelium of the retinal vessels, the pigment epithelium of the retina, and the inner layer of the ciliary epithelium (Rodriguez-Peralta, 1968). Although very small amounts of protein pass these barriers, the entry of low-molecular-weight substances does not seem to be impeded.

The concentrations of the major ions of rabbit vitreous, Na^+ and Cl^-, are similar to those found in aqueous and plasma (Table 6.3). There are, however, species differences in other solutes, although relatively few detailed studies are available. Ascorbic acid is known to be actively secreted by the ciliary epithelium into the posterior aqueous, and from there it is thought to diffuse into the vitreous. Nevertheless, studies on inflamed eyes show a decrease in ascorbate concentration in the aqueous and an increase in the vitreous (McGahan, 1985). Moreover, there is no correlation between ascorbate levels in the aqueous and vitreous from noninflamed control eyes. These findings cast some doubt on the commonly held view that vitreous ascorbate is derived mainly from the posterior aqueous. The concentrations of free amino acids in the vitreous have been measured in rabbits, cattle, and other species and found to be considerably lower than in either the aqueous or the plasma (see review by Balazs and Denlinger, 1984). A recent study corroborates their low level in rat vitreous as well (Heinamaki and Lindfors, 1988). Although small molecules such as amino acids would be expected to diffuse freely throughout the vitreous and hence display homogeneous distribution, several studies have shown that their concentration is lower in the posterior than in the anterior vitreous. This may be the consequence of active posterior transport as well as utilization by the retina.

Table 6.3. Concentrations (mM) of Low-Molecular-Weight Solutes in the Vitreous, Aqueous, and Plasma of Rabbit[a]

Solute	Vitreous	Aqueous[b]	Plasma
Sodium	134	136	143
Potassium	9.5	5.0	5.6
Chloride	104	96	97
Phosphate	0.4	0.6	2
Ascorbate	0.4	1.4	0.04
Glucose	3.0	5.6	5.7
Lactate	12.0	9.9	10.3

[a]Adapted from Gloor (1981), Moorhead (1983), and Balazs and Denlinger (1984).
[b]Analytical values for posterior aqueous except potassium, which is the concentration measured in the anterior aqueous.

6.3. METABOLISM

Our knowledge of the metabolic characteristics of the vitreous and of the dynamic processes controlling the synthesis and maintenance of its major macromolecular components, collagen and hyaluronic acid, is very meager. As described in Section 6.2.1.1, type II vitreous collagen in the chicken is synthesized in early embryonic stages by the retina and later in development by the vitreous. However, there are no biochemical data available on the metabolism of the collagen network in normal adult vitreous. After vitrectomy it is generally assumed that the gel-forming collagen fibrils do not regenerate (Balazs and Denlinger, 1984). Nevertheless, electron microscopic findings suggest that fibroblasts and fibrocytes present in human vitreous cortex at the ora serrata may secrete collagen-like fibrils (Gartner, 1986). The contribution of these cells, which form only about 10% of the total vitreous cell population, to the overall metabolism of vitreous structure remains to be established.

The presence of cells on the surface of mature vitreous has been recognized for more than a century; historical perspectives, as well as descriptions of the distribution and phagocytic activities of these cells, have appeared in several articles (see, for example, Szirmai and Balazs, 1958; Hamburg, 1959). It is generally acknowledged that the cells present in the vitreous cavity during embryonic development are of mesenchymal origin; they are thought to be responsible for the synthesis of the primary vitreous, the hyaloid anteriole, and other components, but after regression of the hyaloid artery, they are not replaced by a "second generation" of similar cells (Swann, 1980). Rather, they die out following atrophy of the vascular supply, and afterward the vitreous is populated by a small number of transient cells. The cells detected in the cortical layer of adult vitreous and called hyalocytes (Bloom and Balazs, 1965) are best defined as macrophages (see discussions by Swann, 1980; Gloor, 1981; Streeten, 1982). Hyalocytes isolated from human vitreous show strong adherence to glass and plastic surfaces; they also display phagocytic properties and in addition have receptors for IgG and complement components (Grabner *et al.*, 1980). There seems little doubt that they are hematogenous in origin, derived from mature mononuclear phagocytes that pass from the retinal vessels into the cortical layer of the vitreous (Swann, 1980).

Crude enzyme extracts from cellular (and other) material sedimenting after high-speed centrifugation of calf vitreous contain transferase activities capable of synthesizing small amounts of oligosaccharide as well as hyaluronic acid (HA), if the latter is present endogenously (Osterlin and Jacobson, 1968a). Transferase activity has also been detected in the extracellular matrix of the vitreous (Osterlin and Jacobson, 1968b), and additional evidence implying enzyme-like activity in acellular portions of the vitreous came from studies using labeled glucose as a precursor (Berman and Gombos, 1969). In this case the glucose appeared to be incorporated into nondialyzable substances tentatively identified as proteins and glycoproteins; very little, if any, turnover of hyaluronic acid could be detected. Little difference in metabolic activity could be found in cellular (cortical) or in acellular (central) regions of the vitreous.

In addition to these studies on HA, proteins, and glycoproteins, lipid metabolism has also been investigated in the vitreous (Reddy *et al.*, 1986). Aspirates of human vitreous and pellets obtained after centrifugation of dog vitreous showed active incorporation of labeled arachidonic acid into glycerolipids. Although arachidonic acid is not metabolized by either the cyclooxygenase or lipoxygenase pathways in the vitreous, it may undergo autooxidation. The site of lipid metabolism, whether cellular or extracellular, was not examined in these studies.

Other approaches to vitreous metabolism have utilized *in vivo* measurements of turnover and synthesis of HA in the owl monkey, an animal with a predominantly liquid vitreous that can be sampled or replaced without causing undue trauma. In early experiments, gel vitreous containing HA was removed and replaced with an equivalent amount of saline; evidence for HA regeneration several months later was reported (see review by Balazs and Denlinger, 1984). With labeled glucosamine used as a precursor of HA, some incorporation could be detected within 24 h but there was little turnover after 72 days (Osterlin, 1968, 1969). Regeneration of HA, if it occurs, is an extremely slow process; 3 weeks after removal of vitreous, there is a net loss of HA, and evidence for its renewal was found in only a limited number of animals (Osterlin, 1969). In fact, other studies using essentially the same technique in the owl monkey showed that after aspiration of vitreous and saline replacement, HA was not regenerated at all for as long as 6 months (Swann and Constable, 1972b). There was, however, a large influx of serum protein in response to the unavoidable tissue injury resulting from aspiration of vitreous samples. Similar results had been obtained previously in the cat; in this case also, replacement of the vitreous with air, HA, or synthetic implants gave evidence for inflammatory responses and degenerative changes, judging from increased levels of protein and loss of HA over and above the small amounts that had been aspirated (Gombos and Berman, 1967). It appears that breakdown of the vitreous matrix results in diffusion of HA out of the eye anteriorly through the aqueous (Balazs and Denlinger, 1984). Because of conflicting results from different laboratories, the question of vitreous regeneration in adult animals has yet to be satisfactorally resolved.

In summary, despite its outward simplicity, it has been difficult to characterize the biochemical mechanisms that may be responsible for the turnover and/or remodeling of the structural components of mature vitreous. Vitreous hyalocytes, like any other macrophages, would be expected to have phagocytic, degradative, and possibly synthetic activities. Whether they are indeed responsible for maintaining the macromolecular composition of the vitreous remains to be established. Although enzyme-like activities have been detected in cellular preparations of the vitreous, observations that acellular regions may also be "metabolically active" raise many questions on vitreous metabolism in general and on the putative role of cortical layer cells in particular on the biosynthesis or turnover of vitreous components.

6.4. AGING

Many changes, both biochemical and structural, are now known to be associated with normal aging of the vitreous, but the process appears to be limited to certain primates. For example, in mammals such as cattle, dogs, sheep, cats, rabbits, and most monkeys, there appear to be few if any physical or rheological changes after the eye has reached maturity (Balazs and Denlinger, 1984). On the other hand, there are important changes in rheological properties, chemical composition, and molecular structure in human and rhesus monkey vitreous. One of the most characteristic changes in humans is a decrease in the volume of gel vitreous and a concomitant increase in liquid vitreous commencing at about the age of 45–50 (Green *et al.*, 1984; Balazs and Denlinger, 1984).

The collagen content does not decrease with age, but because of the decrease in volume of gel vitreous, its concentration increases. The volume decrease is thought to be the result of collapse or contraction (syneresis) of the collagen network. This reorganization of the physical structure of the vitreous is often considered one of the major causes of posterior vitreous detachment (PVD) observed clinically in aging persons as well as in about 30% of postmortem eyes from individuals over the age of 60. A study of 61 postmortem eyes has shown that the incidence and/or degree of PVD is in most cases directly related to an increase in the amount of liquid vitreous, as assessed by either slit-lamp examination or dissection (Larsson and Osterlin, 1985). Another study on a series of normal postmortem eyes using dark-field horizontal slit illumination has confirmed the commonly held view that the vitreous is nearly homogeneous in structure in most individuals until the age of about 30, but during the following decades, vitreous tracts appearing as macroscopic fibers oriented in the anterior–posterior axis develop in many individuals (Sebag, 1987). Morphological and ultrastructural studies on normal adult middle-aged vitreous show that these tracts are fibers and not membranes; they appear to represent aggregates of collagen fibrils that had previously been randomly dispersed throughout the vitreous (Sebag and Balazs, 1989).

The foregoing studies support the view that in most individuals beyond the age of 50 to 60 years, the vitreous volume decreases. Syneresis accompanied by increasing incidence of PVD is a common observation. The sequence of events leading to PVD and the clinical importance of molecular changes in the vitreous preceding (or perhaps accompanying) PVD are only now beginning to be fully appreciated. The development of new techniques such as proton nuclear magnetic resonance (NMR) (Gonzalez *et al.*, 1984) and CD (Armand and Chakrabarti, 1987) should in the future clarify more precisely the physicochemical basis of the molecular reorganization of vitreous structure that occurs in the aging human eye.

Changes in either concentration, size, or conformation of HA may also be involved in the development of syneresis and PVD with advancing age. The concentration of HA increases gradually with age; it is about 10–20% higher in individuals 50 years of age or older than in 20- to 40-year-olds (Berman and Michaelson, 1964; Green *et al.*, 1984). No significant difference in HA concentration could be detected in gel or liquid vitreous in postmortem eyes with either complete or partial PVD (Larsson and Osterlin, 1985); there could, however, be subtle conformational differences in gel versus liquid HA (Armand and Chakrabarti, 1987). The investigation of Larsson and Osterlin (1985) revealed that in a small number of eyes with no PVD, the HA concentration in the total vitreous was higher than in eyes with total PVD. Also of interest was the observation that females have a significantly lower concentration of HA than males.

Studies on rhesus monkeys ranging in age from 1 to 19 years show that this animal is an excellent model of human vitreous with respect to physical state and collagen and HA content (Denlinger *et al.*, 1980). Small age-associated changes in HA concentration in the rhesus monkey appear to be similar to those observed in human vitreous. Biomicroscopic examinations reveal the development of liquid pockets and vitreous tracts with advancing age, the latter similar in many respects to the macroscopic fibers observed in humans (Larsson and Osterlin, 1985; Sebag, 1987; Sebag and Balazs, 1989). There is, however, one important difference: whereas posterior vitreous detachment is common in humans beyond the age of 60, this was not observed in rhesus monkeys of an equivalent age, i.e., animals about 21 years of age.

6.5. REFERENCES

Allen, W. S., Otterbein, E. C., and Wardi, A. H., 1977, Isolation and characterization of the sulfated glycosaminoglycans of the vitreous body, *Biochim. Biophys. Acta* **498:**167–175.

Armand, G., and Chakrabarti, B., 1987, Conformational differences between hyaluronates of gel and liquid human vitreous: Fractionation and circular dichroism studies, *Curr. Eye Res.* **6:**445–450.

Armand, G., and Reyes, M., 1983, A new chromatographic method for the fractionation of hyaluronic acid, *Biochem. Biophys. Res. Commun.* **112:**168–175.

Arsenis, C., Kuettner, K. E., and Eisenstein, R., 1988, Isolation and partial characterization of neutrophil elastase inhibitors from bovine vitreous and aorta, *Curr. Eye Res.* **7:**95–102.

Ayad, S., and Weiss, J. B., 1984, A new look at vitreous-humour collagen, *Biochem. J.* **218:**835–840.

Balazs, E. A., 1961, Molecular morphology of the vitreous body, in: *The Structure of the Eye* (G. K. Smelser, ed.), Academic Press, New York, pp. 293–310.

Balazs, E. A., 1965, Amino sugar-containing macromolecules in the tissues of the eye and ear, in: *The Amino Sugars,* Vol. 2A (E. A. Balazs and R. W. Jeanloz, eds.), Academic Press, New York, pp. 401–460.

Balazs, E. A., and Denlinger, J. L., 1984, The vitreous, in: *The Eye,* Vol. Ia (H. Davson, ed.), Academic Press, New York, pp. 533–589.

Balazs, E. A., Laurent, T. C., and Laurent, U. B. G., 1959, Studies on the structure of the vitreous body, *J. Biol. Chem.* **234:**422–430.

Berman, E. R., 1963, Studies on mucopolysaccharides in ocular tissues. I. Distribution and localization of various molecular species of hyaluronic acid in the bovine vitreous body, *Exp. Eye Res.* **2:**1–11.

Berman, E. R., and Gombos, G. M., 1969, Studies on the incorporation of U-^{14}C-glucose into vitreous polymers *in vitro* and *in vivo*, *Invest. Ophthalmol.* **8:**521–534.

Berman, E. R., and Michaelson, I. C., 1964, The chemical composition of the human vitreous body as related to age and myopia, *Exp. Eye Res.* **3:**9–15.

Berman, E. R., and Voaden, M., 1970, The vitreous body, in: *Biochemistry of the Eye* (C. N. Graymore, ed.), Academic Press, New York, pp. 373–471.

Bloom, G. D., and Balazs, E. A., 1965, An electron microscopic study of hyalocytes, *Exp. Eye Res.* **4:**249–255.

Burke, J. M., 1980, An analysis of rabbit vitreous collagen, *Connective Tissue Res.* **8:**49–52.

Chen, C.-H., and Chen, S. C., 1981, Studies on soluble proteins of vitreous in experimental animals, *Exp. Eye Res.* **32:**381–388.

Denlinger, J. L., Eisner, G., and Balazs, E. A., 1980, Age-related changes in the vitreous and lens of rhesus monkeys (*Macaca mulatta*), *Exp. Eye Res.* **31:**67–79.

Fitch, J. M., Mentzer, A., Mayne, R., and Linsenmayer, T. F., 1988, Acquisition of type IX collagen by the developing avian primary corneal stroma and vitreous, *Dev. Biol.* **128:**396–405.

Gartner, J., 1986, Electron-microscopic study on the fibrillar network and fibrocyte–collagen interactions in the vitreous cortex at the ora serrata of human eyes with special regard to the role of disintegrating cells, *Exp. Eye Res.* **42:**21–33.

Geanon, J. D., Tripathi, B. J., Tripathi, R. C., and Barlow, G. H., 1987, Tissue plasminogen activator in avascular tissues of the eye: A quantitative study of its activity in the cornea, lens, and aqueous and vitreous humors of dog, calf and monkey, *Exp. Eye Res.* **44:**55–63.

Gloor, B. P., 1981, The vitreous, in: *Adler's Physiology of the Eye* (R. A. Moses, ed.), C. V. Mosby, St. Louis, pp. 255–276.

Gombos, G. M., and Berman, E. R., 1967, Chemical and clinical observations on the fate of various vitreous substitutes, *Acta Ophthalmol.* **45:**794–803.

Gonzalez, R. G., Cheng, H.-M., Barnett, P., Aguayo, J., Glaser, B., Rosen, B., Burt, C. T., and Brady, T., 1984, Nuclear magnetic resonance imaging of the vitreous body, *Science* **223:**399–400.

Grabner, G., Boltz, G., and Forster, O., 1980, Macrophage-like properties of human hyalocytes, *Invest. Ophthalmol. Vis. Sci.* **19:**333–340.

Green, K., Balazs, E. A., and Denlinger, J. L., 1984, Aqueous humor and vitreous production, in: *Geriatrics 3* (D. Platt, ed.), Springer-Verlag, New York, pp. 352–372.

Hamburg, A., 1959, Some investigations on the cells of the vitreous body, *Ophthalmologica* **138:**81–107.

Hawkins, K. N., 1986, Contribution of plasma proteins to the vitreous of the rat, *Curr. Eye Res.* **5:**655–663.

Heinamaki, A. A., and Lindfors, A. S. H., 1988, Free amino acids in rat ocular tissues during postnatal development, *Biochem. Int.* **16:**405–412.

Larsson, L., and Osterlin, S., 1985, Posterior vitreous detachment. A combined clinical and physicochemical study, *Albrecht von Graefes Arch. Klin. Exp. Ophthalmol.* **223:**92–95.

Linsenmeyer, T. F., and Little, C. D., 1978, Embryonic neural retina collagen: *In vitro* synthesis of high molecular weight forms of type II plus a new genetic type, *Proc. Natl. Acad. Sci. USA* **75:**3235–3239.

Machemer, R., 1988, Proliferative vitreoretinopathy (PVR): A personal account of its pathogenesis and treatment, *Invest. Ophthalmol. Vis. Sci.* **29:**1771–1782.

McGahan, M. C., 1985, Ascorbic acid levels in aqueous and vitreous humors of the rabbit: Effects of inflammation and ceruloplasmin, *Exp. Eye Res.* **41:**291–298.

Mecham, R. P., Hinek, A., Cleary, E. G., Kucich, U., Lee, S. J., and Rosenbloom, J., 1988, Development of immunoreagents to ciliary zonules that react with protein components of elastic fiber microfibrils and with elastin-producing cells, *Biochem. Biophys. Res. Commun.* **151:**822–826.

Moorehead, L. A., 1983, Vitreous, in: *Biochemistry of the Eye* (R. E. Anderson, ed.), American Academy of Ophthalmology, San Francisco, pp. 145–163.

Moseley, H., and Foulds, W. S., 1982, The movement of xenon-133 from the vitreous to the choroid, *Exp. Eye Res.* **34:**169–179.

Moseley, H., Foulds, W. S., Allan, D., and Kyle, P.M., 1984, Routes of clearance of radioactive water from the rabbit vitreous, *Br. J. Ophthalmol.* **68:**145–151.

Newsome, D. A., Linsenmayer, T. F., and Trelstad, R. L., 1976, Vitreous body collagen. Evidence for a dual origin from the neural retina and hyalocytes, *J. Cell Biol.* **71:**59–67.

Osterlin, S. E., 1968, The synthesis of hyaluronic acid in vitreous. III. *In vivo* metabolism in the owl monkey, *Exp. Eye Res.* **7:**524–533.

Osterlin, S. E., 1969, The synthesis of hyaluronic acid in the vitreous. IV. Regeneration in the owl monkey, *Exp. Eye Res.* **8:**27–34.

Osterlin, S. E., and Jacobson, B., 1968a, The synthesis of hyaluronic acid in vitreous. I. Soluble and particulate transferases in hyalocytes, *Exp. Eye Res.* **7:**497–510.

Osterlin, S. E., and Jacobson, B., 1968b, The synthesis of hyaluronic acid in vitreous. II. The presence of soluble transferase and nucleotide sugar in the acellular vitreous gel, *Exp. Eye Res.* **7:**511–523.

Pirie, A., 1969, The vitreous body, in: *The Eye*, Vol. 1 (H. Davson, ed.), Academic Press, New York, pp. 273–297.

Pirie, A., Schmidt, G., and Waters, J. W., 1948, Ox vitreous humour. I. The residual protein, *Br. J. Ophthalmol.* **32:**321–339.

Reddy, T. S., Birkle, D. L., Packer, A. J., Dobard, P., and Bazan, N. G., 1986, Fatty acid composition and arachidonic acid metabolism in vitreous lipids from canine and human eyes, *Curr. Eye Res.* **5:**441–447.

Rodriguez-Peralta, L. A., 1968, Hematic and fluid barriers of the retina and vitreous body, *J. Comp. Neurol.* **132:**109–124.

Sakai, L. Y., Keene, D. R., and Engvall, E., 1986, Fibrillin, a new 350-kD glycoprotein, is a component of extracellular microfibrils, *J. Cell Biol.* **103:**2499–2509.

Sebag, J., 1987, Age-related changes in human vitreous structure, *Albrecht von Graefes Arch. Klin. Exp. Ophthalmol.* **225:**89–93.

Sebag, J., and Balazs, E. A., 1989, Morphology and ultrastructure of human vitreous fibers, *Invest. Ophthalmol. Vis. Sci.* **30:**1867–1871.

Steinberg, R. H., 1986, Research update: Report from a workshop on cell biology of retinal detachment, *Exp. Eye Res.* **43:**695–706.

Streeten, B. W., 1982, Disorders of the vitreous, in: *Pathobiology of Ocular Disease* (A. Garner and G. K. Klintworth, eds.), Marcel Dekker, New York, pp. 1383–1419.

Streeten, B. W., and Gibson, S. A., 1988, Identification of extractable proteins from the bovine ocular zonule: Major zonular antigens of 32kD and 250kD, *Curr. Eye Res.* **7:**139–146.

Streeten, B. W., Licari, P. A., Marucci, A. A., and Dougherty, R. M., 1981, Immunohistochemical comparison of ocular zonules and the microbifrils of elastic tissue, *Invest. Ophthalmol. Vis. Sci.* **21:**130–135.

Streeten, B. W., Swann, D. A., Licari, P. A., Robinson, M. R., Gibson, S. A., Marsh, N. J., Vergnes, J.-P., and Freeman, I. L., 1983, The protein composition of the ocular zonules, *Invest. Ophthalmol. Vis. Sci.* **24:**119–123.

Swann, D. A., 1980, Chemistry and biology of the vitreous body, *Int. Rev. Exp. Pathol.* **22:**2–64.

Swann, D. A., 1987, Biochemistry of the vitreous, *Bull. Soc. Belge Ophthalmol.* **223:**59–72.

Swann, D. A., and Constable, I. J., 1972a, Vitreous structure. I. Distribution of hyaluronate and protein, *Invest. Ophthalmol.* **11:**159–163.

Swann, D. A., and Constable, I. J., 1972b, Vitreous structure. II. Role of hyaluronate, *Invest. Ophthalmol.* **11:**164–168.

Swann, D. A., and Sotman, S. S., 1980, The chemical composition of bovine vitreous-humour collagen fibres, *Biochem. J.* **185:**545–554.

Swann, D. A., Constable, I. J., and Harper, E., 1972, Vitreous structure. III. Composition of bovine vitreous collagen, *Invest. Ophthalmol.* **11:**735–738.

Swann, D. A., Constable, I. J., and Caulfield, J. B., 1975, Vitreous structure. IV. Chemical composition of the insoluble residual protein fraction from the rabbit vitreous, *Invest. Ophthalmol.* **14:**613–618.

Szirmai, J. A., and Balazs, E. A., 1958, Studies on the structure of the vitreous body, *Arch. Ophthalmol.* **59:**34–48.

Taylor, C. M., and Weiss, J. B., 1985, Partial purification of a 5.7K glycoprotein from bovine vitreous which inhibits both angiogenesis and collagenase activity, *Biochem. Biophys. Res. Commun.* **133:**911–916.

Ueno, N., and Chakrabarti, B., 1988, Monitoring in situ circular dichroism of the intact vitreous: A new approach, *J. Biochem. Biophys. Methods* **15:**349–356.

Ueno, N., Sebag, J., Hirokawa, H., and Chakrabarti, B., 1987, Effects of visible-light irradiation on vitreous structure in the presence of a photosensitizer, *Exp. Eye Res.* **44:**863–870.

van Bockxmeer, F. M., Martin, C. E., and Constable, I. J., 1983, Iron-binding proteins in vitreous humour, *Biochim. Biophys. Acta* **758:**17–23.

van der Rest, M., and Mayne, R., 1988, Type IX collagen proteoglycan from cartilage is covalently cross-linked to type II collagen, *J. Biol. Chem.* **263:**1615–1618.

Wiggert, B., Lee, L., Rodrigues, M., Hess, H., Redmond, T. M., and Chader, G. J., 1986, Immunochemical distribution of interphotoreceptor retinoid-binding protein in selected species, *Invest. Ophthalmol. Vis. Sci.* **27:**1041–1049.

Wright, D. W., and Mayne, R., 1988, Vitreous humor of chicken contains two fibrillar systems: An analysis of their structure, *J. Ultrastruct. Mol. Struct. Res.* **100:**224–234.

Retina 7

7.1. INTRODUCTION: GENERAL COMPOSITION AND METABOLISM

Structurally and functionally, the retina consists of two components: (1) the nonneural retinal pigment epithelium, a single layer of hexagonally shaped cells resting on Bruch's membrane, and (2) the tightly apposed, but not anatomically joined, neural retina (Fig. 7.1). The neural retina is a peripheral extension of the forebrain, with which it has common embryological origins. It consists of six distinct neuronal cell types: photoreceptors (rods and cones), bipolar cells, horizontal cells, amacrine cells, interplexiform cells, and ganglion cells. Its principal glial cell is the Müller cell, whose fibers extend from the vitreous surface (or inner limiting "membrane" of the retina) to beyond the outer limiting "membrane," where they surround the photoreceptor inner segments. In histological cross section, the neural retina is organized into ten distinct layers, including the pigment epithelium, as described in the legend to Fig. 7.1.

Operationally and functionally, the neural retina is usually considered as two regions: the outer (light-detecting) photoreceptor cells and the inner (signal-processing) layers. In man, monkey, cat, dog, mouse, and frog, the photoreceptors receive their blood supply from the choroidal circulation, which accounts for about 70–80% of the oxygen consumed by the neural retina; the inner layers of the retina, whose oxygen utilization is far less than that of the photoreceptors, receive their blood supply from the retinal circulation.

The rod visual cells are located in the peripheral areas of the retina in humans, rats, and other nocturnal animals. The rod cells are receptors for dim light (i.e., in night vision) and do not distinguish colors. Rod photoreceptors are thought to provide scotopic visual activity. Cone visual cells, which provide photopic vision, are shorter than rods and usually cone-shaped; in most (but not all) species they are concentrated principally in the central area of the retina. The cones are specialized for color vision and require higher light intensities to function than rods. The proportion of rods to cones varies considerably among species. The human eye contains 100 to 120 million rods and about 6 million cones. Most rodents have rod-dominant retinas, and at the other extreme, the ground squirrel and the western fence lizard are cone-dominant species (Farber *et al.*, 1981).

Several extensive reviews of both general and specialized aspects of retinal biochemistry have provided much of the background for the general organization of this chapter (Graymore, 1965, 1970; Zinn and Marmor, 1979; Anderson, 1983a,b; Basinger and Hoffman, 1983; Fliesler and Anderson, 1983; Lolley, 1983; Redburn and Hollyfield, 1983; Shichi, 1983; Winkler, 1983a; Reif-Lehrer, 1984; Bazan and Reddy, 1985; Bridges and Adler, 1985; Dowling, 1987; Chader, 1989). Other reviews and monographs on specific topics such as photoreceptor proteins and glycoproteins, neurotransmitters, and phototransduction are cited in appropriate sections.

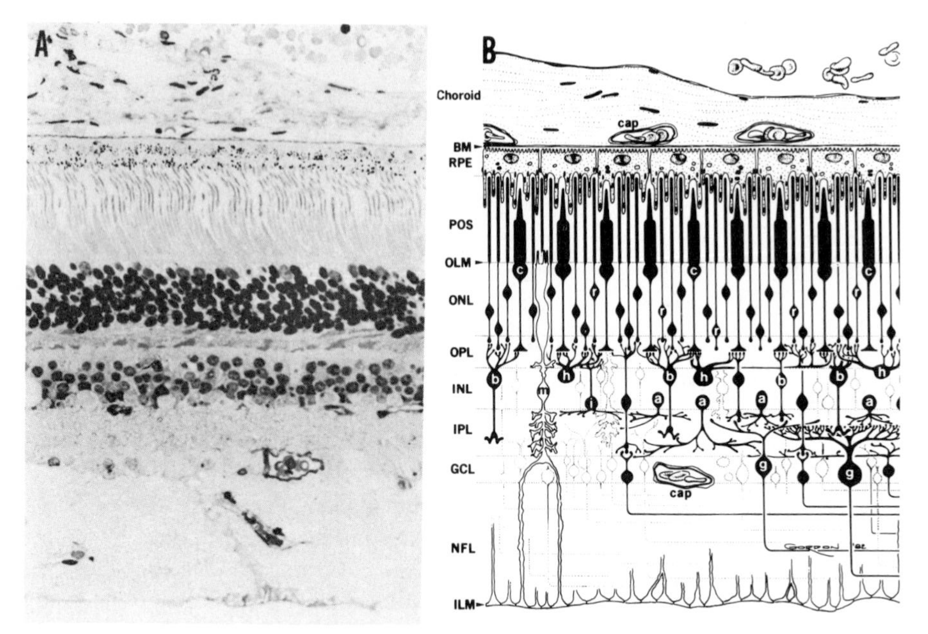

Figure 7.1. Cellular organization of the vertebrate retina. (A) Histological cross-section of a human retina, and (B) a schematic diagram showing the spatial arrangement of the neuronal cell layers. Abbreviations used: a, amacrine cell; b, bipolar cell; c, cone; g, ganglion cell; h, horizontal cell; i, interplexiform cell; r, rod. The cell layers seen histologically consist of: RPE, retinal pigment epithelium; POS, photoreceptor outer segments; OLM, outer limiting "membrane" (a network of Müller cell apical processes in contact with photoreceptor inner segments); ONL, outer nuclear layer (the photoreceptor cell nuclei); OPL, outer plexiform layer (a region of synapses between photoreceptor, bipolar, and horizontal cells); INL, inner nuclear layer; IPL, interplexiform layer (another region of synapses between bipolar, amacrine, and ganglion cells); GCL, ganglion cell layer; NFL, nerve fiber layer (containing the axonal processes of the ganglion cells); ILM, inner limiting "membrane" (the Müller cell processes that terminate on the vitreal surface); M, Müller cell. (From Fliesler and Anderson, 1983.) (Reprinted with permission from *Prog. Lipid Res.* **22,** Steven J. Fliesler and Robert A. Anderson, "Chemistry and metabolism of lipids in the vertebrate retina," © 1983, Pergamon Press PLC.)

Biochemical studies of the nonneural retina, the pigment epithelium, are carried out either on freshly isolated cells or on cultured cells, which appear to be metabolically and functionally viable over several passages. The metabolism of this cell layer has also been studied *in vivo* by following the fate of injected radioactive substances.

The design of *in vitro* experiments on the neural retina is far more complex; with the exception of rod outer segments and the interphotoreceptor matrix, individual cell layers cannot be isolated. Hence, for practical purposes, many studies are carried out on the intact neural retina consisting of the heterogeneous population of neuronal cells shown in Fig. 7.1. In other approaches, outer segments isolated in highly purified form are used extensively, and the "rodless" inner retina remaining after removal of outer segments provides supplementary information. Additionally, studies on subcellular fractions of retinal homogenates are often used, but the fractions isolated (e.g., mitochondria or microsomes) represent organelles derived from all of the cell types present in the retina. Light and electron microscopy and, in increasing use, immunofluorescent and immunocytochemical techniques now provide characterization and localization of specific components. An important *in vivo* approach consists of injecting precursor substances parenterally or intravitreally and following their metabolic fate in the neural retina and in the pigment epithelium.

Thus, different types of information are obtained depending on the starting material, the *in vivo* or *in vitro* design, and, most importantly, lighting conditions and the state of light or dark adaptation. To the extent possible, this chapter is organized taking all of these experimental conditions into account. In addition, the chapter includes, in an abbreviated form, recent findings using the techniques of molecular biology that are now opening new horizons in our understanding of retinal biochemistry.

7.1.1. Intact Neural Retina (Excluding Lipids)

The neural retina is especially suitable as a model for relating biochemical processes such as carbohydrate and protein metabolism to retinal function under physiological conditions. The experimental conditions required to ensure viable tissue preparations for this purpose, using the relatively avascular rabbit retina, have been known for many years through the classic investigations of Ames and co-workers (Ames and Gurian, 1963a,b). Freshly excised rabbit retinas incubated in bicarbonate-buffered medium in an atmosphere of 40% oxygen, 5% CO_2, and 55% nitrogen are metabolically active and responsive to light for periods of at least 8 h (Ames and Nesbett, 1981). These retinas maintain gradients for electrolytes and amino acids, carry on active aerobic and anaerobic glycolysis, maintain protein synthesis, and respond to photic stimulation. Using similar experimental conditions for rat retina, essential relationships between glycolysis and respiration on the one hand and generation of electrical potential on the other have been elucidated (Winkler, 1981, 1983a).

7.1.1.1. Glucose

Short-term incubations of intact neural retina have yielded considerable information on overall pathways of carbohydrate metabolism (see Figs. 3.2 and 5.3 for outlines of cellular glucose metabolism). An exceptionally high rate of both anaerobic and aerobic glycolysis, and active lactate production, have been recognized since Warburg's early experiments in the 1920s. In fact, neural retina has a higher rate of respiration and glucose oxidation than any other tissue examined *in vitro* (Cohen and Noell, 1960, 1965; Graymore, 1970). The retina appears to be similar to most facultative organisms in that it can function either aerobically or anaerobically; these tissues display an unusual sensitivity to the Pasteur effect (Cohen and Noell, 1965; Shichi, 1983). Despite its high rate of respiration, most of the glucose utilized by the retina is converted to lactate (see review by Winkler, 1983a). This phenomenon is present in all of the cell layers but is especially prominent in photoreceptor cells. Although some questions have been raised about the significance of aerobic glycolysis, it is a normal property of the retina, and its presence can be clearly demonstrated under appropriate experimental conditions (Winkler, 1989).

An unusual feature of carbohydrate metabolism in the retina is the higher rate of lactic acid production, oxygen uptake, and glucose consumption in bicarbonate/CO_2 buffer than in phosphate buffer, even though overall metabolic patterns of glucose utilization are similar (Winkler *et al.*, 1977). A comparison of the metabolic activities of adult retinas from various species in bicarbonate and phosphate buffer is given in Table 7.1. It

Table 7.1. Glucose Metabolism of Mammalian Retina[a]

Buffer	Lactic acid produced (μmol/mg dry wt. per hour)		O_2 uptake (μmol/mg dry wt. per hour)	[1 − ^{14}C]glucose/ [6 − ^{14}C]glucose
	Anaerobic	Aerobic		
Bicarbonate	2.01	1.12	1.49	1.16
Phosphate	1.25	0.56	0.48	1.21

[a]Adapted from Winkler (1983a). The values shown are the averages from the following species: rat, rabbit, cattle, ox, and pig.

has been calculated that glucose consumption in the retina in the presence of oxygen is 0.73 μmol/mg dry weight per hour, and about 70% of the glucose utilized aerobically is converted to lactate (Winkler, 1983a). Retinas from rat and rabbit incubated anaerobically in the absence of glucose produce only small amounts of lactic acid during the initial 10–50 min of incubation. Production then ceases, suggesting that glycogen stores, localized mainly if not entirely in the Müller cells, are very limited.

The major substrate for respiration in the retina is glucose, although mannose can also serve as an energy source. Rat retinas incubated aerobically for 30 min with either 5 mM glucose or 5 mM mannose have similar levels of ATP, namely, about 10 μmol/g dry weight (Winkler, 1983a). Under anaerobic conditions and/or in the presence of iodoacetate, the ATP levels decline to about one-fourth of this value. Other substrates that can be utilized for ATP production in rat retina include pyruvate and lactate as well as glutamine and glutamate and tricarboxylic acid intermediates such as malate and succinate.

The foregoing studies were carried out on intact neural retinas, which does not allow the assignment of specific metabolic patterns to individual cell types. This problem can be partially circumvented by studying glucose metabolism in "photoreceptor-poor" retinas, i.e., either degenerate or immature or chemically poisoned retinas. These conditions would be expected to minimize the contribution of the photoreceptors to glucose oxidation, and the difference between metabolic activity in whole and "photoreceptor-poor" retinas should provide an estimation of glucose oxidation by the photoreceptors alone. One such classic study by Cohen and Noell (1965) suggests that glucose oxidation is considerably higher in the photoreceptors than in the remaining retinal layers. Glucose oxidation accounts for about 80% of the oxygen uptake by photoreceptors and only about 55% in the remaining cell layers. Histochemical studies corroborate the high oxidative capacity of the photoreceptor cells (Lowry *et al.*, 1956, 1961).

The ratios of carbon-1- and carbon-6-labeled glucose oxidized to CO_2 by adult mammalian retina (Table 7.1) suggest that under basal conditions the pentose phosphate pathway accounts for only about 23% of the total glucose converted to CO_2 (Winkler, 1983a). Regardless of its relatively low activity, this pathway is partly responsible for maintaining GSH levels of approximately 1.25 μmol/g wet weight in the rat retina (Winkler and Giblin, 1983). Oxidation of glucose through this pathway is dramatically increased in the presence of diamide, an agent that specifically oxidizes GSH to GSSG. This observation suggests that the retina has the capacity to respond to oxidative stress. When the tissue requires additional NADPH as a cofactor for glutathione reductase to regenerate depleted levels of GSH, it responds through increased activity of the pentose phosphate pathway (see Fig. 1.9). Although this pathway has long been considered to be the major source of NADPH in the retina and in other tissues, there is now compelling evidence that other NADPH-generating systems contribute substantially to the intracellular pool of NADPH (Winkler *et al.*, 1986). The activity of the pentose phosphate pathway can be decreased by depleting the retina of glucose stores, but rather unexpectedly, on exposure to diamide, there is an almost complete recovery of GSH in both glucose-repleted and glucose-depleted retinas. The finding that the NADPH required for regeneration of GSH was not coming from the pentose phosphate pathway led to a search for other NADPH-generating systems. Two cytosolic enzymes capable of reducing $NADP^+$ and generating NADPH were detected in rat retina: malic enzyme and isocitrate dehydrogenase. Together they appear to provide at least the same amount of reducing power as the pentose phosphate pathway.

Electrical activity (ERG potentials) of rat retinas remains stable for several hours when incubated in bicarbonate media equilibrated with 95% oxygen/5% CO_2 and containing 5 mM glucose (Winkler *et al.*, 1977; Winkler, 1983a). The ERG potential is dependent on active glucose oxidation since it is quickly abolished either by withdrawal of the oxygen supply or by the addition of iodoacetate to the incubating medium (Winkler, 1981). Under these conditions lactate production declines, and ATP levels are reduced. An adequate supply of GSH is also essential to maintain normal electrical activity in the retina (Winkler and Giblin, 1983). Exposure of rat retinas to diamide results in a 26-fold increase in pentose phosphate pathway activity, a dramatic drop in GSH concentration, and substantial changes in the receptor potential. This interesting effect is reversible by superfusion with diamide-free medium.

The plasma membrane of vertebrate photoreceptors is unusual because, in the resting state in the dark, it is highly permeable to Na^+. There is a constant inward flow of Na^+, but there is no extrusion of Na^+ from the outer segments because they lack a Na^+,K^+-ATPase pump. This enzyme is localized in the inner segments; hence, the Na^+ ions that enter the outer segments of the photoreceptors in the dark through the open Na^+ channels flow to the inner segments, where they are extruded from the cell. This is the basis of the well-known "dark current." Light closes the Na^+ channels and causes hyperpolarization of the photoreceptor plasma membrane. The concentration gradient of Na^+ required for the "dark current" as well as the receptor potential is maintained by a constant supply of ATP generated by the mitochondria in the inner segments. The ouabain-sensitive Na^+,K^+-ATPase is thought to play an important, though possibly indirect, role in the generation of receptor potential through maintenance of the sodium gradient necessary for the light response (Winkler, 1983b).

7.1.1.2. Amino Acids

Neural retina, like brain, contains a rather large pool of neuroactive amino acids. Their concentrations have been measured in a number of animal species (Voaden *et al.*, 1977), and the levels found are summarized in Table 7.2. Measurements have also been made of some of the precursors and metabolites of these amino acids as well as taurine, the most abundant free amino acid in the retina in all species examined. Although its precise physiological function in the neural retina remains obscure, taurine is thought to play a role in maintaining the structural integrity of the photoreceptors (Pasantes-Morales, 1986). Dietary restriction in certain animal species causes a severe retinal degeneration and blindness (see Section 7.9.1.6). Although taurine is released from the retina (probably

Table 7.2. Free Amino Acids in Adult Retina of Various Species

	Species (μmol/g wet wt.)[a]				
Amino acid	Rat[b]	Cat[b]	Rabbit[b]	Pigeon[b] (red spot)	Chicken[c]
Taurine	50	43	52	21	9.60
Glutamate	10.5	6.4	13.2	7.1	2.70
GABA	2.8	2.0	2.7	5.3	3.02
Glycine	2.2	2.0	2.1	4.0	1.49
Aspartate	3.1	2.9	2.2	2.0	0.41
Alanine	n.a.[d]	n.a.	n.a.	n.a.	0.24
Glutamine	5.2	3.6	4.8	8.6	2.13

[a]The values shown are the means (not including SEM).
[b]Adapted from Voaden *et al.* (1977).
[c]Adapted from Reif-Lehrer (1984).
[d]n.a., not analyzed.

from the photoreceptors) in response to light (Schmidt, 1978), it is not known to be a neurotransmitter, and there is no evidence for its involvement in visual excitation. In all cells where it is present, taurine is found as the free amino acid; it is not incorporated into either proteins or any other known macromolecule. Most tissues, including the retina, contain all of the enzymes necessary for endogenous biosynthesis of taurine from cysteine (see review by Pasantes-Morales, 1986). Paradoxically, although more than 60% of the retinal taurine is concentrated in the photoreceptors, only low levels of taurine-synthesizing enzymes are found in this cell layer. Instead, most of the active synthesis takes place in the inner retina. The major source of taurine in the outer retinal layers appears to be through a highly effective active transport system; avid uptake by photoreceptors has been demonstrated both *in vivo* and *in vitro*.

Some of the metabolic and enzymatic interconversions of the neuroactive amino acids—GABA, glycine, glutamate, and aspartate as well as glutamine—are summarized in the present section, and their major function as neurotransmitters in specific neuronal cells of the retina is discussed in Section 7.7.1. The four neuroactive amino acids, as well as others, are synthesized *in situ* through well-known pathways of amino acid metabolism outlined in Fig. 7.2. Glucose metabolized through the tricarboxylic acid cycle and glutamine are important sources of precursor carbon atoms for several four- and five-carbon amino acids. Rat retinas incubated with either labeled glucose or glutamine convert about 20–30% of these precursors to glutamate (Morjaria and Voaden, 1979). Smaller amounts are metabolized to aspartate and GABA. Studies on microdissected retinas show that the greatest incorporation of label into glutamate and aspartate occurs in the photoreceptors. Glutamine is also a source of a relatively stable pool of GABA, which appears to be localized in either amacrine or ganglion cells, depending on the species examined (Voaden *et al.*, 1978).

Glutamate, found in high concentration in the retina of all species examined, is actively metabolized by this tissue (see review by Reif-Lehrer, 1984). Some of the reactions it undergoes, as shown in Fig. 7.2, include transamination with pyruvate and enzymatic reaction with glutamic dehydrogenase to form α-ketoglutarate, a key inter-

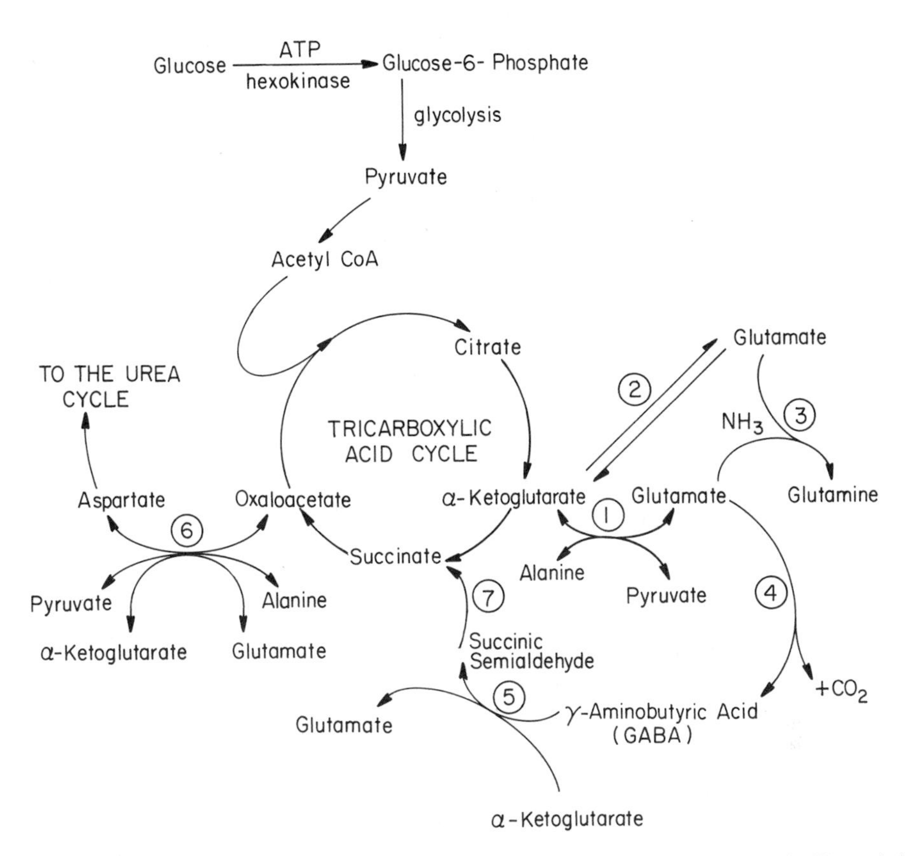

Figure 7.2. Metabolic pathways of amino acids and glucose through the tricarboxylic acid cycle. The encircled numbers correspond to the following enzymes: (1) glutamate transaminase; (2) glutamic dehydrogenase; (3) glutamine synthetase (GS), which also has glutamine transferase (GT) activity; (4) glutamic acid decarboxylase (GAD); (5) GABA aminotransferase; (6) aspartate transaminase; and (7) succinic semialdehyde dehydrogenase.

mediate in the tricarboxylic acid cycle. Glutamate is also decarboxylated to GABA, an inhibitory neurotransmitter substance that reenters the tricarboxylic acid cycle through the GABA "shunt" after transamination with α-ketoglutarate. One of the most important and extensively studied metabolic reactions of glutamate is its conversion to glutamine by glutamine synthetase (GS), an important pathway for the detoxification of ammonia. The enzyme is also called glutamine transferase (GT) since it can catalyze the exchange of the amide group of glutamine for ammonia or hydroxylamine. A highly purified preparation from bovine retina has a M_r of 360,000, and SDS-PAGE analyses show that it is composed of eight 45-kDa subunits (Pahuja *et al.*, 1985). Both activities, GS and GT, are catalyzed by the same enzyme, although they have strikingly different requirements for divalent cations. When the enzyme is measured as GS activity, it shows threefold higher activity with Mn^{2+} over Mg^{2+}, whereas the opposite is true when assayed as GT activity (Pahuja and Reid, 1985).

The neurotoxic properties of glutamate have been recognized for several decades, and recently considerable attention has been focused on its unusual neurological effects because of the widespread use of monosodium glutamate (MSG) as a food additive (see review by Reif-Lehrer, 1984). Glutamate (or MSG) also has striking effects on embryonic and neonatatal retina. For example, subcutaneous injection of MSG into newborn rats or mice causes severe retinal damage, especially to the inner layers. Addition of low concentrations of the L isomer of MSG to cultured embryonic chick retinas has similar effects and, in addition, causes a thinning of the photoreceptor cell layer (Reif-Lehrer *et al.*, 1975). There is also a marked inhibition of cortisol-induced glutamine synthetase activity (measured as glutamine transferase), although the L isomer of glutamic acid has no effect on preformed enzyme (Reif-Lehrer, 1984).

Immunohistochemical studies have shown that both the endogenous and the steroid-induced enzymes are localized exclusively in the cytoplasm of the glial (Müller) cells of chick retina (Norenberg *et al.*, 1980). Moreover, it is evident from both light and electron microscopic studies that the initial lesion, after exposure of embryonic chick retina to 0.1–0.3 mM MSG for 2 h, is in the Müller cells (Casper *et al.*, 1982). Incubation of chick embryos with α-aminoadipic acid, a six-carbon analogue of glutamic acid, results in glial swelling and neuronal necrosis similar to that observed after exposure to glutamate (Casper and Reif-Lehrer, 1983). Neuronal damage is most severe in embryonic retinas incubated with the L isomer, whereas the D,L racemic mixture is somewhat less toxic. In both cases the damage is limited to Müller cell swelling without neuronal necrosis. The increase in GS activity detectable late in embryonic development in the chick occurs only postnatally in rodents (Reif-Lehrer, 1984). In this case, the D isomer of glutamic acid not only inhibits cortisol-induced GS activity but also suppresses already existing enzyme activity.

7.1.1.3. Protein Synthesis

Autoradiographic studies in whole animals as well as biochemical studies of incubated retinas show active turnover and synthesis of protein from labeled precursor amino acids. The renewal of photoreceptor proteins *in vivo*, first demonstrated in the classic studies of Young and co-workers (Young, 1967, 1971; Young and Droz, 1968), is a dynamic process in which outer segment proteins are synthesized in the rough endoplasmic reticulum of the inner segments and transported through the connecting cilium to the base of the outer segments for insertion into disk membranes. Further details of this unique biosynthetic system in the photoreceptors are discussed in Section 7.2.2.1.

The ability of neural retina to incorporate labeled precursor amino acids into tissue protein *in vitro* is also well established. Functionally viable intact rabbit retinas incorporate labeled leucine at a linear rate of about 2 μmol/g dry weight per hour, representing a turnover of 0.55% of the leucine pool in retinal proteins (Parks *et al.*, 1976). This synthesis is balanced by an approximately equal rate of breakdown, the most rapidly metabolizing proteins being those of M_r 33,000–43,000 (Ames *et al.*, 1980a). The total rate of protein synthesis in rabbit retina is 103 nmol/g protein per hour; it is not affected by the functional activity of the tissue, since similar rates of incorporation of labeled leucine occur in darkness or after photic stimulation. The latter causes an increase in ganglion cell firing but no corresponding enhancement of protein synthesis (Ames *et al.*, 1980b). Protein synthesis in the retina is probably not directly related to light *per se*. Thus, studies on frog (*Xenopus laevis*) retina revealed no measurable differences in protein synthesis during any phase of the 12-h alternating light–dark cycle (Hollyfield and Anderson, 1982). Only in animals maintained in continuous darkness for 3 days is there a 40–48%

reduction in incorporation of labeled leucine into proteins present in all cell layers of the retina. This decrease was ascribed to the general drop in metabolic activity of the retina in amphibians maintained in total darkness.

7.1.1.4. Nucleic Acids

The retina of most species contains an unusually high ratio of DNA/RNA, as shown in a number of early studies (see review by Bazan and Reddy, 1985). The values, in milligrams per gram wet weight, for DNA and RNA, respectively, are 11 and 3.5 in mice, 10 and 2.5 in rat, and 8 and 2.25 in the rabbit. Histochemical investigations of frog and rat retinas have localized both DNA and RNA to the nucleus and inner segments of both species (Bok, 1970). Although under normal conditions there is no regeneration of DNA in the mature retina, RNA renewal takes place within 15 min after injection of [^{3}H]cytidine. The label, located by autoradiographic techniques, was concentrated in the rod cell nuclei of both species. After 1 h, the radioactivity is displaced to the inner segment, the site of active protein synthesis in photoreceptor cells. Studies on [^{3}H]uridine incorporation into rod and cone RNA of *Rana pipiens* maintained in diurnal cyclic light (14L:10D) show that the highest rate of incorporation occurs during the middle of the light cycle, and the lowest during the dark cycle (Hollyfield and Basinger, 1980). Also *in vitro*, most of the newly synthesized RNA in rat retinas incubated with labeled cytidine is localized in the nuclei of the photoreceptors, with relatively little being found in the inner retina (Schmidt, 1983b).

7.1.1.5. Cyclic Nucleotides

Cyclic GMP. The two major cyclic nucleotides present in the retina, cAMP and cGMP, are highly compartmentalized in cone-dominant and rod-dominant retinas, respectively (Farber *et al.*, 1980, 1981). The concentration of cGMP in rod-dominant rodent retinas is 100 times greater than in any other neural tissue examined (Ferrendelli *et al.*, 1980; Berger *et al.*, 1980; Farber and Shuster, 1986), and in dark-adapted retinas, the concentration of cGMP is 100-fold higher than that of cAMP (Krishna *et al.*, 1976). Several different lines of evidence corroborate that approximately 90% of the cGMP of the retina is localized in the photoreceptors. Its levels are increased with maturation of the visual cells (Farber and Lolley, 1977), and in human and monkey eyes, the highest concentration of cGMP is in the rod-rich periphery, while the lowest concentration is in the cone-rich central (macular) region (Newsome *et al.*, 1980; Farber *et al.*, 1985). The distribution patterns are particularly striking when represented as three-dimensional plots (Farber *et al.*, 1985; Farber and Shuster, 1986). Studies on biologically fractionated mouse retina also show the localization of about 95% of cGMP in the photoreceptor cells and the relatively even distribution of cAMP throughout the retina (DeVries *et al.*, 1978). In animals with inherited retinal degenerations that result in death of the visual cells (see Section 7.9.1), there are abnormal increases in cGMP content initially because of enzymatic defects in light-activated photoreceptor-specific phosphodiesterase, and with death of the photoreceptors, the cGMP content falls to undetectable levels (Farber and Lolley, 1974, 1977; Aguirre *et al.*, 1978; Woodford *et al.*, 1982).

The concentration of photoreceptor cGMP is modulated by light (Chader *et al.*,

1974; Goridis *et al.*, 1974; Fletcher and Chader, 1976; Krishna *et al.*, 1976; Orr *et al.*, 1976; Farber and Lolley, 1977; Mitzel *et al.*, 1978). Specifically, the major response of either whole retina or isolated photoreceptors or rod outer segments after exposure to light is a decrease in cGMP concentration, the extent depending on the degree of dark adaptation, the intensity of the light used, the type of tissue preparation, the concentration of Ca^{2+}, as well as other experimental and/or environmental conditions (Farber and Shuster, 1986). The reduction in cGMP concentration in the light is not only rapid but is also correlated with a large amplification of the light signal as measured by electrophysiological methods. These observations are now perceived in terms of the central role of cGMP as the internal transmitter in phototransduction in the vertebrate retina (see Section 7.8.2).

Cyclic GMP levels in rod-dominant photoreceptors are controlled by two enzymes. One is guanylate cyclase, a rod outer segment enzyme that is considerably more active than the adenylate cyclase in this cell layer (Berger *et al.*, 1980; Ferrendelli *et al.*, 1980). Guanylate cyclase is a peripherally bound enzyme requiring divalent cations for activation. Early studies using isolated bovine rod outer segments suggested that its activity is inhibited by light (Krishna *et al.*, 1976), but this interpretation has been questioned (Bitensky *et al.*, 1978). There is, however, compelling (though indirect) evidence that guanylate cyclase activity in intact rabbit retinas is increased after illumination (Goldberg *et al.*, 1983). By measuring the insertion of ^{18}O from [^{18}O]water into the phosphoryl of the 5′-nucleotide product of phosphodiesterase activity, it has been shown that a 20-s light impulse greater than 3 log units of intensity increases the metabolic flux of cGMP without changing its steady-state level. This finding implies that not only the hydrolysis of cGMP but also its synthesis by guanylate cyclase is enhanced by light, a provocative hypothesis requiring further investigation. More recent studies have shown that the activity of guanylate cyclase is stimulated five- to 20-fold when Ca^{2+} levels are lowered experimentally from 200 nM to 50 nM; this cooperative feedback may be a key event in restoring the dark current after visual excitation (Koch and Stryer, 1988).

The other enzyme that modulates cGMP concentration in rod-dominant species is the photoreceptor-specific cGMP phosphodiesterase (ROS PDE), an extensively studied enzyme that plays a major role in phototransduction. The apparent K_m of this enzyme from retinas of various species is in the range of 10^{-4}–10^{-5} M (Farber and Shuster, 1986); it shows an absolute dependence on light for its activation (Chader *et al.*, 1974) and displays optimal activity in the presence of divalent cations such as Mg^{2+} or Mn^{2+}. Cyclic GMP phosphodiesterase has been isolated and purified from bovine rod outer segments (Baehr *et al.*, 1979); the properties and mechanism of action of this 170-kDa peripherally bound membrane protein, and its central role in the cGMP cascade in visual transduction, are discussed in Sections 7.2.1.1 and 7.8.2, respectively.

Cyclic AMP. In contrast to cGMP, which is localized in the photoreceptors of rod-dominant retinas, cAMP is evenly distributed throughout the cell layers except for the photoreceptors, which have very low, albeit detectable, levels (Orr *et al.*, 1976). The total content of cAMP in rodent retina is considerably lower than that of cGMP (Ferrendelli *et al.*, 1980); moreover, its concentration in rod outer segments is not affected by bleaching (Fletcher and Chader, 1976; Orr *et al.*, 1976). In contrast, cAMP levels in the outer plexiform and outer nuclear layers are reduced by about 40% after illumination of dark-adapted mouse retinas (DeVries *et al.*, 1978). Cyclic AMP concentration in rod-dominant retinas is modulated not only by light but also by adenylate cyclase, a dopamine-stimulated enzyme localized mainly in the inner retinal layers (see review by Shuster and Farber, 1986). Dopamine receptors have been identified in retinas from vertebrate and nonvertebrate species, and the role of dopamine as a neurotransmitter in the retina is discussed in Section 7.7.2. Other putative neurotransmitters may also modulate the level of cAMP in the retina. For example, both vasoactive intestinal peptide (VIP) and glucagon cause a substantial rise in cAMP levels in cultured Müller cells from embryonic chick retina (Koh *et al.*, 1984).

As in other tissues throughout the body, so also in the retina, cAMP as a second messenger has numerous and diverse biological functions. It may modulate the synthesis and breakdown of glycogen stored in the Müller cells by activation of protein kinase A, analagous to its action in the liver. This cyclic nucleotide also activates other protein kinases in the retina, resulting in the phosphorylation of the 200-kDa chain of bovine retinal myosin (Hesketh *et al.*, 1978). Another cAMP-dependent protein kinase that phosphorylates tyrosine hydroxylase has been detected in rat retina (Iuvone *et al.*, 1982). Tyrosine hydroxylase is activated *in vitro* by a cAMP-dependent phosphorylation that closely resembles the *in vivo* activation resulting from photic stimulation. In both cases there is a decrease in the apparent K_m for synthetic pterin cofactor $6MPH_4$ as well as similar time-dependent inactivation of the enzyme by incubation at 25°. These observations and others support the commonly held view that the *in vivo* (photic) activation of tyrosine hydroxylase is mediated through phosphorylation of the enzyme.

Cone-dominant retinas of the western fence lizard and ground squirrel contain 2.8 and 8.0 times more cAMP than cGMP, respectively (Farber *et al.*, 1980, 1981). Destruction of retinal cones of the ground squirrel by injection of iodoacetate provides an indirect method of studying the distribution of cyclic nucleotides in this cone-dominant species. Measurements before and after iodoacetate treatment show that the loss of cAMP is about four times greater than the loss of cGMP, implying that cAMP is the dominant cyclic nucleotide in ground squirrel (Farber *et al.*, 1981; Farber and Shuster, 1986). Light causes reductions in cAMP levels of dark-adapted western fence lizard and ground squirrel retinas by 30% and 55%, respectively, but has little effect on cGMP levels (Farber and Shuster, 1986). Cyclic AMP appears to be distributed almost equally between the visual cells and the inner retina of cone-dominant retinas as shown by direct dissection as well as by selective destruction of ground squirrel cones by iodoacetic acid.

7.1.1.6. Retinoids and Retinoid-Binding Proteins

The highest concentrations of vitamin A in the body are in the liver, with relatively little being stored in extrahepatic tissues apart from the eye. In the retina, the absolute amounts and stereoisomeric forms of retinol and other retinoids vary with the species and the state of light or dark adaptation. Except for retinoids bound to visual pigment and the small amount (about 2% of the total) associated with the retinoid-binding proteins, most of the vitamin A of the retina is found as retinyl ester in the pigment epithelium (Berman, 1979; Bridges, 1984). The vitamin A levels in frog (Bridges, 1976) and human retina (Bridges *et al.*, 1982) are only about 4% and 15%, respectively, of those present in the pigment epithelium. Whereas about 95–98% of the vitamin A in the pigment epithelium is esterified, human and frog retina contain only 79% and 51%, respectively, in the esterified form. In both retina and pigment epithelium, the major fatty acids esterified to retinol are palmitate and stearate.

In the dark-adapted state, 11-*cis* retinal is bound to opsin in a protonated Schiff-base linkage to the ϵ-amino group of lysine (Lys-296) to form the photosensitive visual pigment rhodopsin (for reviews see Dratz and Hargrave, 1983; Applebury and Hargrave, 1986). Exposure of neural retina (or isolated photoreceptors) to light results in the isomerization of 11-*cis* retinal to all-*trans* retinal, which is released from its binding site in rhodopsin and rapidly reduced to all-*trans* retinol by a membrane-bound retinol dehydrogenase (Futterman, 1963; Bridges, 1977). It is then transferred to the pigment epithelium (Dowling, 1960), where it is esterified and isomerized (see Section 7.4.4.2); retinoid (in all probability 11-*cis* retinal) is then returned to the photoreceptors during dark adaptation. Only one other retinoid, all-*trans* retinoic acid, is known to be present in neural retina. It is the endogenous ligand of cellular retinoic acid-binding protein but, unlike the other retinoids discussed above, has no known role in the visual cycle (Saari *et al.*, 1982).

Cellular retinoid-binding proteins (as distinct from serum retinol-binding protein) are found in the cytosol of a wide variety of cell types; they form noncovalent complexes with specific endogenous retinoids and are thought to function in the stabilization and metabolism of intracellular retinoids. They are of particular importance in the retina, where four distinct retinoid-binding proteins have now been characterized and localized in three anatomic compartments of the tissue: the neural retina, the interphotoreceptor matrix, and the pigment epithelium (for reviews see Bridges, 1984; Bok, 1985; Chader, 1982, 1989). A brief summary of current nomenclature as well as localizations and molecular masses of the retinoid-binding proteins of the retina is given in Table 7.3. This section considers mainly the binding proteins of neural retina, keeping in mind that retinoids are transformed and transferred, both intracellularly and intercellularly, through the various retinal

Table 7.3. Retinoid-Binding Proteins of the Retina[a]

Endogenous retinoid ligand	Name of binding protein	Molecular mass (kDa)	Retinal compartment: Neural retina[b]	Retinal compartment: Interphotoreceptor matrix	Retinal compartment: Pigment epithelium
All-*trans* retinol	CRBP	16.6	■		■
	IRBP	140		■[c]	
11-*cis* retinol }	CRALBP	33;36[d]	■		
11-*cis* retinal }			■		■
All-*trans* retinoic acid	CRABP	15.4[e]	■		

[a]Details and literature references on retinoid-binding proteins in neural retina, interphotoreceptor matrix, and retinal pigment epithelium are given in Sections 7.1.1.6, 7.3.3, and 7.4.4.2, respectively.
[b]CRBP and CRALBP of neural retina are localized in the Müller cell cytoplasm; although less certain, CRABP also appears to be localized in this glial cell (see Bok, 1985).
[c]Also carries about 12% 11-*cis* retinol as endogenous ligand.
[d]Molecular mass from data of Saari and Bredberg (1988a).
[e]Data from Crabb and Saari (1986).

compartments. Several provocative schemes have been published showing possible routes of transfer within the tissue (Chader, 1982; Bridges, 1984; Bok, 1985).

Binding proteins for retinoic acid and retinol were detected more than a decade ago in cytosolic preparations of retina from a variety of animal species (Wiggert and Chader, 1975; Saari and Futterman, 1976; Futterman *et al.*, 1976; Wiggert *et al.*, 1978a,b). The cytosolic proteins from bovine retina that bind all-*trans* retinoic acid (CRABP) and all-*trans* retinol (CRBP) have been partially purified and shown to be of similar M_r (16,300 and 16,600, respectively); they have similar but not identical amino acid compositions, and each binds 1 mol of ligand (Saari *et al.*, 1978a; Crabb and Saari, 1981). The complete amino acid sequence of bovine CRABP reported in later studies by Saari and co-workers gave a molecular mass of 15,460 Da for the protein; moreover, the amino acid sequence of CRABP from bovine retina is identical to CRABP of bovine adrenal gland, implying that the same gene is expressed in both tissues (Crabb and Saari, 1986). Similarly, the CRBPs from various species and tissues, including the retina, appear to be stucturally conserved proteins, as shown by comparative biochemical and immunologic studies on CRBP from bovine retina, and rat and dog liver (Liou *et al.*, 1981). The CRBP of rat, bovine, and human neural retina is found exclusively in the Müller cell cytoplasm (Bok *et al.*, 1984; Bok, 1985); it is also probable that CRABP is localized in this glial cell.

A third retinoid-binding protein, CRALBP, was first detected in the cytosol of bovine retina by Futterman and co-workers (1977). In the native state, this 33-kDa protein contains two endogenous ligands, 11-*cis* retinal and 11-*cis* retinol (Stubbs *et al.*, 1979), present in a ratio of approximately 3:1, respectively (Saari *et al.*, 1982). The binding site is nearly fully saturated by the two ligands (0.9 mol of retinoids/mol of protein) in preparations obtained from frozen bovine retinas. Similar to CRBP (and probably CRABP), the CRALBP of rod-dominant (Bunt-Milam and Saari, 1983) and cone-dominant (Anderson *et al.*, 1986b) neural retina is localized exclusively in the Müller cells. CRALBP complexed to 11-*cis* retinal has many properties characteristic of visual pigments (Saari *et al.*, 1984); the complex shows a bathochromic shift in absorption maximum from 380 to 425 nm and a concomitant decrease in extinction coefficient. Illumination of the complex causes photoisomerization of the bound 11-*cis* retinal to all-*trans* retinal, but the photosensitivity of the complex is only about 4% of that for rhodopsin when measured in bovine homogenates or in eyecup preparations (Saari and Bredberg, 1987). This low photosensitivity suggests that the physiological role of CRALBP is to solubilize 11-*cis* retinal and to protect it from photoisomerization; moreover, CRALBP is highly stereoselective for 11-*cis* retinoids and hence may be associated with the visual cycle. The CRALBPs of bovine retina and pigment epithelium (which contains only 11-*cis* retinal as the endogenous ligand) have been purified to apparent homogeneity (Saari and Bredberg, 1988a). The concentration of CRALBP in adult bovine retina is approximately 3 nmol per eye, about two-thirds being present in the neural retina and the remainder in the pigment epithelium.

7.1.2. Lipids in Neural Retina

Retinal lipids have been extensively investigated, and several detailed reviews and monographs have appeared (Fliesler and Anderson, 1983; Bazan and Reddy, 1985; Birkle and Bazan, 1986). The overall composition and metabolism of lipids in whole retina, and in some subcellular fractions, are discussed in this section; rod outer segment (ROS) lipids are considered separately in Section 7.2.1.2.

7.1.2.1. General Composition

Lipids comprise about 20% of the dry weight of the retina: the lowest content (12%) is found in the outer nuclear layer, and the highest (34%) in the nerve fiber layer; intermediate values (22–25%) are found in the photoreceptors (Lowry *et al.*, 1956). Phospholipids (see Fig. 7.3) constitute the major class of lipids, accounting for about 65–75% of the total. The remainder consist of cholesterol (10–12%), diglycerides, triglycerides, free fatty acids (5–7%), and sphingolipids (2%) (Fliesler and Schroepfer, 1982; Bazan and Reddy, 1985). There is a remarkable similaritiy not only in the various lipid classes but also in phospholipid composition among different species (Table 7.4). Analyses of three principal subcellular fractions of rat, toad, and cattle have revealed that in isolated fractions of nuclei, mitochondria, and microsomes, phosphatidylcholine (PC) and phosphatidylethanolamine (PE) together comprise close to 80% of the total phospholipids (Anderson *et al.*, 1975).

The major saturated fatty acids in the retina are palmitate (16:0) and stearate (18:0). As shown in Table 7.5, PC is especially enriched in palmitic acid, whereas the major saturated fatty acid in PE and PS is stearate (Anderson and Maude, 1972; Fliesler and Anderson, 1983). The retina is unique among body tissues in its unusually high content of

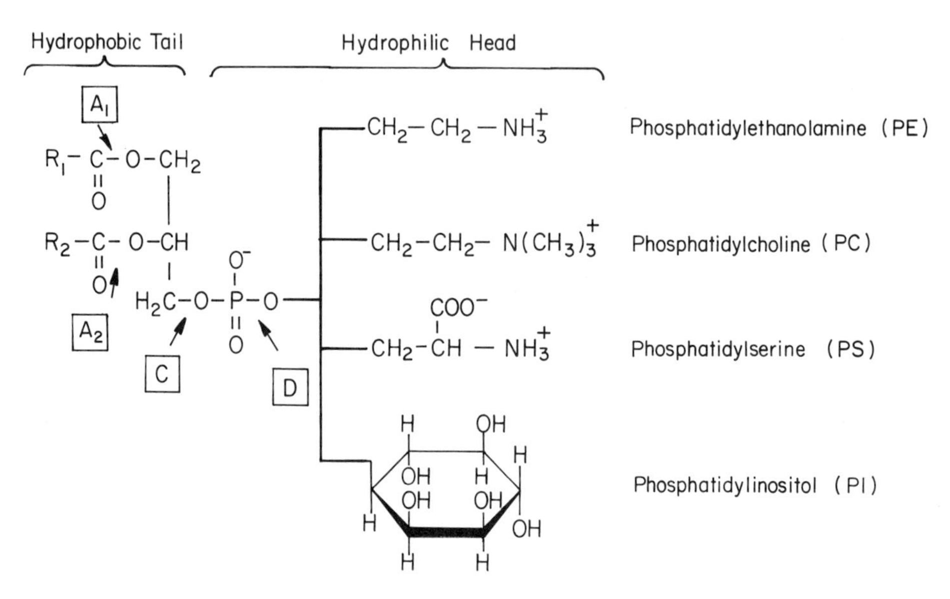

Figure 7.3. Phospholipid structures and sites of enzymatic degradation. The *sn*-1 position of glycerol is occupied mainly by long-chain saturated fatty acids (R_1), and *sn*-2 mainly by unsaturated species (R_2). Other combinations are also possible, as discussed in the text.

Table 7.4. Phospholipid Composition of Vertebrate Retinas[a]

Phospholipid[b]	Cattle[c]	Toad	Human	Rat[d]	Rabbit	Pig	Sheep
PC[e]	43.2	50.1	47.8	43.7	43.9	47.1	46.9
PE[e]	34.1	30.9	31.7	32.3	34.7	32.2	30.8
PS	10.0	9.8	8.6	11.0	7.4	8.1	9.7
PI	5.6	3.5	4.4	4.3	4.3	5.5	5.2
SPH	2.1	2.5	4.3	3.8	4.4	3.7	3.4
LPC	0.2	0.3	0.3	—	2.6	0.3	0.6

[a]Adapted from Fliesler and Anderson (1983). Data are expressed as mol% of total lipid phosphorus.
[b]Abbreviations used are: PC, phosphatidylcholine; PE, phosphatidylethanolamine; PS, phosphatidylserine; PI, phosphatidylinositol; SPH, sphingomyelin; LPC, lysophosphatidylcholine.
[c]Adapted from Anderson *et al.* (1970).
[d]Adapted from Anderson and Maude (1972).
[e]Includes plasmalogens.

unsaturated fatty acids; over 50% of the total fatty acids in most species are unsaturated, and of these, about 60% are polyunsaturated. The photoreceptors are particularly enriched in polyenoic fatty acids, the major species being 22:6(*n*-3). When precautions are taken to avoid autooxidation, maximum values of 22:6(*n*-6) are obtained; the relative concentrations of this fatty acid isolated under argon in the presence of a metal ion chelator are 50.7, 50.9, and 46.2 mol% in bovine, frog, and rat rod outer segment membranes, respectively (Stone *et al.*, 1979a). There is some variability in the fatty acid composition among different animal species, and detailed data on frog, rat, bovine, human, chick, and goldfish retina are available (Fliesler and Anderson, 1983). A comparison of the fatty acid content of the three principal phospholipids, PC, PE, and PS, in whole retina and in ROS of the rat (Table 7.5) shows a relative enrichment of 22:6(*n*-6) in ROS. This is characteristic of virtually all species examined to date.

Table 7.5. Fatty Acid Composition (mol%) of Three Phospholipids (PC, PE, and PS) in Whole Retina and Rod Outer Segments of the Rat[a]

	Whole retina			Rod outer segments		
Fatty acid	PC	PE	PS	PC	PE	PS
14:0	0.3	—	—	0.9	0.2	—
16:0	36.5	8.6	2.2	29.5	5.9	2.5
16:1	—	—	0.5	1.0	0.3	—
18:0	27.9	38.7	37.6	11.7	29.3	29.1
18:1	13.8	1.8	2.2	8.3	3.2	1.9
18:2	0.5	0.1	0.2	0.8	0.6	—
20:2 (*n*−6)	—	—	—	0.8	0.8	0.4
20:3 (*n*−9)	0.8	—	—	—	—	—
20:4 (*n*−6)	3.3	7.6	2.4	3.4	2.9	1.9
22:4 (*n*−6)	0.2	1.1	3.4	—	0.2	—
22:5 (*n*−6)	—	—	—	3.2	1.1	—
22:5 (*n*−3)	—	—	0.8	—	—	—
22:6 (*n*−3)	29.2	42.2	50.0	38.4	54.8	61.8
Other	—	—	0.5	1.4	—	—

[a]Adapted from Bazan and Reddy (1985).

Saturated and unsaturated fatty acids are found mainly in the *sn*-1 and *sn*-2 positions, respectively, of glycerolipids. Although phospholipids are thought to contain 1 mol of each fatty acid at these positions, the heterogeneous nature of phospholipids from most sources, including the retina, suggests a more complex composition. Molecular species of all three major classes (PE, PC, and PE) have been isolated and shown to contain not only saturated–unsaturated fatty acids but also saturated–saturated and unsaturated–unsaturated molecular species (Miljanich *et al.*, 1979).

By use of improved techniques to resolve the various phospholipid subclasses, a group of highly unsaturated molecular species termed "supraenes" have been isolated and partially characterized (Aveldano and Bazan, 1983). The "supraenes" contain more than six double bonds per phospholipid molecule and are present in all of the retinal phospholipid classes, PC, PE, PS, and PI. Among the phospholipids isolated in this study were dipolyunsaturated species containing polyenoic fatty acids, both hexaenes and supraenes, esterified at both the *sn*-1 and *sn*-2 positions of glycerol. In bovine ROS, dipolyunsaturated species of phospholipids comprise about one-third of the total membrane lipid. The highest content is in PS (51%), and the lowest in PI (9%). About 30% of the PC molecules contain dipolyunsaturated species, and further studies using highly purified PC from ROS of bovine retina, as well as PC isolated from rabbit, rat, chicken, toad, and cod retinas, have led to the characterization of a novel series of very long-chain polyunsaturated fatty acids (VLCPUFA) (Aveldano, 1987). They are homologues of polyenoic fatty acids of the *n*-3 and *n*-6 families since they have the characteristics of 24- to 36-carbon tetraenoic (*n*-6), pentaenoic (*n*-3 and *n*-6), and hexaenoic (*n*-3) fatty acids. The long-chain tetraenes belong to the *n*-6 series, hexaenes to the *n*-3 series, and the major pentaenes to the *n*-3 series of fatty acids (Aveldano and Sprecher, 1987).

Cholesterol is the principal neutral lipid of the retina, accounting for about 40% of the total neutral lipids (Fliesler and Schroepfer, 1982; Fliesler and Anderson, 1983). It is mainly, if not entirely, unesterified. Diacylglycerol and triacylglycerol (see Fig. 7.4) account for 10.2% and 7.0%, respectively, of the neutral lipids of toad retina (Bazan and Reddy, 1985; Birkle and Bazan, 1985). The diglycerides of toad are rich in 22:6(*n*-3) but have only small amounts of 20:4(*n*-6), whereas bovine and rabbit retinas show the opposite distributions of these two polyenoic fatty acids.

Neural tissues are relatively rich in glycolipids, but retinal levels are very low (Urban *et al.*, 1980; Fliesler and Anderson, 1983; Bazan and Reddy, 1985). Cerebrosides and sulfatides have been studied only in chicken retina, but gangliosides have been analyzed in a variety of vertebrate species (Table 7.6).

7.1.2.2. Metabolism

Both *in vivo* and *in vitro* studies have demonstrated active biosynthesis of phospholipids and neutral lipids from labeled precursors. Lipid metabolism in intact retina and in subcellular fractions is discussed in this section, and studies on phospholipid biosynthesis in intact ROS and in disk membranes are considered in Section 7.2.2.1. Much of our current understanding of lipid metabolism in the retina is based on the early investigations of Young and co-workers (see review by Young, 1976) and later work by Anderson and co-workers (see review by Fliesler and Anderson, 1983) and Bazan and co-workers (for reviews see Bazan and Reddy, 1985; Birkle and Bazan, 1985, 1986; Bazan *et al.*, 1986).

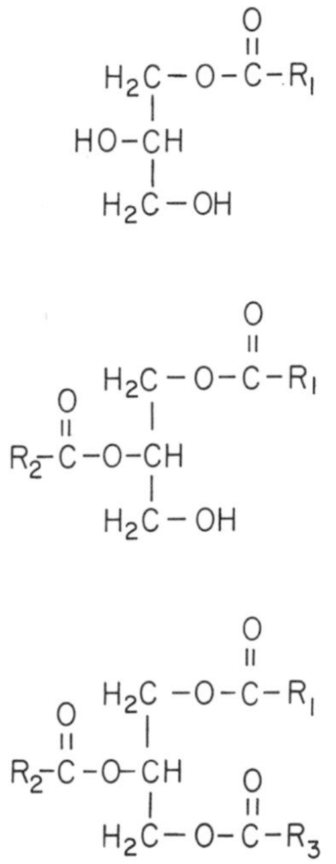

$$\begin{array}{l} \quad\quad\quad\quad O \\ \quad\quad\quad\quad \| \\ H_2C-O-C-R_1 \\ \quad | \\ HO-CH \\ \quad | \\ H_2C-OH \end{array}$$

Monoacylglycerol

$$\begin{array}{l} \quad\quad\quad\quad\quad\quad\quad O \\ \quad\quad\quad\quad\quad\quad\quad \| \\ \quad\quad O \quad H_2C-O-C-R_1 \\ \quad\quad \| \quad\quad | \\ R_2-C-O-CH \\ \quad\quad\quad\quad | \\ \quad\quad\quad H_2C-OH \end{array}$$

Diacylglycerol

$$\begin{array}{l} \quad\quad\quad\quad\quad\quad\quad O \\ \quad\quad\quad\quad\quad\quad\quad \| \\ \quad\quad O \quad H_2C-O-C-R_1 \\ \quad\quad \| \quad\quad | \\ R_2-C-O-CH \quad\quad O \\ \quad\quad\quad\quad | \quad\quad\quad \| \\ \quad\quad\quad H_2C-O-C-R_3 \end{array}$$

Triacylglycerol

Figure 7.4. Neutral lipids: mono-, di-, and triacylglycerols. By convention, glycerol is written as a Fischer projection with carbon 2 on the left side, and the atoms are numbered 1, 2, and 3 from top to bottom. Glycerol numbered by this convention is called *sn*-glycerol (stereospecifically numbered). Naturally occurring phosphoglycerides are all of the *sn*-glycerol 3-phosphate type. R_1, R_2 and R_3 designate fatty acid substituents at carbon atoms 1, 2, and 3, respectively.

Glycerolipid Biosynthesis. Early *in vivo* studies on the time course of the biosynthesis of retinal lipids after injection of [^{14}C]glycerol into toads showed phosphatidic acid (PA) as the initial product formed (Bazan and Bazan, 1976). Di- and triacylglycerols are synthesized later, and there is a high rate of *de novo* synthesis of phosphatidylinositol (PI). Other phospholipids become labeled at a later period. Similar biosynthetic profiles were also found *in vitro* in toad and bovine retinas (Giusto and Bazan, 1979). Subsequent studies have utilized [2-^{3}H]glycerol as a glycerolipid precursor since there is little or no recycling of the label (Birkle and Bazan, 1986). Brief incubation of intact bovine retinas or subcellular fractions with this precursor results in extensive labeling of PA, especially in the microsomal fraction. Labeled glycerol is also rapidly incorporated into hexaenes and dipolyunsaturates (supraenes) (Aveldano *et al.*, 1983), suggesting direct utilization of docohexaenoate (22:6) during the *de novo* synthesis of phospholipids (Bazan and Reddy, 1985).

In addition to *de novo* synthesis, base-exchange reactions are also involved in the formation of retinal phospholipids (Anderson and Kelleher, 1981). Bovine retinal microsomes are able to exchange free serine, choline, and ethanolamine with endogenous phospholipid substrates. The microsomal base-exchange enzymes are Ca^{2+} dependent and show maximum activity at pH 7.5–8. The apparent K_m values of serine, ethanolamine, and choline measured under these experimental conditions were 148, 63, and 616 μM, respectively, and the apparent V_{max} values for the three bases were 24, 32, and 18 nmol/mg protein per hour, respectively. Although the enzymes responsible for these base-exchange reactions can be demonstrated *in vitro*, their physiological relevance is uncertain apart from the biosynthesis of phosphatidylserine.

Other enzymes involved in phospholipid biosynthesis in the retina have also been described. Incorporation of long-chain fatty acids (palmitate, oleate, and linoleate) into retinal phospholipids in the presence of lysophosphatide acceptors strongly suggests the presence of acyltransferase reactions in the retina (Swartz and Mitchell, 1974). The activation requires ATP, CoA, and Mg^{2+}, and is detectable in homogenates and subcellular fractions of bovine retina but not in ROS. Later studies demonstrated an active long-chain acyl-CoA synthetase that converts labeled 20:4 and 22:6 to arachidonoyl-CoA and docosahexaenoyl-CoA, respectively, in the presence of ATP, CoA, and Mg^{2+} in microsomes of human, rat, frog, and bovine retinas (Reddy and Bazan, 1984; Bazan *et al.*, 1986). The rate of synthesis of arachidonoyl-CoA is two to four times higher than that of docosahexaenoyl-CoA. Enzymes involved in phosphatidylcholine (PC) synthesis have also been examined. The incorporation of CDP-choline into PC by homogenates as well as isolated microsomal preparations from bovine retina suggests a role for cytidyl transferase in the *de novo* biosynthesis of PC in the retina (Swartz and Mitchell, 1970). Choline kinase activity has been demonstrated in homogenates (Masland and Mills, 1980) and cytosol (Pu and Anderson, 1983) of rabbit retina.

An important aspect of phospholipid biosynthesis in the retina is the mechanism of insertion of polyunsaturated fatty acids into the molecule. The foregoing discussion has briefly summarized current evidence for (1) *de novo* synthesis (condensation of a fatty acid with lyso derivatives of α-glycerol phosphate to form PA) and (2) acyltransferase reactions (esterification of the *sn*-1 or *sn*-2 positions of endogenous phospholipids). Base-exchange reactions may also be responsible for the formation of phosphatidylserine (PS), as described above. These reactions are localized mainly in microsomal fractions derived from homogenates of whole retina. Phosphatidic acid (PA) is a key intermediate in phospholipid biosynthesis, and incubation of retinal microsomes with labeled 22:6 in the presence or absence of lyso-PA can help to distinguish between *de novo* synthesis and deacylation–reacylation reactions (Bazan *et al.*, 1984; Birkle and Bazan, 1986). The formation of 22:6-labeled PA from [^{14}C]docosahexaenoic acid is greatly enhanced in the presence of lyso-PA, the preferred acceptors being 1-palmitoyl- and 1-oleoyl-*sn*-glycerol 3-phosphate. Newly synthesized 22:6-PA is further metabolized to di- and triacylglycerols and to phospholipids PC and PS. Competition studies using lyso-PA and lyso-PC show that in addition to a deacylation–reacylation cycle, 22:6 is actively incorporated into retinal glycerolipids during the *de novo* synthesis of PA.

Apart from *de novo* synthesis, base-exchange reactions, and deacylation–reacylation, other mechanisms operating in the biosynthesis and/or remodeling of retinal glycerides include (1) elongation and desaturation, (2) decarboxylation, (3) transmethylation, (4) phospholipid transfer, and (5) enzymatic hydrolysis by phospholipases. These topics have been less extensively investigated and are only considered briefly.

Intravitreal injection of ^{14}C-labeled eicosapentaenoic acid, 20:5 (*n*-3), in the rat results in the incorporation, after acylation, of labeled docosapentaenoate, 22:5(*n*-3), and docosahexaenoate, 22:6(*n*-3), into retinal phospholipids (Bazan *et al.*, 1982). The elongation and desaturation products represent 8% and 4%, respectively, of the total radioactivity in phospholipids 3 min after injection.

Decarboxylation of PS to PE has been demonstrated in frog retina following injection of labeled serine (Anderson *et al.*, 1980a,b). The reaction occurs in both the microsomes and the ROS, the former appearing to be more active. Within 4 weeks after injection, the specific activities of PS and PE are nearly equal; hence, decarboxylation of PS may be a major source of PE in ROS. Decarboxylation can also be demonstrated *in vitro*. Studies with bovine retina show that microsomes actively incorporate labeled serine into both PS and PE; the synthesis of PS and its decarboxylation are stimulated by Ca^{2+} (for review see Bazan and Reddy, 1985).

Active transmethylation of PE to PC has been shown in frogs after injection of labeled ethanolamine (Anderson *et al.*, 1980b). Although the reaction takes place in both microsomes and ROS, the ratios of specific activities of PC/PE in the two compartments are very different. The PE derived from decarboxylation of PS in ROS undergoes relatively little transmethylation to PC, although PE arising through *de novo* synthesis from labeled ethanolamine in the inner segments is actively transmethylated. Other findings on PC biosynthesis in ROS reported by Anderson and co-workers (1980c) and other investigators are discussed in Section 7.2.2.1.

Phospholipid transfer proteins that actively exchange PC between liposomes and ROS membranes have been identified in cytosolic fractions of bovine retina (Dudley and Anderson, 1978). Only one such protein has been studied, but there are probably others. Their localization in the retina was not established, but they may play a role in the ROS in interdisk phospholipid transfer.

Phospholipases A_1 and A_2 have been detected in subcellular fractions of bovine retina (Swartz and Mitchell, 1973), but isolated ROS showed no phospholipase activity. In retrospect, this is probably because phospholipase A_2 activity of ROS is best demonstrated under specific conditions that take into account its light-activated, transducin-dependent properties (Jelsema, 1987; Jelsema and Axelrod, 1987).

Inositol Phosphatides (Vertebrate Retina). The biosynthetic pathway for phosphatidylinositol (PI) in the retina is different from that of the three major phospholipids, PE, PC, and PS. Recognition of the "phosphatidylinositol or phosphoinositide effect" in the retina, and particularly its activation by light in the photoreceptors, has led to extensive investigations of PI metabolism in the photoreceptors (see review by Anderson and Brown, 1988 and Section 7.2.2.2). The present discussion is limited to general aspects of inositol phosphatide metabolism in the whole retina.

Phosphatidylinositol (PI) of cattle and toad retinas has an unusually high content of arachidonic acid (AA), representing approximately 44% and 25%, respectively, of the total fatty acids (Aveldano de Caldironi and Bazan, 1980). Extensive labeling of PI from radioactive glycerol in bovine retina (Bazan *et al.*, 1977; Aveldano *et al.*, 1983) and from AA has been shown both *in vitro* and *in vivo* in rat and bovine retina; the synthesis appears to be mediated by deacylation–reacylation reactions (Birkle and Bazan, 1986).

The stimulation of phosphoinositide turnover by α-adrenergic agents, muscarinic cholinergic agonists, light, and other agonists has been discussed in Chapter 1, Section 1.3.2, and a diagram of this membrane transduction system is shown in Fig. 1.6. Many of the intermediates in this pathway have also been identified in the retina; moreover, their turnover can be studied using labeled glycerol (Birkle and Bazan, 1986). The receptor-mediated breakdown of phosphatidylinositol 4,5-bisphosphate (PIP_2) is the key step in the production of two second messengers, inositol 1,4,5-trisphosphate (IP_3) and diacylglycerol (DG). Agonist-stimulated phosphoinositide breakdown has recently been demonstrated in intact rabbit retinas as well as in 3-day-old retinal cultures (Ghazi and Osborne, 1988). Carbachol, norepinephrine, and serotonin stimulate the accumulation of inositol phosphates (principally inositol 1-phosphate) by as much as 172%, 71%, and 51%, respectively. Receptors for all three agonists were identified pharmacologically in

3-day-old cultures that contained both neuronal cells and glia; however, older cultures consisting primarily of Müller cells appear to have only muscarinic cholinergic and α_1-adrenergic receptors but lack serotonin (5-HT_2) receptors. These findings suggest that the 5-HT_2 receptors are linked mainly to phosphoinositide breakdown in neuronal cells of the retina and not to glial components.

Ganglioside and Sterol Biosynthesis. Most of the work on ganglioside biosynthesis was carried out over a decade ago using embryonic chick retinas (Dreyfus *et al.*, 1975, 1977, 1980; Caputto *et al.*, 1980). The activity of enzymes responsible for the activation of carbohydrate precursors of gangliosides is a function of embryonic age; maximum biosynthesis from labeled precursors reaches a peak within 5 days after injection, and maximum biosynthetic activity is found in 10-day-old embryonic retina. The highest specific activity from labeled precursors is found in disialoganglioside GD_3 (see Table 7.6). Because these studies were carried out on total retinas, there is no information on cellular localization of this biosynthetic activity. Gangliosides are catabolized by neuraminidase, a particulate enzyme that hydrolyzes terminal N-acetylneuraminic acid (NANA) residues of endogenous gangliosides in chick retina (Preti *et al.*, 1978) and in ROS membranes isolated from calf retina (Dreyfus *et al.*, 1983).

Cholesterol is the major sterol, albeit minor lipid, component of the retina, comprising about 5–7 mol% of the total membrane lipids (Fliesler and Schroepfer, 1982). Although its biosynthesis and metabolic fate in nonocular tissues have been studied extensively, relatively little is known about either its origin or its turnover in the retina. Studies on *de novo* biosynthesis of cholesterol from labeled mevalonic acid in homogenates of bovine retina (Fliesler and Schroepfer, 1983) and in intact bovine retina *in vitro* (Fliesler and Schroepfer, 1986) have shown that only 1% of the label is incorporated into cholesterol; the major labeled unsaponifiable intermediates are squalene and lanosterol (Fliesler and Schroepfer, 1983; Fliesler and Anderson, 1983). Labeled mevalonic acid is metabolized in a $10{,}000 \times g$ supernatant from bovine retina to a variety of products, principally squalene and unsaponifiable isoprenoid alcohols; as in the studies with intact retina, very little is found in cholesterol. Similar results were obtained from *in vivo* investigations after intraocular injection of labeled mevalonic acid (Fliesler and Anderson, 1983). These findings suggest that although cholesterol precursors and biologically related intermediates are formed in the retina from mevalonic acid, there is nevertheless minimal

Table 7.6. Ganglioside Composition of Vertebrate Retina (% of Total N-Acetylneuraminic Acid Recovered)[a]

Ganglioside	Rat	Rabbit	Cattle	Frog	Mouse
GM_3	6.1	3.8	2.8	0.6	5.0
GM_2	0	0	0.6	0.3	0.6
GM_1	2.7	7.3	2.1	—	3.8
GD_3	36.5	41.8	37.3	1.8	34.2
GD_{1a}	11.8	16.5	13.2	—	11.7
GD_{1b}	17.7	8.2	15.0	12.6	12.1
GT_{1b}	15.3	13.9	22.3	26.0	19.8
GQ_1	9.9	0.9	2.4	4.0	1.8

[a]Modified from Urban *et al.* (1980), Fliesler and Anderson (1983), and Bazan and Reddy (1985).

de novo synthesis of cholesterol from these precursors, either *in vivo* or *in vitro*. The endogenous cholesterol present in the tissue may, like other hydrophobic substances such as retinol, pass the blood–retinal barrier and utilize a carrier protein (possibly interphotoreceptor retinol- binding protein) as a vehicle for deposition in the retina (Fliesler and Schroepfer, 1986).

Pathways in the biosynthesis of cholesterol and dolichol, a lipid intermediate of major importance in the retina, are known to have several common intermediates, the most prominent being isoprenoid acids. A regulatory interrelationship in these biosynthetic pathways has been demonstrated *in vitro* in the retina of *Rana pipiens* (Keller *et al.*, 1988). The endogenous levels of squalene, cholesterol, and dolichol phosphate (Dol-P) are approximately 6, 134, and 0.14 nmol/retina, respectively. Incubation of intact retinas for 4.5 h with ^3H_2O results in the formation of labeled squalene, cholesterol, lathosterol, and methyl sterols. Calculations from these data show that the upper limits of the absolute rates of sterol and dolichol synthesis are 3.4 and 0.022 pmol/h per retina, respectively. This rate of *de novo* synthesis of cholesterol could supply only about one-tenth of the sterol required for the assembly of new disk membranes, a finding that lends further support to the view that cholesterol reaches the retina preformed from exogenous sources.

Dolichols. As described above, there are low but measurable endogenous levels of Dol-P in *Rana pipiens* retina, and the absolute rate of synthesis of this key intermediate is 0.022 pmol/h per retina (Keller *et al.*, 1988). Dolichol compounds are probably not degraded in tissues, and Dol-P does not appear to be dephosphorylated. The biosynthesis of the core region of asparagine-linked glycoproteins and proteoglycans through the dolichol pathway and its inhibition by tunicamycin have been demonstrated in a variety of tissues (Kornfeld and Kornfeld, 1985), and this pathway is the major mechanism involved in the glycosylation of rhodopsin (Plantner *et al.*, 1980; Kean, 1980a; Fliesler and Basinger, 1985; Fliesler *et al.*, 1985, 1986a). The synthesis, assembly, and translocation of this integral membrane glycoprotein from the rod inner segments to the outer segments are discussed in Section 7.2.2.1.

Enzymes required for lipid activation of mannose and glucosamine and their transfer to endogenous Dol-P have been detected in cell-free preparations from bovine, squid, dogfish, and chick retinas (Kean, 1977, 1980a). Incubation of tissue homogenates with GDP-[^{14}C]mannose results in the formation of several labeled lipid-extractable products, the major one being Dol-P-mannose. Dol-P-[^{14}C]Mannose is a highly effective substrate for the transfer of labeled mannose into both oligosaccharide lipids and endogenous glycoproteins. Of interest is the novel observation that GDP-mannose not only acts as a substrate for mannosyltransferase reactions but it also causes a 10- to 15-fold stimulation in the biosynthesis of non-mannose-containing GlcNAc-lipids. This sugar nucleotide specifically enhances the activities of N-acetylglucosaminyl (GlcNAc) transferases in embryonic chick retina (Kean, 1980b). In the presence of Dol-P-mannose, either added exogenously or generated *in situ*, there is a seven- to 15-fold stimulation in the formation of early lipid-linked intermediates in glycoprotein synthesis from Dol-P and UDP-GlcNAc (Kean, 1982). The Dol-P-mannose is not a substrate in these reactions; rather, it functions as an allosteric activator of GlcNAc transferases that are involved in regulating the synthesis of early glucosaminyl polyprenol intermediates. This activation is about four times greater with Mg^{2+} than Mn^{2+} and appears to be specific for the formation of the first intermediate in the dolichol pathway, GlcNAc-P-P-Dol (Kean, 1983).

Phosphatidylglycerol also stimulates the activation of the specific glucosaminyltransferase involved in GlcNAc-P-P-Dol synthesis, and this phospholipid appears to be acting at the same allosteric site as Dol-P-mannose (Kean, 1985). When they are present together, there is a mutual inhibition of stimulation compared to additive effects observed for Dol-P-mannose or phosphatidylglycerol alone. Although these studies using embryonic chick retina have shed considerable light on early events in the formation of oligosaccharide-lipid intermediates in the retina, the use of intact bovine retina has led to the identification of a family of oligosaccharide-lipid intermediates synthesized from four radioactive sugars: mannose, glucose, galactose, and glucosamine (Plantner and Kean, 1988). The major intermediate formed in short-term incubations (less than 15 min) in bovine retina is the large fully glucosylated key intermediate, $Glc_3Man_9GlcNAc_2$-P-P-Dol, which is transferred directly to nascent glycoproteins. After longer periods of incubation of intact retina, the nonglucosylated intermediate $Man_9GlcNAc_2$ is the major compound formed. The accumulation of the nonglucosylated intermediate is not caused by glucose deprivation but rather it is due to the presence of α-glucosidase in the preparations. This enzyme may be acting as a glucosyl transferase:α-glucosidase "shuttle" that functions in modulating the level of oligosaccharide-lipid donors available for the final step in glycoprotein synthesis (Plantner and Kean, 1988).

Enzymatic Oxygenation of Polyunsaturated Fatty Acids. An overview of the metabolism of arachidonic acid (AA) to eicosanoids through three major oxidative pathways has been given in Chapter 1, Section 1.5. Two of these pathways for AA metabolism have now been described in the neural retina: the cyclooxygenase pathway resulting in the production of prostaglandins, prostacyclin, and thromboxane and the lipoxygenase pathway, which produces HETEs and leukotrienes (for reviews see Birkle and Bazan, 1985, 1986; Bazan *et al.*, 1986). In addition to AA, docosahexaenoic acid, 22:6(*n*-3), is also a substrate for lipoxygenation in the retina, as described below.

The synthesis of prostaglandins in rat retina has been demonstrated *in vitro* (Naveh-Floman *et al.*, 1984) and *in vivo* after intravitreal injection of labeled AA (Birkle and Bazan, 1984a). About 25% of the injected AA is incorporated into retinal lipids, and incubation of AA-labeled retinas results in the production of small but detectable amounts of prostaglandins PGE_2, $PGF_{2\alpha}$, PGD_2, and 6-keto-$PGF_{1\alpha}$, as well as thromboxane TXB_2 and 12-HETE. The acylation of labeled AA and its metabolism through the cyclooxygenase and lipoxygenase pathways have also been studied *in vitro* using intact bovine retina (Birkle and Bazan, 1984b). Approximately 6% of the labeled AA is converted to eicosanoids. Endogenous leukotrienes, further products of 5-HETE lipoxygenation, have been demonstrated in frog retina using a radioimmunoassay specific for LTC_4 (Bazan *et al.*, 1987). This leukotriene appears to be concentrated in ROS but is not produced there, as inferred from studies using the Ca^{2+} ionophore A23187.

In the classical view, it is arachidonic acid, 20:4(*n*-6), that is metabolized to eicosanoids through the cyclooxygenase, lipoxygenase, and cytochrome P450 pathways. However, another pathway has also been described in canine retina involving the lipoxygenation of docosahexaenoic acid, 22:6(*n*-3), to docosanoids, higher homologues of the eicosanoids produced by lipoxygenation of AA (Birkle and Bazan, 1986). The enzymes involved in lipoxygenation of docosahexaenoic acid may be different from those catalyzing the lipoxygenation of arachidonic acid, judging from their responses to lipoxygenase inhibitors.

7.2. PHOTORECEPTOR CELLS

The anatomic, functional, and topographical organization of rod and cone visual cells are well understood, and a simplified diagram of their overall structure in the vertebrate retina is given in Fig. 7.5. The outer segments of these cells contain the photosensitive visual pigments involved in the capture of photons and the transduction of light energy into electrical impulses. Signals are transmitted through the synaptic body of the photoreceptor cells to the bipolar cells, which then interact with second-order neurons before being transmitted to the brain. The unique structural design and chemical organization of photoreceptor cells subserve this central role in the visual process.

7.2.1. Chemical Composition

Rod outer segments (ROS) are one of the few cellular fractions of the retina that can be isolated in essentially pure form. The connecting cilium is very fragile, allowing

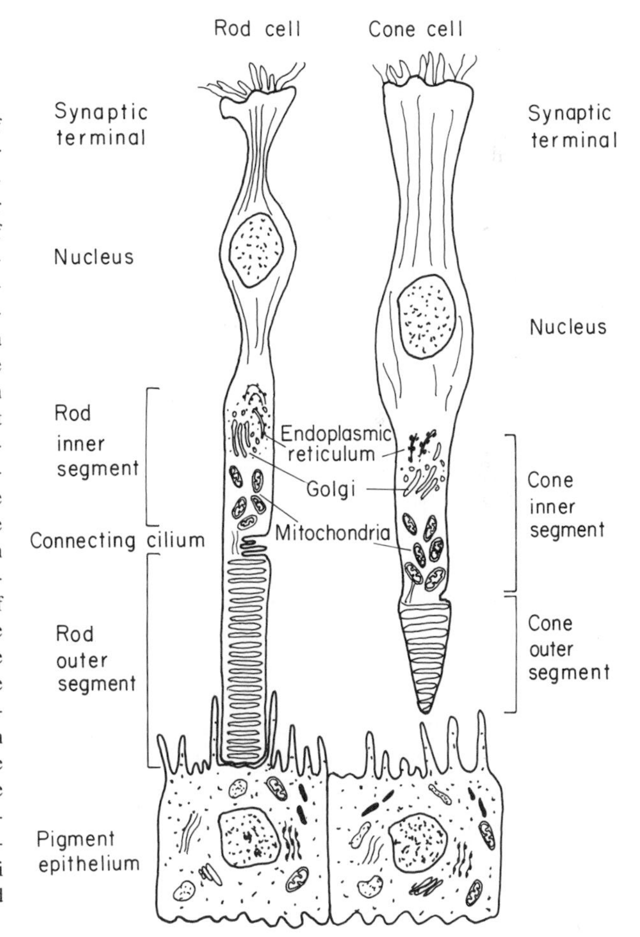

Figure 7.5. Schematic diagram of primate rod and cone cells and their relationship to the underlying pigment epithelial cell layer. The outer segments of the rod cell consist of parallel stacks of flattened disks; although usually considered to be free-floating, there may be weak attachments to each other and to the plasma membrane through thin spectrin-like filaments (see Section 7.2.1.1). In contrast, cone outer segments consist of continuous infoldings of the plasma membrane. The rod outer segments of most species are in close apposition to the apical surface of the pigment epithelium, surrounded in their distal portions by villous processes projecting from the surface of these cells. Cone outer segments are usually not in close apposition to the pigment epithelium but may have weak attachments through "cone matrix sheaths" (see Section 7.3.1). In both types of visual cell, the mitochondria are concentrated in the ellipsoid region of the inner segments; rough and smooth endoplasmic reticulum and the Golgi stacks are localized in the myoid region.

cleavage of ROS from the intact neural retina by simple agitation or mechanical shearing. Many methods have been developed to isolate and purify ROS membranes, all of the techniques utilizing flotation and/or centrifugation in continuous or discontinuous buffered sucrose density gradients (Papermaster and Dreyer, 1974; Papermaster, 1982; R. S. Molday and Molday, 1987). Membranes collecting at the 1.11/1.13 g/ml interface represent highly purified ROS membranes; however, the possibility of contamination with opsin-containing vesicles originating from the inner segment near the connecting cilium should not be overlooked (Fliesler and Anderson, 1983; Lolley, 1983). Although most studies utilize ROS membranes, methods are also available to isolate osmotically intact ROS disk membranes (Nemes and Dratz, 1982) as well as osmotically sealed ROS with intact plasma membranes (Schnetkamp and Daemen, 1982; Zimmerman and Godchaux, 1982; Schnetkamp and Kaupp, 1985). These preparations are especially useful for studying the total protein composition of ROS, as described below. New methods for isolating purified plasma membranes (Kamps *et al.*, 1982; L. L. Molday and Molday, 1987) as well as ROS disk and plasma membranes from retinas of cattle (R. S. Molday and Molday, 1987) and frog (Witt and Bownds, 1987) have also been described.

Rod outer segment (ROS) membranes contain nearly equal proportions, by weight, of protein and lipid. The major protein is rhodopsin, and the dominant lipid class consists of phospholipids, which account for about 80–90% of the ROS lipids in most species.

7.2.1.1. Proteins

Many of the ROS proteins other than integral membrane proteins such as rhodopsin are lost during the extensive washing used in the purification of ROS membranes. Hence, in the past, the rhodopsin content of ROS has usually been estimated as 90–95% of the total ROS protein. A truer picture emerges by studying the protein composition of osmotically intact (sealed) ROS. In one study on frog ROS, 20 major polypeptides in addition to rhodopsin have been detected and quantitated after SDS-PAGE (Hamm and Bownds, 1986). Another 50 peptides were also detectable in small amounts. These studies showed that rhodopsin comprises only about 70% of the total protein in osmotically intact frog ROS; the molar ratios of rhodopsin, transducin, and phosphodiesterase (PDE) in these preparations were estimated to be 100:10:1.

As described above, many methods are available for obtaining purified ROS membranes or osmotically intact ROS or ROS disk and plasma membranes. Moreover, new techniques have been developed to characterize the proteins in these preparations (Table 7.7). Among them, chemical labeling combined with immunocytochemistry and electron microscopy has proved to be a powerful tool for the identification and localization of specific ROS disk and plasma membrane proteins (L. L. Molday and Molday, 1987; R. S. Molday and Molday, 1987). The production of monoclonal antibodies and their use in immunocytochemical labeling of rhodopsin, cone opsin, and other ROS proteins have been reviewed in detail by Molday (1988).

Current information derived from several laboratories suggests a classification of ROS proteins into three functional and/or structural compartments: (1) integral membrane proteins of ROS disks; (2) peripheral and soluble proteins; and (3) plasma membrane proteins (both integral and peripheral). This should be considered only as an operational classification, since rhodopsin is a major integral membrane protein of both ROS disks and plasma membranes. It appears to be the only major protein that is present in both

Table 7.7. Techniques for Isolation and/or Characterization of Rod Outer Segment Proteins[a]

Technique	Reference
Lactoperoxidase-catalyzed radioiodination	Clark and Hall (1982)
Con A-coated polystyrene beads	Kamps *et al.* (1982)
Monoclonal antibodies to ROS disk and plasma membrane proteins	Wong and Molday (1986)
Percoll density gradients	Hamm and Bownds (1986)
Percoll density gradients; lactoperoxidase-catalyzed radioiodination	Witt and Bownds (1987)
Immunogold-dextran	R. S. Molday *et al.* (1987)
Sucrose density gradient; neuraminidase; ricin–gold–dextran labeling; dissociation of plasma membranes from disks by trypsin or low-ionic-strength buffer	L. L. Molday and Molday (1987); R. S. Molday and Molday (1987)

[a]Proteins are resolved by SDS-PAGE and/or identified *in situ*.

membranes and comprises about 90% and 50% of disk and plasma membrane protein, respectively (Molday, 1988). Apart from rhodopsin, however, the protein compositions of these two membranes are strikingly different (R. S. Molday and Molday, 1987; Molday, 1988). An excellent working model showing the probable spatial arrangements and interrelationships of most of the known proteins in vertebrate ROS membranes has appeared in a recent review (Fig. 20 in Molday, 1988).

Integral ROS Disk Membrane Proteins. *Rhodopsin* is the major integral membrane protein of ROS disks (Table 7.8); it is a glycoprotein consisting of a single 40-kDa polypeptide chain, opsin, in combination with 11-*cis* retinal chromophore. Its composition and structure have been studied extensively, and the brief description that follows is based mainly on several detailed reviews and monographs (Hargrave *et al.*, 1980, 1984; Dratz and Hargrave, 1983; Shichi, 1983; Anderson, 1983a; Applebury and Hargrave, 1986; Baehr

Table 7.8. Integral Membrane Proteins of Rod Outer Segment Disks[a]

Protein	Molecular mass (kDa)	Location	Reference
Rhodopsin	40	Lamellar region of disk membrane and plasma membrane	Molday *et al.* (1987); R. S. Molday (1987)
Rim protein (ROS 1.2; or 120-kDa trypsin fragment)	220	Rim of disk membrane	Szuts (1985); Molday and Molday (1979, 1987)
Rim protein	290[b]	Rim of disk membrane and incisures	Papermaster *et al.* (1978a)
Peripherin	33–35[c]	Rim of disk membrane	Molday *et al.* (1987)

[a]Data are on bovine ROS except where indicated.
[b]Frog retina.
[c]In the presence of 2-mercaptoethanol; in its absence, a 67- and 69-kDa doublet is formed (Molday *et al.*, 1987).

and Applebury, 1986; Findlay and Pappin, 1986; Hargrave, 1986; Dohlman *et al.*, 1987). The light-stimulated opsins in photoreceptors form part of a multi-gene family and are structurally and functionally related to the hormone-stimulated β-adrenergic and muscarinic cholinergic receptors found in a wide variety of tissues (see Chapter 1, Section 1.3). The present discussion is limited to vertebrate rhodopsins, particularly bovine ROS rhodopsin, whose complete amino acid sequence was elucidated some years ago (Ovchinnikov *et al.*, 1982; Hargrave *et al.*, 1983).

Bovine and other vertebrate rhodopsins have been sequenced and found to contain 348 amino acids. Primary amino acid sequences deduced from molecular cloning show that all of them share sequence homology and have remarkably similar topography in the membrane. Although there is some variation in molecular weight from various sources, most opsins are structurally and topographically similar; a molecular model showing the organization of bovine opsin in the ROS disk membrane is given in Fig. 7.6. The seven membrane-spanning hydrophobic α-helices, which comprise about half of the molecular

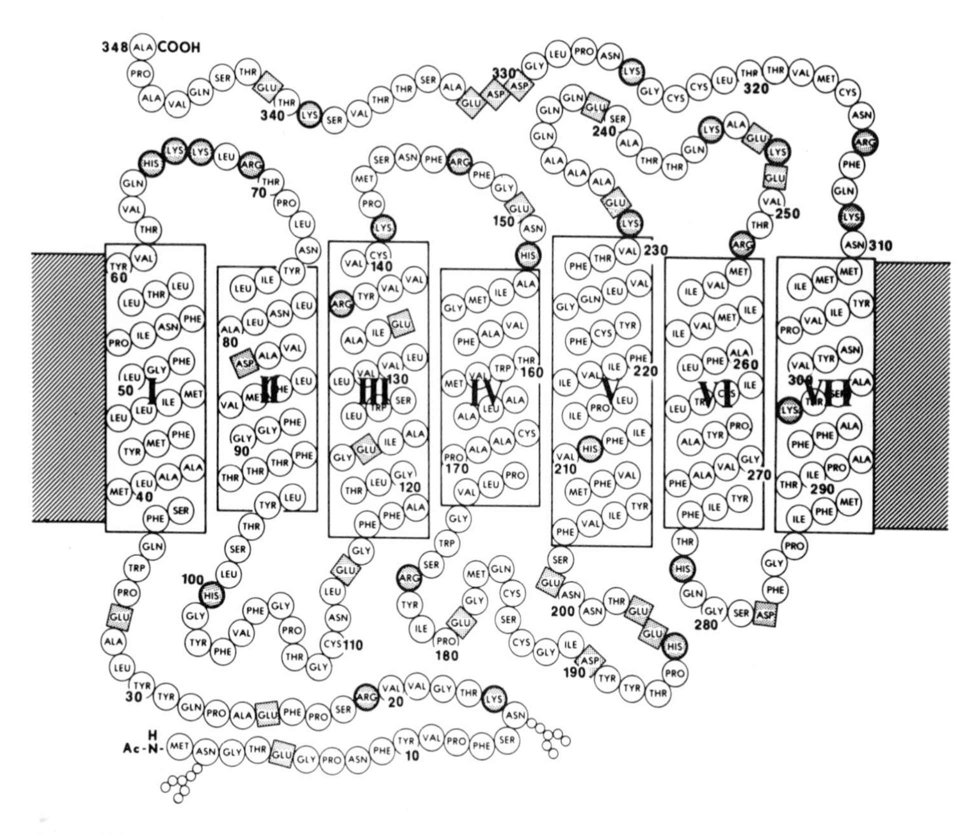

Figure 7.6. Topographical model of bovine opsin in the lipid bilayer membrane of the rod outer segment disk. The amino terminus is exposed at the internal cytoplasmic (intradisk) surface, and the nonpolar polypeptides are shown as seven transmembrane helices. The carboxyl terminus is exposed on the external cytoplasmic surface. The shaded areas at either side represent the low-polarity regions of the lipid bilayer. The binding site for 11-*cis* retinal is located in a pocket on helix VII in Schiff-base linkage to Lys-296. (From Applebury and Hargrave, 1986.) (Reprinted with permission from *Vision Res.* **26,** M.L. Applebury and P.A. Hargrave, "Molecular biology of the visual pigments," © 1986, Pergamon Journals Ltd.)

mass of the molecule (Dratz *et al.*, 1979), are embedded in the lipid bilayer approximately perpendicular to the plane of the membrane. Each helix contains about 21–28 amino acids, which are highly conserved, as discussed below. The helical domains are connected by sequences of predominantly polar amino acids that form loops into both the cytoplasmic and luminal surfaces of the membrane.

The carboxyl-terminal region of rhodopsin is exposed on the cytoplasmic surface of the disk membrane (Dratz *et al.*, 1979). It is rich in hydroxy amino acids, serine and threonine, and provides the sites for phosphorylation (after illumination) by rhodopsin kinase. The 11-*cis* retinal is bound through a protonated Schiff-base linkage to the ϵ-amino group of lysine (Lys-296) on helix VII in a hydrophobic "pocket" near the center of the membrane. The chromophore lies parallel to the plane of the membrane and perpendicular to the helix; incoming light is thought to induce small conformational changes in this transmembrane segment and larger changes on the cytoplasmic surface, where peripheral proteins are located. These events on the cytoplasmic surface initiate the cascade of visual excitation.

The amino-terminal region is exposed on the luminal (or intradiskal) surface; it is the principal site for the glycosylation of rhodopsin, which occurs through two N-linked asparagines, Asn-2 and Asn-15 (Fig. 7.6). The principal Asn-linked oligosaccharide chains consist entirely of short hexasaccharides of mannose (Man) and N-acetyl-glucosamine (GlcNAc) (Fukuda *et al.*, 1979; Liang *et al.*, 1979). There are two oligosaccharides per mol of rhodopsin; the major one (whose structure is shown below) accounts for about 70% of the total oligosaccharide content (Hargrave *et al.*, 1984). Two minor components having one and two additional mannose residues represent about 10% and 20%, respectively, of the rhodopsin oligosaccharides.

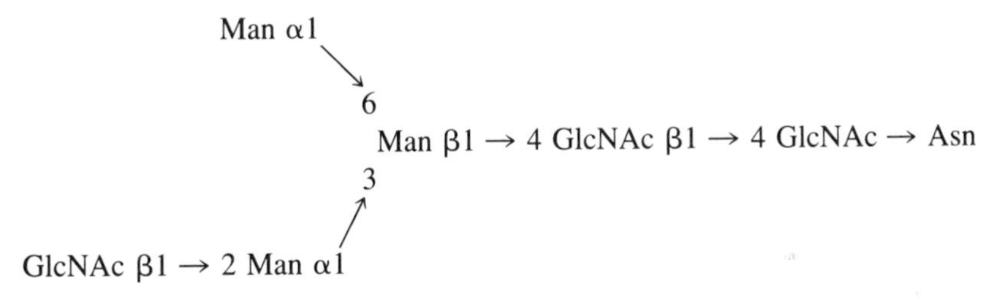

Three main categories of intermediate oligosaccharides have been identified in the series of enzymatic steps leading to the formation of Asn-linked oligosaccharides: high-mannose, complex, and hybrid (Kornfeld and Kornfeld, 1985). In the majority of tissues, they all have a common pentasaccharide core structure because they arise through the same dolichol-linked oligosaccharide precursor. The oligosaccharide $Glc_3Man_9GlNAc_2$ is transferred from this lipid-linked intermediate to nascent polypeptides in the rough endoplasmic reticulum and afterward to the cis Golgi cisternae, a step requiring trimming of glucose residues. Further processing is carried out sequentially in the medial and trans cisternae. It is of interest that the oligosaccharides in rhodopsin are of the hybrid type, intermediate in composition between the initially processed high-mannose type and the final complex type. The localization of this lipid intermediate pathway in the photoreceptor cell and its inhibition by tunicamycin are discussed in Section 7.2.2.1.

Immunochemical and immunocytochemical approaches are yielding valuable information on structural–functional relationships in the rhodopsin molecule, but a detailed discussion of these studies is beyond the scope of the present chapter. However, an in-depth review is now available describing the production and characterization of anti-rhodopsin monoclonal antibodies and their use as probes for studying defined regions of the molecule (Molday, 1988). These monoclonal antibodies are also being used as specific tools for studying the localization and orientation of rhodopsin in the photoreceptor cell membrane.

With the elucidation of the amino acid sequence of bovine rhodopsin through the application of recombinant DNA techniques (Nathans and Hogness, 1983, 1984; Nathans *et al.*, 1986a,b), it is now possible to study the genetic information encoding rhodopsin and other photoreceptor opsins (for reviews see Baehr and Applebury, 1986; Applebury and Hargrave, 1986). Specific oligonucleotide probes for a short amino acid sequence near the C terminus were used to prime reverse transcriptase for the synthesis of opsin cDNA. Opsin clones in cDNA retinal libraries were subsequently identified, and primary structures of at least 11 visual pigments have been deduced. The 6.4-kb bovine gene consists of a 96-bp 5′ untranslated region, a 1044-bp coding region, and a long (approximately 1400-bp) untranslated region. The sequences in the human opsin gene are similar, having four introns interrupting five exons; the deduced amino acid sequence is 348 residues long and is 93.4% homologous to bovine rhodopsin. Studies of visual pigment gene structures from a variety of vertebrate and invertebrate sources have revealed that although the position of the first intron is not conserved between vertebrate opsins and *Drosophila* R1-6 opsin, the other introns are in fact highly conserved. Strongly homologous coding regions have been noted in a wide variety of organisms (Martin *et al.*, 1986), implying that visual pigments of vertebrates, invertebrates, and even single-cell organisms are derived from a common ancestral precursor.

A *220-kDa protein*, another integral ROS disk membrane protein, is localized in the rim region (the far edge) of rod outer segment disks. This region appears to be a specialized domain whose function may be to maintain the structural integrity of the photoreceptor (R. S. Molday *et al.*, 1987). A 290-kDa integral membrane protein was first characterized in the rim and incisures of frog disk membranes (Papermaster *et al.*, 1978a), and short filamentous structures extending between adjacent disks as well as from disks to plasma membranes have been described (Roof and Heuser, 1982). A similar integral membrane protein of M_r 220,000 was detected in both frog and bovine ROS disks and named ROS 1.2 glycoprotein (Molday and Molday, 1979). There is now firm evidence that the 220-kDa concanavalin A-binding rim protein is the second major protein, after rhodopsin, in bovine disk membranes isolated free of plasma membranes (L. L. Molday and Molday, 1987; R. S. Molday and Molday, 1987). Dilute trypsin cleaves the 220-kDa rim protein to a 120-kDa fragment, previously designated ROS 1.2 glycoprotein by Molday and Molday (1979). This transmembrane glycoprotein probably links adjacent disks or may be involved in maintaining the highly curved structure of the rim through interaction with peripherin (described below) (Molday, 1988).

A physiological function for the 220-kDa rim glycoprotein has not been established, but it has several interesting properties. Using suspensions of dark-adapted ROS prepared from frog retina, it has been shown that this protein as well as a 240-kDa polypeptide is phosphorylated after brief illumination (Szuts, 1985). In addition, the 220-kDa rim protein has specific binding sites for ^{32}P-labeled azido-GTP and -ATP (Shuster *et al.*, 1988). Unlabeled guanine nucleotides are the most effective competitive inhibitors of azido-nucleotide binding, whereas divalent cations, especially zinc, stimulate the binding.

Peripherin is a 33- to 35-kDa protein present in significant quantities in pure bovine

ROS disk membranes (Molday *et al.*, 1987; R. S. Molday and Molday, 1987). It is an integral membrane protein that is extracted in detergents together with rhodopsin. It was not detected previously using standard gel-staining methods because it is masked under the broad rhodopsin band. However, two monoclonal antibodies generated against ROS disk proteins have been shown to bind to this protein, which travels as a sharp band at the leading edge of the rhodopsin band in samples treated with 2-mercaptoethanol and subjected to SDS-PAGE. In the absence of reducing agent, it migrates as a doublet with molecular weights of 67,000 and 69,000. Immunogold labeling of ROS sections as well as intact disks distinguishes peripherin from 220-kDa rim protein and shows its localization in the periphery of rod outer segment disks, in contrast to rhodopsin, which occupies the lamellar region (the flat surface) of the disk.

In summary, at least three integral membrane proteins are present in ROS disks. The major one is rhodopsin, localized in the lamellar region of bovine disks. Two additional proteins, the 220-kDa rim-specific glycoprotein and the 33- to 35-kDa protein named peripherin, are located in the rim region of the disk. Other ROS disk membrane proteins detected in silver-stained SDS gels migrate as bands of 43- and 57- to 63-kDa polypeptides but have not yet been identified (R. S. Molday and Molday, 1987).

Peripheral and Soluble (Cytosolic) Proteins. Peripheral membrane proteins represent a major class of proteins in all cells. In contrast to integral membrane proteins, whose hydrophobic domains span the lipid bilayer, peripheral proteins do not interact with the hydrophobic core of the membrane. They are usually bound indirectly, through ionic or other weak interactions, with integral membrane proteins or more directly by interactions with phospholipid polar head groups. Most peripheral proteins can be removed from the membrane by extraction in buffered aqueous solutions.

At least five peripheral and soluble proteins, first noted over a decade ago in osmotically intact bovine ROS (Godchaux and Zimmerman, 1979), have since been extensively investigated, and several are now known to have key roles in phototransduction. The names assigned to them and their molecular masses and localizations are summarized in Table 7.9. Two of these are peripheral proteins, located on the photoreceptor disk surface: transducin (G protein) and phosphodiesterase (PDE) (Fung, 1987). Three others, rhodopsin kinase, 48-kDa protein (S-antigen), and guanylate cyclase, are probably cytosolic proteins.

Transducin, a member of the superfamily of G proteins (see Chapter 1, Section 1.3), is a heterotrimer composed of two functional subunits: T_α (a 40-kDa polypeptide) and $T_{\beta\gamma}$ (T_β and T_γ are 37- and 8-kDa polypeptides, respectively) (Baehr *et al.*, 1982; Fung, 1983). The T_α subunit binds both GTP and GDP, but it is the GTP-bound form that activates phosphodiesterase (PDE) (for reviews see Stryer and Bourne, 1986; Stryer, 1986; Fung, 1987; Liebman *et al.*, 1987; Casey and Gilman, 1988). After illumination, transducin becomes tightly bound to photolyzed rhodopsin (R*) in the ROS membranes; this binding is reversible by the addition of GTP (Kuhn, 1980).

The primary structures of all the transducin subunits have been deduced from molecular cloning and cDNA sequencing. The T_α subunit consists of 350 amino acids and has a molecular weight of 39,971 (Medynski *et al.*, 1985). Sequencing studies have revealed two different T_α subunits, one in bovine rod outer segments (Yatsunami and Khorana, 1985) and the other in cone outer segments (Lerea *et al.*, 1986). More recently, a cDNA clone for $G_{o\alpha}$ has been isolated from a bovine retinal library that exhibits 90% homology with a cDNA clone from rat brain and is also closely related to $G_{o\alpha}$ of bovine

Table 7.9. Peripheral and Soluble Proteins of Bovine Rod Outer Segments[a]

Protein	Molecular mass (kDa)	Location	Reference
Transducin		Disk surface	Fung (1987)
α subunit	40		
β subunit	37		
γ subunit	8		
Phosphodiesterase		Disk surface	Fung (1987)
α subunit	88		
β subunit	84 (86)		
γ subunit	13		
Guanylate cyclase	—	Cytosol and axoneme	Fung (1987)
Rhodopsin kinase	68	Cytosol	Fung (1987)
48-kDa protein (S-antigen)	48	Cytosol	Fung (1987)
33-kDa–$T_{\beta\gamma}$ phospho protein	77	Cytosol	Lee *et al.* (1987, 1988a,b)

[a]Adapted from Fung (1987).

brain and other G proteins of bovine origin (van Meurs *et al.*, 1987). The β subunit of bovine transducin is also highly conserved and may even be identical to other G protein β subunits (Hildebrandt *et al.*, 1985; Fong *et al.*, 1986; Fung, 1987). The amino acid sequence of the 74 residues in the 8-kDa γ subunit of transducin have also been determined (Yatsunami *et al.*, 1985), but it does not share extensive homologies with γ subunits of other G proteins (Hildebrandt *et al.*, 1985). Because the β subunits of transducin have identical sequences to those of other G proteins, it seems possible that it is the γ subunit that confers specificity to the βγ complex in its interaction with the α subunit (Fung, 1987).

The mRNA encoding T_α as well as immunoreactive T_α are both localized solely in the visual cells of rat retina (Brann and Cohen, 1987). The combined use of immunohistochemical techniques and *in situ* hybridization histochemistry shows that whereas T_α mRNA is localized primarily in the inner segments at all times of the day or night, T_α immunoreactivity reaches peak levels in the inner segments during the day and in the outer segments at night. These observations suggest that T_α is synthesized in the inner segments after an increase in T_α mRNA with the onset of light, and the newly synthesized T_α remains in the inner segments until it is transported to the outer segments at night. Further confirmation of this interesting finding is needed; if it is true, then this diurnal cycle follows a time course remarkably similar to that of disk shedding in the rat, which is highest shortly after the onset of morning light (LaVail, 1976). Whether these two processes are related remains to be established, but the diurnal expression of immunoreactive T_α and its circadian or light-regulated translocation in the visual cells of rat may have a physiological role in regulating photoreceptor sensitivity.

Phosphodiesterase, a peripheral membrane heterotrimer that hydrolyzes cGMP to 5′-GMP, plays a key role in the cGMP cascade of phototransduction (for reviews see Stryer, 1986; Liebman *et al.*, 1987; Fung, 1987). Photoreceptor-specific cGMP phosphodiesterase (ROS PDE) isolated and purified from bovine rod outer segments is a 170-kDa protein composed of three subunits (Baehr *et al.*, 1979). The enzymatic activity has been assigned to a core protein consisting of subunits α and β, of M_r 88,000 and 84,000, respectively. The 11- to 13-kDa γ subunit has an important and specific regulatory function in inhibiting the enzyme in the dark (Hurley and Stryer, 1982). Removal of this inhibitory constraint by trypsin degradation of the γ subunit results in a striking increase in PDE catalytic activity. Moreover, addition of purified γ subunit to trypsin-activated ROS PDE results in almost complete inhibition of enzyme activity. In the cGMP cascade, the interaction of ROS PDE with the α subunit of transducin (visual cell G protein) releases the γ subunit of ROS PDE, which removes the enzymatic inhibition. Activation of the enzyme leads to a rapid drop in the concentration of cytosolic cGMP. It is this event that results in closing of the Na^+ channels and hyperpolarization of the plasma membrane (Fesenko *et al.*, 1985; Baehr and Applebury, 1986; Baylor, 1987). The molecular events that precede and follow this crucial step, and the role of cGMP as the second messenger in visual transduction, are discussed in Section 7.8.2.

Guanylate cyclase is a peripherally bound enzyme that catalyses the synthesis of cGMP from GTP. Relatively little is known of factors controlling its activity, although several studies suggest that it is stimulated after lowering of cytosolic Ca^{2+} following illumination. In bovine ROS, guanylate cyclase activity decreases fivefold when the concentration of Ca^{2+} is raised from 20 to 120 nM (Koch and Stryer, 1988). Conversely, its activity increases as the concentration of Ca^{2+} is lowered. Thus, the activity of this enzyme is regulated by a calcium switch (possibly a protein) that senses changes in calcium concentration. This activation of guanylate cyclase, which is controlled by a highly cooperative feedback mechanism, may be a key event in restoring the dark current after visual excitation.

Rhodopsin kinase is a cytosolic enzyme that catalyzes the phosphorylation of light-activated rhodopsin (Shichi and Somers, 1978; Kuhn, 1978). In the reaction, the γ-phosphate of ATP is transferred to threonine and serine residues located in the C-terminal peptide on the cytoplasmic surface of the disk membrane (Hargrave *et al.*, 1980; Miller and Dratz, 1984). The number of phosphorylation sites has been estimated as seven in bovine rhodopsin (Wilden and Kuhn, 1982). This, however, represents an average value for the total population of rhodopsin molecules; under optimum conditions, some individual molecules of bovine and frog rhodopsin may contain 8 or 9 mol of phosphate per mol of rhodopsin.

Phosphorylated rhodopsin has a decreased ability to activate transducin and enhanced ability to bind 48-kDa protein, the latter completing the deactivation of rhodopsin, as described below. Despite the important regulatory role played by rhodopsin kinase, little has been known until recently about its kinetic properties or substrate specificity. The enzyme has now been purified to near homogeneity from bovine ROS and shown to be a single polypeptide of M_r 67,000–70,000 (Hargrave *et al.*, 1988; Palczewski *et al.*, 1988a,b). It has a high specificity and low K_m for each of its substrates, freshly bleached rhodopsin and ATP. The binding of the kinase to rhodopsin and the phosphorylation of rhodopsin appear to be two separate reactions. In addition to serine and threonine residues in the C terminus, rhodopsin kinase also phosphorylates similar residues in the third cytoplasmic loop, the hydrophilic sequence that connects helices V and VI (see Fig. 7.6). Compounds such as cGMP and others whose concentration changes when photoreceptors are exposed to light do not substantially influence the activity of rhodopsin kinase. Purified rhodopsin kinase differs from other protein kinases such as cAMP-dependent kinases and protein kinase C but shares many properties with β-adrenergic receptor kinase (βARK). Rhodopsin kinase may in fact be a member of a family of kinases that phosphorylate (and densensitize) G protein receptors.

Dephosphorylation of phosphoopsin has also been demonstrated in bovine ROS (Palczewski *et al.*, 1989). The activity is similar to that of the catalytic subunit of skeletal muscle protein phosphatase 2A. The phosphatase is only active on highly phosphorylated opsin (4 mol of phosphate per mol of rhodopsin).

The *48-kDa protein* (S-antigen, arrestin) is cytoplasmic in the dark but in the light binds specifically to photoexcited-phosphorylated rhodopsin (R*-P) and quenches its activation of transducin (Wilden *et al.*, 1986). The 48-kDa protein is highly soluble in dark-adapted retinas and represents about 2–7% of the ROS proteins (Pfister *et al.*, 1985). The biochemical identity of 48-kDa protein and S-antigen has been demonstrated by immunological techniques and functional tests. The complete primary sequence of 48-kDa protein deduced from cDNA clones shows that it contains 404 amino acids and has a M_r of 45,275 (Shinohara *et al.*, 1987). It also contains small amounts of carbohydrate, approximately 4 mol of sugar per mol of S-antigen, in N linkage to two Asn residues. Sequence similarities with T_α, the α subunit of transducin, are found in the C-terminal portions of both peptides, including the site of ADP-ribophosphorylation by pertussis toxin. S-Antigen plays an important role in phototransduction, probably by competing with transducin for binding sites on photolyzed-phosphorylated rhodopsin, thus preventing further activation of phosphodiesterase (Fung, 1987).

The *33-kDa–Tβγ phosphoprotein* was first described as a 33-kDa protein that is phosphorylated in a cyclic nucleotide-dependent manner in darkness and dephosphorylated by exposure to light (Lee *et al.*, 1984). The protein kinase that catalyzes the phosphorylation appears to be distinct from rhodopsin kinase (Lee *et al.*, 1981). Purification of the 33-kDa protein from bovine retinas has shown that its native conformation *in situ* is that of a 77-kDa heterotrimer (Lee *et al.*, 1987). Under conditions that prevent dissociation, the 33-kDa protein is complexed to 37-kDa and 10-kDa subunits that are immunologically identical to the β and γ subunits of transducin. Monospecific antibodies for the 33-kDa and β subunits used to localize the complex by immunocytochemical methods show that the 33-kDa–$T_{\beta\gamma}$ phosphoprotein is found exclusively in photoreceptor cells (Lee *et al.*, 1988a). It does not seem probable that the 33-kDa–$T_{\beta\gamma}$ complex plays a direct role in phototransduction; rather, it may represent a branch point from the cGMP cascade that regulates light-modulated metabolic activities in the photoreceptors (Lee *et al.*, 1987).

Rod Outer Segment Plasma Membrane Proteins. There are relatively few studies on ROS plasma membrane proteins because of difficulties in the purification of such a small fraction (about 5%) of the total membranes; moreover, there are few if any suitable markers for identifying them. Recently, however, a novel technique using ricin–gold–dextran particles has been developed for isolating either highly purified plasma membranes (L. L. Molday and Molday, 1987) or disk and plasma membranes (R. S. Molday and Molday, 1987) from bovine ROS. Disruption of the filaments connecting the disks to the plasma membranes (Roof and Heuser, 1982) is an essential step, and when this is done, usually by mild trypsin digestion or exposure to low-ionic-strength buffer, the two membrane types can be separated by a density-gradient fractionation procedure based on the specific labeling of the plasma membranes with ricin–gold–dextran particles. Two proteins of major functional importance are under active investigation using these and closely related techniques for isolating ROS membranes. One of them, the cGMP-gated channel, is now known to be localized exclusively in the plasma membrane (Cook *et al.*, 1989), and the other, the Na^{+}–Ca^{2+} exchanger, is also likely to be localized there (Cook and Kaupp, 1988; R. N. Molday, personal communication). Other plasma membrane-specific proteins are listed in Table 7.10.

The *cGMP-sensitive channel* has been identified using modifications of the novel and powerful patch-clamp suction-pipet technique developed in recent years for measuring ion channels in other tissues. The key step involves formation of an ultrahigh-resistance seal between a glass micropipet and a small patch of plasma membrane. Currents may be recorded through a patch of membrane while it is still attached to the cell (cell-attached patch) or after detachment from the cell (excised patch). A commonly used variant of the patch-clamp technique involves a pulse of suction or voltage that ruptures the patch after seal formation, leaving the interior of the pipet continuous with the cell cytoplasm. The composition of bathing solutions can be precisely controlled, and membrane conductance measured under a variety of experimental conditions.

The first breakthrough in identifying the cGMP-sensitive channel came with the demonstration that application of cGMP to the intracellular side of excised ROS patches activates a conductance having permeability properties similar to the light-sensitive conductance (Fesenko *et al.*, 1985). Membrane conductance has also been measured in truncated rod outer segments; the ROS are drawn into the suction pipet, and the inner segment and basal portions of the outer segment outside the pipet are broken off. This leaves an open-ended outer segment whose internal space can be dialyzed with cGMP and other solutions (Yau and Nakatani, 1985). These experiments showed conclusively that cGMP activates a light-suppressable conductance. These and other findings (see review by Fung, 1987) leave little doubt that the light-sensitive channel of the ROS membrane is controlled solely by cGMP, and the cGMP-gated channel reported by several investigators

Table 7.10. Plasma Membrane Proteins of Bovine Rod Outer Segments

Protein	Molecular mass (kDa)	Reference
cGMP-sensitive channel	63[a] 52;54 39	Cook *et al.* (1987, 1989); R. Molday and Molday (1987); Matesik and Liebman (1987)
Rhodopsin	40	R. Molday and Molday (1987)
Proteins; neuraminidase-sensitive glycoproteins	110 160 226;230[b]	Kamps *et al.* (1982); Clark and Hall (1982); L. Molday and Molday (1987); R. Molday and Molday (1987)
Spectrin-like protein[c]	240	Roof and Heuser (1982); Molday and Molday (1979); Wong and Molday (1986)
Other proteins	24[d] 13[d] 38	Witt and Bownds (1987); R. Molday and Molday (1987)

[a]The 63-kDa cGMP-sensitive channel is localized almost exclusively, if not entirely, in the plasma membrane (Cook *et al.*, 1989).
[b]It is highly probable that the 226;230 glycoprotein is the Na^{+}–Ca^{2+} exchanger (see text). This 220-kDa polypeptide has been purified to homogeneity from ROS disks (Cook and Kaupp, 1988).
[c]Identified as filaments connecting the inner plasma membrane to the disk membranes.
[d]Frog retina.

using excised patches is identical to the light-sensitive conductance measured in intact rods (Zimmerman *et al.*, 1985). This conductance is activated by the cooperative binding of at least three cGMP molecules per channel (Haynes *et al.*, 1986; Zimmerman and Baylor, 1986; Karpen *et al.*, 1988). Other experiments using the patch-clamp technique also provide direct evidence that cGMP acts as an internal transmitter in keeping the channels open in darkness; they are closed in light as the concentration of cGMP in the ROS drops (Matthews, 1987).

An important question in membrane conductance, whether mediation is through a pore (or channel) or a shuttle-type carrier, has been resolved by the isolation of the cGMP-sensitive channel itself, reported variously as a 39-, 52-, or 63-kDa protein (see Table 7.10). A 39-kDa protein isolated from bovine ROS membranes stripped of peripheral proteins mediates cGMP-dependent cation fluxes in reconstructed systems (Matesic and Liebman, 1987). Using similar techniques, a highly purified channel preparation isolated by Cook *et al.* (1987) had an apparent molecular mass of 63 kDa. Later studies using ricin–gold–dextran particles for the purification of plasma membranes and monoclonal antibody to the 63-kDa cGMP-gated channel show that it is localized exclusively in the plasma membrane (Cook *et al.*, 1989). Although several laboratories had shown that the channel protein is present in ROS disk membranes, this was an artifact arising from fusion of plasma membrane components during permeabilization of photoreceptor cells. The cGMP channel comprises about 7% of the total protein of bovine ROS plasma membranes. Previous investigations using purified plasma membranes suggest that a major 52- to 54-kDa protein is probably the same as the 63-kDa protein, since it cross-reacts with purified 63-kDa cGMP-sensitive channel protein (Cook *et al.*, 1987; Molday, 1988). The possibility has also been raised that the native channel consists of both 39- and 63-kDa subunits that reassemble after detergent extraction into active preparations having slightly different molecular characteristics (Applebury, 1987).

In addition to rhodopsin, proteins of M_r 230,000 and 110,000 were detected some years ago in ROS plasma membranes (Kamps *et al.*, 1982; Clark and Hall, 1982). Although the membranes were highly purified, suitable markers were not available at the time. More recent investigations demonstrate the presence of ricin-binding glycoproteins on the surface of neuraminidase-treated ROS (L. L. Molday and Molday, 1987; R. S. Molday and Molday, 1987; Molday, 1988). These 110- and 230-kDa glycoproteins are specific to ROS plasma membranes. One of them, the 230-kDa glycoprotein, has been isolated and purified from bovine ROS; after reconstitution into liposomes, it was identified as the Na^{+}–Ca^{2+} exchanger (Cook and Kaupp, 1988). This 220-kDa glycoprotein is distinct from rim protein (of similar M_r) but appears to be identical to the 230-kDa neuramindase-sensitive glycoprotein localized in ROS plasma membranes (R. S. Molday, personal communication).

Another important component of ROS plasma membranes is the 240-kDa spectrin-like protein (Table 7.10). This protein exhibits immunologic cross-reactivity with cytoskeletal proteins spectrin and fodrin and may represent a unique type of cytoskeletal system that lines the inner surface of the ROS plasma membrane (Wong and Molday, 1986; Molday, 1988). Spectrin-like protein is thought to be anchored to the plasma membrane through the 54-kDa (or 63-kDa) plasma membrane channel protein, a view supported by the finding these proteins copurify (Cook *et al.*, 1987). Direct interactions between ion channels and cytoskeletal networks have been demonstrated in red blood cells and may also be present in ROS (Molday, 1988).

7.2.1.2. Lipids

An important property of biomembranes in general, and photoreceptor membranes in particular, is their fluidity. The viscosity of ROS membranes has been estimated as approximately 2 poise, which is similar to that of olive oil (Fliesler and Anderson, 1983; Shichi, 1983). Viscosity is inversely related to fluidity, and comparison with other biomembranes places retinal disks in the lower range of viscosity and hence in the higher range of fluidity. As a consequence, the translational freedom of rhodopsin in the disk membrane is high, and it has been estimated that there are approximately 100,000 collisions per second between molecules (Shichi, 1983). One of the factors contributing to the high fluidity of the disk membrane is the low ratio of cholesterol/phospholipid, which is in the range of 0.09–0.11 in ROS membranes of most vertebrates except goldfish (Fliesler and Anderson, 1983). Another factor is the high content of polyunsaturated fatty acids. Despite many studies on the molecular organization of lipids in the disk bilayer membrane, their precise locations and orientations are still uncertain (see review by Fliesler and Anderson, 1983).

The lipid composition of ROS membranes is similar in most vertebrates, and analyses of mammalian species as well as amphibia and fish have been reported from several laboratories. Some of these analytical data are summarized in Table 7.11. Phospholipids account for about 85–90 mol% of the total lipids; the remainder consist of neutral lipids (Fliesler and Anderson, 1983). The two principal phospholipids are PC and PE, and PS is the next most abundant. Although only small amounts of PI are present in ROS membranes, its turnover is much faster than that of other phospholipids (Anderson *et al.*, 1980d). A unique metabolism for PI in ROS membranes is now recognized and best explained in terms of receptor-stimulated polyphosphoinositide (PI) breakdown or the "phosphoinositide effect" (see Section 7.2.2.2).

The major neutral lipids of frog and rat ROS membranes are cholesterol, free fatty acids (FFA), and 1,2-diglycerides (Wiegand and Anderson, 1983). Cholesterol is present only as the free sterol; it comprises about 5–10 mol% of the total lipid and about 50–60 mol% of the neutral lipids (Table 7.11). The major free fatty acids in frog ROS are

Table 7.11. Lipid Composition of Vertebrate Rod Outer Segments[a]

	Species			
Lipid[b]	Rat	Cattle	Human	Frog
Phospholipids				
PC	27.5	32–36	32.3	36.4
PE	38.1	37–43	37.6	29.0
PS	12.8	12–14	12.0	10.8
PI	0.9	2	—	1.9
SPH	0.7	1–4	1.8	1.6
Neutral lipids				
Cholesterol	7.3	9.1	9.8	4.9
FFA	3.5	6.8	2.1	2.2
Diglycerides	0.5	0.9	0.9	2.0

[a]Adapted from Fliesler and Anderson (1983), Wiegand and Anderson (1983), and Shichi (1983). Data are expressed as mol%.
[b]The following abbreviations are used: PC, phosphatidylcholine; PE, phosphatidylethanolamine; PS, phosphatidylserine; PI, phosphatidylinositol; SPH, sphingomyelin; FFA, free fatty acids.

22:6(*n*-3), 16:0, and 18:0, whereas rat ROS contain mainly saturated FFA, 16:0 and 18:0, with smaller amounts of unsaturated FFA. The diglycerides of ROS membranes consist almost entirely of two molecular species; in the frog, C-38 and C-40 are the dominant species, whereas in rat ROS, the most abundant are the C-36 and C-38 types.

Purified ROS membranes from rat, bovine, and frog retinas have an unusually high content of polyunsaturated acids, accounting for about 50–60 mol% of the total fatty acids; docosahexaenoate, 22:6(*n*-3), comprises approximately 80 mol% of these polyunsaturates (Fliesler and Anderson, 1983). The major saturated fatty acid in PE and PS of rod-dominant retinas is 18:0, whereas in PC, the major species is palmitate (16:0). The fatty acid content of PI differs from that of the major phospholipids in that it contains relatively more 18:0 and 20:4 (*n*-6) and less 22:6. A complete tabulation of the total fatty acid content of PE, PC, and PS in frog, rat, cow, human, chick, and goldfish has been published by Fliesler and Anderson (1983).

The possibility of altering the composition of polyunsaturated fatty acids by dietary manipulations has been examined by several investigators. In one study, the content of 22:6(*n*-3) was somewhat lower in whole retinas and in ROS of rats maintained on diets deficient in antioxidants (α-tocopherol and selenium) than in controls, but the differences were not highly statistically significant (Farnsworth *et al.*, 1979). An age-dependent decrease in 22:6(*n*-3) and increase in 22:5(*n*-6) has been observed in rats reared on synthetic diets with or without vitamin A supplementation (Organisciak *et al.*, 1986). The changes were attributed to lack of dietary *n*-3 fatty acids. Linolenic acid, 18:3(*n*-3), is a dietary precursor of 22:6(*n*-3), but there are no remarkable changes in the pool of polyunsaturated fatty acids in the retina or in ROS membranes of rats placed on fat-free diets from weaning (Anderson and Maude, 1972). However, if rats are maintained on fat-free diets and this regimen is continued for first-generation offspring, there is a 50% decrease in the content of 22:6(*n*-3) in ROS membranes, and concomitant abnormalities in electroretinographic (ERG) responses to light (Benolken *et al.*, 1973).

More recent studies on rhesus monkeys provide additional evidence supporting the notion that 22:6(*n*-3) in ROS membranes may have a function in vision (Neuringer *et al.*, 1984). In these experiments, semipurified diets low in *n*-3 fatty acids were fed to adult female monkeys as well as to their infants from birth. By 12 weeks of age the content of 22:6(*n*-3) in plasma phospholipids of deficient infants had dropped to 6% of control values. During the same period, there was a significant (50%) impairment in the development of visual acuity in deprived infant monkeys compared to normal controls. The visual loss was attributed to biochemical changes in the retina and/or brain resulting from an inadequate supply of 22:6(*n*-3) during the development of the visual system.

Environmental lighting conditions have a profound influence on both phospholipid content and rhodopsin packing densities in ROS membranes of the albino rat (Organisciak and Noell, 1977). The opsin content is lower and the phospholipid/opsin ratio significantly higher in rats maintained in weak cyclic light (12L:12D) than in those maintained in continuous darkness for at least 2 weeks. Further detailed studies on rhodopsin packing density and lipid composition of ROS membranes in albino rats raised in dim (5 lx), bright (300 lx), or intense (800 lx) cyclic illumination corroborated these findings and also demonstrated striking differences between the three groups (Penn and Anderson, 1987). With increasing illuminance, the mole percentage of 22:6(*n*-3) decreases by about 70%, while the level of the next most abundant polyunsaturate in ROS, 20:4(*n*-6), is nearly three times higher. There is a striking threefold increase in the mole percentage of cholesterol and a concomitant decrease in rhodopsin packing density with increasing illumination. The ROS are significantly reduced in length in both inferior and superior regions of the retina in rats reared in cyclic 800-lx illumination. The biochemical and morphological changes observed under different lighting conditions were thought to reflect adaptative protection from potentially damaging effects of bright or intense cyclic illumination.

7.2.2. Metabolism (Excluding Phototransduction)

The photoreceptor cell is a highly compartmentalized and polarized organelle; phototransduction occurs in the rod and cone outer segments, whereas the enzymatic machinery necessary to maintain viable outer segments is located in the inner segments (see Fig. 7.5). Cellular components and metabolites are exchanged and/or transported between the inner and outer segments through the narrow connecting cilium. The photoreceptor cells do not undergo mitosis; instead, they are replaced throughout life by a unique process of renewal: new disk membranes are assembled at the base of the outer segment, and "old" disk membranes are shed at the distal end of the outer segment and phagocytized by the pigment epithelium (see review by Young, 1976). This spatially and temporally balanced process maintains the rod outer segment at a constant length, and some of the factors controlling this dynamic turnover are discussed below. Two major types of metabolic renewal are now recognized in the photoreceptor cells: membrane replacement and molecular replacement. Opsin, the major integral membrane protein of ROS disks, is renewed mainly if not entirely by membrane replacement, whereas phospholipids appear to be renewed by both mechanisms.

7.2.2.1. Renewal of Outer Segments

The dynamic process of rod outer segment membrane renewal was elucidated more than two decades ago by Young and co-workers (Young, 1967, 1971; Young and Droz, 1968). These classic studies using light and electron microscopy and autoradiographic techniques revealed that injected radioactive amino acids are initially incorporated into proteins in the rough endoplasmic reticulum (RER) of the inner segment; the newly synthesized protein is afterward transferred through the Golgi complex to the base of the outer segment, where it is assembled into the proximal basal disks (Fig. 7.7). The radioactivity is displaced distally as discrete bands along the length of the rod outer segment, and finally small packets of terminal disks are shed and rapidly phagocytized by the pigment epithelium. The major protein synthesized is visual pigment (rhodopsin), and once it is assembled into the disk structure, it remains as an integral component until it is lost by shedding at the distal tip (Hall *et al.*, 1969). Complete renewal of integral membrane proteins in mammalian ROS occurs over a period of about 9 to 10 days.

The renewal mechanism in cone outer segments follows a different pattern; randomized, in contrast to discrete, labeling is observed after injection of radioactive amino acids *in vivo* (see reviews by Young, 1976; Anderson, 1983b) or after a 1-h incubation *in vitro* (Fliesler and Anderson, 1983). This correlates with structural differences between the rod and cone outer segments: the latter consist of continuous infoldings of the plasma membrane, and the newly assembled protein is free to diffuse throughout the lipid bilayer. Most studies of membrane biogenesis have been carried out on rod outer segments (as discussed below), and relatively few on renewal mechanisms in cone outer segments.

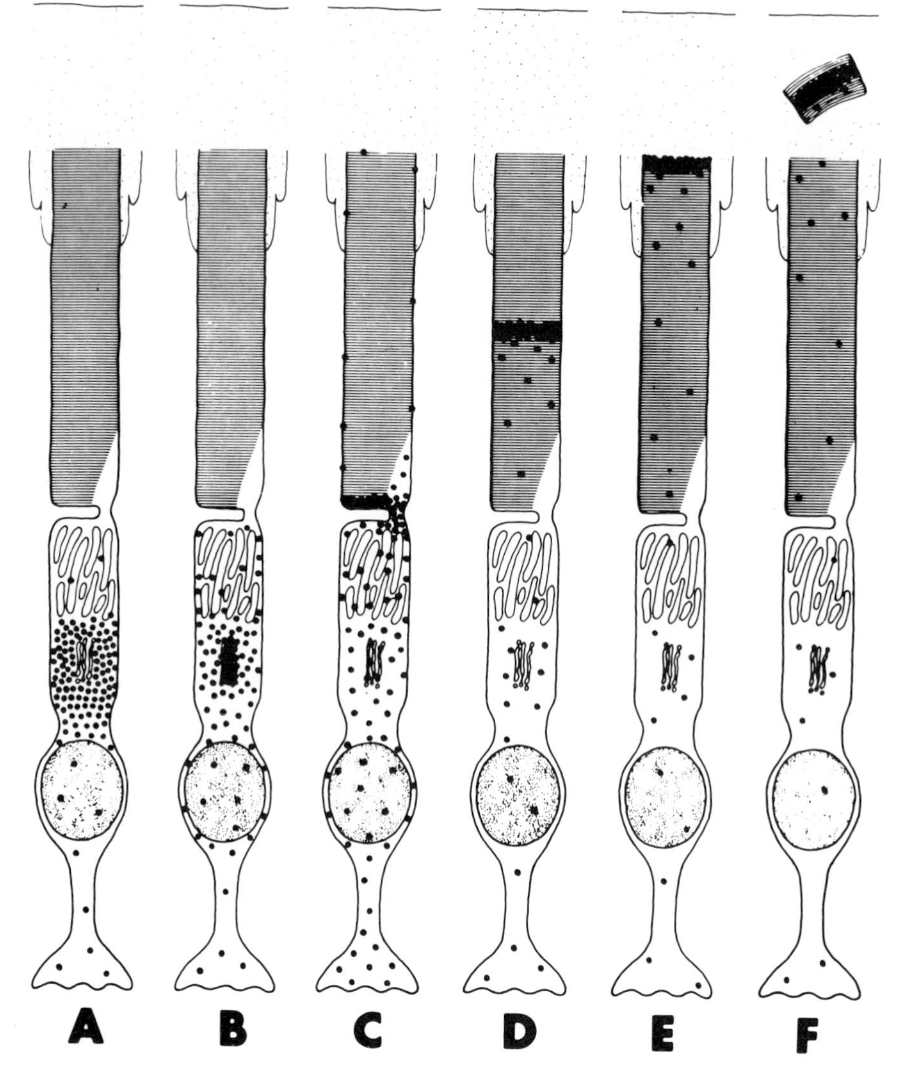

Figure 7.7. Diagram illustrating protein renewal in vertebrate rod outer segments. Autoradiographic studies show that newly synthesized protein is first concentrated in the rough endoplasmic reticulum and shortly afterward passes into the Golgi complex. After posttranslational glycosylation, the protein is transferred vectorially in vesicles through the inner segment to the connecting cilium, where it is incorporated into newly formed disk membranes. They are displaced distally, and, on reaching the distal tip, packets of outer segment disks are shed and phagocytized by the pigment epithelium. (From Young, 1976.)

However, fucose has been shown to be a useful marker for distinguishing differences in renewal patterns between the two types of photoreceptors (Bunt and Klock, 1980). Intravitreal injection of radioactive fucose in a wide variety of animal species shows diffuse labeling after 24 h in the cone outer segments and only a light surface label (as well as a basal band in some species) in rod outer segments. In further studies using goldfish, which have prominent cone outer segments, a fucose-containing integral membrane glycoprotein of M_r 33,000 was identified (Bunt and Saari, 1982). It is localized specifically to cone outer segments and is distinct from rod outer segment rhodopsin, which does not contain fucose.

Some of the early events in cone outer segment membrane renewal have been elucidated by following the fate of intraocularly injected labeled fucose into cone-dominant squirrels (Anderson *et al.*, 1986a). These studies show that the initial labeling pattern is not random; 30 min after injection there is heavy labeling in the Golgi complex, and after 1.5 h the label begins to appear in the periciliary region and basal portion of the cone outer segment. Fucose-labeled molecules persist in this region for about 12 h, and only afterward is longitudinal diffusion observed throughout the cone outer segment. If this component diffused freely along the infoldings of the plasma membrane, then a more rapid distribution would be expected, suggesting the possibility of "barriers" to diffusion at the lamellar rims of cone outer segments.

Although the renewal patterns of lipids and proteins appear to be similar in the early stages after injection of labeled precursor, the overall mechanism of lipid renewal is different since it involves both membrane replacement and molecular replacement (Bibb and Young, 1974a,b). Shortly after injection of labeled glycerol in the frog, the rough endoplasmic reticulum (RER) and nuclear envelope become heavily labeled, indicating that these are the primary sites for the *de novo* synthesis of ROS lipids (Mercurio and Holtzman, 1982). Within 1–4 h, the label appears at the base of the outer segment, but afterward, most of it is distributed diffusely throughout the outer segment. Using labeled fatty acids as precursors, there is little radioactivity in the inner segments; rather, most of the label is concentrated in the pigment epithelium shortly after injection, and some label is present throughout the outer segments. These findings suggest that fatty acids may be incorporated directly into ROS phospholipids through acyl exchange reactions.

In summary, autoradiographic techniques show that the biosynthesis and renewal of ROS lipids and proteins are similar in one respect: label is found initially in the RER and afterward appears at the base of the outer segment following injection of precursors such as glycerol or amino acids. There are, however, important differences. Whereas passage through the Golgi is obligatory for the posttranslational modifications involved in the biosynthesis of proteins such as opsin, phospholipid biosynthesis appears to bypass the Golgi apparatus. Moreover, the ultimate fate of lipids and proteins is different: phospholipids not only diffuse freely between individual disk membranes, they are also acylated and deacylated *in situ* in the ROS membrane (Fliesler and Anderson, 1983); in contrast, opsin is inserted into the disk membrane at the base of the photoreceptor and remains as an integral component of the disk until it is lost by phagocytosis.

Proteins and Glycoproteins. The synthesis of disk membrane proteins, particularly opsin, has been extensively investigated, and several comprehensive reviews have appeared (Papermaster and Schneider, 1982; Bok, 1985; Besharse, 1986). In the overall sequence of events, opsin is synthesized and inserted into a membrane in the rough endoplasmic reticulum (RER) and is afterward transferred to the Golgi apparatus for posttranslational modifications. As shown in Fig. 7.8, newly synthesized opsin is then transported vectorially in post-Golgi intracellular transport vesicles through the densely packed mitochondria in the ellipsoid region to the base of the connecting cilium (Papermaster *et al.*, 1978b). The combined use of EM autoradiographic and immunocytochemical techniques provides direct evidence that specific vesicles are involved in the vectorial transport of newly synthesized opsin from the Golgi apparatus to the periciliary ridge

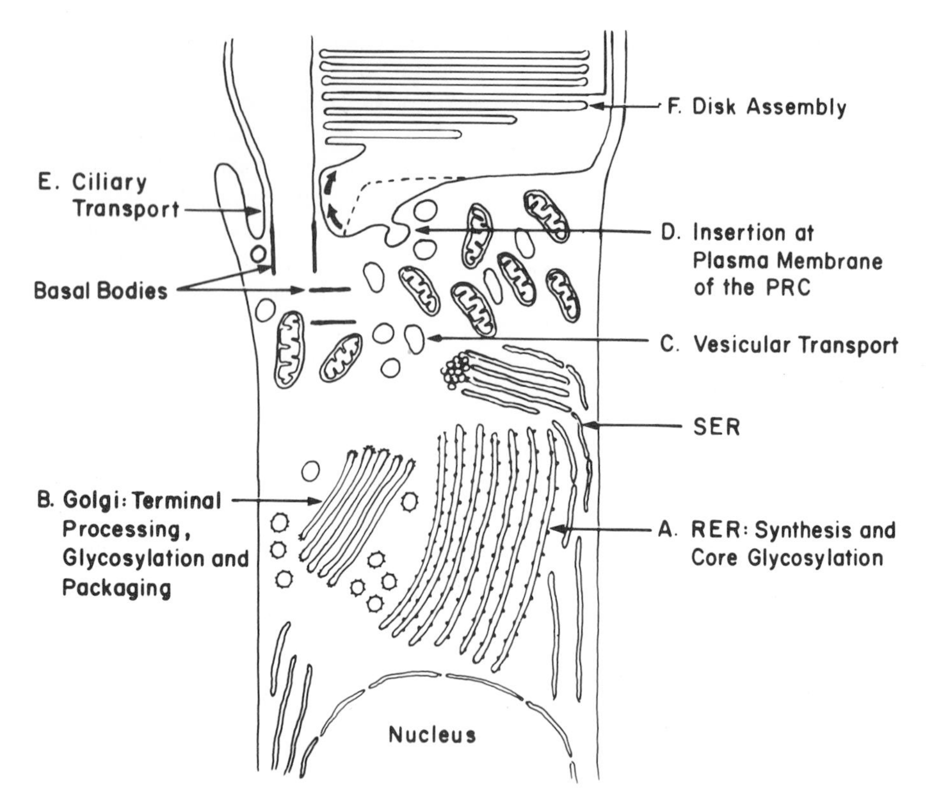

Figure 7.8. Biosynthesis and transport of opsin in *X. laevis* from its site of synthesis in the rough endoplasmic reticulum to the base of the outer segment. The diagram illustrates the vectorial transport of newly synthesized opsin in post-Golgi vesicles through the mitochondia-rich ellipsoid region; the vesicles apparently fuse with grooves of the periciliary ridge complex near the base of the connecting cilium. The opsin then travels along the ciliary plasma membrane and is then incorporated into newly forming disks. (From Papermaster *et al.*, 1985.)

complex (Papermaster *et al.*, 1985). The plasma membranes of the inner segment, connecting cilium, and outer segment are continuous, and the opsin-laden vesicles move along these membranes prior to insertion into newly formed disks.

The earliest form of opsin synthesized has been characterized as a nonglycosylated primary translation product of opsin mRNA generated in a wheat germ cell-free system (Schechter *et al.*, 1979; Papermaster *et al.*, 1980). However, unlike many other proteins that are either secreted or inserted into membranes, the immunoprecipitable translation product of opsin mRNA lacks a transient hydrophobic N-terminal signal peptide; instead, it has the same N-terminal sequence as mature opsin. Lacking this signal peptide, opsin is thought to be integrated into the lipid bilayer by cotranslational coupling of glycosylation and asymmetric insertion, a process mediated entirely by specific insertion sequences (Goldman and Blobel, 1981). In further studies, two out of four theoretically possible signal sequences for the insertion of opsin into microsomal membranes have been localized by deletion of selected segments from a bovine opsin cDNA clone and subsequent analyses of transcripts produced in a cell-free system (Friedlander and Blobel, 1985). These two signal sequences have been localized to the first transmembrane segment and the N-terminal portion of the sixth transmembrane segment of opsin (see Fig. 7.6). A provocative model of possible locations of alternating signal and stop transfer sequences within the opsin molecule highlights the various points at which this integration process can be tested (Friedlander and Blobel, 1985).

Autoradiographic studies on the glycosylation of rhodopsin using radioactive glucosamine have shown sequential incorporation of label into the RER and the Golgi apparatus (see review by Bok, 1985). Opsin is initially glycosylated cotranslationally with mannose and N-acetylglucosamine in the RER, and terminal glycosylation with a single amino sugar residue occurs posttranslationally in the Golgi apparatus (Besharse, 1986). Passage of opsin through the Golgi apparatus is an obligatory step for its incorporation into ROS disk membranes, as shown by studies using the ionophore monensin, which disrupts the transit of secretory materials through the Golgi apparatus (Matheke *et al.*, 1983, 1984).

With the elucidation of the structure of rhodopsin's major oligosaccharides and their N-linked attachments to Asn-2 and Asn-15 residues located near the N terminus, it is now recognized that glycosylation of opsin occurs through the lipid intermediate pathway (Kean, 1980a; Plantner *et al.*, 1980). Many details of this pathway have been elucidated in other tissues (see review by Kornfeld and Kornfeld, 1985); in the overall process, mannose, N-acetylglucosamine, and glucose are utilized for the assembly of the major Dol-P-linked oligosaccharide intermediate ($Glc_3Man_9GlcNAc_2$). It is from this carrier that the mannose-rich oligosaccharide is transferred to Asn receptor sites in the nascent polypeptides synthesized in the RER. Processing to form complex oligosaccharides involves "trimming" of all the glucose residues and some of the mannose residues; the newly synthesized glycoprotein is transported to the cis Golgi cisternae, and in subsequent passage through the medial and trans Golgi, further processing by sequential addition of specific sugars results in the formation of either high-mannose, complex, or hybrid oligosaccharides. Rhodopsin is considered to be a hybrid type, intermediate in structure between the high-mannose and complex types (Fukuda *et al.*, 1979; Liang *et al.*, 1979).

Tunicamycin inhibits the synthesis of the first intermediate in lipid-linked oligosaccharide synthesis, Dol-P-P-GlcNAc, and in a wide variety of tissues it effectively blocks protein glycosylation. In bovine retina, tunicamycin strongly inhibits the incorporation of labeled mannose and glucosamine into rhodopsin, although the incorporation of labeled methionine is only slightly depressed (Kean, 1980a; Plantner *et al.*, 1980). Under the conditions of these experiments, glycosylation did not appear to be an absolute requirement for insertion of rhodopsin into ROS disk membranes.

Tunicamycin also causes a marked reduction in the incorporation of labeled mannose and glucosamine into photoreceptors of human retina, as shown by light microscopic autoradiography (Fliesler *et al.*, 1984). Dual-label experiments with various sugars and amino acid precursors have shown that tunicamycin selectively inhibits the incorporation of mannose (but not other sugars) into retinal glycoproteins, one of which had an electrophoretic mobility similar to opsin. Other dual-label experiments show that tunicamycin causes a marked inhibition of mannose incorporation into TCA-precipitable materials from whole retina but has relatively little effect on leucine incorporation (Fliesler and Basinger, 1985). However, examination of ROS membranes of retinas that had been incubated with labeled leucine alone in the presence of tunicamycin showed a 95% inhibition of incorporation of radioactivity into TCA-precipitable material compared to controls. SDS-PAGE analyses of retinal proteins after incubation with labeled methionine in the presence of tunicamycin showed a virtual disappearance of the 37-kDa opsin band and a concomitant appearance of a prominent 32-kDa band identified as nonglycosylated opsin. No labeled bands, either 32 or 37 kDa, were detected in isolated ROS. These and other studies (Fliesler *et al.*, 1985) provide compelling evidence that even though non-glycosylated opsin can be transported from its site of synthesis in the RER to the base of the outer segments, it cannot be utilized for the normal assembly of ROS disks. In place of disk-like evaginations of outer segment plasmalemma, tunicamycin-treated rod cells accumulate vesicles of abnormal structure in the extracellular space between the inner and outer segments that cannot be used for the assembly of normal disks.

The functional significance of small, truncated oligosaccharides in the native opsin molecule is unknown, but apparently their size *per se* does not directly influence normal disk morphogenesis (Fliesler *et al.*, 1986a,b). Incubation of frog retinas with castanospermine, an inhibitor of α-glucosidase I, blocks the trimming of glucose from Asn-linked $Glc_3Man_9GlcNAc_2$ and results in a two- to threefold increase in the incorporation of labeled mannose into TCA-precipitable proteins of whole retina and into opsin. The apparent M_r of opsin synthesized in the presence of castanospermine is increased by about 2500 over that of normal opsin. This opsin with abnormally large oligosaccharides is, nevertheless, effectively incorporated into newly formed disks of normal morphological appearance. Thus, although glycosylation of opsin appears to be required for normal disk membrane morphogenesis, the exact structural features of the N-linked oligosaccharide may not be stringently specified for this role (Fliesler, 1988).

Another covalent modification in the synthesis of opsin is acylation by palmitic acid. Freshly prepared native rhodopsin contains one to two bound fatty acid residues per molecule, and several studies show that the acylation events occur at two different intracellular sites and by different mechanisms. Rat (St. Jules and O'Brien, 1986) and bovine (O'Brien and Zatz, 1984; O'Brien *et al.*, 1987) retinas and ROS membranes incorporate labeled palmitate into rhodopsin both *in vivo* and *in vitro*. Palmitate is esterified to both newly synthesized and mature rhodopsin. In the former case the site of acylation is the RER, and the reaction occurs prior to the transfer of nascent polypeptides of opsin to the Golgi apparatus. This is the primary site of acylation; the secondary acylation of rhodopsin occurs in the rod outer segments, possibly nonenzymatically through formation of a thioester bond.

The spatial and temporal site of attachment of 11-*cis* retinal to the apoprotein opsin has been somewhat controversial (see review by Bok, 1985), and until recently the chromophore was thought to be added to opsin after its transport to the ROS. There is, however, compelling evidence that in the rat, 11-*cis* retinal is attached to opsin shortly after its translation in the RER and before it is transported to the ROS (St. Jules *et al.*, 1989). Support for the view that opsin acquires its chromophore while still in the RER derives from the isolation of labeled rhodopsin in an RER-enriched fraction within an hour after intravitreal injection of either labeled amino acids or retinol. The RER fraction had only minimal (1–3%) contamination from ROS, and the relatively short time interval would preclude significant transport of opsin to newly assembled disk membranes and subsequent acquisition of chromophore.

Phospholipids (Excluding Phosphatidylinositol). Opsin remains as an integral component of the photoreceptor disk until it is lost by shedding, but lipids are randomized throughout the outer segment during their synthesis, and their turnover cannot be accounted for by loss from the distal tips alone. Autoradiographic studies on the fate of labeled glycerol or long-chain fatty acids injected into frogs show that the radioactivity is initially concentrated in the endoplasmic reticulum, and after several hours the outer segments become diffusely labeled with newly synthesized phospholipids (Bibb and Young, 1974a,b). The rough endoplasmic reticulum (RER) is the primary site of lipid synthesis (Mercurio and Holtzman, 1982). Biochemical studies on the kinetics of phospholipid biosynthesis using labeled glycerol as a precursor have shown that the half-life of phosphatidylserine (PS) in frog ROS is 23 days (Anderson *et al.*, 1980a); phosphatidylethanolamine (PE) and phosphatidylcholine (PC) have half-lives of 18–19 days (Anderson *et al.*, 1980b,c). Parallel studies of protein synthesis showed that the turnover of these phospholipids is nearly double that of protein. These findings suggest that, unlike integral membrane proteins such as opsin that are renewed solely by membrane replacement, lipids are renewed both by membrane replacement and molecular replacement (Fliesler and Anderson, 1983).

Phosphatidylcholine (PC), one of the most abundant ROS phospholipids, is thought to be formed by three biosynthetic routes: methylation of phosphatidylethanolamine (Anderson *et al.*, 1980b), base-exchange reaction, and *de novo* microsomal synthesis (Pu and Anderson, 1983). In the latter, often called the Kennedy pathway, free choline is activated in two stages: the first is phosphorylation by ATP through the action of choline kinase to form phosphorylcholine; the second step involves its reaction with CTP to form CDP-choline. In this form choline is transferred to 1,2-diacylglycerol to form PC. Photoreceptors maintain a relatively large pool of choline; they have high-affinity receptors that can accumulate choline from low extracellular concentrations and subsequently convert it to phosphorylcholine through the action of an endogenous choline kinase (Masland and Mills, 1980; Pu and Anderson, 1983). Incubation of rabbit retinas with hemicholinium-3, a choline analogue that inhibits the high-affinity uptake of choline, causes a selective photoreceptor degeneration without affecting other cell layers (Pu and Anderson, 1983; Pu and Masland, 1984). Hemicholinium-3 has no effect either on protein synthesis or on the formation of PC from glycerol. However, the synthesis of PC from choline is severely inhibited in its presence because of the reduced utilization, uptake, and subsequent phosphorylation and incorporation into PC.

The rapid turnover and dispersion of phospholipids throughout the outer segments may be through remodeling reactions mediated by phospholipid-exchange proteins (Dudley and Anderson, 1978) or by acyl exchange. The latter has been demonstrated *in vivo* in the rat by following the fate of intravitreally injected labeled palmitate, arachidonate, and glycerol into the major phospholipids of the ROS membranes (Wetzel and O'Brien, 1986). The incorporation of these labeled fatty acids precedes the peak of glycerol incorporation (which takes place in the inner segments) by several hours. Both fatty acids are rapidly incorporated into ROS phospholipids, and kinetic studies suggest their extensive reutilization in specific phospholipids. Remodeling by fatty acyl exchange thus appears to play a major role in the initial incorporation of both palmitate and arachidonate into phospholipids of rat ROS.

Active acylation–deacylation reactions have also been demonstrated using isolated bovine ROS membrane and photoreceptor disks (Giusto *et al.*, 1986). These preparations incorporate labeled docosahexaenoate (22:6) into phospholipids, mainly PC, when incubated in the presence of CoA, ATP, and Mg^{2+}. The reaction is not dependent on biosynthetic processes localized in the inner segment, since it takes place in isolated ROS membranes or disks free of inner segment organelles such as microsomes. Thus, two enzymes required for the biosynthesis of PC and other phospholipids from labeled docosahexaenoic acid appear to be present in bovine ROS membranes: acyl-CoA synthetase and docosahexaenoyl-CoA:lyso-PC acyltransferase. About two-thirds of the label in the newly synthesized PC was found in dipolyunsaturated species (see Section 7.1.2.1) and the remainder in hexaenes.

Recent evidence suggests that both acylation and deacylation in bovine ROS take place mainly at the *sn*-2 position of the phospholipids (Zimmerman and Keys, 1988). This is based on calculations showing that in 11 days, 99%, 95%, and 51% of the fatty acids in the *sn*-2 position of PC, PS, and PE, respectively, could be exchanged once, assuming no interconversions by base exchange. Prooxidizing conditions (exclusion of dithioerythritol) and exposure to light stimulate both the acylation and deacylation of ROS phospholipids, and preferential hydrolysis of polyenoates was noted. The latter finding is of particular interest in view of recent reports on the light-stimulated activation of phospholipase A_2 and its regulation by the βγ subunits of transducin (Jelsema, 1987; Jelsema and Axelrod, 1987).

Studies on the biosynthesis and turnover of molecular species of PC, PE, and PS in frog ROS after intravitreal injection of [2-^{3}H]glycerol show that phospholipids synthesized in the inner segment are rapidly incorporated into the outer segment; their specific activities increase for 6–8 days and then decline exponentially (Louie *et al.*, 1988). Time studies on changing specific radioactivities of the docosahexaenoate species in PC and PE indicated considerable acyl exchange within the same phospholipid class; however, these reactions are not responsible for the initial incorporation of 22:6, which occurs prior to delivery of phospholipids to the ROS. The highest specific radioactivity was found in the dipolyenoic species 22:6-22:6 from PC. Other data show that the molecular species composition of PC and PE is determined by several factors such as rates of biosynthesis, turnover, and interconversion. In contrast, the steady-state composition of two major molecular species of PS in ROS is determined mainly by their rates of synthesis; PS does not undergo extensive remodeling once it is synthesized.

Monensin, the Na^+,H^+ ionophore that disrupts the Golgi apparatus and inhibits the release of secretory proteins, is a highly effective agent for demonstrating partial dissociation of phospholipid and protein synthesis in the ROS (Matheke and Holtzman, 1984; Fliesler and Basinger, 1987). Incubation of frog retinas with 50 nM monensin causes two- and threefold increases in specific activities of newly synthesized ROS phospholipids from labeled glycerol and palmitate, respectively. Under these conditions, retinal protein synthesis is maintained at about 90% of normal level, but the newly synthesized proteins are not transported and assembled into new ROS disks. These interesting experiments show that even though newly synthesized lipids and proteins originate in the same organelles (the rough endoplasmic reticulum and nuclear envelope), they have independent routes of transport to the base of the outer segment before assembly into disk membranes. Whereas transport of opsin for posttranslational processing through the Golgi apparatus is an obligatory step, newly synthesized glycerolipids may bypass the Golgi apparatus (Mercurio and Holtzman, 1982; Matheke and Holtzman, 1984; Fliesler and Basinger, 1987).

The biosynthesis of phospholipids has been studied extensively, but little is known about their possible enzymatic breakdown in photoreceptors. Phospholipase A_1 and A_2 activities have been described in bovine retina (Swartz and Mitchell, 1973), and a specific phospholipase C involved in polyphosphoinositide (PI) turnover in ROS is now well established (see Section 7.2.2.2). More recently, the observation that light activates phospholipase A_2 in dark-adapted bovine ROS by a transducin-dependent mechanism suggests hitherto unknown pathways for phospholipid turnover in photoreceptors (Jelsema, 1987). The reaction was measured by the liberation of arachidonic acid from phosphatidylcholine (PC), and the involvement of transducin was demonstrated in several ways: (1) the ability of the hydrolysis-resistant GTP analogue GTPγS to mimic the light activation of phospholipase A_2 in dark-adapted ROS membranes; (2) the inhibitory effects of both pertussis and cholera toxin; (3) the decrease in phospholipase A_2 activation following removal (by hypotonic washing) of transducin from the membranes; and (4) a partial restoration of activity by supplying exogenous transducin to transducin-depleted membranes. Moreover, phospholipase A_2 activity may be under dual regulation by two different forms or pools of transducin, since addition of GTPγS together with exposure to light inhibits the light activation. This appears to be the first report of a G protein that is coupled to two effector systems, namely, the light activation of both phospholipiase A_2 and of cGMP phosphodiesterase. Whereas the latter is mediated by the dissociated α subunit (Fig. 7.9), the light activation of phospholipase A_2 in dark-adapted bovine ROS is thought to be mediated through the dissociated $\beta\gamma$ subunits (Jelsema and Axelrod, 1987; Axelrod *et al.*, 1988). The mechanism for the transducin-mediated, light-induced increase in phospholipase A_2 activity appears to involve interaction of the $\beta\gamma$ subunits with a non-G

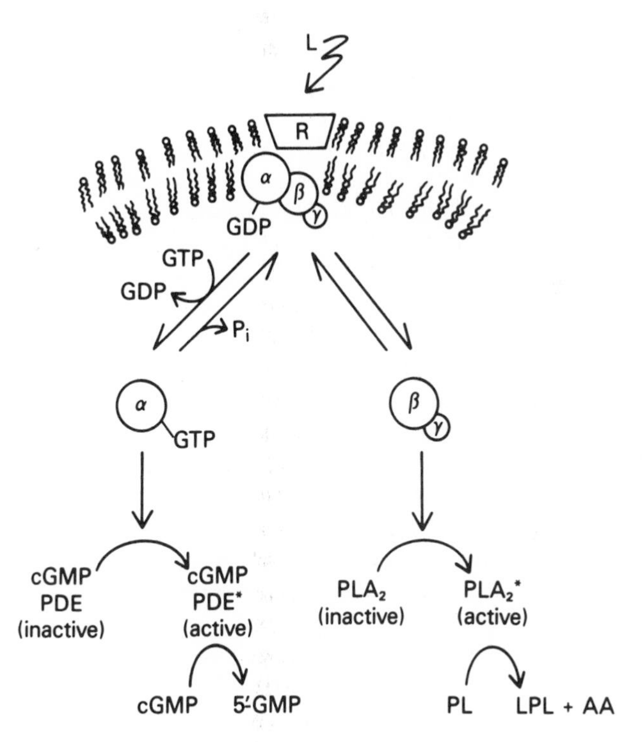

Figure 7.9. Schematic diagram showing transducin-dependent light activation of bovine ROS phospholipase A_2. (From Jelsema and Axelrod, 1987.)

protein inhibitor of phospholipase A_2, resulting in disinhibition of the enzyme (Jelsema, 1989).

Disk Shedding. The periodic detachment of ROS disks from the apical portions of the photoreceptors, their phagocytosis by the pigment epithelium, and subsequent formation of phagosomes within these cells (Fig. 7.7) were first described two decades ago by Young and Bok (1969). At first thought to be a random event, disk shedding in photoreceptors of frog (Basinger *et al.*, 1976) and rat (LaVail, 1976) was later shown to follow a circadian rhythm, with maximum shedding occurring after the onset of light. In the rat, the burst of disk shedding appears to be light-entrained, since it also takes place in continuous darkness in a cyclic manner without the onset of light. In contrast, cone shedding in some but not all species takes place after the onset of darkness (for reviews see Young, 1976; Bok, 1985). The molecular basis of disk shedding is under active investigation, and there is still some uncertainty about whether it is initiated locally in the photoreceptor cell or whether the retinal pigment epithelium also plays an active role. In amphibians, disk detachment may result from events occurring in both the outer segments and the pigment epithelium (see review by Besharse, 1986).

Disk shedding is not controlled by humoral factors, nor is the pineal essential (Tamai *et al.*, 1978); rather, initiation of shedding in the frog (Hollyfield and Basinger, 1978) and rat (Teirstein *et al.*, 1980) resides within the eye. In these classic experiments, animals entrained in a 12-h light–dark cycle had one eye occluded and were then exposed to constant light. The patched eye did not respond to changes in ambient lighting conditions, whereas the open eye did. Thus, each eye contains an endogenous circadian oscillator that controls rhythmic disk shedding independently of the other eye. This has also been demonstrated in 12-h cyclic-light-entrained eyes of the frog *in vitro* (Flannery and Fisher, 1984). In this case, frogs were placed in constant darkness after cyclic light entrainment, and at various intervals, the eyes were removed and the pattern of disk shedding examined in culture medium. Under these conditions, the eyes continued rhythmic shedding of disks at the approximate time expected from previous entrainment. Taken together, these and other experiments suggest local production in the eye of substances that control diurnal rhythms.

Although the factors mediating the intrinsic control of rhythmic disk shedding of photoreceptor tips have not been precisely defined, there is compelling evidence that the pineal hormone melatonin may play an important role. Autoradiographic and biochemical studies have demonstrated saturable melatonin binding in melanosomes of the retinal pigment epithelium–choroid and in the outer plexiform layer of *Rana pipiens* (Wiechmann *et al.*, 1986). As shown in Fig. 7.10, melatonin is synthesized from serotonin in two enzymatic steps. Both enzyme activities, serotonin N-acetyltransferase (NAT) and hydroxyindole O-methyltransferase (HIOMT), have been detected in the retina. Moreover, the activity of NAT and the levels of melatonin in the retinas of chick (Hamm and Menaker, 1980) and frog (Iuvone and Besharse, 1983) fluctuate in a circadian pattern. The highest levels of melatonin and the peak activity of NAT are found during subjective darkness. The NAT activity in frog retina can be entrained by light and dark, and its dark-induced increase is completely blocked by inhibitors of protein synthesis, puromycin and cycloheximide. These observations suggest the possibility of enzyme induction during darkness.

The development of an *in vitro* system in which disk shedding can be measured in

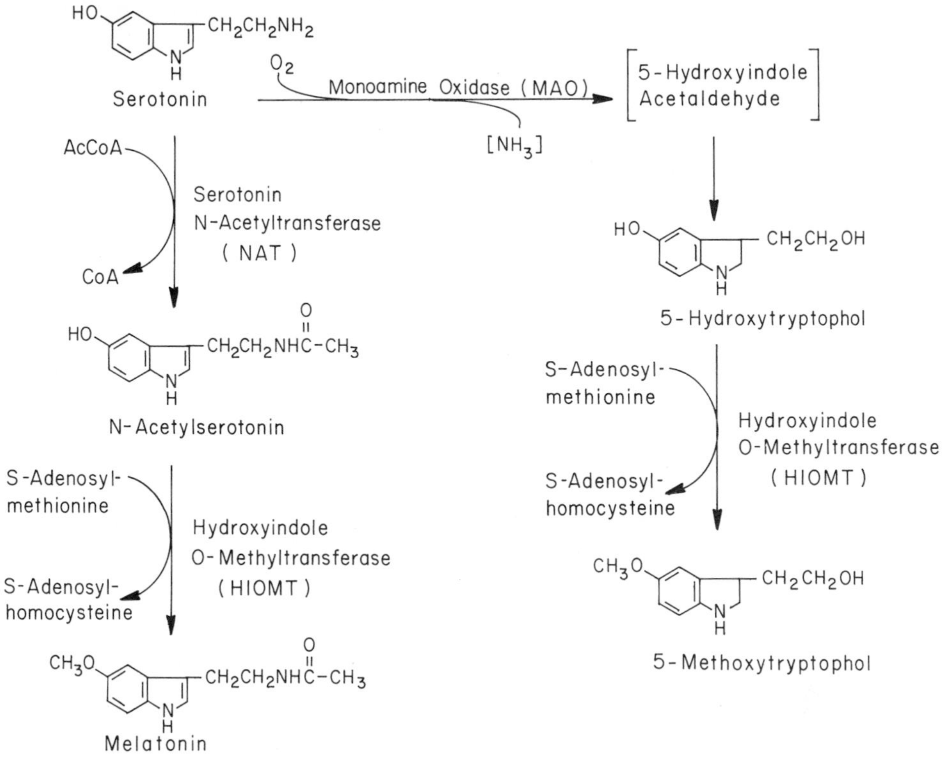

Figure 7.10. Enzymatic conversion of serotonin to melatonin and to 5-methoxytryptophol in the pineal and in the retina. The first step in the enzymatic synthesis of melatonin from serotonin is acylation of the terminal amino group; this step is mediated by NAT, an enzyme that is under diurnal control. The final synthesis of the two methoxyindoles is catalyzed by HIOMT, an enzyme that shows no circadian regulation.

response to various environmental or chemical agents allows a more precise evaluation of the role of NAT and melatonin in this process. Eye cups of *Xenopus laevis* can be maintained in a viable state and responsive to light for several days in culture medium containing a mixture of 14 amino acids and a relatively high ($\geq$30 mM) concentration of bicarbonate (Besharse and Dunis, 1983a). In these preparations, the cyclic activity of NAT persists during the normal light–dark cycle as well as in constant darkness or after phase reversal of the lighting regimen, implying involvement of a local circadian oscillator in the regulation of NAT activity (Besharse and Iuvone, 1983). The induction of NAT may be regulated by cAMP, since incubation of eye cups in the light in the presence of agents that increase the cellular levels of cAMP (dibutyryl cAMP, IBMX, or forskolin) increases the NAT activity to levels similar to those present in darkness (Besharse *et al.*, 1984; Iuvone and Besharse, 1986a). This stimulation does not require calcium, in contrast to the nocturnal increase of NAT activity, which shows an absolute dependence on Ca^{2+} (Iuvone and Besharse, 1986b). Antagonists of voltage-sensitive calcium channels inhibit the dark-dependent increase of NAT activity. Thus, Ca^{2+}-dependent channels, possibly localized in the inner segments and tonically open in darkness, could supply Ca^{2+} to systems such as calmodulin-dependent adenylate cyclase. This is consistent with the enhanced nocturnal NAT activity described above.

Further studies using eye cup preparations from *X. laevis* have established a direct relationship between methoxyindole metabolism and photoreceptor disk shedding (Besharse and Dunis, 1983b). Both melatonin and 5-methoxytryptophol (see Fig. 7.10) activate disk shedding in eye cup preparations in a manner that mimics the light-evoked stimulation *in vivo*. Another interesting finding is that excitatory amino acids also induce massive disk shedding in *Xenopus* eye cups *in vitro* (Greenberger and Besharse, 1985). Millimolar concentrations of L-aspartate and L-glutamate as well as micromolar levels of kainic acid (a glutamate analogue) cause a three- to fivefold greater stimulation of disk shedding than light-evoked shedding. This effect is not related to the neurotoxicity of the amino acids, since nonneurotoxic amino acids have a similar effect. The mechanism of hypershedding in the presence of excitatory amino acids is unclear, but recent evidence supports the view that excitatory amino acids exert their effect on disk shedding by binding to a receptor of the kainate type located in postreceptoral neurons (Besharse *et al.*, 1988).

Combined use of an antibody to HIOMT and immunocytochemistry have demonstrated melatonin-synthesizing enzyme in the photoreceptors of bovine, rat, and human retinas and in the inner retina of some species (Wiechmann *et al.*, 1985). The techniques used in this study showed labeling of the entire retina with the exception of the outer segments, suggesting possible cytosolic localization of HIOMT. The enzyme is also present in pinealocytes of bovine and human pineal glands obtained during the light period. SDS-PAGE analyses of bovine pineal HIOMT show that it is composed of two subunits of M_r 39,000 and 25,000. However, only the 25-kDa subunit could be detected in retinal extracts using the HIOMT antibody. Based on sequence homology and immunologic identity, the 25-kDa peptide bears strong resemblance to a portion of the 39-kDa protein and hence may represent either a true subunit of the larger peptide or one of its degradation products.

It is also possible that Mn^{2+}-dependent pyrimidine 5′-nucleotidase (MDPNase) plays a role in disk shedding and/or phagocytosis by the retinal pigment epithelium (RPE) (Irons, 1987a). Rats were reared in cyclic illumination (12L:12D), and MDPNase activity was measured cytochemically in retinal slices at various times during the lighting cycle. In retinas fixed prior to the beginning of the shedding period, heavy cytochemical staining was present on the surfaces of the RPE villous processes and on the surfaces of ROS in contact with these processes. During the shedding peak, staining in the apical processes is diminished, while labeling becomes more intense in the ROS tips and in ROS phagosomes. After the shedding period, little staining is observed in either the ROS tips or in the apical processes of the RPE. Thus, the redistribution of this enzyme activity in relation to cyclic light appears to correlate closely with disk shedding. These findings were not artifacts resulting from diffusion of the reaction product from RPE lysosomes to the ROS, since a similar diurnal pattern is present in isolated ROS (Irons, 1987b). In this case, the reaction product arises from MDPNase intrinsic to the ROS that appears to be localized in the rim regions and in narrow zones in the disk interiors.

Further characterization of MDPNase in extracts of rat ROS has corroborated cytochemical findings (Irons and O'Brien, 1987). The hydrolysis of cytidine 5′-monophosphate is stimulated five- to sixfold in the presence of Mn^{2+} and is strongly inhibited by NaF. At pH 5.0–5.5, most of the MDPNase activity is Mn^{2+} dependent, whereas between pH 7 and 8, the activity is mainly Mn^{2+} independent. The latter in all probability is an enzyme (or group of enzymes) similar to the 5′-nucleotidases described in bovine

ROS (Fukui and Shichi, 1981, 1982). No clear-cut diurnal variations in MDPNase activity could be detected by biochemical assay, possibly because small differences would have been masked by the fairly broad range of specific activities of the extracted enzyme.

The foregoing discussion suggests that many factors, principally enzymatic, are involved in the regulation of disk shedding in photoreceptors. However, a definitive conclusion on the relative importance of each in this complex process has not yet been reached.

7.2.2.2. Phosphoinositide Turnover and Protein Kinase C

Phosphatidylinositol (PI) comprises less than 2% of the total ROS membrane lipids (see Table 7.11), yet its rate of synthesis from labeled glycerol and arachidonic acid in bovine or frog retinas incubated *in vitro* is considerably higher than that of other phospholipids (Bazan and Bazan, 1977; Bazan *et al.*, 1977). The unique metabolism of PI has also been noted *in vivo* in *Rana pipiens* after injection of radioactive precursors (Anderson *et al.*, 1980d). The specific activity of PI in the ROS after injection of labeled glycerol is ten times higher than that of other phospholipids, and one particularly active pool of PI has a half-life of 3.0–3.5 days, a turnover about seven times greater than that of PC or PE. Other studies on PI metabolism in ROS have recently been reviewed by Anderson and Brown (1988) and can be summarized briefly as follows: kinases that phosphorylate endogenous PI to PIP and PIP to PIP_2, as well as DG kinase (see Fig. 1.6), have been detected in bovine ROS (Giusto and Ilincheta de Boschero, 1986). However, PI synthesis from DG via PA → CDP-DG and PI synthetase does not take place in washed frog ROS (Anderson and Brown, 1988). These preparations have the enzymatic capabilities to metabolize PI to PA through breakdown of PIP_2 and production of DG, but they do not have the capacity to regenerate PI from PA. Hence, it appears that, in this species, key intermediates in the recycling of PI from PA must be supplied from the inner segments.

The observations that light enhances the incorporation of glycerol (Bazan and Bazan, 1977) and inositol (Anderson and Hollyfield, 1981) into retinal PI and PA pointed to the presence, in the retina, of a receptor-mediated turnover of PI in response to specific external stimuli, in this case light (see review by Anderson and Brown, 1988). The "phosphoinositide effect," extensively investigated in a wide wariety of tissues, has in recent years become firmly established for both vertebrate and invertebrate retinas. The receptor-stimulated activation of phospholipase C and the breakdown of PIP_2 (phosphatiylinositol 4,5-bisphosphate) to produce two second messengers, IP_3 (inositol 1,4,5-trisphosphate) and DG (diacylglycerol), is now thought to modulate a wide range of physiological responses in photoreceptor cells.

In *Xenopus laevis*, light stimulates the incorporation of labeled inositol and phosphate into PI and other phosphoinositides, the reponse being localized in the horizontal cells of the retina (Anderson and Hollyfield, 1981; Anderson *et al.*, 1983). Since light is absorbed in the outer segments, and the response (enhanced turnover of phosphoinositides) appears in the horizontal cells, it has been suggested that blocking of neurotransmitter release may stimulate the PI effect in the horizontal cells. Of several neurotransmitters tested that could modulate this effect in *X. laevis*, only two appear to be involved in regulating the PI response: glycine and acetylcholine (Anderson and Hollyfield, 1984). Glycine inhibits the light-stimulated increase in phosphoinositide turnover, whereas acetylcholine, one of the most widely studied and best known direct activators of PIP_2 breakdown, enhances this turnover.

Light-activated breakdown of PIP_2, probably mediated by PIP_2-specific phospholipase C, has been detected in the retina of *Rana pipiens* (Ghalayini and Anderson, 1984). This was shown by the marked decrease in ROS membrane-labeled PIP_2 after brief (5–15 sec) illumination; under these conditions light-induced breakdown of other phosphoinositides was not observed. Similar findings on the light-stimulated breakdown of PIP_2 have been reported in isolated ROS preparations from frog (Hayashi and Amakawa, 1985) and bovine retinas (Millar and Hawthorne, 1985). The major product of PIP_2 breakdown, IP_3, is rapidly produced in toad ROS after brief (millisecond) flashes of bright light (Brown *et al.*, 1987); moreover, pressure injection of IP_3 into dark-adapted salamander ROS induces a reversible hyperpolarization of the rod membranes (Waloga and Anderson, 1985). These interesting observations raise the possibility that PIP_2 breakdown may play a role in phototransduction in some vertebrate retinas, in a manner possibly analogous to light- and IP_3-induced excitation and adaptation in invertebrate retinas (see Section 7.8.3).

Further studies on phospholipase C in ROS membranes of *R. pipiens* show that the major activity of this enzyme is toward PIP_2 and PIP (phosphatidylinositol 4-phosphate); PI (phosphatidylinositol) is hydrolyzed at only 2% of the rate of PIP_2 (Tarver and Anderson, 1988). It was also noted that the relative rates of PIP_2 and PIP hydrolysis are highly sensitive to the concentration of Ca^{2+} in the assay medium. There is now compelling evidence that phospholipase C activity in ROS is under tonic inhibitory control by a pertussis toxin-sensitive inhibitory G protein (Jelsema, 1989). The transducin-mediated, light-induced increase in phospholipase C activity is thought to occur by preventing the action of this inhibitory protein, resulting in disinhibition of the enzyme activity (Jelsema, 1989).

The formation of cytidine triphosphate (CTP) and cytidine diphosphate (CDP) in rat photoreceptor cells is also stimulated by light (Schmidt, 1983a,b). Since CDP-diacylglycerol and inositol are the immediate precursors of newly synthesized PI (see Fig. 1.6), these findings suggest a role for cytidine nucleotides in the light-enhanced formation of PI. Exogenous cytidine and inositol appear to enhance not only the synthesis but also the turnover of light-stimulated PI formation in the photoreceptors. Light-stimulated formation of PI from glycerol and phosphate precursors has also been demonstrated, and this stimulation is not simply a consequence of changes in the size and/or turnover of the precursor pool; rather, it represents a true stimulatory effect of light on PI synthesis and turnover in the rat retina (Schmidt, 1983c). These findings have been confirmed by Anderson *et al.*, (1985a). Light does not, however, increase *de novo* synthesis of retinal lipids, even though there is enhanced incorporation of labeled glycerol into all phospholipids and diglycerides in retinas incubated in the light, compared to those incubated in darkness. The possibility of light-dependent deacylation–reacylation of PI, mediated by a phospholipase A_2 that selectively removes arachidonic acid, was raised in these studies. The subsequent finding that light activates phospholipase A_2 of bovine ROS through a transducin-dependent mechanism (Jelsema, 1987) confirms this interesting prediction.

Combined immunocytochemical and image-analysis techniques have localized the light-regulated breakdown and synthesis of PIP_2 to the rod outer segments of rat retina (N. Das *et al.*, 1986, 1987). Dark-adapted retinas show intense staining of the outer segments and less intense staining of the inner segments. After a light flash of only 1 ms, there is a marked decrease in PIP_2 content, and fully bleached retinas show little or no staining. Dark adaptation of the flash-bleached eye for 5 min is sufficient to restore the PIP_2 levels to those of the fully dark-adapted state. These observations imply that the rapid hydrolysis of PIP_2 may be directly correlated with the photoactivation of rhodopsin and that light-induced hydrolysis of PIP_2 may be fast enough to function in the visual process in rodents and possibly other vertebrates.

Light-mediated breakdown of PIP_2 generates two second messengers, inositol 1,4,5-trisphosphate (IP_3) and diacylglycerol (DG). The latter activates Ca^{2+}/phospholipid-dependent protein kinase C; however, activation of this enzyme is also coupled to the tumor-producing phorbol ester receptor (see Fig. 1.6). Protein kinase C is a ubiquitous enzyme with a broad spectrum of biological activities. Several distinct types, both soluble and membrane-bound, have been characterized in nonocular tissues and have also been detected in soluble extracts of dark-adapted bovine ROS (Kapoor and Chader, 1984). In the presence of Ca^{2+} and phosphatidylserine, this enzyme phosphorylates several endogenous retinal proteins of molecular weights ranging from 11,000 to 95,000. Other studies have shown that purified protein kinase C, in the presence of calcium, binds tightly to ROS membranes and phosphorylates serine and threonine residues of rhodopsin near the C terminus (Kelleher and Johnson, 1986). The phosphorylation is not dependent on illumination and requires Ca^{2+}, in contrast to rhodopsin kinase, which phosphorylates only bleached rhodopsin in a Ca^{2+}-independent manner.

It is difficult to exclude the possibility that, in addition to the putative phosphorylation of rhodopsin by protein kinase C, small amounts of bleached rhodopsin unavoidably present in dark-adapted bovine ROS could have provided phosphorylation sites for rhodopsin kinase as well. This question has been approached by investigating ROS phosphorylation in fully dark-adapted rats, with all procedures being carried out using an infrared image converter (Kapoor *et al.*, 1987). After intravitreal injection of labeled phosphate, retinas were incubated either in the light or in total darkness; in addition, the effects of two agents known to activate protein kinase C (phorbol esters and a diacylglyceride, 1-oleoyl-2-acetylglycerol) were examined. In the light, rhodopsin as well as 80- and 65-kDa proteins and lower-M_r species of crude ROS are phosphorylated. No phosphorylation is observed in ROS prepared in total darkness, whereas variable low phosphorylation occurs in ROS prepared under dim red light. Neither phorbol esters nor diacylglycerol induce rhodopsin phosphorylation, but they do promote phosphorylation of other proteins, mainly 65- and 80-kDa species. The similarity in responses of light, phorbol ester, and diacylglycerol in stimulating the phosphorylation of proteins other than rhodopsin suggests a possible link between light-induced PIP_2 breakdown and phosphorylation of ROS proteins by protein kinase C.

Purified protein kinase C also phosphorylates serine residues of the GDP-bound α subunit of transducin isolated from bovine ROS (Zick *et al.*, 1986). Moreover, tyrosine residues of this subunit are also phosphorylated by insulin receptor tyrosine kinase activity. Although these experiments were carried out in a cell-free system, it is not unreasonable to extrapolate the findings to intact ROS. The multisite phosphorylation of the α subunit of transducin appears to be under hormonal control mediated through insulin receptors as well as by the light- and/or ligand-stimulated breakdown of PIP_2 that activates protein kinase C.

7.2.2.3. Exocytosis (Interphotoreceptor Matrix) and Endocytosis

Photoreceptors exhibit both endocytotic and exocytotic (secretory) activities that may be mediated by the inner segment cytoplasmic vesicles involved in opsin transport. Interphotoreceptor retinol-binding protein (IRBP) is the most abundant protein in the interphotoreceptor matrix; it is a glycolipoprotein of M_r 135,000—144,000 (depending on the species) and is secreted by the retina. Labeled IRBP is detectable in the medium after incubation of intact monkey neural retina with either labeled glucosamine or amino acids (Wiggert *et al.*, 1984; Fong *et al.*, 1984a). It is not present in the medium of retinas exposed to monensin, an ionophore that disrupts the Golgi apparatus and blocks the release of proteins destined for secretion (Wiggert *et al.*, 1984). Quantitative autoradiographic analyses of human retina incubated with labeled fucose show that although all retinal layers incorporate fucose, the most intense label is found in the rod inner segments (Hollyfield *et al.*, 1985b). A 4-h chase results in the loss of fucose-labeled macromolecules, mainly from the rod photoreceptors, and a concomitant appearance of IRBP in the medium.

Thus, the rod photoreceptors appear to be primarily responsible for both the synthesis and secretion of IRBP. Curiously, however, later immunocytochemical studies using rabbit anti-bovine IRBP failed to detect IRBP in photoreceptor inner segments of frog, human, or bovine retinas (Schneider *et al.*, 1986). No labeling was observed either in the Golgi or in vesicles near the connecting cilium, putative sites of IRBP synthesis. Nevertheless, later studies using rabbit anti-monkey IRBP as the primary antibody showed small amounts of IRBP in the rod inner segments of the monkey retina (Rodrigues *et al.*, 1987b). Compelling evidence that rod photoreceptors of both bovine and monkey retinas contain the synthetic machinery for IRBP synthesis has come from studies utilizing labeled cDNA probes and *in situ* hybridization techniques (van Veen *et al.*, 1986). Substantial amounts of IRBP mRNA were observed in rod neurons, but considerably less in other areas of the retina. Also of interest was the abundance of IRBP mRNA in pinealocytes.

Intact bovine retinas incubated with labeled fucose, glucosamine, or leucine synthesize and secrete IRBP into the medium (Fong *et al.*, 1984b). It is the only major labeled protein found in the medium under these experimental conditions. That photoreceptor cells are the source of IRBP in the interphotoreceptor matrix of the rat retina has been deduced from comparative studies of RCS and normal (congenic) strains (Gonzalez-Fernandez *et al.*, 1984, 1985). The IRBP identified by immunocytochemical techniques and immunoassay is present in the interphotoreceptor matrix of normal rat retina as early as the fifth postnatal day, and the level continues to increase for several months thereafter. In the degenerate RCS retina, although detectable early in development, IRBP levels decline sharply as the photoreceptor cells degenerate.

Photoreceptors also display endocytotic activity. For example, bovine retinas *in vitro* internalize colloidal gold particles coated with IRBP but show no activity toward particles coated with ovalbumin, a glycoprotein not found in the interphotoreceptor matrix (Hollyfield *et al.*, 1985a). The label was identified in small multivesicular bodies in both rod and cone inner segments; these organelles also show acid phosphatase reaction product, suggesting this as a possible site of IRBP degradation. Frog retinas also internalize horseradish peroxidase (HRP) in a time-dependent manner for periods up to 4 h *in vitro* (Hollyfield and Rayborn, 1987). The HRP reaction product is concentrated in a population of endocytotic vesicles, possibly distinct from opsin-carrying vesicles, located in the ellipsoid region of the inner segment. Thus, as shown by studies using HRP and collidal gold coated particles, the inner segment may play an important role in internalizing and degrading physiological components of the interphotoreceptor matrix such as IRBP.

7.2.2.4. Insulin and IGF-I Receptors

Insulin-like growth factor (IGF-I) is a 70-kDa polypeptide with extensive (49%) structural homology to insulin; both hormones exert their biological effects by binding to specific receptors that have similar properties but are nevertheless distinctly different membrane glycoproteins in most cell types. The receptors for insulin and IGF-I are heterotetramers composed of two 135-kDa α subunits and two 90- to 95-kDa β subunits; they are, however, products of distinct genes located on separate chromosomes (Ullrich *et al.*, 1986). On binding to the α subunit of the receptors, IGF-I and insulin stimulate a tyrosine-specific protein kinase activity that results in autophosphorylation of the β subunit as well as phosphorylation of exogenously added substrates.

Although insulin and IGF-I receptors have long been recognized in nonneural tissues, they have only recently been detected in neural tissues and in the retina. The observation that the GDP-bound α subunit of transducin can serve as a substrate for insulin receptor tyrosine-specific protein kinase in a cell-free system (Zick *et al.*, 1986) led to the characterization of both insulin and IGF-I receptors in ROS membranes (Zick *et al.*, 1987). The IGF-I receptors are detectable by the high affinity and specific binding of [^{125}I]IGF-I to bovine ROS membranes. The binding capacity of [^{125}I]insulin to ROS membranes is less than 5% of that shown by IGF-I. Partially purified tyrosine kinase activity from ROS membranes phosphorylates its own 95-kDa β subunits as well as transducin. Moreover, labeled IGF-I, but not insulin, can be specifically cross-linked to the 135-kDa α subunit of the receptor. These and other findings suggest that in bovine ROS membranes, both ligands may stimulate a single pool of tyrosine kinases, and that insulin acts through binding to the IGF-I receptor. In spite of its lower binding capacity, insulin appears to be more potent than IGF-I in coupling ligand binding with the activation of tyrosine kinase.

Crude membrane preparations of bovine retina have one IGF-I and two insulin binding sites (Waldbillig and Chader, 1988). One of the insulin binding sites has a higher affinity for insulin than for IGF-I, whereas the second one binds both hormones with equal affinity. Receptor solubilization and purification do not affect the insulin receptors' affinity for insulin but do increase this site's affinity for IGF-I. Thus, purified "insulin receptors" have a higher affinity for IGF-I than for insulin. Membranes isloated from whole bovine retina (Waldbillig *et al.*, 1987a) or from purified ROS (Waldbillig *et al.*, 1987b) have two subpopulations of insulin receptors differing in the size of the α subunit. One of them, a 133-kDa protein that migrates on SDS-PAGE as a diffuse band, is similar in size to the α subunit of bovine liver receptor, whereas the second 120-kDa protein is characteristic of bovine brain receptor. The smaller subunit is relatively insensitive to neuraminidase and is less soluble in Triton X-100 than the larger species. The heterogeneity is not a consequence of vascular contamination, since cross-linking studies on insulin receptors in bovine vascular and nonvascular tissue reveal 125- and 116-kDa subunit sizes, respectively (Im *et al.*, 1986). The latter investigations did not, however, detect the 133-kDa species in nonvascular retinal membranes, possibly because of its loss during the extraction procedure.

7.2.3. Lipid Peroxidation, Free Radicals, Light Damage, and Protective Mechanisms

Photoreceptor membranes, with their high concentration of polyunsaturated fatty acids (PUFAs), are potential targets for lipid peroxidation. The retina is chronically exposed to light, and its well-developed circulation provides a rich source of oxygen for the dense concentration of energy-consuming mitochondria in the inner segments. Together, these are major predisposing factors for *in situ* production of free radicals with the potential of initiating peroxidative chain reactions in the PUFAs of the disk membranes (Ham *et al.*, 1984). Light-induced free radical oxidation of ROS membrane lipids (Kagan *et al.*, 1973) and extensive morphological changes resulting from exposure of isolated ROS disks to air (Farnsworth and Dratz, 1976) are a few of the early observations on the damaging effects of light and oxygen (Feeney and Berman, 1976) on the retina. A brief description of free radical generation in nonocular tissues and of protective mechanisms that have evolved to prevent or minimize their toxicity is given in Chapter 1, Section 1.4. During the past decade, these topics have been investigated extensively in ocular tissues, and several detailed and provocative overviews of free radical-induced biochemical and structural changes in the photoreceptors have appeared (Noell, 1980; Wiegand *et al.*, 1984; Handelman and Dratz, 1986).

Lipid peroxidation and retinal damage resulting from oxidative stress, antioxidant-deficient diets, and constant illumination can be demonstrated under a variety of experimental conditions. Iron toxicity is a well-known clinical problem; penetration of a foreign body containing iron through the vitreous to the retina results in extensive retinal degeneration, often leading to loss of vision (Handelman and Dratz, 1986). It has been shown in model systems that highly reactive hydroxyl radical (OH·) can be produced by a Fenton-type reaction involving ferrous iron (Fe^{2+}) and H_2O_2 (see Fig. 1.9), and there is indirect evidence supporting the view that this reaction may also occur in biological systems. Incubation of frog retinas with Fe^{2+} plus ascorbate decreases the magnitude of the ERG waves generated by the retina in comparison to controls that had been preincubated with antioxidants (α-tocopherol or BHT) (Shvedova *et al.*, 1979). Intravitreal injection of 1 mM ferrous sulfate into the frog eye results in an extinct ERG and extensive disruption of ROS membranes within 4 h (Anderson *et al.*, 1984); after 8 days, the photoreceptor cell layer is obliterated, although other retinal layers are spared. Analyses of fatty acids 24 h after injection show a decrease in 22:6(*n*-3) and an increase in hydroperoxide levels. These *in vivo* findings, suggesting a causal relationship between peroxidation of PUFAs and specific photoreceptor cell degeneration, have been further clarified by *in vitro* investigations (Anderson *et al.*, 1985b). Incubation of ROS isolated from frog retinas with graded amounts of Fe^{2+} in the presence of ascorbate shows a concentration-dependent production of lipid peroxides, measured as conjugated dienes, and decreased regenerability of rhodopsin from 11-*cis* retinal in ROS having the greatest increase in hydroperoxides. Preincubation with chelators of iron (EDTA, EGTA, and DTPA) inhibits lipid peroxidation, and in their presence essentially complete regeneration of rhodopsin was observed.

Light damage to the retina has long been recognized, although the mechanisms involved are not completely understood. Five types of light-induced retinal degenerations in vertebrate retinas are described in detail by both Lawwill (1982) and Handelman and Dratz (1986). These classifications are based on the wavelength, intensity, and duration of light exposure used in experimental animals. Because of space limitations, the present discussion is limited to biochemical changes in the retina resulting from exposure to constant illumination administered under defined conditions.

The initial site of injury after exposure of experimental animals, usually albino rats, to illumination of several days' duration is the outer segment, and measurements of action spectra strongly suggest that the light damage is rhodopsin mediated (Williams and Howell, 1983). The extent of injury is best assessed by integration of entire retinal sections and not just severely damaged localized areas such as the superior retina. In support of this concept, conditions that lower the level of rhodopsin in the retina, such as restriction of dietary vitamin A, seem to minimize the light-induced photoreceptor damage (see review by Handelman and Dratz, 1986).

If free radicals are involved in these retinal degenerations, then vitamin E deficiency should exacerbate the damage caused by exposure to constant light. It has been reported that constant illumination is more injurious to the retinas of vitamin E-deficient rats than to retinas of animals receiving a normal diet (Kagan *et al.*, 1981). Other studies, however, gave no clear evidence of increased retinal damage in vitamin E-deficient albino rats after prolonged light exposure (Stone *et al.*, 1979b). Some of the injury observed may have been caused by the dietary deficiency itself, prior to light exposure. Further investigations on the putative role of vitamin E in light-induced retinal degeneration show that vitamin E levels in rat ROS do not change appreciably, and may even increase, during light stress. The vitamin E levels in ROS of cyclic-light-reared albino rats exposed to constant illumination of 200–250 ft-cd (foot-candles) for 24 h are only slightly (8–10%) lower than those in controls not exposed to light (Hunt *et al.*, 1984). However, after constant illumination, the vitamin E levels of ROS from cyclic-light-reared rats are 40% higher than those in dark-reared rats.

Further studies using pigmented rats with dilated pupils show that exposure to 10–20 ft-cd of light for periods up to 5 days causes a progressive degeneration of photoreceptor cells and concomitant loss of docosahexaenoic acid, 22:6(*n*-3) (Wiegand *et al.*, 1986). During the same period, there is a significant increase in vitamin E levels compared to nonilluminated controls. These findings suggest that vitamin E is utilized as a free radical scavenger during light stress but may be regenerated by interaction with another antioxidant, ascorbic acid. The levels of ascorbic acid in rat retina decrease during constant illumination (Organisciak *et al.*, 1984), and other investigations have shown that rats supplemented with ascorbic acid sustain less photic injury than unsupplemented controls (Z.-Y. Li *et al.*, 1985).

The intensity and duration of light exposure are important parameters in demonstrating lipid peroxidation in ROS membranes or in light-induced retinal degeneration. Exposure of either dark-reared or cyclic-light-reared albino rats to 300–400 ft-cd of illumination for 1 h does not result in detectable accumulation of hydroperoxides in ROS membranes of either group of animals (Organisciak *et al.*, 1983). However, constant illumination of 100–125 ft-cd for 3 days causes degeneration of the photoreceptors of albino rats, a reduction in levels of 22:6(*n*-3), and evidence for an increase in lipid hydroperoxides in isolated ROS (Wiegand *et al.*, 1983; Anderson *et al.*, 1984). Thus, prolonged exposure to light of moderate intensity light may promote peroxidation of ROS lipids and possibly initiate retinal degeneration.

Light history, intensity, and duration of exposure all influence the susceptibility of the albino rat retina to light-induced injury. Comparisons of antioxidant levels in retinas of rats reared in cyclic light of 5, 300, or 800 lx (0.4, 24, or 64 ft-cd, respectively) show increased amounts of vitamin E, ascorbic acid, and glutathione as well as higher activities of glutathione peroxidase and glutathione S-transferase in rats reared in 800-lx cyclic illumination than at lower light intensities (Penn *et al.*, 1987). After exposure to 2000 lx for 24 h, animals raised in 800-lx cyclic light showed no retinal damage, whereas those raised in 5-lx cyclic illumination had an almost complete loss of photoreceptor cells. Other studies have shown a significant decrease in 22:6(n-3) in ROS of rats raised in 800-lx cyclic light (Penn and Anderson, 1987); thus, both factors (increase in antioxidant capability and decrease in peroxidizable substrate) contribute to increased protection from oxidative insult to the retina from constant illumination. Further details of specific antioxidants present in the retina are discussed in the following sections.

7.2.3.1. Ascorbic Acid

Ascorbic acid (see Fig. 5.7) is present at relatively high concentration in the retina of most species. Premature infant human retinas have higher levels, by a factor of 35–50%, than mature human retinas (Nielsen *et al.*, 1988). In pigmented guinea pig, the neural retina contains 22 mg/dl of ascorbic acid, mainly in the reduced form; the concentration in the pigment epithelium–choroid is considerably lower (7 mg/dl) and consists principally of the oxidized (dehydro) form (Woodford *et al.*, 1983). After photic stimulation there is a decrease in the proportion of reduced ascorbic acid in the neural retina and an increase in the oxidized form in the pigment epithelium. A similar decrease in ascorbic acid levels after light exposure has been reported for albino rats (Organisciak *et al.*, 1984). Moreover, ascorbic acid is remarkably effective in protecting the retina from light-induced injury (Organisciak *et al.*, 1985; Z.-Y. Li *et al.*, 1985). Injection before, but not after, light exposure prevents the depletion of docosahexaenoic acid, reduces the loss of rhodopsin, and minimizes the damage caused by photic stimulation. An important role for ascorbate may be in the regeneration of reduced vitamin E, the form in which it is most effective as a free radical scavenger.

7.2.3.2. Glutathione Peroxidase and Glutathione S-Transferases

Glutathione (GSH) is thought to play an important role in defense against oxidative insult to the retina (Winkler and Giblin, 1983). Freshly excised rat retina contains 1.2 μmole of reduced glutathione per gram wet weight and no detectable amounts of oxidized glutathione (GSSG). Three enzymes appear to be involved in controlling the redox state of glutathione and the hydroperoxide levels in the retina: glutathione peroxidase, glutathione reductase, and glutathione S-transferases.

In most tissues, glutathione peroxidase activity is expressed by two enzyme proteins, one a selenium (Se)-containing enzyme and the other a nonselenium enzyme (Singh *et al.*, 1984; Handelman and Dratz, 1986). The Se-dependent enzyme (GSH I) is active in reducing both H_2O_2 and lipid hydroperoxides in the presence of GSH. The Se-independent enzyme (GSH II) reduces only lipid hydroperoxides; in many tissues, its activity is expressed by glutathione S-transferases. The latter comprise a group of multifunctional enzymes that not only reduce organic hydroperoxides (Stone and Dratz, 1982) but also detoxify electrophilic xenobiotics through conjugation to GSH. This detoxifying activity has been demonstrated in both rat (Stone and Dratz, 1980) and bovine (Saneto *et al.*, 1982a) retina, where it exists in two isoenzymic forms.

Selenium-dependent glutathione peroxidase (GSH I) does not appear to be present in bovine retina; all of the measurable glutathione peroxidase activity is associated with Se-independent glutathione S-transferase enzymes (Saneto *et al.*, 1982a). Measurements of enzymes involved in the mercapturic acid pathway for detoxification of xenobiotics (see Fig. 4.2) show high activities in all ocular tissues, and especially in bovine retina (Saneto *et al.*, 1982b). Three major classes of glutathione S-transferases, designated α, μ, and π, have recently been delineated in human and other mammalian tissues. Immunological studies and amino acid sequencing of the amino terminus of bovine retinal isoenzymes show significant primary structural homology with the μ and π classes of human enzymes (Ahmad *et al.*, 1988). Both of the retinal enzymes are heterodimers consisting 23.5- and 24.5-kDa subunits.

The Se and non-Se glutathione peroxidase activities in rat retinas have been measured using either H_2O_2 or cumene hydroperoxide as a substrate; H_2O_2 reflects the Se-dependent activity, whereas cumene hydroperoxide measures both (Stone and Dratz, 1982). High activities were found for both enzymes, and, not unexpectedly, in rats on diets deficient in Se and vitamin E, the detectable level of Se glutathione peroxidase (GSH I) is only about one-tenth of that in animals fed a supplemented diet. A recent study on GSH-dependent enzyme activities in whole retinas from rat and rabbit and in isolated sealed ROS confirms previous findings (Naash and Anderson, 1989). Of the three enzymes measured, the Se-dependent glutathione peroxidase is the major activity in ROS of both rat and rabbit. There are, however, species differences in glutathione S-transferase; e.g., the activities are strikingly higher in rat ROS than in rabbit ROS. Glutathione peroxidase activities have also been examined in frog retinas; pH optima as well as other parameters were studied to determine optimum conditions for assaying this enzyme activity *in vitro* (Naash and Anderson, 1984).

Human retina has two forms of glutathione S-transferase activity (anionic and cationic); however, neither of them express GSH II activity (Singh *et al.*, 1984). In contrast to bovine retina, all of the glutathione peroxidase activity of human retina is the Se-dependent GSH I type. Other studies have shown that the specific activity of glutathione peroxidase is higher in premature infants than in mature human retinas, although age-related changes in glutathione S-transferase activities were not observed (Naash *et al.*, 1988).

In summary, all species of retina examined to date appear to be protected from peroxidative damage under normal conditions. They contain glutathione enzymes active in reducing both H_2O_2 and lipid hydroperoxides at the sites of their formation. The most important of these sites is the ROS, and under conditions of stress such as constant illumination, the activities of at least two of these enzymes, glutathione peroxidase and glutathione S-transferase, are considerably elevated (Penn *et al.*, 1987).

7.2.3.3. Superoxide Dismutase and Catalase

Retinal superoxide dismutase (SOD) activity has been examined in several animal species using biochemical, immunochemical, and immunohistochemical methods. Although all studies show that retinal SOD is the cytosolic Cu-Zn enzyme, there are nevertheless discrepancies in its localization and also in its relationship to the erythrocyte enzyme.

On the basis of the nitroblue tetrazolium (NBT) photoreduction assay, the Cu-Zn enzyme was reported to be highly concentrated in ROS of frog and bovine retinas (Hall and Hall, 1975), yet immunohistochemical studies localize the enzyme to the inner segments and the plexiform layers, with essentially no reaction product detectable in the outer segments (Rao *et al.*, 1985). Inhibition by cyanide identifies the retinal enzyme as the cytosolic Cu-Zn type (Hall and Hall, 1975; Crouch *et al.*, 1978); mitochondrial Mn SOD appears to be absent from retina. This seems rather curious in view of the dense population of mitochondria in the inner segments. The SOD activity in rabbit retina (4.8 units/mg protein) is about one-third of that present in the iris or ciliary body (Bhuyan and Bhuyan, 1978).

The Cu-Zn SOD of bovine retina has the same electrophoretic mobility as erythrocyte SOD (Hall and Hall, 1975, Crouch *et al.*, 1978). However, later studies showed that although the retinal and erythrocyte enzymes have similar amino acid composition, there are differences in tryptic peptide maps and immunoreactivity (Bensinger *et al.*, 1982). On the other hand, the SOD enzymes of canine and rat retina and erythrocytes appear to be identical (Harlan and Crouch, 1984). The possibility that species differences could account for these conflicting results requires further investigation.

Very low levels of catalase have been detected biochemically in rabbit retina (Bhuyan and Bhuyan, 1977). Immunocytochemical studies of rat and bovine retina show a distribution similar to that of SOD; both enzymes are localized mainly in the inner segments (Atalla *et al.*, 1987).

7.2.3.4. Vitamin E (α-Tocopherol)

Standard dietary sources are usually sufficient to supply the nutritional requirements for this lipid-soluble vitamin whose antioxidant properties have long been recognized. As a free radical scavenger, it is thought to react directly with lipid free radicals (ROO·) and block the propagation of autooxidative chain reactions involved in lipid peroxidation (Fig. 7.11). As discussed above, α-tocopherol appears to be regenerated by ascorbic acid in the retina under conditions of light stress.

That vitamin E is essential for maintaining the structural integrity of the photoreceptors postnatally may be gleaned from the striking observation that vitamin E deficiency in the monkey causes extensive disruption of the photoreceptors, particularly those in the macular area (Hayes, 1974). Subsequent studies on vitamin E-deficient rats have shown specific degeneration of the ROS apical tips, with little damage occurring in the basal regions (see review by Handelman and Dratz, 1986). Even more remarkable are the changes observed in the retinal pigment epithelium (RPE) of vitamin E-deficient rats, the principal one being a striking accumulation of lipofuscin (Katz *et al.*, 1978, 1982). In fact, the RPE may be the primary site of autooxidative changes resulting from vitamin E deficiency; damage to adjacent apical ROS tips may be secondary.

As an antioxidant and free radical scavenger localized in the lipid matrix of the cell, α-tocopherol impedes the autooxidation of membrane PUFAs and protects cells and tissues from free radical damage. A large number of clinical disorders are thought to be associated with, or caused by, antioxidant deficiency that is expressed as an inability to respond to oxidative stress (Lubin and Machlin, 1982; Berman, 1987). In particular,

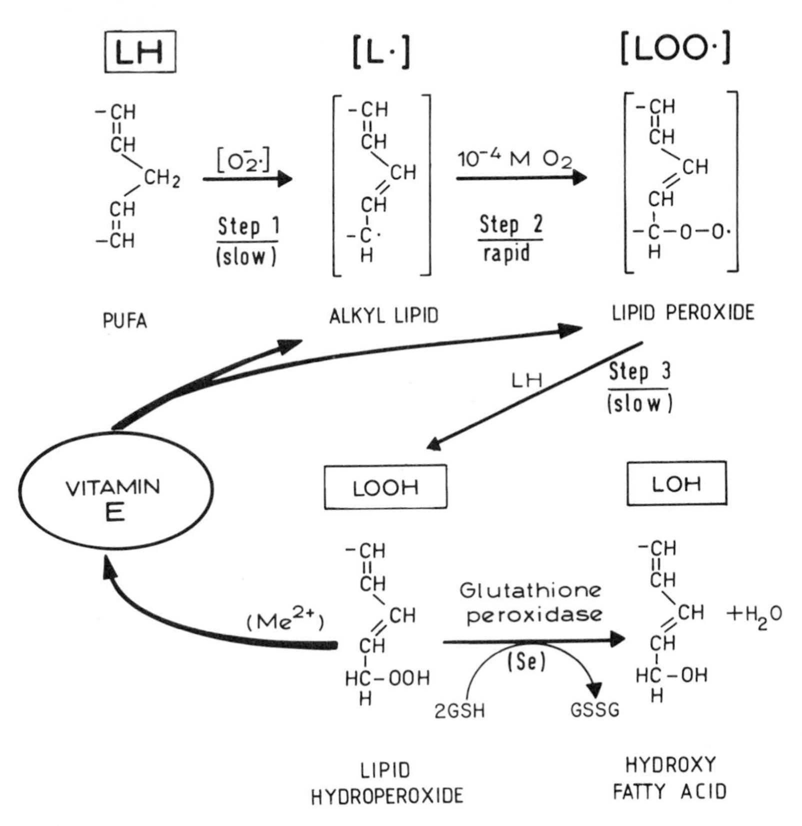

Figure 7.11. Diagram suggesting a possible site of action of vitamin E (α-tocopherol) in the interruption of autooxidation of polyunsaturated fatty acids (PUFAs) in photoreceptor membranes.

vitamin E appears to play an important role in the developing retina, since formation of abnormal gap junctions between spindle cells can be suppressed by early supplementation of preterm infants with the vitamin (Kretzer *et al.*, 1984). A relative deficiency of vitamin E in the retina of premature human infants of <27 weeks' gestational age is thought to be one of the major factors leading to retinopathy of prematurity (ROP) (Nielsen *et al.*, 1988). The precise cause of the abnormal proliferation of retinal vessels in ROP is poorly understood, but one possibility is that exposure of the premature retina to a sudden increase in oxygen tension triggers autooxidative reactions in the retina; the premature termination of vascular development following abnormal proliferation of blood vessels may be an adaptive response to oxidative stress, an attempt to minimize autooxidative reactions in the tissue (Katz and Robison, 1988).

The clinical importance of an adequate supply of vitamin E in the retina has prompted several studies on its endogenous content both in neural retina and in the RPE of humans as well as several animal species. Variations in diet and age contribute to the rather large range of values reported. The vitamin E content of cattle ROS is considerably higher than that in the rest of the retina, and the absolute amount depends on dietary sources. The ROS of cattle fed green winter rations contain about 1.5 μg vitamin E/mg protein, whereas little or none is detectable in those fed summer rations lacking vitamin E (Farnsworth and Dratz, 1976; Handelman and Dratz, 1986).

Measurements of ocular vitamin E in human donor eyes show that the highest levels are in the retina and pigment epithelium (RPE)–choroid (2.8 and 3.3 mg/100 g wet wt., respectively) and the lowest in the iris–ciliary body (0.4 mg/100 g wet wt.) (Alvarez *et al.*, 1987). The summed values for the vitamin E content of retina plus RPE–choroid of 70- to 80-year-old individuals are approximately 35% higher than those measured in 30- to 40-year-olds; these differences were not considered significant. However, by analyzing the neural retina and RPE (without attached choroid) separately, a distinct tendency toward higher levels of vitamin E in older individuals can be detected (Organisciak *et al.*, 1987). This was especially evident in the pigment epithelium, where vitamin E levels in the 80-year-old age group were about three to four times higher than those in 20-year-old individuals. There are smaller increases with age in the neural retina. Since the RPE has a higher concentration of vitamin E than the neural retina (see also Stephens *et al.*, 1988), age-related changes would be masked if neural retina is analyzed together with the RPE–choroid, as in the studies of Alvarez *et al.* (1987).

In albino rats on standard laboratory rations, vitamin E levels in the retina increase in an age-dependent manner from weanling (18–20 days) until 60 days (Hunt *et al.*, 1984). At this age, the content is about 0.45 nmol/retina, and similar to cattle, the levels are about three times higher in the ROS than in whole retina. As in human eyes (Organisciak *et al.*, 1987), in rats on standard laboratory diets, there are marked increases in the α-tocopherol content of the RPE during senescence (Katz and Robison, 1987). Measurements of α-tocopherol levels in the RPE of rats ranging in age from 12 to 32 months show increases of 21% and 48% in the dark-adapted and light-adapted states, respectively. New techniques have been reported that utilize a highly sensitive gas chromatographic–mass spectrometric procedure for analyzing the vitamin E content of microdissected rat, rabbit, and cat retinas (Stephens *et al.*, 1988). Confirming the findings of Organisciak *et al.* (1987), the highest content in all species examined was in the pigment epithelium, followed by outer segments of the photoreceptor cells. On a dry weight basis, the vitamin E content of the ciliary body and iris of cat and rabbit is about one-half of that in whole retina, but in the rat, it is 50% higher. In rats maintained on vitamin E-deficient diets, the vitamin E levels of neural retina and RPE decline relatively little compared to the iris and ciliary body. The relative resistance of the neural retina and pigment epithelium to dietary depletion of vitamin E may explain some of the ambiguous results described above on light-induced retinal degeneration in vitamin E-deficient animals.

7.2.3.5. Macular Pigments (Carotenoids)

The nonbleaching yellow pigments in the primate macula are concentrated mainly in the fiber layers (receptor axon and inner plexiform layers) of the retina (Snodderly *et al.*, 1984). Although their function is not completely understood, they are thought to attenuate the blue light reaching the cone-rich fovea and thus prevent chromatic aberration and glare; the pigments may also filter out potentially phototoxic short-wave (blue) visible light and may quench singlet oxygen generated in the retina (Kirschfeld, 1982; Handelman and Dratz, 1986). In this context, it is interesting to note that macular pigments are derivatives of carotenoids whose antioxidant activity is especially effective at low partial pressures of oxygen such as those found in the neural layers of the retina (Burton and Ingold, 1984). In monkeys maintained on xanthophyll-deficient diets for periods of 2–5 years, the plasma levels of xanthophyll fall to undetectable levels, and macular yellow

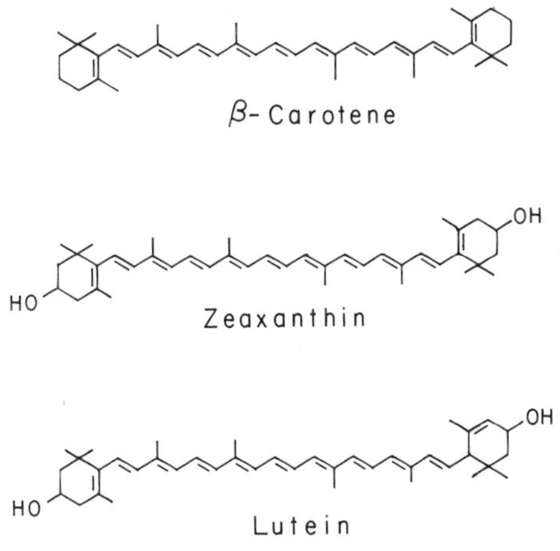

Figure 7.12. Carotenoid pigments of primate macula and their structural relationship to β-carotene.

pigmentation is absent from most experimental animals (Malinow *et al.*, 1980). The major macular anomaly observed in rhesus macaques was an increase of hyperfluorescence in fluorescein angiograms; possible impairment of visual function was not examined.

The yellow pigments have been characterized by high-pressure liquid chromatography, and their structures established (Bone *et al.*, 1985; Handelman and Dratz, 1986; Handelman *et al.*, 1988). As shown in Fig. 7.12, the macular pigments consist of a mixture of two isomeric dihydroxycarotenes, zeaxanthin and lutein. Between 20 and 250 ng of macular pigments are present in the human retina, and the amount may correlate with age in some individuals but not in others (Handelman *et al.*, 1988). β-Carotene is probably not an endogenous component of the retina; the trace amounts detected in this study probably represented blood contamination. Zeaxanthin is concentrated in the macular region, whereas lutein is found throughout the retina. Substantial amounts of both carotenoids are present in infant and in prenatal retinas (Bone *et al.*, 1988). Zeaxanthin is the dominant pigment in most individuals; moreover, in agreement with the findings of Handelman *et al.* (1988), the ratio of zeaxanthin/lutein is greatest in the central area and lowest in the periphery. This study showed no age-related changes in either of the macular pigments, nor could any changes in pigment density be detected as a function of age in 50 volunteers ranging in age from 10 to 90 years (Werner *et al.*, 1987).

7.2.3.6. Melanin

Melanins are high-molecular-weight complex polymers whose composition and structure are poorly understood. They are synthesized by successive oxidative polymerization of either dihydroxyphenylalanine (DOPA) to form brown-black eumelanin or cysteinyl-DOPA to form red-brown pheomelanin. Melanin can act as either a prooxidant or an antioxidant, and the reduced form is a highly effective free radical trap; studies in model systems show that it is both a singlet oxygen quencher and a scavenger for superoxide anion (see review by Handelman and Dratz, 1986).

Although there is some evidence suggesting that in the rat the melanosomes of the retinal pigment epithelium (RPE) may play a role in protecting photoreceptors from light damage, this does not seem to be the case in mice. Studies using mouse chimeras with alternating pigmented and nonpigmented RPE cells show no relationship between the extent of photoreceptor degeneration after constant illumination and the pigmentation phenotype in the immediately underlying RPE (LaVail and Gorrin, 1987). Other independent studies with chimeric mice gave similar results; although light-induced injury to the retina is less extensive in chimeras with more pigmented cells, no exact correlation between retinal damage and pigmentation could be detected (Sanyal and Zeilmaker, 1988). Taken together, these findings leave some doubt regarding a biochemical role for RPE melanin in protecting the photoreceptors from light-induced injury.

7.3. INTERPHOTORECEPTOR MATRIX

7.3.1. General Description and Structure

The interphotoreceptor matrix (IPM) is defined operationally as the mixture of macromolecules (proteins, glycoproteins, and proteoglycans) that fill the closed space lying between the retinal pigment epithelium (RPE), the photoreceptors, and the external limiting membrane (Fig. 7.13). The "mucosubstances" in this unique extracellular matrix of the retina were first detected in the late 1950s (see review by Feeney-Burns, 1985) and later characterized by histochemical (Fine and Zimmerman, 1963; Feeney, 1973b), biochemical (Berman and Bach, 1968), electron microscopic (Rohlich, 1970), and autoradiographic (Ocumpaugh and Young, 1966; Feeney, 1973a) techniques. The volumes of human and bovine IPM have been estimated at about 10 and 100 μl, respectively (see Adler and Severin, 1981). There is now firm evidence that the IPM consists of both soluble and insoluble glycoconjugates, the latter present as matrix sheaths surrounding the cone outer segments (L. V. Johnson *et al.*, 1985; Hageman and Johnson, 1986) and the rod outer segments in some species.

Owing to the strategic location of the IPM between the photoreceptors and the RPE, considerable attention has been focused in recent years on its chemical composition, physical properties, origin, and biological functions (see Bridges and Adler, 1985; Hewitt, 1986). Among the functions ascribed to the IPM, those concerned with the normal development and maintenance of the visual process appear to be the most important. Interphotoreceptor retinol-binding protein (IRBP), the major glycoprotein of the IPM, is thought to act as a transport vehicle for all-*trans* retinol in the visual cycle. Other possible functions of the IPM that require further clarification include transport of nutrients and metabolites as well as promotion or maintenance of adhesion between the photoreceptor cells and the pigment epithelium (Marmor, 1989; Yao and Marmor, 1989). It has also been speculated that the IPM may supply certain "recognition" factors involved in the phagocytosis of outer segments.

Early ultrastructural studies of the IPM emphasized its amorphous composition (Rohlich, 1970), but improved fixation techniques and the use of specific stains such as Ruthenium red and Alcian blue clearly delineated polysaccharide-rich regions on the surface of the photoreceptors as well as numerous strands throughout the matrix (Feeney, 1973b; Feeney-Burns, 1985). Other studies using liquid helium for rapid freezing of the

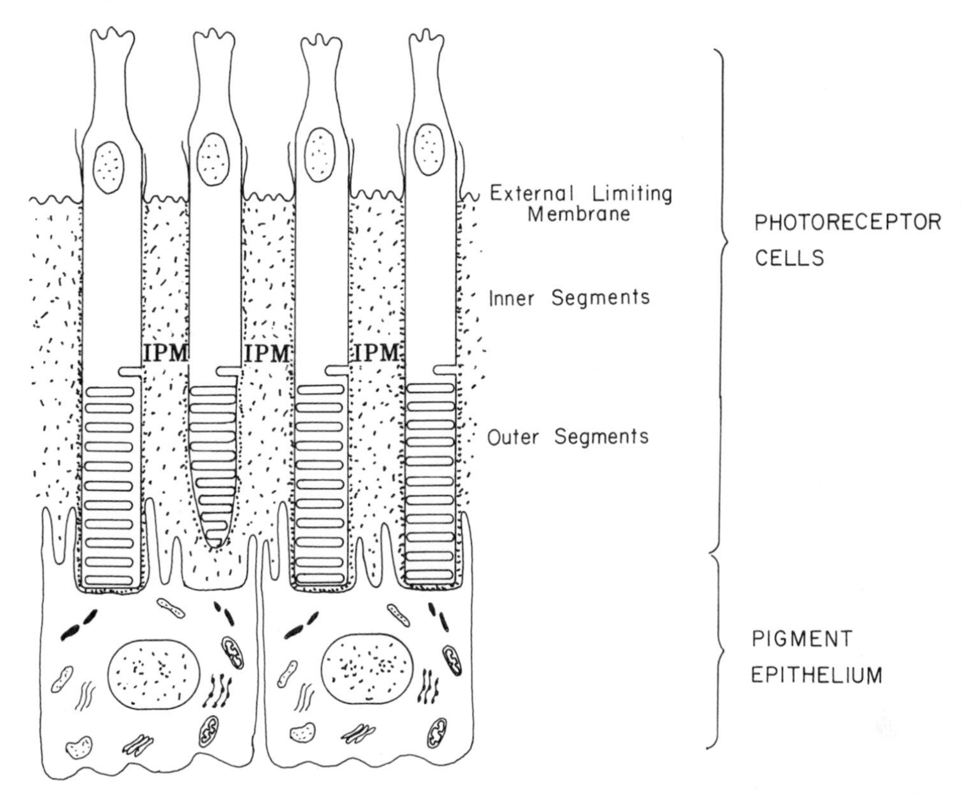

Figure 7.13. Schematic diagram showing the interphotoreceptor matrix (IPM) lying between the apical surface of the retinal pigment epithelium and the external limiting membrane. Soluble proteoglycans and glycoproteins are dispersed throughout the IPM, and insoluble glycoconjugates are concentrated in specialized domains that form sheaths surrounding the photoreceptor cells. Further details are given in the text.

IPM of frog retina have revealed a three-dimensional filamentous network distributed evenly throughout the matrix (Yamada, 1985). Since neither fibronectin nor collagen has been detected in the IPM (Adler and Klucznik, 1982), the chemical nature of these filaments is unclear, but the possibility of fixation artifacts cannot be entirely excluded.

It is now recognized that the IPM, as examined in a variety of species, is not homogeneous; rather, it is a highly organized extracellular matrix with specialized regional domains. The major glycoprotein of the IPM, IRBP, is preferentially localized on the apical surface of the RPE (Bunt-Milam and Saari, 1983). Moreover, histochemical and immunocytochemical techniques demonstrate a striking regional heterogeneity in the distribution of both glycosaminoglycans (Porrello and LaVail, 1986a,b) and glycoproteins (L. V. Johnson *et al.*, 1985; Hageman and Johnson, 1986).

The finding that carbohydrate-specific lectins bind to receptor sites on ROS plasma membranes of frog, cattle, monkey, rabbit, fish, and human retinas (Nir and Hall, 1979; Bridges, 1981; Uehara *et al.*, 1983; Blanks and Johnson, 1984) provided strong evidence for specific glycoconjugates associated with these cell surfaces. Of the lectins tested, soybean agglutinin (SBA) and particularly peanut agglutinin (PNA) showed preferential binding to cone outer segments. Peanut agglutinin, which is specific for the galactose β1,3 N-acetylgalactosamine sequence in glycoconjugates, shows little if any reaction with rod photoreceptors (Blanks and Johnson, 1984). Later studies provided evidence that most of the PNA binding is associated with cylindrical domains termed "cone matrix sheaths" extending from the external limiting membrane and ensheathing both the inner and outer segments of the cone photoreceptors (L. V. Johnson *et al.*, 1985, 1986; Hageman and Johnson, 1986). The domains, isolated intact by physical dissociation, consist of trypsin-sensitive glycoproteins of M_r ranging from 30,000 to 88,000. Judging from the behavior of the PNA-binding glycoconjugates toward proteolytic and glycosidic enzymes *in situ*, they appear to be protease-sensitive glycoconjugates containing O-glycosidic-linked oligosaccharides (Johnson and Hageman, 1987). Further studies on trypsin and urea extracts from porcine retinas have resulted in the tentative identification of very high-molecular-weight proteoglycans or proteoglycan aggregates as major components of the cone matrix sheaths (Hageman *et al.*, 1989).

Chondroitinases have a highly disruptive effect on the morphology of the cone matrix sheaths, and the use of anti-proteoglycan antibodies directed against several proteoglycans suggests that chondroitin 6-sulfate is a major component of monkey, pig, and human cone sheaths (Hageman and Johnson, 1987). The cone matrix sheath of human retina has been visualized histochemically with Cuprolinic blue, a phthalocyanin-like dye that stains anionic sites on sulfated proteoglycans when used according to the critical electrolyte concentration. Staining of human retina with Cuprolinic blue reveals spherical electron-dense particles approximately 40 nm in diameter that form elongated structures in the cone matrix sheaths; they appear to extend as far as the apical surface of the retinal pigment epithelium (Varner *et al.*, 1987). Unlike frog, rabbit, and human retina, the mouse retina does not seem to have a cone matrix sheath; instead, cytochemical studies using Cuprolinic blue staining have revealed two different types of sulfated glycosaminoglycan filaments in the IPM (Tawara *et al.*, 1988). Large filaments appear to form a complex network concentrated mainly on the surface of the outer segments and the outer portions of the inner segments of both rods and cones. Smaller filaments are found on the surface of the inner segments. The filaments are insoluble and closely associated with the plasma membranes of the photoreceptors; incubation with chondroitinase AC or ABC abolishes the Cuprolinic blue staining of both types of filament.

The foregoing discussion has considered mainly the specialized IPM matrices surrounding the cone photoreceptors, but recent evidence points to the existence of binding domains in rod photoreceptors as well. Studies using fluorescein-labeled (FITC) lectins have shown that FITC-WGA stains cylindrical domains of the IPM surrounding the rod inner and outer segments of monkey retina (Sameshima *et al.*, 1987). Wheat germ agglutinin (WGA) recognizes sialyl and N-acetylglucosamine residues, both of which appear to be present on the ROS plasma membrane surfaces. The source of the sialic acid binding sites may be sialic acid-containing glycoconjugates (Cohen and Nir, 1987), whereas rhodopsin probably supplies the N-acetylglucosamine binding sites.

These studies using FITC-WGA (Sameshima *et al.*, 1987), as well as others, corroborate previous findings suggesting the presence of glycoconjugates in the cell coat of rodent photoreceptors (Feeney, 1973a,b). Some of the cell coat sloughed off into the IPM space may correspond to the lectin-binding glycoconjugates described by Sameshima *et al.* (1987). It has been postulated (Varner *et al.*, 1987; Sameshima *et al.*, 1987) that the cone-associated and rod-associated matrix domains may play a role in adhesion between the photoreceptors and the retinal pigment epithelium (RPE), a notion arising from early studies on the glycosaminoglycans of the IPM (Berman, 1969; Zauberman and Berman, 1969). Since both rod and cone sheaths appear to reach as far as the RPE, they may be able to form noncovalent electrostatic or ionic bonds with charged groups of sialoglycoconjugates located on the apical surface of the pigment epithelium (Cohen and Nir, 1983). That such putative interactions could provide adhesive forces between the two cell layers is supported by the observation that exposing the rabbit retina to either testicular hyaluronidase or neuraminidase reduces within 2 min the peeling force required to separate the neural retina from the pigment epithelium (Yoon and Marmor, 1988). Recent studies using other specific enzymes lend further support to the view that the cone matrix sheaths and sialyl residues in the pigment epithelium may promote retinal adhesiveness (Yao and Marmor, 1989).

7.3.2. Glycosaminoglycans, Proteins, and Glycoproteins (Excluding IRBP)

7.3.2.1. Glycosaminoglycans

Whereas early investigations utilized mainly histochemical methods for identifying the soluble "mucosubstances" of the IPM, biochemical characterization of specific components became possible with the development of methods for rinsing or extracting the matrix from the retina (Berman and Bach, 1968; Berman, 1982). Agitation of intact bovine retinas in buffered saline releases about 90% of the soluble IPM; the IPM adhering to the surface of the RPE is also accessible by rinsing the surface of the eye cup after removal of the neural retina. The crude preparations obtained in this way are centrifuged, and the supernatants provide starting material for purification and characterization of either glycosaminoglycans (GAGs) (Berman and Bach, 1968; Berman, 1985) or proteins (Adler and Severin, 1981; Adler and Evans, 1985a).

The soluble GAGs of bovine IPM, which account for about 2% of the macromolecules of the matrix, were isolated and characterized many years ago (Berman and Bach, 1968; Bach and Berman, 1971a,b). Approximately 60% of the total population consists of three species of chondroitin sulfate. Of these, more than half are nonsulfated or undersulfated, and only about 16% appear to be fully sulfated (at the C-6 position). Hyaluronic acid comprises about 14% of the GAGs in bovine IPM. In addition, low-molecular-weight sialoglycans (not necessarily GAGs) have also been detected in bovine IPM. They probably represent fragments of larger sialoglycoconjugates that were later detected histochemically (Porrello and LaVail, 1986b). The GAG composition of human IPM is similar to that of bovine tissue, although fully sulfated chondroitin sulfate appears to be absent in extracts of postmortem human retina (Edwards, 1982a).

The alkali lability of serine and threonine in purified chondroitin sulfate from bovine IPM is indicative of O-glycosidic linkages to a peptide core (Bach and Berman, 1971b). Hence, in the native state, the GAGs of bovine IPM, like those in other extracellular matrices, are present as proteoglycans. This has been confirmed in the rat retina using monoclonal antibodies directed against various chondroitin sulfate proteoglycans (Porrello and LaVail, 1986a,b; Hageman and Johnson, 1987). Of all the monoclonal antibodies tested, immunofluorescent labeling was found only for unsulfated chondroitin and chondroitin 6-sulfate; neither 4-sulfated chondroitin sulfate nor dermatan sulfate could be detected in the IPM of rat retina (Porrello and LaVail, 1986a,b). These immunocytological studies, as well as previous histochemical findings (LaVail *et al.*, 1981, 1985), suggested that both species of chondroitin sulfate are heavily concentrated in the basal inner segment/outer segment zone. However, immunogold cytochemistry has localized a significant portion of chondroitin 6-sulfate to the RPE apical surface as well (Porrello *et al.*, 1989). The presence of major proteoglycan(s) on the apical surface of the RPE lends further support to the view that these substances have a variety of important roles in photoreceptor–RPE interactions such as promoting adhesion and/or facilitating exchange of nutrients.

The major sites of synthesis of IPM GAGs and/or proteoglycans have not been established with certainty. Early autoradiographic studies in the rat using labeled sulfate suggested that they may be synthesized and secreted by the inner segments (Ocumpaugh and Young, 1966), and recent biochemical and autoradiographic studies on the mouse retina *in vitro* corroborate active synthesis and secretion of sulfated proteoglycans by the neural retina (Landers *et al.*, 1989).

Nevertheless, other studies using a variety of experimental approaches suggest that the pigment epithelium may be an important source IPM GAGs and/or proteoglycans. For example, freshly harvested bovine RPE cells incubated with labeled glucose synthesize several glycoconjugates, at least one of which has the characteristics of a proteoglycan (Berman, 1964). Cultured human RPE cells also synthesize and secrete GAGs tentatively identified as chondroitin sulfate, dermatan sulfate, and hyaluronic acid (Edwards, 1982a,b). However, the GAG profile expressed by cultured RPE cells differs substantially from that of native human IPM not only in specific components but also in the relative proportions of individual GAGs. Other studies using cultured human RPE cells show a similar pattern of GAG synthesis and secretion into the culture medium (Yue and Fishman, 1985). In cultured feline RPE cells, the major GAG secreted is chondroitin sulfate (Stramm, 1987). When compared to freshly isolated feline RPE cells, the profile of newly synthesized GAGs in cultured cells is similar but still not identical to the endogenous GAGs present in the IPM.

In a different approach to determine the origin of GAGs in the IPM, Tawara and coworkers (1989) have followed GAG synthesis in developing mouse using Cupromeronic blue to stain and localize chondroitin sulfate-type proteoglycans. Three types of filaments were observed in the IPM of developing mouse retina; these filaments were not detected in either photoreceptor cytoplasm or in Müller cells but were found only in the RPE cells during all stages of development. Type A filaments were especially prominent in the Golgi apparatus and in cytoplasmic vesicles of the RPE. The proteoglycans of developing mouse IPM form an extracellular meshwork that surrounds the photoreceptors during their elongation to adult length. The fact that they are elaborated into the IPM a few days prior to outer segment development again suggests the RPE, and not the photoreceptor cell, as their site of synthesis.

7.3.2.2. Proteins and Glycoproteins (Excluding IRBP)

The protein profile of bovine IPM on SDS-PAGE is distinctly different from that of homogenates derived from adjacent retinal tissues (Adler and Severin, 1981; Adler and Evans, 1985a). Comparison of IPM proteins with those of soluble extracts of neural retina, photoreceptors, and RPE show that the most prominent band in the IPM preparation is a 140-kDa glycoprotein; this protein, later named interphotoreceptor retinol-binding protein (IRBP), is discussed in Section 7.3.3. About ten other proteins are also observed on SDS-PAGE in crude preparations of IPM obtained by rinsing or stirring the retina. Although some are unique to the IPM, others undoubtedly originate by diffusion or leakage from adjacent retinal compartments. Fewer proteins are present in IPM obtained by washing or cannulating the interphotoreceptor space *in situ*, a technique that in the monkey eye causes little or no injury to either the neural retina or the RPE (Pfeffer *et al.*, 1983). The major protein in IPM obtained by this nontraumatic method is IRBP, which accounts for close to 70% of the total protein of the matrix. Other retinoid-binding proteins originally thought to be present in the IPM probably originated through leakage from adjacent retinal cells.

Apart from IRBP, only a few endogenous IPM proteins or glycoproteins have been characterized. One is a 31-kDa glycoprotein that is secreted into the medium of cultured human RPE cells and also appears to be a component of native IPM (Edwards *et al.*, 1987b). A number of lectin-binding glycoproteins have been found in bovine IPM using two-dimensional gel electrophoresis combined with Western blotting (Uehara *et al.*, 1986). They are of lower M_r than IRBP, and many of them show altered binding characteristics after neuraminidase digestion.

Several enzyme activities have been reported in the IPM. One is a cGMP phosphodiesterase having catalytic properties similar to those of ROS PDE but differing in location, solubility characteristics, and subunit size from the photoreceptor enzyme (Barbehenn *et al.*, 1985). This cGMP phosphodiesterase has been isolated and purified from bovine and monkey IPM. Like ROS PDE, the IPM enzyme is activated by both protamine and trypsin, and is highly specific for cGMP. Moreover, the two proteins are immunologically related as shown by cross-reactivities to a monoclonal antibody (ROS-1) prepared to ROS PDE. The function of IPM PDE has not been established, but Barbehenn *et al.* (1985) speculate that it may regulate the extracellular concentration of cGMP in the RPE–photoreceptor complex.

Another enzyme detected in bovine IPM, lysosomal acid protease, represents close to 10% of the total activity present in the neural retina–IPM–RPE complex (Adler and Martin, 1982/1983). The highest specific activity was found on the apical surface of the RPE. Although some of the enzyme activity may have originated from damaged RPE cells, later studies using cultured human RPE support the view that RPE cells may in fact actively secrete lysosomal acid hydrolases (Wilcox, 1987). The extracellular release of β-N-acetylglucosaminidase, α-mannosidase, and β-glucuronidase is enhanced 120% in the presence of mannose 6-phosphate. In other cell types, and possibly in the RPE, mannose 6-phosphate receptors play key roles both in the intracellular transport of newly synthesized acid hydrolases and in their uptake from the extracellular space. The studies of Wilcox (1987) show that protein synthesis is necessary for the extracellular release of β-N-acetylglucosaminidase; moreover, this enzyme was shown to be specifically bound to a mannose 6-phosphate receptor present on the RPE cell surface.

7.3.3. Interphotoreceptor Retinol-Binding Protein

Absorption of light by the photoreceptors is the first event in the visual process in the vertebrate retina; after isomerization and reduction, all-*trans* retinol migrates into the retinal pigment epithelium (RPE). Retinoid is returned to the photoreceptors during dark adaptation. The existence of a transport vehicle for retinol between these two cell layers has long been suspected since, in the free form, retinol is unstable, cytotoxic, and insoluble in aqueous media. The retinol carrier (IRBP), later shown to be localized exclusively in the IPM, was in fact initially detected as a 7 S retinol-binding protein in supernatant fractions of retinal homogenates (Wiggert *et al.*, 1978b). In later studies, direct isolation of the IPM by rinsing or soaking excised neural retina led to a more precise localization and identification of IRBP by three independent groups of investigators (Adler and Martin, 1982; Lai *et al.*, 1982; Liou *et al.*, 1982). Chader (1989) has recently presented an in-depth analysis of the biochemical, molecular biological, and physiological properties of IRBP; this highly recommended review also includes current information on its gene structure as well as the putative role of IRBP in retinal degenerations and in experimental autoimmune uveitis.

7.3.3.1. Chemical Characterization

The IRBP from bovine IPM was first isolated by concanavalin A-Sepharose chromatography and shown to be a glycoprotein rich in mannose and containing N-glycosidically linked carbohydrate chains (Adler and Klucznik, 1982). It has an apparent M_r of about 250,000 on gel filtration (Liou *et al.*, 1982; Adler and Martin, 1982; Fong *et al.*, 1984b; Saari *et al.*, 1985), but on SDS-PAGE, it migrates as a 140- to 145-kDa polypeptide. When measured by sedimentation equilibrium, the molecular masses of bovine (Saari *et al.*, 1985; Adler *et al.*, 1985) and monkey (Wiggert and Chader, 1985) IRBP are 131.7 or 133 kDa and 106 kDa, respectively. The frictional ratios of IRBP from these two sources are 1.64 and 1.59, respectively, corresponding to an axial ratio of about 8 or 11:1. Thus, IRBP is an elongated, asymmetric molecule; this, plus the fact that it is glycosylated, would account for the anomalously high M_r values obtained by gel filtration. All studies show that IRBP is a 140- to 146-kDa monomer and not a dimer as suspected in early investigations. Some of these data are summarized in Table 7.12.

Bovine IRBP isolated from dark-adapted eyes carries both 11-*cis* retinol and all-*trans* retinol as endogenous ligands, corresponding to 6–7% saturation of binding sites (Liou *et al.*, 1982; Fong *et al.*, 1984b; Saari *et al.*, 1985). Illumination of the eye causes a four- to fivefold increase in the amount of all-*trans* retinol bound to IRBP but does not affect the binding of 11-*cis* retinol (Lai *et al.*, 1982; Adler and Martin, 1982; Liou *et al.*, 1982; Adler and Evans, 1985b; Saari *et al.*, 1985). Thus, only about 30% of IRBP's binding sites are saturated with endogenous retinoid in the light-adapted eye (Bridges *et al.*, 1984; Adler and Evans, 1985a). Full saturation can be achieved by incubation of IRBP with labeled exogenous all-*trans* retinol, resulting in a ligand:protein ratio of 2 mol of retinol per mol of IRBP (Fong *et al.*, 1984b; Saari *et al.*, 1985). Dark-adapted IRBP binds 11-*cis* retinal, suggesting a role for this protein in returning the aldehyde from the RPE to the photoreceptors for regeneration of rhodopsin (Adler and Evans, 1985a). The apparent K_d for all-*trans* retinol is relatively high, 1.3×10^{-6} M, which implies a loose attachment of ligand and hence the possibility that the retinoid can be readily transferred to other binding proteins.

These findings have led to the notion that IRBP serves as a buffer system and/or a transport vehicle, or shuttle, for the translocation of retinoids between the photoreceptors and the RPE during operation of the visual cycle (see review by Chader, 1989). However, experiments using model membrane systems *in vitro* have shown that all-*trans* retinol is transferred rapidly and spontaneously between liposomes and ROS membranes via the aqueous phase, and IRBP retards rather than facilitates this transfer (Ho *et al.*, 1989). The

Table 7.12. Some Chemical and Physicochemical Properties of Interphotoreceptor Retinol-Binding Protein[a]

Property	IRBP	
	Bovine	Monkey
Molecular weight		
SDS-PAGE	140–145 k	146 k
Sediment. equil.	131.7 or 133 k	106 k
Sedimentation coefficient	5.73 S	5.4 S
Stokes radius	55 Å	56 Å
Frictional ratio	1.64	1.59
Axial ratio	8:1	11:1
Endogenous ligand	all-*trans* Retinol[b]	all-*trans* Retinol
No. of binding sites	2	—
Fraction of sites occupied		
Dark	5%	—
Light	30%	—
Apparent K_d	1.3×10^{-6} M	—
Fluorescence max		
Emission (nm)	470 (350 exc.)	470 (340 exc.)
Excitation (nm)	333 (470 emiss.)	333 (470 emiss.)
Carbohydrate	8.4%	18%

[a]Further details and literature references are given in the text.
[b]Also contains about 12% 11-*cis* retinol (Bridges *et al.*, 1984; Saari *et al.*, 1985).

kinetics of retinol transfer between single unilamellar vesicles was also studied, and the overall results suggest the following two-step mechanism: first, retinol partitions between liposomes (or membranes) and the aqueous phase, and second, the aqueous "soluble" retinol diffuses spontaneously to the acceptor liposome (or membrane).

These experiments imply that even though a major function of IRBP is to bind retinoids and possibly act as a buffer during extensive bleaching, it was not thought to facilitate the transport of bound retinoids through the IPM. Nevertheless, under "simulated" physiological conditions *in vitro* using toad eye cup preparations devoid of neural retina, IRBP was shown to enhance the rate of retinol esterification in the RPE in a time- and concentration-dependent manner (Okajima *et al.*, 1989). These findings imply that IRBP could function in the visual cycle as a carrier of retinol to the pigment epithelium. Not only does IRBP effectively mediate the delivery of labeled retinol to the RPE for esterification, but it also plays an active role in the bleaching and regeneration of rhodopsin through its ability to transport 11-*cis* retinal from the pigment epithelium to the photoreceptor cells (G. J. Chader, personal communication).

The IPM of bovine retina contains about 3–4 nmol of IRBP per eye (Fong *et al.*, 1984b; Adler and Evans, 1985b), and its major (if not only) site of synthesis and secretion in the rod-dominant (van Veen *et al.*, 1986) and cone-dominant (Anderson *et al.*, 1986b) retina is the photoreceptor cell (see Section 7.2.2.3). Although present throughout the IPM, the highest concentration of IRBP is on the apical surface of the RPE (Bunt-Milam and Saari, 1983). However, IRBP is not limited to the retina. Using a newly developed ELISA combined with immunocytochemical techniques, IRBP has been detected in the vitreous and aqueous humors and in the pineal gland of monkey (Wiggert *et al.*, 1986).

Bovine IRBP contains 8.4% carbohydrate consisting of sialic acid, N-acetylglucosamine, and neutral sugars (galactose, mannose and fucose) that appear to be N-linked biantennary complex structures (Fong *et al.*, 1984a,b, 1985). The glycoprotein contains one major type of oligosaccharide of M_r approximately 2500. It displays microheterogeneity possibly because of differing numbers of sialic acid and fucose residues. There are four or possibly five N-linked oligosaccharide chains per mol of bovine IRBP. However, glycosylation is not required for its secretion by bovine retina, as shown in experiments using labeled fucose and leucine as IRBP precursors. Even after preincubation with tunicamycin, castanospermine, or swainsonin, well-known inhibitors of glycoprotein synthesis and processing, labeled IRBP is still secreted into the incubating medium.

The chemical and physicochemical properties of highly purified IRBP prepared from monkey retina (Wiggert and Chader, 1985; Redmond *et al.*, 1985) are similar, though not identical, to those of bovine IRBP (Table 7.12). Both are asymmetric monomers whose molecular weights are considerably lower when measured by sedimentation equilibrium than by SDS-PAGE. Monkey IRBP also contains endogenous covalently and noncovalently linked fatty acids, the latter accounting for about two-thirds of the total bound fatty acids (Bazan *et al.*, 1985). The fatty acids are present in relatively high concentration, 6.51 mol per mol of protein, and consist mainly of palmitic (35%), oleic (29%), and stearic (21%) acids; IRBP is now often referred to as a glycolipoprotein.

The amino acid compositions of purified bovine and monkey IRBP are similar, with about 50% of the residues consisting of hydrophobic amino acids (Fong *et al.*, 1984b; Wiggert and Chader, 1985). The N-terminal residues of bovine (Bridges *et al.*, 1984; Saari *et al.*, 1985) and monkey (Redmond *et al.*, 1985) IRBP show extensive sequence homology, and subsequent studies suggest that although IRBP is a relatively well-conserved protein, there are species differences (Redmond *et al.*, 1986). Primate IRBPs (human and monkey) have "N" and "N + 5" terminal sequences that are qualitatively identical. However, bovine IRBP lacks the five-amino acid extension. In addition, analyses of peptide fragments produced by limited proteolysis also reveal similarities between human and monkey IRBP on the one hand but significant differences when compared to bovine IRBP.

The molecular biology of IRBP has been reviewed in detail by Chader (1989); hence, only a brief summary of these findings is given here. In an early study, bovine retinal cDNA libraries were probed with antibodies to purified IRBP and then hybridized to yield several cDNA clones; the deduced amino acid sequence from one of the clones was identical to the sequence of a tryptic peptide obtained from authentic IRBP (Barrett *et al.*, 1985). In other investigations, a 3400-nucleotide cDNA segment has been isolated that encodes the amino acid sequence of eight bovine IRBP tryptic peptides, including one that appears to be glycosylated (Liou *et al.*, 1986). More recently, genomic clones have been obtained that contain full-length copies of the gene (Borst and Nickerson, 1988). The gene is surprisingly small, about 12 kb, and sequence analysis of one of the gene clones revealed a short five-amino-acid sequence between the signal sequence and the N terminus, confirming the observations of Redmond *et al.* (1986) on the presence of an "N + 5" terminal sequence in primates and its absence in bovine IRBP.

In other studies, the complete 1264-amino-acid sequence of human IRBP has been deduced; the protein was reported to have a molecular mass of 136,600 Da, and except for two short segments at the C and N termini, the entire molecule consists of four contiguous repeat sequences, each containing about 300 amino acids with 33–38% identity (Fong and Bridges, 1988). There is 87% identity between human and cattle IRBP, implying that gene duplications had occurred 600–800 million years ago, before the emergence of vertebrates. A later study by Liou *et al.* (1989) using the same cDNA clone has, however, revealed certain discrepancies in the deduced amino acid sequence of human IRBP reported by Fong and Bridges (1988). While confirming that the approximately 11-kb human IRBP gene contains four exons interrupted by three introns, the deduced amino acid sequence of Liou *et al.* (1989) predicts a protein of 1230 residues with a calculated M_r of 133,000. The discrepancy is in part due to an additional C-terminal peptide sequence at position 1228—1233 in the data of Fong and Bridges (1988), giving a sequence containing 29 additional amino acids. Three additional amino acids were also deduced at position 982—1010 by Fong and Bridges (1988) which do not match the data of Liou *et al.* (1989) or Borst and co-workers (1988, 1989). Maps of the human IRBP gene have been published (Liou *et al.*, 1989; Chader, 1989), and in addition the structural gene for human IRBP is reported to be located on chromosome 10p11.2→q11.2 (Chin *et al.*, 1988).

7.3.3.2. Experimental Autoimmune Uveitis

Immunization with microgram quantities of purified monkey IRBP causes uveoretinitis (EAU) and pinealitis (EAP) in Lewis rats (Gery *et al.*, 1986) but not in guinea pigs (Vistica *et al.*, 1987). The uveitis induced in Lewis rats resembles in some respects that provoked by S-antigen, but there are nevertheless differences in intensity, time of onset, and duration of the inflammatory reaction. IRBP is more uveitogenic than S-antigen at low doses, but the reverse is true at higher doses (Fox *et al.*, 1987). Moreover, various inbred strains of rats show differing susceptibility to EAU induced by the two antigens. For example, Wistar Furth and RCS-rdy$^+$ rats, like guinea pigs, show little ocular inflammation after immunization with IRBP.

As in Lewis rats, immunization of rabbits with bovine IRBP induces a uveitis resembling that caused by S-antigen (Eisenfeld *et al.*, 1987). Detailed investigation of the retinal pathology shows that the immunization causes extensive photoreceptor degeneration and probably a breakdown in the outer blood–retinal barrier. Bovine IRBP at doses as low as 10 μg/kg body weight also induces severe uveitis in monkeys, but monkey IRBP administered at levels ten times higher causes no inflammatory response (Hirose *et al.*, 1987). When tested using a newly developed highly sensitive culture procedure, blood lymphocytes from most human uveitis patients, as well as a large proportion of normal healthy volunteers, react against both S-antigen and IRBP (Hirose *et al.*, 1988).

Three cyanogen bromide fragments of bovine IRBP, obtained after reduction and S-carboxymethylation of the purified protein, produce EAU when injected into Lewis rats (Redmond *et al.*, 1988). Two of the peptides are localized in the C-terminal portion of the IRBP molecule, and the third in the N-terminal region. Part of the genomic DNA-deduced sequence of one of them (CB*-47) localized near the N-terminal portion of the molecule, as well as the entire sequence of another (CB-71) localized near the C terminal, have been determined and shown to display extensive homology. These findings imply internal gene duplication events in the evolution of the IRBP molecule. The authors propose that bovine IRBP contains multiple homologous uveitogenic sites; S-antigen is reported to have two uveitogenic sites (Donoso *et al.*, 1986), but the uveitogenic sequences in IRBP and S-antigen are very different. In further studies to identify the immunopathogenic epitopes of IRBP, two closely related synthetic peptides (designated R4 and R9) derived from the highly uveitogenic cyanogen bromide cleavage fragment CB-71 were tested and found to induce mild EAU and EAP in immunized Lewis rats (Sanui *et al.*, 1988). Contrary to prevailing theories on immunogenicity, neither of these peptides was predicted to form amphipathic helixes. The inflammatory changes induced by R4 and R9 resemble human uveitis more closely than the changes induced by either S-antigen or IRBP; therefore, these peptides may provide a better model for studying the etiology of human autoimmune disorders than the intact parent molecules.

7.3.3.3. Subretinal Fluid

An early study of subretinal fluid (SRF) from two patients with rhegmatogenous retinal detachments failed to reveal IRBP in samples analyzed by SDS-PAGE (Adler and Martin, 1982). However, with more sensitive Western blotting, IRBP has been detected in seven out of eight SRF samples (Bridges *et al.*, 1986a). When calculated as nanograms per microliter of fluid, the concentration of IRBP is only about 5–19% of normal human IPM. These low concentrations do not appear to be related to the duration of the detachment, but rather to dilution of the samples in the subretinal space. Measurements of IRBP in a large series of SRF samples using a newly developed ELISA revealed highly variable concentrations ranging from 1 to 176 μg of IRBP per milligram protein (Newsome *et al.*, 1988a). In these studies the concentrations appeared to be related to the duration of the detachment, with the highest levels being found in detachments of short duration.

In contrast to rhegmatogenous retinal detachments, IRBP is not detectable in SRF samples from patients with retinopathy of prematurity (Bridges *et al.*, 1986a; Newsome *et al.*, 1988a). Since IRBP is detectable in the subretinal space as early as the 20th week in preterm human infants and continues to accumulate as the photoreceptors develop (A. T. Johnson *et al.*, 1985), the absence of IRBP in patients with retinopathy of prematurity implies impaired synthesis, possibly because of metabolically nonviable photoreceptors.

7.4. RETINAL PIGMENT EPITHELIUM

7.4.1. General Description

The retinal pigment epithelium (RPE), a single layer of cuboidal cells lying between Bruch's membrane and the photoreceptors, is an essential component of the visual system. As shown in Fig. 7.14, the apical RPE cell surface faces the photoreceptors, and its villous processes interdigitate with the outer segments. The basal surface, with its numerous infoldings, is functionally linked to the choroid via Bruch's membrane. The cells are joined to one another by tight junctions that, together with the endothelium of the choriocapillaris, effectively exclude the exchange of potentially toxic substances between

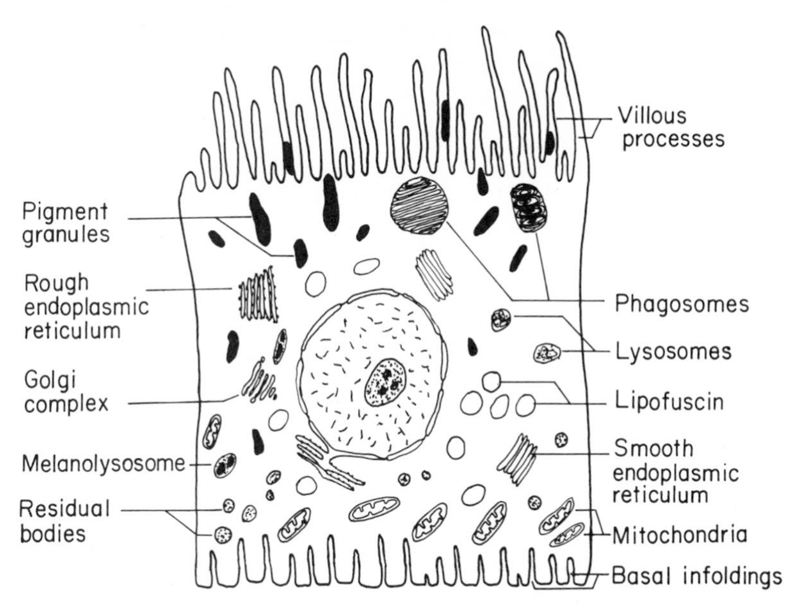

Figure 7.14. Diagram of a retinal pigment epithelial (RPE) cell. Note the villous processes projecting from the apical surface and the extensive infoldings on the basal surface of this highly polarized cell. Melanin granules are more numerous in the apical region, whereas lipofuscin particles are more abundant in the central and basal areas. The organelles shown include those present in all cell types (nucleus, mitochondria, lysosomes, rough and smooth endoplasmic reticulum, and Golgi). In addition, the RPE contains specialized organelles such as melanin granules, phagosomes, and phagolysosomes.

the choroidal circulation and the neural retina. Thus, the RPE constitutes an important structural and functional part of the outer blood–retinal barrier.

The RPE cells are polarized not only morphologically but also biochemically. For example, Na^+,K^+-ATPase is localized on the apical surface (Ostwald and Steinberg, 1980), whereas receptors for serum retinol-binding protein (RBP) are found on the basal and lateral plasma membranes of the RPE of rat (Bok and Heller, 1976) and human (Pfeffer *et al.*, 1986) retina. Even more striking is the dual localization of proteoglycan-synthesizing systems in cultured human RPE (Hewitt and Newsome, 1988). The proteoglycans synthesized on the apical surface that are secreted into the medium differ substantially, in both size and GAG composition, from those deposited basally onto the substratum. These topics are discussed in greater detail in appropriate sections below.

The retinal pigment epithelium displays a wide variety of activities, including phagocytosis, uptake and storage of retinoids, secretion of IPM and basement membrane components, and transport of substances between the photoreceptors and the choriocapillaris. Many of the known functions of the RPE are summarized briefly in Table 7.13, and considerably more information on all aspects of the RPE in health and disease is available in a full text edited by Zinn and Marmor (1979). Despite the large amount of information available from a variety of disciplinary approaches, the present discussion, based in part on a number of reviews and monographs (Berman, 1979; Young and Bok, 1979; Basinger and Hoffman, 1983; Clark, 1986), is limited to the biochemistry, metabolism, and biology of the RPE.

Table 7.13. Major Functions of the Retinal Pigment Epithelium

Biochemical
Phagocytosis of rod and cone outer segments.
Enzymatic (and nonenzymatic) breakdown of phagosomes to residual bodies.
Uptake and enzymatic conversions of retinoids for use in the visual cycle.
Synthesis of extracellular matrices.
GAGs and proteins for secretion into the IPM.
GAGs and collagen for deposition into Bruch's membrane.
Detoxification of drugs.
Synthesis of melanin.
Physiological
Exchange of nutrients and metabolites between the photoreceptors and the choriocapillaris.
Transport of fluids sclerally, creating a force thought to contribute to adhesion between the neural retina and the RPE.
Maintenance of photoreceptor microenvironment.
Physical and optical
Promotion of retinal adhesion.
Maintenance of the outer blood–retinal barrier.
Control of light scatter and absorption of light energy.

7.4.2. Methods for Isolation and Culture

Whereas adequate amounts of neural retina are available for biochemical study in virtually all species, the paucity of tissue in the single layer of RPE cells has in the past hindered the study of RPE biochemistry. The majority of investigations were limited to conventional histochemical, electron microscopic, and autoradiographic techniques. Now, however, several types of tissue preparation are available for direct biochemical study: fresh suspensions of RPE cells, organ or cell cultures, and explants. Each has its advantages and, as discussed below, a number of limitations as well.

Retinal pigment epithelial cells can be isolated from bovine eyes fairly easily: the anterior segment and neural retina are removed, buffered sucrose is added to the exposed eyecup, and the cells are loosened from their attachments to Bruch's membrane with a soft brush and then harvested by aspiration (Berman, 1964, 1971; Feeney-Burns and Berman, 1982). Although this procedure is rapid and convenient, it is now recognized that because of damage to the villous processes while peeling away the neural retina as well as rupture of RPE–Bruch's membrane attachments during the brushout, both the apical and basal plasma membranes are injured, and cytosolic components are lost (Saari *et al.*, 1977). The cytosol can, however, be recovered, and the cells purified either by repeated washing in buffered sucrose (Feeney-Burns and Berman, 1982) or by centrifugation on a Ficoll density gradient (Salceda, 1986).

Injury to RPE cells can be minimized prior to their removal by trypsinization of either the whole eye (Edwards, 1977) or the exposed optic cup (Edwards, 1982a) or by combined hyaluronidase and collagenase treatment plus trypsin (Mayerson *et al.*, 1985). Sheets of cells or single cells obtained in this way can be cultured to confluent monolayers that are viable for several months. Their morphology is well preserved, and, most importantly, they retain many of their metabolic and functional characteristics. They have active phagocytic capabilities (Hall, 1978), and cultured RPE cells from rats with inherited retinal dystrophy (RCS rats) express the same phagocytic defect in culture (Edwards and Szamier, 1977) as they do *in vivo* (Bok and Hall, 1971).

Since the first successful culture of rat RPE, cells from other species have also been grown to confluency and subcultured; these include human (Flood *et al.*, 1980), embryonic chick (Israel *et al.*, 1980), bovine (Basu *et al.*, 1983), and cat (Stramm *et al.*, 1983) RPE. In all cases the cells have normal morphological characteristics and appear to be well differentiated. Further improvements in cell culture technique involve the use of laminin-coated microporous filter supports; rat RPE cells become rapidly confluent, form tight junctions, and maintain their apical–basal orientation in the secretion of extracellular matrix (Heth *et al.*, 1987).

Cultured RPE thus provides a convenient source of tissue for studying a wide range of biochemical and other functional properties. Nevertheless, cultured RPE cells lack at least two activities present in freshly prepared tissue, or in RPE cells *in vivo*. The first is a failure to maintain their pigmentation. The pigment granules become diluted by cell division, probably the consequence of a failure to synthesize melanin. Another observation reported from many laboratories is the inability of cultured RPE cells to store or metabolize retinoids. Within 48 h of being placed in culture, the content of cellular vitamin A is reduced to about 20% of its initial value (Flood *et al.*, 1983). Changes in the composition of retinyl esters in primary and subcultures are further evidence of biochemical alterations in retinoid metabolism in cultured RPE cells. It may, however, be possible to partially restore retinyl ester-synthesizing activity in human RPE cells by enriching the culture medium with hormones and other defined components (Edwards *et al.*, 1987a).

In summary, despite certain inherent limitations using either freshly prepared RPE cells or cultured preparations, these two types of tissue preparation are widely used today for studying RPE biochemistry.

7.4.3. Chemical Composition

The overall composition of bovine RPE cells prepared by the brushing-out procedure and purified until free of red blood cells and rod outer segments is shown in Table 7.14. Although the relative amounts of the principal components are very similar to those of neural retina, there are major differences in protein profiles, lipid classes, and fatty acid composition.

7.4.3.1. Proteins: Cytoskeletal Proteins and Plasma Membranes

Actin, a prominent component of the cytoskeletal system of the RPE, was first identified morphologically more than a decade ago in the apical processes of both rat (Burnside, 1976) and primate (Burnside and Laties, 1976) RPE. More recent studies on the cytoskeletal organization of chick RPE using immunofluorescent techniques have revealed two different groupings of microfilaments: one is composed of actin, myosin, α-actinin, and vimentin, which forms a circumferential band near the zonula adherens; the other, localized in the apical processes, consists of myosin, vinculin, α-actinin, and fodrin (Philp and Nachmias, 1985). In addition to their *in situ* identification, these proteins were also identified by SDS-PAGE and immunoblots.

High-resolution two-dimensional gel electrophoresis is now being successfully ap-

Table 7.14. Chemical Composition of Bovine Retinal Pigment Epithelium and Neural Retina[a]

Component	Pigment epithelium	Neural retina
Water	81.4	80.6
Protein	8.1	7.7
Total lipid	3.00	2.74
Phospholipids	1.59	1.67
DNA	0.72	0.79

[a]Adapted from Berman *et al.*, 1974. Data are expressed as percentage wet weight.

plied to the separation of labeled RPE proteins according to their isoelectric points and molecular weights. This powerful technique has revealed a broad array of several hundred proteins in cultured RPE. The question of which proteins are common to RPE and other cell types and which are unique to the RPE is under active investigation. The first report using this technique showed that incubation of human RPE cell cultures with [^{35}S]methionine results in the labeling of most of the major proteins of the cell; two-dimensional gel electrophoresis and fluorography revealed at least 200 different proteins, one of which was a 43-kDa polypeptide tentatively identified as actin (Haley *et al.*, 1983). This method depicts mainly the acidic proteins synthesized by the cell, but later studies identified nearly 850 proteins, including both acidic and basic proteins (Haley and Gouras, 1984). More than 100 of the proteins were identified and catalogued by comparison with highly purified protein standards. Quantitation of individual ^{35}S-labeled proteins and glycoproteins, the latter detected by labeling with tritiated sugars, showed considerable variability from gel to gel and from donor to donor.

Specific receptors in plasma membranes play key roles in many essential cellular functions. However, isolation of plasma membranes by techniques established for other cell types is not generally applicable to the RPE. The small fraction of the total membranes that form the plasma membrane of cells in general and the limited amount of material available in RPE cells in particular have precluded the isolation of pure RPE plasma membranes. Nevertheless, plasma membrane-enriched fractions of rat, bovine, and human RPE have been prepared through the development of several novel techniques.

In one study, the apical and basal cell surfaces of cultured rat RPE cells were labeled by lactoperoxidase-catalyzed radioiodination (Clark *et al.*, 1984). Differential detergent extraction followed by SDS-PAGE revealed three major ^{125}I-labeled proteins of molecular mass 152, 138, and 123 kDa. These proteins were presumed to be derived from the plasmalemma. Independent cell fractionations showed that they copurified with plasma membrane marker enzymes; moreover, all three appear to be glycoproteins. In an extension of these studies, plasma membrane proteins were characterized by two-dimensional gel electrophoresis at three levels: (1) total proteins, visualized by silver staining; (2) glycoproteins, detected by autoradiography of fucose- or glucosamine-labeled proteins; and (3) cell surface proteins, labeled by lactoperoxidase-catalyzed radioiodination (Clark and Hall, 1986). A maximum of 102 different plasma membrane proteins were detected by this procedure; 38 proteins carried a fucose label, and 40 were labeled with ^{125}I. In other studies, wheat germ agglutinin (WGA)-conjugated Sepharose beads have been used to remove selectively the villous processes of the apical plasma membranes of rat RPE

(Cooper *et al.*, 1987). Separation of the proteins in this subfraction by SDS-PAGE and characterization by comparative lectin binding in Western blots reveals an array of glycoproteins. A 175-kDa glycoprotein is prominent in normal (Long–Evans) rats but is not detectable in RCS rats; this glycoprotein may be the same "underglycosylated" 183-kDa polypeptide detected in the dystrophic RCS membrane by Clark and Hall (1986).

Plasma membrane-enriched fractions of bovine RPE have been isolated by gradient centrifugation of glass-bead-bound, collagenase-treated cells (Braunagel *et al.*, 1985). The fractions were enriched in three commonly used plasma membrane markers: Na^+,K^+-ATPase, 5′-nucleotidase, and alkaline phosphodiesterase. Little mitochondrial or lysosomal contamination was noted, but the preparations did contain endoplasmic reticulum. More than 20 proteins were detected by SDS-PAGE, although there appeared to be a selective loss of both low- and high-molecular-weight species. Plasma membrane-enriched fractions have also been isolated from cultured human RPE by discontinuous gradient centrifugation of cells labeled with either iodine, methionine, or sugars (Ishizaki *et al.*, 1987). Two-dimensional electrophoresis of the plasma membrane-enriched fractions revealed approximately 180 proteins, several of which were glycoproteins with molecular weights ranging from 43,000 to 139,000.

7.4.3.2. Lipids

The total lipid content of bovine RPE is 3% of wet weight, and about half of the lipids are phospholipids. Although similar in total content to neural retina, the compositions of RPE phospholipids (Table 7.15) and especially their fatty acid profiles are strikingly different (Anderson *et al.*, 1976; Braunagel *et al.*, 1985; Batey *et al.*, 1986). The two major phospholipid classes of ROS and RPE are the same, consisting of phosphatidylethanolamine (PE) and phosphatidylcholine (PC); however, the latter forms a much higher proportion of the total phospholipids in RPE than in ROS. Differences in fatty acid composition of the phospholipids are even more striking (Berman *et al.*, 1974; Anderson *et al.*, 1976; Braunagel *et al.*, 1985, 1988). About 50–60 mol% of the total fatty acids in ROS phospholipids are polyunsaturates, the principal species being

Table 7.15. Lipid Classes in Whole Bovine and Rat RPE Cells and in Plasma Membrane-Enriched Fractions from Bovine and Rat RPE

	Whole cells				Plasma membrane-enriched fractions
Phospholipid[a]	Bovine[b]	Bovine[c]	Rat[d]	Rat[e]	from rat RPE[d]
PC	47.6	39.9	36.2	44.6	44.0
PE	32.4	25.8	26.6	23.2	25.3
PS	4.0	5.4	11.0	5.8	4.4
PI	6.2	8.6	11.2	5.7	3.1
SPH	6.0	11.8	14.9	14.2	20.4

[a]Abbreviations used: PC, phosphatidylcholine; PE, phosphatidylethanolamine; PS, phosphatidylserine; PI, phosphatidylinositol; SPH, sphingomyelin.
[b]Adapted from Anderson *et al.* (1976). Data are expressed as mol%.
[c]Adapted from Braunagel *et al.* (1985). Data are expressed as mol%.
[d]Data for 12- to 14-day-old rats adapted from Batey *et al.* (1986) and expressed as mol%.
[e]Adapted from Braunagel *et al.* (1988). Data are expressed as percentage of lipid phosphorus.

Table 7.16. Major Fatty Acids of Bovine Retinal Pigment Epithelium (mol%)

	Whole cells		Individual phospholipids[c]			
Fatty acid	*a*	*b*	PC	PE	PS	PI
16:0	30.7	34.4	61.3	40.1	18.3	17.9
16:1	2.6	2.8	1.9	0.3	0.7	0.8
18:0	14.9	15.4	7.0	20.1	40.5	38.0
18:1	13.1	12.5	11.2	6.1	13.9	5.9
18:2(*n*-6)	14.9	9.2	10.2	2.6	2.6	1.8
20:4(*n*-6)	16.6	12.1	2.5	11.4	7.3	25.1
22:6(*n*-3)	Trace	6.1	2.7	7.5	11.7	3.6

[a]Adapted from Berman *et al.* (1974).
[b]Adapted from Braunagel *et al.* (1985).
[c]Adapted from Anderson *et al.* (1976).

22:6(*n*-3). In contrast, all of the phospholipid classes in bovine RPE (Table 7.16) and rat RPE (Table 7.17) contain significantly high proportions of saturated fatty acids. In addition, there are relatively high levels of arachidonic acid, 20:4(*n*-6), and only small amounts of 22:6(*n*-3) in RPE phospholipids. The levels of 20:4 are higher in rat than in bovine RPE; phosphatidylinositol of bovine RPE is especially enriched in arachidonic acid.

Plasma membrane-enriched fractions of bovine (Braunagel *et al.*, 1985) and rat (Braunagel *et al.*, 1988) RPE have been prepared by a novel technique utilizing binding of glass microbeads to intact RPE cells, removal of the cells, and subsequent density gradient centrifugation. The fractions isolated from bovine RPE contain about 1 mol of cholesterol per mol of lipid phosphorus, 39 mol% of palmitate (16:0), and only about 2.3 mol% of docosahexaenoate (22:6). The ratio of PC/PE in the membranes is about 2.5:1, in contrast to ROS where this ratio is about 0.8. The relatively high proportion of cholesterol and low content of 22:6 suggests that RPE plasma membranes are less fluid than those of ROS. Plasma membrane-enriched fractions of rat RPE are similarly enriched in cholesterol and PC, and also have very low levels ($< 3\%$) of 22:6.

Table 7.17. Major Fatty Acids of Rat Retinal Pigment Epithelium (mol%)

	Whole cells		
Fatty acid	*a*	*b*	Plasma membrane-enriched fractions[a]
16:0	21.9	17.6	33.3
16:1	1.6	—	2.8
18:0	22.9	24.2	24.6
18:1	14.6	7.2	14.5
18:2(*n*-6)	6.9	10.3	3.9
20:4(*n*-6)	19.1	23.8	12.8
22:6(*n*-3)	5.9	10.7	2.9

[a]Adapted from Braunagel *et al.* (1988).
[b]Adapted from Batey *et al.* (1986).

Lipid inclusion bodies containing vitamin A have been observed in the RPE of frog (Bridges, 1975) and rabbit (Alvarez *et al.*, 1981), but apart from hypervitaminotic A mice (Robison and Kuwabara, 1977) and aging rats (Katz and Robison, 1984), oil droplets are not commonly observed in rodents. Although lipid droplets have been difficult to demonstrate in rat RPE in the basal state, there is now evidence of an increase in sudanophilic droplets that correlates with the clearance of phagosomes (Baker *et al.*, 1986). These droplets, isolated by centrifugal flotation of RPE after addition of hexane, contain mainly triglycerides and only small amounts ($< 5\%$) of retinyl esters. Their principal fatty acids are palmitic (16:0), oleic (18:1), and linoleic (18:2). Docosahexaenoic acid (22:6), so characteristic of ROS, and stearic acid (18:0) are not present in the oil droplets. On the basis of this unusual fatty acid composition, the authors speculate that although the fatty acids in the droplets are hydrolytic products of ROS membranes in the phagosome, they are not stored in the RPE; rather, two of them (22:6 and 18:0) are selectively cleared from the RPE and recycled to the retina.

7.4.4. Metabolism

7.4.4.1. Carbohydrate and Energy Metabolism

The abundance of mitochondria in the RPE suggests active oxidative pathways, and histochemical evidence supports this view (see review by Young and Bok, 1979). There are, however, only a few direct biochemical studies on glucose metabolism in pure preparations of RPE cells. An early report showed that in developing embryonic chick pigment epithelium, glucose is metabolized mainly if not entirely by glycolysis and tricarboxylic acid oxidation to CO_2 (Masterson *et al.*, 1978). Carbon-1- and carbon-6-labeled glucose were oxidized at equal rates, suggesting an absence of pentose phosphate pathway in developing chick RPE. In contrast, cultured RPE from 2-week-old hatched chicks shows considerable pentose phosphate activity; nevertheless, oxidation of glucose through this pathway does not appear to be required for active phagocytosis (Masterson and Chader, 1981a). Most of the energy requirements for phagocytosis are provided by oxidation of glucose via the tricarboxylic acid cycle.

Cultured chick (Masterson and Chader, 1981b) and freshly prepared frog RPE cells (Salceda, 1986) accumulate C-1-labeled glucose from the medium at comparable rates and oxidize it to CO_2. The uptake requires sodium but is only partially sensitive to ouabain. The recent application of ^{31}P-NMR spectroscopy to studying intact viable cultured human RPE is an important step forward in understanding the bioenergetics of this cell layer (Miceli *et al.*, 1987). Through this technique, steady-state concentrations of intracellular metabolites can be determined with great precision. The major metabolites found in human RPE are phosphorylethanolamine and phosphorylcholine, substances whose function in these cells is not known. Other major intermediates identified by this technique include ATP, Pi, and phosphocreatine. Significant amounts of UDP-N-acetylglucosamine and UTP are also present, which is not surprising in view of their role as intermediates in the biosynthesis of proteoglycans and glycoproteins (see Fig. 1.4).

7.4.4.2. Vitamin A and Cellular Retinoid-Binding Proteins

The RPE plays a major role in the uptake, storage, and mobilization of vitamin A for use in the visual cycle. This single layer of cells is the major storage site for vitamin A in the visual system. Its concentration in this cell layer is higher than in any tissue of the body except liver, and RPE stores are difficult to deplete even in vitamin A-deprived animals. The unique metabolism and turnover of RPE retinoids have been extensively studied, but the present discussion can include only a brief summary of their uptake, transport, metabolism, and mobilization in the photoreceptor–pigment epithelium complex. Several review articles are highly recommended for in-depth coverage of this topic (Bridges, 1984; Bok, 1985).

Retinoids are delivered to the RPE through both the basal and apical surfaces, the former deriving from the circulation and the latter through operation of the visual cycle. Functional receptors for serum retinol-binding protein (RBP) are localized on the basolateral membranes of rat RPE (Bok and Heller, 1976) and can also be demonstrated in cultured human cells (Pfeffer *et al.*, 1986). Labeled all-*trans* retinol bound to its physiological carrier, RBP, is internalized by both primary and subcultured human RPE. Nevertheless, retinol can also be internalized by cultured human RPE cells when bound to a nonspecific carrier such as bovine serum albumin (Flood *et al.*, 1983). In both cases, the retinol is rapidly converted to retinyl esters by freshly isolated cells. The ester formed when retinol is supplied in the physiological form bound to RBP is retinyl palmitate, which is the major endogenous ester in human RPE cells (Pfeffer *et al.*, 1986).

The mechanism of retinol internalization is still under investigation. In model systems nonesterified retinoids undergo rapid intermembranous bulk transfer in the absence of exchange proteins (Rando and Bangerter, 1982), and by analogy, it is generally assumed that in biological systems, retinoids, like other lipids, pass freely though the lipid bilayer. In RPE cells, cytosolic RBP is thought to be a necessary component in receptor-mediated internalization of retinol (Pfeffer *et al.*, 1986), but there is also evidence that in bovine RPE, a plasma membrane- associated CRPB may mediate both the uptake and the esterification of retinol bound to RBP (Ottonello *et al.*, 1987). The plasma membrane-enriched fractions examined in these studies also displayed retinyl ester hydrolase activity in the presence of apo-CRBP as retinol acceptor. Thus, the uptake, esterification, and hydrolysis of retinol appear to be functionally coupled in this cell-free plasma membrane-enriched system from bovine RPE.

The amount as well as the storage site(s) of RPE vitamin A vary widely among different species, and the absolute level depends on the state of light or dark adaptation. Dark-adapted rats store little or no vitamin A in the RPE (Dowling, 1960; Zimmerman, 1974; Alvarez *et al.*, 1981); however, in light-adapted rat retinas, virtually all of the vitamin A in the RPE is esterified and consists of all-*trans* isomers of retinyl palmitate and retinyl stearate. Senescent rats store considerably more retinyl ester in the dark-adapted state than young animals, and the ratio of palmitate/stearate esters increases significantly with age in light-adapted animals (Katz *et al.*, 1987a). In vitamin A-deprived rats, this ratio is decreased, and there is a concomitant reduction in total retinyl ester content, as would be expected (Katz *et al.*, 1987b). In most other tissues vitamin E deficiency lowers the intracellular levels of vitamin A, presumably as a result of its autooxidative destruction, but in the RPE, the retinyl ester content increases in vitamin E-deficient rats. These and other findings suggest that vitamin E not only modulates the uptake and/or storage of vitamin A in the RPE–photoreceptor complex but may also regulate the proportion of palmitate or stearate esterified.

In human RPE, more than 98% of the vitamin A is esterified. The esters consist mainly of all-*trans* isomers of palmitate and stearate as well as small amounts of all-*trans* oleate; some 11-*cis* esters are also detectable (Bridges *et al.*, 1982; Flood *et al.*, 1983). Freshly isolated cells contain from 1.0 to 4.0 pg vitamin A per cell, or approximately 8 nmol/eye. The vitamin A stores are rapidly lost in cultured cells, as discussed below.

Not only in human and rat RPE but also in all other species examined, 95% or more of the retinol is esterified. In the frog and rabbit, the ester is stored mainly, if not entirely, as oil droplets dispersed throughout the cytoplasm (Bridges, 1975, 1976; Alvarez *et al.*, 1981). By contrast, the major site of retinyl ester storage in bovine RPE is in the microsomes (Berman *et al.*, 1979). Only small but measurable amounts can be detected in fractions that would correspond to oil droplets *in situ*.

Cytosolic binding proteins play major roles in the transport and metabolism of retinoids in the RPE. The two that have been detected to date are cytosolic retinol-binding protein (CRBP) and cytosolic retinal-binding protein (CRALBP); the former carries all-*trans* retinol, and the latter 11-*cis* retinal, as endogenous ligands (see Table 7.3). CRBP, first recognized as a 2 S retinol-binding protein in chick RPE (Wiggert and Chader, 1975), is also present in a variety of other vertebrate RPE cytosols. High concentrations are found in bovine RPE (Saari *et al.*, 1977), which contains three to five times more CRBP than neural retina (Saari *et al.*, 1984). In the native state, approximately 95% of the binding sites are saturated with all-*trans* retinol; highly purified CRBP from bovine RPE contains no 11-*cis* retinol (Saari *et al.*, 1982). Immunocytochemical studies show that CRBP is evenly distributed throughout bovine and rat cytosol (Bok *et al.*, 1984); it has also been detected in the nucleus of rat RPE in association with euchromatin. As in many other cells (see Section 1.7.3), and in the RPE as well, CRBP mediates the delivery of retinol to the nucleus where it may modify gene expression. This is an important additional role of CRBP in the RPE, apart from intracellular translocation and metabolism of retinol.

The CRALBP of bovine RPE, a 36-kDa protein, carries 11-*cis* retinal as the only endogenous ligand (Saari *et al.*, 1982). In the native state, approximately 95% of the binding sites are saturated with this ligand. The protein has been purified to homogeneity, and, taking into account losses during isolation, bovine RPE was found to contain about 1 nmol of CRALBP per eye (Saari and Bredberg, 1988a).

Several metabolic transformations of retinoids have been elucidated in the RPE in recent years: (1) esterification of all-*trans* and 11-*cis* retinol, (2) hydrolysis of retinyl esters, (3) oxidoreduction of 11-*cis* retinol and 11-*cis* retinal, and (4) isomerization of all-*trans* retinol to 11-*cis* retinol. These enzymatic interconversions are shown diagrammatically in Fig. 7.15. The enzymes are membrane-bound, yet many or perhaps all of the reactions appear to be regulated by cytosolic retinoid-binding proteins acting as substrate carriers.

Esterification of All-*trans* and 11-*cis* Retinol. Given the continuous uptake of all-*trans* retinol from the circulation through the basal surface of the RPE and the entrance of additional amounts through the apical surface during rhodopsin bleaching, rapid and efficient mechanisms for the storage and stereoisomeric transformation of all-*trans* retinol would be expected. As discussed above, only a small portion of RPE retinol is bound to CRBP; most of the retinol of the RPE is esterified to palmitate and/or stearate (Bridges *et al.*, 1982), and the principal storage site is in the microsomes in some species and oil droplets in others.

Active esterification of retinol by bovine RPE was reported many years ago (Krinsky,

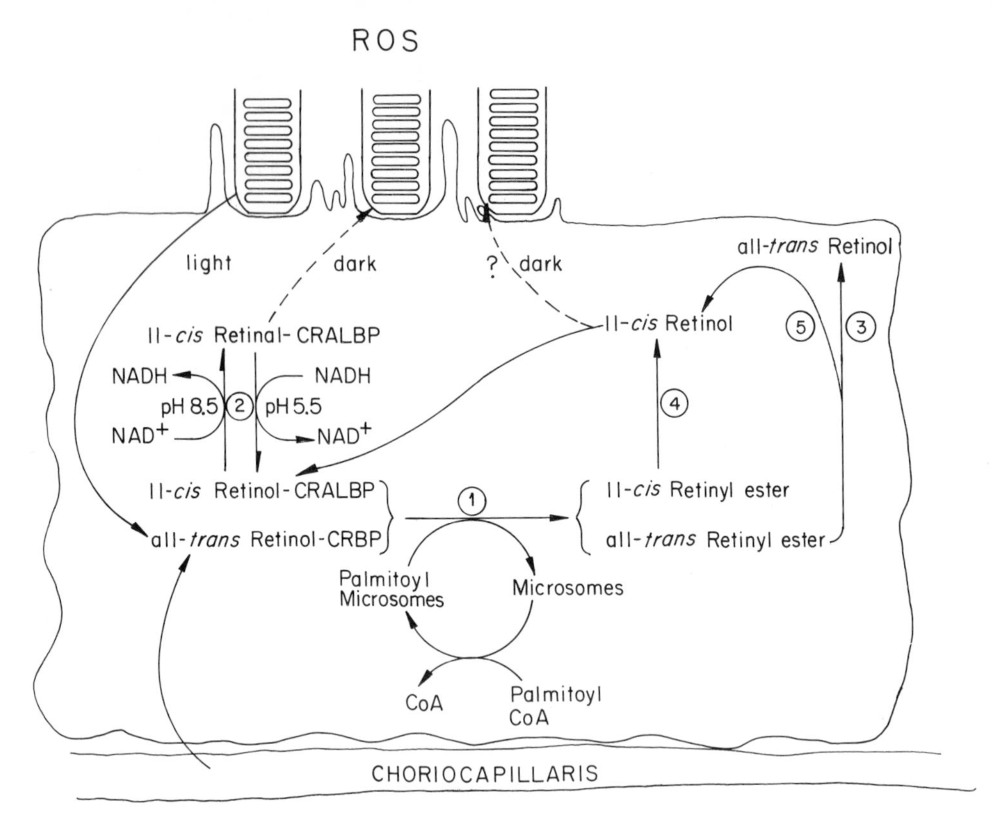

Figure 7.15. Enzymatic interconversions of retinoids in the RPE, showing the cyclic nature of uptake, storage, and mobilization of vitamin A for use in the visual cycle. The enzymes designated by encircled numbers are (1) retinyl ester synthetase, or lecithin:retinol acyltransferase (LRAT) (Saari and Bredberg, 1988b, 1989), (2) 11-*cis* retinol dehydrogenase, (3) all-*trans* retinyl ester hydrolase, (4) 11-*cis* retinyl ester hydrolase, and (5) retinol isomerase. As described in the text, all of the enzymes are associated with particulate fractions of the cell, although their activities appear to be modulated directly or indirectly by cytosolic retinoid-binding proteins.

1958), and later studies of this reaction show that the highest esterifying activity is in the microsomes, with an optimum pH of 7.5 (Berman *et al.*, 1980). All-*trans* retinol, either free or complexed to CRBP, is rapidly esterified by RPE microsomes, but the bound form is three times more effective as a substrate. Addition of exogenous acyl donors does not enhance either the rate or the extent of esterification of all-*trans* retinol. Similar observations were reported in later studies on frog and human RPE (Fong *et al.*, 1983; Bridges, 1984). Both palmitate and stearate esters were identified as the principal reaction products. Microsomes from bovine RPE are also active in esterifying 11-*cis* retinol complexed to CRALBP, the rate of ester formation being five times faster than that of all-*trans* retinol complexed to CRBP (Saari *et al.*, 1984). The esters formed are firmly bound to the microsomes.

Retinol esterification in the RPE has two unusual features: first, it is extremely active (Berman *et al.*, 1980), the apparent V_{max} being about 200 times higher than that of hydrolysis (Berman *et al.*, 1985); second, the microsomal esterification does not require an exogenous acyl donor, implying that there are sufficient stores of endogenous fatty acyl-CoA to support the rapid rate of esterification. This, however, is not the case since bovine RPE microsomes contain virtually no endogenous fatty acyl-CoA; hence, the esterification is not mediated through this donor (Saari and Bredberg, 1988b). As shown in Fig. 7.15, labeled palmitate incubated with RPE microsomes appears to be converted to an "activated" acylated microsomal component; it is through this intermediate that retinol is esterified by an enzyme activity that transfers palmitate first from palmitoyl-CoA to a microsomal (lipid) acceptor and then to retinol. Further studies have shown that phosphatidylcholine (PC) synthesized *in situ* from [1-^{14}C]palmitoyllyso-PC and fatty acyl-CoA serves as a highly effective acyl donor for retinyl ester synthesis (Saari and Bredberg, 1989). This acyl transfer from the 1-position of PC to retinol in RPE microsomes is mediated by a lecithin:retinol acyltransferase (LRAT), and it appears likely that PC may be a major endogenous acyl donor in CoA-independent retinol esterification in RPE microsomes.

In contrast to the active esterification of either free or bound retinol in freshly prepared tissue, the expression of this activity in cultured human RPE cells is drastically modified (Flood *et al.*, 1983; Bridges *et al.*, 1986b). There is a sharp reduction in the rate of esterification of all-*trans* retinol; moreover, esterifying activity cannot be enhanced or restored by supplementation with palmitoyl-CoA. The possibility that the microsomal LRAT activity described above is somehow impaired in cultured RPE cells should be considered. The loss of esterifying activity is accompanied by, or leads to, a virtually complete depletion of endogenous stores of vitamin A in cultured RPE cells. The normal levels of vitamin A are reduced by 80% after 48 h in culture, and after subculturing the levels of vitamin A are less than 1% of those found in freshly isolated cells. The stores of cellular retinoid-binding proteins, CRBP and CRALBP, are also rapidly depleted in human cultured RPE. Although the reasons for these declines in vitamin A metabolism were not established, the use of defined culture medium containing hormones and other components can restore the esterifying activity to about 40% of the normal level (Edwards *et al.*, 1987a).

Hydrolysis of Retinyl Esters. Labeled retinyl ester synthesized *in vitro* by bovine RPE microsomes has been used as a physiological substrate to examine hydrolytic activity in various subcellular fractions (Berman *et al.*, 1985). Maximum hydrolysis occurs in the lysosomal fraction at pH 4.5–5.0, but other fatty acyl ester lipids such as cholesteryl oleate and triolein are also actively hydrolyzed by bovine RPE. The behavior of these enzyme activities toward detergents and their characteristic pH profiles suggest that they are distinct from the retinyl ester-degrading activity. The rate of lysosomal hydrolysis of retinyl ester is only about 1/200 of the rate of microsomal synthesis, implying a possible mechanism for the buildup of retinyl ester stores in the RPE. Homogenates of freshly isolated RPE cells from human postmortem eyes also hydrolyze exogenous labeled retinyl palmitate; the 11-*cis* ester is hydrolyzed 20 times faster than the all-*trans* isomer (Blaner *et al.*, 1987). Differential behavior toward detergents as well as different intracellular localizations suggest the presence of (at least) two different enzymes capable of hydrolyzing the 11-*cis* and all-*trans* isomers of retinyl palmitate in human RPE.

Oxidoreduction of 11-*cis* Retinol and 11-*cis* Retinal. An 11-*cis* retinol dehydrogenase of high specific activity was reported many years ago in bovine RPE microsomes (Zimmerman *et al.*, 1975; Zimmerman, 1976). This enzyme and the recently described retinol isomerase (described below) are two key RPE enzymes possibly operating in concert to convert all-*trans* retinol to 11-*cis* retinal for recycling to the photoreceptors. The 11-*cis* retinol dehydrogenase interconverts 11-*cis* isomers of retinol and retinal with $NAD^+/NADH$ as the preferred cofactor. Using CRALBP as substrate carrier for 11-*cis* retinal, the aldehyde is nearly quantitavely reduced to 11-*cis* retinol by bovine RPE microsomes at pH 5.5 in the presence of NADH (Saari and Bredberg, 1982). The same enzyme oxidizes CRALBP-bound 11-*cis* retinol to 11-*cis* retinal at pH 8.5 in the presence of NAD^+. However, the reaction is incomplete because of competing esterification of the alcohol, which is very active at this pH.

Isomerization of all-*trans* Retinol to 11-*cis* Retinol. The reactions described above are stereospecific for either all-*trans* or 11-*cis* retinoids. Isomerizing enzyme(s) in the RPE that convert all-*trans* isomers to 11-*cis* forms have long been suspected, but only recently has this key reaction, the "missing link" in the visual cycle, been demonstrated unequivocally. Rando, Bernstein, and co-workers (Bernstein *et al.*, 1987a,b; Fulton and Rando, 1987) have characterized an enzymatic isomerizing system, assayed in the absence of light, in particulate fractions of frog and bovine RPE that transforms exogenous labeled all-*trans* retinol to 11-*cis* retinol and its palmitate ester. Bovine serum albumin must be used as the retinol carrier, since ethanol, a solvent commonly used for studying retinoid metabolism *in vitro*, is inhibitory at concentrations greater than 2%.

Further studies by the above investigators show that the isomerization, which is an endergonic process, is localized in membranes obtained after high-speed centrifugation of RPE homogenates from which low-speed debris had been previously discarded. These membranes are devoid of redox activity [$NAD(P)^+/NAD(P)H$] and hence do not interconvert retinol and retinal through the dehydrogenase system. The membranes do, however, contain microsomes and not unexpectedly show active retinyl ester-synthesizing activity. Rando and co-workers demonstrated that the esterification of all-*trans* retinol is functionally coupled to the isomerizing enzyme. The endergonic isomerization in this multicomponent system utilizes the metabolic energy stored in the esterification to drive the isomerization reaction (Deigner *et al.*, 1989). That the reaction is reversible was shown by the use of chiral retinols, the latter being required because of the spontaneous isomerization of 11-*cis* to all-*trans* retinol. This isomerization occurs with concomitant inversion of the stereochemistry at the C-15 prochiral methylene hydroxyl group (Law and Rando, 1988). There is strong experimental evidence showing that all-*trans* retinol is first esterified to an active (all-*trans*) ester, which is then isomerized to form 11-*cis* retinol through a mechanism that couples the free energy of hydrolysis of a C–O bond in the retinyl ester to the isomerization reaction. This appears to be mediated by the chemical energy of the acyl groupings in RPE membrane phospholipids (see Saari and Bredberg, 1988b).

The fate of 11-*cis* retinol generated by the retinol isomerase enzyme has yet to be clarified, but there appear to be two possibilities (Fig. 7.15): (1) direct transfer to the ROS for rhodopsin regeneration, which is not likely, or (2) oxidation in the RPE to 11-*cis* retinal by 11-*cis* retinol dehydrogenase, the probable form in which this isomer is returned to the ROS for rhodopsin regeneration.

7.4.4.3. Production of Extracellular Matrix

The highly polarized RPE cell secretes two specialized extracellular matrices (ECMs), the interphotoreceptor matrix (IPM) on its apical surface and a unique basement membrane, Bruch's membrane, on its basal surface. The role of the RPE in the synthesis of IPM components (proteoglycans, glycoproteins, and proteins) by the RPE is discussed in Section 7.3.2; this section deals mainly with the biochemistry and biosynthesis of the basal ECM. Bruch's membrane consists ultrastructurally of five components: RPE basement membrane, inner collagenous zone, elastin layer, outer collagenous zone, and choriocapillaris basement membrane. This structure acts as a molecular sieve for nutrients passing between the choriocapillaris and the inner retina; it also provides support and attachment for the RPE cells. Further details on the cell biology and morphology of this unique ECM are available in a recent review (Hewitt, 1986).

The first definitive biochemical identification of Bruch's membrane proteoglycans was done on tissue microdissected from monkey eyes (Robey and Newsome, 1983). Native Bruch's membrane contains about 60% heparan sulfate proteoglycan and smaller amounts of chondroitin and/or dermatan sulfate and hyaluronic acid. Eye organ cultures incubated with labeled precursors synthesize the same proteoglycans but in different proportions. Newly developed techniques for studying GAG synthesis in cultured feline RPE provide additional tools for comparing the proteoglycans secreted on the apical surface with those retained on the cell layer. The former are secreted into the medium in cultured cells and probably into the IPM *in vivo*, whereas the GAGs retained on the cell surface may represent basement membrane extracellular matrix (ECM) proteoglycans (Stramm, 1987). Their GAG profiles are strikingly different: chondroitin sulfate is the major GAG secreted into the medium, whereas heparan sulfate predominates in the cell layer.

Human cultured RPE cells also exhibit a distinct polarity with respect to the proteoglycans released apically and basally (Hewitt and Newsone, 1988). Similar to feline cells, the predominant proteoglycans secreted into the medium by human RPE consist of a mixture of chondroitin sulfates with varying degrees of sulfation, suggesting close identity with native IPM proteoglycans (Berman and Bach, 1968; Bach and Berman, 1971a). In the studies of Hewitt and Newsome (1988), the cell membrane-associated proteoglycans were further "fractionated" according to their differential solubilities. Separate analyses of cell-associated and EDTA-extractable populations of proteoglycans show significant differences in both size distribution and GAG composition, implying a selective localization of proteoglycans in distinct physiological or environmental domains of the native cell membrane.

The foregoing discussion considered the proteoglycans of the RPE extracellular matrix, but this ECM also contains at least three other major components: type IV collagen, laminin, and fibronectin. Type IV collagen was identified by standard biochemical procedures as a major component of newly synthesized ECM secreted basally by cultured feline RPE cells (Li *et al.*, 1984). This finding was later corroborated and extended using immunohistochemical techniques (Campochiaro *et al.*, 1986). Cultured human RPE cells produce not only type IV collagen but also small amounts of types I, II, and III collagen as well as fibronectin and laminin. Integrin and fibronectin are localized exclusively in the basement membrane of both chick and rat RPE (Philp and Nachmias, 1987). Contrary to a previous report (Pino, 1986), the immunofluorescence techniques used in this study failed to confirm the presence of fibronectin on the apical surface of the RPE. Further detailed studies on the composition of the ECM elaborated by cultured human and monkey RPE cells suggest an age-related synthesis of type I collagen, its detectability increasing with the number of subcultures and with increasing donor age

(Newsome *et al.*, 1988b). Type II collagen does not appear to be synthesized by cultured primate RPE, although other characteristic ECM components such as laminin, fibronectin, and heparan sulfate proteoglycan have been positively identified in primate cultures. Future studies of the chemical composition and the supramolecular structure and organization of this unique ECM are needed in order to clarify the important physiological and biological functions ascribed to it.

7.4.4.4. Transport of Ions, Solutes, and Fluid

The RPE plays an important role in controlling both the fluid volume of the subretinal space and the ionic milieu in which the photoreceptors are embedded. This is achieved through solute-linked active transport of dissolved solutes from the subretinal space across the RPE to the choroid. The pumping force(s) generated by this active outward transport of fluid and ions may help promote normal adhesion between the neural retina and the RPE.

Ouabain-sensitive Na^+,K^+-ATPase activity in frog RPE has been demonstrated in whole-cell preparations and in membrane-enriched fractions; those derived from the apical membranes contain 75–100% of the total activity found in the fractions (Ostwald and Steinberg, 1980). This activity is also expressed in cultured human RPE cells, and measurements of specific binding of labeled ouabain in confluent and sparse cultures show approximately 1.1×10^6 and 6.6×10^6 sites per cell, respectively (Jaffe *et al.*, 1989). In addition to ion fluxes, the RPE also mediates a net flux of taurine from the retina to the choroid; inhibition by ouabain suggests that taurine is cotransported with Na^+ through the gradient produced by the Na^+,K^+-ATPase pump (Ostwald and Steinberg, 1981). Two acidic amino acids, aspartate and glutamate, are also transported across the RPE membrane in the direction of retina to choroid (Pautler and Tengerdy, 1986). The unidirectional net transport of both amino acids is Na^+ dependent and ouabain inhibitable, adding further evidence for the existence of an active ATPase transport system that is probably localized on the apical surface. Thus, Na^+, taurine, and certain amino acids are actively transported by the RPE in the same direction (symport) against concentration gradients utilizing the energy stored in the transmembrane gradient of Na^+.

Glucose uptake in cultured chick RPE is stereospecific and does not require metabolic energy (Masterson and Chader, 1981b). Two glucose analogues, 3-O-methylglucose and 2-deoxy-D-glucose, use the same carrier system, and kinetic data suggest that D-glucose enters the cell by facilitated diffusion. Zinc uptake in human cultured RPE is also mediated by facilitated diffusion (Newsome and Rothman, 1987). On the other hand, ascorbate accumulation in cultured feline RPE cells appears to be an energy-dependent process, sensitive to both sodium ion depletion and ouabain (Khatami *et al.*, 1986a).

Fluid transport is coupled to the active transport of ions, amino acids, and organic solutes, and recent studies have shed considerable light on the mechanism of net movement of fluid from the subretinal space across the RPE to the choroid. The net rate of fluid transported in isolated frog RPE–choroid is 4–6 $\mu l/cm^2 \cdot h$, a value similar to that found in other epithelia (Miller *et al.*, 1982; Hughes *et al.*, 1984). This fluid transport is completely inhibited by 2,4-dinitrophenol, an uncoupler of mitochondrial oxidative phosphorylation, and is reduced by 70% in the absence of bicarbonate. Active HCO_3^- transport appears to be the dominant driving force for fluid absorption across the RPE. Ouabain has little effect, suggesting that RPE fluid transport is not directly coupled to the active

Na^+,K^+-ATPase pump located on the apical surface of the cell. Most importantly, it was shown that elevating the intracellular concentration of cAMP causes a rapid decrease (84% within 5 min) in fluid absorption. This dramatic effect of cAMP was attributed to stimulation of a secretory flux of Na^+ and Cl^-. As discussed in the following section, the intracellular concentration of cAMP in RPE cells may be under the control of the adenylate cyclase system.

7.4.4.5. Adrenergic Receptors, Adenylate Cyclase, and Phosphoinositide Turnover

Recent findings leave little doubt that the RPE is a target tissue for circulating hormones and neurotransmitters. However, the precise nature of the receptor-stimulated membrane tranduction system(s) and the intracellular second messengers generated by them are still being clarified. Primate RPE contains low but detectable levels of cAMP (Newsome *et al.*, 1980), and similar endogenous (unstimulated) levels are probably present in the RPE of other species. Cyclic AMP levels in the RPE can be modulated by a variety of agonists, suggesting that these cells contain a hormone-sensitive adenylate cyclase system similar to that found in other cell types.

Intracellular cAMP production in cultured embryonic chick RPE is stimulated by catecholamines in the following order of potency: L-isoproterenol > (-)epinephrine ≥ (-)norepinephrine > dopamine (Koh and Chader, 1984a). As shown in Fig. 1.5, this response is typical for β_2-adrenergic agents that interact with the stimulatory β-adrenergic receptor. The signal is transduced through interaction with a G protein, in this case G_s, which results in the activation of adenylate cyclase and production of cAMP. The stimulation of cAMP production by isoproterenol is blocked by propranolol and hydroxybenzylpindolol (HYP), which are selective β-adrenergic antagonists (Koh and Chader, 1984a). Other agonists that stimulate cAMP production in cultured chick RPE include vasoactive intestinal peptide (VIP), glucagon, and prostaglandin PGE_1 (Koh and Chader, 1984b). The production of cAMP in response to VIP is especially pronounced, with a 100-fold increase of intracellular cAMP being measurable within 3 min.

Adenylate cyclase has been isolated and partially characterized in membranous fractions of cultured human RPE (Friedman *et al.*, 1987). As in other tissues, the RPE enzyme can be activated at three different sites: (1) through cell-surface receptors that appear to be of the β_2-adrenergic type, (2) through direct stimulation of the α subunit of G_s regulatory protein, and (3) by direct stimulation of the catalytic site of adenylate cyclase with forskolin (see Fig. 1.5). At least two RPE cell functions may be regulated by the adenylate cyclase system *in vivo*: fluid transport and phagocytosis.

In addition to β_2-adrenergic receptors that regulate intracellular levels of cAMP in the RPE, there is also evidence for α_1-adrenergic receptors that act independently of the adenylate cyclase system (Frambach *et al.*, 1988). This was shown by studying the effects of epinephrine and other specific adrenergic agonists and antagonists on the short-circuit current in mounted rabbit RPE–choroid–sclera explants. The epinephrine receptor is not of the β-adrenergic type, since it is neither stimulated by β-adrenergic agonists such as isoproterenol nor blocked by propranolol, a β-adrenergic antagonist. The second messenger responses for α_1-adrenergic receptors are coupled to polyphosphoinositide (PI) breakdown and Ca^{2+} mobilization (see Chapter 1, Section 1.3.2). Of the two second messengers generated by the phospholipase C-stimulated hydrolysis of PIP_2, one of them, DG, is involved in the Ca^{2+}–phospholipid activation of protein kinase C. This enzyme can also be activated by phorbol esters, tumor-promoting substances that mimic the action of DG, and such an effect has been shown in cultured rat RPE cells (Heth and Schmidt, 1988). Several proteins are phosphorylated beyond the basal level in response to phorbol esters; one in particular, an 80-kDa polypeptide widely used as a specific marker for protein kinase C, showed a marked increase in phosphorylation after 15 min exposure of cultured rat RPE cells to phorbol ester.

Another interesting study has addressed the question of whether RPE cells have the potential of generating superoxide anion (O_2^-) in response to a phagocytic challenge (Dorey *et al.*, 1989). It was found that latex beads stimulate the release of O_2^- in cultured porcine RPE, and dioctanoylglycerol, a synthetic cell-permeating activator of protein kinase C, elicits a similar response. Further investigations are needed to define more clearly the receptor-stimulated membrane transduction systems and the role of second messengers in modulating biochemical activities in the RPE.

7.4.4.6. Drug-Metabolizing Enzymes

Cytochrome P450 hemoproteins comprise a multi-gene family of microsomal proteins that serve as terminal acceptors for NADPH-dependent mixed-function oxidases. These microsomal drug-metabolizing systems catalyze the monooxygenation of physiological substances (e.g., steroids, fatty acids, and arachidonic acid) as well as a wide variety of xenobiotics. In addition to their major localization in liver, microsomal drug-metabolizing systems are also present in extrahepatic tissues including the eye (see review by Shichi, 1984).

The identification of cytochrome P450, the hemoprotein component of microsomal aryl hydrocarbon hydroxylase, provided the first evidence for the presence of drug-metabolizing enzymes in the RPE (Shichi, 1969). One of the unusual features of drug-metabolizing systems is the inducibility of several forms of cytochrome P450 by aromatic hydrocarbons. The ocular aryl hydrocarbon hydroxylase activity of C57BL/6N mice, but not of DBA/2N mice, is induced by β-naphthoflavone and other polycyclic hydrocarbons (Shichi *et al.*, 1975). The activity is localized mainly in the RPE, and the inducibility in the eye appears to be under the same genetic control as that of liver microsomes. The hydroxylase activity can also be induced in cultured chick RPE by treatment with 1,2-benz[a]anthracene; as in liver, the RPE activity is found mainly in the microsomes (Shichi *et al.*, 1976).

Drug-metabolizing enzymes in the RPE may play an important role in protecting the outer retina from potentially harmful substances in the uveal circulation. For example, most aromatic hydrocarbons in the posterior uvea will be removed by adsorption on the choroidal melanin granules, but if the amounts exceed the capacity of this barrier, they will enter the RPE (Shichi, 1984). Studies with bovine RPE microsomes have revealed active enzyme systems capable of metabolizing and detoxifying a variety of drugs and chemicals (Das and Shichi, 1981). Among the enzymes identified were aryl hydrocarbon hydroxylase and UDP-glucuronyl transferase; several enzymes involved in mercapturic acid synthesis were also detected in this study. In addition to aryl hydrocarbon hydroxylase, two other cytochrome P450-dependent drug-metabolizing enzymes have been identified in bovine RPE microsomes: 7-ethoxycoumarin *O*-deethylase and benzphetamine demethylase (Schwartzman *et al.*, 1987). In comparison to other ocular tissues, the RPE and ciliary body possess the highest activities of cytochrome P450-dependent monooxygenases in the eye. The activity of heme oxygenase, a key enzyme in the regulation of free heme, is higher in RPE microsomes than in any other ocular tissue. Labeled arachidonic acid is also oxidized by bovine RPE microsomes, mainly via the cyclooxygenase pathway; relatively little is oxidized through the cytochrome P450-dependent monooxygenases system.

7.4.4.7. Melanin

Melanin granules are the hallmark of RPE cells in virtually all vertebrate species. The other characteristic pigment of the RPE, lipofuscin, was first characterized in human RPE but is now known to be present in the RPE of all species. The formation of lipofuscin and its relationship to the phagolysosomal system (Feeney, 1973c) are described in Section 7.4.5. The present discussion on RPE melanin draws heavily from the most extensive review available on melanin and other pigments in the RPE (Feeney-Burns, 1980); detailed descriptions and literature references up to 1978 on the embryological origins, biosynthesis, assembly, and metabolic fate of the RPE pigments are provided in this review. Space limitations permit the inclusion of only selected highlights of this broad and complex topic.

Melanin granules are localized mainly in the apical region of the RPE cell (see Fig. 7.14). They vary not only in size but also in shape; spherical, ellipsoid, and elongated forms have been observed in all RPE cells. Melanin granules are synthesized *in utero* and slowly mature during the first few decades of life. There is at the same time a gradual age-related decrease in both the number of melanin granules (Feeney-Burns *et al* ., 1984) and the melanin content of human RPE (Schmidt and Peisch, 1986).

Ocular tissues contain two populations of melanin-producing cells: (1) melanocytes, the characteristic uveal tract cells; and (2) melanin-containing neuroepithelial cells such as the RPE, ciliary body, and iris. Melanin, an insoluble high-molecular-weight polymer composed of oxidized derivatives of tyrosine, is stored in unique organelles called melanosomes. Melanogenesis, known to occur in four stages, has been extensively studied in a variety of cell types (see review by Feeney-Burns, 1980), but discussion of this complex biological process is beyond the scope of this chapter. Although some of the steps in RPE melanogenesis have been clarified using ultrastructural, cytochemical, and autoradiographic techniques, there is a paucity of biochemical data on RPE melanin.

Tyrosinase, the key enzyme in melanin biosynthesis, has been detected in embryonic as well as adult RPE. The first step in melanin biogenesis involves the oxidation of L-tyrosine to 3,4-dihydroxyphenylalanine (DOPA), and the same enzyme catalyzes the oxidation of DOPA to a complex series of intermediates resulting in the synthesis of two types of melanin, pheomelanins and eumelanins (Fig. 7.16). Both of these types appear to be present in cattle RPE (Dryja *et al*., 1979). Whereas autoradiographic studies on the conversion of labeled tyrosine to melanin provide indirect evidence for tyrosinase activity, the first direct demonstration of this enzyme in RPE was reported using suspensions of bovine RPE (Dryja *et al*., 1978). Enzyme activity was measured as radioactivity released from tritiated tyrosine after 30 min incubation with RPE cells. The specific activity of the enzyme in RPE is severalfold lower than in iris or choroid. Tyrosinase activity has also been measured in primary cultures and subcultures of bovine melanotic and amelanotic cells; both cell types have enzyme activity, although that of melanotic cells is considerably higher (Basu *et al*., 1983).

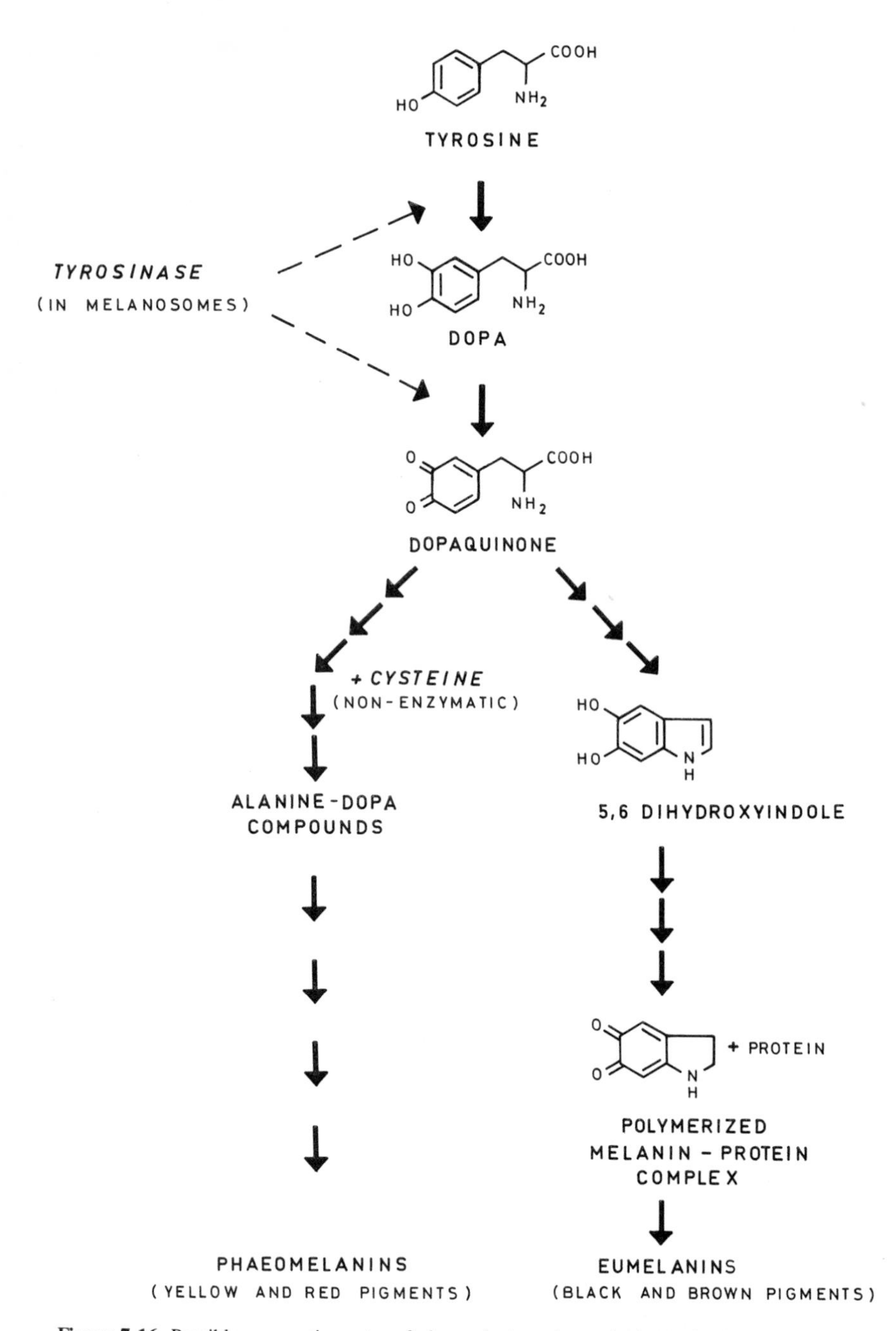

Figure 7.16. Possible enzymatic routes of pheomelanin and eumelanin synthesis from tyrosine.

Computer-assisted morphometric analyses of RPE from human donor eyes show a gradual age-related decrease in the number of "pure" melanin granules and a concomitant increase in the number of complex melanin granules: melanolipofuscin and melanolysosomes (Feeney-Burns *et al.*, 1984). The loss of melanin may be due to lysosomal digestion resulting from the direct fusion of melanin granules with acid hydrolase-containing lysosomes as well as possible reorganization and fusion with secondary lysosomes that are then incorporated into the phagolysosomal system (Feeney, 1978). The age-related loss of melanin has been confirmed by both *in situ* microspectrophotometric analyses (Weiter *et al.*, 1986) and direct spectrophotometric measurements at 260 nm of borohydride-solubilized melanin (Schmidt and Peisch, 1986). In samples obtained from a series of normal human donors, values as high as 95 μg/mg RPE dry weight were found in the 14- to 40-year age group. However from age 40, the melanin concentrations decrease in an almost linear manner to about one-third of those found in the youngest age group. Regional variations were also noted. The concentration of solubilized melanin is highest in the peripheral area and lowest in the macula, confirming previous morphometric analyses (Feeney-Burns *et al.*, 1984).

7.4.5. Phagocytosis, Phagolysosomal System, and Lipofuscin

New membranes are continually added at the base of the rod outer segments (ROS), yet the length of the photoreceptor cell remains relatively constant because of synchronized shedding of terminal packets of ROS containing approximately 8–30 disks (Young, 1971). The disks are phagocytized by the RPE and appear in the RPE cytoplasm as membranous inclusion bodies termed phagosomes (Young and Bok, 1969). Since each photoreceptor cell renews its outer segments about every 10 days, each RPE cell must phagocytize about 10% of the adjacent rod outer segments each day (Young, 1971). If RPE cells phagocytize an average of 80–90 disks per day from about 24–44 apposed photoreceptors, then each cell would ingest and degrade about 2000–4000 ROS disks each day! To accomplish such a gigantic task, the RPE cell has developed a unique catabolic machinery, the phagolysosomal system (Feeney, 1973c). This degradative system is so powerful that about 50% of the phagolysosomes formed in the cell after a phagocytic challenge are digested within 1–2 h (Hall and Abrams, 1987). Thus, phagocytosis and degradation of apical ROS remnants is one of the major functions of the RPE cell (for reviews see Bok and Young, 1979; Clark, 1986). It is essential for normal maintenance of the RPE–photoreceptor complex and of the visual process. An impairment in phagocytosis, as in the RCS rat (see Section 7.9.1.1), leads to retinal degeneration and blindness.

7.4.5.1. Binding and Uptake

Some aspects of disk shedding are discussed in Section 7.2.2.1, and the prevailing view is that shedding is probably initiated and/or controlled by active processes residing in both the RPE cell and the apical outer segment (Bok, 1985; Besharse, 1986). Whatever the biological mechanism(s) may be, once the terminal packet of ROS disks is released, it must become rapidly bound to the apical surface of the RPE prior to ingestion. To study the phagocytic process itself, a variety of particle types (e.g., purified ROS, latex spheres, and carbon particles) have been used in an effort to understand both the dynamics and the specificity of binding and internalization. Studies during the 1970s (reviewed by Clark, 1986) leave the impression that RPE cells possess two types of phagocytic capability: (1) a nonspecific type recognizable as a slow (two to three particles per hour) internalization of synthetic particles such as latex and (2) a specific type that is relatively rapid with the potential of phagocytizing several hundred ROS tips within 1–2 h. More recent studies leave little doubt that ROS are the preferred substrates in comparison with red blood cells, algae, bacteria, or yeast (Mayerson and Hall, 1986).

The characterization of rhodopsin's N-terminal carbohydrate moieties allowed a direct approach to the question of whether externally oriented sugars on the surface of the ROS plasma membrane may provide the "signal" required for their recognition and binding by the RPE. An early proposal (O'Brien, 1976) that a galactose-modified rhodopsin at the tip of the ROS may be acting as a recognition ligand stimulated considerable research into the possible role of rhodopsin in the binding and phagocytosis of ROS tips by RPE cells. Although there is indirect evidence supporting this possibility (Clark, 1986), nevertheless four recent investigations have failed to demonstrate a role for rhodopsin in ROS binding to RPE cells. In one study, bovine ROS disk membranes were frozen and thawed in order to expose intraluminally oriented carbohydrate groups and were then incubated with cultured embryonic chick RPE in the absence and presence of high concentrations of various monosaccharides (Lentrichia *et al.*, 1987). Competition studies in the presence of monosaccharides, as well as comparisons of the rate of binding of exposed (frozen–thawed) and inaccessible (native) ROS membranes, show that rhodopsin's carbohydrates, N-acetylglucosamine and mannose, are not involved in interactions with the RPE. There is no interference with binding in the presence of excess monosaccharides; moreover, there is little if any difference in the binding of membranes whose sugars are exposed compared to those whose sugars are inaccessible. In addition, enzymatic galactosylation of rhodopsin does not enhance the binding of ROS disks to RPE cells. Similar findings have been reported using rhodopsin liposomes as the target membranes for phagocytosis in cultured embryonic chick RPE (Shirakawa *et al.*, 1987).

Further evidence that rhodopsin is not the ligand for ROS binding to RPE is the inability of a soluble tryptic glycopeptide derived from rhodopsin to inhibit phagocytosis of ROS in cultured chick or cat RPE (Philp *et al.*, 1988). In other studies using cultured bovine RPE, neither the 2–39 N-terminal rhodopsin glycopeptide nor a synthetic 1–16 rhodopsin analogue is effective in inhibiting the binding and phagocytosis of ^{125}I-labeled ROS (Laird and Molday, 1988). Frozen and thawed disk membranes, whose carbohydrate residues are exposed, are similarly ineffective. Taken together, it appears that an ROS membrane component other than rhodopsin, possibly one of the newly-described plasma membrane-specific glycoproteins (see Section 7.2.1.1), provides the recognition ligand(s) involved in binding and phagocytosis of ROS by the RPE.

Other factors within the RPE cell, or possibly a glycoconjugate on its apical surface, may also play a role in phagocytosis. Cytoskeletal components such as actin filaments (Burnside, 1976; Burnside and Laties, 1976) appear to be involved in the ingestion phase of phagocytosis, although their distribution does not change during phagocytosis of ROS (Chaitin and Hall, 1983a). The binding and ingestion phases of phagocytosis can be separated by cooling RPE cells to 17°C, a temperature at which maximum binding and minimum ingestion occurs (Hall and Abrams, 1987). Kinetic studies at this temperature show that the binding is saturable, suggesting that there are a limited number of sites for binding ROS on the RPE cell surface.

Considerable attention has been focused recently on the identification of putative receptors for ROS on the RPE cell surface. Several glycoproteins have been detected in plasma membrane-enriched fractions of RPE cells, and some of them may play a role in phagocytosis. One interesting study has shown that removal of cell surface glycoproteins by proteolytic digestion of cultured rat RPE cells is correlated with a loss in phagocytic capability without, however, affecting the viability of the cells (Colley *et al.*, 1987). Two glycoproteins of M_r 160,000 and 214,000 were consistently removed by all of the proteases used in this study, but the possible role of these glycoproteins in phagocytosis remains to be established.

Mannose 6-phosphate receptors in a variety of cell types function in the intracellular translocation of newly synthesized lysosomal enzymes through the Golgi to the lysosomes. About 80% of the receptor is inside the cell; the remainder, on the cell surface, mediates the uptake by pinocytosis of secreted phosphorylated lysosomal enzymes for delivery to lysosomes inside the cell. Studies with RPE explants from rat retinas show that soluble [^{125}I]mannosidase is internalized by RPE cells, a process that is mediated by phosphomannan receptors located on the apical surface of the cell (Tarnowski *et al.*, 1988a). Using the same system, these investigators showed that the mannose-specific receptors can function both in the binding and pinocytosis of soluble ligands such as [^{125}I]mannose-BSA and in the phagocytosis of particulate mannan-coated latex beads (Tarnowski *et al.*, 1988b).

Cyclic nucleotides, particularly cAMP, may also play a role in inhibiting (or stimulating) phagocytosis by RPE cells. Despite our limited knowledge, it is not unreasonable to assume that intracellular levels of cAMP in the RPE are modulated by the adrenergic receptors and adenylate cyclase recently described in these cells (see Section 7.4.4.5). There is evidence suggesting that cAMP may be acting as a second messenger in controlling the rate of phagocytosis by the RPE. For example, the internalization of latex spheres by cultured chick RPE is inhibited by about 50% in the presence of very low concentrations (10^{-11} M) of cAMP added to the incubation medium (Ogino *et al.*, 1983). Other studies on the effect of drug-induced elevations of intracellular cAMP in cultured rat RPE also show that increased cAMP levels are associated with reduced phagocytosis of ROS (Edwards and Flaherty, 1986). Papaverine, cholera toxin, and isoproterenol increase intracellular cAMP levels by different mechanisms, and the effects on phagocytosis differ both quantitatively and temporally. Short-term increases in cAMP levels are associated with reduced phagocytosis; however, short-term reductions in cAMP concentration stimulate phagocytosis. Long-term exposure to cAMP also results in increased phagocytosis.

7.4.5.2. Phagosome Degradation and Lysosomal Enzymes

The 2000–4000 rod and cone disks phagocytized daily by each RPE cell are rapidly and efficiently degraded; the mechanism must undoubtedly involve sequential degradation of the internalized membranes by highly specialized enzyme systems in the RPE. Attempts to understand the processes involved have followed two general lines of investigation: (1) characterization of RPE lysosomal enzymes, measurement of their activities toward synthetic and physiological substrate(s), and identification of degradation products; (2) characterization of the steps leading to the formation of "age pigment" (lipofuscin) through the phagolysosomal system.

In virtually all mammalian cells, lysosomal acid hydrolases play a key role in the degradation of macromolecular substances such as glycoconjugates and proteins. The RPE is no exception, but the process is far more complex than in other cells. The major substrate in RPE cells is the rhodopsin- and phospholipid-containing phagosome, yet the mechanism of its degradation remains obscure. Acid phosphatase has been detected in rat RPE by electron microscopic cytochemistry (Ishikawa and Yamada, 1970), and two other acid hydrolases, N-acetyl-β-glucosaminidase and β-galactosidase, were shown to be present in lysosome-enriched fractions of bovine RPE (Berman, 1971). Several other glycosidases, lipases, and proteases were later detected using similar techniques: cathepsin D (Hayasaka *et al.*, 1975), acid lipase (Rothman *et al.*, 1976; Hayasaka *et al.*, 1977), arylsulfatases (Hara *et al.*, 1979), phospholipases (Berman, 1979; Zimmerman *et al.*, 1983), α-mannosidase, and others (Lentrichia *et al.*, 1978).

The activities of N-acetyl-β-glucosaminidase and α-mannosidase in the studies described above were measured using synthetic substrates, but these enzymes do not appear to be active toward the native substrate, rhodopsin. Measurements of the relative proportions of lysosomal enzymes in bovine RPE suggest that two in particular, a cathepsin D-like proteinase activity and phospholipase A, may be functioning in concert in the degradation of ROS membranes (Zimmerman *et al.*, 1983). A newly described method for measuring the specific activity of acid hydrolases (mainly of lysosomal origin) in RPE cells cultured on microtiter plates is highly sensitive and has many potential applications for directly examining the role of acid hydrolases in phagosome degradation (Cabral *et al.*, 1988).

Two populations of lysosomes have been separated by density gradient centrifugation; the lighter one, in which phagolysosomes were identified, was especially enriched in acid lipase (Rothman *et al.*, 1976). Cathepsin D activity also has a bimodal distribution, but in contrast to acid lipase, this enzyme appears to be localized mainly in the denser lysosomal population (Regan *et al.*, 1980). Bovine RPE lysosomes are as effective as purified cathepsin D in degrading the rhodopsin in ROS membranes, again suggesting an important role for this protease in phagosome degradation. A recent interesting study has shown that in freshly isolated bovine RPE cells, the activity of cathepsin D is significantly higher in the area centralis than in the peripheral regions (Burke and Twining, 1988). These differences are not expressed in cultured cells, suggesting that stimulatory modulators may be present in this region *in vivo* but are lost during culture of RPE cells.

In a more direct approach to the question of phagosomal degradation, highly purified rhodopsin radiolabeled in its sugar moieties has been used as a substrate, and its degradation (release of labeled glycopeptide) followed after 30 min incubation with bovine RPE extracts (Hara *et al.*, 1983). The pH optimum for this reaction is at 3.5, and the only product detectable is a 9-kDa glycopeptide. The reaction appears to be catalyzed by a specific rhodopsin-cleaving enzyme activity that, however, does not copurify with BSA-cleaving enzyme (cathepsin D). The oligosaccharide groups of the cleavage product are identical to those found in the native N-terminal region of rhodopsin, implying that the carbohydrate structure is completely conserved during the degradation of rhodopsin (Kean *et al.*, 1983). This interesting observation further suggests that RPE acid hydrolases such as α-mannosidase and N-acetyl-β-glucosaminidase, whose activities are readily detectable using synthetic substrates, do not act on native rhodopsin.

7.4.5.3. Phagolysosomal System and Lipofuscin Formation

Enzyme cytochemical and ultrastructural studies have revealed the presence of a highly developed phagolysosomal system in human RPE (Feeney, 1973c, 1978). The major processes include (1) fusion of primary lysosomes with phagosomes to form phagolysosomes, (2) fusion of lysosomes with lipofuscin granules (mainly in older eyes), and (3) fusion of melanin with either lipofuscin or lysosomes to form melanolipofuscin or melanolysosomes, respectively. In humans under 10 years of age and in animals such as the frog or rat, phagosomes are rapidly degraded through the following sequence of events: formation of the phagolysosome, fall in pH with resulting degradation of its major macromolecular components, diffusion of low-molecular-weight products of acid hydrolase activity out of the phagolysosome, and finally formation of a small compact residual body (Feeney-Burns and Eldred, 1983). However, the process is different in older humans. The phagolysosomes described above are not completely digested; instead, there are frequent fusions of phagolysosomes with preexisting lipofuscin as well as continued fusion of multiple lipofuscin granules with primary lysosomes. The time sequence is obscure, but the final result is an age-related accumulation of autofluorescent "age pigment" or lipofuscin bodies in the human eye (Feeney, 1978; Wing *et al.*, 1978).

In the mechanism recently postulated for the formation of lipofuscin in primate RPE (Fig. 7.17), opsin-containing membranes of outer segments are initially incorporated into the phagolysosomal system by fusion of the phagosome with a primary lysosome. To follow later stages of phagosome breakdown, opsin has been used as a "marker," and its possible fate in the formation of lipofuscin examined by the use of monoclonal and polyclonal antibodies to human and bovine rhodopsin (Feeney-Burns *et al.*, 1988). These results were negative in the sense that although ROS, phagosomes, and secondary lysosomes showed heavy immunoreactive labeling, there was no detectable antibody label in either lipofuscin or melanolipofuscin granules. The monoclonal antibodies used were specific for the N and C terminals of rhodopsin; hence, it seems that opsin fragments that may be present in the lipofuscin granule lack the epitopes that recognize these opsin antibodies. They may have been lost by prior hydrolytic cleavage, and possible hydrophobic domains of opsin that remain in the lipofuscin granule are not reactive.

Apart from age-related accumulation of lipofuscin, deficiencies in dietary antioxidants also lead to a buildup of fluorescent pigment in the RPE (Katz *et al.*, 1978, 1982). In rats fed diets high in polyunsaturated fatty acids (PUFAs) and low in α-tocopherol, selenium, and other antioxidants, there is a dramatic accumulation of yellow autofluorescent pigments with spectral properties similar to those of lipofuscin. It has been argued that these pigments may have arisen from excessive peroxidation of PUFAs derived from phagocytized ROS membranes. Vitamin E is a powerful antioxidant, thought to act as a free radical sink in the termination of autooxidative chain reactions involving lipid peroxide free radicals (see Fig. 7.11). In its absence this unstable free radical could be broken down to small reactive fragments such as malondialdehyde, which in turn become crosslinked with various primary amines to form Schiff bases.

One theory holds that such Schiff base intermediates may be responsible for the autofluorescence of lipofuscin granules in human RPE, especially in older individuals (Feeney-Burns *et al.*, 1980). Uncorrected fluorescence spectra of partially purified chloroform–methanol extracts of human RPE from individuals over 40 years of age show a blue emission peak at about 430 nm; moreover, the excitation and emission spectra are similar to those observed in lipofuscin pigments isolated from other tissues. However, using a fluorescence microscope to avoid instrumental bias, corrected spectra show that the major fluorescent components of intact human RPE lipofuscin granules are golden-yellow fluorophores emitting as a broad peak from 540 to 640 nm when excited at 366 nm

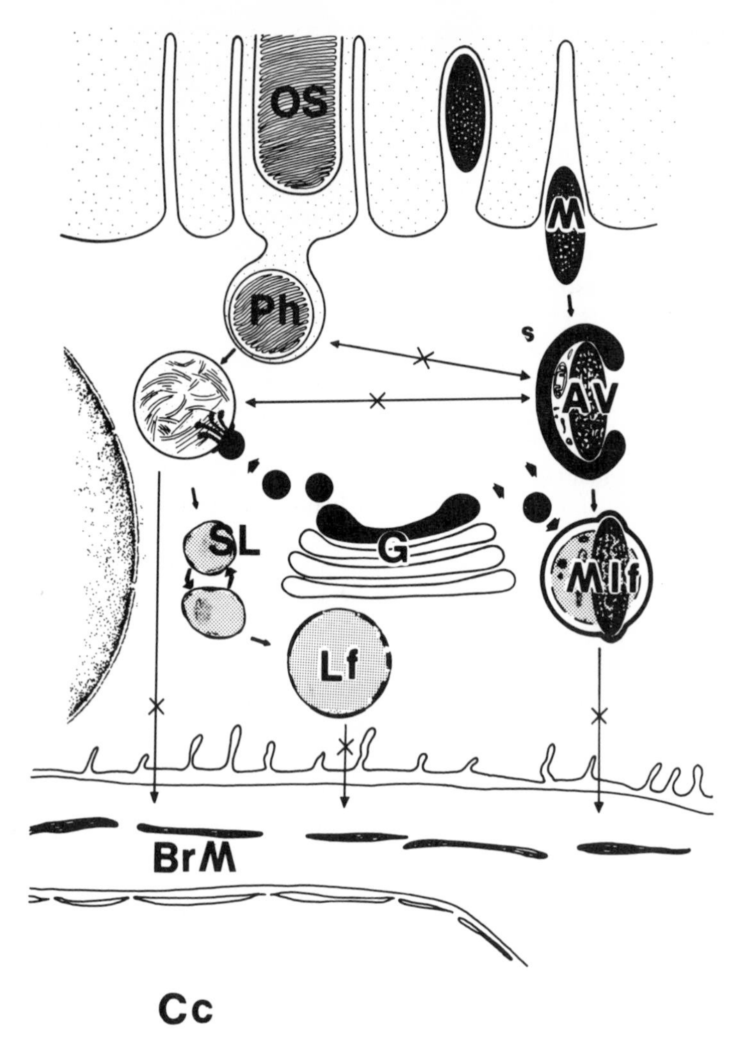

Figure 7.17. Lipofuscin formation in primate retinal pigment epithelium. The abbreviations are: OS, outer segments; Ph, phagosome; SL, secondary lysosome; G, Golgi cisternae; M, melanin granule; AV, autophagic vacuole; Lf, lipofuscin; MIf, melanolipofuscin; BrM, Bruch's membrane; and Cc, choriocapillaris. (From Feeney-Burns *et al.*, 1988.)

(Eldred *et al.*, 1982). Since imine-conjugated Schiff bases fluoresce with an emission ranging from 430 to 480 nm, they do not contribute to the yellow-emitting fluorophores, the major fluorescent pigments present in lipofuscin granules. Other studies have shown that the age-related pigment (lipofuscin) and pigments that accumulate in the RPE of vitamin E-deficient rats have identical fluorescence emission spectra (Katz *et al.*, 1984). The corrected emission spectrum in rats is somewhat narrower (590–650 nm) than that of human lipofuscin.

A major problem in all of these studies is the insolubility of lipofuscin fluorophores, and those that are extracted may not represent all of the fluorescent pigments present in the lipofuscin granule. A recent modification of extraction and purification procedures has led to the isolation and partial characterization of ten fluorescent fractions in human RPE from adult donor eyes (Eldred and Katz, 1988). Based on their spectral properties, they have been grouped into four categories: (1) two green-emitting fluorophores tentatively identified as retinol and retinyl palmitate, (2) a golden-yellow-emitting fluorophore, (3) at least three yellow-green-emitting fluorophores, and (4) at least three orange-red-emitting fluorophores. The spectral analyses reported in this study exclude Schiff base structures generated from the autooxidation of PUFAs as major fluorophores in lipofuscin. Instead, there is now compelling evidence that substances derived from the oxidation of vitamin A metabolites are involved in the formation of certain of the RPE lipofuscin pigments.

Dietary restriction of retinyl palmitate leads to a substantial reduction in the deposition of lipofuscin in rat RPE; the relative magnitude of this vitamin A effect is greater in vitamin E-supplemented rats, suggesting that the action of vitamin E is more complex than that of a simple antioxidant (Katz *et al.*, 1986a). It has also been noted that the outer segment debris accumulating in the subretinal space of young RCS rats on normal diets has spectral properties similar to those of RPE lipofuscin that accumulates during normal senescence (Katz *et al.*, 1986b). Thus, photoreceptors appear to provide the major precursor(s) for lipofuscin fluorophores, and further experiments to identify them have utilized vitamin A-deprived RCS rats (Katz *et al.*, 1987c). Vitamin A deficiency in these rats reduces the autofluorescence of outer segment debris, suggesting that vitamin A contributes directly to the formation of lipofuscin in the RPE. One fluorophore in particular extracted from these vitamin A-deficient RCS rat retinas differs substantially in TLC mobility from the major age-dependent pigment that accumulates in the RPE of normal rats. Further studies will undoubdtedly reveal more precisely the involvement of vitamin A-dependent photoreceptor fluorophore(s) in RPE lipofuscin formation.

Drusen or "senile warts" are scattered excrescences commonly found beneath the RPE cells near the vortex veins (equator) in humans beyond the age of 30. Drusen in the macular region develop several decades later and are important clinical harbingers of age-related macular degeneration, a common disease affecting central vision. Drusen vary in their morphology, content, and location in or on Bruch's membrane; nonetheless, the RPE is generally recognized as the main source of the various components, whether particulate or "hyaline" in appearance. Apoptosis, or budding off, of the RPE cytoplasm appears to be a common event causing deposition and gradual accumulation of granular and fibrillar debris within Bruch's membrane of human and primate eyes (Ishibashi *et al.*, 1986). Additionally, secretion of basement membrane-like materials by the RPE produces hyaline mounds in Bruch's membrane and/or linear deposits of collagen-like material internal to Bruch's membrane. These materials modify the normal relationships between the RPE and Bruch's membrane and appear to compound other aging changes, particularly in the macular region.

The question of whether the shedding of RPE cytoplasmic buds introduces active lysosomal enzymes into Bruch's membrane was examined by enzyme cytochemistry (Feeney-Burns *et al.*, 1987). These studies failed to reveal either acid phosphatase or arylsulfatase B in granular drusen or in Bruch's membrane in the eyes of 30- to 70-year-old individuals. Only in aged eyes (80 years old) are there detectable levels of lysosomal enzymes. Thus, exocytosis of RPE organelles such as lysosomes into the basal extracellular matrix is not a normal process, although it may occur in elderly individuals. Other studies using specific antibodies to extracellular matrix components point to marked heterogeneity in drusen composition (Newsome *et al.*, 1987). Diffuse types of drusen often contain fibronectin, but in general there appears to be no specific relationship between abnormal deposits and immunoreactive substances in Bruch's membrane. There is, however, an age-related increase in type I collagen reactivity in Bruch's membrane.

7.5. METABOLISM OF THE MICROVASCULATURE

Two characteristic cell types, endothelial cells and pericytes (mural cells), line the retinal capillary lumen. The endothelial cells, which form a continuous monolayer throughout the lumen, are joined to one another by zonulae occludentes. These tight encircling complexes probably provide the physical basis for the inner blood–retinal barrier. Basement membrane, which acts as a filter in controlling the flow of solutes, forms enveloping sheaths surrounding the capillaries and provides structural support for the endothelial cells and pericytes, the latter lying external to the endothelial lining. Selective loss of pericytes is the hallmark of human diabetic retinopathy; observed histochemically for several decades, the biochemical basis of this lesion remains unknown. Extensive morphological, biological, ultrastructural, metabolic, and biochemical investigations have been carried out on retinal microvessels, but only the latter two areas of research are included in this section.

7.5.1. Glucose Transport, Insulin, and Aldose Reductase

The mechanism of glucose transport in the retinal microvasculature has been studied in several laboratories, and there is general agreement that D-glucose and its nonmetabolizable analogue, 3-O-methyl-D-glucose (3-O-MG) are transported by facilitated diffusion both in cultured endothelial cells (Betz *et al.*, 1983) and in cultured pericytes (W. Li *et al.*, 1985). The transport is stereospecific for D-glucose and under the conditions used in these studies, the uptake appears to be insulin independent. Kinetic studies show that sugar transport is a saturable, sodium- and energy-independent carrier-mediated process.

Hyperglycemia *per se* is thought to play a role in the development of diabetic microangiopathy, possibly through specific effects on the transport of important physiological substances such as inositol and ascorbate. In the case of inositol, which is transported in capillary pericytes by a carrier-mediated, sodium-dependent, ouabain-sensitive mechanism, high concentrations of metabolizable sugars such as D-glucose or D-galactose inhibit inositol uptake; the inhibition is significantly reversed by the aldose reductase inhibitor sorbinil (Li *et al.*, 1986). This suggests that sugar alcohols generated by the polyol pathway may in some way interfere with the transport of inositol. The transport of ascorbate, which is an insulin-insensitive process, is also impaired by high glucose concentrations (Khatami *et al.*, 1986b). Ascorbate uptake is mediated by facilitated diffusion and appears to use the same carrier as D-glucose or 3-O-MG. Kinetic studies of ascorbate transport in pericytes show that inhibition by 3-O-MG is noncompetitive.

7.5.1.1. Insulin Receptors

Specific high-affinity receptors for insulin have been characterized in cultured pericytes and endothelial cells obtained from calf microvessels (King *et al.*, 1983). The 125-kDa subunit of the bovine microvessel insulin receptor obtained following cross-linking of labeled insulin to its receptor has a mobility on SDS-PAGE identical to that of insulin-sensitive tissues such as liver (Haskell *et al.*, 1984). In the light of these findings, the effect of insulin on glucose uptake in the retina has been reassessed. An insulin-stimulated increase in the transport of 2-deoxy-D-glucose has been clearly demonstrated in cultured bovine retinal endothelial cells that had been preincubated with insulin for 90 min in serum-free medium (Allen and Gerritsen, 1986). Because of the long preincubation required, the insulin response in these cells was thought to reflect an induction of new glucose carriers rather than activation of existing ones, an interesting hypothesis requiring further study.

In addition to insulin receptors, bovine retinal capillary endothelial cells and pericytes also contain specific receptors for insulin-like growth factor I (IGF-I) and multiplication-stimulating activity (MSA; IGF-II). Studies on cultured pericytes show that insulin can bind both to its own receptors at high affinity and to IGF-I receptors at low affinity (King *et al.*, 1985). In pericytes as well as in endothelial cells, insulin and IGF-I and -II probably mediate their biological effects through the same pathway, possibly even using the same receptors. Insulin receptors have also been detected in the nonvascular portions of bovine retina and ROS membranes (see Section 7.2.2.4), but they exhibit different molecular properties from those of retinal microvessels (Im *et al.*, 1986). On a weight basis, the insulin receptor density of retinal microvessels is about six times greater than that of nonvascular tissue. Triton X-102 solubilization of bovine microvessels followed by purification, cross-linking with labeled insulin, and SDS-PAGE under reducing conditions reveals a 125-kDa α subunit in microvessels identical in size to that of the liver insulin receptor. The α subunit isolated under the same conditions from nonvascular retina is a 116-kDa peptide.

7.5.1.2. Aldose Reductase

Immunoreactive aldose reductase (AR) has been clearly demonstrated in pericytes but not in endothelial cells of trypsin-digested human (Akagi *et al.*, 1983; Hohman *et al.*, 1989) and canine (Akagi *et al.*, 1986) retinas. This key enzyme in the polyol pathway converts glucose and galactose to sorbitol and dulcitol (galactitol), respectively, in the presence of NADPH (see Fig. 5.8). Given the presence of AR in dog retinal microvessels, it is not surprising that isolated canine microvessels are able to convert galactose to dulcitol, albeit at a rate severalfold slower than in lens epithelium (Kern and Engerman, 1985). The AR inhibitor sorbinil inhibits the reduction of galactose to dulcitol in purified preparations of canine microvessels. Sorbinil and other AR inhibitors are also effective in inhibiting the reduction of the more physiological substrate, glucose, using purified enzyme isolated from human retinas (Poulsom, 1987). These studies were carried out on AR derived from whole retina, and further data would be needed to relate these findings to the microvasculature.

Another characteristic feature of diabetic retinopathy is retinal capillary basement membrane thickening. Its pathogenesis is unknown, but there is indirect evidence linking basement membrane thickening to an accumulation of polyols. Although this has not yet been demonstrated in diabetic animals, retinal capillary basement membrane thickening can be induced in rats fed galactose-rich diets (Robison *et al.*, 1983; Frank *et al.*, 1983). The lesions produced are similar to those found in diabetic retinopathy and can be prevented by including sorbinil in the diet of the galactosemic rats. The basement membrane thickening in response to galactose-rich diets progresses with time; the thickening is 73% greater in rats maintained on 50% galactose diets for 88 weeks than for 28 weeks (Robison *et al.*, 1988). The basement membrane thickening in galactose-fed rats can be diminished, though not completely prevented, by another AR inhibitor, tolrestat.

7.5.2. Production of Extracellular Matrix

Heparan sulfate is a small but functionally important component of most basement membranes, where it is thought to serve as a permeability barrier to low-molecular-weight anions. This glycosaminoglycan (GAG) has been detected in basement membranes isolated from bovine retinal microvessels, and it appears to be the only GAG present (Kennedy *et al.*, 1986a). However, cultured bovine pericytes incubated with labeled sulfate or glucosamine synthesize primarily chondroitin sulfate and relatively little heparan sulfate; endothelial cells synthesize both GAGs. These alterations in phenotypic expression may be more apparent than real, since no distinction was made between GAGs secreted into the medium and those associated with either the cell layer or the extracellular matrix. In other studies using cultured bovine capillary pericytes, GAGs synthesized from radioactive precursors were extracted separately from the medium and from the cell layer (Stramm *et al.*, 1987). The GAG profiles in the two preparations were found to be distinctly different. Heparan sulfate was the principal component in the cell-layer-associated fraction, whereas those secreted into the medium consisted of both heparan sulfate and chondroitin sulfate in approximately equal proportions.

The collagen composition of bovine retinal microvascular basement membrane appears to be similar to that of other extracellular matrices and consists primarily of type IV collagen. However, rather considerable amounts of type I and small amounts of type III collagen have also been detected by immunocytochemical and biochemical methods in highly purified basement membrane preparations that were considered free of interstitial tissue contamination (Kennedy *et al.*, 1986b). Rather unexpectedly, cultured retinal pericytes and endothelial cells were found to produce mainly type I collagen polypeptides together with only small and variable amounts of type IV collagen. The reasons for alterations in collagen expression by cultured microvascular cells are not clear but may be related to conditions under which the cells were grown or possibly to the production of a collagenase capable of degrading newly formed type IV collagen.

7.5.3. Adrenergic Receptors and Phosphoinositide Turnover

Purified preparations of bovine retinal vessels contain α_1-, α_2-, β_1-, and β_2-adrenergic receptors, as shown by the kinetic behavior of specific agonists for each receptor type (Forster *et al.*, 1987; Ferrari-Dileo, 1988). Although the binding sites are few in number, they are of high affinity and hence capable of responding to low concentrations of adrenergic agonists. Retinal vessels lack autonomic innervation, but as in many noninnervated tissues throughout the body, the receptors may be activated by circulating hormones and neurotransmitters (see Chapter 1, Section 1.3). In other tissues, these receptors are coupled to two major membrane transduction systems, adenylate cyclase and polyphosphoinositide (PI) hydrolysis, but the biological responses modulated by the adrenergic receptors in retinal vessels have not yet been established.

Although there is as yet no definitive information on receptor-stimulated breakdown of phosphatidylinositol 4,5-bisphosphate (PIP_2) in retinal microvessels, two of the second messengers generated by its breakdown, IP_3 and DG, may regulate the proliferation of bovine capillary pericytes (Li *et al.*, 1987a). The incorporation of tritiated thymidine into cellular DNA by permeabilized bovine pericytes is enhanced severalfold in the presence of increasing concentrations of either IP_3 or DG. Moreover, these two second messengers of PIP_2 hydrolysis act synergistically in stimulating pericyte proliferation. This striking effect appears to be mediated through the mobilization of intracellular Ca^{2+}; IP_3 may initiate the response, while DG appears to function in prolonging it. High concentrations of glucose inhibit inositol transport in cultured pericytes (Li *et al.*, 1986), and further studies suggest a possible link between loss of pericyte viability and decrease in intracellular inositol. Incubation of labeled inositol with retinal microvessels for 60 h results in the synthesis of labeled inositol phospholipids, the major component being phosphatidylinositol (PI) (Li *et al.*, 1987b). However, high (30 mM) concentrations of glucose in the incubation medium decrease this labeling, probably by inhibiting the transport of inositol and thus limiting the supply of free inositol required for PI synthesis.

7.5.4. Regulation of Microvessel Caliber

The function of pericytes in the retinal microvasculature is unclear, but they are believed to be the major contractile elements of the capillary wall. There is abundant evidence supporting the view that pericytes are a modified type of smooth muscle cell. A major contractile protein isolated from cultured bovine capillary pericytes has been partially characterized as a 46-kDa protein having the same electrophoretic mobility as muscle actin (Chan *et al.*, 1986). Using a modified method to detect small quantities of functional actin, it was shown that the purified actin isolated from cultures of bovine pericytes could activate skeletal muscle myosin Mg^{2+}-ATPase. These findings imply that pericytes have the capacity to generate contractile forces through their cytoskeletal system. The role of actin in this process has been shown more directly in sequential photographs of cultured permeabilized bovine microvascular pericytes exposed to Mg^{2+}-ATP (Das *et al.*, 1988). The contraction of these Triton X-100-treated cells is dose-dependent and occurs within a matter of minutes. This rapid and specific response suggests that the actin-containing microfilaments in retinal pericytes could play a role *in vivo* in regulating microvessel caliber.

Eicosanoids have a wide range of biological effects in vascular and nonvascular systems throughout the body. When released in small quantities under a variety of physiological and/or experimental conditions, they promote vasodilation, inhibit platelet aggregation, and regulate inflammatory and immune responses. Recent experiments have demonstrated eicosanoid production *in vitro* by cultured bovine pericytes after incubation with labeled arachidonic acid (Hudes *et al.*, 1988). The only eicosanoid detected was 6-keto-$PGF_{1\alpha}$, the major stable metabolite of prostacyclin (PGI_2). This substance in particular is a potent vasodilator and has profound effects on blood flow in tissues such as myocardium and in cerebral capillaries. It has been suggested that loss of pericytes, as in the early stages of diabetic retinopathy, may deprive the microvessels of both contractile (actin) and vasodilatory (prostacyclin) controls that are essential for the regulation of microvessel caliber and stability.

7.6. GLIA (MÜLLER CELLS)

Müller cells are the predominant glial cell type in the vertebrate retina. They traverse the retina from the vitreous surface (or inner limiting "membrane" of the retina) to the outer limiting "membrane," the latter consisting of junctional complexes made between the Müller cell processes and the photoreceptor inner segments. Müller cells provide mechanical support for the retina, and one of their major physiological functions is the generation of the b-wave in the electroretinogram (Dowling, 1987). In recent years, several observations suggest that this glial cell may also play an unexpectedly important role in the metabolism of neural retina. For example, immunoreactive staining for aldose reductase is especially intense in Müller cells of human and rat retina, and it has been suggested that impaired polyol metabolism could cause osmotic fluxes leading to structural and/or functional changes in the retina (Akagi *et al.*, 1984). Indeed, thickening of the inner limiting membrane can be induced in rats by long-term galactose feeding but is effectively prevented by the aldose reductase inhibitor tolrestat (Nagata and Robison, 1987).

7.6.1. Retinoid-Binding Proteins

Three retinoid-binding proteins, cytosolic retinol-binding protein (CRBP), cytosolic retinal-binding protein (CRALBP), and cytosolic retinoic acid binding protein (CRABP), have been characterized biochemically in neural retina (see Table 7.3). Because of their important role in the visual cycle, it had been assumed that they would be localized in the photoreceptors. Surprisingly, however, immunocytochemical studies of vertebrate retinas show that all three retinoid-binding proteins, CRALBP (Bunt-Milam and Saari, 1983), CRBP (Bok *et al.*, 1984; Anderson *et al.*, 1986b), and CRABP (Bok, 1985), are localized in the Müller cells and not in the photoreceptors. The full significance of these unusual findings and the specific role of these proteins in retinoid metabolism in the retina are important topics for future study.

7.6.2. Glial Fibrillary Acidic Protein

Glial fibrillary acidic protein (GFAP) is a 47- to 50-kDa cytoskeletal protein found in neural tissue; it is normally localized to astrocytes and usually occurs as a characteristic 10-nm-diameter intermediate filament (IF). GFAP has also been detected by immunocytochemical methods in normal retinal Müller cells and retinal astrocytes of several vertebtate species. What has aroused considerable interest is the dramatic increase in GFAP immunoreactivity in Müller cells in response to experimentally induced injury or genetically related photoreceptor degeneration, as in the RCS rat (Eisenfeld *et al.*, 1984). Müller cells of cat retina normally contain GFAP-immunoreactive material in the end feet near the vitreoretinal border and in the cytoplasm, but following experimental retinal detachment there is a large accumulation throughout the whole cell (Erickson *et al.*,

1987). Studies on Abyssinian cats as well as on mice with *rd* (retinal degeneration) and *rds* (retinal degeneration slow) genes show a general increase in GFAP immunoreactivity that parallels the course of photoreceptor degeneration both temporally and, in most cases, topographically (Ekstrom *et al.*, 1988). Given the broad spectrum of injuries and genetic disorders examined, it appears that GFAP expression in Müller cells is an early reaction to a variety of metabolic insults in photoreceptor cells.

A panel of antibodies to five known glial proteins has been used to compare, by immunocytochemical methods, Müller cell-specific proteins and astrocyte proteins of cat retina (Lewis *et al.*, 1988a). Whereas Müller cells are labeled by all of the five antibodies [carbonic anhydrase C, α-crystallin, GFAP, cytosol retinal-binding protein (CRALBP) and glutamine synthetase], astrocytes are not labeled by either anti-CRALBP or anti-glutamine synthetase. Thus, there is strong evidence for differences in expression of specific proteins in these two types of glial cells. Müller cells cultured from cat retina continue to express all five of the proteins, even though the cells do not retain their characteristic *in vivo* morphology (Lewis *et al.*, 1988b). The expression of these proteins provides a specific marker for positive identification of Müller cells and exclusion of other cell types such as astrocytes in these cultures. This is the first reported account showing that CRALBP can be expressed in cultured cells. The availability of pure lines of viable Müller cells provides an important new tool for future metabolic and biochemical studies of these unique retinal glial cells.

7.6.3. Insulin Synthesis (Insulin-Specific mRNA)

Immunoreactive insulin-like activity has been detected in several extrapancreatic tissues, including brain and other neural tissue, and it is now recognized that the insulin gene is expressed in a variety of cell types. Insulin-like activity has also been found in glial cells of mouse and human retina (Das *et al.*, 1984). However, in view of the presence of insulin binding sites throughout the retina, the insulin activity in glial cells could reflect uptake and storage of the pancreatic hormone from the circulation. Alternatively, as in other extrapancreatic cells, it could reflect local synthesis. The latter possibility has been convincingly demonstrated in cultured rat retinal glial cells using a tritiated insulin cDNA probe (A. Das *et al.*, 1987). *In situ* hybridization has revealed that Müller cells in particular contain the mRNA necessary for *de novo* synthesis of insulin or a closely homologous peptide. The function of locally produced insulin has not been established, but possible roles as a neurotransmitter or neuromodulator have been suggested. In addition, it may function in regulating glucose metabolism and glycogen turnover in the Müller cell.

7.7. NEUROTRANSMITTERS

Neurotransmitters and neuromodulators are chemical messengers that mediate intercellular communication in neural tissue. Neurotransmitters are generally defined as endogenous substances that elicit rapid electrophysiological responses of short duration. Neuromodulators are also endogenous substances released from neurons, but their effects are distinct from those of neurotransmitters. Neuromodulators appear to act postsynaptically by activating intracellular enzyme systems that control a variety of cellular functions; the actions of most neuromodulators are relatively long-lived (Iuvone, 1986; Dowling, 1987). At least 15 putative neurotransmitters and neuromodulators have been described in the retina, and their action on the six major classes of neuronal cells is briefly summarized in Table 7.18.

Neurotransmitters are all low-molecular-weight substances that can be classified under three general groupings: amino acids, biogenic amines, and neuropeptides. Studies of their specific actions on retinal neurons utilize a broad range of multidisciplinary approaches: biochemical, pharmacological, immunologic, physiological, anatomic, and ultrastructural. A review of the massive amount of information on neurotransmitters that has appeared during the past few years is beyond the scope of a general text on ocular biochemistry. Hence, only a brief outline of selected biochemical aspects are presented in this section, much of the material being drawn from the complete text by Dowling (1987) as well as from two recent comprehensive overviews (Iuvone, 1986; Massey and Redburn, 1987). Other extensive reviews of transmitters and modulators in the retina have appeared on specific topics and are recommended for in-depth coverage as well as literature references (Redburn and Hollyfield, 1983; Osborne, 1984; Dowling, 1986; Masland and Tauchi, 1986; Hutchins, 1987; Redburn and Madtes, 1987).

Several criteria are used to establish that a given substance is a neurotransmitter at a

Table 7.18. Neurotransmitters in the Retina and Their Probable Localizations[a]

Putative transmitter	Neuronal cell					
	Photoreceptor	Horizontal	Bipolar	Amacrine	Interplexiform	Ganglion
Amino acids						
Aspartate	■		±			±
Glutamate	■		±			±
GABA		■[b]		■	±	±
Glycine				■	±	±
Biogenic amines						
Acetylcholine				■		±
Dopamine		■[c]		■	■	
Serotonin[d]	±			±		
Epinephrine				±		
Neuropeptides[e]						
VIP[f]				±		±
Substance P				±	±	±
Enkephalin				±		
Glucagon				±		
Somatostatin				±	±	±
Neurotensin				±		
Neuropeptide Y				±	±	±
Cholecystokinin				±		

[a]Adapted from Bazan and Reddy (1985), Iuvone (1986), and Dowling (1987). The solid squares indicate strong evidence, and ± indicates weak or inconclusive evidence.
[b]In fish, amphibia, and birds but probably not in mammals.
[c]Dopamine is a neuromodulator in horizontal cells of some fish.
[d]Evidence contradictory; may be a neuromodulator or melatonin precursor.
[e]At least 30 peptides are considered as putative neurotransmitters or neuromodulators in brain; eight of them have been identified (to date) in retina.
[f]Abbreviation: VIP, vasoactive intestinal peptide.

particular synapse: (1) the substance must be detectable in sufficient quantities by histochemical, biochemical, or immunocytochemical methods; (2) there should be enzymatic mechanisms for its synthesis from endogenous precursors in presynaptic terminals; (3) it must be released in response to photic, electrical, or other stimulation; additionally, there must be an appropriate postsynaptic electrical response when a putative transmitter is applied exogenously; and (4) there must be efficient mechanisms for the rapid termination of its action. Acetylcholine and neuropeptides are inactivated enzymatically, but for most other transmitters, termination is achieved by high-affinity reuptake.

7.7.1. Amino Acids

General pathways for the enzymatic synthesis of aspartate and glutamate are shown in Fig. 7.2. Both of these acidic amino acids are present in high concentrations in the retina, and it is generally recognized that they fulfill many of the criteria for neurotransmitters in photoreceptor cells of most mammalian species (Iuvone, 1986). They can function in both rod and cone photoreceptors, although there is considerable species variation in their uptake, localization, and release in rod cells compared to cone cells. Whether aspartate and glutamate serve as functional transmitters in other neurons is less certain, since only some of the criteria have been met in a limited number of species. Glutamate and possibly aspartate have been considered as putative transmitters in bipolar cells of cat and mudpuppy and in ganglion cells of pigeon and certain vertebrates (Iuvone, 1986; Massey and Redburn, 1987).

GABA is not only one of the most important inhibitory transmitters in the vertebrate retina, but it is also one of the earliest transmitters expressed during the development of the retina (Redburn and Madtes, 1987). Three classes of GABAergic neurons (amacrine cells, H_1 horizontal cells, and interplexiform cells) have been known for several years, and more recently, there is immunocytochemical evidence suggesting a role for GABA as a neurotransmitter in ganglion cells in the rabbit as well (Yu *et al.*, 1988). Adult retinas contain high levels of GABA; high-affinity uptake as well as the GABA-synthesizing enzyme glutamic acid decarboxylase have been clearly established in H_1 horizontal cells and in amacrine cells of many animal species (for reviews see Iuvone, 1986; Massey and Redburn, 1987). GABAergic neurons have been identified in H_1 horizontal cells of fish, amphibia, and birds but do not appear to be present in mammalian horizontal cells. By contrast, all vertebrate retinas including mammals contain GABAergic amacrine cells. Thus, there is strong evidence implicating GABA as a neurotransmitter in horizontal and amacrine cells. There is, in addition, limited evidence for a similar role in interplexiform cells of the cat retina.

The highest concentrations of glycine in the retina are found in amacrine cells and in the interplexiform and ganglion cell layers. Glycine-accumulating amacrine cells have been clearly demonstrated in all species examined, and there is considerable evidence for its rapid release in amacrine cells; moreover, the presence of receptors and high-affinity uptake systems has been shown in frog, goldfish, and other vertebrate retinas. Taken together, the gross criteria for establishing a transmitter role for glycine in the retina have been satisfied (Massey and Redburn, 1987). There are also studies suggesting a transmitter role for glycine in interplexiform and possibly ganglion cells, although some of the data are in conflict. Some bipolar cells accumulate glycine, but for many reasons they are not considered to be glycinergic neurons (Massey and Redburn, 1987).

7.7.2. Biogenic Amines

Acetylcholine was one of the earliest neuroactive substances identified in the retina, and there are no doubts today that it is a major transmitter in the vertebrate retina (Hutchins, 1987). The cholinergic neurons of the rabbit retina consist of both amacrine and displaced amacrine cells, the latter comprising about one-third or more of the ganglion cell layer neurons (see review by Masland and Tauchi, 1986). Morphological and immunocytochemical studies have established a characteristic starburst appearance for cholinergic amacrine cells in rabbit as well as most other vertebrate retinas. The cholinergic system receives an inhibitory input from GABA amacrine cells, and colocalization of GABA and acetylcholine in amacrine cells of the rabbit retina demonstrated by immunohistochemical staining establishes that cholinergic cells are a major subpopulation of GABA-immunoreactive amacrine and displaced amacrine cells (Brecha *et al.*, 1988). The presence and release of both an inhibitory and an excitatory transmitter from the same cell suggest a mechanism for the formation of complex receptor-field properties of ganglion cells.

Localized in both amacrine and interplexiform cells, dopamine is probably the best understood neurotransmitter in the retina. The presence of dopamine-synthesizing enzyme (tyrosine hydroxylase), the release of dopamine by light stimulation, and the presence of appropriate receptors and enzymes for its degradation leave little doubt that dopamine functions as a neurotransmitter in these cells (Iuvone, 1986; Massey and Redburn, 1987). There are, however, species variations; cat, rat, and frog retinas contain both of the dopamine cell types, but other species contain either dopamine amacrines or dopamine interplexiform cells.

Dopamine also functions as a neuromodulator; extensive studies (mainly on fish retinas) show that a major action of dopamine is to stimulate the accumulation of cAMP in horizontal cells. Dopamine receptors are localized on the horizontal cell membranes, and their activation stimulates the adenylate cyclase system, causing a rise in the intracellular level of cAMP. This important cyclic nucleotide acts as a second messenger in the control of important intracellular processes such as phosphorylation. In the case of horizontal cells, both dopamine and cAMP decrease gap junctional conductance (see review by Dowling, 1986).

Photic stimulation and consequent depolarization of the dopamine-containing neurons activate tyrosine hydroxylase by mechanisms involving its phosphorylation (see review by Iuvone, 1986). This activation can also be elicited by incubation of cell suspensions of dissociated rat retinas with the 8-bromo cAMP, a membrane-permeable analogue of cAMP. The activation produced by photic stimulation *in vivo* is comparable to that produced by cAMP-dependent protein phosphorylation *in vitro*. There is clearly a molecular link between membrane depolarization and activation of cAMP-dependent protein kinase.

Recent experiments suggest that the action of cAMP as the second messenger in modulating the voltage-dependent ion channels in retinal horizontal cells could be through phosphorylation of the 26-kDa gap junction protein (Lasater, 1987). Although not shown directly, this would be consistent with observations made in other tissues. There is additional evidence that dopamine utilizes cAMP as a second messenger to modulate gap junctional conductance between horizontal cells: (1) forskolin, which activates adenylate cyclase directly, mimics the effects of dopamine; (2) phosphodiesterase inhibitors such as

IBMX, which cause a buildup of intracellular cAMP, also mimic the effects of dopamine; and (3) direct injection of cAMP into cultured horizontal cells causes their rapid uncoupling (see discussion by Lasater, 1987). Thus, the major effect of dopamine and cAMP in decreasing gap junctional conductance between horizontal cells now seems firmly established.

Indoleamine-accumulating amacrine cells have been observed in virtually every retina, although there are considerable species variations; moreover, there is some uncertainty as to whether they are all true serotonergic systems (see review by Massey and Redburn, 1987). Serotonin is often considered as a putative neurotransmitter, but this has not yet been firmly established in spite of its endogenous localization and high-affinity uptake by 5-HT_1 and 5-HT_2 receptors in some species (see review by Osborne, 1984). Its concentration is high in cold-blooded vertebrates, intermediate in birds, and very low in mammals. Much of the conflicting data on the possible role of serotonin as a neurotransmitter in the retina are best appreciated by considering the serotonin (or indoleamine) system(s) in three different groups of animals (Massey and Redburn, 1987). There is strong evidence that one group consisting of amphibian, teleost, and avian retinas contains true serotonergic systems. A second group of animals consisting of cat, rabbit, cebus monkey, cow, and pig has little endogenous serotonin. However, a subset of amacrine cells accumulate exogenously applied serotonin and serotonin analogues; moreover, several studies provide evidence for three types of serotonin receptors in rabbit and cat retinas. A third group of animals do not accumulate either serotonin or other indoleamines in their amacrine cells, yet labeled serotonin is accumulated by photoreceptor terminals in at least one of the species, the rat. The other species in this group that appear to have indoleamine systems localized in retinal photoreceptors are man, monkey, guinea pig, and mouse.

In the rat retina, the accumulation of serotonin is greater in the light than in darkness, and the endogenous concentration of serotonin is increased in constant light and decreased in constant darkness. These animals were killed during the daytime, and hence the levels measured may have reflected the effects of prolonged darkness. In the chicken retina, a clear diurnal variation in serotonin content has been demonstrated, with the highest concentration of serotonin being found at night (Ehinger and Rose, 1988). Serotonin is a precursor of melatonin (see Fig. 7.10), and both melatonin content and melatonin-synthesizing enzyme exhibit circadian rhythm, with the highest levels appearing at night. These findings further support the view that both serotonin and melatonin metabolism in the retina are under circadian control; apart from the role of these indoleamines as putative neurotransmitters, they also play a key role in disk shedding (see Section 7.2.2.1).

Two other catecholamines, epinephrine and norepinephrine, are present in rat and bovine retinas, and they appear to be localized in a subclass of amacrine cells. Binding sites for labeled norepinephrine have been reported, and photic stimulation causes a transient increase in the concentration of epinephrine in the rat retina. However, a role for these two catecholamines as putative transmitters, or possibly neuromodulators, in the retina remains to be established (Iuvone, 1986).

7.7.3. Neuropeptides

Thirty or more peptides that appear to function as neurotransmitters or neuromodulators have been characterized in brain and other neural tissue, and in recent years at least eight of them have also been detected in the retina (Table 7.18). Their localization in the retina, mainly in amacrine and displaced amacrine cells, has been deduced entirely from immunohistochemical reactivity to antibodies against either purified or synthetic peptides (see reviews by Iuvone, 1986; Dowling, 1987; Massey and Redburn, 1987). There are certain pitfalls in this methodology, since different antibodies to a specific peptide may label not only the same peptide but also others with closely related structures. In view of their very low concentrations in the retina, it has been diffult to isolate and characterize any of the retinal neuropeptides biochemically. Neuropeptide immunoreactivity has been demonstrated in a variety of vertebrate retinas, and recently in human retina as well (Tornqvist and Ehinger, 1988). In addition to amacrine cells, histochemical reactivity to several neuropeptides was also detected in the interplexiform and ganglion cell layers.

A frequent observation in several fish and avian species is the colocalization of neuropeptides either with another peptide or with a conventional neurotransmitter. This is not unexpected in view of the large number of possible neurotransmitters, especially in amacrine cells. Enkephalin coexists with GABA in catfish and chick retinas, and glycine colocalizes with neurotensin in turtle retina and with neurotensin, enkephalin, and somatostatin in chick retina (see reviews by Massey and Redburn, 1987; Dowling, 1987). There are also instances of colocalization of neuropeptides, for example, neurotensin and enkephalin in chick amacrine cells, and substance P and neurotensin in goldfish retina.

Little is known about the function of neuropeptides in the retina, either singly or in combination with other peptides or neurotransmitters. In general, their physiological effects appear to be slow-acting and long-term, although there are exceptions. Some years ago it was reported that vasoactive intestinal peptide (VIP) modulates cAMP levels in horizontal cells, and, like dopamine, VIP appears to stimulate adenylate cyclase activity in the rabbit retina. Experimental approaches to the study of neuropeptides pose many technical difficulties, but their potential importance in retinal physiology justifies further investigations aimed at clarifying their functions in this tissue.

7.8. VISUAL EXCITATION

Classical studies of visual excitation and phototransduction have utilized a broad spectrum of scientific disciplines that include biophysical, biochemical, and electrophysiological approaches; in recent years immunologic, immunohistochemical, and recombinant DNA techniques have come into increasing use. Several hundred original papers on nearly every aspect of visual transduction have been published during the past few years, attesting to the enormous interest in this challenging field. Eighteen (or possibly more) reviews, monographs, and texts have appeared recently, and most of the material in this section is drawn from them (Anderson and Rapp, 1983; Lolley, 1983; Shichi, 1983; Chabre, 1985; Schwartz, 1985; Baehr and Applebury, 1986; Hargrave, 1986; Lamb, 1986; Stryer, 1986; Stryer and Bourne, 1986; Pugh and Cobbs, 1986; Baylor, 1987; Fung, 1987; Gold and Nakamura, 1987; Hurley, 1987; Liebman *et al.*, 1987; Owen, 1987; Pugh, 1987). These reviews are highly recommended for the wealth of detailed information contained in them. They not only integrate the recent information that has been generated in this rapidly growing field, but they also provide valuable insights into new concepts that are constantly emerging. The discussions that follow are intended as a brief overview of this vast topic, and space limitations allow only a limited number of individual literature citations.

7.8.1. Bleaching of Photopigments: cGMP as the Chemical Messenger

The photoreceptor outer segment has only one role, visual transduction, the conversion of light energy into electrical signals (Stryer, 1986). The molecules in vertebrate rod (and probably also cone) photoreceptors that mediate visual excitation are present in a uniquely organized highly concentrated form in the outer segment, separate from the energy-generating machinery in the inner segment. The initial photochemical event in transduction is the absorption of a single photon of light by the chromophore of rhodopsin; this results in a very rapid isomerization of 11-*cis* retinal to all-*trans* retinal. The formation of the first intermediate, bathorhodopsin, takes place in picoseconds, but subsequent conformational changes in rhodopsin are somewhat slower and light-independent. The products formed during this isomerization are detectable by their characteristic spectra (Fig. 7.18). Because the decays are so rapid at physiological temperature, spectral

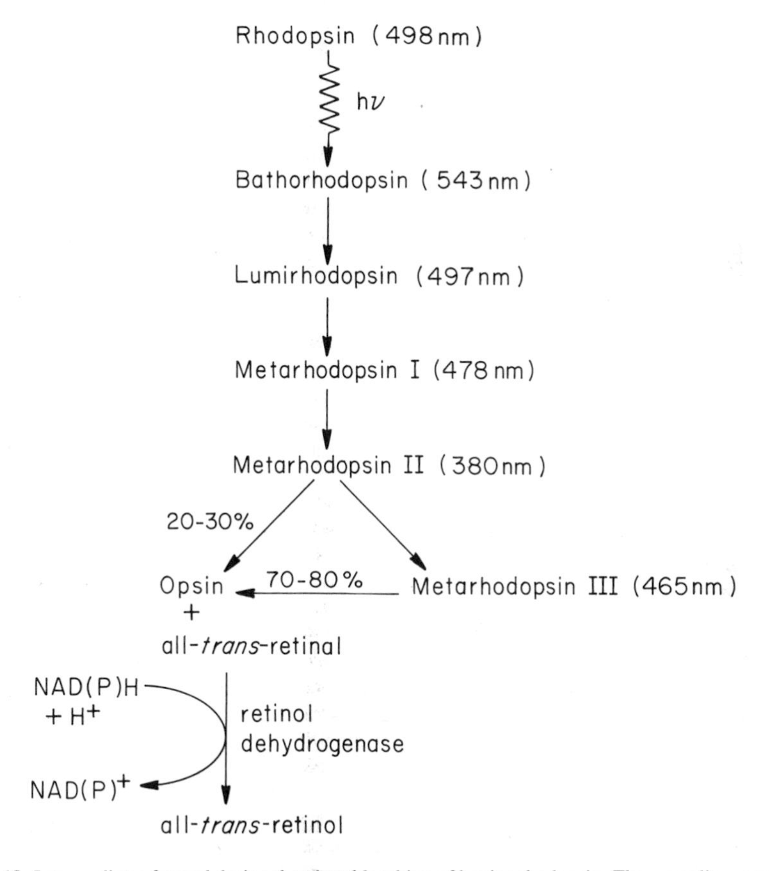

Figure 7.18. Intermediates formed during the photobleaching of bovine rhodopsin. The wavy line represents the photochemical reaction, the absorption of a photon by the 11-*cis* retinal chromophore. The other reactions designated by straight lines are thermal (dark) reactions. The isomerization of the chromophore and hydrolysis of the Schiff base occur in discrete steps. The intermediates have characteristic absorption spectra (measurable at low temperatures) whose maxima are shown in parentheses. The reactions are rapid at physiological temperatures, ranging from picoseconds for the light-catalyzed formation of bathorhodopsin to milliseconds or seconds for the decays of metarhodopsin I and II, respectively.

changes resulting from the initial conformational changes in rhodopsin during bleaching can be studied only at low temperatures, where the lifetimes of the intermediates are relatively long.

It has long been suspected that visual excitation occurs prior to the breakdown of metarhodopsin II, and recent work points to metarhodopsin II as the biochemically activated form of rhodopsin (R*). Only this intermediate is capable of activating transducin (G protein), as discussed in Section 7.8.2. In the final step of rhodopsin bleaching, metarhodopsin II is converted to opsin and all-*trans* retinal directly as well as indirectly, through metarhodopsin III. The all-*trans* retinal formed during bleaching is reduced to vitamin A (all-*trans* retinol) and transported to the pigment epithelium, where it is rapidly esterified and isomerized to 11-*cis* retinol. The visual cycle is completed when this stereoisomeric form of vitamin A is returned in the dark to the photoreceptors for rhodopsin regeneration.

Visual excitation is triggered by activated rhodopsin (R*) formed during the process of photoisomerization. The conformational change of a single molecule of rhodopsin leads to the closure of hundreds of sodium channels in the photoreceptor plasma membrane and a block in the entry of more than 10^6 Na^+ (Stryer, 1986). The resulting hyperpolarization spreads by passive propagation to the synaptic terminal, causing a diminished rate of transmitter release (see reviews by Baylor, 1987; Dowling, 1987; Owen, 1987).

Clearly a chemical messenger is required to carry the signal from activated rhodopsin (R*) on the disk membrane to the plasma membrane, the site of hyperpolarization. Both Ca^{2+} and cGMP have been considered as diffusible transmitter candididates since the early 1970s, and despite intensive efforts by investigators supporting one "school" or the other, the controversy has only been resolved recently. It is now established that cGMP is the chemical messenger in phototransduction. Not only have the enzymatic mechanisms in the cGMP cascade been elucidated, but light-mediated hydrolysis of cGMP with the resultant drop in cGMP concentration is required to terminate the cGMP-induced current (see review by Owen, 1987).

The most compelling evidence that the cGMP-sensitive current and the light-sensitive conductances are one and the same has come from the work of Yau and Nakatani (1985), Fesenko *et al.* (1985) and others using the patch-clamp technique (see reviews by Pugh and Cobbs, 1986; Owen, 1987; Baylor, 1987). These experiments show conclusively that conductance through the light-sensitive channel is controlled by cGMP. In the dark, an active Ca^{2+}-modulated guanylate cyclase maintains high levels of rod outer segment cGMP; under these conditions, three molecules of cGMP are bound to each ion channel, thus keeping the channel open for the influx of Na^+. Studies of the gating kinetics of this cGMP-activated cation channel suggest a simple model in which activation involves three sequential cGMP binding steps; the channel is optimized for rapid responses to changes in cytoplasmic cGMP concentration (Karpen *et al.*, 1988). Light closes the channel by activating the cGMP cascade that leads to the rapid hydrolysis of cGMP (see review by Liebman *et al.*, 1987). A drop in cGMP concentration to <10 μM causes a rapid closure of the light-sensitive channels in the rod cell plasma membrane. The entry of NA^+ is blocked, thereby causing hyperpolarization of the outer segment plasma membrane. The discovery of cGMP-gated conductance as a new class of ion channel represents a major step forward in our understanding of phototransduction (see review by Gold and Nakamura, 1987).

Given these findings on cGMP as the unequivocal transmitter in vertebrate phototransduction, the role of Ca^{2+} is now perceived in terms of a negative feedback (see reviews by Pugh, 1987; Baylor, 1987). It is known that Ca^{2+} enters the photoreceptor through open light-sensitive channels, and these ions are extruded by a powerful $Na^+(K^+)-Ca^{2+}$ exchanger (antiporter). The classical view is that the exchanger transports three Na^+ inward for each Ca^{2+} that is ejected outward. However, there is evidence supporting the view that the gradient of K^+ also contributes to the extrusion of Ca^{2+}. Recovery of the dark state requires the resynthesis of cGMP, which is catalyzed by guanylate cyclase. Recent studies have shown that this enzyme is strongly stimulated when the Ca^{2+} level is lowered after illumination; retinal rod guanylate cyclase activity has recently been shown to be regulated by a highly cooperative calcium feedback control that responds to changes in cytosolic $[Ca^{2+}]$ (Koch and Stryer, 1988). The activation of guanylate cyclase by light-induced lowering of $[Ca^{2+}]$ is thought to be a key event in restoring the dark current after excitation.

7.8.2. Phototransduction in Vertebrate Rods: G Protein (Transducin) and the cGMP Cascade

Within picoseconds after the absorption of a photon by rhodopsin, the chromophore is isomerized, and the protein portion of rhodopsin becomes activated (see reviews by Stryer and Bourne, 1986; Stryer, 1986; Hurley, 1987; Liebman *et al.*, 1987; Fung, 1987). The photoexcitation of rhodopsin triggers a rapid chain of molecular events involving protein–protein interactions, enzymatic reactions, and changes in ion-gated conductance. A diagram of the sequential interactions of the various components in this cascade is shown in Fig. 7.19. The events are mediated specifically by activated rhodopsin (R*), an integral membrane protein, acting in concert with several peripheral proteins: transducin (G protein), cGMP phosphodiesterase (PDE), rhodopsin kinase, 48-kDa protein, guanylate cyclase, and the cGMP-sensitive channel. These proteins have been studied extensively, and their properties are described in Section 7.2.1.1. The individual reactions in this transduction sequence are briefly summarized below.

Light-induced formation of metarhodopsin II (R*) creates a binding site for transducin (T), the specific G protein of rod outer segment disks. The molecular basis of this interaction has recently been examined using synthetic peptides corresponding to two regions near the C terminus of the α subunit of transducin (T_α) (Hamm *et al.*, 1988). Use of these probes demonstrated that regions from amino acids 311 to 329 and from 340 to 350 mimic the effects of transducin on rhodopsin conformation and showed that these peptides bind to and stabilize its activated conformation. The initial complex formed with R* is transducin-GDP. This transient binding alters the conformation of transducin and exposes a pocket on its α subunit, which allows GTP to be exchanged for GDP, a process that is considerably accelerated by Mg^{2+}. As in other G protein systems, the binding of GTP to the α subunit lowers the affinity between R* and transducin and between the α and $\beta\gamma$ subunits of transducin. The T_α-GTP and $T_{\beta\gamma}$ are released from the disk membrane, and the R* is available to recycle additional transducin molecules. A single molecule of R* can, during its lifetime, catalyze the binding of GTP to several hundred α subunits of transducin.

It is the T_α-GTP complex, dissociated from R*, that activates cGMP phosphodiesterase (PDE); the $T_{\beta\gamma}$ subunit does not participate in this process. The mechanism of T_α-GTP activation of PDE is probably through cancellation of the inhibitory restraint of the PDE γ subunit. This results in an increase in PDE activity from about 50 to more than 1000 mol of cGMP hydrolyzed per second per mol of enzyme. Thus, the signal amplification in this cascade is achieved in two stages: the first is the formation of about 500 molecules of T_α-GTP from one R* which leads to the activation of a similar number of

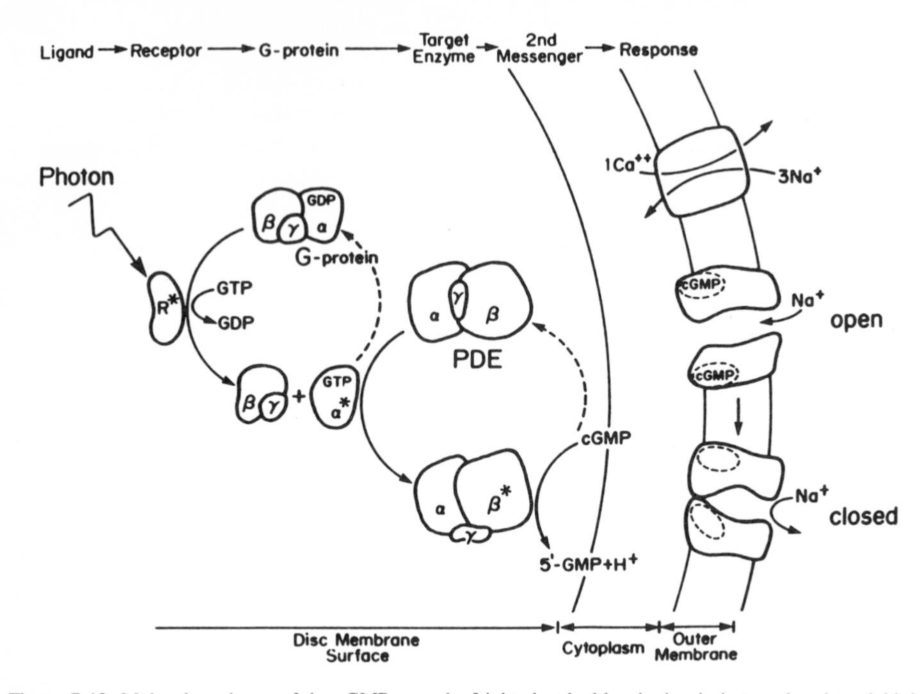

Figure 7.19. Molecular scheme of the cGMP cascade. Light absorbed by rhodopsin is transduced to yield the activated receptor, R*, which triggers the cascade by catalyzing the exchange of GTP for GDP bound to the α subunit of transducin (G protein; G_t; T). In the presence of GTP, the α subunit dissociates from the βγ subunit and activates PDE by removal of an inhibitory γ subunit. One R* catalyzes the activation of several hundred molecules of transducin, and each activated PDE hydrolyzes close to 1000 molecules of cGMP per second. As the level of cGMP falls, it dissociates from the cGMP-gated ion channel in the plasma membrane. This closes the Na^+–Ca^{2+} channel (or exchanger) and causes hyperpolarization of the plasma membrane. An additional 48-kDa protein (not shown in the diagram) interacts with phosphorylated rhodopsin to inactivate the cascade. (From Baehr and Applebury, 1986.)

PDE molecules, and the second is the hydrolysis of several thousand cGMP molecules per second by the activated enzyme. This hydrolysis leads to a fall in cGMP levels, closure of the Na^+ channels, and hyperpolarization of the plasma membrane, as described in Section 7.8.1. The levels of cGMP are restored by the action of guanylate cyclase.

The photoactivation of PDE would continue indefinitely, and the levels of cGMP would be constantly depressed, if there were not mechanisms for inactivation of R*, termination of the cGMP cascade, and restoration of the dark state. Although not completely understood, there appear to be (at least) two mechanisms involved.

Phosphorylation of R* at multiple serine and threonine residues by rhodopsin kinase is followed by the subsequent binding of 48-kDa protein to the phosphorylated region(s) of R*. Phosphorylation itself does not prevent binding of transducin to R*; 48-kDa protein is also required. The 48-kDa protein is believed to act by competing with transducin for binding sites on the phosphorylated R*; the binding of 48-kDa protein blocks the capacity of R* to further catalyze transducin (and PDE) activation (Wilden *et al.*, 1986). Thus, both rhodopsin kinase and 48-kDa protein acting sequentially are essential for the inactivation of PDE. ATP is also known to quench light-activated PDE activity, presumably through the ATP-requiring protein kinase.

Slow hydrolysis of GTP to GDP is mediated by the intrinsic GTPase activity of the α subunit of transducin. The GTPase activity of transducin has a turnover rate of about 30 s, which is too slow to account for the 4-s ATP-induced deactivation of PDE (Fung, 1987; Hurley, 1987). If the deactivation of R* by phosphorylation (described above) were the only effect of ATP, then activated transducin would be present for the lifetime of bound GTP (about 30 s). It is therefore assumed that there are other (as yet unknown) mechanisms that inactivate transducin.

In summary, visual transduction in vertebrate rod outer segments is a complex process initiated by photon capture by rhodopsin, the major integral membrane protein of photoreceptor disks. This results in rapid closing of the Na^+ channels and hyperpolarization of the plasma membrane. It is cGMP and not Ca^{2+} that carries the signal from the rod outer segment disk to the plasma membrane. Phototransduction is perceived as a cascade of events in which T_α-GTP couples the light-activated form of rhodopsin (R*) to the stimulation of cGMP phosphodiesterase. Hydrolysis of cGMP causes a drop in cGMP levels and a closing of the Na^+ channels in the plasma membrane; this leads to hyperpolarization of the membrane and transmission of the visual signal. Termination of the cascade appears to involve several mechanisms: ATP-dependent quenching of PDE, sequential phosphorylation of R* by rhodopsin kinase and binding by 48-kDa protein, and slow hydrolysis of GTP by T_α-GTPase.

7.8.3. Phosphoinositide Turnover and G Protein in Invertebrate Retinas

Despite some similarities to the vertebrate retina, phototransduction in the invertebrate retina is nevertheless distinctly different. A major difference is that invertebrate photoreceptors use inositol 1,4,5-trisphosphate (IP_3) as the internal messenger in phototransduction in contrast to vertebrate receptors, which utilize cGMP. As described below, the intracellular level of IP_3 in a variety of invertebrate retinas is regulated by the light-induced breakdown of phosphatidylinositol 4,5-bisphosphate (PIP_2); moreover, IP_3 administered by pressure injection or other means into invertebrate photoreceptors evokes rapid and specific electrophysiological responses (see reviews by Anderson and Brown, 1988; Selinger and Minke, 1988). A description of receptor-stimulated phosphoinositide (PI) breakdown, a major transmembrane signaling system in a wide variety of tissues, is given in Chapter 1, Section 1.3.2, and the overall process of PI turnover is shown schematically in Fig. 1.6.

Early studies on the *Limulus* eye showed that pressure injection of IP_3 into rhabdomeral lobes isolated from dark-adapted ventral photoreceptors mimics the effect of light in inducing both excitation and adaptation (Fein *et al.*, 1984). Phosphoinositide turnover is increased by illumination, and a concomitant rise in intracellular levels of IP_3 correlates with depolarization responses by the cell (Brown *et al.*, 1984). In the squid retina, a bright light flash causes a rapid and large (240%) increase in IP_3 when compared with dark levels, and, similar to *Limulus*, the light-induced rise in IP_3 correlates with excitation and adaptation responses (Szuts *et al.*, 1986).

The physiological effects of IP_3 in *Limulus* ventral photoreceptors appear to be mediated through mobilization of intracellular stores of Ca^{2+}, the rise being sufficiently rapid to induce light adaptation (Brown and Rubin, 1984; Payne *et al.*, 1986). Increased intracellular levels of Ca^{2+} also excite ventral photoreceptors of *Limulus*; nevertheless, another as yet unidentified messenger may also be required for this excitation (see review by Fein, 1986).

Recent attention in invertebrate phototransduction has focused on the coupling of photoexcited rhodopsin to the activation of G protein and stimulation of phospholipase C. The G protein of vertebrate retina, transducin, has been thoroughly characterized, and a similar light-regulated protein has also been described in squid (Vandenberg and Montal, 1984; Saibil and Michel-Villaz, 1984), *Musca* flies (Blumenfeld *et al.*, 1985), and octopus (Tsuda *et al.*, 1986). The binding of the hydrolysis-resistant GTP analogue Gpp(NH)p as well as GTPase activity in squid photoreceptor membranes are enhanced tenfold and three- to fourfold, respectively, after illumination (Vandenberg and Montal, 1984). These findings, together with cholera toxin-catalyzed labeling of a 44-kDa protein, provide strong evidence for G protein involvement in phototransduction in squid photoreceptors. Moreover, the first stage of interaction between activated rhodopsin (R*) and a 46-kDa protein that bears strong sequence homology with the α subunit of bovine transducin has also been demonstrated in squid photoreceptors (Saibil and Michel-Villaz, 1984). The 41-kDa α subunit of octopus photoreceptor G protein is specifically ADP-ribosylated by pertussis toxin, and a 36-kDa protein from octopus cross-reacts with anti-bovine $T_{\beta\gamma}$; since the β subunits of G proteins are broadly distributed in nature and are highly conserved, it appears that the 36-kDa protein of octopus photoreceptors corresponds to vertebrate T_{β} (Tsuda *et al.*, 1986). The effector enzyme activated by G protein of *Musca* and *Drosophila* flies (Devary *et al.*, 1987) and squid (Baer and Saibil, 1988) is phospholipase C, which stimulates the hydrolysis of PIP_2; this results in a rapid generation of IP_3 and depolarization of the cell.

In summary, the first steps in phototransduction in vertebrate and invertebrate photoreceptors are similar and consist of activation of G protein by photoexcited rhodopsin. However, the G proteins are coupled to different effector enzymes: in vertebrates, G protein (transducin) activates phosphodiesterase, leading to the cGMP cascade; in invertebrates, G protein mediates the activation of phospholipase C, leading to the breakdown of PIP_2, production of IP_3, release of Ca^{2+}, and stimulation of electrical responses.

7.9. RETINAL DEGENERATIONS

Retinal degeneration, either acquired or inherited, is a major cause of visual impairment and blindness in humans. Considerable progress has been made in understanding clinical disorders such as retinitis pigmentosa and gyrate atrophy (see Section 7.9.2), but at the same time in recent years increasing attention has been focused on animal models with inherited or inducible retinal degenerations. Although their specific human counterparts have not yet been established, animal models provide an abundant source of material for studying the molecular basis of retinal degeneration; to this end, histological, cell biological, ultrastructural, immunocytochemical, biochemical, and molecular genetic techniques are being used to maximum advantage. Five animal models with inherited retinal degeneration and two with taurine-depletion-induced degeneration have been chosen for discussion. Retinal degenerations have also been observed in the miniature poodle, Norwegian elk hound, and Abyssinian cat, but biochemical defects have not yet been established.

7.9.1. Animal Models

Of the 10 or 20 well-characterized retinal degenerations in animals, there is considerable genetic and biochemical information on five in particular. One is the RCS rat, in which the major genetic defect is expressed in the pigment epithelium; in the others, two rat models and two canine models, the mutant genes have specific effects on early postnatal development of the neural retina. Their common denominator is defective cGMP metabolism in the photoreceptor cell. Several excellent reviews, symposia, and texts on animal models provide detailed information on this topic (LaVail, 1979, 1981; Chader *et al.*, 1985, 1987; Farber and Shuster, 1986; Hollyfield *et al.*, 1987; Sanyal, 1987).

7.9.1.1. RCS (Royal College of Surgeons) Rat

One of the earliest recognized animal models of photoreceptor degeneration, this dystrophy is inherited as an autosomal recessive trait. The RCS rats used today in most laboratories are inbred descendents of the original strain reared at the Royal College of Surgeons (see review by LaVail, 1981). The morphological hallmark of the RCS rat retinal dystrophy is the accumulation of disordered lamellar debris in the subretinal space beginning at about the 12th postnatal day (Dowling and Sidman, 1962), the consequence of impaired phagocytosis of ROS apical disks by the retinal pigment epithelium (RPE) (Dowling and Sidman, 1962; Herron *et al.*, 1969; Bok and Hall, 1971). Visual cell death occurs after postnatal day 60. Although the cause of the retinal degeneration is still unknown, a classic study using chimeric rats showed that the RCS gene resides in the retinal pigment epithelium (RPE) and not in the neural retina (Mullen and LaVail, 1976). Histological and ultrastructural changes occurring during the course of the retinal degeneration in the RCS rat have been discussed in detail by LaVail (1979).

The impaired phagocytic capability of the RPE demonstrated *in vivo* is also expressed in cultured cells (Edwards and Szamier, 1977); moreover, the use of double immunofluorescent staining to distinguish between externally bound ROS and internalized ROS shows that although binding is normal, the ingestion phase is impaired in dystrophic RPE cells (Chaitin and Hall, 1983b). This defect cannot be attributed to abnormalities in the cytoskeletal system of RPE cells from dystrophic retinas, since both microfilaments (actin) (Chaitin and Hall, 1983a) and microtubules (tubulin) (Irons and Kalnins, 1984) appear to be functionally normal.

Abnormalities in ligand–receptor interactions and/or transmembrane signaling may be involved in the phagocytic defect in the RCS rat (see review by Clark, 1986). Techniques such as lectin binding and labeling of plasma membrane molecules with radioactive sugar precursors have been used to detect and partially characterize several plasma membrane glycoproteins in normal RPE (see Section 7.4.3.1). Application of these techniques to RCS plasma membranes suggests abnormalities in the structure and/or composition of certain cell surface glycoconjugates. For example, incomplete glycosylation of a major 183-kDa plasma membrane glycoprotein has been observed in cultured dystrophic RPE cells (Clark and Hall, 1986). This may be the same protein as the 175-kDa glycoprotein that binds Con A and WGA in normal RPE but binds little or no WGA in dystrophic RPE (Cooper *et al.*, 1987). An 86-kDa glycoprotein in dystrophic RPE shows similar diminished binding to Con A and WGA under conditions where its normal counterpart avidly binds both of these lectins. Although sialoglycoconjugates have been considered as putative receptor sites in normal RPE plasma membrane, a recent study has failed to detect sialic acid binding sites in either normal or dystrophic RPE (McLaughlin and Boykins, 1987). Abnormalities in cell surface receptors that recognize both soluble and particulate mannose ligands have also been found in dystrophic RPE (Tarnowski *et al.*, 1988b). Thus, a wide array of plasma membrane defects are now being detected in the dystrophic RPE, but it is not yet clear which, if any, is directly related to, or primarily responsible for, the impaired phagocytic capability of the RPE cell.

Other biochemical and structural abnormalities have also been noted in the dystrophic RPE. For example, the relative amount of phosphatidylethanolamine in RPE membranes of 9- to 14-day-old RCS rats is significantly higher than that of controls (Batey *et al.*, 1986). Other studies have shown that plasma membrane-enriched fractions from dystrophic RPE have higher levels of docosahexaenoic acid (22:6) and lower levels of arachidonic acid (20:4) than normal rats (Braunagel *et al.*, 1988). Freeze–fracture ultrastructural studies of the dystrophic rat show a breakdown in the tight junctions between RPE cells, a change that is associated with an increased formation of filipin–and digitonin–sterol complexes (Caldwell, 1987). These findings implicate alterations in cholesterol metabolism in RPE membranes of the dystrophic rat and possibly also in the neural retina (Organisciak *et al.*, 1982). The increased RPE cholesterol may be derived from the lamellar debris that accumulates in the interphotoreceptor space of the RCS rat.

An abnormal distribution of proteoglycans and/or other glycoconjugates in the interphotoreceptor matrix (IPM) of the RCS retina has been detected histochemically by the use of cationic stains such as Alcian blue or colloidal iron (LaVail *et al.*, 1981, 1985; Porrello and LaVail, 1986a; Porrello *et al.*, 1986) as well as immunogold (Porrello *et al.*, 1989). The major differences between RCS and control rats are (1) absence of histochemical staining on the apical surface of dystrophic RPE in the posterior region of the retina and (2) a concomitant increase in stainable material in the basal region of the IPM (at the junction of the inner and outer segments) and on the apical surface of RPE cells in the far peripheral areas. These changes occur 6–8 days before signs of photoreceptor degeneration are detectable. The IPM glycoconjugates have been identified histochemically before and after digestion with specific GAG-degrading enzymes (Porrello *et al.*, 1986) and by the use of monoclonal antibodies directed against chondroitin sulfate proteoglycans (Porrello and LaVail, 1986a). The excessive GAGs that accumulate in the basal IPM of the RCS rat consist mainly of chondroitin 6-sulfate proteoglycan. Chondroitin sulfates are not found in the peripheral apical zone, a region that contains sialoglycoconjugates consisting principally of interphotoreceptor retinol-binding protein (IRBP); the level of this glycolipoprotein reaches a peak in the RCS rat at about the 18th (Eisenfeld *et al.*, 1985) or 22nd (Gonzalez-Fernandez *et al.*, 1985) postnatal day but is not detectable after postnatal day 40.

These findings correlate with the spatial and temporal course of the retinal degeneration and suggest that the lamellar debris that accumulates in the subretinal space may partially block IRBP from access to the RPE apical surface, thus preventing efficient retinol transport in the IPM (Gonzalez-Fernandez *et al.*, 1985). In this context, the defect in retinol-esterifying enzyme activity in the RPE of RCS rats (Berman *et al.*, 1981) could provide a potential source of excess retinol and/or lack of substrate for isomerizing enzyme, and the absence of functional IRBP could exacerbate or hasten the photoreceptor degeneration.

It has long been recognized that photoreceptor degeneration is slower in RCS rats reared in darkness than in cyclic illumination. However, dark rearing only delays the time at which visual cell death occurs but does not prevent it. Recently a promising new strategy has been developed that may restore functional activity in the RCS retina. Transplantation of pigmented RPE cells from healthy rats into the subretinal space of 26-day-old pink-eyed RCS rats appears to "rescue" the photoreceptor cells near the site of transplantation from degeneration (Li and Turner, 1988; Sheedlo *et al.*, 1989). When examined at day 60, a time at which most of the visual cells in untreated RCS retina would have degenerated, large areas of intact outer and inner segments were observed in the grafted regions. The "rescued" retinas show normal distribution of, and immunostaining for, Na^+,K^+-ATPase and opsin 3 months after transplantation. Studies using a somewhat similar technique show that phagocytosis is restored 48 h after transplantation; phagosomes as well as melanin granules are present in the transplanted areas but are absent in other regions of the host RPE (Lopez *et al.*, 1989). Further studies on the functional status of the RPE–photoreceptor complex are needed for a more complete evaluation of these interesting observations.

7.9.1.2. The *rd* (Retinal Dystrophy) Mouse

The *rd* retinal degeneration is inherited as an autosomal recessive trait; the mutant gene, located on chromosome 5, is expressed in the photoreceptor cells early in postnatal development (for reviews see Farber and Shuster, 1986; Sanyal, 1987; Chader *et al.*, 1987). Whereas the eyes of normal mice are open and the retina functional by about the third postnatal week, in homozygous *rd* mutant mice, photoreceptors begin to form but do not reach maturity. The earliest signs of degeneration are at postnatal day 8, when vacuolar inclusions suddenly appear in the inner segments; pycnotic nuclei begin to accumulate by about day 10, and shortly afterward there is a massive loss of rod photoreceptor cells resulting in reduction of the outer nuclear layer. By day 20, all of the rods have degenerated; although the cones survive at this stage, they later degenerate, albeit at a much slower rate than rod photoreceptors.

Elevated levels of cGMP are present in the *rd* retina from about the sixth to the eighth postnatal days, before there are overt signs of visual cell degeneration (Farber and Lolley, 1974). Peak levels of cGMP are reached at about postnatal day 14, but afterward there is a steep decline, and by day 20, when the visual cells have disappeared, cGMP is barely detectable. The initial rise in cGMP correlates with a deficiency of rod-specific cGMP phosphodiesterase (PDE) activity (Farber and Lolley, 1976). As discussed in Section 7.8.2, the activation of PDE in normal vertebrate rods is coupled to photolyzed rhodopsin (R*) through G protein (transducin), and cGMP is the chemical messenger in the phototransduction cascade. The PDE of the *rd* mutant is only barely activated by light, in contrast to the normal enzyme.

Various components in the activation of PDE in the *rd* mutant have been examined; for example, rhodopsin phosphorylation is substantially reduced *in vitro*, but this may not be the principal site of the lesion in *rd* mutants (Shuster and Farber, 1986). In other studies, a series of cDNA probes have been used to compare corresponding mRNAs in 8-

to 11-day-old *rd* mutants and in normal age-matched controls (Bowes and Farber, 1987). Probes for transducin (α, β, and γ subunits), 48-kDa protein, and opsin show that there are no differences in either size or intensity of the mRNA transcripts in *rd* and control mice. Moreover, use of a C-terminal-derived opsin probe showed no differences in mRNAs hybridized by *rd* and control retinas, suggesting that the C-terminal phosphorylation sites of opsin in the *rd* mutant are normal. Immunologic studies also show that the transducin subunits in the *rd* mutant are of normal composition (Navon *et al.*, 1987).

Further studies on the expression of mRNAs coding for opsin, 48-kDa protein, and the α, β, and γ subunits of transducin during retinal development show that all are detectable before postnatal day 8 in both *rd* and normal retinas (Bowes *et al.*, 1988). The levels of these transcripts increase in both the normal and *rd* mutant until about the 11th or 12th postnatal day, but afterward in the *rd* mutant there is a sharp decline that parallels the photoreceptor cell degeneration. In contrast, the levels of mRNA transcripts in control retinas continue to increase until maturity.

The investigations described above did not reveal any obvious abnormalities in the three major proteins involved in the light-activated cascade that controls PDE activity. There is, however, increasing evidence suggesting that abnormalities in structure and/or amount of photoreceptor-specific cGMP PDE comprise the principal biochemical lesion in the *rd* retina. The PDE complex in retina is a heterotrimer composed of three subunits, α, β and γ. Removal of the 13-kDa γ subunit by interaction with T_{α}-GTP is the crucial step in the activation of PDE. The solubilized enzyme from frog or bovine ROS shows very little activity, but it can be activated by various agents. Among those tested, histone is particularly effective in activating rat, bovine, toad, and C57 murine ROS PDE; moreover, histone activation appears to be a unique characteristic of the photoreceptor enzyme in normal retinas (Lee *et al.*, 1985). However, histone does not activate the PDE of 7- to 12-day-old *rd* mutants whose visual cells are still viable. Other findings pointing to an absence or abnormality in the PDE complex in the *rd* retina are failure to cross-react with antibody to PDE and abnormal subunit size and/or structure. Further studies using polyclonal antibodies raised against the PDE complex of bovine ROS show that although PDE-immunoreactive polypeptides are present in small amounts in immature *rd* photoreceptors, they form a complex of M_r 105,000 instead of the normal 170-kDa heterotrimeric protein (Lee *et al.*, 1988b). The abnormal complex is derived from affected rod photoreceptors and not from cones, and may represent a pool of precursors that fail to form the complete heterotrimer.

Use of a newly developed highly sensitive radioimmunoassay to detect PDE shows that the enzyme is synthesized at similar rates in *rd* and control retinas for about the first 6–8 postnatal days (Farber *et al.*, 1988). Commencing at about the 10th to 12th postnatal days, the critical period in photoreceptor development, PDE concentration in the *rd* retina declines to an undetectable level while that in controls rises sharply to maximum concentrations. The reasons for similar rates of PDE synthesis in normal and *rd* retina, but lower content in the *rd* mutant, are not clear but may be related to abnormalities in composition and/or structure of the enzyme. Cloning, sequencing, and mapping of the inhibitory γ subunit of mouse cGMP PDE shows 91% homology with the coding regions of the bovine retinal enzyme; moreover, the deduced amino acid sequence of this subunit from *rd* retina has 100% homology with the normal mouse subunit and 97.7% homology with the bovine subunit (Tuteja *et al.*, 1989). In other studies, the cDNA for the γ subunit of mouse cGMP PDE has been mapped to chromosome 11 (Danciger *et al.*, 1989).

However, since linkage analyses have assigned the genetic locus of *rd* degeneration to chromosome 5, the molecular defect in this mutant mouse is not in the γ subunit of cGMP PDE; it may reside in either the α or β subunit of the enzyme, whose chromosomal localizations have not yet been established.

7.9.1.3. The *rds* (Slow Retinal Dystrophy) Mouse

The *rds* gene is located on chromosome 17, and the most distinctive phenotypic feature in mutant rats homozygous for this gene is the absence of photoreceptor outer segments throughout the postnatal period (for reviews see Sanyal, 1987; Chader *et al.*, 1987). In the most extensively studied inbred strain of *rds* mice, designated 020/A, a connecting cilium that protrudes from the inner segments is present, but disk structures typical of outer segments never form. Other retinal layers appear to be normal for about the first 2–3 postnatal weeks; this is followed by a slowly progressive loss of photoreceptors and significantly reduced thickness of the outer nuclear layer. By about 1 year of age, virtually all of the photoreceptor cells have disappeared.

Despite the absence of outer segments in the *rds* retina, a light-induced reduction in the levels of cyclic nucleotides has been demonstrated in 21-day-old mice (Cohen, 1983). The concentrations of cGMP in light-adapted retinas of 7- and 20- to 25-day-old *rds* mice are consistently low when compared with normal controls; moreover, the activities of both cGMP PDE and guanylate cyclase are greatly reduced (Fletcher *et al.*, 1986). Of interest is the observation that in congenic mice doubly homozygous for both the *rd* and *rds* genes (*rd*/*rd*,*rds*/*rds*), there is an initial rise in cGMP in the early postnatal period (as in the *rd* mutant) followed by a slower reduction that parallels the gradual loss of the outer nuclear layer (Chader *et al.*, 1987). In view of the absence of outer segments in the *rd*/*rd*,*rds*/*rds* retina, the unexpected finding of elevated cGMP levels suggests that outer segments are not necessary for the production of high levels of cGMP; therefore, in the doubly homozygous mutant, this cyclic nucleotide must be originating from other regions of the photoreceptor cell.

The *rds* retina lacks outer segments, yet photoreceptors do contain small amounts of photosensitive pigment, and an electrical response of low amplitude is evocable by light. Thus, the components of phototransduction appear to be present, albeit at greatly reduced levels. Small amounts of opsin, not detectable by conventional techniques (Cohen, 1983), can be demonstrated by immunocytochemical assays in the plasma membranes of both the inner segments and the connecting cilium (Nir and Papermaster, 1986; Jansen *et al.*, 1987; Usukura and Bok, 1987). The opsin destined for outer segment formation appears to be retained in the inner segment plasma membranes of the *rds* mutant. Opsin levels reach a maximum at about the 15th postnatal day but then decline to undetectable levels by 30 days. High concentrations of immunoreactive opsin in the distal ends of the ciliary protrusions and at receptor ends of surviving cells suggest that this may be the site of phototransduction in the rudimentary retinas of the *rds* mutant (Jansen *et al.*, 1987).

In the normal retina, interphotoreceptor retinoid-binding protein (IRBP) is synthesized in the photoreceptor cell and secreted into the interphotoreceptor matrix (IPM), as discussed in Section 7.2.2.3. This retinoid-transport glycolipoprotein is present at early stages of retinal development, and its secretion appears to commence after outer segment elongation. It is therefore of interest that in the *rds* mutants and in various genotypes of *rd* and *rds* mice, as in normal mice, IRBP is detectable intracellularly by immunocytochemical methods (van Veen *et al.*, 1988). However, in *rds* mutants, its secretion into the IPM, although normal in early stages, is significantly impaired in later stages of the photoreceptor degeneration. Recent studies on colloidal iron staining of the IPM of chimeric *rds* mice (consisting of mutant *rds,rds* and normal genotypes) show that the IPM in regions adjacent to *rds,rds* photoreceptors stains more intensely than that of the normal genotype (Sanyal *et al.*, 1988). These and other findings on toluidine blue staining patterns in the IPM suggest that the *rds* gene is expressed only within the photoreceptor cells.

7.9.1.4. Irish Setter

Rod–cone dysplasia in Irish setter dogs is a hereditary degeneration characterized by failure of photoreceptor cell differentiation from about the 16th postnatal day. The photoreceptors remain undeveloped and ultimately degenerate by adulthood. There is a tenfold increase in cGMP content in the retinas of affected dogs that is associated with a marked deficiency of cGMP PDE activity (Aguirre *et al.*, 1978); these changes precede morphological signs of retinal degeneration (Aguirre *et al.*, 1982). Unlike the *rd* mutation in mice, the phosphodiesterase complex of the dystrophic retina of the Irish setter appears normal by several criteria, including histone activation (Lee *et al.*, 1985). Nevertheless, the cGMP PDE of the Irish setter appears to be catalytically inactive, or possibly it is not activated *in situ* through the normal light-induced cGMP cascade.

Not only is there a general rise in cGMP concentration in the photoreceptor cells of the Irish setter, but there is also a particularly striking abnormality in its distribution (Barbehenn *et al.*, 1988). The levels of cGMP are 29 times higher in the outer plexiform layer of the Irish setter retina than in age-matched (20-day-old) control retinas. At a later stage, there is a shift of unusually high cGMP levels to the outer nuclear layer. At all ages examined, guanylate cyclase activity is normal. These findings raise the possibility that abnormally high local concentrations of cGMP in the inner retina during the early crucial stages of neuronal differentiation could adversely affect later development of normal photoreceptor cells in the Irish setter.

7.9.1.5. Collie

Retinal degeneration in a strain of collies is characterized by functional abnormalities in the ERG, recognizable by about the second postnatal week. Histological and ultrastructural studies reveal abnormal photoreceptor cells at this age. Most of the photoreceptor nuclei and outer segments disappear by about 6 weeks of age, and blindness follows a few months later.

The levels of cGMP are about ten times higher in the retinas of collies affected with this dysplasia than in age-matched controls (Woodford *et al.*, 1982). The elevated levels persist from about the 15th to the 45th postnatal days; at this time, most of the photoreceptor cells have disappeared, and the levels of cGMP drop sharply, as in the *rd* mouse mutant. However, cGMP PDE activity in retinal homogenates of the affected collie is only about 25% lower than that of normal collie retina. Ultrastructural histochemical studies show that the abnormally low PDE activity is associated mainly with degenerating outer segments. In contrast, considerable enzyme activity is detectable in normal-appearing outer segments. Although elevated cGMP levels are present in both the collie and the Irish setter retinas, the retinopathies in the two canine models are not of common etiology, since PDE activity does not seem to be impaired in the collie retina.

7.9.1.6. Taurine Deficiency in Cat and Rat

Taurine, the most abundant amino acid in the retina in all species examined (Table 7.2), is actively synthesized from precursor amino acids in the inner retina (see review by Pasantes-Morales, 1986). Its highest concentration is in the photoreceptors. Curiously, however, this pool of taurine does not arise from the inner retina; rather, taurine is avidly transported into the retina from exogenous sources, mainly if not entirely from the blood. Taurine uptake is mediated by specific high-affinity receptors localized in the photoreceptor layer (Pasantes-Morales, 1986). A high and constant pool of taurine is essential for maintaining the structural integrity of the retina, and if taurine is depleted (as described below), a rapid photoreceptor degeneration ensues. Taurine is present in the retina as the free amino acid; it is not incorporated into proteins. Although its physiological function has yet to be established, a role as a neurotransmitter has been excluded by several criteria.

Two models of retinal degeneration are associated with taurine deficiency: one can be induced in cats (Hayes *et al.*, 1975) and in monkeys (Neuringer and Sturman, 1987) maintained on taurine-free diets, and the other is inducible in rats treated with guanidinoethyl sulfonate (GES), a structural analogue of taurine that inhibits taurine uptake by the retina and causes a 50–70% drop in its concentration. The retinal degenerations are similar; there is an initial loss of ERG amplitude, and ultrastructural studies show disruption of photoreceptor structure and retinal dysfunction leading to blindness.

A light-stimulated release of taurine from the retina that appears to be localized in the photoreceptors has been described in several species (Schmidt, 1978; Pasantes-Morales, 1986). Recent work suggests that environmental light accelerates the photoreceptor degeneration in taurine-deficient animals, although the findings are somewhat ambiguous. For example, ERG changes are less prominent in taurine-depleted albino rats reared in darkness than in those maintained in dim cyclic light, but pigmented rats in which taurine levels had been reduced by 63% after GES treatment are no more susceptible to light-induced photoreceptor damage than non-GES-treated controls (Rapp *et al.*, 1987). Further studies on taurine-depleted albino rats exposed to either dim (2 lx) or bright (300 lx) cyclic light show that both loss of photoreceptor cells and reduction in ERG response are greater in animals reared under brighter lighting conditions (Rapp *et al.*, 1988). The effect appears to be a synergistic one, and the results imply that reduced levels of taurine in the retina increase their susceptibility to light damage. Other studies on GES-treated albino rats reared either in total darkness or in cyclic light of approximately 160 lx showed that ERG changes were somewhat more pronounced in the cyclic-light-reared animals (Quesada *et al.*, 1988). These and other findings support the view that taurine plays a role in maintaining the structural integrity of the photoreceptors, and if this amino acid is depleted, even physiological levels of illumination may exacerbate the photoreceptor degeneration.

7.9.2. Human Disorders

7.9.2.1. Retinitis Pigmentosa

Retinitis pigmentosa (RP) is a generic term applied to a heterogeneous group of inherited disorders characterized initially by night blindness and later by progressive loss of peripheral visual fields. These functional changes parallel a generalized retinal degeneration that differentially affects the rod and cone photoreceptors. In most cases rod photoreceptors are the first affected, whereas cone cells often survive for long periods. Migration of pigment cells and macrophages into the inner retinal layers is a common occurrence. The onset and rate of degeneration in specific regions of the retina are highly variable among affected individuals.

Although pigmentary retinopathy may be a component of certain systemic or multisystem syndromes, most attention is currently being focused on RP localized to the eye. These types of RP may be transmitted as autosomal dominant, autosomal recessive, or X-linked traits. In general, the autosomal dominant and X-linked modes of inheritance are readily identified, but autosomal recessive inheritance can be ambiguous unless there is parental consanguinity. A certain percentage of presumed autosomal recessive cases are probably sporadic (or simplex), but there is considerable controversy on this question. Retinitis pigmentosa (RP) is in fact now regarded as a family of disorders with similar, though not identical, clinical signs. This implies that a wide range of biochemical defects could give rise to a similar clinical picture. Two general approaches are being used to gain some insight into specific biochemical changes in various forms of RP: one involves postmortem analyses of donor eyes from RP patients, and the other, analyses of blood samples from RP patients.

The glycosaminoglycans (GAGs) secreted into the medium by cultured retinal pigment epithelium (RPE) from an RP patient appear to be similar in size and composition to those of normal controls (Edwards, 1982b), although there may be differences in their relative rates of synthesis (Yue and Fishman, 1985). The proteoglycans of Bruch's membrane from a patient with autosomal dominant RP were larger than those from normal donor eyes and, in addition, contained a higher proportion of heparan sulfate proteoglycan (Hewitt and Newsome, 1985). The RPE is a highly polarized cell, and proteoglycans secreted into the medium probably correspond to interphotoreceptor matrix substances *in vivo*; however, those adhering to the cell membrane represent a different pool of GAGs. The cell-associated proteoglycans in cultured RPE cells from RP donor eyes differ in both molecular size and GAG composition from normal controls (Hewitt and Newsome, 1988). Although not considered to be a primary defect in RP or a change that would be expected in all forms of RP, alterations in cell-associated proteoglycans of RPE cells could contribute to the altered physiological responses of the retina and RPE in retinitis pigmentosa.

Interphotoreceptor retinoid-binding protein (IRBP) is virtually absent in donor eyes of autosomal dominant RP (Rodrigues *et al.*, 1985) and autosomal recessive RP (Bridges *et al.*, 1985a). Both cases examined were in advanced stages of degeneration, although some photoreceptors may still have been present. Whether the absence of IRBP is a primary or secondary event in the pathogenesis of RP remains to be established.

In the most comprehensive study reported to date on RP donor eyes, immunochemical methods were used to evaluate the levels of opsin, α-transducin (T_{α}), 48-kDa protein, cathepsin D, and IRBP in neural retina and in the IPM of RP donor eyes and in 22 normal controls (Schmidt *et al.*, 1988). Homogenates from five RP retinas had less than normal amounts of immunoreactive opsin, 48-kDa protein, T_α, and IRBP, and IPM from six other RP retinas had little or no detectable IRBP compared to normal controls. In contrast, cathepsin D, a predominantly RPE-localized enzyme, was present at comparable levels in both RP and control retinas. These findings imply that the photoreceptor-specific and IPM-specific proteins detected in RP retinas, although present in reduced amounts, probably originate from surviving photoreceptors. However, whether these proteins or the cGMP enzyme cascade are fully functional in the RP retina is still an open question.

In an advanced case of autosomal dominant RP, the cGMP levels were about the same as normal in the macular area but considerably lower in midzone and peripheral areas (Rodrigues *et al.*, 1985). Other studies have been made on cGMP levels and cGMP PDE activity throughout the retina of a 17-year-old RP donor (Farber *et al.*, 1987). Levels of cGMP were generally lower than those in an age-matched control in all areas of the retina, but in those regions where photoreceptor cells could still be detected, higher than normal levels of cGMP were present. There were also striking decreases in PDE activity throughout the RP retina, findings similar in some respects to the abnormalities found in *rd* mice.

Normal photoreceptor outer segments have an unusually high content of polyunsaturated fatty acids, and diets deficient in docosahexaenoic acid (22:6) or its precursors lead to visual impairment in experimental animals (see Section 7.2.1.2). The possibility of defects in the transport and/or metabolism of polyunsaturated fatty acids has been examined in several types of human RP. In one study, the plasma levels of 22:6 in four out of five X-linked RP patients were considerably lower than those in nonaffected family members, and in three of the families, reduced levels of arachidonic acid (20:4) were also found (Converse *et al.*, 1987). Similarly, in Usher's syndrome, the levels of these two unsaturated fatty acids in plasma phospholipids are lower than in normal controls (Bazan and Scott, 1987). Other studies have been performed on the fatty acids of total plasma lipids in four patient groups: controls, dominant RP, X-linked RP, and simplex RP (Anderson *et al.*, 1987). Statistical analyses of the measured values showed that the ratio of 22:6/16:0 is consistently lower in the dominant RP group than in controls. There were, however, no differences in the ratio of 20:4/16:0 among the four groups. Differences in plasma levels of polyunsaturated fatty acids of RP patients compared to normal controls did not appear to be related to dietary habits or life style. Therefore, the possibility of abnormalities in the metabolism of polyunsaturated fatty acids in some forms of RP is an attractive hypothesis requiring further investigation.

7.9.2.2. Gyrate Atrophy

Gyrate atrophy of the choroid and retina (GA), a rare inborn error of metabolism transmitted as an autosomal recessive trait, has been recognized as a disease entity for more than a century (see review by Kaiser-Kupfer and Valle, 1986). The characteristic clinical signs in GA are myopia and night blindness, which develop during the first decade of life, and posterior subcapsular cataracts, which appear a decade or two later. The name GA derives from the ophthalmoscopically unique chorioretinal atrophy that is recognizable in childhood; the disorder leads to blindness in most patients, although there is considerable variability in the rate of progression of the dystrophy.

The finding of an association between hyperornithinemia and GA (Takki and Simell,

1974) later led to the discovery of the basic enzymatic defect, a deficiency of the mitochondrial matrix enzyme ornithine-δ-aminotransferase (OAT) (Valle *et al.*, 1977). Ornithine plays a key role in the urea cycle (Fig. 7.20), and most pathways of its metabolism are well known. One of the enzymes, OAT (reaction 5 in Fig. 7.20), is a pyridoxal phosphate-requiring Ω-transaminase that catalyzes the reversible interconversion of ornithine and α-ketoglutarate to Δ′-pyrroline-5-carboxylate (P5C) and glutamate. The equilibrium is determined by the concentrations of the reacting species in various tissues. This enzyme is a 180-kDa oligomeric polypeptide consisting of four to six identical subunits; it has been purified to homogeneity from various tissues including human liver.

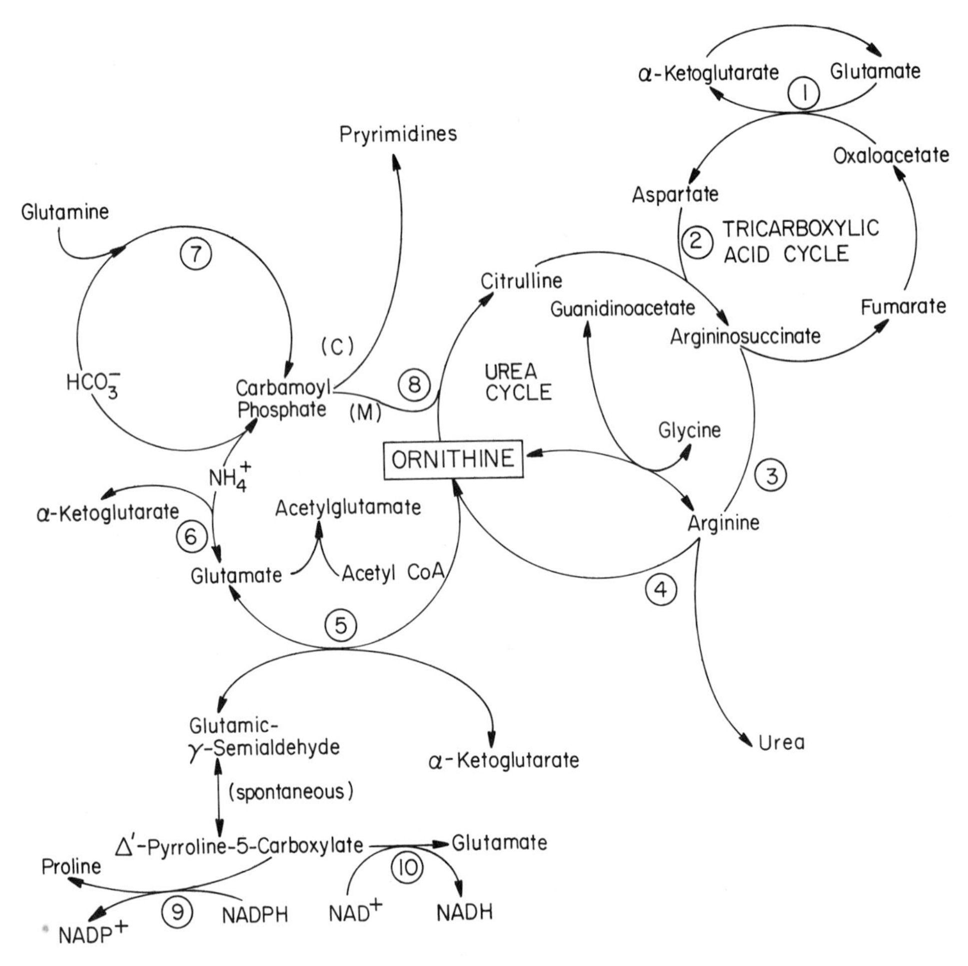

Figure 7.20. Principal reactions in the urea cycle and in ornithine metabolism. The enzymes shown in encircled numbers are: (1) glutamate–oxaloacetate transaminase; (2) argininosuccinate synthetase; (3) argininosuccinate lyase; (4) arginase; (5) ornithine aminotransferase (OAT); (6) glutamate dehydrogenase; (7) carbamoyl phosphate synthetase II; (8) ornithine carbamoyl transferase; (9) Δ′-pyrroline carboxylate reductase; and (10) Δ′-pyrroline carboxylate dehydrogenase. C and M designate cytosolic and mitochondrial carbamoyl phosphate, respectively. Three of the enzymes in the urea cycle (2, 3, and 4) are cytosolic, and two (7 and 8) are localized in the mitochondrial matrix.

The defect in OAT activity is expressed in many tissues and cell types from GA patients, including cultured skin fibroblasts. However, the deficiency is often not a total one, since varying levels of enzyme activity have been detected in about one-third of the GA patients examined. Moreover, the activity *in vitro* can in some cases be enhanced by the addition of high concentrations of pyridoxal phosphate (vitamin B_6) (see review by Kaiser-Kupfer and Valle, 1986). Pharmacological doses of the vitamin to GA patients have resulted in lowering of plasma ornithine levels as well as improvements in electrophysiological responses in the retina in some cases. Arginine-restricted diets are also beneficial, but the effectiveness of these therapies in arresting the progression of the dystrophy remains to be established. As additional cases of GA are reported, clinical heterogeneity has become more apparent. This is now best explained in terms of the OAT gene sequences and the amount(s) of OAT protein detectable in fibroblasts of GA patients.

By means of a cDNA clone for the mRNA encoding human OAT, the OAT gene sequences have been mapped, using a panel of human–mouse somatic cell hybrids and *in situ* hybridization, to two human chromosomes, 10q26 and Xp11.2 (Barrett *et al.*, 1987). As described below, it is probably the OAT gene on chromosome 10 that is abnormal in GA; the significance of OAT-like sequences on the X chromosome and their relationship to the OAT deficiency in GA, an autosomal recessive disorder, are unclear. However, the results imply the existence of a gene family consisting of multiple copies of the OAT gene. Also of interest was the finding that the X chromosome gene sequences map to the same region as a marker closely related to X-linked retinitis pigmentosa. In other studies, the OAT cDNA together with anti-human OAT antibody have been used as probes to examine the OAT gene, mRNA, and protein in 14 GA patients (Inana *et al.*, 1988). Defects in all three were found in one of the patients, the first real demonstration of the OAT defect in GA at the gene level. However, the other cases showed normal OAT gene sequences and mRNA but reduced and variable amounts of immunoreactive OAT protein. These findings raise the possibility that most cases of GA involve a point mutation rather than gross deletions or rearrangements. Moreover, the clinical heterogeneity with respect to levels of detectable OAT activity in tissues of GA patients, responsiveness to vitamin B_6 therapy, and rate of progression to blindness are expressions of the variable amounts of OAT protein produced by each patient.

Further studies using human OAT cDNA have more precisely delineated the mutation(s) occurring in GA. Mitchell and co-workers (1988) coupled the techniques of ribonuclease A protection and polymerase chain reaction (PCR) to define the point mutation(s) in two GA sibling pairs of Lebanese Maronite descent. They showed that in these patients, but not in 12 others of different ethnic backgrounds, there is a probe–target mismatch at the 5′ end of the first coding exon. Sequence analyses showed that the initiator codon in these patients is changed from AT*G* (methionine) to AT*A*, a point mutation that would truncate the OAT enzyme by 138 amino acids and eliminate the entire mitochondrial leader sequence as well as 113 amino acids of the mature polypeptide. Another study in which the cDNAs for OAT have been cloned and sequenced from two GA patients, one B6 responsive and the other nonresponsive, provides direct proof that molecular defects in the gene are responsible for both forms of GA (Ramesh *et al.*, 1988). Both cases represent missense mutations in the gene that potentially alter the conformation of the mature protein. The B6-responsive phenotype is associated with a transition mutation in which valine-332 in the OAT gene is replaced with methionine. Lysine-292 is the critical residue that binds to pyridoxal phosphate, and the replacement of valine with methionine at residue 332 may interfere with, but does not completely abolish, the binding of the vitamin to the enzyme. These mutations, as well as those in the Lebanese GA patients, suggest that the clinical and biochemical heterogeneity in GA is the consequence of diverse mutations at the OAT locus.

More recent investigations may even open the way to therapeutic possibilities. A mammalian expression clone containing human OAT cDNA (pHOAT) corresponding to the OAT gene on chromosome 10 was prepared, transfected, and shown to be expressed in NIH3T3 cells and in Chinese hamster ovary cells (Hotta and Inana, 1989). The synthesis of human OAT mRNAs and active enzyme was demonstrated in both cell types. These findings thus demonstrate that the OAT gene on chromosome 10 is the functional OAT gene responsible for the expression of active OAT enzyme, and replacement therapy in GA patients is not an unrealistic possibility for the future.

7.10. RETINOBLASTOMA: BIOCHEMICAL STUDIES ON CELL LINE Y-79

Retinoblastoma is a childhood tumor of either hereditary or nonhereditary origin; the locus for both types is on chromosome 13q14. Cell lines have been established from these tumors, the most thoroughly investigated being the Y-79 type. These cells are of neuroectodermal origin and relatively undifferentiated. Until recently most biochemical studies had been carried out using suspension cultures of Y-79 cells; however, under defined culture conditions and in the presence of appropriate substrata, Y-79 cells can be made to attach to the substratum and to grow as monolayers (see Section 7.10.4). Treatment with specific modifying agents shows that Y-79 cells have the characteristics of multipotential cells capable of differentiating into neurons, photoreceptors, glia, and pigment epithelial cells (Kyritsis *et al.*, 1986a, 1987b; Chader, 1987).

7.10.1. Retinoid-Binding Proteins

Receptors (or binding proteins) for exogenous labeled retinol and retinoic acid have been detected in soluble extracts of Y-79 cells (Wiggert *et al.*, 1977; Saari *et al.*, 1978b). Their properties are similar to those of cytosol retinol-binding protein (CRBP) and cytosol retinoic acid-binding protein (CRABP) found in bovine and other mammalian retina cytosols (see Table 7.3). Recent studies on surgically removed retinoblastomas and on several tumor cell lines including the Y-79 cell line showed considerable variability among specimens in the binding of either labeled retinol or retinoic acid (Fong *et al.*, 1988). The average levels of CRBP in tumor cells are similar to those found in cytosols of adult human retinas, but the levels of CRABP are considerably lower and often not detectable. Cellular retinal-binding protein (CRALBP), as measured immunochemically, was not detectable in any of the retinoblastoma tumors.

Interphotoreceptor retinol-binding protein (IRBP) is synthesized and secreted by retinal photoreceptors; it is the major glycoprotein of the interphotoreceptor matrix (IPM) in a wide variety of vertebrate species (see Section 7.3.3). This glycolipoprotein has been detected by immunochemical methods in retinoblastomas (Bridges *et al.*, 1985b; Fong *et al.*, 1988). It is found mainly in the medium of dispersed undifferentiated tumor cells, with very little being present in the cell cytosol. These findings, as well as measurements of IRBP mRNA transcripts in tumor cell lines, point to active synthesis and secretion of

IRBP by fresh tumor cells. Other studies on surgically removed retinoblastomas confirm the presence of IRBP and, moreover, suggest a correlation between the degree of tumor differentiation and the amount of immunocytochemically detectable IRBP (Rodrigues *et al.*, 1987a). In contrast to fresh tumors, little if any IRBP is secreted by the Y-79 cell line, either in dispersed cells (Fong *et al.*, 1988) or in those grown as monolayers on poly-D-lysine-coated plastic dishes (Kyritsis *et al.*, 1985). However, synthesis and secretion of IRBP into the medium can be stimulated by pretreatment of Y-79 cells with butyrate, an agent with potent activity in inducing partial differentiation of subpopulations of Y-79 cells.

7.10.2. Insulin and IGF-I Receptors

The first step in the action of insulin in tissues is binding to its membrane receptor. Hence, their characterization is important. Insulin and IGF-I receptors are present in both neural retina and in its microvasculature; these receptors, corresponding to approximately 125,000 binding sites per cell, have also been detected in cultured Y-79 cells (Yorek *et al.*, 1985). Unlike insulin receptors in some neural tissues, those in Y-79 cell lines appear to be down-regulated after prolonged exposure to insulin. Further kinetic studies of cultured Y-79 cells using [^{125}I]insulin show that they are true insulin receptors; negative cooperativity was not found; therefore, a two-site model has been proposed (Saviolakis *et al.*, 1986). The first has a high affinity, and the second a low affinity, for insulin. The latter could include sites for IGF-I. Additional studies have confirmed the two-site binding system for insulin and a one-site, one-affinity system for IGF-I in retinoblastoma cells (Yorek *et al.*, 1987). Both insulin and IGF-I enhance the uptake of glycine in cultured suspensions of Y-79 cells by a factor of 25–50% but have no effect on the accumulation of other amino acids or other putative neurotransmitters. The function of insulin and IGF-I receptors in Y-79 cells is not known, but since they are multipotential cells, the presence of these receptors may be an expression of their neuronal characteristics.

7.10.3. Cyclic-AMP-Dependent Protein Kinases

Cyclic-AMP-dependent protein kinases are ubiquitous in mammalian cells and tissues, where they play key roles in the regulation of cell metabolism and in gene expression. The enzymes are tetramers composed of two catalytic (C) and two regulatory (R) subunits; two isotypes of cAMP-dependent protein kinases are distinguishable on the basis of differing regulatory subunits, RI and RII. In normal developing retina, the type I enzyme predominates in fetal tissue, and only the type II enzyme is present in mature retina (see Gentleman *et al.*, 1989). The tissue-specific effects of cAMP thus appear to be regulated by the enzyme isotype present.

Suspensions of Y-79 retinoblastoma cells contain both type I and type II cAMP-dependent protein kinases and, in addition, significant amounts of free type I regulatory subunit (RI) (Gentleman *et al.*, 1989). Studies of poly(A)$^+$-enriched mRNA from Y-79 cells showed both a 4.5-kb RI transcript and a 6.0-kb RII transcript, the former being typical of normal retina. The ratio of R to C subunits in normal tissues is 1:1 but may be higher in certain transformed cell lines. These authors speculate that the suppression of translation of the RI transcript is lost in Y-79 cells, thus leading to the synthesis of RI dimers in excess of catalytic (C) subunits, an imbalance that could contribute to the unregulated growth of these cells.

7.10.4. Attached (Monolayer) Y-79 Cell Cultures

The Y-79 cells established from human retinoblastoma are customarily grown in suspension cultures. Serum is required to promote rapid growth, but the cells are poorly differentiated both morphologically and biochemically. However, culture conditions can be manipulated by (1) plating on adhesive substrata such as poly-D-lysine and either laminin or fibronectin, (2) using defined culture media, and (3) treating with agents such as butyrate or dibutyryl cAMP (db-cAMP) (for reviews see Chader, 1987; Kyritsis *et al.*, 1987b). Under these conditions or a specific combination of them, Y-79 cells adhere to the substratum and grow as monolayers. There is compelling evidence that these conditions promote partial differentiation of subpopulations of Y-79 cells into cells with morphological and/or biochemical characteristics of neurons, photoreceptors, glia, and pigment epithelial cells.

Undifferentiated Y-79 cells synthesize very little IRBP, but if they are plated on poly-D-lysine-coated dishes and cultured in Eagle's MEM, the cells grow as monolayers, and butyrate causes a dramatic increase in the synthesis of IRBP (Kyritsis *et al.*, 1985). Further evidence that Y-79 cells have the potential for partial differentiation into photoreceptor-like cells is the marked increase of immunoreactive hydroxyindole-O-methyltransferase (HIOMT) in cells grown in attachment culture (Kyritsis *et al.*, 1987a). This enzyme, which catalyzes the final step in the synthesis of melatonin (see Fig. 7.10), is localized mainly in retinal photoreceptors. Although it is barely detectable in Y-79 cells growing in suspension cultures, at least 80% of cells grown in attachment culture have immunologically detectable amounts of HIOMT.

Firm attachment and morphological differentiation of a large percentage of initially seeded cells can be achieved by plating in serum-free medium containing laminin; replacement of laminin with either butyrate or db-cAMP promotes the differentiation of a subpopulation of Y-79 cells into pigment epithelial-like cells (Kyritsis *et al.*, 1986a). Cells treated in this manner have little or no immunoreactive IRBP or opsin, specific markers for photoreceptors. Other studies have shown that laminin, in preference to fibronectin, is a most effective agent in promoting attachment and inducing morphological changes in Y-79 cells (Campbell and Chader, 1988). Laminin is not synthesized by Y-79 cells, but it is present in the inner limiting membrane of mature retina. Thus, laminin may have an important role in neuronal differentiation, not only in Y-79 cells but also in the normal retina *in vivo*.

The effects of agents such as butyrate, db-cAMP, 8-bromo cAMP, and retinoic acid on attachment cultures of Y-79 cells are most strikingly illustrated by the changes they induce in mRNA-translatable proteins (Kapoor *et al.*, 1985). Butyrate in particular induces an intracellular translocation of type II cAMP-dependent protein kinase, the enzyme being measured by immunofluorescent localization of its regulatory subunit, RII (Kyritsis *et al.*, 1986b). In the exponential phase of monolayer growth of Y-79 cells, type II cAMP-dependent protein kinase is localized in Golgi bodies of the cell cytoplasm, but after 3–5 days' exposure to butyrate, the enzyme is translocated to the cell nucleus in many—though not all—of the Y-79 cells. Further studies are needed to clarify the mechanism(s) by which butyrate and other agents are able to promote differentiation and alter gene expression in cultures of attached Y-79 cells.

7.11. REFERENCES

Adler, A. J., and Evans, C. D., 1985a, Proteins of the bovine interphotoreceptor matrix: Retinoid binding and other functions, in: *The Interphotoreceptor Matrix in Health and Disease* (C. D. Bridges and A. J. Adler, eds.), Alan R. Liss, New York, pp. 65–88.

Adler, A. J., and Evans, C. D., 1985b, Some functional characteristics of purified bovine interphotoreceptor retinol-binding protein, *Invest. Ophthalmol. Vis. Sci.* **26:**273–282.

Adler, A. J., and Klucznik, K. M., 1982, Proteins and glycoproteins of the bovine interphotoreceptor matrix: Composition and fractionation, *Exp. Eye Res.* **34:**423–434.

Adler, A. J., and Martin, K. J., 1982, Retinol-binding proteins in bovine interphotoreceptor matrix, *Biochem. Biophys. Res. Commun.* **108:**1601–1608.

Adler, A. J., and Martin, K. J., 1982/1983, Lysosomal enzymes in the interphotoreceptor matrix: Acid protease, *Curr. Eye Res.* **2:**359–366.

Adler, A. J., and Severin, K. M., 1981, Proteins of the bovine interphotoreceptor matrix: Tissues of origin, *Exp. Eye Res.* **32:**755–769.

Adler, A. J., Evans, C. D., and Stafford, W. F. III, 1985, Molecular properties of bovine interphotoreceptor retinol-binding protein, *J. Biol. Chem.* **260:**4850–4855.

Aguirre, G., Farber, D., Lolley, R., Fletcher, R. T., and Chader, G. J., 1978, Rod–cone dysplasia in Irish setters: A defect in cyclic GMP metabolism in visual cells, *Science* **201:**1133–1134.

Aguirre, G., Farber, D., Lolley, R., O'Brien, P., Alligood, J., Fletcher, R. T., and Chader, G., 1982, Retinal degenerations in the dog. III. Abnormal cyclic nucleotide metabolism in rod–cone dysplasia, *Exp. Eye Res.* **35:**625–642.

Ahmad, H., Singh, S. V., Medh, R. D., Ansari, G. A. S., Kurosky, A., and Awasthi, Y. C., 1988, Differential expression of α, μ and π classes of isozymes of glutathione S-transferase in bovine lens, cornea, and retina, *Arch. Biochem. Biophys.* **266:**416–426.

Akagi, Y., Kador, P. F., Kuwabara, T., and Kinoshita, J. H., 1983, Aldose reductase localization in human retinal mural cells, *Invest. Ophthalmol. Vis. Sci.* **24:**1516–1519.

Akagi, Y., Yajima, Y., Kador, P. F., Kuwabara, T., and Kinoshita, J. H., 1984, Localization of aldose reductase in the human eye, *Diabetes* **33:**562–566.

Akagi, Y., Terubayashi, H., Millen, J., Kador, P. F., and Kinoshita, J. H., 1986, Aldose reductase localization in dog retinal mural cells, *Curr. Eye Res.* **5:**883–886.

Allen, L. A., and Gerritsen, M. E., 1986, Regulation of hexose transport in cultured bovine retinal microvessel endothelium by insulin, *Exp. Eye Res.* **43:**679–686.

Alvarez, R. A., Bridges, C. D. B., and Fong, S.-L., 1981, High-pressure liquid chromatography of fatty acid esters of retinol isomers. Analysis of retinyl esters stored in the eye, *Invest. Ophthalmol. Vis. Sci.* **20:**304–312.

Alvarez, R. A., Liou, G. I., Fong, S.-L., and Bridges, C. D. B., 1987, Levels of α- and γ-tocopherol in human eyes: Evaluation of the possible role of IRBP in intraocular α-tocopherol transport, *Am. J. Clin. Nutr.* **46:**481–487.

Ames, A., III, and Gurian, B. S., 1963a, Effects of glucose and oxygen deprivation on function of isolated mammalian retina, *J. Neurophysiol.* **26:**617–634.

Ames, A., III, and Gurian, B. S., 1963b, Electrical recordings from isolated mammalian retina mounted as a membrane, *Arch. Ophthalmol.* **70:**837–841.

Ames, A., III, and Nesbett, F. B., 1981, *In vitro* retina as an experimental model of the central nervous system, *J. Neurochem.* **37:**867–877.

Ames, A., III, Parks, J. M., and Nesbett, F. B., 1980a, Protein turnover in retina, *J. Neurochem.* **35:**131–142.

Ames, A., III, Parks, J. M., and Nesbett, F. B., 1980b, Synthesis and degradation of retinal proteins in darkness and during photic stimulation, *J. Neurochem.* **35:**143–148.

Anderson, D. H., Fisher, S. K., and Breding, D. J., 1986a, A concentration of fucosylated glycoconjugates at the base of cone outer segments: Quantitative electron microscope autoradiography, *Exp. Eye Res.* **42:**267–283.

Anderson, D. H., Neitz, J., Saari, J. C., Kaska, D. D., Fenwick, J., Jacobs, G. H., and Fisher, S. K., 1986b, Retinoid-binding proteins in cone-dominant retinas, *Invest. Ophthalmol. Vis. Sci.* **27:**1015–1026.

Anderson, R. E., 1983a, Chemistry of photoreceptor outer segments, in: *Biochemistry of the Eye* (R. E. Anderson, ed.), American Academy of Ophthalmology, San Francisco, pp. 164–170.

Anderson, R. E., 1983b, Synthesis and turnover of photoreceptor outer segments, in: *Biochemistry of the Eye*, (R. E. Anderson, ed.), American Academy of Ophthalmology, San Francisco, pp. 171–177.

Anderson, R. E., and Brown, J. E., 1988, Phosphoinositides in the retina, *Prog. Retinal Res.* **8:**211–228.

Anderson, R. E., and Hollyfield, J. G., 1981, Light stimulates the incorporation of inositol into phosphatidylinositol in the retina, *Biochim. Biophys. Acta* **665:**619–622.

Anderson, R. E., and Hollyfield, J. G., 1984, Inositol incorporation into phosphoinositides in retinal horizontal cells of *Xenopus laevis*: Enhancement by acetylcholine, inhibition by glycine, *J. Cell Biol.* **99:**686–691.

Anderson, R. E., and Kelleher, P. A., 1981, Biosynthesis of retinal phospholipids by base exchange reactions, *Exp. Eye Res.* **32:**729–736.

Anderson, R. E., and Maude, M. B., 1972, Lipids of ocular tissues. VIII. The effects of essential fatty acid deficiency on the phospholipids of the photoreceptor membranes of rat retina, *Arch. Biochem. Biophys.* **151:**270–276.

Anderson, R. E., and Rapp, L. M., 1983, Visual pigment dynamics and the vitamin A cycle, in: *Biochemistry of the Eye* (R. E. Anderson, ed.), American Academy of Ophthalmology, San Francisco, pp. 189–195.

Anderson, R. E., Feldman, L. S., and Feldman, G. L., 1970, Lipids of ocular tissues. II. The phospholipids of mature bovine and rabbit whole retina, *Biochim. Biophys. Acta* **202:**367–373.

Anderson, R. E., Maude, M. B., and Zimmerman, W., 1975, Lipids of ocular tissues. X. Lipid composition of subcellular fractions of bovine retina, *Vision Res.* **15:**1087–1090.

Anderson, R. E., Lissandrello, P. M., Maude, M. B., and Matthes, M. T., 1976, Lipids of bovine retinal pigment epithelium, *Exp. Eye Res.* **23:**149–157.

Anderson, R. E., Kelleher, P. A., Maude, M. B., and Maida, T. M., 1980a, Synthesis and turnover of lipid and protein components of frog retinal rod outer segments, *Neurochem. Int.* **1:**29–42.

Anderson, R. E., Kelleher, P. A., and Maude, M. B., 1980b, Metabolism of phosphatidylethanolamine in the frog retina, *Biochim. Biophys. Acta* **620:**227–235.

Anderson, R. E., Maude, M. B., Kelleher, P. A., Maida, T. M., and Basinger, S. F., 1980c, Metabolism of phosphatidylcholine in the frog retina, *Biochim. Biophys. Acta* **620:**212–226.

Anderson, R. E., Maude, M. B., and Kelleher, P. A., 1980d, Metabolism of phosphatidylinositol in the frog retina, *Biochim. Biophys. Acta* **620:**236–246.

Anderson, R. E., Maude, M. B., Kelleher, P. A., Rayborn, M. E., and Hollyfield, J. G., 1983, Phosphoinositide metabolism in the retina: Localization to horizontal cells and regulation by light and divalent cations, *J. Neurochem.* **41:**764–771.

Anderson, R. E., Rapp, L. M., and Wiegand, R. D., 1984, Lipid peroxidation and retinal degeneration, *Curr. Eye Res.* **3:**223–227.

Anderson, R. E., Maude, M. B., Pu, G. A.-W., and Hollyfield, J. G., 1985a, Effect of light on the metabolism of lipids in the rat retina, *J. Neurochem.* **44:**773–778.

Anderson, R. E., Maude, M. B., and Nielsen, J. C., 1985b, Effect of lipid peroxidation on rhodopsin regeneration, *Curr. Eye Res.* **4:**65–71.

Anderson, R. E., Maude, M. B., Lewis, R. A., Newsome, D. A., and Fishman, G. A., 1987, Abnormal plasma levels of polyunsaturated fatty acid in autosomal dominant retinitis pigmentosa, *Exp. Eye Res.* **44:**155–159.

Applebury, M. L., 1987, Biochemical puzzles about the cyclic- GMP-dependent channel, *Nature* **326:**546–547.

Applebury, M. L., and Hargrave, P. A., 1986, Molecular biology of the visual pigments, *Vision Res.* **26:**1881–1895.

Atalla, L., Fernandez, M. A., and Rao, N. A., 1987, Immunohistochemical localization of catalase in ocular tissue, *Curr. Eye Res.* **6:**1181–1187.

Aveldano, M. I., 1987, A novel group of very long chain polyenoic fatty acids in dipolyunsaturated phosphatidylcholines from vertebrate retina, *J. Biol. Chem.* **262:**1172–1179.

Aveldano de Caldironi, M. I., and Bazan, N. G., 1980, Composition and biosynthesis of molecular species of retina phosphoglycerides, *Neurochem. Int.* **1:**381–392.

Aveldano, M. I., and Bazan, N. G., 1983, Molecular species of phosphatidylcholine, -ethanolamine, -serine, and -inositol in microsomal and photoreceptor membranes of bovine retina, *J. Lipid Res.* **24:**620–627.

Aveldano, M. I., and Sprecher, H., 1987, Very long chain (C_{24} to $/_{36}$) polyenoic fatty acids of the *n*-3 and *n*-6 series in dipolyunsaturated phosphatidylcholines from bovine retina, *J. Biol. Chem.* **262:**1180–1186.

Aveldano, M. I., Pasquare de Garcia, S. J., and Bazan, N. G., 1983, Biosynthesis of molecular species of inositol, choline, serine, and ethanolamine glycerophospholipids in the bovine retina, *J. Lipid Res.* **24:**628–638.

Axelrod, J., Burch, R. M., and Jelsema, C. L., 1988, Receptor- mediated activation of phospholipase A_2 via GTP-binding proteins: Arachidonic acid and its metabolites as second messengers, *Trends Neurosci.* **11:**117–123.

Bach, G., and Berman, E. R., 1971a, Amino sugar-containing compounds of the retina. I. Isolation and identification, *Biochim. Biophys. Acta* **252:**453–461.

Bach, G., and Berman, E. R., 1971b, Amino sugar-containing compounds of the retina. II. Structural studies, *Biochim. Biophys. Acta* **252:**462–471.

Baehr, W., and Applebury, M. L., 1986, Exploring visual transduction with recombinant DNA techniques, *Trends Neurosci.* **9:**198–203.

Baehr, W., Devlin, M. J., and Applebury, M. L., 1979, Isolation and characterization of cGMP phosphodiesterase from bovine rod outer segments, *J. Biol. Chem.* **254:**11669–11677.

Baehr, W., Morita, E. A., Swanson, R. J., and Applebury, M. L., 1982, Characterization of bovine rod outer segment G-protein, *J. Biol. Chem.* **257:**6452–6460.

Baer, K. M., and Saibil, H. R., 1988, Light- and GTP-activated hydrolysis of phosphatidylinositol bisphosphate in squid photoreceptor membranes, *J. Biol. Chem.* **263:**17–20.

Baker, B. N., Moriya, M., Maude, M. B., Anderson, R. E., and Williams, T. P., 1986, Oil droplets of the retinal epithelium of the rat, *Exp. Eye Res.* **42:**547–557.

Barbehenn, E. K., Wiggert, B., Lee, L., Kapoor, C. L., Zonnenberg, B. A., Redmond, T. M., Passonneau, J. V., and Chader, G. J., 1985, Extracellular cGMP phosphodiesterase related to the rod outer segment phosphodiesterase isolated from bovine and monkey retinas, *Biochemistry* **24:**1309–1316.

Barbehenn, E., Gagnon, C., Noelker, D., Aguirre, G., and Chader, G. J., 1988, Inherited rod–cone dysplasia: Abnormal distribution of cyclic GMP in visual cells of affected Irish setters, *Exp. Eye Res.* **46:**149–159.

Barrett, D. J., Redmond, T. M., Wiggert, B., Oprian, D. D., Chader, G. J., and Nickerson, J. M., 1985, cDNA clones encoding bovine interphotoreceptor retinoid binding protein, *Biochem. Biophys. Res. Commun.* **131:**1086–1093.

Barrett, D. J., Bateman, J. B., Sparkes, R. S., Mohandas, T., Klisak, I., and Inana, G., 1987, Chromosomal localization of human ornithine aminotransferase gene sequences to 10q26 and Xp11.2, *Invest. Ophthalmol. Vis. Sci.* **28:**1037–1042.

Basinger, S. F., and Hoffman, R. T., 1983, Biochemistry of the pigment epithelium, in: *Biochemistry of the Eye* (R. E. Anderson, ed.), American Academy of Ophthalmology, San Francisco, pp. 256–264.

Basinger, S., Hoffman, R., and Matthes, M., 1976, Photoreceptor shedding is initiated by light in the frog retina, *Science* **194:**1074–1076.

Basu, P. K., Sarkar, P., Menon, I., Carre, F., and Persad, S., 1983, Bovine retinal pigment epithelial cells cultured *in vitro*: Growth characteristics, morphology, chromosomes, phagocytosis ability, tyrosinase activity and effect of freezing, *Exp. Eye Res.* **36:**671–683.

Batey, D. W., Mead, J. F., and Eckhert, C. D., 1986, Lipids of the retinal pigment epithelium in RCS dystrophic and normal rats, *Exp. Eye Res.* **43:**751–757.

Baylor, D. A., 1987, Photoreceptor signals and vision, *Invest. Ophthalmol. Vis. Sci.* **28:**34–49.

Bazan, H. E. P., and Bazan, N. G., 1976, Phospholipid composition and [^{14}C]glycerol incorporation into glycerolipids of toad retina and brain, *J. Neurochem.* **27:**1051–1057.

Bazan, H. E. P., and Bazan, N. G., 1977, Effects of temperature, ionic environment, and light flashes on the glycerolipid neosynthesis in the toad retina, in: *Function and Biosynthesis of Lipids* (N. G. Bazan, R. R. Brenner, and N. M. Giusto, eds.), Plenum Press, New York, pp. 489–495.

Bazan, H. E. P., Careaga, M. M., Sprecher, H., and Bazan, N. G., 1982, Chain elongation and desaturation of eicosapentaenoate to docosahexaenoate and phosholipid labeling in the rat retina in vivo, *Biochim. Biophys. Acta* **712:**123–128.

Bazan, H. E. P., Sprecher, H., and Bazan, N. G., 1984, *De novo* biosynthesis of docosahexaenoyl-phosphatidic acid in bovine retinal microsomes, *Biochim. Biophys. Acta* **796:**11–19.

Bazan, N. G., and Reddy, T. S., 1985, Retina, in: *Handbook of Neurochemistry*, Vol. 8 (A. Lajtha, ed.), Plenum Press, pp. 507–575. Bazan, N. G., and Scott, B. L., 1987, Docosahexaenoic acid metabolism and inherited retinal degeneration, in: *Degenerative Retinal Disorders: Clinical and Laboratory Investigations* (J. G. Hollyfield, R. E. Anderson, and M. M. LaVail, eds.), Alan R. Liss, New York, pp. 103–118.

Bazan, N. G., Ilincheta de Boschero, M. G., and Giusto, N. M., 1977, Neobiosynthesis of phosphatidylinositol and of other glycerolipids in the entire cattle retina, *Adv. Exp. Med. Biol.* **83:**377–388.

Bazan, N. G., Reddy, T. S., Redmond, T. M., Wiggert, B., and Chader, G. J., 1985, Endogenous fatty acids are covalently and noncovalently bound to interphotoreceptor retinoid-binding protein in the monkey retina, *J. Biol. Chem.* **260:**13677–13680.

Bazan, N. G., Reddy, T. S., Bazan, H. E. P., and Birkle, D. L., 1986, Metabolism of arachidonic and docosahexaenoic acids in the retina, *Prog. Lipid Res.* **25:**595–606.

Bazan, N. G., Bazan, H. E. P., Birkle, D. L., and Rossowska, M., 1987, Synthesis of leukotrienes in frog retina and retinal pigment epithelium, *J. Neurosci. Res.* **18:**591–596.

Benolken, R. M., Anderson, R. E., and Wheeler, T. G., 1973, Membrane fatty acids associated with the electrical response in visual excitation, *Science* **182:**1253–1254.

Bensinger, R. E., Crabb, J. W., and Johnson, C. M., 1982, Purification and properties of superoxide dismutase from bovine retina, *Exp. Eye Res.* **34:**623–634.

Berger, S. J., DeVries, G. W., Carter, J. G., Schulz, D. W., Passonneau, P. N., Lowry, O. H., and Ferrendelli, J. A., 1980, The distribution of the components of the cyclic GMP cycle in retina, *J. Biol. Chem.* **255:**3128–3133.

Berman, E. R., 1964, The biosynthesis of mucopolysaccharides and glycoproteins in pigment epithelial cells of bovine retina, *Biochim. Biophys. Acta* **83:**371–373.

Berman, E. R., 1969, Mucopolysaccharides (glycosaminoglycans) of the retina: Identification, distribution and possible biological role, *Mod. Probl. Ophthalmol.* **8:**5–31.

Berman, E. R., 1971, Acid hydrolases of the retinal pigment epithelium, *Invest. Ophthalmol.* **10:**64–68.

Berman, E. R., 1979, Biochemistry of the retinal pigment epithelium, in: *The Retinal Pigment Epithelium* (K. M. Zinn and M. F. Marmor, eds.), Harvard University Press, Cambridge, pp. 83–102.

Berman, E. R., 1982, Isolation and characterization of the interphotoreceptor matrix, in: *Methods in Enzymology*, Vol. 81 (L. Packer, ed.), Academic Press, New York, pp. 77–85.

Berman, E. R., 1985, An overview of the biochemistry of the interphotoreceptor matrix, in: *The Interphotoreceptor Matrix in Health and Disease* (C. D. Bridges and A. J. Adler, eds.), Alan R. Liss, New York, pp. 47–64.

Berman, E. R., 1987, Does vitamin E have a protective role in the retina as an anti-oxidant and free radical scavenger? in: *Ocular Circulation and Neovascularization* (D. BenEzra, J. Ryan, B. M. Glaser, and R. P. Murphy, eds.), Martinus Nijhoff/Dr. W. Junk, Dordrecht, pp. 163–167.

Berman, E. R., and Bach, G., 1968, The acid mucopolysaccharides of cattle retina, *Biochem. J.* **108:**75–88.

Berman, E. R., Schwell, H., and Feeney, L., 1974, The retinal pigment epithelium. Chemical composition and structure, *Invest. Ophthalmol.* **13:**675–687.

Berman, E. R., Segal, N., and Feeney, L., 1979, Subcellular distribution of free and esterified forms of vitamin A in the pigment epithelium of the retina and in liver, *Biochim. Biophys. Acta* **572:**167–177.

Berman, E. R., Horowitz, J., Segal, N., Fisher, S., and Feeney-Burns, L., 1980, Enzymatic esterification of vitamin A in the pigment epithelium of bovine retina, *Biochim. Biophys. Acta* **630:**36–46.

Berman, E. R., Segal, N., Photiou, S., Rothman, H., and Feeney-Burns, L., 1981, Inherited retinal dystrophy in RCS rats: A deficiency in vitamin A esterification in pigment epithelium, *Nature* **293:**217–220.

Berman, E. R., Segal, N., Rothman, H., and Weiner, A., 1985, Retinyl ester hydrolase of bovine retina and pigment epithelium: Comparisons to the rat liver enzyme, *Curr. Eye Res.* **4:**867–876.

Bernstein, P. S., Law, W. C., and Rando, R. R., 1987a, Isomerization of all-*trans*-retinoids to 11-*cis*-retinoids *in vitro*, *Proc. Natl. Acad. Sci. USA* **84:**1849–1853.

Bernstein, P. S., Law, W. C., and Rando, R. R., 1987b, Biochemical characterization of the retinoid isomerase system of the eye, *J. Biol. Chem.* **262:**16848–16857.

Besharse, J. C., 1986, Photosensitive membrane turnover: Differentiated membrane domains and cell–cell interaction, in: *The Retina. A Model for Cell Biology Studies*, Part I (R. Adler and D. Farber, eds.), Academic Press, Orlando, FL, pp. 297–352.

Besharse, J. C., and Dunis, D. A., 1983a, Rod photoreceptor disc shedding in eye cups: Relationship to bicarbonate and amino acids, *Exp. Eye Res.* **36:**567–580.

Besharse, J. C., and Dunis, D. A., 1983b, Methoxyindoles and photoreceptor metabolism: Activation of rod shedding, *Science* **219:**1341–1343.

Besharse, J. C., and Iuvone, P. M., 1983, Circadian clock in *Xenopus* eye controlling retinal serotonin N-acetyltransferase, *Nature* **305:**133–135.

Besharse, J. C., Dunis, D. A., and Iuvone, P. M., 1984, Regulation and possible role of serotonin *N*-acetyltransferase in the retina, *Fed. Proc.* **43:**2704–2708.

Besharse, J. C., Iuvone, P. M., and Pierce, M. E., 1988, Regulation of rhythmic photoreceptor metabolism: A role for post-receptoral neurons, *Prog. Retinal Res.* **7:**21–61.

Betz, A. L., Bowman, P. D., and Goldstein, G. W., 1983, Hexose transport in microvascular endothelial cells cultured from bovine retina, *Exp. Eye Res.* **36:**269–277.

Bhuyan, K. C., and Bhuyan, D. K., 1977, Regulation of hydrogen peroxide in eye humors. Effect of 3-amino-1H-1,2,4-triazole on catalase and glutathione peroxidase of rabbit eye, *Biochim. Biophys. Acta* **497:**641–651.

Bhuyan, K. C., and Bhuyan, D. K., 1978, Superoxide dismutase of the eye. Relative functions of superoxide dismutase and catalase in protecting the ocular lens from oxidative damage, *Biochim. Biophys. Acta* **542:**28–38.

Bibb, C., and Young, R. W., 1974a, Renewal of fatty acids in the membranes of visual cell outer segments, *J. Cell Biol.* **61:**327–343.

Bibb, C., and Young, R. W., 1974b, Renewal of glycerol in the visual cells and pigment epithelium of the frog retina, *J. Cell Biol.* **62:**378–389.

Birkle, D. L., and Bazan, N. G., 1984a, Effect of K^+ depolarization on the synthesis of prostaglandins and hydroxyeicosatetra(5,8,11,14)enoic acids (HETE) in the rat retina, *Biochim. Biophys. Acta* **795:**564–573.

Birkle, D. L., and Bazan, N. G., 1984b, Lipoxygenase- and cyclooxygenase-reaction products and incorporation into glycerolipids of radiolabeled arachidonic acid in the bovine retina, *Prostaglandins* **27:**203–216.

Birkle, D. L., and Bazan, N. G., 1985, Metabolism of arachidonic acid in the central nervous system. The enzymatic cyclooxygenation and lipoxygenation of arachidonic acid in the mammalian retina, in: *Phospholipids in the Nervous System*, Vol. 2: *Physiological Roles* (L. A. Horrocks, J. N. Kanfer, and G. Porcellati, eds.), Raven Press, New York, pp. 193–208.

Birkle, D. L., and Bazan, N. G., 1986, The arachidonic acid cascade and phospholipid and docosahexaenoic acid metabolism in the retina, *Prog. Retinal Res.* **5:**309–335.

Bitensky, M. W., Wheeler, G. L., Aloni, B., Vetury, S., and Matuo, Y., 1978, Light- and GTP-activated photoreceptor phosphodiesterase: Regulation by a light-activated GTPase and identification of rhodopsin as the phosphodiesterase binding site, *Adv. Cyclic Nucleotide Res.* **9:**553–572.

Blaner, W. S., Das, S. R., Gouras, P., and Floor, M. T., 1987, Hydrolysis of 11-*cis*- and all-*trans*-retinyl palmitate by homogenates of human retinal epithelial cells, *J. Biol. Chem.* **262:**53–58.

Blanks, J. C., and Johnson, L. V., 1984, Specific binding of peanut lectin to a class of retinal photoreceptor cells, *Invest. Ophthalmol. Vis. Sci.* **25:**546–557.

Blumenfeld, A., Erusalimsky, J., Heichal, O., Selinger, Z., and Minke, B., 1985, Light-activated guanosinetriphosphatase in *Musca* eye membranes resembles the prolonged depolarizing afterpotential in photoreceptor cells, *Proc. Natl. Acad. Sci. USA* **82:**7116–7120.

Bok, D., 1970, The distribution and renewal of RNA in retinal rods, *Invest. Ophthalmol.* **9:**516–523.

Bok, D., 1985, Retinal photoreceptor–pigment epithelium interactions, *Invest. Ophthalmol. Vis. Sci.* **26:**1659–1694.

Bok, D., and Hall, M. O., 1971, The role of the pigment epithelium in the etiology of inherited retinal dystrophy in the rat, *J. Cell Biol.* **49:**664–682.

Bok, D., and Heller, J., 1976, Transport of retinol from the blood to the retina: An autoradiographic study of the pigment epithelial cell surface receptor for plasma retinol-binding protein, *Exp. Eye Res.* **22:**395–402.

Bok, D., and Young, R. W., 1979, Phagocytic properties of the retinal pigment epithelium, in: *The Retinal Pigment Epithelium* (K. M. Zinn and M. F. Marmor, eds.), Harvard University Press, Cambridge, pp. 148–174.

Bok, D., Ong, D. E., and Chytil, F., 1984, Immunocytochemical localization of cellular retinol binding protein in the rat retina, *Invest. Ophthalmol. Vis. Sci.* **25:**877–883.

Bone, R. A., Landrum, J. T., and Tarsis, S. L., 1985, Preliminary identification of the human macular pigment, *Vision Res.* **25:**1531–1535.

Bone, R. A., Landrum, J. T., Fernandez, L., and Tarsis, S. L., 1988, Analysis of the macular pigment by HPLC: Retinal distribution and age study, *Invest. Ophthalmol. Vis. Sci.* **29:**843–849.

Borst, D. E., and Nickerson, J. M., 1988, The isolation of a gene encoding interphotoreceptor retinoid-binding protein, *Exp. Eye Res.* **47:**825–838.

Borst, D. E., Redmond, T. M., Elser, J. E., Gonda, M. A., Wiggert, B., Chader, G. J., and Nickerson, J. M., 1989, Interphotoreceptor retinoid-binding protein. Gene characterization, protein repeat structure, and its evolution, *J. Biol. Chem.* **264:**1115–1123.

Bowes, C., and Farber, D. B., 1987, mRNAs coding for proteins of the cGMP cascade in the degenerative retina of the *rd* mouse, *Exp. Eye Res.* **45:**467–480.

Bowes, C., van Veen, T., and Farber, D. B., 1988, Opsin, G-protein and 48-kDa protein in normal and *rd* mouse retinas: Developmental expression of mRNAS and proteins and light/dark cycling of mRNAs, *Exp. Eye Res.* **47**:369–390.

Brann, M. R., and Cohen, L. V., 1987, Diurnal expression of transducin mRNA and translocation of transducin in rods of rat retina, *Science* **235**:585–587.

Braunagel, S. C., Organisciak, D. T., and Wang, H.-M., 1985, Isolation of plasma membranes from the bovine retinal pigment epithelium, *Biochim. Biophys. Acta* **813**:183–194.

Braunagel, S. C., Organisciak, D. T., and Wang, H.-M., 1988, Characterization of pigment epithelial cell plasma membranes from normal and dystrophic rats, *Invest. Ophthalmol. Vis. Sci.* **29**:1066–1075.

Brecha, N., Johnson, D., Peichl, L., and Wassle, H., 1988, Cholinergic amacrine cells of the rabbit retina contain glutamate decarboxylase and γ-aminobutyrate immunoreactivity, *Proc. Natl. Acad. Sci. USA* **85**:6187–6191.

Bridges, C. D. B., 1975, Storage, distribution and utilization of vitamins A in the eyes of adult amphibians and their tadpoles, *Vision Res.* **15**:1311–1323.

Bridges, C. D. B., 1976, Vitamin A and the role of the pigment epithelium during bleaching and regeneration of rhodopsin in the frog eye, *Exp. Eye Res.* **22**:435–455.

Bridges, C. D. B., 1977, Rhodopsin regeneration in rod outer segments: Utilization of 11-*cis* retinal and retinol, *Exp. Eye Res.* **24**:571–580.

Bridges, C. D. B., 1981, Lectin receptors of rods and cones: Visualization by fluorescent label, *Invest. Ophthalmol. Vis. Sci.* **20**:8–16.

Bridges, C. D. B., 1984, Retinoids in photosensitive systems, in: *The Retinoids*, Vol. 2 (M. B. Sporn, A. B. Roberts, and D. S. Goodman, eds.), Academic Press, Orlando, FL pp. 125–176.

Bridges, C. D., and Adler, A. J. (eds.), 1985, *The Interphotoreceptor Matrix in Health and Disease*, Alan R. Liss, New York.

Bridges, C. D. B., Alvarez, R. A., and Fong, S.-L., 1982, Vitamin A in human eyes: Amount, distribution, and composition, *Invest. Ophthalmol. Vis. Sci.* **22**:706–714.

Bridges, C. D. B., Alvarez, R. A., Fong, S.-L., Gonzalez-Fernandez, F., Lam, D. M. K., and Liou, G. I., 1984, Visual cycle in the mammalian eye. Retinoid-binding proteins and the distribution of 11-*cis* retinoids, *Vision Res.* **24**:1581–1594.

Bridges, C. D. B., O'Gorman, S., Fong, S.-L., Alvarez, R. A., and Berson, E., 1985a, Vitamin A and interstitial retinol-binding protein in an eye with recessive retinitis pigmentosa, *Invest. Ophthalmol. Vis. Sci.* **26**:684–691.

Bridges, C. D. B., Fong, S.-L., Landers, R. A., Liou, G. I., and Font, R. L., 1985b, Interstitial retinol-binding protein (IRBP) in retinoblastoma, *Neurochem. Int.* **7**:875–881.

Bridges, C. D. B., Price, J., Landers, R. A., Fong, S.-L., Liou, G. I., Hong, B.-S., and Tsin, A. T. C., 1986a, Interstitial retinol-binding protein (IRBP) in subretinal fluid, *Invest. Ophthalmol. Vis. Sci.* **27**:1027–1030.

Bridges, C. D. B., Oka, M. S., Fong, S.-L., Liou, G. I., and Alvarez, R. A., 1986b, Retinoid-binding proteins and retinol esterification in cultured retinal pigment epithelium cells, *Neurochem. Int.* **8**:527–534.

Brown, J. E., and Rubin, L. J., 1984, A direct demonstration that inositol-*tris*phosphate induces an increase in intracellular calcium in *limulus* photoreceptors, *Biochem. Biophys. Res. Commun.* **125**:1137–1142.

Brown, J. E., Rubin, L. J., Ghalayini, A. J., Tarver, A. P., Irvine, R. F., Berridge, M. J., and Anderson, R. E., 1984, *myo*-nositol polyphosphate may be a messenger for visual excitation in *limulus* photoreceptors, *Nature* **311**:160–163.

Brown, J. E., Blazynski, C., and Cohen, A. I., 1987, Light induces a rapid and transient increase in inositol-trisphosphate in toad rod outer segments, *Biochem. Biophys Res. Commun.* **146**:1392–1396.

Bunt, A. H., and Klock, I. B., 1980, Comparative study of ^{3}H-fucose incorporation into vertebrate photoreceptor outer segments, *Vision. Res.* **20**:739–747.

Bunt, A. H., and Saari, J. C., 1982, Fucosylated protein of retinal cone photoreceptor outer segments: Morphological and biochemical analyses, *J. Cell Biol.* **92**:269–276.

Bunt-Milam, A. H., and Saari, J. C., 1983, Immunocytochemical localization of two retinoid-binding proteins in vertebrate retina, *J. Cell Biol.* **97**:703–712.

Burke, J. M., and Twining, S. S., 1988, Regional comparisons of cathepsin D activity in bovine retinal pigment epithelium, *Invest. Ophthalmol. Vis. Sci.* **29**:1789–1793.

Burnside, B., and Laties, A. M., 1976, Actin filaments in apical projections of the primate pigmented epithelial cell, *Invest. Ophthalmol.* **15**:570–575.

Burnside, M. B., 1976, Possible roles of microtubules and actin filaments in retinal pigmented epithelium, *Exp. Eye Res.* **23**:257–275.

Burton, G. W., and Ingold, K. U., 1984, β-Carotene: An unusual type of lipid antioxidant, *Science* **224:**569–573.

Cabral, L., Unger, W., Boulton, M., and Marshall, J., 1988, A microsystem to assay lysosomal enzyme activities in cultured retinal pigment epithelial cells, *Curr. Eye Res.* **7:**1097–1104.

Caldwell, R. B., 1987, Filipin and digitonin studies of cell membrane changes during junction breakdown in the dystrophic rat retinal pigment epithelium, *Curr. Eye Res.* **6:**515–526.

Campbell, M. A., and Chader, G. J., 1988, Effects of laminin on attachment, growth and differentiation of cultured Y-79 retinoblastoma cells, *Invest. Ophthalmol. Vis. Sci.* **29:**1517–1522.

Campochiaro, P. A., Jerdan, J. A., and Glaser, B. M., 1986, The extracellular matrix of human retinal pigment epithelial cells *in vivo* and its synthesis *in vitro*, *Invest. Ophthalmol. Vis. Sci.* **27:**1615–1621.

Caputto, R., de Maccioni, A. H. R., Maccioni, H. J. F., Caputto, B. L., and Landa, C. A., 1980, The gangliosides of the chicken retina and optic tectum, *Neurochem. Int.* **1:**43–57.

Casey, P. J., and Gilman, A. G., 1988, G protein involvement in receptor–effector coupling, *J. Biol. Chem.* **263:**2577–2580.

Casper, D. S., and Reif-Lehrer, L., 1983, Effects of alpha-aminoadipate isomers on the morphology of the isolated chick embryo retina, *Invest. Ophthalmol. Vis. Sci.* **24:**1480–1488.

Casper, D. S., Trelstad, R. L., and Reif-Lehrer, L., 1982, Glutamate-induced cellular injury in isolated chick embryo retina: Müller cell localization of initial effects, *J. Comp. Neurol.* **209:**79–90.

Chabre, M., 1985, Trigger and amplification mechanisms in visual phototransduction, *Annu. Rev. Biophys. Chem.* **14:**331–360.

Chader, G. J., 1982, Retinoids in ocular tissues: Binding proteins, transport, and mechanism of action, in: *Cell Biology of the Eye* (D. S. McDevitt, ed.), Academic Press, New York, pp. 377–433.

Chader, G. J., 1987, Multipotential differentiation of human Y-79 retinoblastoma cells in attachment culture, *Cell Differ.* **20:**209–216.

Chader, G. J., 1989, Interphotoreceptor retinoid-binding protein (IRBP): A model protein for molecular biological and clinically relevant studies, *Invest. Ophthalmol. Vis. Sci.* **30:**7–22.

Chader, G. J., Herz, L. R., and Fletcher, R. T., 1974, Light activation of phosphodiesterase activity in retinal rod outer segments, *Biochim. Biophys. Acta* **347:**491–493.

Chader, G. J., Fletcher, R. T., Sanyal, S., and Aguirre, G. D., 1985, A review of the role of cyclic GMP in neurological mutants with photoreceptor dysplasia, *Curr. Eye Res.* **4:**811–819.

Chader, G. J., Fletcher, R. T., Barbehenn, E., Aguirre, G., and Sanyal, S., 1987, Studies on abnormal cyclic CMP metabolism in animal models of retinal degeneration: Genetic relationships and cellular compartmentalization, in: *Degenerative Retinal Disorders: Clinical and Laboratory Investigations* (J. G. Hollyfield, R. E. Anderson, and M. M. LaVail, eds.), Alan R. Liss, New York, pp. 289–307.

Chaitin, M. H., and Hall, M. O., 1983a, The distribution of actin in cultured normal and dystrophic rat pigment epithelial cells during the phagocytosis of rod outer segments, *Invest. Ophthalmol. Vis. Sci.* **24:**821–831.

Chaitin, M. H., and Hall, M. O., 1983b, Defective ingestion of rod outer segments by cultured dystrophic rat pigment epithelial cells, *Invest. Ophthalmol. Vis. Sci.* **24:**812–820.

Chan, L. S., Li, W., Khatami, M., and Rockey, J. H., 1986, Actin in cultured bovine retinal capillary pericytes: Morphological and functional correlation, *Exp. Eye Res.* **43:**41–54.

Chin, K. S., Mathew, C. G. P., Fong, S. L., Bridges, C. D. B., and Ponder, B. A. J., 1988, Styl RFLP recognised by a human IRBP cDNA localised to chromosome 10, *Nucleic Acids Res.* **16:**1645.

Clark, V. M., 1986, The cell biology of the retinal pigment epithelium, in: *The Retina. A Model for Cell Biology Studies*, Part II (R. Adler and D. Farber, eds.), Academic Press, Orlando, FL, pp. 129–168.

Clark, V. M., and Hall, M. O., 1982, Labeling of bovine rod outer segment surface proteins with ^{125}I, *Exp. Eye Res.* **34:**847–859.

Clark, V. M., and Hall, M. O., 1986, RPE cell surface proteins in normal and dystrophic rats, *Invest. Ophthalmol. Vis. Sci.* **27:**136–144.

Clark, V. M., Hall, M. O., Mayerson, P. L., and Schechter, C., 1984, Identification of some plasma membrane proteins of cultured rat pigment epithelial cells by labeling with ^{125}I, *Exp. Eye Res.* **39:**611–628.

Cohen, A. I., 1983, Some cytological and initial biochemical observations on photoreceptors in retinas of *rds* mice, *Invest. Ophthalmol. Vis. Sci.* **24:**832–843.

Cohen, D., and Nir, I., 1983, Cytochemical evaluation of anionic sites on the surface of cultured pigment epithelium cells from normal and dystrophic RCS rats, *Exp. Eye Res.* **37:**575–582.

Cohen, D., and Nir, I., 1987, Cytochemical characterization of sialoglyconjugates on rat photoreceptor cell surface, *Invest. Ophthalmol. Vis. Sci.* **28:**640–645.

Cohen, L. H., and Noell, W. K., 1960, Glucose catabolism of rabbit retina before and after development of visual function, *J. Neurochem.* **5:**253–276.

Cohen, L. H., and Noell, W. K., 1965, Relationships between visual function and metabolism, in: *Biochemistry of the Retina* (C. N. Graymore, ed.), Academic Press, London, pp. 36–50.

Colley, N. J., Clark, V. M., and Hall, M. O., 1987, Surface modification of retinal pigment epithelial cells: Effects on phagocytosis and glycoprotein composition, *Exp. Eye Res.* **44:**377–392.

Converse, C. A., McLachlan, T., Bow, A. C., Packard, C. J., and Shepherd, J., 1987, Lipid metabolism in retinitis pigmentosa, in: *Degenerative Retinal Disorders: Clinical and Laboratory Investigations* (J. G. Hollyfield, R. E. Anderson, and M. M. LaVail, eds.), Alan R. Liss, New York, pp. 93–101.

Cook, N. J., and Kaupp, U. B., 1988, Solubilization, purification, and reconstitution of the sodium–calcium exchanger from bovine retinal rod outer segments, *J. Biol. Chem.* **263:**11382–11388.

Cook, N. J., Hanke, W., and Kaupp, U. B., 1987, Identification, purification, and functional reconstitution of the cyclic GMP-dependent channel from rod photoreceptors, *Proc. Natl. Acad. Sci. USA* **84:**585–589.

Cook, N. J., Molday, L. L., Reid, D., Kaupp, U. B., and Molday, R. S., 1989, The cGMP-gated channel of bovine rod photoreceptors is localized exclusively in the plasma membrane, *J. Biol. Chem.* **264:**6996–6999.

Cooper, N. G. F., Tarnowski, B. I., and McLaughlin, B. J., 1987, Lectin-affinity isolation of microvillous membranes from the pigmented epithelium of rat retina, *Curr. Eye Res.* **6:**969–979.

Crabb, J. W., and Saari, J. C., 1981, N-Terminal sequence homology among retinoid-binding proteins from bovine retina, *FEBS Lett.* **130:**15–17.

Crabb, J. W., and Saari, J. C., 1986, The complete amino acid sequence of the cellular retinoic acid-binding protein from bovine retina, *Biochem. Int.* **12:**391–395.

Crouch, R., Priest, D. G., and Duke, E. J., 1978, Superoxide dismutase activities of bovine ocular tissues, *Exp. Eye Res.* **27:**503–509.

Danciger, M., Tuteja, N., Kozak, C. A., and Farber, D. B., 1989, The gene for the γ-subunit of retinal cGMP-phosphodiesterase is on mouse chromosome 11, *Exp. Eye Res.* **48:**303–308.

Das, A., Pansky, B., Budd, G. C., and Kollarits, C. R., 1984, Immunocytochemistry of mouse and human retina with antisera to insulin and S-100 protein, *Curr. Eye Res.* **3:**1397–1403.

Das, A., Pansky, B., and Budd, G. C., 1987, Demonstration of insulin-specific mRNA in cultured rat retinal glial cells, *Invest. Ophthalmol. Vis. Sci.* **28:**1800–1810.

Das, A., Frank, R. N., Weber, M. L., Kennedy, A., Reidy, C. A., and Mancini, M. A., 1988, ATP causes retinal pericytes to contract *in vitro*, *Exp. Eye Res.* **46:**349–362.

Das, N. D., and Shichi, H., 1981, Enzymes of mercapturate synthesis and other drug-metabolizing reactions—specific localization in the eye, *Exp. Eye Res.* **33:**525–533.

Das, N. D., Yoshioka, T., Samuelson, D., and Shichi, H., 1986, Immunocytochemical localization of phosphatidylinositol-4,5- bisphosphate in dark- and light-adapted rat retinas, *Cell Struct. Funct.* **11:**53–63.

Das, N. D., Yoshioka, T., Samuelson, D., Cohen, R. J., and Shichi, H., 1987, Immunochemical evidence for the light-regulated modulation of phosphatidylinositol-4,5-bisphosphate in rat photoreceptor cells, *Cell Struct. Funct.* **12:**471–481.

Deigner, P. S., Law, W. C., Canada, F. J., and Rando, R. R., 1989, Membranes as the energy source in the endergonic transformation of vitamin A to 11-*cis*-retinol, *Science* **244:**968–971.

Devary, O., Heichal, O., Blumenfeld, A., Cassel, D., Suss, E., Barash, S., Rubinstein, C. T., Minke, B., and Selinger, Z., 1987, Coupling of photoexcited rhodopsin to inositol phospholipid hydrolysis in fly photoreceptors, *Proc. Natl. Acad. Sci. USA* **84:**6939–6943.

DeVries, G. W., Cohen, A. I., Hall, I. A., and Ferrendelli, J. A., 1978, Cyclic nucleotide levels in normal and biologically fractionated mouse retina: Effects of light and dark adaptation, *J. Neurochem.* **31:**1345–1351.

Dohlman, H. G., Caron, M. G., and Lefkowitz, R. J., 1987, A family of receptors coupled to guanine nucleotide regulatory proteins, *Biochemistry* **26:**2657–2664.

Donoso, L. A., Merryman, C. F., Shinohara, T., Dietzschold, B., Wistow, G., Craft, C., Morley, W., and Henry, R. T., 1986, S-antigen: Identification of the MAbA9-C6 monoclonal antibody binding site and the uveitopathogenic sites, *Curr. Eye Res.* **5:**995–1004.

Dorey, C. K., Khouri, G. G., Syniuta, L. A., Curran, S. A., and Weiter, J. J., 1989, Superoxide production by porcine retinal pigment epithelium *in vitro*, *Invest. Ophthalmol. Vis. Sci.* **30:**1047–1054.

Dowling, J. E., 1960, Chemistry of visual adaptation in the rat, *Nature* **188:**114–118.

Dowling, J. E., 1986, Dopamine: A retinal neuromodulator? *Trends Neurosci.* **9:**236–240.

Dowling, J. E., 1987, *The Retina. An Approachable Part of the Brain*, Belknap Press/Harvard University Press, Cambridge.
Dowling, J. E., and Sidman, R. L., 1962, Inherited retinal dystrophy in the rat, *J. Cell Biol.* **14:**73–109.
Dratz, E. A., and Hargrave, P. A., 1983, The structure of rhodopsin and the rod outer segment disk membrane, *Trends Biochem. Sci.* **8:**128–131.
Dratz, E. A., Miljanich, G. P., Nemes, P. P., Gaw, J. E., and Schwartz, S., 1979, The structure of rhodopsin and its disposition in the rod outer segment disk membrane, *Photochem. Photobiol.* **29:**661–670.
Dreyfus, H., Urban, P. F., Edel-Harth, S., and Mandel, P., 1975, Developmental patterns of gangliosides and phospholipids in chick retina and brain, *J. Neurochem.* **25:**245–250.
Dreyfus, H., Edel-Harth, S., Urban, P. F., Neskovic, N., and Mandel, P., 1977, Enzymatic synthesis of lactosylceramide by a galactosyltransferase from developing chicken retina, *Exp. Eye Res.* **25:**1–7.
Dreyfus, H., Harth, S., Yusufi, A. N. K., Urban, P. F., and Mandel, P., 1980, Sialyltransferase activities in two neuronal models: Retina and cultures of isolated neurons, in: *Structure and Function of Gangliosides* (L. Svennerholm, P. Mandel, H. Dreyfus, and P.-F. Urban, eds.), Plenum Press, New York, pp. 227–237.
Dreyfus, H., Preti, A., Harth, S., Pellicone, C., and Virmaux, N., 1983, Neuraminidase in calf retinal outer segment membranes, *J. Neurochem.* **40:**184–188.
Dryja, T. P., O'Neil-Dryja, M., Pawelek, J. M., and Albert, D. M., 1978, Demonstration of tyrosinase in the adult bovine uveal tract and retinal pigment epithelium, *Invest. Ophthalmol. Vis. Sci.* **17:**511–514.
Dryja, T. P., O'Neil-Dryja, M., and Albert, D. M., 1979, Elemental analysis of melanins from bovine hair, iris, choroid, and retinal pigment epithelium, *Invest. Ophthalmol. Vis. Sci.* **18:** 231–236.
Dudley, P. A., and Anderson, R. E., 1978, Phospholipid transfer protein from bovine retina with high activity towards retinal rod disc membranes, *FEBS Lett.* **95:**57–60.
Edwards, R. B., 1977, Culture of rat retinal pigment epithelium, *In Vitro* **13:**301–304.
Edwards, R. B., 1982a, Culture of mammalian retinal pigment epithelium and neural retina, in: *Methods in Enzymology*, Vol. 81 (L. Packer, ed.), Academic Press, New York, pp. 39–45.
Edwards, R. B., 1982b, Glycosaminoglycan synthesis by cultured human retinal pigmented epithelium from normal postmortem donors and a postmortem donor with retinitis pigmentosa, *Invest. Ophthalmol. Vis. Sci.* **23:**435–446.
Edwards, R. B., and Flaherty, P. M., 1986, Association of changes in intracellular cyclic AMP with changes in phagocytosis in cultured rat pigment epithelium, *Curr. Eye Res.* **5:**19–26.
Edwards, R. B., and Szamier, R. B., 1977, Defective phagocytosis of isolated rod outer segments by RCS retinal pigment epithelium in culture, *Science* **197:**1001–1003.
Edwards, R. B., Adler, A. J., and Southwick, R. E., 1987a, Maintenance of retinyl ester synthetase activity in cultured human retinal pigment epithelial cells, *Exp. Eye Res.* **45:**187–190.
Edwards, R. B., Brandt, J. T., and Hardenbergh, G. S., 1987b, A 31,000-dalton protein released by cultured human retinal pigment epithelium, *Invest. Ophthalmol. Vis. Sci.* **28:**1213–1218.
Ehinger, B., and Rose, B., 1988, Diurnal variation in chick retinal 5-hydroxytryptamine, *Exp. Eye Res.* **46:**819–821.
Eisenfeld, A. J., Bunt-Milam, A. H., and Sarthy, P. V., 1984, Muller cell expression of glial fibrillary acidic protein after genetic and experimental photoreceptor degeneration in the rat retina, *Invest. Ophthalmol. Vis. Sci.* **25:**1321–1328.
Eisenfeld, A. J., Bunt-Milam, A. H., and Saari, J. C., 1985, Immunocytochemical localization of interphotoreceptor retinoid-binding protein in developing normal and RCS rat retinas, *Invest. Ophthalmol. Vis. Sci.* **26:**775–778.
Eisenfeld, A. J., Bunt-Milam, A. H., and Saari, J. C., 1987, Uveoretinitis in rabbits following immunization with interphotoreceptor retinoid-binding protein, *Exp. Eye Res.* **44:**425–438.
Ekstrom, P., Sanyal, S., Narfstrom, K., Chader, G. J., and van Veen, T., 1988, Accumulation of glial fibrillary acidic protein in Müller radial glia during retinal degeneration, *Invest. Ophthalmol. Vis. Sci.* **29:**1363–1371.
Eldred, G. E., and Katz, M. L., 1988, Fluorophores of the human retinal pigment epithelium: Separation and spectral characterization, *Exp. Eye Res.* **47:**71–86.
Eldred, G. E., Miller, G. V., Stark, W. S., and Feeney-Burns, L., 1982, Lipofuscin: Resolution of discrepant fluorescence data, *Science* **216:**757–759.
Erickson, P. A., Fisher, S. K., Guerin, C. J., Anderson, D. H., and Kaska, D. D., 1987, Glial fibrillary acidic protein increases in Müller cells after retinal detachment, *Exp. Eye Res.* **44:**37–48.

Farber, D. B., and Lolley, R. N., 1974, Cyclic guanosine monophosphate: Elevation in degenerating photoreceptor cells of the C3H mouse retina, *Science* **186:**449–451.

Farber, D. B., and Lolley, R. N., 1976, Enzymatic basis for cyclic GMP accumulation in degenerative photoreceptor cells of mouse retina, *J. Cyclic Nucleotide Res.* **2:**139–148.

Farber, D. B., and Lolley, R. N., 1977, Light-induced reduction in cyclic GMP of retinal photoreceptor cells *in vivo*: Abnormalities in the degenerative diseases of RCS rats and *rd* mice, *J. Neurochem.* **28:**1089–1095.

Farber, D. B., and Shuster, T. A., 1986, Cyclic nucleotides in retinal function and degeneration, in: *The Retina. A Model for Cell Biology Studies*, Part I (R. Adler and D. Farber, eds.), Academic Press, Orlando, FL, pp. 239–296.

Farber, D. B., Chase, D. G., and Lolley, R. N., 1980, Cyclic nucleotides in rod- and cone-dominant retinas, *Neurochem. Int.* **1:**327–336.

Farber, D. B., Souza, D. W., Chase, D. G., and Lolley, R. N., 1981, Cyclic nucleotides of cone-dominant retinas, *Invest. Ophthalmol. Vis. Sci.* **20:**24–31.

Farber, D. B., Flannery, J. G., Lolley, R. N., and Bok, D., 1985, Distribution patterns of photoreceptors, protein, and cyclic nucleotides in the human retina, *Invest. Ophthalmol. Vis. Sci.* **26:**1558–1568.

Farber, D. B., Flannery, J. G., Bird, A. C., Shuster, T., and Bok, D., 1987, Histopathological and biochemical studies on donor eyes affected with retinitis pigmentosa, in: *Degenerative Retinal Disorders: Clinical and Laboratory Investigations* (J. G. Hollyfield, R. E. Anderson, and M. M. LaVail, eds.), Alan R. Liss, New York, pp. 53–67.

Farber, D. B., Park, S., and Yamashita, C., 1988, Cyclic GMP-phosphodiesterase of *rd* retina: Biosynthesis and content, *Exp. Eye Res.* **46:**363–374.

Farnsworth, C. C., and Dratz, E. A., 1976, Oxidative damage of retinal rod outer segment membranes and the role of vitamin E, *Biochim. Biophys. Acta* **443:**556–570.

Farnsworth, C. C., Stone, W. L., and Dratz, E. A., 1979, Effects of vitamin E and selenium deficiency on the fatty acid composition of rat retinal tissues, *Biochim. Biophys. Acta* **552:**281–293.

Feeney, L., 1973a, Synthesis of interphotoreceptor matrix. I. Autoradiography of ^{3}H-fucose incorporation, *Invest. Ophthalmol.* **12:**739–751.

Feeney, L., 1973b, The interphotoreceptor space. II. Histochemistry of the matrix, *Dev. Biol.* **32:**115–128.

Feeney, L., 1973c, The phagolysosomal system of the pigment epithelium. A key to retinal disease, *Invest. Ophthalmol.* **12:**635–638.

Feeney, L., 1978, Lipofuscin and melanin of human retinal pigment epithelium. Fluorescence, enzyme cytochemical, and ultrastructural studies, *Invest. Ophthalmol. Vis. Sci.* **17:** 583–600.

Feeney, L., and Berman, E. R., 1976, Oxygen toxicity: Membrane damage by free radicals, *Invest. Ophthalmol.* **15:**789–792.

Feeney-Burns, L., 1980, The pigments of the retinal pigment epithelium, in: *Current Topics in Eye Research*, Vol. 2 (J. A. Zadunaisky and H. Davson, eds.), Academic Press, New York, pp. 119–178.

Feeney-Burns, L., 1985, The early years of research, in: *The Interphotoreceptor Matrix in Health and Disease* (C. D. Bridges and A. J. Adler, eds.), Alan R. Liss, New York, pp. 3–23.

Feeney-Burns, L., and Berman, E. R., 1982, Methods for isolating and fractionating pigment epithelial cells, in: *Methods in Enzymology*, Vol. 81 (L. Packer, ed.), Academic Press, New York, pp. 95–110.

Feeney-Burns, L., and Eldred, G. E., 1983, The fate of the phagosome: Conversion to "age pigment" and impact in human retinal pigment epithelium, *Trans. Ophthalmol. Soc. U.K.* **103:**416–421.

Feeney-Burns, L., Berman, E. R., and Rothman, H., 1980, Lipofuscin of human retinal pigment epithelium, *Am. J. Ophthalmol.* **90:**783–791.

Feeney-Burns, L., Hilderbrand, E. S., and Eldridge, S., 1984, Aging human RPE: Morphometric analysis of macular, equatorial, and peripheral cells, *Invest. Ophthalmol. Vis. Sci.* **25:**195–200.

Feeney-Burns, L., Gao, C.-L., and Tidwell, M., 1987, Lysosomal enzyme cytochemistry of human RPE, Bruch's membrane and drusen, *Invest. Ophthalmol. Vis. Sci.* **28:**1138–1147.

Feeney-Burns, L., Gao, C.-L., and Berman, E. R., 1988, The fate of immunoreactive opsin following phagocytosis by pigment epithelium in human and monkey retinas, *Invest. Ophthalmol. Vis. Sci.* **29:**708–719.

Fein, A., 1986, Excitation and adaptation of *Limulus* photoreceptors by light and inositol 1,4,5-trisphosphate, *Trends Neurosci.* **9:**110–114.

Fein, A., Payne, R., Corson, D. W., Berridge, M. J., and Irvine, R. F., 1984, Photoreceptor excitation and adaptation by inositol 1,4,5-trisphosphate, *Nature* **311:**157–160.

Ferrari-Dileo, G., 1988, $Beta_1$ and $Beta_2$ adrenergic binding sites in bovine retina and retinal blood vessels, *Invest. Ophthalmol. Vis. Sci.* **29:**695–699.

Ferrendelli, J. A., De Vries, G. W., Cohen, A. I., and Lowry, O. H., 1980, Localization and roles of cyclic nucleotide systems in retina, *Neurochem. Int.* **1:**311–326.

Fesenko, E. E., Kolesnikov, S. S., and Lyubarsky, A. L., 1985, Induction by cyclic GMP of cationic conductance in plasma membrane of retinal rod outer segment, *Nature* **313:**310–311.

Findlay, J. B. C., and Pappin, D. J. C., 1986, The opsin family of proteins, *Biochem. J.* **238:**625–642.

Fine, B. S., and Zimmerman, L. E., 1963, Observations on the rod and cone layer of the human retina, *Invest. Ophthalmol.* **2:**446–459.

Flannery, J. G., and Fisher, S. K., 1984, Circadian disc shedding in *Xenopus* retina *in vitro*, *Invest. Ophthalmol. Vis. Sci.* **25:**229–232.

Fletcher, R. T., and Chader, G. J., 1976, Cyclic GMP: Control of concentration by light in retinal photoreceptors, *Biochem. Biophys. Res. Commun.* **70:**1297–1302.

Fletcher, R. T., Sanyal, S., Krishna, G., Aguirre, G., and Chader, G. J., 1986, Genetic expression of cyclic GMP phosphodiesterase activity defines abnormal photoreceptor differentiation in neurological mutants of inherited retinal degeneration, *J. Neurochem.* **46:**1240–1245.

Fliesler, S. J., 1988, Retinal rod outer segment membrane assembly: Studies with inhibitors of enzymes involved in N-linked oligosacchraride biosynthesis and processing, in: *Proceedings of the Kagoshima International Symposium on Glycoconjugates in Medicine*, pp. 316–323.

Fliesler, S. J., and Anderson, R. E., 1983, Chemistry and metabolism of lipids in the vertebrate retina, *Prog. Lipid Res.* **22:**79–131.

Fliesler, S. J., and Basinger, S. F., 1985, Tunicamycin blocks the incorporation of opsin into retinal rod outer segment membranes, *Proc. Natl. Acad. Sci. USA* **82:**1116–1120.

Fliesler, S. J., and Basinger, S. F., 1987, Monensin stimulates glycerolipid incorporation into rod outer segment membranes, *J. Biol. Chem.* **262:**17516–17523.

Fliesler, S. J., and Schroepfer, G. J., Jr., 1982, Sterol composition of bovine retinal rod outer segment membranes and whole retinas, *Biochim. Biophys. Acta* **711:**138–148.

Fliesler, S. J., and Schroepfer, G. J., Jr., 1983, Metabolism of mevalonic acid in cell-free homogenates of bovine retinas. Formation of novel isoprenoid acids, *J. Biol. Chem.* **258:**15062–15070.

Fliesler, S. J., and Schroepfer, G. J., Jr., 1986, *In vitro* metabolism of mevalonic acid in the bovine retina, *J. Neurochem.* **46:**448–460.

Fliesler, S. J., Tabor, G. A., and Hollyfield, J. G., 1984, Glycoprotein synthesis in the human retina: Localization of the lipid intermediate pathway, *Exp. Eye Res.* **39:**153–173.

Fliesler, S. J., Rayborn, M. E., and Hollyfield, J. G., 1985, Membrane morphogenesis in retinal rod outer segments: Inhibition by tunicamycin, *J. Cell Biol.* **100:**574–587.

Fliesler, S. J., Rayborn, M. E., and Hollyfield, J. G., 1986a, Inhibition of oligosaccharide processing and membrane morphogenesis in retinal rod photoreceptor cells, *Proc. Natl. Acad. Sci. USA* **83:**6435–6439.

Fliesler, S. J., Rayborn, M. E., and Hollyfield, J. G., 1986b, Protein-bound carbohydrate involvement in plasma membrane assembly: The retinal rod photoreceptor cell as a model, in: *Protein–Carbohydrate Interactions in Biological Systems* (D. L. Lark, ed.), Academic Press, London, pp. 191–205.

Flood, M. T., Gouras, P., and Kjeldbye, H., 1980, Growth characteristics and ultrastructure of human retinal pigment epithelium in vitro, *Invest. Ophthalmol. Vis. Sci.* **19:**1309–1320.

Flood, M. T., Bridges, C. D. B., Alvarez, R. A., Blaner, W. S., and Gouras, P., 1983, Vitamin A utilization in human retinal pigment epithelial cells *in vitro*, *Invest. Ophthalmol. Vis. Sci.* **24:**1227–1235.

Fong, H. K .W., Hurley, J. B., Hopkins, R. S., Miake-Lye, R., Johnson, M. S., Doolittle, R. F., and Simon, M. I., 1986, Repetitive segmental structure of the transducin β subunit: Homology with the CDC4 gene and identification of related mRNAs, *Proc. Natl. Acad. Sci. USA* **83:**2162–2166.

Fong, S.-L., and Bridges, C. D. B., 1988, Internal quadruplication in the structure of human interstitial retinol-binding protein deduced from its cloned cDNA, *J. Biol. Chem.* **263:**15330–15334.

Fong, S.-L., Bridges, C. D. B., and Alvarez, R. A., 1983, Utilization of exogenous retinol by frog pigment epithelium, *Vision Res.* **23:**47–52.

Fong, S.-L., Liou, G. I., Landers, R. A., Alvarez, R. A., Gonzalez-Fernandez, F., Glazebrook, P. A., Lam, D. M. K., and Bridges, C. D. B., 1984a, Characterization, localization, and biosynthesis of an interstitial retinol-binding glycoprotein in the human eye, *J. Neurochem.* **42:**1667–1676.

Fong, S.-L., Liou, G. I., Landers, R. A., Alvarez, R. A., and Bridges, C. D. B., 1984b, Purification and characterization of a retinol-binding glycoprotein synthesized and secreted by bovine neural retina, *J. Biol. Chem.* **259:**6534–6542.

Fong, S.-L., Irimura, T., Landers, R. A., and Bridges, C. D. B., 1985, The carbohydrate of bovine interstitial retinol-binding protein, in: *The Interphotoreceptor Matrix in Health and Disease* (C. D. Bridges and A. J. Adler, eds.), Alan R. Liss, New York, pp. 111–128.

Fong, S.-L., Balakier, H., Canton, M., Bridges, C. D. B., and Gallie, B., 1988, Retinoid-binding proteins in retinoblastoma tumors, *Cancer Res.* **48:**1124–1128.

Forster, B. A., Ferrari-Dileo, G., and Anderson, D. R., 1987, Adrenergic $alpha_1$ and $alpha_2$ binding sites are present in bovine retinal blood vessels, *Invest. Ophthalmol. Vis. Sci.* **28:**1741–1746.

Fox, G. M., Kuwabara, T., Wiggert, B., Redmond, T. M., Hess, H. H., Chader, G. J., and Gery, I., 1987, Experimental autoimmune uveoretinitis (EAU) induced by retinal interphotoreceptor retinoid-binding protein (IRBP): Differences between EAU induced by IRBP and by S-antigen, *Clin. Immunol. Immunopathol.* **43:**256–264.

Frambach, D. A., Valentine, J. L., and Weiter, J. J., 1988, Alpha-1 adrenergic receptors on rabbit retinal pigment epithelium, *Invest. Ophthalmol. Vis. Sci.* **29:**737–741.

Frank, R. N., Keirn, R. J., Kennedy, A., and Frank, K. W., 1983, Galactose-induced retinal capillary basement membrane thickening: Prevention by sorbinil, *Invest. Ophthalmol. Vis. Sci.* **24:**1519–1524.

Friedlander, M., and Blobel, G., 1985, Bovine opsin has more than one signal sequence, *Nature* **318:**338–343.

Friedman, Z., Hackett, S. F., and Campochiaro, P. A., 1987, Characterization of adenylate cyclase in human retinal pigment epithelial cells *in vitro*, *Exp. Eye Res.* **44:**471–479.

Fukuda, M. N., Papermaster, D. S., and Hargrave, P. A., 1979, Rhodopsin carbohydrate. Structure of small oligosaccharides attached at two sites near the NH_2 terminus, *J. Biol. Chem.* **254:**8201–8207.

Fukui, H., and Shichi, H., 1981, 5′-Nucleotidases of retinal rod membranes, *Arch. Biochem. Biophys.* **212:**78–87.

Fukui, H., and Shichi, H., 1982, Soluble 5′-nucleotidase: Purification and reversible binding to photoreceptor membranes, *Biochemistry* **21:**3677–3681.

Fulton, B. S., and Rando, R. R., 1987, Biosynthesis of 11-*cis*- retinoids and retinyl esters by bovine pigment epithelium membranes, *Biochemistry* **26:**7938–7945.

Fung, B. K.-K., 1983, Characterization of transducin from bovine retinal rod outer segments. I. Separation and reconstitution of the subunits, *J. Biol. Chem.* **258:**10495–10502.

Fung, B. K.-K., 1987, Transducin: Structure, function and role in phototransduction, *Prog. Retinal Res.* **6:**151–177.

Futterman, S., 1963, Metabolism of the retina. III. The role of reduced triphosphopyridine nucleotide in the visual cycle, *J. Biol. Chem.* **238:**1145–1150.

Futterman, S., Saari, J. C., and Swanson, D. E., 1976, Retinol and retinoic acid-binding proteins in bovine retina: Aspects of binding specificity, *Exp. Eye Res.* **22:**419–424.

Futterman, S., Saari, J. C., and Blair, S., 1977, Occurrence of a binding protein for 11-*cis*-retinal in retina, *J. Biol. Chem.* **252:**3267–3271.

Gentleman, S., Russell, P., Hemmings, B. A., and Chader, G. J., 1989, Abnormal expression of the RI subunit of cyclic AMP-dependent protein kinase in Y-79 retinoblastoma cells, *Exp. Eye Res.* **48:**497–507.

Gery, I., Wiggert, B., Redmond, T. M., Kuwabara, T., Crawford, M. A., Vistica, B. P., and Chader, G. J., 1986, Uveoretinitis and pinealitis induced by immunization with interphotoreceptor retinoid-binding protein, *Invest. Ophthalmol. Vis. Sci.* **27:**1296–1300.

Ghalayini, A., and Anderson, R. E., 1984, Phosphatidylinositol 4,5-bisphosphate: Light-mediated breakdown in the vertebrate retina, *Biochem. Biophys. Res. Commun.* **124:**503–506.

Ghazi, H., and Osborne, N. N., 1988, Agonist-induced stimulation of inositol phosphates in primary rabbit retinal cultures, *J. Neurochem.* **50:**1851–1858.

Giusto, N. M., and Bazan, N. G., 1979, Phospholipids and acylglycerols biosynthesis and $^{14}CO_2$ production from [^{14}C]glycerol in the bovine retina: The effects of incubation time, oxygen and glucose, *Exp. Eye Res.* **29:**155–168.

Giusto, N. M., and Ilincheta de Boschero, M. G., 1986, Synthesis of polyphosphoinositides in vertebrate photoreceptor membranes, *Biochim. Biophys. Acta* **877:**440–446.

Giusto, N. M., Ilincheta de Boschero, M. G., Sprecher, H., and Aveldano, M. I., 1986, Active labeling of phosphatidylcholines by [1-^{14}C]docosahexaenoate in isolated photoreceptor membranes, *Biochim. Biophys. Acta* **860:**137–148.

Godchaux, W., III, and Zimmerman, W. F., 1979, Soluble proteins of intact bovine rod cell outer segments, *Exp. Eye Res.* **28:**483–500.

Gold, G. H., and Nakamura, T., 1987, Cyclic nucleotide-gated conductances: A new class of ion channels mediates visual and olfactory transduction, *Trends Pharmacol. Sci.* **8:**312–316.

Goldberg, N. D., Ames, A., III, Gander, J. E., and Walseth, T. F., 1983, Magnitude of increase in retinal cGMP metabolic flux determined by ^{18}O incorporation into nucleotide α-phosphoryls corresponds with intensity of photic stimulation, *J. Biol. Chem.* **258:**9213–9219.

Goldman, B. M., and Blobel, G., 1981, *In vitro* biosynthesis, core glycosylation and membrane integration of opsin, *J. Cell Biol.* **90:**236–242.

Gonzalez-Fernandez, F., Landers, R. A., Glazebrook, P. A., Fong, S.-L., Liou, G. I., Lam, D. M. K., and Bridges, C. D. B., 1984, An extracellular retinol-binding glycoprotein in the eyes of mutant rats with retinal dystrophy: Development, localization, and biosynthesis, *J. Cell Biol.* **99:**2092–2098.

Gonzalez-Fernandez, F., Fong, S.-L., Liou, G. I., and Bridges, C. D. B., 1985, Interstitial retinol-binding protein (IRBP) in the RCS rat: Effect of dark-rearing, *Invest. Ophthalmol. Vis. Sci.* **26:**1381–1385.

Goridis, C., Virmaux, N., Cailla, H. L., and Delaage, M. A., 1974, Rapid, light-induced changes of retinal cyclic GMP levels, *FEBS Lett.* **49:**167–169.

Graymore, C. N. (ed.), 1965, *Biochemistry of the Retina*, Academic Press, London.

Graymore, C. N., 1970, Biochemistry of the retina, in: *Biochemistry of the Eye* (C. N. Graymore, ed.), Academic Press, London, pp. 645–735.

Greenberger, L. M., and Besharse, J. C., 1985, Stimulation of photoreceptor disc shedding and pigment epithelial phagocytosis by glutamate, aspartate, and other amino acids, *J. Comp. Neurol.* **239:**361–372.

Hageman, G. S., and Johnson, L. V., 1986, Biochemical characterization of the major peanut-agglutinin-binding glycoproteins in vertebrate retinae, *J. Comp. Neurol.* **249:** 499–510.

Hageman, G. S., and Johnson, L. V., 1987, Chondroitin 6-sulfate glycosaminoglycan is a major constituent of primate cone photoreceptor matrix sheaths, *Curr. Eye Res.* **6:**639–646.

Hageman, G. S., Hewitt, A. T., Kirchoff, M., and Johnson, L. V., 1989, Selective extration and characterization of cone matrix sheath-specific molecules, *Invest. Ophthalmol. Vis. Sci. Suppl.* **30:**489.

Haley, J. E., and Gouras, P., 1984, Two-dimensional electrophoresis of proteins from cultured human retinal-pigment epithelial cells: Internal references, cataloging, and glycoproteins, *Clin. Chem.* **30:**1906–1913.

Haley, J. E., Flood, M. T., Gouras, P., and Kjeldbye, H. M., 1983, Proteins from human retinal pigment epithelial cells: Evidence that a major protein is actin, *Invest. Ophthalmol. Vis. Sci.* **24:**803–811.

Hall, M. O., 1978, Phagocytosis of light- and dark-adapted rod outer segments by cultured pigment epithelium, *Science* **202:** 526–528.

Hall, M. O., and Abrams, T., 1987, Kinetic studies of rod outer segment binding and ingestion by cultured rat RPE cells, *Exp. Eye Res.* **45:**907–922.

Hall, M., and Hall, D. O., 1975, Superoxide dismutase of bovine and frog rod outer segments, *Biochem. Biophys. Res. Commun.* **67:**1199–1204.

Hall, M. O., Bok, D., and Bacharach, A. D. E., 1969, Biosynthesis and assembly of the rod outer segment membrane system. Formation and fate of visual pigment in the frog retina, *J. Mol. Biol.* **45:**397–406.

Ham, W. T., Jr., Mueller, H. A., Ruffolo, J. J., Jr., Millen, J. E., Cleary, S. F., Guerry, R. K., and Guerry, D. III, 1984, Basic mechanisms underlying the production of photochemical lesions in the mammalian retina, *Curr. Eye Res.* **3:**165–174.

Hamm, H. E., and Bownds, M. D., 1986, Protein complement of rod outer segments of frog retina, *Biochemistry* **25:**4512–4523.

Hamm, H. E., and Menaker, M., 1980, Retinal rhythms in chicks: Circadian variation in melatonin and serotonin *N*-acetyltransferase activity, *Proc. Natl. Acad. Sci. USA* **77:**4998–5002.

Hamm, H. E., Deretic, D., Arendt, A., Hargrave, P. A., Koenig, B., and Hofmann, K. P., 1988, Site of G protein binding to rhodopsin mapped with synthetic peptides from α subunit, *Science* **241:**832–835.

Handelman, G. J., and Dratz, E. A., 1986, The role of antioxidants in the retina and retinal pigment epithelium and the nature of prooxidant-induced damage, *Adv. Free Radical Biol. Med.* **2:**1-89.

Handelman, G. J., Dratz, E. A., Reay, C. C., and van Kuijk, F. J. G. M., 1988, Carotenoids in the human macula and whole retina, *Invest. Ophthalmol. Vis. Sci.* **29:**850–855.

Hara, S., Hayasaka, S., and Mizuno, K., 1979, Distribution and some properties of lysosomal arylsulfatases in the bovine eye, *Exp. Eye Res.* **28:**641–650.

Hara, S., Plantner, J. J., and Kean, E. L., 1983, The enzymatic cleavage of rhodopsin by the retinal pigment epithelium. I. Enzyme preparation, properties and kinetics: Characterization of the glycopeptide product, *Exp. Eye Res.* **36:**799–816.

Hargrave, P. A., 1986, Molecular dynamics of the rod cell, in: *The Retina. A model for Cell Biology Studies*, Part I (R. Adler and D. Farber, eds.), Academic Press, Orlando, FL, pp. 207–237.

Hargrave, P. A., Fong, S.-L., McDowell, J. H., Mas, M. T., Curtis, D. R., Wang, J. K., Juszczak, E., and

Smith, D. P., 1980, The partial primary structure of bovine rhodopsin and its topography in the retinal rod cell disc membrane, *Neurochem. Int.* **1**:231–244.

Hargrave, P. A., McDowell, J. H., Curtis, D. R., Wang, J. K., Juszczak, E., Fong, S.-L., Rao, J. K. M., and Argos, P., 1983, The structure of bovine rhodopsin, *Biophys. Struct. Mech.* **9**: 235–244.

Hargrave, P. A., McDowell, J. H., Feldmann, R. J., Atkinson, P. H.,Rao, J. K. M., and Argos, P., 1984, Rhodopsin's protein and carbohydrate structure: Selected aspects, *Vision Res.* **24**:1487–1499.

Hargrave, P. A., Palczewski, K., Arendt, A., Adamus, G., and McDowell, J. H., 1988, Rhodopsin and its kinase, in: *Molecular Biology of the Eye: Genes, Vision, and Ocular Disease* (J. Piatigorsky, T. Shinohara, and P. S. Zelenka, eds.), Alan R. Liss, New York, pp. 35–44.

Harlan, E. W., and Crouch, R., 1984, Canine retinal superoxide dismutase: Identity with the erythrocyte enzyme, *Curr. Eye Res.* **3**:1455–1459.

Haskell, J. F., Meezan, E., and Pillion, D. J., 1984, Identification and characterization of the insulin receptor of bovine retinal microvessels, *Endocrinology* **115**:698–704.

Hayasaka, S., Hara, S., and Mizuno, K., 1975, Distribution and some properties of cathepsin D in the retinal pigment epithelium, *Exp. Eye Res.* **21**:307–313.

Hayasaka, S., Hara, S., Takaku, Y., and Mizuno, K., 1977, Distribution of acid lipase in the bovine retinal pigment epithelium, *Exp. Eye Res.* **21**:307–313.

Hayashi, F., and Amakawa, T., 1985, Light-mediated breakdown of phosphatidylinositol 4,5-bisphosphate in isolated rod outer segments of frog photoreceptor, *Biochem. Biophys. Res. Commun.* **128**:954–959.

Hayes, K. C., 1974, Retinal degeneration in monkeys induced by deficiencies of vitamin E or A, *Invest. Ophthalmol.* **13**:499–510.

Hayes, K. C., Carey, R. E., and Schmidt, S. Y., 1975, Retinal degeneration associated with taurine deficiency in the cat, *Science* **188**:949–951.

Haynes, L. W., Kay, A. R., and Yau, K.-W., 1986, Single cyclic GMP-activated channel activity in excised patches of rod outer segment membrane, *Nature* **321**:66–70.

Herron, W. L., Riegel, B. W., Myers, O. E., and Rubin, M. L., 1969, Retinal dystrophy in the rat - A pigment epithelial disease, *Invest. Ophthalmol.* **8**:595–604.

Hesketh, J. E., Virmaux, N., and Mandel, P., 1978, Evidence for a cyclic nucleotide-dependant phosphorylation of retinal myosin, *FEBS Lett.* **94**:357–360.

Heth, C. A., and Schmidt, S. Y., 1988, Protein phosphorylation in cultured rat RPE. Effects of protein kinase C activation, *Invest. Ophthamol. Vis. Sci.* **29**:1794–1799.

Heth, C. A., Yankauckas, M. A., Adamian, M., and Edwards, R. B., 1987, Characterization of retinal pigment epithelial cells cultured on microporous filters, *Curr. Eye Res.* **6**:1007–1019.

Hewitt, A. T., 1986, Extracellular matrix molecules: Their importance in the structure and function of the retina, in: *The Retina. A Model for Cell Biology Studies*, Part II (R. Adler and D. Farber, eds.), Academic Press, Orlando, FL, pp. 169–214.

Hewitt, A. T., and Newsome, D. A., 1985, Altered synthesis of Bruch's membrane proteoglycans associated with dominant retinitis pigmentosa, *Curr. Eye Res.* **4**:169–174.

Hewitt, A. T., and Newsome, D. A., 1988, Altered proteoglycans in cultured human retinitis pigmentosa retinal pigment epithelium, *Invest. Ophthalmol. Vis. Sci.* **29**:720–726.

Hildebrandt, J. D., Codina, J., Rosenthal, W., Birnbaumer, L., Neer, E. J., Yamazaki, A., and Bitensky, M. W., 1985, Characterization by two-dimensional peptide mapping of the γ subunits of N_s and N_i, the regulatory proteins of adenylyl cyclase, and of transducin, the guanine nucleotide-binding protein of rod outer segments of the eye, *J. Biol. Chem.* **260**:14867–14872.

Hirose, S., Wiggert, B., Redmond, T. M., Kuwabara, T., Nussenblatt, R. B., Chader, G. J., and Gery, I., 1987, Uveitis induced in primates by IRBP: Humoral and cellular immune responses, *Exp. Eye Res.* **45**:695–702.

Hirose, S., Tanaka, T., Nussenblatt, R. B., Palestine, A. G., Wiggert, B., Redmond, T. M., Chader, G. J., and Gery, I., 1988, Lymphocyte responses to retinal-specific antigens in uveitis patients and healthy subjects, *Curr. Eye Res.* **7**:393–402.

Ho, M.-T. P., Massey, J. B., Pownall, H. J., Anderson, R. E., and Hollyfield, J. G., 1989, Mechanism of vitamin A movement between rod outer segments, interphotoreceptor retinoid-binding protein, and liposomes, *J. Biol. Chem.* **264**:928–935.

Hohman, T. C., Nishimura, C., and Robison, W. G., Jr., 1989, Aldose reductase and polyol in cultured pericytes of human retinal capillaries, *Exp. Eye Res.* **48**:55–60.

Hollyfield, J. G., and Anderson, R. E., 1982, Retinal protein synthesis in relationship to environmental lighting, *Invest. Ophthalmol. Vis. Sci.* **23**:631–639.

Hollyfield, J. G., and Basinger, S. F., 1978, Photoreceptor shedding can be initiated within the eye, *Nature* **274:**794–796.

Hollyfield, J. G., and Basinger, S. F., 1980, RNA metabolism in the retina in relation to cyclic lighting, *Vision Res.* **20:**1151–1155.

Hollyfield, J. G., and Rayborn, M. E., 1987, Endocytosis in the inner segment of rod photoreceptors: Analysis of *Xenopus laevis* retinas using horseradish peroxidase, *Exp. Eye Res.* **45:**703–719.

Hollyfield, J. G., Varner, H. H., Rayborn, M. E., Liou, G. I., and Bridges, C. D., 1985a, Endocytosis and degradation of interstitial retinol-binding protein: Differential capabilities of cells that border the interphotoreceptor matrix, *J. Cell Biol.* **100:**1676–1681.

Hollyfield, J. G., Fliesler, S. J., Rayborn, M. E., Fong, S.-L., Landers, R. A., and Bridges, C. D. B., 1985b, Synthesis and secretion of interstitial retinol-binding protein by the human retina, *Invest. Ophthalmol. Vis. Sci.* **26:**58–67.

Hollyfield, J. G., Anderson, R. E., and LaVail, M.M. (eds.), 1987, *Degenerative Retinal Disorders. Clinical and Laboratory Investigations*, Alan R. Liss, New York.

Hotta, Y., and Inana, G., 1989, Gene transfer and expression of human ornithine aminotransferase, *Invest. Ophthalmol. Vis. Sci.* **30:**1024–1031.

Hudes, G. R., Li, W., Rockey, J. H., and White, P., 1988, Prostacyclin is the major prostaglandin synthesized by bovine retinal capillary pericytes in culture, *Invest. Ophthalmol. Vis. Sci.* **29:**1511–1516.

Hughes, B. A., Miller, S. S., and Machen, T. E., 1984, Effects of cyclic AMP on fluid absorption and ion transport across frog retinal pigment epithelium, *J. Gen. Physiol.* **83:**875–899.

Hunt, D. F., Organisciak, D. T., Wang, H. M., and Wu, R. L. C., 1984, α-Tocopherol in the developing rat retina: A high pressure liquid chromatographic analysis, *Curr. Eye Res.* **3:**1281–1288.

Hurley, J. B., 1987, Molecular properties of the cGMP cascade of vertebrate photoreceptors, *Annu. Rev. Physiol.* **49:**793–812.

Hurley, J. B., and Stryer, L., 1982, Purification and characterization of the γ regulatory subunit of the cyclic GMP phosphodiesterase from retinal rod outer segments, *J. Biol. Chem.* **257:**11094–11099.

Hutchins, J. B., 1987, Review: Acetylcholine as a neurotransmitter in the vertebrate retina, *Exp. Eye Res.* **45:**1–38.

Im, J. H., Pillion, D. J., and Meezan, E., 1986, Comparison of insulin receptors from bovine retinal blood vessels and nonvascular retinal tissue, *Invest. Ophthalmol. Vis. Sci.* **27:**1681–1690.

Inana, G., Hotta, Y., Zintz, C., Takki, K., Weleber, R. G., Kennaway, N. G., Nakayasu, K., Nakajima, A., and Shiono, T., 1988, Expression defect of ornithine aminotransferase gene in gyrate atrophy, *Invest. Ophthalmol. Vis. Sci.* **29:**1001–1005.

Irons, M. J., 1987a, Redistribution of Mn^{++}-dependent pyrimidine 5′-nucleotidase (MDPNase) activity during shedding and phagocytosis, *Invest. Ophthalmol. Vis. Sci.* **28:**83–91.

Irons, M. J., 1987b, Cytochemical localization of Mn^{2+}-dependent pyrimidine 5′-nucleotidase activity in isolated rod outer segments, *Exp. Eye Res.* **44:**789–803.

Irons, M. J., and Kalnins, V. I., 1984, Distribution of macrotubules in cultured RPE cells from normal and dystrophic RCS rats, *Invest. Ophthalmol. Vis. Sci.* **25:**434–439.

Irons, M. J., and O'Brien, P. J., 1987, Biochemical evidence for Mn^{2+}-dependent 5′-nucleotidase activity in isolated rod outer segments, *Exp. Eye Res.* **45:**813–821.

Ishibashi, T., Sorgente, N., Patterson, R., and Ryan, S. J., 1986, Pathogenesis of Drusen in the primate, *Invest. Ophthalmol. Vis. Sci.* **27:**184–193.

Ishikawa, T., and Yamada, E., 1970, The degradation of the photoreceptor outer segment within the pigment epithelial cell of rat retina, *J. Electron Microsc.* **19:**85–99.

Ishizaki, H., Haley, J. E., Gouras, P., Liang, J. T., and Kjeldbye, H. M., 1987, Isolation and characterization of plasma membrane proteins of cultured human retinal pigment epithelium, *Exp. Eye Res.* **44:**1–16.

Israel, P., Masterson, E., Goldman, A. I., Wiggert, B., and Chader, G. J., 1980, Retinal pigment epithelial cell differentiation *in vitro*, *Invest. Ophthalmol. Vis. Sci.* **19:**720–727.

Iuvone, P. M., 1986, Neurotransmitters and neuromodulators in the retina: Regulation, interactions, and cellular effects, in: *The Retina. A Model for Cell Biology Studies*, Part II (R. Adler and D. Farber, eds.), Academic Press, Orlando, FL, pp. 1–72.

Iuvone, P. M., and Besharse, J. C., 1983, Regulation of indoleamine N-acetyltransferase activity in the retina: Effects of light and dark, protein synthesis inhibitors and cyclic nucleotide analogs, *Brain. Res.* **273:**111–119.

Iuvone, P. M., and Besharse, J. C., 1986a, Cyclic AMP stimulates serotonin N-acetyltransferase activity in *Xenopus* retina *in vitro*, *J. Neurochem.* **46:**33–39.

Iuvone, P. M., and Besharse, J. C., 1986b, Involvement of calcium in the regulation of serotonin N-acetyltransferase in retina, *J. Neurochem.* **46:**82–88.

Iuvone, P. M., Rauch, A. L., Marshburn, P. B., Glass, D. B., and Neff, H. N., 1982, Activation of retinal tyrosine hydroxylase *in vitro* by cyclic AMP-dependent protein kinase: Characterization and comparison to activation *in vivo* by photic stimulation, *J. Neurochem.* **39:**1632–1640.

Jaffe, G. J., Burke, J. M., and Geroski, D. H., 1989, Ouabain-sensitive Na^+-K^+ ATPase pumps in cultured human retinal pigment epithelium, *Exp. Eye Res.* **48:**61–68.

Jansen, H. G., Sanyal, S., De Grip, W. J., and Schalken, J. J., 1987, Development and degeneration of retina in *rds* mutant mice: Ultraimmunohistochemical localization of opsin, *Exp. Eye Res.* **44:**347–361.

Jelsema, C. L., 1987, Light activation of phospholipase A_2 in rod outer segments of bovine retina and its modulation by GTP-binding proteins, *J. Biol. Chem.* **262:**163–168.

Jelsema, C. L., 1989, Regulation of phospholipase A_2 and phospholipase C in rod outer segments of bovine retina involves a common CTP-binding protein but different mechanisms of action, *Neurol. Neurobiol.* **49:**25–47.

Jelsema, C. L., and Axelrod, J., 1987, Stimulation of phospholipase A_2 activity in bovine rod outer segments by the $\beta\gamma$ subunits of transducin and its inhibition by the α subunit, *Proc. Natl. Acad. Sci. USA* **84:**3623–3627.

Johnson, A. T., Kretzer, F. L., Hittner, H. M., Glazebrook, P. A., Bridges, C. D. B., and Lam, D. M. K., 1985, Development of the subretinal space in the preterm human eye: Ultrastructural and immunocytochemical studies, *J. Comp. Neurol.* **233:**497–505.

Johnson, L. V., and Hageman, G. S., 1987, Enzymatic characterization of peanut agglutinin-binding components in the retinal interphotoreceptor matrix, *Exp. Eye Res.* **44:**553–565.

Johnson, L. V., Hageman, G. S., and Blanks, J. C., 1985, Restricted extracellular matrix domains ensheath cone photoreceptors in vertebrate retinae, in: *The Interphotoreceptor Matrix in Health and Disease* (C. D. Bridges and A. J. Adler, eds.), Alan R. Liss, New York, pp. 33–44.

Johnson, L. V., Hageman, G. S., and Blanks, J. C., 1986, Interphotoreceptor matrix domains ensheath vertebrate cone photoreceptor cells, *Invest. Ophthalmol. Vis. Sci.* **27:**129–135.

Kagan, V. E., Shvedova, A. A., Novikov, K. N., and Kozlov, Y. P., 1973, Light-induced free radical oxidation of membrane lipids in photoreceptors of frog retina, *Biochim. Biophys. Acta* **330:**76–79.

Kagan, V. E., Kuliev, I. Y., Spirichev, V. B., Shvedova, A. A., and Kozlov, Y. P., 1981, Accumulation of lipid peroxidation products and depression of retinal electrical activity in vitamin E-deficient rats exposed to high-intensity light, *Bull. Exp. Biol. Med.* **91:**165–167.

Kaiser-Kupfer, M. I., and Valle, D. L., 1986, Clinical, biochemical and therapeutic aspects of gyrate atrophy, *Prog. Retinal Res.* **6:**179–206.

Kamps, K. M. P., de Grip, W. J., and Daemen, F. J. M., 1982, Use of a density modification technique for isolation of the plasma membrane of rod outer segments, *Biochim. Biophys, Acta* **687:**296–302.

Kapoor, C. L., and Chader, G. J., 1984, Endogenous phosphorylation of retinal photoreceptor outer segment proteins by calcium phospholipid-dependent protein kinase, *Biochem. Biophys. Res. Comm.* **122:**1397–1403.

Kapoor, C. L., Kyritsis, A. P., and Chader, G. J., 1985, Alteration in gene expression at the onset of human Y-79 retinoblastoma cell differentiation, *Neurochem. Int.* **7:**285–294.

Kapoor, C. L., O'Brien, P. J., and Chader, G. J., 1987, Phorbol ester- and light-induced endogenous phosphorylation of rat rod outer-segment proteins, *Exp. Eye Res.* **45:**545–556.

Karpen, J. W., Zimmerman, A. L., Stryer, L., and Baylor, D. A., 1988, Gating kinetics of the cyclic-GMP-activated channel of retinal rods: Flash photolysis and voltage-jump studies, *Proc. Natl. Acad. Sci. USA* **85:**1287–1291.

Katz, M. L., and Robison, W. G., Jr., 1984, Age-related changes in the retinal pigment epithelium of pigmented rats, *Exp. Eye Res.* **38:**137–151.

Katz, M. L., and Robison, W. G., Jr., 1987, Light and aging effects on vitamin E in the retina and retinal pigment epithelium, *Vision Res.* **27:**1875–1879.

Katz, M. L., and Robison, W. G., Jr., 1988, Autoxidative damage to the retina: Potential role in retinopathy of prematurity, *Birth Defects* **24:**237–248.

Katz, M. L., Stone, W. L., and Dratz, E. A., 1978, Fluorescent pigment accumulation in retinal pigment epithelium of antioxidant-deficient rats, *Invest. Ophthalmol. Vis. Sci.* **17:**1049–1058.

Katz, M. L., Parker, K. R., Handelman, G. J., Bramel, T. L., and Dratz, E. A., 1982, Effects of antioxidant nutrient deficiency on the retina and retinal pigment epithelium of albino rats: A light and electron microscopic study, *Exp. Eye Res.* **34:**339–369.

Katz, M. L., Robison, W. G., Jr., Herrmann, R. K., Groome, A. B., and Bieri, J. G., 1984, Lipofuscin accumulation resulting from senescence and vitamin E deficiency: Spectral properties and tissue distribution, *Mech. Ageing Dev.* **25:**149–159.

Katz, M. L., Drea, C. M., and Robison, W. G., Jr., 1986a, Relationship between dietary retinol and lipofuscin in the retinal pigment epithelium, *Mech. Ageing Dev.* **35:**291–305.

Katz, M. L., Drea, C. M., Eldred, G. E., Hess, H. H., and Robison, W. G., Jr., 1986b, Influence of early photoreceptor degeneration on lipofuscin in the retinal pigment epithelium, *Exp. Eye Res.* **43:**561–573.

Katz, M. L., Drea, C. M., and Robison, W. G., Jr., 1987a, Age-related alterations in vitamin A metabolism in the rat retina, *Exp. Eye Res.* **44:**939–949.

Katz, M. L., Drea, C. M., and Robison, W. G., Jr., 1987b, Dietary vitamins A and E influence retinyl ester composition and content of the retinal pigment epithelium, *Biochim. Biophys. Acta* **924:**432–441.

Katz, M. L., Eldred, G. E., and Robison, W. G., Jr., 1987c, Lipofuscin autofluorescence: Evidence for vitamin A involvement in the retina, *Mech. Ageing Dev.* **39:**81–90.

Kean, E. L., 1977, GDP-mannose-polyprenyl phosphate mannosyltransferases of the retina, *J. Biol. Chem.* **252:**5622–5629.

Kean, E. L., 1980a, The lipid intermediate pathway in the retina for the activation of carbohydrates involved in the glycosylation of rhodopsin, *Neurochem. Int.* **1:**59–68.

Kean, E. L., 1980b, Stimulation by GDP-mannose of the biosynthesis of N-acetylglucosaminylpyrophosphoryl polyprenols by the retina, *J. Biol. Chem.* **255:**1921–1927.

Kean, E. L., 1982, Activation by dolichol phosphate-mannose of the biosynthesis of N-acetylglucosaminylpyrophosphoryl polyprenols by the retina, *J. Biol. Chem.* **257:**7952–7954.

Kean, E. L., 1983, Influence of metal ions on the biosynthesis of N-acetylglucosaminyl polyprenols by the retina, *Biochim. Biophys. Acta* **750:**268–273.

Kean, E. L., 1985, Stimulation by dolichol phosphate-mannose and phospholipids of the biosynthesis of N-acetylglucosaminylpyrophosphoryl dolichol, *J. Biol. Chem.* **260:**12561–12571.

Kean, E. L., Hara, S., Mizoguchi, A., Matsumoto, A., and Kobata, A., 1983, The enzymatic cleavage of rhodopsin by the retinal pigment epithelium. II. The carbohydrate composition of the glycopeptide cleavage product, *Exp. Eye Res.* **36:**817–825.

Kelleher, D. J., and Johnson, G. L., 1986, Phosphorylation of rhodopsin by protein kinase C *in vitro*, *J. Biol. Chem.* **261:**4749–4757.

Keller, R. K., Fliesler, S. J., and Nellis, S. W., 1988, Isoprenoid biosynthesis in the retina, *J. Biol. Chem.* **263:**2250–2254.

Kennedy, A., Frank, R. N., and Mancini, M. A., 1986a, *In vitro* production of glycosaminoglycans by retinal microvesel cells and lens epithelium, *Invest. Ophthalmol. Vis. Sci.* **27:**746–754.

Kennedy, A., Frank, R. N., Mancini, M. A., and Lande, M., 1986b, Collagens of the retinal microvascular basement membrane and of retinal microvascular cells *in vitro*, *Exp. Eye Res.* **42:**177–199.

Kern, T. S., and Engerman, R. L., 1985, Hexitrol production by canine retinal microvessels, *Invest. Ophthalmol. Vis. Sci.* **26:**382–384.

Khatami, M., Stramm, L. E., and Rockey, J. H., 1986a, Ascorbate transport in cultured cat retinal pigment epithelial cells, *Exp. Eye Res.* **43:**607–615.

Khatami, M., Li, W., and Rockey, J. H., 1986b, Kinetics of ascorbate transport by cultured retinal capillary pericytes, *Invest. Ophthalmol. Vis. Sci.* **27:**1665–1671.

King, G. L., Buzney, S. M., Kahn, C. R., Hetu, N., Buchwald, S., Macdonald, S. G., and Rand, L. I., 1983, Differential responsiveness to insulin of endothelial and support cells from micro- and macrovessels, *J. Clin. Invest.* **71:**974–979.

King, G. L., Goodman, A. D., Buzney, S., Moses, A., and Kahn, C. R., 1985, Receptors and growth-promoting effects of insulin and insulinlike growth factors on cells from bovine retinal capillaries and aorta, *J. Clin. Invest.* **75:**1028–1036.

Kirschfeld, K., 1982, Carotenoid pigments: Their possible role in protecting against photooxidation in eyes and photoreceptor cells, *Proc. R. Soc. Lond.* **216:**71–85.

Koch, K.-W., and Stryer, L., 1988, Highly cooperative feedback control of retinal rod guanylate cyclase by calcium ions, *Nature* **334:**64–66.

Koh, S.-W. M., and Chader, G. J., 1984a, Retinal pigment epithelium in culture demonstrates a distinct β-adrenergic receptor, *Exp. Eye Res.* **38:**7–13.

Koh, S.-W. M., and Chader, G. J., 1984b, Agonist effects on the intracellular cyclic AMP concentration of retinal pigment epithelial cells in culture, *J. Neurochem.* **42:**287–289.

Koh, S.-W. M., Kyritsis, A., and Chader, G. J., 1984, Interaction of neuropeptides and cultured glial (Müller) cells of the chick retina: Elevation of intracellular cyclic AMP by vasoactive intestinal peptide and glucagon, *J. Neurochem.* **43:** 199–203.

Kornfeld, R., and Kornfeld, S., 1985, Assembly of asparagine-linked oligosaccharides, *Annu. Rev. Biochem.* **54:** 631–664.

Kretzer, F. L., Mehta, R. S., Johnson, A. T., Hunter, D. G., Brown, E. S., and Hittner, H. M., 1984, Vitamin E protects against retinopathy of prematurity through action on spindle cells, *Nature* **309:**793–795.

Krinsky, N. I., 1958, The enzymatic esterification of vitamin A, *J. Biol. Chem.* **232:**881–894.

Krishna, G., Krishnan, N., Fletcher, R. T., and Chader, G. J., 1976, Effects of light on cyclic GMP metabolism in retinal photoreceptors, *J. Neurochem.* **27:**717–722.

Kuhn, H., 1978, Light-regulated binding of rhodopsin kinase and other proteins to cattle photoreceptor membranes, *Biochemistry* **17:**4389–4395.

Kuhn, H., 1980, Light- and GTP-regulated interaction of GTPase and other proteins with bovine photoreceptor membranes, *Nature* **283:**587–589.

Kyritsis, A. P., Wiggert, B., Lee, L., and Chader, G. J., 1985, Butyrate enhances the synthesis of interphotoreceptor retinoid-binding protein (IRBP) by Y-79 human retinoblastoma cells, *J. Cell. Physiol.* **124:**233–239.

Kyritsis, A. P., Tsokos, M., Triche, T. J., and Chader, G. J., 1986a, Retinoblastoma: A primitive tumor with multipotential characteristics, *Invest. Ophthalmol. Vis. Sci.* **27:**1760–1764.

Kyritsis, A. P., Kapoor, C. L., and Chader, G. J., 1986b, Distribution of the regulatory subunit of type II cAMP-dependent protein kinase in Y-79 retinoblastoma cells, *Invest. Ophthalmol. Vis. Sci.* **27:**1420–1423.

Kyritsis, A. P., Wiechmann, A. F., Bok, D., and Chader, G. J., 1987a, Hydroxyindole-*O*-methyltransferase in Y-79 human retinoblastoma cells: Effect of cell attachment, *J. Neurochem.* **48:**1612–1616.

Kyritsis, A. P., Tsokos, M., and Chader, G. J., 1987b, Behavior of human retinoblastoma cells in tissue culture, *Prog. Retinal Res.* **6:**245–274.

Lai, Y. L., Wiggert, B., Liu, Y. P., and Chader, G. J., 1982, Interphotoreceptor retinol-binding proteins: Possible transport vehicles between compartments of the retina, *Nature* **298:**848–849.

Laird, D. W., and Molday, R. S., 1988, Evidence against the role of rhodopsin in rod outer segment binding to RPE cells, *Invest. Ophthalmol. Vis. Sci.* **29:**419–428.

Lamb, T. D., 1986, Transduction in vertebrate photoreceptors: The roles of cyclic GMP and calcium, *Trends Neurosci.* **9:**224–228.

Landers, R. A., Varner, H. H., Tawara, A., Gay, C. A., Rayborn, M. E., and Hollyfield, J. G., 1989, Retinal contributions to chondroitin sulfate proteoglycans present in the mouse interphotoreceptor matrix, *Invest. Ophthalmol. Vis. Sci. Suppl.* **30:**489.

Lasater, E. M., 1987, Retinal horizontal cell gap junctional conductance is modulated by dopamine through a cyclic AMP-dependent protein kinase, *Proc. Natl. Acad. Sci. USA* **84:**7319–7323.

LaVail, M. M., 1976, Rod outer segment disk shedding in rat retina: Relationship to cyclic lighting, *Science* **194:**1071–1074.

LaVail, M. M., 1979, The retinal pigment epithelium in mice and rats with inherited retinal degeneration, in: *The Retinal Pigment Epithelium* (K. M. Zinn and M. F. Marmor, eds.), Harvard University Press, Cambridge, pp. 357–380.

LaVail, M. M., 1981, Analysis of neurological mutants with inherited retinal degeneration, *Invest. Ophthalmol. Vis. Sci.* **21:**638–656.

LaVail, M. M., and Gorrin, G. M., 1987, Protection from light damage by ocular pigmentation: Analysis using experimental chimeras and translocation mice, *Exp. Eye Res.* **44:**877–889.

LaVail, M. M., Pinto, L. H., and Yasumura, D., 1981, The interphotoreceptor matrix in rats with inherited retinal dystrophy, *Invest. Ophthalmol. Vis. Sci.* **21:**658–668.

LaVail, M. M., Yasumura, D., and Porrello, K., 1985, Histochemical analysis of the interphotoreceptor matrix in hereditary retinal degeneration, in: *The Interphotoreceptor Matrix in Health and Disease* (C. D. Bridges and A. J. Adler, eds.), Alan R. Liss, New York, pp. 179–193.

Law, W. C., and Rando, R. R., 1988, Stereochemical inversion at C-15 accompanies the enzymatic isomerization of *all-trans-* to 11-*cis*-retinoids, *Biochemistry* **27:**4147–4152.

Lawwill, T., 1982, Three major pathologic processes caused by light in the primate retina: A search for mechanisms, *Trans. Am. Ophthalmol. Soc.* **80:**517–579.

Lee, R. H., Brown, B. M., and Lolley, R. N., 1981, Protein kinases of retinal rod outer segments: Identification and partial characterization of cyclic nucleotide dependent protein kinase and rhodopsin kinase, *Biochemistry* **20:**7532–7538.

Lee, R. H., Brown, B. M., and Lolley, R. N., 1984, Light-induced dephosphorylation of a 33K protein in rod outer segments of rat retina, *Biochemistry* **23:**1972–1977.

Lee, R. H., Lieberman, B. S., Hurwitz, R. L., and Lolley, R. N., 1985, Phosphodiesterase-probes show distinct defects in *rd* mice and Irish setter dog disorders, *Invest. Ophthalmol. Vis. Sci.* **26:**1569–1579.

Lee, R. H., Lieberman, B. S., and Lolley, R. N., 1987, A novel complex from bovine visual cells of a 33000-dalton phosphoprotein with β- and γ-transducin: Purification and subunit structure, *Biochemistry* **26:**3983–3990.

Lee, R. H., Whelan, J. P., Lolley, R. N., and McGinnis, J. F., 1988a, The photoreceptor-specific 33 kDa phosphoprotein of mammalian retina: Generation of monospecific antibodies and localization by immunocytochemistry, *Exp. Eye Res* **46:**829–840.

Lee, R. H., Navon, S. E., Brown, B. M., Fung, B. K.-K., and Lolley, R. N., 1988b, Characterization of a phosphodiesterase- immunoreactive polypeptide from rod photoreceptors of developing *rd* mouse retinas, *Invest. Ophthalmol. Vis. Sci.* **29:**1021–1027.

Lentrichia, B. B., Bruner, W. E., and Kean, E. L., 1978, Glycosidases of the retinal pigment epithelium, *Invest. Ophthalmol. Vis. Sci.* **17:**884–895.

Lentrichia, B. B., Itoh, Y., Plantner, J. J., and Kean, E. L., 1987, The influence of carbohydrates on the binding of rod outer-segment (ROS) disc membranes and intact ROS by the cells of the retinal pigment epithelium of the embryonic chick, *Exp. Eye Res.* **44:**127–142.

Lerea, C. L., Somers. D. E., Hurley, J. B., Klock, I. B., and Bunt-Milam, A. H., 1986, Identification of specific transducin α subunits in retinal rod and cone photoreceptors, *Science* **234:**77–80.

Lewis, G. P., Erickson, P. A., Kaska, D. D., and Fisher, S. K., 1988a, An immunocytochemical comparison of Müller cells and astrocytes in the cat retina, *Exp. Eye Res.* **47:**839–853.

Lewis, G. P., Kaska, D. D., Vaughan, D. K., and Fisher, S. K., 1988b, An immunocytochemical study of cat retinal Müller cells in culture, *Exp. Eye Res.* **47:**855–868.

Li, L., and Turner, J. E., 1988, Inherited retinal dystrophy in the RCS rat: Prevention of photoreceptor degeneration by pigment epithelial cell transplantation, *Exp. Eye Res.* **47:**911–917.

Li, W., Stramm, L. E., Aguirre, G. D., and Rockey, J. H., 1984, Extracellular matrix production by cat retinal pigment epithelium *in vitro*: Characterization of type IV collagen synthesis, *Exp. Eye Res.* **38:**291–304.

Li, W., Chan, L. S., Khatami, M., and Rockey, J. H., 1985, Characterization of glucose transport by bovine retinal capillary pericytes in culture, *Exp. Eye Res.* **41:**191–199.

Li, W., Chan, L. S., Khatami, M., and Rockey, J. H., 1986, Non-competitive inhibition of *myo*-inositol transport in cultured bovine retinal capillary pericytes by glucose and reversal by sorbinil, *Biochim. Biophys. Acta* **857:**198–208.

Li, W., Hu, T. S., Stramm, L. E., Rockey, J. H., and Liu, S. L., 1987a, Synergistic activation of retinal capillary pericyte proliferation in culture by inositol triphosphate and diacylglycerol, *Exp. Eye Res.* **44:**29–35.

Li, W., Zhou, Q., Qin, M., and Hu, T.-S., 1987b, Reduction of inositol triphosphate in retinal microvessels by glucose and restimulation by *myo*-inositol, *Exp. Eye Res.* **45:**517–524.

Li, Z.-Y., Tso, M. O. M., Wang, H.-M., and Organisciak, D. T., 1985, Amelioration of photic injury in rat retina by ascorbic acid: A histopathologic study, *Invest. Ophthalmol. Vis. Sci* . **26:**1589–1598.

Liang, C.-J., Yamashita, K., Muellenberg, C. G., Shichi, H., and Kobata, A., 1979, Structure of the carbohydrate moieties of bovine rhodopsin, *J. Biol. Chem.* **254:**6414–6418.

Liebman, P. A., Parker, K. R., and Dratz, E. A., 1987, The molecular mechanism of visual excitation and its relation to the structure and composition of the rod outer segment, *Annu. Rev. Physiol.* **49:**765–791.

Liou, G. I., Fong, S.-L., and Bridges, C. D. B., 1981, Comparison of cytosol retinol binding proteins from bovine retina, dog liver, and rat liver, *J. Biol. Chem.* **256:**3153–3155.

Liou, G. I., Bridges, C. D. B., Fong, S.-L., Alvarez, R. A., and Gozalez-Fernandez, F., 1982, Vitamin A transport between retina and pigment epithelium—An interstitial protein carrying endogenous retinol (interstitial retinol-binding protein), *Vision Res.* **22:**1457–1467.

Liou, G. I., Fong, S.-L., Beattie, W. G., Cook, R. G., Leone, J., Landers, R. A., Alvarez, R. A., Wang, C., Li, Y., and Bridges, C. D. B., 1986, Bovine interstitial retinol-binding protein (IRBP)—Isolation and sequence analysis of cDNA clones, characterization and *in vitro* translation of mRNA, *Vision Res.* **26:**1645–1653.

Liou, G. I., Ma, D.-P., Yang, Y.-W., Geng, L., Zhu, C., and Baehr, W. 1989, Human interstitial retinoid-binding protein. Gene structure and primary sequence, *J. Biol. Chem.* **264:**8200-8206.

Lolley, R. N., 1983, Metabolism of retinal rod outer segments, in: *Biochemistry of the Eye* (R. E. Anderson, ed.), American Academy of Ophthalmology, San Francisco, pp. 178–188.

Lopez, R., Gouras, P., Kjeldbye, H., Sullivan, B., Reppucci, V., Brittis, M., Wapner, F., and Goluboff, E., 1989, Transplanted retinal pigment epithelium modifies the retinal degeneration in the RCS rat, *Invest. Ophthalmol. Vis. Sci.* **30:**586–588.

Louie, K., Wiegand, R. D., and Anderson, R. E., 1988, Docosahexaenoate-containing molecular species of glycerophospholipids from frog retinal rod outer segments show different rates of biosynthesis and turnover, *Biochemistry* **27:**9014–9020.

Lowry, O. H., Roberts, N. R., and Lewis, C., 1956, The quantitative histochemistry of the retina, *J. Biol. Chem.* **220:**879–892.

Lowry, O. H., Roberts, N. R., Schultz, D. W., Clow, J. E., and Clark, J. R., 1961, Quantitative histochemistry of retina. II. Enzymes of glucose metabolism, *J. Biol. Chem.* **236:**2813–2820.

Lubin, B., and Machlin, L. J. (eds.), 1982, *Vitamin E: Biochemical, Hematological and Clinical Aspects, Annals of the New York Academy of Sciences*, Vol. 393, The New York Academy of Sciences, New York.

Malinow, M. R., Feeney-Burns, L., Peterson, L. H., Klein, M. L., and Neuringer, M., 1980, Diet-related macular anomalies in monkeys, *Invest. Ophthalmol. Vis. Sci.* **19:**857–869.

Marmor, M. F., 1989, Mechanisms of normal retinal adhesion, in: *Retina* (S. J. Ryan, ed.), C. V. Mosby, St. Louis, pp. 71–87.

Martin, R. L., Wood, C., Baehr, W., and Applebury, M. L., 1986, Visual pigment homologies revealed by DNA hybridization, *Science* **232:**1266–1269.

Masland, R. H., and Mills, J. W., 1980, Choline accumulation by photoreceptor cells of the rabbit retina, *Proc. Natl. Acad. Sci. USA* **77:**1671–1675.

Masland, R. H., and Tauchi, M., 1986, The cholinergic amacrine cell, *Trends Neurosci.* **9:**218–223.

Massey, S. C., and Redburn, D. A., 1987, Transmitter circuits in the vertebrate retina, *Prog. Neurobiol.* **28:**55–96.

Masterson, E., and Chader, G. J., 1981a, Pigment epithelial cells in culture. Metabolic pathways required for phagocytosis, *Invest. Ophthalmol. Vis. Sci.* **20:**1–7.

Masterson, E., and Chader, G. J., 1981b, Characterization of glucose transport by cultured chick pigmented epithelium, *Exp. Eye Res.* **32:**279–289.

Masterson, E., Israel, P., and Chader, G. J., 1978, Pentose shunt activity in developing chick retina and pigment epithelium: A switch in biochemical expression in cultured pigment epithelial cells, *Exp. Eye Res.* **27:**409–416.

Matesic, D., and Liebman, P. A., 1987, cGMP-dependent cation channel of retinal rod outer segments, *Nature* **326:**600–603.

Matheke, M. L., and Holtzman, E., 1984, The effects of monensin and of puromycin on transport of membrane components in the frog retinal photoreceptor. II. Electron microscopic autoradiography of proteins and glycerolipids, *J. Neurosci.* **4:**1093–1103.

Matheke, M. L., Kalff, M., and Holtzman, E., 1983, Effects of monensin on photoreceptors of isolated frog retinas, *Tissue Cell* **15:**509–513.

Matheke, M. L., Fliesler, S. J., Basinger, S. F., and Holtzman, E., 1984, The effects of monensin on transport of membrane components in the frog retinal photoreceptor. I. Light microscopic autoradiography and biochemical analysis, *J. Neurosci.* **4:**1086–1092.

Matthews, G., 1987, Single-channel recordings demonstrate that cGMP opens the light-sensitive ion channel of the rod photoreceptor, *Proc. Natl. Acad. Sci. USA* **84:**299–302.

Mayerson, P. L., and Hall, M. O., 1986, Rat retinal pigment epithelial cells show specificity of phagocytosis *in vitro*, *J. Cell Biol.* **103:**299–308.

Mayerson, P. L., Hall, M. O., Clark, V., and Abrams, T., 1985, An improved method for isolation and culture of rat retinal pigment epithelial cells, *Invest. Ophthalmol. Vis. Sci.* **26:**1599–1609.

McLaughlin, B. J., and Boykins, L. G., 1987, Examination of sialic acid binding on dystrophic and normal retinal pigment epithelium, *Exp. Eye Res.* **44:**439–450.

Medynski, D. C., Sullivan, K., Smith, D., Van Dop, C., Chang, F.-H., Fung, B. K.-K., Seeburg, P. H., and Bourne, H. R., 1985, Amino acid sequence of the α subunit of transducin deduced from the cDNA sequence, *Proc. Natl. Acad. Sci. USA* **82:**4311–4315.

Mercurio, A. M., and Holtzman, E., 1982, Ultrastructural localization of glycerolipid synthesis in rod cells of the isolated frog retina, *J. Neurocytology* **11:**295–322.

Miceli, M. V., Kan, L.-S., and Newsome, D. A., 1987, Phosphorus-31 NMR spectroscopy of cultured human retinal pigmented epithelial cells, *Invest. Ophthalmol. Vis. Sci.* **28:**70–75.

Miljanich, G. P., Sklar, L. A., White, D. L., and Dratz, E. A., 1979, Disaturated and dipolyunsaturated phospholipids in the bovine retinal rod outer segment disk membrane, *Biochim. Biophys. Acta* **552:**294–306.

Millar, F. A., and Hawthorne, J. N., 1985, Polyphosphoinositide metabolism in response to light stimulation of retinal rod outer segments, *Biochem. Soc. Trans.* **13:**185–186.

Miller, J. L., and Dratz, E. A., 1984, Phosphorylation at sites near rhodopsin's carboxyl-terminus reglates light initiated cGMP hydrolysis, *Vision Res.* **24:**1509–1521.

Miller, S. S., Hughes, B. A., and Machen, T. E., 1982, Fluid transport across retinal pigment epithelium is inhibited by cyclic AMP, *Proc. Natl. Acad. Sci. USA* **79:**2111–2115.

Mitchell, G. A., Brody, L. C., Looney, J., Steel, G., Suchanek, M., Dowling, C., Der Kaloustian, V., Kaiser-Kupfer, M., and Valle, D., 1988, An initiator codon mutation in ornithine-δ- aminotransferase causing gyrate atrophy of the choroid and retina, *J. Clin, Invest.* **81:**630–633.

Mitzel, D. L., Hall, I. A., DeVries, G. W., Cohen, A. I., and Ferrendelli, J. A., 1978, Comparison of cyclic nucleotide and energy metabolism of intact mouse retina *in situ* and *in vitro*, *Exp. Eye Res.* **27:**27–37.

Molday, L. L., and Molday, R. S., 1987, Glycoproteins specific for the retinal rod outer segment plasma membrane, *Biochim. Biophys. Acta* **897:**335–340.

Molday, R. S., 1988, Monoclonal antibodies to rhodopsin and other proteins of rod outer segments, *Prog. Retinal Res.* **8:** 173–209.

Molday, R. S., and Molday, L. L., 1979, Identification and characterization of multiple forms of rhodopsin and minor proteins in frog and bovine rod outer segment disc membranes, *J. Biol. Chem.* **254:**4653–4660.

Molday, R. S., and Molday, L. L., 1987, Differences in the protein composition of bovine retinal rod outer segment disk and plasma membranes isolated by a ricin–gold–dextran density perturbation method, *J. Cell Biol.* **105:**2589–2601.

Molday, R. S., Hicks, D., and Molday, L., 1987, Peripherin. A rim-specific membrane protein of rod outer segment discs, *Invest. Ophthalmol. Vis. Sci.* **28:**50–61.

Morjaria, B., and Voaden, M. J., 1979, The formation of glutamate, aspartate and GABA in the rat retina; glucose and glutamine as precursors, *J. Neurochem.* **33:**541–551.

Mullen, R. J., and LaVail, M. M., 1976, Inherited retinal dystrophy: Primary defect in pigment epithelium determined with experimental rat chimeras, *Science* **192:**799–801.

Naash, M. I., and Anderson, R. E., 1984, Characterization of glutathione peroxidase in frog retina, *Curr. Eye Res.* **3:**1299–1304.

Naash, M. I., and Anderson, R. E., 1989, Glutathione-dependent enzymes in intact rod outer segments, *Exp. Eye Res.* **48:**309–318.

Naash, M. I., Nielsen, J. C., and Anderson, R. E., 1988, Regional distribution of glutathione peroxidase and glutathione S-transferase in adult and premature human retinas, *Invest. Ophthalmol. Vis. Sci.* **29:**149–152.

Nagata, M., and Robison, W. G., Jr., 1987, Aldose reductase, diabetes, and thickening of the retinal inner limiting membrane, *Invest. Ophthalmol. Vis. Sci.* **28:**1867–1869.

Nathans, J., and Hogness, D. S., 1983, Isolation, sequence analysis, and intron–exon arrangement of the gene encoding bovine rhodopsin, *Cell* **34:**807–814.

Nathans, J., and Hogness, D. S., 1984, Isolation and nucleotide sequence of the gene encoding human rhodopsin, *Proc. Natl. Acad. Sci. USA* **81:**4851–4855.

Nathans, J., Thomas, D., and Hogness, D. S., 1986a, Molecular genetics of human color vision: The genes encoding blue, green, and red pigments, *Science* **232:**193–202.

Nathans, J., Piantanida, T. P., Eddy, R. L., Shows, T. B., and Hogness, D. S., 1986b, Molecular genetics of inherited variation in human color vision, *Science* **232:**203–210.

Naveh-Floman, N., Weissman, C., and Belkin, M., 1984, Arachidonic acid metabolism by retinas of rats with streptozotocin-induced diabetes, *Curr. Eye Res.* **3:**1135–1139.

Navon, S. E., Lee, R. H., Lolley, R. N., and Fung, B. K.-K., 1987, Immunological determination of transducin content in retinas exhibiting inherited degeneration, *Exp. Eye Res.* **44:**115–125.

Nemes, P. P., and Dratz, E. A., 1982, Preparation of isolated osmotically intact bovine rod outer segment disk membranes, in: *Methods in Enzymology*, Vol. 81 (L. Packer, ed.), Academic Press, New York, pp. 116–123.

Neuringer, M., and Sturman, J., 1987, Visual acuity loss in rhesus monkey infants fed a taurine-free human infant formula, *J. Neurosci. Res.* **18:**597–601.

Neuringer, M., Connor, W. E., Van Petten, C., and Barstad, L., 1984, Dietary omega-3 fatty acid deficiency and visual loss in infant rhesus monkeys, *J. Clin. Invest.* **73:**272–276.

Newsome, D. A., and Rothman, R. J., 1987, Zinc uptake *in vitro* by human retinal pigment epithelium, *Invest. Ophthalmol. Vis. Sci.* **28:**1795–1799.

Newsome, D. A., Fletcher, R. T., and Chader, G. J., 1980, Cyclic nucleotides vary by area in the retina and pigmented epithelium of the human and monkey, *Invest. Ophthalmol. Vis. Sci.* **19:**864–869.

Newsome, D. A., Hewitt, A. T., Huh, W., Robey, P. G., and Hassell, J. R., 1987, Detection of specific extracellular matrix molecules in Drusen, Bruch's membrane, and ciliary body, *Am. J. Ophthalmol.* **104:**373–381.

Newsome, D. A., Chader, G. J., Wiggert, B., Yeo, J. H., Welch, R. B., Blacharski, P. A., and Charles, S. T., 1988a, Interphotoreceptor retinoid-binding protein levels in subretinal fluid from rhegmatogenous retinal detachment and retinopathy of prematurity, *Arch. Ophthalmol.* **106:**106–110.

Newsome, D. A., Pfeffer, B. A., Hewitt, A. T., Robey, P. G., and Hassell, J. R., 1988b, Detection of extracellular matrix molecules synthesized *in vitro* by monkey and human retinal pigment epithelium: Influence of donor age and multiple passages, *Exp. Eye Res.* **46:**305–321.

Nielsen, J. C., Naash, M. I., and Anderson, R. E., 1988, The regional distribution of vitamins E and C in mature and premature human retinas, *Invest. Ophthalmol. Vis. Sci.* **29:**22–26.

Nir, I. and Hall, M. O., 1979, Ultrastructural localization of lectin binding sites on the surface of retinal photoreceptors and pigment epithelium, *Exp. Eye Res.* **29:**181–194.

Nir, I., and Papermaster, D. S., 1986, Immunocytochemical localization of opsin in the inner segment and ciliary plasma membrane of photoreceptors in retinas of *rds* mutant mice, *Invest. Ophthalmol. Vis. Sci.* **27:**836–840.

Noell, W. K., 1980, Possible mechanisms of photoreceptor damage by light in mammalian eyes, *Vision Res.* **20:**1163–1171.

Norenberg, M. D., Dutt, K., and Reif-Lehrer, L., 1980, Glutamine synthetase localization in cortisol-induced chick embryo retinas, *J. Cell Biol.* **84:**803–807.

O'Brien, P. J., 1976, Rhodopsin as a glycoprotein: A possible role for the oligosaccharide in phagocytosis, *Exp. Eye Res.* **23:**127–137.

O'Brien, P. J., and Zatz, M., 1984, Acylation of bovine rhodopsin by [^{3}H]palmitic acid, *J. Biol. Chem.* **259:**5054–5057.

O'Brien, P. J., St. Jules, R. S., Reddy, T. S., Bazan, N. G., and Zatz, M., 1987, Acylation of disc membrane rhodopsin may be nonenzymatic, *J. Biol. Chem.* **262:**5210–5215.

Ocumpaugh, D. E., and Young, R. W., 1966, Distribution and synthesis of sulfated mucopolysaccharides in the retina of the rat, *Invest. Ophthalmol.* **5:**196–203.

Ogino, N., Matsumura, M., Shirakawa, H., and Tsukahara, I., 1983, Phagocytic activity of cultured retinal pigment epithelial cells from chick embryo: Inhibition by melatonin and cyclic AMP, and its reversal taurine and cyclic GMP, *Ophthalmic Res.* **15:**72–89.

Okajima, T.-I. L., Pepperberg, D. R., Ripps, H., Wiggert, B., and Chader, G. J., 1989, Interphotoreceptor retinoid-binding protein: Role in delivery of retinol to the pigment epithelium, *Exp. Eye Res.* **49:**629–644.

Organisciak, D. T., and Noell, W. K., 1977, The rod outer segment phospholipid/opsin ratio of rats maintained in darkness or cyclic light, *Invest. Ophthalmol. Vis. Sci.* **16:**188–190.

Organisciak, D. T., Wang, H.-M., and Kou, A. L., 1982, Rod outer segment lipid-opsin ratios in the developing normal and retinal dystrophic rat, *Exp. Eye Res.* **34:**401–412.

Organisciak, D. T., Favreau, P., and Wang, H. M., 1983, The enzymatic estimation of organic hydroperoxides in the rat retina, *Exp. Eye Res.* **36:**337–349.

Organisciak, D. T., Wang, H. M., and Kou, A. L., 1984, Ascorbate and glutathione levels in the developing normal and dystrophic rat retina: Effect of intense light exposure, *Curr. Eye Res.* **3:**257–267.

Organisciak, D. T., Wang, H.-M., Li, Z.-Y., and Tso, M. O. M., 1985, The protective effect of ascorbate in retinal light damage of rats, *Invest. Ophthalmol. Vis. Sci.* **26:**1580–1588.

Organisciak, D. T., Wang, H.-M., Noell, W. K., Plantner, J. J., and Kean, E. L., 1986, Rod outer segment lipids in vitamin A-adequate and -deficient rats, *Exp. Eye Res.* **42:**73–82.

Organisciak, D. T., Berman, E. R., Wang, H.-M., and Feeney-Burns, L., 1987, Vitamin E in human neural retina and retinal pigment epithelium: Effect of age, *Curr. Eye Res.* **6:**1051–1055.

Orr, H. T., Lowry, O. H., Cohen, A. I., and Ferrendelli, J. A., 1976, Distribution of 3′:5′-cyclic AMP and 3′:5′-cyclic GMP in rabbit retina *in vivo*: Selective effects of dark and light adaptation and ischemia, *Proc. Natl. Acad. Sci. USA* **73:**4442–4445.

Osborne, N. N., 1984, Indoleamines in the eye with special reference to the serotonergic neurons of the retina, *Prog. Retinal Res.* **3:**61–103.

Ostwald, T. J., and Steinberg, R. H., 1980, Localization of frog retinal pigment epithelium Na^{+}-K^{+} ATPase, *Exp. Eye Res.* **31**:351–360.

Ostwald, T. J., and Steinberg, R. H., 1981, Transmembrane components of taurine flux across frog retinal pigment epithelium, *Curr. Eye Res.* **1**:437–443.

Ottonello, S., Petrucec, S., and Maraini, G., 1987, Vitamin A uptake from retinol-binding protein in a cell-free system from pigment epithelial cells of bovine retina, *J. Biol. Chem.* **262**:3975–3981.

Ovchinnikov, Y. A., Abdulaev, N. G., and Feigina, M. Y., 1982, The complete amino acid sequence of visual rhodopsin, *Bioorg. Khim.* **8**:1011–1014.

Owen, W. G., 1987, Ionic conductances in rod photoreceptors, *Annu. Rev. Physiol.* **49**:743–764.

Pahuja, S. L., and Reid, T. W., 1985, Bovine retinal glutamine synthetase 2. Regulation and properties on the basis of glutamine synthetase and glutamyl transferase reactions, *Exp. Eye Res.* **40**:75–83.

Pahuja, S. L., Mullins, B. T., and Reid, T. W., 1985, Bovine retinal glutamine synthetase 1. Purification, characterization and immunological properties, *Exp. Eye Res.* **40**:61–74.

Palczewski, K., McDowell, J. H., and Hargrave, P. A., 1988a, Rhodopsin kinase: Substrate specificity and factors that influence activity, *Biochemistry* **27**:2306–2313.

Palczewski, K., McDowell, J. H., and Hargrave, P. A., 1988b, Purification and characterization of rhodopsin kinase, *J. Biol. Chem.* **263**:14067–14073.

Palczewski, K., Hargrave, P. A., McDowell, J. H., and Ingebritsen, T. S., 1989, The catalytic subunit of phosphatase 2A dephosphorylates phosphoopsin, *Biochemistry* **28**:415–419.

Papermaster, D. S., 1982, Preparation of retinal rod outer segments, in: *Methods in Enzymology*, Vol. 81 (L. Packer, ed.), Academic Press, New York, pp. 48–52.

Papermaster, D. S., and Dreyer, W. J., 1974, Rhodopsin content in the outer segment membranes of bovine and frog retinal rods, *Biochemistry* **13**:2438–2444.

Papermaster, D. S., and Schneider, B. G., 1982, Biosynthesis and morphogenesis of outer segment membranes in vertebrate photoreceptor cells, in: *Cell Biology of the Eye* (D. S. McDevitt, ed.), Academic Press, New York, pp. 475–531.

Papermaster, D. S., Schneider, B. G., Zorn, M. A., and Kraehenbuhl, J. P., 1978a, Immunocytochemical localization of a large intrinsic membrane protein to the incisures and margins of frog rod outer segment disks, *J. Cell Biol.* **78**:415–425.

Papermaster, D. S., Schneider, B. G., Zorn, M. A., and Kraehenbuhl, J. P., 1978b, Immunocytochemical localization of opsin in outer segments and Golgi zones of frog photoreceptor cells, *J. Cell Biol.* **77**:196–210.

Papermaster, D. S., Burstein, Y., and Schechter, I., 1980, Opsin mRNA isolation from bovine retina and partial sequence of the *in vitro* translation product, *Ann. N.Y. Acad. Sci.* **343**:347–355.

Papermaster, D. S., Schneider, B. G., and Besharse, J. C., 1985, Vesicular transport of newly synthesized opsin from the Golgi apparatus toward the rod outer segment, *Invest. Ophthalmol. Vis. Sci.* **26**:1386–1404.

Parks, J. M., Ames, A., III, and Nesbett, F. B., 1976, Protein synthesis in central nervous tissue: Studies on retina *in vitro*, *J. Neurochem.* **27**:987–997.

Pasantes-Morales, H., 1986, Current concepts on the role of taurine in the retina, *Prog. Retinal Res.* **5**:207–229.

Pautler, E. L., and Tengerdy, C., 1986, Transport of acidic amino acids by the bovine pigment epithelium, *Exp. Eye Res.* **43**:207–214.

Payne, R., Corson, D. W., Fein, A., and Berridge, M. J., 1986, Excitation and adaptation of *Limulus* ventral photoreceptors by inositol 1,4,5 trisphosphate result from a rise in intracellular calcium, *J. Gen. Physiol.* **88**:127–142.

Penn, J. S., and Anderson, R. E., 1987, Effect of light history on rod outer-segment membrane composition in the rat, *Exp. Eye Res.* **44**:767–778.

Penn, J. S., Naash, M. I., and Anderson, R. E., 1987, Effect of light history on retinal antioxidants and light damage susceptibility in the rat, *Exp. Eye Res.* **44**:779–788.

Pfeffer, B., Wiggert, B., Lee, L., Zonnenberg, B., Newsome, D., and Chader, G. J., 1983, The presence of a soluble interphotoreceptor retinol-binding protein (IRBP) in the retinal interphotoreceptor space, *J. Cell. Physiol.* **117**:333–341.

Pfeffer, B. A., Clark, V. M., Flannery, J. G., and Bok, D., 1986, Membrane receptors for retinol-binding protein in cultured human retinal pigment epithelium, *Invest. Ophthalmol. Vis. Sci.* **27**:1031–1040.

Pfister, C., Chabre, M., Plouet, J., Tuyen, V. V., De Kozak, Y., Faure, J. P., and Kuhn, H., 1985, Retinal S antigen identified as the 48K protein regulating light-dependent phosphodiesterase in rods, *Science* **228**:891–893.

Philp, N. J., and Nachmias, V. T., 1985, Components of the cytoskeleton in the retinal pigmented epithelium of the chick, *J. Cell Biol.* **101:**358–362.

Philp, N. J., and Nachmias, V. T., 1987, Polarized distribution of integrin and fibronectin in retinal pigment epithelium, *Invest. Ophthalmol. Vis. Sci.* **28:**1275–1280.

Philp, N. J., Nachmias, V. T., Lee, D., Stramm, L., and Buzdygon, B., 1988, Is rhodopsin the ligand for receptor-mediated phagocytosis of rod outer segments by retinal pigment epithelium? *Exp. Eye Res.* **46:**21–28.

Pino, R. M., 1986, Immunocytochemical localization of fibronectin to the retinal pigment epithelium of the rat, *Invest. Ophthalmol. Vis. Sci.* **27:**840–844.

Plantner, J. J., and Kean, E. L., 1988, The dolichol pathway in the retina: Oligosaccharide—lipid biosynthesis, *Exp. Eye Res.* **46:**785–800.

Plantner, J. J., Poncz, L., and Kean, E. L., 1980, Effect of tunicamycin on the glycosylation of rhodopsin, *Arch. Biochem. Biophys.* **201:**527–532.

Porrello, K., and LaVail. M. M., 1986a, Immunocytochemical localization of chondroitin sulfates in the interphotoreceptor matrix of the normal and dystrophic rat retina, *Curr. Eye Res.* **5:**981–993.

Porrello, K., and LaVail, M. M., 1986b, Histochemical demonstration of spatial heterogeneity in the interphotoreceptor matrix of the rat retina, *Invest. Ophthalmol. Vis. Sci.* **27:**1577–1586.

Porrello, K., Yasumura, D., and LaVail, M. M., 1986, The interphotoreceptor matrix in RCS rats: Histochemical analysis and correlation with the rate of retinal degeneration, *Exp. Eye Res.* **43:**413–429.

Porrello, K., Yasumura, D., and LaVail, M. M., 1989, Immunogold localization of chondroitin 6-sulfate in the interphotoreceptor matrix of normal and RCS rats, *Invest. Ophthalmol. Vis. Sci.* **30:**638–651.

Poulsom, R., 1987, Inhibition of aldose reductase from human retina, *Curr. Eye Res.* **6:**427–432.

Preti, A., Fiorilli, A., Dreyfus, H., Harth, S., Urban, P. F., and Mandel, P., 1978, Developmental studies of the catabolism of sialocompounds by endogenous neuraminidase in the chick retina, *Exp. Eye Res.* **26:**621–628.

Pu, G. A-W., and Anderson, R. E., 1983, Alteration of retinal choline metabolism in an experimental model for photoreceptor cell degeneration, *Invest. Ophthalmol. Vis. Sci.* **24:**288–293.

Pu, G. A-W., and Masland, R. H., 1984, Biochemical interruption of membrane phospholipid renewal in retinal photoreceptor cells, *J. Neurosci.* **4:**1559–1576.

Pugh, E. N., Jr., 1987, The nature and identity of the internal excitational transmitter of vertebrate phototransduction, *Annu. Rev. Physiol.* **49:**715–741.

Pugh, E. N., Jr., and Cobbs, W. H., 1986, Visual transduction in vertebrate rods and cones: A tale of two transmitters, calcium and cyclic GMP, *Vision. Res.* **26:**1613–1643.

Quesada, O., Picones, A., and Pasantes-Morales, H., 1988, Effect of light deprivation on the ERG responses of taurine-deficient rats, *Exp. Eye Res.* **46:**13–20.

Ramesh, V., McClatchey, A. I., Ramesh, N., Benoit, L. A., Berson, E. L., Shih, V. E., and Gusella, J. F., 1988, Molecular basis of ornithine aminotransferase deficiency in B-6-responsive and -nonresponsive forms of gyrate atrophy, *Proc. Natl. Acad. Sci. USA* **85:**3777–3780.

Rando, R. R., and Bangerter, F. W., 1982, The rapid intermembraneous transfer of retinoids, *Biochem. Biophys. Res. Comm.* **104:**430–436.

Rao, N. A., Thaete, L. G., Delmage, J. M., and Sevanian, A., 1985, Superoxide dismutase in ocular structures, *Invest. Ophthalmol. Vis. Sci.* **26:**1778–1781.

Rapp, L. M., Thum, L. A., Tarver, A. P., and Wiegand, R. D., 1987, Retinal changes associated with taurine depletion in pigmented rats, in: *Degenerative Retinal Disorders: Clinical and Laboratory Investigations* (J. C. Hollyfield, R. E. Anderson, and M. M. LaVail, eds.), Alan R. Liss, New York, pp. 485–495.

Rapp, L. M., Thum, L. A., and Anderson, R. E., 1988, Synergism between environmental lighting and taurine depletion in causing photoreceptor cell degeneration, *Exp. Eye Res.* **46:**229–238.

Redburn, D. A., and Hollyfield, J., 1983, Chemical transmission in retina, in: *Biochemistry of the Eye* (R. E. Anderson, ed.), American Academy of Ophthalmology, San Francisco, pp. 196–226.

Redburn, D. A., and Madtes, P, Jr., 1987, GABA - Its role and development in retina, in: *Prog. Retinal Res.* **6:**69–84.

Reddy, T. S., and Bazan, N. G., 1984, Synthesis of arachidonoyl coenzyme A and docosahexaenoyl coenzyme A in retina, *Curr. Eye Res.* **3:**1225–1232.

Redmond, T. M., Wiggert, B., Robey, F. A., Nguyen, N. Y., Lewis, M. S., Lee, L., and Chader, G. J., 1985, Isolation and characterization of monkey interphotoreceptor retinoid-binding protein, a unique extracellular matrix component of the retina, *Biochemistry* **24:**787–793.

Redmond, T. M., Wiggert, B., Robey, F. A., and Chader, G. J., 1986, Interspecies conservation of structure of interphotoreceptor retinoid-binding protein, *Biochem. J.* **240:**19–26.

Redmond, T. M., Sanui, H., Nickerson, J. M., Borst, D. E., Wiggert, B., Kuwabara, T., and Gery, I., 1988, Cyanogen bromide fragments of bovine interphotoreceptor retinoid-binding protein induce experimental autoimmune uveoretinitis in Lewis rats, *Curr. Eye Res.* **7:**375–385.

Regan, C. M., de Grip, W. J., Daemen, F. J. M., and Bonting, S. L., 1980, Degradation of rhodopsin by a lysosomal fraction of retinal pigment epithelium: Biochemical aspects of the visual process. XLI, *Exp. Eye Res.* **30:**183–191.

Reif-Lehrer, L., 1984, Glutamate metabolism in the retina: A larger perspective, in: *Current Topics in Eye Research*, Vol. 4 (J. A. Zadunaisky and H. Davson, eds.), Academic Press, Orlando, FL, pp. 1–95.

Reif-Lehrer, L., Bergenthal, J., and Hanninen, L., 1975, Effects of monosodium glutamate on chick embryo retina in culture, *Invest. Ophthalmol.* **14:**114–124.

Robey, P. G., and Newsome, D. A., 1983, Biosynthesis of proteoglycans present in primate Bruch's membrane, *Invest. Ophthalmol. Vis. Sci.* **24:**898–905.

Robison, W. G., Jr., and Kuwabara, T., 1977, Vitamin A storage and peroxisomes in retinal pigment epithelium and liver, *Invest. Ophthalmol. Vis. Sci.* **16:**1110–1116.

Robison, W. G., Jr., Kador, P. F., and Kinoshita, J. H., 1983, Retinal capillaries: Basement membrane thickening by galactosemia prevented with aldose reductase inhibitor, *Science* **221:**1177–1179.

Robison, W. G., Jr., Nagata, M., and Kinoshita, J. H., 1988, Aldose reductase and retinal capillary basement membrane thickening, *Exp. Eye Res.* **46:**343–348.

Rodrigues, M. M., Wiggert, B., Hackett, J., Lee, L., Fletcher, R. T., and Chader, G. J., 1985, Dominantly inherited retinitis pigmentosa. Ultrastructure and biochemical analysis, *Ophthalmology* **92:**1165–1172.

Rodrigues, M. M., Wiggert, B., Shields, J., Donoso, L., Bardenstein, D., Katz, N., Friendly, D., and Chader, G. J., 1987a, Retinoblastoma. Immunohistochemistry and cell differentiation, *Ophthalmology* **94:**378–387.

Rodrigues, M. M., Hackett, J., Wiggert, B., Gery, I., Spiegel, A., Krishna, G., Stein, P., and Chader, G. J., 1987b, Immunoelectron microscopic localization of photoreceptor-specific markers in the monkey retina, *Curr. Eye Res.* **6:**369–380.

Rohlich, P., 1970, The interphotoreceptor matrix: Electron microscopic and histochemical observations on the vertebrate retina, *Exp. Eye Res.* **10:**80–96.

Roof, D. J., and Heuser, J. E., 1982, Surfaces of rod photoreceptor disk membranes: Integral membrane components, *J. Cell Biol.* **95:**487–500.

Rothman, H., Feeney, L., and Berman, E. R., 1976, The retinal pigment epithelium. Analytical subcellular fractionation with special reference to acid lipase, *Exp. Eye Res.* **22:**519–532.

Saari, J. C., and Bredberg, L., 1982, Enzymatic reduction of 11-*cis* retinal bound to cellular retinal-binding protein, *Biochim. Biophys. Acta* **716:**266–272.

Saari, J. C., and Bredberg, D. L., 1987, Photochemistry and stereoselectivity of cellular retinaldehyde-binding protein from bovine retina, *J. Biol. Chem.* **262:**7618–7622.

Saari, J. C., and Bredberg, D. L., 1988a, Purification of cellular retinaldehyde-binding protein from bovine retina and retinal pigment epithelium, *Exp. Eye Res.* **46:**569–578.

Saari, J. C., and Bredberg, D. L., 1988b, CoA- and non-CoA- dependent retinol esterification in retinal pigment epithelium, *J. Biol. Chem.* **263:**8084–8090.

Saari, J. C., and Bredberg, D. L., 1989, Lecithin:retinol acyltransferase in retinal pigment epithelial microsomes, *J. Biol. Chem.* **264:**8636–8640.

Saari, J. C., and Futterman, S., 1976, Separable binding proteins for retinoic acid and retinol in bovine retina, *Biochim. Biophys. Acta.* **444:**789–793.

Saari, J. C., Bunt, A. H., Futterman, S., and Berman, E. R., 1977, Localization of cellular retinol-binding protein in bovine retina and retinal pigment epithelium, with a consideration of the pigment epithelium isolation technique, *Invest. Ophthalmol. Vis. Sci.* **16:**797–806.

Saari, J. C., Futterman, S., and Bredberg, L., 1978a, Cellular retinol- and retinoic acid-binding proteins of bovine retina. Purification and properties, *J. Biol. Chem.* **253:**6432–6436.

Saari, J. C., Futterman, S., Stubbs, G. W., Heffernan, J. T., Bredberg, L., Chan, K. Y., and Albert, D. M., 1978b, Cellular retinol- and retinoic acid-binding proteins in transformed mammalian cells, *Invest. Ophthalmol. Vis. Sci.* **17:**988–992.

Saari, J. C., Bredberg, L., and Garwin, G. G., 1982, Identification of the endogenous retinoids associated with three cellular retinoid-binding proteins from bovine retina and retinal pigment epithelium, *J. Biol. Chem.* **257:**13329–13333.

Saari, J. C., Bunt-Milam, A. H., Bredberg, D.L ., and Garwin, G. G., 1984, Properties and immunocytochemical localization of three retinoid-binding proteins from bovine retina, *Vision Res.* **24:**1595–1603.

Saari, J. C., Teller, D. C., Crabb, J. W., and Bredberg, L., 1985, Properties of an interphotoreceptor retinoid-binding protein from bovine retina, *J. Biol. Chem.* **260:**195–201.

Saibil, H. R., and Michel-Villaz, M., 1984, Squid rhodopsin and GTP-binding protein crossreact with vertebrate photoreceptor enzymes, *Proc. Natl. Acad. Sci. USA* **81:**5111–5115.

Salceda, R., 1986, Isolation and biochemical characterization of frog retinal pigment epithelium cells, *Invest. Ophthalmol. Vis. Sci.* **27:**1172–1176.

Sameshima, M., Uehara, F., and Ohba, N., 1987, Specialization of the interphotoreceptor matrices around cone and rod photoreceptor cells in the monkey retina, as revealed by lectin cytochemistry, *Exp. Eye Res.* **45:**845–863.

Saneto, R. P., Awasthi, Y. C., and Srivastava, S. K., 1982a, Glutathione S-transferases of the bovine retina, *Biochem. J.* **205:**213–217.

Saneto, R. P., Awasthi, Y. C., and Srivastava, S. K., 1982b, Mercapturic acid pathway enzymes in bovine ocular lens, cornea, retina and retinal pigmented epithelium, *Exp. Eye Res.* **34:**107–111.

Sanui, H., Redmond, T. M., Hu, L.-H., Kuwabara, T., Margalit, H., Cornette, J. L., Wiggert, B., Chader, G. J., and Gery, I., 1988, Synthetic peptides derived from IRBP induce EAU in Lewis rats, *Curr. Eye Res.* **7:**727–735.

Sanyal, S., 1987, Cellular site of expression and genetic interaction of the *rd* and the *rds* loci in the retina of the mouse, in: *Degenerative Retinal Disorders: Clinical and Laboratory Investigations* (J. G. Hollyfield, R. E. Anderson, and M. M. LaVail, eds.), Alan R. Liss, New York, pp. 175–194.

Sanyal, S., and Zeilmaker, G. H., 1988, Retinal damage by constant light in chimaeric mice: Implications for the protective role of melanin, *Exp. Eye Res.* **46:**731–743.

Sanyal, S., Hawkins, R. K., and Zeilmaker, G. H., 1988, Development and degeneration of retina in *rds* mutant mice: Analysis of interphotoreceptor matrix staining in chimaeric retina, *Curr. Eye Res.* **7:**1183–1190.

Saviolakis, G. A., Kyritsis, A. P., and Chader, G. J., 1986, Human Y-79 retinoblastoma cells exhibit specific insulin receptors, *J. Neurochem.* **47:**70–76.

Schechter, I., Burstein, Y., Zemell, R., Ziv, E., Kantor, F., and Papermaster, D. S., 1979, Messenger RNA of opsin from bovine retina: Isolation and partial sequence of the *in vitro* translation product, *Proc. Natl. Acad. Sci. USA* **76:**2654–2658.

Schmidt, S. Y., 1978, Taurine fluxes in isolated cat and rat retinas: Effects of illumination, *Exp. Eye Res.* **26:**529–537.

Schmidt, S. Y., 1983a, Light- and cytidine-dependent phosphatidylinositol synthesis in photoreceptor cells of the rat, *J. Cell Biol.* **97:**832–837.

Schmidt, S.Y., 1983b, Cytidine metabolism in photoreceptor cells of the rat, *J. Cell Biol.* **97:**824–831.

Schmidt, S. Y., 1983c, Phosphatidylinositol synthesis and phosphorylation are enhanced by light in rat retinas, *J. Biol. Chem.* **258:**6863–6868.

Schmidt, S. Y., and Peisch, R. D., 1986, Melanin concentration in normal human retinal pigment epithelium, *Invest. Ophthalmol. Vis. Sci.* **27:**1063–1067.

Schmidt, S. Y., Heth, C. A., Edwards, R. B., Brandt, J. T., Adler, A. J., Spiegel, A., Shichi, H., and Berson, E. L., 1988, Identification of proteins in retinas and IPM from eyes with retinitis pigmentosa, *Invest. Ophthalmol. Vis. Sci.* **29:**1585–1593.

Schneider, B. G., Papermaster, D. S., Liou, G. I., Fong, S.-L., and Bridges, C. D., 1986, Electron microscopic immunocytochemistry of interstitial retinol-binding protein in vertebrate retinas, *Invest. Ophthalmol. Vis. Sci.* **27:**679–688.

Schnetkamp, P. P. M., and Daemen, F. J. M., 1982, Isolation and characterization of osmotically sealed bovine rod outer segments, in: *Methods in Enzymology*, Vol. 81 (L. Packer, ed.), Academic Press, New York, pp. 110–123.

Schnetkamp, P. P. M., and Kaupp, U. B., 1985, Calcium–hydrogen exchange in isolated bovine rod outer segments, *Biochemistry* **24:**723–727.

Schwartz, E. A., 1985, Phototransduction in vertebrate rods, *Annu. Rev. Neurosci.* **8:**339–367.

Schwartzman, M. L., Masferrer, J., Dunn, M. W., McGiff, J. C., and Abraham, N. G., 1987, Cytochrome P450, drug metabolizing enzymes and arachidonic acid metabolism in bovine ocular tissues, *Curr. Eye Res.* **6:**623–630.

Selinger, Z., and Minke, B., 1988, Inositol lipid cascade of vision studied in mutant flies, *Cold Spring Harbor Symp. Quant. Biol.* **53**:333–341.
Sheedlo, H. J., Li, L., and Turner, J. E., 1989, Functional and structural characteristics of photoreceptor cells rescued in RPE- cell grafted retinas of RCS dystrophic rats, *Exp. Eye Res.* **48**:841–854.
Shichi, H., 1969, Microsomal electron transfer system of bovine retinal pigment epithelium, *Exp. Eye Res.* **8**:60–68.
Shichi, H., 1983, *Biochemistry of Vision*, Academic Press, New York.
Shichi, H., 1984, Biotransformation and drug metabolism, in: *Handbook of Pharmacology*, Vol. 69, *Pharmacology of the Eye* (M. L. Sears, ed.), Springer-Verlag, Berlin, pp. 117–148.
Shichi, H., and Somers, R. L., 1978, Light-dependent phosphorylation of rhodopsin. Purification and properties of rhodopsin kinase, *J. Biol. Chem.* **253**:7040–7046.
Shichi, H., Atlas, S. A., and Nebert, D. W., 1975, Genetically regulated aryl hydrocarbon hydroxylase induction in the eye: Possible significance of the drug-metabolizing enzyme system for the retinal pigmented epithelium–choroid, *Exp. Eye Res.* **21**:557–567.
Shichi, H., Tsunematsu, Y., and Nebert, D. W., 1976, Aryl hydrocarbon hydroxylase induction in retinal pigmented epithelium: Possible association of genetic differences in a drug-metabolizing enzyme system with retinal degeneration, *Exp. Eye Res.* **23**:165–176.
Shinohara, T., Dietzschold, B., Craft, C. M., Wistow, G., Early, J. J., Donoso, L. A., Horwitz, J., and Tao, R., 1987, Primary and secondary structure of bovine retinal S antigen (48-kDa protein), *Proc. Natl. Acad. Sci. USA* **84**:6975–6979.
Shirakawa, H., Ishiguro, S.-I., Itoh, Y., Plantner, J. J., and Kean, E. L., 1987, Are sugars involved in the binding of rhodopsin-membranes by the retinal pigment epithelium? *Invest. Ophthalmol. Vis. Sci.* **28**:628–632.
Shuster, T. A., and Farber, D. B., 1986, Rhodopsin phosphorylation in developing normal and degenerative mouse retinas, *Invest. Ophthalmol. Vis. Sci.* **27**:264–268.
Shuster, T. A., Nagy, A. K., and Farber, D. B., 1988, Nucleotide binding to the rod outer segment rim protein, *Exp. Eye Res.* **46**:647–655.
Shvedova, A. A., Sidorov, A. S., Novikov, K. N., Galushchenko, I. V., and Kagan, V. E., 1979, Lipid peroxidation and electric activity of the retina, *Vision Res.* **19**:49–55.
Singh, S. V., Dao, D. D., Srivastava, S. K., and Awasthi, Y. C., 1984, Purification and characterization of glutathione S-transferases in human retina, *Curr. Eye Res.* **3**:1273–1280.
Snodderly, D. M., Brown, P. K., Delori, F. C., and Auran, J. D., 1984, The macular pigment. I. Absorbance spectra, localization, and discrimination from other yellow pigments in primate retinas, *Invest. Ophthalmol. Vis. Sci.* **25**:660–673.
Stephens, R. J., Negi, D. S., Short, S. M., van Kuijk, F. J. G. M., Dratz, E. A., and Thomas, D. W., 1988, Vitamin E distribution in ocular tissues following long-term dietary depletion and supplementation as determined by microdissection and gas chromatography–mass spectrometry, *Exp. Eye Res.* **47**:237–245.
St. Jules, R. S., and O'Brien, P. J., 1986, The acylation of rat rhodopsin *in* vitro and *in vivo*, *Exp. Eye Res.* **43**:929–940.
St. Jules, R. S., Wallingford, J. C., Smith, S. B., and O'Brien, P. J., 1989, Addition of the chromophore to rat rhodopsin is an early post translational event, *Exp. Eye Res.* **48**:653–665.
Stone, W. L., and Dratz, E. A., 1980, Increased glutathione-S- transferase activity in antioxidant-deficient rats, *Biochim. Biophys. Acta* **631**:503–506.
Stone, W. L., and Dratz, E. A., 1982, Selenium and non-selenium glutathione peroxidase activities in selected ocular and non-ocular rat tissues, *Exp. Eye Res.* **35**:405–412.
Stone, W. L., Farnsworth, C. C., and Dratz, E. A., 1979a, A reinvestigation of the fatty acid content of bovine, rat and frog retinal rod outer segments, *Exp. Eye Res.* **28**:387–397.
Stone, W. L., Katz, M. L., Lurie, M., Marmor, M. F., and Dratz, E. A., 1979b, Effects of dietary vitamin E and selenium on light damage to the rat retina, *Photochem. Photobiol.* **29**:725–730.
Stramm, L. E., 1987, Synthesis and secretion of glycosaminoglycans in cultured retinal pigment epithelium, *Invest. Ophthalmol. Vis. Sci.* **28**:618–627.
Stramm, L. E., Haskins, M. E., McGovern, M. M., and Aguirre, G. D., 1983, Tissue culture of cat retinal pigment epithelium, *Exp. Eye Res.* **36**:91–101.
Stramm, L. E., Li., W., Aguirre, G. D., and Rockey, J. H., 1987, Glycosaminoglycan synthesis and secretion by bovine retinal capillary pericytes in culture, *Exp. Eye Res.* **44**:17–28.
Stryer, L., 1986, Cyclic GMP cascade of vision, *Annu. Rev. Neurosci.* **9**:87–119.

Stryer, L., and Bourne, H. R., 1986, G Proteins: A family of signal transducers, *Annu. Rev. Cell Biol.* **2**:391–419.

Stubbs, G. W., Saari. J. C., and Futterman, S., 1979, 11-*cis*-Retinal-binding protein from bovine retina. Isolation and partial characterization, *J. Biol. Chem.* **254**:8529–8533.

Swartz, J. G., and Mitchell, J. E., 1970, Biosynthesis of retinal phospholipids: Incorporation of radioactivity from labeled phosphorylcholine and cytidine diphosphate choline, *J. Lipid Res.* **11**:544–550.

Swartz, J. G., and Mitchell, J. E., 1973, Phospholipase activity of retina and pigment epithelium, *Biochemistry* **12**:5273–5278.

Swartz, J. G., and Mitchell, J. E., 1974, Acyl transfer reactions of retina, *Biochemistry*, **13**:5053–5059.

Szuts, E. Z., 1985, Light stimulates phosphorylation of two large membrane proteins in frog photoreceptors, *Biochemistry* **24**:4176–4184.

Szuts, E. Z., Wood, S. F., Reid, M. S., and Fein, A., 1986, Light stimulates the rapid formation of inositol trisphosphate in squid retinas, *Biochem. J.* **240**:929–932.

Takki, K., and Simell, O., 1974, Genetic aspects in gyrate atrophy of the choroid and retina with hyperornithinaemia, *Br. J. Ophthalmol.* **58**:907–916.

Tamai, M., Teirstein, P., Goldman, A., O'Brien, P., and Chader, G. J., 1978, The pineal gland does not control rod outer segment shedding and phagocytosis in the rat retina and pigment epithelium, *Invest. Ophthalmol. Vis. Sci.* **17**:558–562.

Tarnowski, B. I., Shepherd, V. L., and McLaughlin, B. J., 1988a, Mannose 6-phosphate receptors on the plasma membrane on rat retinal pigment epithelial cells, *Invest. Ophthalmol. Vis. Sci.* **29**:291–297.

Tarnowski, B. I., Shepherd, V. L., and McLaughlin, B. J., 1988b, Expression of mannose receptors for pinocytosis and phagocytosis on rat retinal pigment epithelium, *Invest. Ophthalmol. Vis. Sci.* **29**:742–748.

Tarver, A. P., and Anderson, R. E., 1988, Phospholipase C activity and substrate specificity in frog photoreceptors, *Exp. Eye Res.* **46**:29–35.

Tawara, A., Varner, H. H., and Hollyfield, J. G., 1988, Proteoglycans in the mouse interphotorecetor matrix. I. Histochemical studies using Cuprolinic blue, *Exp. Eye Res.* **46**:689–704.

Tawara, A., Varner, H. H., and Hollyfield, J. G., 1989, Proteoglycans in the mouse interphotoreceptor matrix. II. Origin and development of proteoglycans, *Exp. Eye Res.* **48**: 815–839.

Teirstein, P. S., Goldman, A. I., and O'Brien, P. J., 1980, Evidence for both local and central regulation of rat rod outer segment disc shedding, *Invest. Ophthalmol. Vis. Sci.* **19**:1268–1273.

Tornqvist, K., and Ehinger, B., 1988, Peptide immunoreactive neurons in the human retina, *Invest. Ophthalmol. Vis. Sci.* **29**:680–686.

Tsuda, M., Tsuda, T., Terayama, Y., Fukada, Y., Akino, T., Yamanaka, G., Stryer, L., Katada, T., Ui, M., and Ebrey, T., 1986, Kinship of cephalopod photoreceptor G-protein with vertebrate transducin, *FEBS Lett.* **198**:5–10.

Tuteja, N., Tuteja, R., and Farber, D. B., 1989, Cloning and sequencing of the γ-subunit of retinal cyclic-GMP phosphodiesterase from *rd* mouse, *Exp. Eye Res.* **48**:863–872.

Uehara, F., Sameshima, M., Muramatsu, T., and Ohba, N., 1983, Localization of fluorescence-labeled lectin binding sites on photoreceptor cells of the monkey retina, *Exp. Eye Res.* **36**: 113–123.

Uehara, F., Muramatsu, T., and Ohba, N., 1986, Two-dimensional gel electrophoretic analysis of lectin receptors in the bovine interphotoreceptor matrix, *Exp. Eye Res.* **43**:227–234.

Ullrich, A., Gray, A., Tam, A. W., Yang-Feng, T., Tsubokawa, M., Collins, C., Henzel, W., Le Bon, T., Kathuria, S., Chen, E., Jacobs, S., Francke, U., Ramachandran, J., and Fujita-Yamaguchi, Y., 1986, Insulin-like growth factor I receptor primary structure: Comparison with insulin receptor suggests structural determinants that define functional specificity, *EMBO J.* **5**:2503–2512.

Urban, P. F., Harth, S., Freysz, L., and Dreyfus, H., 1980, Brain and retinal ganglioside composition from different species determined by TLC and HPTLC, in: *Structure and Function of Gangliosides* (L. Svennerholm, P. Mandel, H. Dreyfus, and P. F. Urban, eds.), Plenum Press, New York, pp. 149–157.

Usukura, J., and Bok, D., 1987, Changes in the localization and content of opsin during retinal development in the *rds* mutant mouse: Immunocytochemistry and inmmunoassay, *Exp. Eye Res.* **45**:501–515.

Valle, D., Kaiser-Kupfer, M. I., and Del Valle, L. A., 1977, Gyrate atrophy of the choroid and retina: Deficiency of ornithine aminotransferase in transformed lymphocytes, *Proc. Natl. Acad. Sci. USA* **74**:5159–5161.

Vandenberg, C. A., and Montal, M., 1984, Light-regulated biochemical events in invertebrate photoreceptors. I. Light- activated guanosinetriphosphatase, guanine nucleotide binding, and cholera toxin catalyzed labeling of squid photoreceptor membranes, *Biochemistry* **23**:2339–2347.

van Meurs, K. P., Angus, C. W., Lavu, S., Kung, H.-F., Czarnecki, S. K., Moss, J., and Vaughan, M., 1987, Deduced amino acid sequence of bovine retinal $G_{o\alpha}$: Similarities to other guanine nucleotide-binding proteins, *Proc. Natl. Acad. Sci. USA* **84:**3107–3111.

van Veen, T., Katial, A., Shinohara, T., Barrett, D. J., Wiggert, B., Chader, G. J., and Nickerson, J. M., 1986, Retinal photoreceptor neurons and pinealocytes accumulate mRNA for interphotoreceptor retinoid-binding protein (IRBP), *FEBS Lett.* **208:**133–137.

van Veen, T., Ekstrom, P., Wiggert, B., Lee, L., Hirose, Y., Sanyal, S., and Chader, G. J., 1988, A developmental study of interphotoreceptor retinoid-binding protein (IRBP) in single and double homozygous *rd* and *rds* mutant mouse retinae, *Exp. Eye Res.* **47:**291–305.

Varner, H. H., Rayborn, M. E., Osterfeld, A. M., and Hollyfield, J. G., 1987, Localization of proteoglycan within the extracellular matrix sheath of cone photoreceptors, *Exp. Eye Res.* **44:**633–642.

Vistica, B. P., Usui, M., Kuwabara, T., Wiggert, B., Lee, L., Redmond, T. M., Chader, G. J., and Gery, I., 1987, IRBP from bovine retina is poorly uveitogenic in guinea pigs and is identical to A-antigen, *Curr. Eye Res.* **6:**409–417.

Voaden, M. J., Lake, N., Marshall, J., and Morjaria, B., 1977, Studies on the distribution of taurine and other neuroactive amino acids in the retina, *Exp. Eye Res.* **25:**249–257.

Voaden, M. J., Lake, N., Marshall, J., and Morjaria, B., 1978, The utilization of glutamine by the retina: An autoradiographic and metabolic study, *J. Neurochem.* **31:**1069–1076.

Waldbilling, R. J., and Chader, G. J., 1988, Anomalous insulin-binding activity in the bovine neural retina: A possible mechanism for regulation of receptor binding specificity, *Biochem. Biophys. Res. Commun.* **151:**1105–1112.

Waldbilling, R. J., Fletcher, R. T., Chader, G. J., Rajagopalan, S., Rodrigues, M., and LeRoith, D., 1987a, Retinal insulin receptors. 1. Structural heterogeneity and functional characterization, *Exp. Eye Res.* **45:**823–835.

Waldbilling, R. J., Fletcher, R. T., Chader, G. J., Rajagopalan, S., Rodrigues, M., and LeRoith, D., 1987b, Retinal insulin receptors. 2. Characterization and insulin-induced tyrosine kinase activity in bovine retinal rod outer segments, *Exp. Eye Res.* **45:**837–844.

Waloga, G., and Anderson, R. E., 1985, Effects of inositol 1,4,5-trisphosphate injections into salamander rods, *Biochem. Biophys. Res. Commun.* **126:**59–62.

Weiter, J. J., Delori, F. C., Wing, G. L., and Fitch, K. A., 1986, Retinal pigment epithelial lipofuscin and melanin and choroidal melanin in human eyes, *Invest. Ophthalmol. Vis. Sci.* **27:**145–152.

Werner, J. S., Donnelly, S. K., and Kliegl, R., 1987, Aging and human macular pigment density, *Vision Res.* **27:**257–268.

Wetzel, M. G., and O'Brien, P. J., 1986, Turnover of palmitate, arachidonate and glycerol in phospholipids of rat rod outer segments, *Exp. Eye Res.* **43:**941–954.

Wiechmann, A. F., Bok, D., and Horwitz, J., 1985, Localization of hydroxyindole-O-methyltransferase in the mammalian pineal gland and retina, *Invest. Ophthalmol. Vis. Sci.* **26:**253–265.

Wiechmann, A. F., Bok, D., and Horwitz, J., 1986, Melatonin-binding in the frog retina: Autoradiographic and biochemical analysis, *Invest. Ophthalmol. Vis. Sci.* **27:**153–163.

Wiegand, R. D., and Anderson, R. E., 1983, Neutral lipids of frog and rat rod outer segments, *Exp. Eye Res.* **36:**389–396.

Wiegand, R. D., Giusto, N. M., Rapp, L. M., and Anderson, R. E., 1983, Evidence for rod outer segment lipid peroxidation following constant illumination of the rat retina, *Invest. Ophthalmol. Vis. Sci.* **24:**1433–1435.

Wiegand, R. D., Jose, J. G., Rapp, L. M., and Anderson, R. E., 1984, Free radicals and damage to ocular tissues, in: *Free Radicals in Molecular Bilogy, Aging and Disease* (D. Armstrong, R. S. Sohol, R. G. Cutler, and T. F. Slater, eds.), Raven Press, New York, pp. 317–353.

Wiegand, R. D., Joel, C. D., Rapp, L. M., Nielsen, J. C., Maude, M. B., and Anderson, R. E., 1986, Polyunsaturated fatty acids and vitamin E in rat rod outer segments during light damage, *Invest. Ophthalmol. Vis. Sci.* **27:**727–733.

Wiggert, B. O., and Chader, G. J., 1975, A receptor for retinol in the developing retina and pigment epithelium, *Exp. Eye Res.* **21:**143–151.

Wiggert, B., and Chader, G. J., 1985, Monkey interphotoreceptor retinoid-binding protein (IRBP): Isolation, characterization and synthesis, in: *The Interphotoreceptor Matrix in Health and Disease* (C. D. Bridges and A. J. Adler, eds.), Alan R. Liss, New York, pp. 89–110.

Wiggert, B., Russell, P., Lewis, M., and Chader, G. J., 1977, Differential binding to soluble nuclear receptors and effects on cell viability of retinol and retinoic acid in cultured retinoblastoma cells, *Biochem. Biophys. Res. Commun.* **79:**218–225.

Wiggert, B., Bergsma, D. R., Helmsen, R., and Chader, G. J., 1978a, Vitamin A receptors. Retinoic acid binding in ocular tissues, *Biochem. J.* **169:**87–94.

Wiggert, B., Mizukawa, A., Kuwabara, T., and Chader, G. J., 1978b, Vitamin A receptors: Multiple species in retina and brain and possible compartmentalization in retinal photoreceptors, *J. Neurochem.* **30:**653–659.

Wiggert, B., Lee, L., O'Brien, P. J., and Chader, G. J., 1984, Synthesis of interphotoreceptor retinoid-binding protein (IRBP) by monkey retina in organ culture: Effect of monensin, *Biochem. Biophys. Res. Commun.* **118:**789–796.

Wiggert, B., Lee, L., Rodrigues, M., Hess, H., Redmond, T. M., and Chader, G. J., 1986, Immunochemical distribution of interphotoreceptor retinoid-binding protein in selected species, *Invest. Ophthalmol. Vis. Sci.* **27:**1041–1049.

Wilcox, D. K., 1987, Extracellular release of acid hydrolases from cultured retinal pigmented epithelium, *Invest. Ophthalmol. Vis. Sci.* **28:**76–82.

Wilden, U., and Kuhn, H., 1982, Light-dependent phosphorylation of rhodopsin: Number of phosphorylation sites, *Biochemistry* **21:**3014–3022.

Wilden, U., Hall, S. W., and Kuhn, H., 1986, Phosphodiesterase activation by photoexcited rhodopsin is quenched when rhodopsin is phosphorylated and binds the intrinsic 48-kDa protein of rod outer segments, *Proc. Natl. Acad. Sci. USA* **83:**1174–1178.

Williams, T. P., and Howell, W. L., 1983, Action spectrum of retinal light-damage in albino rats, *Invest. Ophthalmol. Vis. Sci.* **24:**285–287.

Wing, G. L., Blanchard, G. C., and Weiter, J. J., 1978, The topography and age relationship of lipofuscin concentration in the retinal pigment epithelium, *Invest. Ophthalmol. Vis. Sci.* **17:**601–607.

Winkler, B. S., 1981, Glycolytic and oxidative metabolism in relation to retinal function, *J. Gen. Physiol.* **77:**667–692.

Winkler, B. S., 1983a, The intermediary metabolism of the retina: Biochemical and functional aspects, in: *Biochemistry of the Eye* (R. E. Anderson, ed.), American Academy of Ophthalmology, San Francisco, pp. 227–242.

Winkler, B. S., 1983b, Relative inhibitory effects of ATP depletion, ouabain and calcium on retinal photoreceptors, *Exp. Eye Res.* **36:**581–594.

Winkler, B. S., 1989, Retinal aerobic glycolysis revisited, *Invest. Ophthalmol. Vis. Sci.* **30:**1023.

Winkler, B. S., and Giblin, F. J., 1983, Glutathione oxidation in retina: Effects on biochemical and electrical activities, *Exp. Eye Res.* **36:**287–297.

Winkler, B. S., Simson, V., and Benner, J., 1977, Importance of bicarbonate in retinal function, *Invest. Ophthalmol. Vis. Sci.* **16:**766–768.

Winkler, B. S., DeSantis, N., and Solomon, F., 1986, Multiple NADPH-producing pathways control glutathione (GSH) content in retina, *Exp. Eye Res.* **43:**829–847.

Witt, P. L., and Bownds, M. D., 1987, Identification of frog photoreceptor plasma and disk membrane proteins by radioiodination, *Biochemistry* **26:**1769–1776.

Wong, S., and Molday, R. S., 1986, A spectrin-like protein in retinal rod outer segments, *Biochemistry* **25:**6294–6300.

Woodford, B. J., Liu, Y., Fletcher, R. T., Chader, G. J., Farber, D. B., Santos-Anderson, R., and Tso, M. O. M., 1982, Cyclic nucleotide metabolism in inherited retinopathy in collies: A biochemical and histochemical study, *Exp. Eye Res.* **34:**703–714.

Woodford, B. J., Tso, M. O. M., and Lam, K.-W., 1983, Reduced and oxidized ascorbates in guinea pig retina under normal and light-exposed conditions, *Invest. Ophthalmol. Vis. Sci.* **24:**862–867.

Yamada, E., 1985, Morphology of the interphotoreceptor matrix as revealed by rapid freezing technique, in: *The Interphotoreceptor Matrix in Health and Disease* (C. D. Bridges and A. J. Adler, eds.), Alan R. Liss, New York, pp. 25–32.

Yao, X.-Y., and Marmor, M. F., 1989, Interphotoreceptor matrix plays an important role in retinal adhesiveness, *Invest. Ophthalmol. Vis. Sci. Suppl.* **30:**240.

Yatsunami, K., and Khorana, H. G., 1985, GTPase of bovine rod outer segments: The amino acid sequence of the α subunit as derived from the cDNA sequence, *Proc. Natl. Acad. Sci. USA* **82:**4316–4320.

Yatsunami, K., Pandya, B. V., Oprian, D. D., and Khorana, H. G., 1985, cDNA-derived amino acid sequence of the γ subunit of GTPase from bovine rod outer segments, *Proc. Natl. Acad. Sci. USA* **82:**1936–1940.

Yau, K.-W., and Nakatani, K., 1985, Light-suppressible, cyclic GMP-sensitive conductance in the plasma membrane of a truncated rod outer segment, *Nature* **317:**252–255.

Yoon, Y. H., and Marmor, M. F., 1988, Effects on retinal adhesion of temperature, cyclic AMP, cytochalasin, and enzymes, *Invest. Ophthalmol. Vis. Sci.* **29:**910–914.

Yorek, M. A., Spector, A. A., and Ginsberg, B. H., 1985, Characterization of an insulin receptor in human Y-79 retinoblastoma cells, *J. Neurochem.* **45:**1590–1595.

Yorek, M. A., Dunlap, J. A., and Ginsberg, B. H., 1987, Amino acid and putative neurotransmitter transport in human Y79 retinoblastoma cells, *J. Biol. Chem.* **262:**10986–10993.

Young, R. W., 1967, The renewal of photoreceptor cell outer segments, *J. Cell Biol.* **33:**61–72.

Young, R. W., 1971, The renewal of rod and cone outer segments in the rhesus monkey, *J. Cell Biol.* **49:**303–318.

Young, R. W., 1976, Visual cells and the concept of renewal, *Invest. Ophthalmol.* **15:**700–725.

Young, R. W., and Bok, D., 1969, Participation of the retinal pigment epithelium in the rod outer segment renewal process, *J. Cell Biol.* **42:**392–403.

Young, R. W., and Bok, D., 1979, Metabolism of the retinal pigment epithelium, in: *The Retinal Pigment Epithelium* (K. M. Zinn and M. F. Marmor, eds.), Harvard University Press, Cambridge, pp. 103–123.

Young, W. Y., and Droz, B., 1968, The renewal of protein in retinal rods and cones, *J. Cell Biol.* **39:**169–184.

Yu, B. C.-Y., Watt, C. B., Lam, D. M. K., and Fry, K. R., 1988, GABAergic ganglion cells in the rabbit retina, *Brain Res.* **439:**376–382.

Yue, B. Y. J. T. and Fishman, G. A., 1985, Synthetic activities of cultured retinal pigment epithelial cells from a patient with retinitis pigmentosa, *Arch. Ophthalmol.* **103:**1563–1566.

Zauberman, H., and Berman, E. R., 1969, Measurement of adhesive forces between the sensory retina and the pigment epithelium, *Exp. Eye Res.* **8:**276–283.

Zick, Y., Sagi-Eisenberg, R., Pines, M., Gierschik, P., and Spiegel, A. M., 1986, Multisite phosphorylation of the α subunit of transducin by the insulin receptor kinase and protein kinase C, *Proc. Natl. Acad. Sci. USA* **83:**9294–9297.

Zick, Y., Spiegel, A. M., and Sagi-Eisenberg, R., 1987, Insulin-like growth factor I receptors in retinal rod outer segments, *J. Biol. Chem.* **262:**10259–10264.

Zimmerman, A. L., and Baylor, D. A., 1986, Cyclic GMP-sensitive conductance of retinal rods consists of aqueous pores, *Nature* **321:**70–72.

Zimmerman, A. L., Yamanaka, G., Eckstein, F., Baylor, D. A. and Stryer, L., 1985, Interaction of hydrolysis-resistant analogs of cyclic GMP with the phosphodiesterase and light-sensitive channel of retinal rod outer segments, *Proc. Natl. Acad. Sci. USA* **82:**8813–8817.

Zimmerman, W. F., 1974, The distribution and proportions of vitamin A compounds during the visual cycle in the rat, *Vision Res.* **14:**795–802.

Zimmerman, W. F., 1976, Subcellular distribution of 11-*cis* retinol dehydrogenase activity in bovine pigment epithelium, *Exp. Eye Res.* **23:**159–164.

Zimmerman, W. F., and Godchaux, W. III, 1982, Preparation and characterization of sealed bovine rod cell outer segments, in: *Methods in Enzymology*, Vol. 81 (L. Packer, ed.), Academic Press, New York, pp. 52–57.

Zimmerman, W. F., and Keys, S., 1988, Acylation and deacylation of phospholipids in isolated bovine rod outer segments, *Exp. Eye Res.* **47:**247–260.

Zimmerman, W. F., Lion, F., Daemen, F. J. M., and Bonting, S. L., 1975, Biochemical aspects of the visual process. XXX. Distribution of stereospecific retinol dehydrogenase activities in subcellullar fractions of bovine retina and pigment epithelium, *Exp. Eye Res.* **21:**325–332.

Zimmerman, W. F., Godchaux, W. III, and Belkin, M., 1983, The relative proportions of lysosomal enzyme activities in bovine retinal pigment epithelium, *Exp. Eye Res.* **36:**151–158.

Zinn, K. M., and Marmor, M. F. (eds.), 1979, *The Retinal Pigment Epithelium*, Harvard University Press, Cambridge.

Index